WORLD HEALTH ORGANIZATION

INTERNATIONAL AGENCY FOR RESEARCH ON CANCER

IARC MONOGRAPHS

ON THE

EVALUATION OF CARCINOGENIC RISKS TO HUMANS

Some Traditional Herbal Medicines, Some Mycotoxins, Naphthalene and Styrene

VOLUME 82

This publication represents the views and expert opinions of an IARC Working Group on the Evaluation of Carcinogenic Risks to Humans, which met in Lyon,

12–19 February 2002

2002

IARC MONOGRAPHS

In 1969, the International Agency for Research on Cancer (IARC) initiated a programme on the evaluation of the carcinogenic risk of chemicals to humans involving the production of critically evaluated monographs on individual chemicals. The programme was subsequently expanded to include evaluations of carcinogenic risks associated with exposures to complex mixtures, life-style factors and biological and physical agents, as well as those in specific occupations.

The objective of the programme is to elaborate and publish in the form of monographs critical reviews of data on carcinogenicity for agents to which humans are known to be exposed and on specific exposure situations; to evaluate these data in terms of human risk with the help of international working groups of experts in chemical carcinogenesis and related fields; and to indicate where additional research efforts are needed.

The lists of IARC evaluations are regularly updated and are available on Internet: http://monographs.iarc.fr/

This project was supported by Cooperative Agreement 5 UO1 CA33193 awarded by the United States National Cancer Institute, Department of Health and Human Services, and was funded in part by the European Commission, Directorate-General EMPL (Employment, and Social Affairs), Health, Safety and Hygiene at Work Unit. Additional support has been provided since 1993 by the United States National Institute of Environmental Health Sciences.

Distributed by IARC*Press* (Fax: +33 4 72 73 83 02; E-mail: press@iarc.fr) and by the World Health Organization Marketing and Dissemination (MDI), 1211 Geneva 27 (Fax: +41 22 791 4857; E-mail: publications@who.int)

IARC Library Cataloguing in Publication Data

Some traditional herbal medicines, some mycotoxins, naphthalene and styrene/
IARC Working Group on the Evaluation of Carcinogenic Risks to Humans
(2002 : Lyon, France)

(IARC monographs on the evaluation of carcinogenic risks to humans ; 82)

1. Traditional herbal medicines – congresses 2. Mycotoxins – congresses
3. Naphthalene – congresses 4. Styrene – congresses I. IARC Working
Group on the Evaluation of Carcinogenic Risks to Humans II. Series

ISBN 92 832 1282 7 (NLM Classification: W1)
ISSN 1017-1606

PRINTED IN FRANCE

CONTENTS

NOTE TO THE READER

The term 'carcinogenic risk' in the *IARC Monographs* series is taken to mean the probability that exposure to an agent will lead to cancer in humans.

Inclusion of an agent in the *Monographs* does not imply that it is a carcinogen, only that the published data have been examined. Equally, the fact that an agent has not yet been evaluated in a monograph does not mean that it is not carcinogenic.

The evaluations of carcinogenic risk are made by international working groups of independent scientists and are qualitative in nature. No recommendation is given for regulation or legislation.

Anyone who is aware of published data that may alter the evaluation of the carcinogenic risk of an agent to humans is encouraged to make this information available to the Unit of Carcinogen Identification and Evaluation, International Agency for Research on Cancer, 150 cours Albert Thomas, 69372 Lyon Cedex 08, France, in order that the agent may be considered for re-evaluation by a future Working Group.

Although every effort is made to prepare the monographs as accurately as possible, mistakes may occur. Readers are requested to communicate any errors to the Unit of Carcinogen Identification and Evaluation, so that corrections can be reported in future volumes.

IARC WORKING GROUP ON THE EVALUATION OF CARCINOGENIC RISKS TO HUMANS: SOME TRADITIONAL HERBAL MEDICINES, SOME MYCOTOXINS, NAPHTHALENE AND STYRENE

Lyon, 12–19 February 2002

LIST OF PARTICIPANTS

Members

Ahti Anttila, Finnish Cancer Registry, Institute for Statistical and Epidemiological Cancer Research, Liisankatu 21 B, 00170 Helsinki, Finland

Ramesh V. Bhat, National Institute of Nutrition, Indian Council of Medical Research, Jamai-Osmania PO, Hyderabad-500 007 AP, India

James A. Bond, Chemico-Biological Interactions, Toxcon, 5505 Frenchmans Creek, Durham, NC 27713, USA

Susan J. Borghoff, CIIT Centers for Health Research, 6 Davis Drive, Box 12137, Research Triangle Park, NC 27709-2127, USA

F. Xavier Bosch, Epidemiology Unit and Cancer Registry, Catalan Institute of Oncology, Av. Gran via s/n, Km. 2.7, 08907 L'Hospitalet del Llobregat, Spain

Gary P. Carlson, School of Health Sciences, 1338 Civil Engineering Building, Purdue University, West Lafayette, IN 47907-1338, USA

Marcel Castegnaro, Les Collanges, 07240 Saint-Jean-Chambre, France

George Cruzan, ToxWorks, 1153 Roadstown Road, Bridgeton, NJ 08302-6640, USA

Wentzel C.A. Gelderblom, Programme on Mycotoxins and Experimental Carcinogenesis, Medical Research Council (MRC), PO Box 19070, Tygerberg, South Africa 7505

Ulla Hass, Institute of Food Safety and Toxicology, Mørkhøj Bygade 19, 2860 Søborg, Denmark

Sara H. Henry, 5100 Paint Branch Parkway, College Park, MD 20740-3835, USA

Ronald A. Herbert, Laboratory of Experimental Pathology, National Institute of Environmental Health Sciences, PO Box 12233, Mail Drop B3-08, Research Triangle Park, NC 27709-2233, USA

Marc Jackson, Integrated Laboratory Systems, Inc., PO Box 13501, Research Triangle Park, NC 27709, USA

Tingliang Jiang, Department of Pharmacology, Institute of Chinese Materia Medica, China Academy of Traditional Chinese Medicine, 18 Beixincang, Dongzhimen Nei, Beijing 100700, People's Republic of China

A. Douglas Kinghorn, Department of Medicinal Chemistry and Pharmacognosy (M/C 781), College of Pharmacy, University of Illinois at Chicago, 833 S. Wood St, Chicago, IL 60612, USA

Siegfried Knasmüller, University of Vienna, Institute of Tumour Biology–Cancer Research, Borschkegasse 8a, 1090 Vienna, Austria

Len Levy, MRC Institute for Environment and Health, University of Leicester, 94 Regent Road, Leicester LE1 7DD, United Kingdom (*Chairman*)

Douglas McGregor, 102 rue Duguesclin, 69006 Lyon, France

J. David Miller, College of Natural Sciences, Department of Chemistry, 228 Steacie Boulevard, Carleton University, 1125 Colonel By Drive, K1S 5B6 Ottawa, Canada

Hideki Mori, Department of Pathology, Gifu University School of Medicine, 40 Tsukasa-Machi, Gifu 500-8705, Japan

Steve S. Olin, ILSI Risk Science Institute, One Thomas Circle, NW, 9th Floor, Washington, DC 20005-5802, USA

Douglas L. Park, Division of Natural Products (HFS-345), Center for Food Safety and Applied Nutrition, Food and Drug Administration, 5100 Paint Branch Parkway, College Park, MD 20740-3, USA

John I. Pitt, Commonwealth Scientific and Industrial Research Organization (CSIRO), Food Science Australia, PO Box 52, 15 Julius Avenue, North Ryde, NSW 2113, Australia

Ronald T. Riley, United States Department of Agriculture/ARS, Toxicology and Mycotoxin Research Unit, Richard B. Russell Agricultural Research Center, Room 361, PO Box 5677, 950 College Station Road, Athens, GA 30604-5677, USA

Heinz H. Schmeiser, Division of Molecular Toxicology, German Cancer Research Center, Im Neuenheimer Feld 280, 69120 Heidelberg, Germany

Jack Siemiatycki, Armand-Frappier INRS-Institute, Québec University, 531 Boulevard des Prairies, Laval, Québec H7V 1B7, Canada

Jean-Louis Vanherweghem, Department of Nephrology, Erasmus Hospital, Université Libre de Bruxelles, Route de Lennik 808, 1070 Brussels, Belgium

Christopher P. Wild, Molecular Epidemiology Unit, Epidemiology and Health Services Research, School of Medicine, Algernon Firth Building, University of Leeds, Leeds LS2 9JT, United Kingdom

Takumi Yoshizawa, Department of Biochemistry and Food Science, Faculty of Agriculture, Kagawa University, Miki-cho, Kagawa 761-0795, Japan

Representatives/Observers

Observer, representing the American Chemistry Council

Alan R. Buckpitt, Department of Molecular Biosciences, School of Veterinary Medicine, Haring Hall, One Shields Avenue, University of California, Davis, CA 95616, USA

Observer, representing the US National Institutes of Health
Paul M. Coates, Office of Dietary Supplements, National Institutes of Health, 31 Center Drive, Room 1B29, Bethesda, MD 20892-2086, USA

Observer, representing the National Center for Environmental Assessment, US Environmental Protection Agency
Gary L. Foureman, EPA (MD-52), National Center for Environmental Assessment, Research Triangle Park, NC 27711, USA

Observer, representing the European Centre for Ecotoxicology and Toxicology of Chemicals
Trevor Green, Investigative Toxicology Section, Syngenta, Alderley Park, Macclesfield, Cheshire SK10 4TJ, United Kingdom

Observer, representing the US National Toxicology Program
C. William Jameson, National Toxicology Program, National Institute of Environmental Health Sciences, 79 Alexander Drive, Building 4401, PO Box 12233 (MD EC-14), Research Triangle Park, NC 27709, USA

Observer, representing the US National Cancer Institute
David Longfellow, Chemical and Physical Carcinogenesis Branch, Division of Cancer Biology, National Cancer Institute, 6130 Executive Boulevard, Suite 5000 MSC7368, Rockville, MD 20892

IARC Secretariat
Robert Baan, Unit of Carcinogen Identification and Evaluation (*Responsible Officer*)
Michael Bird[1]
John Cheney (*Editor*)
Silvia Franceschi, Unit of Field and Intervention Studies
Marlin Friesen, Unit of Nutrition and Cancer
Yann Grosse, Unit of Carcinogen Identification and Evaluation
Ted Junghans[2]
Nikolai Napalkov[3]
Christiane Partensky, Unit of Carcinogen Identification and Evaluation (*co-Editor*)
Jerry Rice, Unit of Carcinogen Identification and Evaluation (*Head of Programme*)

---

[1] Present address: Toxicology Division, ExxonMobil Biomedical Sciences, Inc., 1545 Route 22 East, PO Box 971, Annandale, NJ 08801-0971, USA
[2] Technical Resources International Inc., 6500 Rock Spring Drive, Suite 650, Bethesda, MD 20817-1197, USA
[3] Present address: Director Emeritus, Petrov Institute of Oncology, Pesochny-2, 197758 St Petersburg, Russian Federation

Vikash Sewram, Unit of Environmental Cancer Epidemiology
Leslie Stayner[1], Unit of Carcinogen Identification and Evaluation
Kurt Straif, Unit of Carcinogen Identification and Evaluation
Eero Suonio, Unit of Carcinogen Identification and Evaluation

Technical assistance
Sandrine Egraz
Brigitte Kajo
Martine Lézère
Jane Mitchell
Elspeth Perez

[1] Present address: Risk Evaluation Branch, Education and Information Division, Robert A. Taft Labs, C-15, 4676 Columbia Parkway, Cincinnati, OH 45226, USA

PREAMBLE

IARC MONOGRAPHS PROGRAMME ON THE EVALUATION OF CARCINOGENIC RISKS TO HUMANS

PREAMBLE

1. BACKGROUND

In 1969, the International Agency for Research on Cancer (IARC) initiated a programme to evaluate the carcinogenic risk of chemicals to humans and to produce monographs on individual chemicals. The *Monographs* programme has since been expanded to include consideration of exposures to complex mixtures of chemicals (which occur, for example, in some occupations and as a result of human habits) and of exposures to other agents, such as radiation and viruses. With Supplement 6 (IARC, 1987a), the title of the series was modified from *IARC Monographs on the Evaluation of the Carcinogenic Risk of Chemicals to Humans* to *IARC Monographs on the Evaluation of Carcinogenic Risks to Humans*, in order to reflect the widened scope of the programme.

The criteria established in 1971 to evaluate carcinogenic risk to humans were adopted by the working groups whose deliberations resulted in the first 16 volumes of the *IARC Monographs series*. Those criteria were subsequently updated by further ad-hoc working groups (IARC, 1977, 1978, 1979, 1982, 1983, 1987b, 1988, 1991a; Vainio *et al.*, 1992).

2. OBJECTIVE AND SCOPE

The objective of the programme is to prepare, with the help of international working groups of experts, and to publish in the form of monographs, critical reviews and evaluations of evidence on the carcinogenicity of a wide range of human exposures. The *Monographs* may also indicate where additional research efforts are needed.

The *Monographs* represent the first step in carcinogenic risk assessment, which involves examination of all relevant information in order to assess the strength of the available evidence that certain exposures could alter the incidence of cancer in humans. The second step is quantitative risk estimation. Detailed, quantitative evaluations of epidemiological data may be made in the *Monographs*, but without extrapolation beyond the range of the data available. Quantitative extrapolation from experimental data to the human situation is not undertaken.

The term 'carcinogen' is used in these monographs to denote an exposure that is capable of increasing the incidence of malignant neoplasms; the induction of benign neoplasms may in some circumstances (see p. 19) contribute to the judgement that the exposure is carcinogenic. The terms 'neoplasm' and 'tumour' are used interchangeably.

Some epidemiological and experimental studies indicate that different agents may act at different stages in the carcinogenic process, and several mechanisms may be involved. The aim of the *Monographs* has been, from their inception, to evaluate evidence of carcinogenicity at any stage in the carcinogenesis process, independently of the underlying mechanisms. Information on mechanisms may, however, be used in making the overall evaluation (IARC, 1991a; Vainio *et al.*, 1992; see also pp. 25–27).

The *Monographs* may assist national and international authorities in making risk assessments and in formulating decisions concerning any necessary preventive measures. The evaluations of IARC working groups are scientific, qualitative judgements about the evidence for or against carcinogenicity provided by the available data. These evaluations represent only one part of the body of information on which regulatory measures may be based. Other components of regulatory decisions vary from one situation to another and from country to country, responding to different socioeconomic and national priorities. **Therefore, no recommendation is given with regard to regulation or legislation, which are the responsibility of individual governments and/or other international organizations.**

The *IARC Monographs* are recognized as an authoritative source of information on the carcinogenicity of a wide range of human exposures. A survey of users in 1988 indicated that the *Monographs* are consulted by various agencies in 57 countries. About 2500 copies of each volume are printed, for distribution to governments, regulatory bodies and interested scientists. The Monographs are also available from IARC*Press* in Lyon and via the Marketing and Dissemination (MDI) of the World Health Organization in Geneva.

3. SELECTION OF TOPICS FOR MONOGRAPHS

Topics are selected on the basis of two main criteria: (a) there is evidence of human exposure, and (b) there is some evidence or suspicion of carcinogenicity. The term 'agent' is used to include individual chemical compounds, groups of related chemical compounds, physical agents (such as radiation) and biological factors (such as viruses). Exposures to mixtures of agents may occur in occupational exposures and as a result of personal and cultural habits (like smoking and dietary practices). Chemical analogues and compounds with biological or physical characteristics similar to those of suspected carcinogens may also be considered, even in the absence of data on a possible carcinogenic effect in humans or experimental animals.

The scientific literature is surveyed for published data relevant to an assessment of carcinogenicity. The IARC information bulletins on agents being tested for carcinogenicity (IARC, 1973–1996) and directories of on-going research in cancer epidemiology (IARC, 1976–1996) often indicate exposures that may be scheduled for future meetings. Ad-hoc working groups convened by IARC in 1984, 1989, 1991, 1993 and

1998 gave recommendations as to which agents should be evaluated in the IARC Monographs series (IARC, 1984, 1989, 1991b, 1993, 1998a,b).

As significant new data on subjects on which monographs have already been prepared become available, re-evaluations are made at subsequent meetings, and revised monographs are published.

4. DATA FOR MONOGRAPHS

The *Monographs* do not necessarily cite all the literature concerning the subject of an evaluation. Only those data considered by the Working Group to be relevant to making the evaluation are included.

With regard to biological and epidemiological data, only reports that have been published or accepted for publication in the openly available scientific literature are reviewed by the working groups. In certain instances, government agency reports that have undergone peer review and are widely available are considered. Exceptions may be made on an ad-hoc basis to include unpublished reports that are in their final form and publicly available, if their inclusion is considered pertinent to making a final evaluation (see pp. 25–27). In the sections on chemical and physical properties, on analysis, on production and use and on occurrence, unpublished sources of information may be used.

5. THE WORKING GROUP

Reviews and evaluations are formulated by a working group of experts. The tasks of the group are: (i) to ascertain that all appropriate data have been collected; (ii) to select the data relevant for the evaluation on the basis of scientific merit; (iii) to prepare accurate summaries of the data to enable the reader to follow the reasoning of the Working Group; (iv) to evaluate the results of epidemiological and experimental studies on cancer; (v) to evaluate data relevant to the understanding of mechanism of action; and (vi) to make an overall evaluation of the carcinogenicity of the exposure to humans.

Working Group participants who contributed to the considerations and evaluations within a particular volume are listed, with their addresses, at the beginning of each publication. Each participant who is a member of a working group serves as an individual scientist and not as a representative of any organization, government or industry. In addition, nominees of national and international agencies and industrial associations may be invited as observers.

6. WORKING PROCEDURES

Approximately one year in advance of a meeting of a working group, the topics of the monographs are announced and participants are selected by IARC staff in consultation with other experts. Subsequently, relevant biological and epidemiological data are collected by the Carcinogen Identification and Evaluation Unit of IARC from recognized sources of information on carcinogenesis, including data storage and retrieval systems such as MEDLINE and TOXLINE.

For chemicals and some complex mixtures, the major collection of data and the preparation of first drafts of the sections on chemical and physical properties, on analysis, on production and use and on occurrence are carried out under a separate contract funded by the United States National Cancer Institute. Representatives from industrial associations may assist in the preparation of sections on production and use. Information on production and trade is obtained from governmental and trade publications and, in some cases, by direct contact with industries. Separate production data on some agents may not be available because their publication could disclose confidential information. Information on uses may be obtained from published sources but is often complemented by direct contact with manufacturers. Efforts are made to supplement this information with data from other national and international sources.

Six months before the meeting, the material obtained is sent to meeting participants, or is used by IARC staff, to prepare sections for the first drafts of monographs. The first drafts are compiled by IARC staff and sent before the meeting to all participants of the Working Group for review.

The Working Group meets in Lyon for seven to eight days to discuss and finalize the texts of the monographs and to formulate the evaluations. After the meeting, the master copy of each monograph is verified by consulting the original literature, edited and prepared for publication. The aim is to publish monographs within six months of the Working Group meeting.

The available studies are summarized by the Working Group, with particular regard to the qualitative aspects discussed below. In general, numerical findings are indicated as they appear in the original report; units are converted when necessary for easier comparison. The Working Group may conduct additional analyses of the published data and use them in their assessment of the evidence; the results of such supplementary analyses are given in square brackets. When an important aspect of a study, directly impinging on its interpretation, should be brought to the attention of the reader, a comment is given in square brackets.

7. EXPOSURE DATA

Sections that indicate the extent of past and present human exposure, the sources of exposure, the people most likely to be exposed and the factors that contribute to the exposure are included at the beginning of each monograph.

Most monographs on individual chemicals, groups of chemicals or complex mixtures include sections on chemical and physical data, on analysis, on production and use and on occurrence. In monographs on, for example, physical agents, occupational exposures and cultural habits, other sections may be included, such as: historical perspectives, description of an industry or habit, chemistry of the complex mixture or taxonomy. Monographs on biological agents have sections on structure and biology, methods of detection, epidemiology of infection and clinical disease other than cancer.

For chemical exposures, the Chemical Abstracts Services Registry Number, the latest Chemical Abstracts primary name and the IUPAC systematic name are recorded; other synonyms are given, but the list is not necessarily comprehensive. For biological agents, taxonomy and structure are described, and the degree of variability is given, when applicable.

Information on chemical and physical properties and, in particular, data relevant to identification, occurrence and biological activity are included. For biological agents, mode of replication, life cycle, target cells, persistence and latency and host response are given. A description of technical products of chemicals includes trade names, relevant specifications and available information on composition and impurities. Some of the trade names given may be those of mixtures in which the agent being evaluated is only one of the ingredients.

The purpose of the section on analysis or detection is to give the reader an overview of current methods, with emphasis on those widely used for regulatory purposes. Methods for monitoring human exposure are also given, when available. No critical evaluation or recommendation of any of the methods is meant or implied. The IARC published a series of volumes, *Environmental Carcinogens: Methods of Analysis and Exposure Measurement* (IARC, 1978–93), that describe validated methods for analysing a wide variety of chemicals and mixtures. For biological agents, methods of detection and exposure assessment are described, including their sensitivity, specificity and reproducibility.

The dates of first synthesis and of first commercial production of a chemical or mixture are provided; for agents which do not occur naturally, this information may allow a reasonable estimate to be made of the date before which no human exposure to the agent could have occurred. The dates of first reported occurrence of an exposure are also provided. In addition, methods of synthesis used in past and present commercial production and different methods of production which may give rise to different impurities are described.

Data on production, international trade and uses are obtained for representative regions, which usually include Europe, Japan and the United States of America. It should not, however, be inferred that those areas or nations are necessarily the sole or major sources or users of the agent. Some identified uses may not be current or major applications, and the coverage is not necessarily comprehensive. In the case of drugs, mention of their therapeutic uses does not necessarily represent current practice, nor does it imply judgement as to their therapeutic efficacy.

Information on the occurrence of an agent or mixture in the environment is obtained from data derived from the monitoring and surveillance of levels in occupational environments, air, water, soil, foods and animal and human tissues. When available, data on the generation, persistence and bioaccumulation of the agent are also included. In the case of mixtures, industries, occupations or processes, information is given about all agents present. For processes, industries and occupations, a historical description is also given, noting variations in chemical composition, physical properties and levels of occupational exposure with time and place. For biological agents, the epidemiology of infection is described.

Statements concerning regulations and guidelines (e.g., pesticide registrations, maximal levels permitted in foods, occupational exposure limits) are included for some countries as indications of potential exposures, but they may not reflect the most recent situation, since such limits are continuously reviewed and modified. The absence of information on regulatory status for a country should not be taken to imply that that country does not have regulations with regard to the exposure. For biological agents, legislation and control, including vaccines and therapy, are described.

8. STUDIES OF CANCER IN HUMANS

(*a*) *Types of studies considered*

Three types of epidemiological studies of cancer contribute to the assessment of carcinogenicity in humans — cohort studies, case–control studies and correlation (or ecological) studies. Rarely, results from randomized trials may be available. Case series and case reports of cancer in humans may also be reviewed.

Cohort and case–control studies relate the exposures under study to the occurrence of cancer in individuals and provide an estimate of relative risk (ratio of incidence or mortality in those exposed to incidence or mortality in those not exposed) as the main measure of association.

In correlation studies, the units of investigation are usually whole populations (e.g. in particular geographical areas or at particular times), and cancer frequency is related to a summary measure of the exposure of the population to the agent, mixture or exposure circumstance under study. Because individual exposure is not documented, however, a causal relationship is less easy to infer from correlation studies than from cohort and case–control studies. Case reports generally arise from a suspicion, based on clinical experience, that the concurrence of two events — that is, a particular exposure and occurrence of a cancer — has happened rather more frequently than would be expected by chance. Case reports usually lack complete ascertainment of cases in any population, definition or enumeration of the population at risk and estimation of the expected number of cases in the absence of exposure. The uncertainties surrounding interpretation of case reports and correlation studies make them inadequate, except in rare instances, to form the sole basis for inferring a causal relationship. When taken together with case–control and cohort studies, however, relevant case reports or correlation studies may add materially to the judgement that a causal relationship is present.

Epidemiological studies of benign neoplasms, presumed preneoplastic lesions and other end-points thought to be relevant to cancer are also reviewed by working groups. They may, in some instances, strengthen inferences drawn from studies of cancer itself.

(*b*) *Quality of studies considered*

The Monographs are not intended to summarize all published studies. Those that are judged to be inadequate or irrelevant to the evaluation are generally omitted. They may be mentioned briefly, particularly when the information is considered to be a useful supplement to that in other reports or when they provide the only data available. Their inclusion does not imply acceptance of the adequacy of the study design or of the analysis and interpretation of the results, and limitations are clearly outlined in square brackets at the end of the study description.

It is necessary to take into account the possible roles of bias, confounding and chance in the interpretation of epidemiological studies. By 'bias' is meant the operation of factors in study design or execution that lead erroneously to a stronger or weaker association than in fact exists between disease and an agent, mixture or exposure circumstance. By 'confounding' is meant a situation in which the relationship with disease is made to appear stronger or weaker than it truly is as a result of an association between the apparent causal factor and another factor that is associated with either an increase or decrease in the incidence of the disease. In evaluating the extent to which these factors have been minimized in an individual study, working groups consider a number of aspects of design and analysis as described in the report of the study. Most of these considerations apply equally to case–control, cohort and correlation studies. Lack of clarity of any of these aspects in the reporting of a study can decrease its credibility and the weight given to it in the final evaluation of the exposure.

Firstly, the study population, disease (or diseases) and exposure should have been well defined by the authors. Cases of disease in the study population should have been identified in a way that was independent of the exposure of interest, and exposure should have been assessed in a way that was not related to disease status.

Secondly, the authors should have taken account in the study design and analysis of other variables that can influence the risk of disease and may have been related to the exposure of interest. Potential confounding by such variables should have been dealt with either in the design of the study, such as by matching, or in the analysis, by statistical adjustment. In cohort studies, comparisons with local rates of disease may be more appropriate than those with national rates. Internal comparisons of disease frequency among individuals at different levels of exposure should also have been made in the study.

Thirdly, the authors should have reported the basic data on which the conclusions are founded, even if sophisticated statistical analyses were employed. At the very least, they should have given the numbers of exposed and unexposed cases and controls in a case–control study and the numbers of cases observed and expected in a cohort study. Further tabulations by time since exposure began and other temporal factors are also important. In a cohort study, data on all cancer sites and all causes of death should have been given, to reveal the possibility of reporting bias. In a case–control study, the effects of investigated factors other than the exposure of interest should have been reported.

Finally, the statistical methods used to obtain estimates of relative risk, absolute rates of cancer, confidence intervals and significance tests, and to adjust for confounding should have been clearly stated by the authors. The methods used should preferably have been the generally accepted techniques that have been refined since the mid-1970s. These methods have been reviewed for case–control studies (Breslow & Day, 1980) and for cohort studies (Breslow & Day, 1987).

(*c*) *Inferences about mechanism of action*

Detailed analyses of both relative and absolute risks in relation to temporal variables, such as age at first exposure, time since first exposure, duration of exposure, cumulative exposure and time since exposure ceased, are reviewed and summarized when available. The analysis of temporal relationships can be useful in formulating models of carcinogenesis. In particular, such analyses may suggest whether a carcinogen acts early or late in the process of carcinogenesis, although at best they allow only indirect inferences about the mechanism of action. Special attention is given to measurements of biological markers of carcinogen exposure or action, such as DNA or protein adducts, as well as markers of early steps in the carcinogenic process, such as proto-oncogene mutation, when these are incorporated into epidemiological studies focused on cancer incidence or mortality. Such measurements may allow inferences to be made about putative mechanisms of action (IARC, 1991a; Vainio *et al.*, 1992).

(*d*) *Criteria for causality*

After the individual epidemiological studies of cancer have been summarized and the quality assessed, a judgement is made concerning the strength of evidence that the agent, mixture or exposure circumstance in question is carcinogenic for humans. In making its judgement, the Working Group considers several criteria for causality. A strong association (a large relative risk) is more likely to indicate causality than a weak association, although it is recognized that relative risks of small magnitude do not imply lack of causality and may be important if the disease is common. Associations that are replicated in several studies of the same design or using different epidemiological approaches or under different circumstances of exposure are more likely to represent a causal relationship than isolated observations from single studies. If there are inconsistent results among investigations, possible reasons are sought (such as differences in amount of exposure), and results of studies judged to be of high quality are given more weight than those of studies judged to be methodologically less sound. When suspicion of carcinogenicity arises largely from a single study, these data are not combined with those from later studies in any subsequent reassessment of the strength of the evidence.

If the risk of the disease in question increases with the amount of exposure, this is considered to be a strong indication of causality, although absence of a graded response is not necessarily evidence against a causal relationship. Demonstration of a decline in risk after cessation of or reduction in exposure in individuals or in whole populations also supports a causal interpretation of the findings.

Although a carcinogen may act upon more than one target, the specificity of an association (an increased occurrence of cancer at one anatomical site or of one morphological type) adds plausibility to a causal relationship, particularly when excess cancer occurrence is limited to one morphological type within the same organ.

Although rarely available, results from randomized trials showing different rates among exposed and unexposed individuals provide particularly strong evidence for causality.

When several epidemiological studies show little or no indication of an association between an exposure and cancer, the judgement may be made that, in the aggregate, they show evidence of lack of carcinogenicity. Such a judgement requires first of all that the studies giving rise to it meet, to a sufficient degree, the standards of design and analysis described above. Specifically, the possibility that bias, confounding or misclassification of exposure or outcome could explain the observed results should be considered and excluded with reasonable certainty. In addition, all studies that are judged to be methodologically sound should be consistent with a relative risk of unity for any observed level of exposure and, when considered together, should provide a pooled estimate of relative risk which is at or near unity and has a narrow confidence interval, due to sufficient population size. Moreover, no individual study nor the pooled results of all the studies should show any consistent tendency for the relative risk of cancer to increase with increasing level of exposure. It is important to note that evidence of lack of carcinogenicity obtained in this way from several epidemiological studies can apply only to the type(s) of cancer studied and to dose levels and intervals between first exposure and observation of disease that are the same as or less than those observed in all the studies. Experience with human cancer indicates that, in some cases, the period from first exposure to the development of clinical cancer is seldom less than 20 years; studies with latent periods substantially shorter than 30 years cannot provide evidence for lack of carcinogenicity.

9. STUDIES OF CANCER IN EXPERIMENTAL ANIMALS

All known human carcinogens that have been studied adequately in experimental animals have produced positive results in one or more animal species (Wilbourn *et al.*, 1986; Tomatis *et al.*, 1989). For several agents (aflatoxins, 4-aminobiphenyl, azathioprine, betel quid with tobacco, bischloromethyl ether and chloromethyl methyl ether (technical grade), chlorambucil, chlornaphazine, ciclosporin, coal-tar pitches, coal-tars, combined oral contraceptives, cyclophosphamide, diethylstilboestrol, melphalan, 8-methoxypsoralen plus ultraviolet A radiation, mustard gas, myleran, 2-naphthylamine, nonsteroidal estrogens, estrogen replacement therapy/steroidal estrogens, solar radiation, thiotepa and vinyl chloride), carcinogenicity in experimental animals was established or highly suspected before epidemiological studies confirmed their carcinogenicity in humans (Vainio *et al.*, 1995). Although this association cannot establish that all agents and mixtures that cause cancer in experimental animals also cause cancer in humans, nevertheless, **in the absence of adequate data on humans, it is biologically plausible and prudent to regard agents and mixtures for which there is *sufficient evidence* (see p. 24) of carcinogenicity in experimental animals as if they presented a carcinogenic risk to humans**. The possibility that a given agent may cause cancer through a species-specific mechanism which does not operate in humans (see p. 27) should also be taken into consideration.

The nature and extent of impurities or contaminants present in the chemical or mixture being evaluated are given when available. Animal strain, sex, numbers per group, age at start of treatment and survival are reported.

Other types of studies summarized include: experiments in which the agent or mixture was administered in conjunction with known carcinogens or factors that modify carcinogenic effects; studies in which the end-point was not cancer but a defined precancerous lesion; and experiments on the carcinogenicity of known metabolites and derivatives.

For experimental studies of mixtures, consideration is given to the possibility of changes in the physicochemical properties of the test substance during collection, storage, extraction, concentration and delivery. Chemical and toxicological interactions of the components of mixtures may result in nonlinear dose–response relationships.

An assessment is made as to the relevance to human exposure of samples tested in experimental animals, which may involve consideration of: (i) physical and chemical characteristics, (ii) constituent substances that indicate the presence of a class of substances, (iii) the results of tests for genetic and related effects, including studies on DNA adduct formation, proto-oncogene mutation and expression and suppressor gene inactivation. The relevance of results obtained, for example, with animal viruses analogous to the virus being evaluated in the monograph must also be considered. They may provide biological and mechanistic information relevant to the understanding of the process of carcinogenesis in humans and may strengthen the plausibility of a conclusion that the biological agent under evaluation is carcinogenic in humans.

(*a*) *Qualitative aspects*

An assessment of carcinogenicity involves several considerations of qualitative importance, including (i) the experimental conditions under which the test was performed, including route and schedule of exposure, species, strain, sex, age, duration of follow-up; (ii) the consistency of the results, for example, across species and target organ(s); (iii) the spectrum of neoplastic response, from preneoplastic lesions and benign tumours to malignant neoplasms; and (iv) the possible role of modifying factors.

As mentioned earlier (p. 11), the *Monographs* are not intended to summarize all published studies. Those studies in experimental animals that are inadequate (e.g., too short a duration, too few animals, poor survival; see below) or are judged irrelevant to the evaluation are generally omitted. Guidelines for conducting adequate long-term carcinogenicity experiments have been outlined (e.g. Montesano *et al.*, 1986).

Considerations of importance to the Working Group in the interpretation and evaluation of a particular study include: (i) how clearly the agent was defined and, in the case of mixtures, how adequately the sample characterization was reported; (ii) whether the dose was adequately monitored, particularly in inhalation experiments; (iii) whether the doses and duration of treatment were appropriate and whether the survival of treated animals was similar to that of controls; (iv) whether there were adequate numbers of animals per group; (v) whether animals of each sex were used; (vi) whether animals were allocated randomly to groups; (vii) whether the duration of observation was adequate; and (viii) whether the data were adequately reported. If available, recent data on the incidence of specific tumours in historical controls, as well as in concurrent controls, should be taken into account in the evaluation of tumour response.

When benign tumours occur together with and originate from the same cell type in an organ or tissue as malignant tumours in a particular study and appear to represent a stage in the progression to malignancy, it may be valid to combine them in assessing tumour incidence (Huff *et al.*, 1989). The occurrence of lesions presumed to be preneoplastic may in certain instances aid in assessing the biological plausibility of any neoplastic response observed. If an agent or mixture induces only benign neoplasms that appear to be end-points that do not readily progress to malignancy, it should nevertheless be suspected of being a carcinogen and requires further investigation.

(*b*) *Quantitative aspects*

The probability that tumours will occur may depend on the species, sex, strain and age of the animal, the dose of the carcinogen and the route and length of exposure. Evidence of an increased incidence of neoplasms with increased level of exposure strengthens the inference of a causal association between the exposure and the development of neoplasms.

The form of the dose–response relationship can vary widely, depending on the particular agent under study and the target organ. Both DNA damage and increased cell division are important aspects of carcinogenesis, and cell proliferation is a strong determinant of dose–response relationships for some carcinogens (Cohen & Ellwein, 1990). Since many chemicals require metabolic activation before being converted into their reactive intermediates, both metabolic and pharmacokinetic aspects are important in determining the dose–response pattern. Saturation of steps such as absorption, activation, inactivation and elimination may produce nonlinearity in the dose–response relationship, as could saturation of processes such as DNA repair (Hoel *et al.*, 1983; Gart *et al.*, 1986).

(c) *Statistical analysis of long-term experiments in animals*

Factors considered by the Working Group include the adequacy of the information given for each treatment group: (i) the number of animals studied and the number examined histologically, (ii) the number of animals with a given tumour type and (iii) length of survival. The statistical methods used should be clearly stated and should be the generally accepted techniques refined for this purpose (Peto *et al.*, 1980; Gart *et al.*, 1986). When there is no difference in survival between control and treatment groups, the Working Group usually compares the proportions of animals developing each tumour type in each of the groups. Otherwise, consideration is given as to whether or not appropriate adjustments have been made for differences in survival. These adjustments can include: comparisons of the proportions of tumour-bearing animals among the effective number of animals (alive at the time the first tumour is discovered), in the case where most differences in survival occur before tumours appear; life-table methods, when tumours are visible or when they may be considered 'fatal' because mortality rapidly follows tumour development; and the Mantel-Haenszel test or logistic regression, when occult tumours do not affect the animals' risk of dying but are 'incidental' findings at autopsy.

In practice, classifying tumours as fatal or incidental may be difficult. Several survival-adjusted methods have been developed that do not require this distinction (Gart *et al.*, 1986), although they have not been fully evaluated.

10. OTHER DATA RELEVANT TO AN EVALUATION OF CARCINOGENICITY AND ITS MECHANISMS

In coming to an overall evaluation of carcinogenicity in humans (see pp. 25–27), the Working Group also considers related data. The nature of the information selected for the summary depends on the agent being considered.

For chemicals and complex mixtures of chemicals such as those in some occupational situations or involving cultural habits (e.g. tobacco smoking), the other data considered to be relevant are divided into those on absorption, distribution, metabolism and excretion; toxic effects; reproductive and developmental effects; and genetic and related effects.

Concise information is given on absorption, distribution (including placental transfer) and excretion in both humans and experimental animals. Kinetic factors that may affect the dose–response relationship, such as saturation of uptake, protein binding, metabolic activation, detoxification and DNA repair processes, are mentioned. Studies that indicate the metabolic fate of the agent in humans and in experimental animals are summarized briefly, and comparisons of data on humans and on animals are made when possible. Comparative information on the relationship between exposure and the dose that reaches the target site may be of particular importance for extrapolation between species. Data are given on acute and chronic toxic effects (other than cancer), such as organ toxicity, increased cell proliferation, immunotoxicity and endocrine effects. The presence and toxicological significance of cellular receptors is described. Effects on reproduction, teratogenicity, fetotoxicity and embryotoxicity are also summarized briefly.

Tests of genetic and related effects are described in view of the relevance of gene mutation and chromosomal damage to carcinogenesis (Vainio *et al.*, 1992; McGregor *et al.*, 1999). The adequacy of the reporting of sample characterization is considered and, where necessary, commented upon; with regard to complex mixtures, such comments are similar to those described for animal carcinogenicity tests on p. 18. The available data are interpreted critically by phylogenetic group according to the end-points detected, which may include DNA damage, gene mutation, sister chromatid exchange, micronucleus formation, chromosomal aberrations, aneuploidy and cell transformation. The concentrations employed are given, and mention is made of whether use of an exogenous metabolic system *in vitro* affected the test result. These data are given as listings of test systems, data and references. The data on genetic and related effects presented in the *Monographs* are also available in the form of genetic activity profiles (GAP) prepared in collaboration with the United States Environmental Protection Agency (EPA) (see also Waters *et al.*, 1987) using software for personal computers that are Microsoft Windows® compatible. The EPA/IARC GAP software and database may be downloaded free of charge from *www.epa.gov/gapdb*.

Positive results in tests using prokaryotes, lower eukaryotes, plants, insects and cultured mammalian cells suggest that genetic and related effects could occur in mammals. Results from such tests may also give information about the types of genetic effect produced and about the involvement of metabolic activation. Some end-points described are clearly genetic in nature (e.g., gene mutations and chromosomal aberrations), while others are to a greater or lesser degree associated with genetic effects (e.g. unscheduled DNA synthesis). In-vitro tests for tumour-promoting activity and for cell transformation may be sensitive to changes that are not necessarily the result of genetic alterations but that may have specific relevance to the process of carcinogenesis. A critical appraisal of these tests has been published (Montesano *et al.*, 1986).

Genetic or other activity detected in experimental mammals and humans is regarded as being of greater relevance than that in other organisms. The demonstration that an agent or mixture can induce gene and chromosomal mutations in whole mammals indicates that it may have carcinogenic activity, although this activity may not be detectably expressed in any or all species. Relative potency in tests for mutagenicity and related effects is not a reliable indicator of carcinogenic potency. Negative results in tests for mutagenicity in selected tissues from animals treated *in vivo* provide less weight, partly because they do not exclude the possibility of an effect in tissues other than those examined. Moreover, negative results in short-term tests with genetic end-points cannot be considered to provide evidence to rule out carcinogenicity of agents or mixtures that act through other mechanisms (e.g. receptor-mediated effects, cellular toxicity with regenerative proliferation, peroxisome proliferation) (Vainio *et al.*, 1992). Factors that may lead to misleading results in short-term tests have been discussed in detail elsewhere (Montesano *et al.*, 1986).

When available, data relevant to mechanisms of carcinogenesis that do not involve structural changes at the level of the gene are also described.

The adequacy of epidemiological studies of reproductive outcome and genetic and related effects in humans is evaluated by the same criteria as are applied to epidemiological studies of cancer.

Structure–activity relationships that may be relevant to an evaluation of the carcinogenicity of an agent are also described.

For biological agents — viruses, bacteria and parasites — other data relevant to carcinogenicity include descriptions of the pathology of infection, molecular biology (integration and expression of viruses, and any genetic alterations seen in human tumours) and other observations, which might include cellular and tissue responses to infection, immune response and the presence of tumour markers.

11. SUMMARY OF DATA REPORTED

In this section, the relevant epidemiological and experimental data are summarized. Only reports, other than in abstract form, that meet the criteria outlined on p. 11 are considered for evaluating carcinogenicity. Inadequate studies are generally not summarized: such studies are usually identified by a square-bracketed comment in the preceding text.

(*a*) *Exposure*

Human exposure to chemicals and complex mixtures is summarized on the basis of elements such as production, use, occurrence in the environment and determinations in human tissues and body fluids. Quantitative data are given when available. Exposure to biological agents is described in terms of transmission and prevalence of infection.

(*b*) *Carcinogenicity in humans*

Results of epidemiological studies that are considered to be pertinent to an assessment of human carcinogenicity are summarized. When relevant, case reports and correlation studies are also summarized.

(*c*) *Carcinogenicity in experimental animals*

Data relevant to an evaluation of carcinogenicity in animals are summarized. For each animal species and route of administration, it is stated whether an increased incidence of neoplasms or preneoplastic lesions was observed, and the tumour sites are indicated. If the agent or mixture produced tumours after prenatal exposure or in single-dose experiments, this is also indicated. Negative findings are also summarized. Dose–response and other quantitative data may be given when available.

(*d*) *Other data relevant to an evaluation of carcinogenicity and its mechanisms*

Data on biological effects in humans that are of particular relevance are summarized. These may include toxicological, kinetic and metabolic considerations and evidence of DNA binding, persistence of DNA lesions or genetic damage in exposed humans. Toxicological information, such as that on cytotoxicity and regeneration, receptor binding and hormonal and immunological effects, and data on kinetics and metabolism in experimental animals are given when considered relevant to the possible mechanism of the carcinogenic action of the agent. The results of tests for genetic and related effects are summarized for whole mammals, cultured mammalian cells and nonmammalian systems.

When available, comparisons of such data for humans and for animals, and particularly animals that have developed cancer, are described.

Structure–activity relationships are mentioned when relevant.

For the agent, mixture or exposure circumstance being evaluated, the available data on end-points or other phenomena relevant to mechanisms of carcinogenesis from studies in humans, experimental animals and tissue and cell test systems are summarized within one or more of the following descriptive dimensions:

(i) Evidence of genotoxicity (structural changes at the level of the gene): for example, structure–activity considerations, adduct formation, mutagenicity (effect on specific genes), chromosomal mutation/aneuploidy

(ii) Evidence of effects on the expression of relevant genes (functional changes at the intracellular level): for example, alterations to the structure or quantity of the product of a proto-oncogene or tumour-suppressor gene, alterations to metabolic activation/inactivation/DNA repair

(iii) Evidence of relevant effects on cell behaviour (morphological or behavioural changes at the cellular or tissue level): for example, induction of mitogenesis, compensatory cell proliferation, preneoplasia and hyperplasia, survival of premalignant or malignant cells (immortalization, immunosuppression), effects on metastatic potential

(iv) Evidence from dose and time relationships of carcinogenic effects and interactions between agents: for example, early/late stage, as inferred from epidemiological studies; initiation/promotion/progression/malignant conversion, as defined in animal carcinogenicity experiments; toxicokinetics

These dimensions are not mutually exclusive, and an agent may fall within more than one of them. Thus, for example, the action of an agent on the expression of relevant genes could be summarized under both the first and second dimensions, even if it were known with reasonable certainty that those effects resulted from genotoxicity.

12. EVALUATION

Evaluations of the strength of the evidence for carcinogenicity arising from human and experimental animal data are made, using standard terms.

It is recognized that the criteria for these evaluations, described below, cannot encompass all of the factors that may be relevant to an evaluation of carcinogenicity. In considering all of the relevant scientific data, the Working Group may assign the agent, mixture or exposure circumstance to a higher or lower category than a strict interpretation of these criteria would indicate.

(*a*) *Degrees of evidence for carcinogenicity in humans and in experimental animals and supporting evidence*

These categories refer only to the strength of the evidence that an exposure is carcinogenic and not to the extent of its carcinogenic activity (potency) nor to the mechanisms involved. A classification may change as new information becomes available.

An evaluation of degree of evidence, whether for a single agent or a mixture, is limited to the materials tested, as defined physically, chemically or biologically. When the agents evaluated are considered by the Working Group to be sufficiently closely related, they may be grouped together for the purpose of a single evaluation of degree of evidence.

(i) *Carcinogenicity in humans*

The applicability of an evaluation of the carcinogenicity of a mixture, process, occupation or industry on the basis of evidence from epidemiological studies depends on the variability over time and place of the mixtures, processes, occupations and industries. The Working Group seeks to identify the specific exposure, process or activity which is considered most likely to be responsible for any excess risk. The evaluation is focused as narrowly as the available data on exposure and other aspects permit.

The evidence relevant to carcinogenicity from studies in humans is classified into one of the following categories:

Sufficient evidence of carcinogenicity: The Working Group considers that a causal relationship has been established between exposure to the agent, mixture or exposure circumstance and human cancer. That is, a positive relationship has been observed between the exposure and cancer in studies in which chance, bias and confounding could be ruled out with reasonable confidence.

Limited evidence of carcinogenicity: A positive association has been observed between exposure to the agent, mixture or exposure circumstance and cancer for which a causal interpretation is considered by the Working Group to be credible, but chance, bias or confounding could not be ruled out with reasonable confidence.

Inadequate evidence of carcinogenicity: The available studies are of insufficient quality, consistency or statistical power to permit a conclusion regarding the presence or absence of a causal association between exposure and cancer, or no data on cancer in humans are available.

Evidence suggesting lack of carcinogenicity: There are several adequate studies covering the full range of levels of exposure that human beings are known to encounter, which are mutually consistent in not showing a positive association between exposure to the agent, mixture or exposure circumstance and any studied cancer at any observed level of exposure. A conclusion of 'evidence suggesting lack of carcinogenicity' is inevitably limited to the cancer sites, conditions and levels of exposure and length of observation covered by the available studies. In addition, the possibility of a very small risk at the levels of exposure studied can never be excluded.

In some instances, the above categories may be used to classify the degree of evidence related to carcinogenicity in specific organs or tissues.

(ii) *Carcinogenicity in experimental animals*

The evidence relevant to carcinogenicity in experimental animals is classified into one of the following categories:

Sufficient evidence of carcinogenicity: The Working Group considers that a causal relationship has been established between the agent or mixture and an increased incidence of malignant neoplasms or of an appropriate combination of benign and malignant neoplasms in (a) two or more species of animals or (b) in two or more independent studies in one species carried out at different times or in different laboratories or under different protocols.

Exceptionally, a single study in one species might be considered to provide sufficient evidence of carcinogenicity when malignant neoplasms occur to an unusual degree with regard to incidence, site, type of tumour or age at onset.

Limited evidence of carcinogenicity: The data suggest a carcinogenic effect but are limited for making a definitive evaluation because, e.g. (a) the evidence of carcinogenicity is restricted to a single experiment; or (b) there are unresolved questions regarding the adequacy of the design, conduct or interpretation of the study; or (c) the agent or mixture increases the incidence only of benign neoplasms or lesions of uncertain neoplastic potential, or of certain neoplasms which may occur spontaneously in high incidences in certain strains.

Inadequate evidence of carcinogenicity: The studies cannot be interpreted as showing either the presence or absence of a carcinogenic effect because of major qualitative or quantitative limitations, or no data on cancer in experimental animals are available.

Evidence suggesting lack of carcinogenicity: Adequate studies involving at least two species are available which show that, within the limits of the tests used, the agent or mixture is not carcinogenic. A conclusion of evidence suggesting lack of carcinogenicity is inevitably limited to the species, tumour sites and levels of exposure studied.

(*b*) *Other data relevant to the evaluation of carcinogenicity and its mechanisms*

Other evidence judged to be relevant to an evaluation of carcinogenicity and of sufficient importance to affect the overall evaluation is then described. This may include data on preneoplastic lesions, tumour pathology, genetic and related effects, structure–activity relationships, metabolism and pharmacokinetics, physicochemical parameters and analogous biological agents.

Data relevant to mechanisms of the carcinogenic action are also evaluated. The strength of the evidence that any carcinogenic effect observed is due to a particular mechanism is assessed, using terms such as weak, moderate or strong. Then, the Working Group assesses if that particular mechanism is likely to be operative in humans. The strongest indications that a particular mechanism operates in humans come from data on humans or biological specimens obtained from exposed humans. The data may be considered to be especially relevant if they show that the agent in question has caused changes in exposed humans that are on the causal pathway to carcinogenesis. Such data may, however, never become available, because it is at least conceivable that certain compounds may be kept from human use solely on the basis of evidence of their toxicity and/or carcinogenicity in experimental systems.

For complex exposures, including occupational and industrial exposures, the chemical composition and the potential contribution of carcinogens known to be present are considered by the Working Group in its overall evaluation of human carcinogenicity. The Working Group also determines the extent to which the materials tested in experimental systems are related to those to which humans are exposed.

(c) Overall evaluation

Finally, the body of evidence is considered as a whole, in order to reach an overall evaluation of the carcinogenicity to humans of an agent, mixture or circumstance of exposure.

An evaluation may be made for a group of chemical compounds that have been evaluated by the Working Group. In addition, when supporting data indicate that other, related compounds for which there is no direct evidence of capacity to induce cancer in humans or in animals may also be carcinogenic, a statement describing the rationale for this conclusion is added to the evaluation narrative; an additional evaluation may be made for this broader group of compounds if the strength of the evidence warrants it.

The agent, mixture or exposure circumstance is described according to the wording of one of the following categories, and the designated group is given. The categorization of an agent, mixture or exposure circumstance is a matter of scientific judgement, reflecting the strength of the evidence derived from studies in humans and in experimental animals and from other relevant data.

Group 1 — The agent (mixture) is carcinogenic to humans.
The exposure circumstance entails exposures that are carcinogenic to humans.

This category is used when there is *sufficient evidence* of carcinogenicity in humans. Exceptionally, an agent (mixture) may be placed in this category when evidence of carcinogenicity in humans is less than sufficient but there is *sufficient evidence* of carcinogenicity in experimental animals and strong evidence in exposed humans that the agent (mixture) acts through a relevant mechanism of carcinogenicity.

Group 2

This category includes agents, mixtures and exposure circumstances for which, at one extreme, the degree of evidence of carcinogenicity in humans is almost sufficient, as well as those for which, at the other extreme, there are no human data but for which there is evidence of carcinogenicity in experimental animals. Agents, mixtures and exposure circumstances are assigned to either group 2A (probably carcinogenic to humans) or group 2B (possibly carcinogenic to humans) on the basis of epidemiological and experimental evidence of carcinogenicity and other relevant data.

Group 2A — The agent (mixture) is probably carcinogenic to humans.
The exposure circumstance entails exposures that are probably carcinogenic to humans.

This category is used when there is *limited evidence* of carcinogenicity in humans and *sufficient evidence* of carcinogenicity in experimental animals. In some cases, an agent (mixture) may be classified in this category when there is *inadequate evidence* of carcinogenicity in humans, *sufficient evidence* of carcinogenicity in experimental animals and strong evidence that the carcinogenesis is mediated by a mechanism that also operates in humans. Exceptionally, an agent, mixture or exposure circumstance may be classified in this category solely on the basis of *limited evidence* of carcinogenicity in humans.

Group 2B — The agent (mixture) is possibly carcinogenic to humans.
The exposure circumstance entails exposures that are possibly carcinogenic to humans.

This category is used for agents, mixtures and exposure circumstances for which there is *limited evidence* of carcinogenicity in humans and less than *sufficient evidence* of carcinogenicity in experimental animals. It may also be used when there is *inadequate evidence* of carcinogenicity in humans but there is *sufficient evidence* of carcinogenicity in experimental animals. In some instances, an agent, mixture or exposure circumstance for which there is *inadequate evidence* of carcinogenicity in humans but *limited evidence* of carcinogenicity in experimental animals together with supporting evidence from other relevant data may be placed in this group.

Group 3 — The agent (mixture or exposure circumstance) is not classifiable as to its carcinogenicity to humans.

This category is used most commonly for agents, mixtures and exposure circumstances for which the *evidence of carcinogenicity* is *inadequate* in humans and *inadequate* or *limited* in experimental animals.

Exceptionally, agents (mixtures) for which the *evidence of carcinogenicity* is *inadequate* in humans but *sufficient* in experimental animals may be placed in this category when there is strong evidence that the mechanism of carcinogenicity in experimental animals does not operate in humans.

Agents, mixtures and exposure circumstances that do not fall into any other group are also placed in this category.

Group 4 — The agent (mixture) is probably not carcinogenic to humans.

This category is used for agents or mixtures for which there is *evidence suggesting lack of carcinogenicity* in humans and in experimental animals. In some instances, agents or mixtures for which there is *inadequate evidence* of carcinogenicity in humans but *evidence suggesting lack of carcinogenicity* in experimental animals, consistently and strongly supported by a broad range of other relevant data, may be classified in this group.

13. REFERENCES

Breslow, N.E. & Day, N.E. (1980) *Statistical Methods in Cancer Research*, Vol. 1, *The Analysis of Case–Control Studies* (IARC Scientific Publications No. 32), Lyon, IARC*Press*

Breslow, N.E. & Day, N.E. (1987) *Statistical Methods in Cancer Research*, Vol. 2, *The Design and Analysis of Cohort Studies* (IARC Scientific Publications No. 82), Lyon, IARC*Press*

Cohen, S.M. & Ellwein, L.B. (1990) Cell proliferation in carcinogenesis. *Science*, **249**, 1007–1011

Gart, J.J., Krewski, D., Lee, P.N., Tarone, R.E. & Wahrendorf, J. (1986) *Statistical Methods in Cancer Research*, Vol. 3, *The Design and Analysis of Long-term Animal Experiments* (IARC Scientific Publications No. 79), Lyon, IARC*Press*

Hoel, D.G., Kaplan, N.L. & Anderson, M.W. (1983) Implication of nonlinear kinetics on risk estimation in carcinogenesis. *Science*, **219**, 1032–1037

Huff, J.E., Eustis, S.L. & Haseman, J.K. (1989) Occurrence and relevance of chemically induced benign neoplasms in long-term carcinogenicity studies. *Cancer Metastasis Rev.*, **8**, 1–21

IARC (1973–1996) *Information Bulletin on the Survey of Chemicals Being Tested for Carcinogenicity/Directory of Agents Being Tested for Carcinogenicity*, Numbers 1–17, Lyon, IARC*Press*

IARC (1976–1996), Lyon, IARC*Press*

Directory of On-going Research in Cancer Epidemiology 1976. Edited by C.S. Muir & G. Wagner

Directory of On-going Research in Cancer Epidemiology 1977 (IARC Scientific Publications No. 17). Edited by C.S. Muir & G. Wagner

Directory of On-going Research in Cancer Epidemiology 1978 (IARC Scientific Publications No. 26). Edited by C.S. Muir & G. Wagner

Directory of On-going Research in Cancer Epidemiology 1979 (IARC Scientific Publications No. 28). Edited by C.S. Muir & G. Wagner

Directory of On-going Research in Cancer Epidemiology 1980 (IARC Scientific Publications No. 35). Edited by C.S. Muir & G. Wagner

Directory of On-going Research in Cancer Epidemiology 1981 (IARC Scientific Publications No. 38). Edited by C.S. Muir & G. Wagner

Directory of On-going Research in Cancer Epidemiology 1982 (IARC Scientific Publications No. 46). Edited by C.S. Muir & G. Wagner

Directory of On-going Research in Cancer Epidemiology 1983 (IARC Scientific Publications No. 50). Edited by C.S. Muir & G. Wagner

Directory of On-going Research in Cancer Epidemiology 1984 (IARC Scientific Publications No. 62). Edited by C.S. Muir & G. Wagner

Directory of On-going Research in Cancer Epidemiology 1985 (IARC Scientific Publications No. 69). Edited by C.S. Muir & G. Wagner

Directory of On-going Research in Cancer Epidemiology 1986 (IARC Scientific Publications No. 80). Edited by C.S. Muir & G. Wagner

Directory of On-going Research in Cancer Epidemiology 1987 (IARC Scientific Publications No. 86). Edited by D.M. Parkin & J. Wahrendorf

Directory of On-going Research in Cancer Epidemiology 1988 (IARC Scientific Publications No. 93). Edited by M. Coleman & J. Wahrendorf

Directory of On-going Research in Cancer Epidemiology 1989/90 (IARC Scientific Publications No. 101). Edited by M. Coleman & J. Wahrendorf

Directory of On-going Research in Cancer Epidemiology 1991 (IARC Scientific Publications No.110). Edited by M. Coleman & J. Wahrendorf

Directory of On-going Research in Cancer Epidemiology 1992 (IARC Scientific Publications No. 117). Edited by M. Coleman, J. Wahrendorf & E. Démaret

Directory of On-going Research in Cancer Epidemiology 1994 (IARC Scientific Publications No. 130). Edited by R. Sankaranarayanan, J. Wahrendorf & E. Démaret

Directory of On-going Research in Cancer Epidemiology 1996 (IARC Scientific Publications No. 137). Edited by R. Sankaranarayanan, J. Wahrendorf & E. Démaret

IARC (1977) *IARC Monographs Programme on the Evaluation of the Carcinogenic Risk of Chemicals to Humans*. Preamble (IARC intern. tech. Rep. No. 77/002)

IARC (1978) *Chemicals with* Sufficient Evidence *of Carcinogenicity in Experimental Animals* — IARC Monographs *Volumes 1–17* (IARC intern. tech. Rep. No. 78/003)

IARC (1978–1993) *Environmental Carcinogens. Methods of Analysis and Exposure Measurement*, Lyon, IARC*Press*

Vol. 1. Analysis of Volatile Nitrosamines in Food (IARC Scientific Publications No. 18). Edited by R. Preussmann, M. Castegnaro, E.A. Walker & A.E. Wasserman (1978)

Vol. 2. Methods for the Measurement of Vinyl Chloride in Poly(vinyl chloride), Air, Water and Foodstuffs (IARC Scientific Publications No. 22). Edited by D.C.M. Squirrell & W. Thain (1978)

Vol. 3. Analysis of Polycyclic Aromatic Hydrocarbons in Environmental Samples (IARC Scientific Publications No. 29). Edited by M. Castegnaro, P. Bogovski, H. Kunte & E.A. Walker (1979)

Vol. 4. Some Aromatic Amines and Azo Dyes in the General and Industrial Environment (IARC Scientific Publications No. 40). Edited by L. Fishbein, M. Castegnaro, I.K. O'Neill & H. Bartsch (1981)

Vol. 5. Some Mycotoxins (IARC Scientific Publications No. 44). Edited by L. Stoloff, M. Castegnaro, P. Scott, I.K. O'Neill & H. Bartsch (1983)

Vol. 6. N-Nitroso Compounds (IARC Scientific Publications No. 45). Edited by R. Preussmann, I.K. O'Neill, G. Eisenbrand, B. Spiegelhalder & H. Bartsch (1983)

Vol. 7. Some Volatile Halogenated Hydrocarbons (IARC Scientific Publications No. 68). Edited by L. Fishbein & I.K. O'Neill (1985)

Vol. 8. Some Metals: As, Be, Cd, Cr, Ni, Pb, Se, Zn (IARC Scientific Publications No. 71). Edited by I.K. O'Neill, P. Schuller & L. Fishbein (1986)

Vol. 9. Passive Smoking (IARC Scientific Publications No. 81). Edited by I.K. O'Neill, K.D. Brunnemann, B. Dodet & D. Hoffmann (1987)

*Vol. 10. Benzene and Alkylated Benzenes (*IARC Scientific Publications No. 85). Edited by L. Fishbein & I.K. O'Neill (1988)

Vol. 11. Polychlorinated Dioxins and Dibenzofurans (IARC Scientific Publications No. 108). Edited by C. Rappe, H.R. Buser, B. Dodet & I.K. O'Neill (1991)

Vol. 12. Indoor Air (IARC Scientific Publications No. 109). Edited by B. Seifert, H. van de Wiel, B. Dodet & I.K. O'Neill (1993)

IARC (1979) *Criteria to Select Chemicals for* IARC Monographs (IARC intern. tech. Rep. No. 79/003)

IARC (1982) *IARC Monographs on the Evaluation of the Carcinogenic Risk of Chemicals to Humans*, Supplement 4, *Chemicals, Industrial Processes and Industries Associated with Cancer in Humans* (IARC Monographs, Volumes 1 to 29), Lyon, IARC*Press*

IARC (1983) *Approaches to Classifying Chemical Carcinogens According to Mechanism of Action* (IARC intern. tech. Rep. No. 83/001)

IARC (1984) *Chemicals and Exposures to Complex Mixtures Recommended for Evaluation in IARC Monographs and Chemicals and Complex Mixtures Recommended for Long-term Carcinogenicity Testing* (IARC intern. tech. Rep. No. 84/002)

IARC (1987a) *IARC Monographs on the Evaluation of Carcinogenic Risks to Humans*, Supplement 6, *Genetic and Related Effects: An Updating of Selected* IARC Monographs *from Volumes 1 to 42*, Lyon, IARC*Press*

IARC (1987b) *IARC Monographs on the Evaluation of Carcinogenic Risks to Humans*, Supplement 7, *Overall Evaluations of Carcinogenicity: An Updating of* IARC Monographs *Volumes 1 to 42*, Lyon, IARC*Press*

IARC (1988) *Report of an IARC Working Group to Review the Approaches and Processes Used to Evaluate the Carcinogenicity of Mixtures and Groups of Chemicals* (IARC intern. tech. Rep. No. 88/002)

IARC (1989) *Chemicals, Groups of Chemicals, Mixtures and Exposure Circumstances to be Evaluated in Future IARC Monographs, Report of an ad hoc Working Group* (IARC intern. tech. Rep. No. 89/004)

IARC (1991a) *A Consensus Report of an IARC Monographs Working Group on the Use of Mechanisms of Carcinogenesis in Risk Identification* (IARC intern. tech. Rep. No. 91/002)

IARC (1991b) *Report of an ad-hoc* IARC Monographs *Advisory Group on Viruses and Other Biological Agents Such as Parasites* (IARC intern. tech. Rep. No. 91/001)

IARC (1993) *Chemicals, Groups of Chemicals, Complex Mixtures, Physical and Biological Agents and Exposure Circumstances to be Evaluated in Future* IARC Monographs, *Report of an ad-hoc Working Group* (IARC intern. Rep. No. 93/005)

IARC (1998a) *Report of an ad-hoc* IARC Monographs *Advisory Group on Physical Agents* (IARC Internal Report No. 98/002)

IARC (1998b) *Report of an ad-hoc* IARC Monographs *Advisory Group on Priorities for Future Evaluations* (IARC Internal Report No. 98/004)

McGregor, D.B., Rice, J.M. & Venitt, S., eds (1999) *The Use of Short and Medium-term Tests for Carcinogens and Data on Genetic Effects in Carcinogenic Hazard Evaluation* (IARC Scientific Publications No. 146), Lyon, IARC*Press*

Montesano, R., Bartsch, H., Vainio, H., Wilbourn, J. & Yamasaki, H., eds (1986) *Long-term and Short-term Assays for Carcinogenesis — A Critical Appraisal* (IARC Scientific Publications No. 83), Lyon, IARC*Press*

Peto, R., Pike, M.C., Day, N.E., Gray, R.G., Lee, P.N., Parish, S., Peto, J., Richards, S. & Wahrendorf, J. (1980) Guidelines for simple, sensitive significance tests for carcinogenic effects in long-term animal experiments. In: *IARC Monographs on the Evaluation of the Carcinogenic Risk of Chemicals to Humans*, Supplement 2, *Long-term and Short-term Screening Assays for Carcinogens: A Critical Appraisal*, Lyon, IARC*Press*, pp. 311–426

Tomatis, L., Aitio, A., Wilbourn, J. & Shuker, L. (1989) Human carcinogens so far identified. *Jpn. J. Cancer Res.*, **80**, 795–807

Vainio, H., Magee, P.N., McGregor, D.B. & McMichael, A.J., eds (1992) *Mechanisms of Carcinogenesis in Risk Identification* (IARC Scientific Publications No. 116), Lyon, IARC*Press*

Vainio, H., Wilbourn, J.D., Sasco, A.J., Partensky, C., Gaudin, N., Heseltine, E. & Eragne, I. (1995) Identification of human carcinogenic risk in IARC Monographs. *Bull. Cancer*, **82**, 339–348 (in French)

Waters, M.D., Stack, H.F., Brady, A.L., Lohman, P.H.M., Haroun, L. & Vainio, H. (1987) Appendix 1. Activity profiles for genetic and related tests. In: *IARC Monographs on the Evaluation of Carcinogenic Risks to Humans*, Suppl. 6, *Genetic and Related Effects: An Updating of Selected IARC Monographs from Volumes 1 to 42*, Lyon, IARC*Press*, pp. 687–696

Wilbourn, J., Haroun, L., Heseltine, E., Kaldor, J., Partensky, C. & Vainio, H. (1986) Response of experimental animals to human carcinogens: an analysis based upon the IARC Monographs Programme. *Carcinogenesis*, **7**, 1853–1863

GENERAL REMARKS ON THE SUBSTANCES CONSIDERED

This volume of the *IARC Monographs* considers some traditional herbal medicines, including extracts from certain plants of the genera *Aristolochia, Rubia, Morinda* and *Senecio;* some mycotoxins, specifically aflatoxins and fumonisin B_1; and two industrial chemicals, naphthalene and styrene. Of these, the *Monographs* have previously evaluated several of the pyrrolizidine alkaloids that occur in certain species of *Senecio*, *Crotalaria* and other plant genera, including riddelliine (IARC, 1976, 1987); various mycotoxins, including the aflatoxins (IARC, 1993a) and the family of mycotoxins to which fumonisin B_1 belongs (IARC, 1993b); and styrene (IARC, 1994). These previous evaluations are summarized in Table 1.

Table 1. Previous *IARC Monographs* evaluations of substances considered[a]

Agent	Degree of evidence		Overall evaluation	Volume, year
	Human	Animal		
Riddelliine	ND	I	3	**10**, 1976; **S7**, 1987
Aflatoxins, naturally occurring mixtures of	S	S	1	**56**, 1993
Aflatoxin B_1	S	S		
Aflatoxin B_2		L		
Aflatoxin G_1		S		
Aflatoxin G_2		I		
Aflatoxin M_1	I	S	2B	**56**, 1993
Toxins derived from *Fusarium moniliforme* (now known as *F. verticillioides*)	I	S	2B	**56**, 1993
Fumonisin B_1		L		
Fumonisin B_2		I		
Fusarin C		L		
Styrene	I	L	2B[b]	**60**, 1994

[a] Abbreviations: ND, no data; I, inadequate; S7, Supplement 7 (IARC, 1987); S, sufficient; L, limited

[b] The evaluation was upgraded taking into consideration other relevant data on genetic and related effects.

Since these previous reviews, new data have become available, and these have been incorporated into the *Monographs* and considered in the evaluations. The existing Group 1 evaluation of naturally occurring aflatoxins was reaffirmed.

Traditional herbal medicines

Traditional herbal medicines encompass an extremely diverse group of preparations, and originate from many different cultures. Many herbal medicines have emerged from healing traditions around the world. Digitalis (from the dried leaf of *Digitalis purpurea* L.) and quinine (from the bark of the cinchona tree, *Cinchona pubescens* Vahl) are well-known examples of valuable therapeutic products of botanical origin. Some herbal products in current use in many parts of the world, such as ginseng (e.g., from *Panax ginseng* C.A. Mey) and valerian (e.g., from *Valeriana officinalis* L.), have long standing for their modest efficacy and few side-effects. Some, however, such as ephedra (e.g., from *Ephedra sinica* Stapf) have been imported from traditional healing systems and then used for indications (weight loss, athletic performance enhancement) never contemplated in the traditions from which they emerged.

Rather few data on possible carcinogenic hazards of any of these substances have been collected until recently. Previous *IARC Monographs* have reviewed a small number of food plants (cycad nuts, bracken fern) and some natural products that occur in these and other plants for which there were published data on carcinogenicity in experimental animals (Table 2). The only substance listed in Table 2 that has been used for medicinal purposes is dantron (1,8-dihydroxyanthraquinone), which was once widely used as a stimulant laxative. The naturally occurring glycoside derivatives of 1,8-dihydroxyanthraquinone are the pharmacologically active constituents of a herbal purgative (laxative) preparation, senna, which is obtained from the dried leaflets or seed pods of the subtropical shrubs, *Cassia (Senna) acutifolia* and *Cassia (Senna) angustifolia* (Brunton, 1996). The cathartic properties of senna have been known for centuries, and were already described in Arabic writings in the ninth century. No data were available to previous working groups on cancer risk in humans exposed to any of the substances in Table 2, except bracken fern. Several of these compounds, however, including dantron, are potent carcinogens in experimental animals.

It is clear from these examples that some natural products derived from plants are carcinogenic, including certain compounds that are present in, and may be the active ingredients of, some traditional herbal medicines. It is a reasonable inference that others may be also. Data relating to three categories of natural products used in herbal medicines are reviewed in separate sections of this volume.

The first section considers the *Aristolochia* species and some of their chemical constituents. Roots of plants of the genus *Aristolochia* have recently been imported from China and sold in Europe in powdered form, to be taken by mouth in capsules as an aid to body weight reduction. Nephrotoxicity and urothelial carcinomas have occurred in individuals who consumed these products. Carcinogenic risks associated with these

Table 2. Previous *IARC Monographs* evaluations of edible plants and of plant-derived substances occurring naturally in food or used for medicinal purposes

Agent	Degree of evidence[c]		Overall evaluation	Volume, year
	Human	Animal		
Bracken fern[a]	I	S	2B	**40**, 1986; **S7**, 1987
Shikimic acid	ND	I	3	**40**, 1986; **S7**, 1987
Ptaquiloside	ND	L	3	**40**, 1986; **S7**, 1987
Carrageenan[a]				
Native	ND	I	3	**31**, 1983; **S7**, 1987
Degraded	ND	S	2B	**31**, 1983; **S7**, 1987
Cycasin[a]	ND	S	2B	**10**, 1976; **S7**, 1987
Safrole[a]	ND	S	2B	**10**, 1976; **S7**, 1987
Dantron[b] (1,8-Dihydroxyanthraquinone)	ND	S	2B	**50**, 1990

Abbreviations: I, inadequate; S, sufficient; S7, Supplement 7 (IARC, 1987); ND, no data; L, limited

[a] Substances occurring naturally in food

[b] Substance used for medicinal purposes

preparations are evaluated for the first time. The carcinogenicity of aristolochic acids (nitrophenanthrene compounds that are natural products of this genus of plants) is also evaluated for the first time in this volume.

The second section deals with some anthraquinone derivatives that are structurally related to the previously evaluated 1,8-dihydroxyanthraquinone. These compounds occur naturally in certain herbaceous plants that have been used in some traditional Oriental medicinal preparations.

The final section of the herbal medicines monograph concerns toxic pyrrolizidine alkaloids including riddelliine that occur in several widely distributed genera of wild plants including *Senecio*, *Crotalaria* and some others. In western North America these plants co-exist with edible forage plants and may be ingested by livestock. The alkaloids may contaminate foods derived from these animals, including meat and milk, and may be found in honey (National Toxicology Program, 2001). Riddelliine occurs in the plant *Senecio longilobus*, which has been used as a herbal tea called 'gordolobo yerba' by members of the Mexican-American community in the south-western region of the USA, and has been linked with acute hepatic veno-occlusive disease (Stillman *et al.*, 1977; Segall & Molyneux, 1978). Pyrrolizidine alkaloids were previously reviewed by the *IARC Monographs* programme (IARC, 1976). Experimental carcinogenicity data on riddelliine available at that time were considered insufficient to evaluate the carcinogenicity of this compound, and there were no data on cancer risks in humans who might have been exposed to it either in food or as a traditional herbal medicine. Riddelliine was subsequently placed in Group 3, *not classifiable as to its carcinogenicity to humans* (IARC, 1987) (Table 1). A recent bioassay of riddelliine for carcinogenicity conducted by the US National Toxicology Program provides important new data on this compound (National Toxicology Program, 2001) and is the basis for its re-evaluation in this volume.

Mycotoxins

Aflatoxins comprise four secondary metabolites of toxins produced by a number of species of *Aspergillus* of which *A. flavus* and *A. parasiticus* are the most common. These fungi and their toxins contaminate maize and peanuts (groundnuts) along with other commodities in the field and when improperly stored. Naturally occurring aflatoxins and aflatoxin B_1 were previously reviewed and evaluated in *IARC Monographs* Volume 56 (IARC, 1993a). Naturally occurring aflatoxins (as a group) were evaluated as *carcinogenic to humans* (Group 1). Aflatoxin M_1, the metabolite of aflatoxin B_1 found in the milk of lactating mammals, was classified in Group 2B as *possibly carcinogenic to humans* (IARC, 1993a) (Table 1). An update of the scientific literature on these substances is provided in this volume. The existing Group 1 evaluation of naturally occurring aflatoxins was reaffirmed; an update was undertaken because of the concurrent re-evaluation of another mycotoxin, fumonisin B_1.

Mention is also made in this monograph of carcinogenicity studies of materials containing aflatoxins that have been treated with ammonia by a number of methods to reduce aflatoxin content. These processes result in a number of known and unknown reaction products which are not evaluated as such in this monograph; the elimination of the carcinogenicity of aflatoxin B_1 in ammoniated feed as well as the reduction of the mutagenicity and carcinogenicity due to aflatoxin M_1 in milk from dairy cows fed ammoniated feed was noted. This monograph also contains an annex on the causes and occurrence of aflatoxin contamination and on the management of this problem.

The fungi *Fusarium verticillioides* (Sacc.) Nirenburg (formerly known as *Fusarium moniliforme* Sheldon) and *F. proliferatum* are maize endophytes. Fumonisins are toxins produced by these fungi. Under environmental conditions that result in stress of the plant, the maize disease 'Fusarium kernel rot' occurs and the crop may become contaminated with fumonisins. Experimental carcinogenicity data on these toxins that were cited in the previous *IARC Monographs* review (IARC, 1993b) were from studies in which crude mixtures, rather than purified individual compounds, were fed to experimental animals, and the resulting evaluation was for 'toxins derived from *Fusarium moniliforme*' (Table 1) (IARC, 1993b). New bioassays to assess the carcinogenicity of purified fumonisin B_1 in experimental animals have been published since the previous *IARC Monographs* evaluation. These bioassays, as well as numerous studies on the toxicity and mechanisms of action of fumonisin B_1 are reviewed and evaluated in the present volume.

Naphthalene and styrene

Naphthalene was originally planned for inclusion in *IARC Monographs* Volume 77, *Some Industrial Chemicals* (2000), but was withdrawn from consideration at that time because the *Monographs* programme became aware of new carcinogenicity studies of this compound that were nearing completion. These studies have now been published, and their results are included in the present evaluation. In addition, there are extensive new data that contribute to understanding the mechanisms of carcinogenicity of this compound.

Styrene is an important industrial chemical and a major intermediate in the manufacture of both synthetic rubber and certain plastics. It was previously evaluated by the *IARC Monographs* programme in 1994 (IARC, 1994) (see Table 1). At that time it was classified in Group 2B as *possibly carcinogenic to humans,* on the basis of *limited evidence* for carcinogenicity in experimental animals that was supported by an extensive set of other relevant data, including biomarkers of exposure and of effect. Recently, new results have been published on carcinogenicity of styrene in experimental animals by inhalation and on mechanistic aspects, which are included in the present re-evaluation.

References

Brunton, L.L. (1996) Agents affecting gastrointestinal water flux and motility; emesis and anti-emetics; bile acids and pancreatic enzymes. In: Hardman, J.G., Limbird, L.E., Molinoff, P.B., Ruddon, R.W. & Gilman, A.G., eds, *Goodman & Gilman's The Pharmacological Basis of Therapeutics*, 9th Ed., New York, McGraw-Hill, pp. 922–923

IARC (1976) *IARC Monographs on the Evaluation of Carcinogenic Risk of Chemicals to Man*, Vol. 56, *Some Naturally Occurring Substances*, Lyon, IARC*Press*, pp. 313–317

IARC (1987) *IARC Monographs on the Evaluation of Carcinogenic Risks to Humans*, Suppl. 7, *Overall Evaluations of Carcinogenicity: An Updating of* IARC Monographs *Volumes 1 to 42*, Lyon, IARC*Press*, pp. 59, 61, 71, 135–136

IARC (1993a) *IARC Monographs on the Evaluation of Carcinogenic Risks to Humans*, Vol. 56, *Some Naturally Occurring Substances: Food Items and Constituents, Heterocyclic Amines and Mycotoxins*, Lyon, IARC*Press*, pp. 245–395

IARC (1993b) *IARC Monographs on the Evaluation of Carcinogenic Risks to Humans*, Vol. 56, *Some Naturally Occurring Substances: Food Items and Constitutents, Heterocyclic Amines and Mycotoxins*, Lyon, IARC*Press*, pp. 445–466

IARC (1994) *IARC Monographs on the Evaluation of Carcinogenic Risks to Humans*, Vol. 60, *Some Industrial Chemicals*, Lyon, IARC*Press*, pp. 233–320

National Toxicology Program (2001) *NTP Technical Report on the Toxicology and Carcinogenesis Studies of Riddelliine (CAS No. 23246-96-0) in F344/N Rats and B6C3F$_1$ Mice (Gavage Studies)* (NTP Technical Report TR 508), Research Triangle Park, NC

Segall, H.J. & Molyneux, R.J. (1978) Identification of pyrrolizidine alkaloids (*Senecio longilobus*). *Res. Commun. Chem. Pathol. Pharmacol.*, **19**, 545–548

Stillman, A.E., Huxtable, R., Consroe, P., Kohnen, P. & Smith, S. (1977) Hepatic veno-occlusive disease due to pyrrolizidine (Senecio) poisoning in Arizona. *Gastroenterology*, **73**, 349–352

THE MONOGRAPHS

SOME TRADITIONAL HERBAL MEDICINES

A. INTRODUCTION

1. History of Use of Traditional Herbal Medicines

By definition, 'traditional' use of herbal medicines implies substantial historical use, and this is certainly true for many products that are available as 'traditional herbal medicines'. In many developing countries, a large proportion of the population relies on traditional practitioners and their armamentarium of medicinal plants in order to meet health care needs. Although modern medicine may exist side-by-side with such traditional practice, herbal medicines have often maintained their popularity for historical and cultural reasons. Such products have become more widely available commercially, especially in developed countries. In this modern setting, ingredients are sometimes marketed for uses that were never contemplated in the traditional healing systems from which they emerged. An example is the use of ephedra (= Ma huang) for weight loss or athletic performance enhancement (Shaw, 1998). While in some countries, herbal medicines are subject to rigorous manufacturing standards, this is not so everywhere. In Germany, for example, where herbal products are sold as 'phytomedicines', they are subject to the same criteria for efficacy, safety and quality as are other drug products. In the USA, by contrast, most herbal products in the marketplace are marketed and regulated as dietary supplements, a product category that does not require pre-approval of products on the basis of any of these criteria. These matters are covered extensively in Section 3 below.

1.1 The role of herbal medicines in traditional healing

The pharmacological treatment of disease began long ago with the use of herbs (Schulz *et al.*, 2001). Methods of folk healing throughout the world commonly used herbs as part of their tradition. Some of these traditions are briefly described below, providing some examples of the array of important healing practices around the world that used herbs for this purpose.

1.1.1 *Traditional Chinese medicine*

Traditional Chinese medicine has been used by Chinese people from ancient times. Although animal and mineral materials have been used, the primary source of remedies is botanical. Of the more than 12 000 items used by traditional healers, about 500 are in common use (Li, 2000). Botanical products are used only after some kind of processing, which may include, for example, stir-frying or soaking in vinegar or wine. In clinical practice, traditional diagnosis may be followed by the prescription of a complex and often individualized remedy.

Traditional Chinese medicine is still in common use in China. More than half the population regularly uses traditional remedies, with the highest prevalence of use in rural areas. About 5000 traditional remedies are available in China; they account for approximately one fifth of the entire Chinese pharmaceutical market (Li, 2000).

1.1.2 *Japanese traditional medicine*

Many herbal remedies found their way from China into the Japanese systems of traditional healing. Herbs native to Japan were classified in the first pharmacopoeia of Japanese traditional medicine in the ninth century (Saito, 2000).

1.1.3 *Indian traditional medicine*

Ayurveda is a medical system primarily practised in India that has been known for nearly 5000 years. It includes diet and herbal remedies, while emphasizing the body, mind and spirit in disease prevention and treatment (Morgan, 2002).

1.2 Introduction of traditional herbal medicines into Europe, the USA and other developed countries

The desire to capture the wisdom of traditional healing systems has led to a resurgence of interest in herbal medicines (Tyler, 2000), particularly in Europe and North America, where herbal products have been incorporated into so-called 'alternative', 'complementary', 'holistic' or 'integrative' medical systems.

During the latter part of the twentieth century, increasing interest in self-care resulted in an enormous growth in popularity of traditional healing modalities, including the use of herbal remedies; this has been particularly true in the USA. Consumers have reported positive attitudes towards these products, in large part because they believe them to be of 'natural' rather than 'synthetic' origin, they believe that such products are more likely to be safe than are drugs, they are considered part of a healthy lifestyle, and they can help to avoid unnecessary contact with conventional 'western' medicine.

While centuries of use in traditional settings can be used as testimony that a particular herbal ingredient is effective or safe, several problems must be addressed as these ingredients are incorporated into modern practice.

One problem is that ingredients once used for symptomatic management in traditional healing are now used in developed countries as part of health promotion or disease prevention strategies; thus, acute treatment has been replaced by chronic exposure (e.g., herbal products used for weight loss, Allison *et al.*, 2001). This means that a statement about 'thousands of years of evidence that a product is safe' may not be valid for the way the product is now being used. This does not expressly mean that an ingredient is unsafe; it does mean that safety in the modern context cannot be assumed.

A second problem is that efficacy and effectiveness have rarely been demonstrated using modern scientific investigations. An evidence-based approach to this issue has only recently been implemented, and the results reveal that for most herbal products, considerable gaps in knowledge need to be remedied before one can be convinced about their efficacy.

One of the most difficult issues to contend with in translating traditional herbal practices into conventional 'western' medicine is the individualization of prescriptions containing multiple herbal and other ingredients. There is little incentive for standardization of products for a mass market, when the intention has been to provide an individual prescription. To the small grower or the traditionally trained herbalist, standardization means understanding the growth conditions, the time of harvesting, the manner of extraction or other preparation of material so that a reliable (albeit small amount of) active ingredient can be offered to people. To the manufacturer or distributor of large quantities that will be sold in a supermarket or a health food store, standardization refers to industrial production under defined conditions, using so-called Good Manufacturing Practices (GMP) (Food & Drug Administration, 2002) akin to those used for drug production.

In the USA, there is both small-scale and large-scale production of herbal products and there can be wide variation in their content and quality in the marketplace. Regulations in the USA do not yet require that dietary supplement manufacturers adhere to standard manufacturing practices, and so quality is not guaranteed (see Section 3). The public becomes discouraged by reports that products taken from store shelves do not consistently contain the ingredients — or in the amounts — that are claimed on the label.

For herbal products in common use, evidence of efficacy may be based upon traditional use, testimonials, clinical studies, both controlled and uncontrolled, and randomized, double-blind, placebo-controlled trials. For the most part, however, there is a lack of systematic clinical studies to support claims.

Safety of some herbal ingredients has been recently called into question, in part because of the identification of adverse events associated with their use and, increasingly, because of the demonstration of clinically relevant interactions between herbs and prescription drugs.

Adverse events (stroke, heart attacks, heart-rate irregularities, liver toxicity, seizures, psychoses and death) associated with use of ephedra for weight loss, body-building effects and increased energy or kava-kava (also known as kawa), widely used in Europe and increasingly in Canada to treat anxiety, nervousness, insomnia, pain and muscle tension, for example, have caused some countries to issue regulations restricting or banning these products (e.g. Health Canada Online, 2002a,b). Only a few herbs in common use have been suspected of causing cancer. These include *Aristolochia*, *Rubia tinctorum*, *Morinda officinalis* and *Senecio riddellii*, as discussed in detail below.

2. Use of Traditional Herbal Medicines in Developed Countries

2.1 Origin, type and botanical data

Plants and their secondary metabolite constituents have a long history of use in modern 'western' medicine and in certain systems of traditional medicine, and are the sources of important drugs such as atropine, codeine, digoxin, morphine, quinine and vincristine.

Use of herbal medicines in developed countries has expanded sharply in the latter half of the twentieth century. Monographs on selected herbs are available from a number of sources, including the European Scientific Cooperative on Phytotherapy (ESCOP, 1999), German Commission E (Blumenthal *et al.*, 1998) and the World Health Organization (WHO, 1999). The WHO monographs, for example, describe the herb itself by a number of criteria (including synonyms and vernacular names) and the herb part commonly used, its geographical distribution, tests used to identify and characterize the herb (including macroscopic and microscopic examination and purity testing), the active principles (when known), dosage forms and dosing, medicinal uses, pharmacology, contra-indications and adverse reactions. Other resources that provide detailed information about herbal products in current use include the Natural Medicines Comprehensive Database (Jellin, 2002) and NAPRALERT (NAtural PRoducts ALERT) (2001). Information about other available databases has been published by Bhat (1995).

2.2 Medicinal applications, beneficial effects and active components

In some cases, the active principles of plant-derived products have been isolated and characterized, and their mechanisms of action are understood (e.g., ephedrine alkaloids in some species of *Ephedra*). For many, however, including virtually all of the most common products in the marketplace, such information is incomplete or unavailable. This is in large part due to the complexity of herbal and botanical preparations; they are not pure compounds. It is also a function of the traditionally-held belief that the synergistic combination of several active principles in some herbal preparations is responsible for their beneficial effects.

2.3 Trends in use

Data on the global nutrition products industry, in which herbal and botanical products are often included, are given in Table 1.

Sales of dietary supplement products, including herbal and botanical supplements, in the USA increased dramatically during the 1990s, stimulated in the latter part of the

Table 1. The global nutrition products industry in 1999, including herbal and botanical products (in millions of US $)

Country	Vitamins/ minerals	Herbs/ botanicals	Sports, meal replacement, homeopathy, specialty	Natural[a] foods	Natural personal care	Functional foods[b]	Total
USA	7 070	4 070	4 320	9 470	3 590	16 080	44 520
Europe	5 670	6 690	2 510	8 280	3 660	15 390	42 200
Japan	3 200	2 340	1 280	2 410	2 090	11 830	23 150
Canada	510	380	250	700	330	1 500	3 670
Asia	1 490	3 170	970	710	880	1 450	8 670
Latin America	690	260	250	460	250	360	2 270
Australia and New Zealand	300	190	90	340	140	540	1 600
Eastern Europe and Russian Federation	350	220	250	180	40	269	1 300
Middle East	180	90	60	70	30	140	570
Africa	160	80	70	80	10	120	520
Total global	19 260	17 490	9 960	22 700	11 020	47 670	128 470

From Nutrition Business Journal (2000), derived from a number of sources. Totals may not add up due to rounding.

[a] Natural foods: foods grown or marketed with a focus on the perceived benefits of 'foods derived from natural sources' and that are, to varying degrees, free of pesticides, additives, preservatives, and refined ingredients

[b] Functional foods: foods fortified with added or concentrated ingredients to improve health and/or performance

decade by the Dietary Supplements Health and Education Act of 1994 (DSHEA) (Tyler, 2000). This pattern of growth has been replicated elsewhere in the world (Table 2), although more recently, sales of herbal products have apparently experienced a decline.

In the European Union (EU), in general, herbal products for which therapeutic claims are made must be marketed and regulated as drugs, while those that do not make such claims may be found in the food or cosmetic categories. Attempts are at present being made to harmonize the scientific and regulatory criteria that govern the marketing of herbal products (AESGP, 1998).

Table 2. Trends in the global nutrition products industry, 1997–2000 (in millions of US $)

	1997	1998	1999	2000
Vitamins/minerals	18 000	18 870	19 620	20 440
Herbs/botanicals	15 990	16 980	17 490	18 070
Sports, meal replacement, homeopathy, specialty	8 760	9 310	9 960	10 710
Natural foods[a]	16 690	19 910	22 700	25 420
Natural personal care	9 620	10 280	11 020	11 850
Functional foods[b]	40 320	43 940	47 670	51 480
Total	109 380	119 290	128 470	137 980

From Nutrition Business Journal (2000), derived from a number of sources
[a] Natural foods: foods grown or marketed with a focus on the perceived benefits of 'foods derived from natural sources' and that are, to varying degrees, free of pesticides, additives, preservatives, and refined ingredients
[b] Functional foods: foods fortified with added or concentrated ingredients to improve health and/or performance

In 1994, when the Dietary Supplements Health and Education Act (DSHEA) was passed in the USA, approximately 50% of the adult population of the country was reported to use dietary supplements and sales of all products combined were approximately $4 billion. This category of products includes vitamins, minerals and a variety of other ingredients; herbal products accounted for about one quarter of those sales. In 2000, the last year for which comparable data are available, again 50% of the adult population reported use of dietary supplements, and sales were close to $15 billion; herbals accounted for nearly one third of those sales. Table 3 identifies some trends in herbal supplement use in the USA from 1997 to 2000.

In the 1990s, the USA saw the growth of government organizations concerned with dietary supplements, such as the National Institutes of Health (NIH) National Center for Complementary and Alternative Medicine and Office of Dietary Supplements, and the National Cancer Institute (NCI) Chemoprevention Program of the Division of Cancer Prevention and Control. Organizations involved with dietary supplements such as the

Table 3. Ten top-selling herbs in the USA, 1997–2000 (in millions of US $)[a]

	1997	1998	1999	2000
Combination herbs[b]	1 659	1 762	1 740	1 821
Ginkgo biloba	227	300	298	248
Echinacea[c]	203	208	214	210
Garlic (*Allium sativum*)	216	198	176	174
Ginseng[d]	228	217	192	173
St John's wort (*Hypericum perforatum*)	100	308	233	170
Saw palmetto (*Serenoa repens*)	86	105	117	131
Soy (soya)	NA	NA	36	61
Valerian (*Valeriana officinalis*)	30	41	57	58
Kava-kava	22	44	70	53
Total herbal supplements	NA	NA	4 070	4 130

NA, not available
[a] From Nutrition Business Journal (2001) and Schulz *et al.* (2001). US consumer sales via all channels (includes all retail channels, direct sales, multilevel marketing, mail order and practitioner sales)
[b] Combination herbs include products sold for weight management, athletic performance enhancement or energy enhancement and often include mixtures of several herbal extracts, as well as single-compound ingredients. Others that have appeared in the top 10 list in earlier years, but not in 2000, include: goldenseal (*Hydrastis canadensis*), cranberry, bilberry (European blueberry), aloe (see monograph on *Rubia tinctorum*, *Morinda officinalis* and anthraquinones in this volume).
[c] Two types of coneflower preparation can be recommended and prescribed today: alcoholic extracts made from the root of the pale purple coneflower (*Echinacea pallida*) and juices expressed from the fresh aerial parts of the purple coneflower (*Echinacea purpurea*). It is noteworthy that until about 1990, the root of *Echinacea pallida* appears to have been regularly confused with that of the species *Echinacea angustifolia*.
[d] *Panax ginseng* is cultivated in Asia; *panax quinquefolius* is cultivated in the USA.

American Nutraceutical Association and the Foundation for Innovative Medicine, as well as industry trade associations such as the American Herbal Products Association, the Consumer Healthcare Products Association, the National Natural Foods Association, the Utah Natural Products Alliance and the Council for Responsible Nutrition have been expanding during the 1990s.

In Canada, herbal use has also increased. Berger (2001) noted, in summarizing the results of a 2001 survey of 2500 persons, 15 years of age and older, that herbal remedies were used by 38% of respondents, up from 28% in 1999. A survey in 1998 of the most popular remedies reported in Canada is given in Table 4.

In 1994, the European herbal medicine market was worth over £1.8 billion [US$ 2.8 billion] at retail selling prices. Although the UK market was smaller than that of Germany (in 1994 it was £88 million, compared with £1400 million), it had one of the highest forecast growth rates in Europe (Shaw, 1998).

Table 4. Top 10 most popular herbal remedies in Canada[a]

Herb	% who use among herbal users	% of users in general population
Echinacea	54	19
Garlic (*Allium sativum*)	52	18
Ginseng[b]	42	15
Camomile (*Chamomilla recutita*)[c]	38	13
Ginkgo biloba	20	7
Evening primrose (*Oenothera biennis*)	20	7
Devil's claw (*Harpagoghytum procumbens*)	17	6
St John's wort (*Hypericum perforatum*)	17	6
Tea tree oil (*Melaleuca alternifolia*)	15	5
Valerian (*Valeriana officinalis*)	13	5

From Non-Prescription Drug Manufacturers Association of Canada (1998), Sibbald (1999) and Schultz *et al.* (2001)
[a] From a survey of 6849 adults in April 1998
[b] See Table 3.
[c] Reported previously as *Matricaria chamomilla* (WHO, 1999)

The European market for herbal medicinal products was estimated to be worth $5.6 billion at public price level in 1995 (AESGP, 1998).

3. Awareness, Control, Regulation and Legislation on Use

3.1 WHO guidelines for herbal medicines

In 1992, the WHO Regional Office for the Western Pacific invited a group of experts to develop criteria and general principles to guide research work on evaluating herbal medicines (WHO, 1993). This group recognized the importance of herbal medicines to the health of many people throughout the world, stating: 'A few herbal medicines have withstood scientific testing, but others are used simply for traditional reasons to protect, restore, or improve health. Most herbal medicines still need to be studied scientifically, although the experience obtained from their traditional use over the years should not be ignored. As there is not enough evidence produced by common scientific approaches to answer questions of safety and efficacy about most of the herbal medicines now in use, the rational use and further development of herbal medicines will be supported by further appropriate scientific studies of these products, and thus the development of criteria for such studies'.

The document covered such topics as developing protocols for clinical trials using herbal medicines, evaluating herbal medicine research, guidelines for quality specifications of plant materials and preparations, and guidelines for pharmacodynamic and general pharmacological studies of herbal medicines and for toxicity investigations of herbal medicines.

WHO has also issued Guidelines for the Assessment of Herbal Medicines (WHO, 1996). These guidelines defined the basic criteria for the evaluation of quality, safety and efficacy of herbal medicines with the goal of assisting national regulatory authorities, scientific organizations and manufacturers in assessing documentation, submissions and dossiers in respect of such products. It was recommended that such assessments take into account long-term use in the country (over at least several decades), any description in the medical and pharmaceutical literature or similar sources or documentation of knowledge on the application of a herbal medicine, and marketing authorizations for similar products. Although prolonged and apparently uneventful use of a substance usually offers testimony of its safety, investigation of the potential toxicity of naturally occurring substances may reveal previously unsuspected problems. It was also recommended that regulatory authorities have the authority to respond promptly to new information on toxicity by withdrawing or limiting the licences of registered products containing suspect substances, or by reclassifying the substances to limit their use to medical prescription. The guidelines stressed the need for assessment of efficacy including the determination of pharmacological and clinical effects of the active ingredients, and labelling which includes a quantitative list of active ingredient(s), dosage, and contra-indications.

3.2 The European Union

The Association Européenne des Spécialités Pharmaceutiques Grand Public (Association of the European Self-Medication Industry; AESGP) has carried out a study for the European Commission on herbal medicinal products in the European Union (EU). The following summary is taken from this report (AESGP, 1998).

The importance of herbal medicinal products varies from one country to another. These products are not a homogeneous group. In general, they are either fully licensed medicinal products with efficacy proven by clinical studies or by references to published scientific literature (in accordance with Article 4.8 a (ii) of Council Directive 65/65/EEC) (European Commission, 1965) or are available as products with a more or less simplified proof of efficacy according to their national use. Many Member States have these two categories, but there are major discrepancies between the Member States in the classification of individual herbal drug preparations and products into one of these categories as well as in the requirements for obtaining a marketing authorization.

3.2.1 *Definition of herbal medicinal products*

According to Council Directive 65/65/EEC (European Commission, 1965), which has been implemented in national law in all Member States, medicinal products require prior marketing approval before gaining access to the market. In almost all Member States, herbal medicinal products are considered as medicinal products, and are, in principle, subject to the general regulations for medicines as laid down in the various national medicine laws. In many cases, a specific definition of herbal medicinal products is available, which is in line with the EU Guideline 'Quality of Herbal Medicinal Products'. This includes plants, parts of plants and their preparations, mostly presented with therapeutic or prophylactic claims. Different categories of medicinal products containing plant preparations exist or are in the process of being created. For instance, draft legislation in Spain includes the definitions 'herbal medicinal products' and 'phytotraditional products'. The latter are not considered as 'pharmaceutical specialties' and are therefore not classified as herbal medicinal products.

3.2.2 *Classification of herbal products*

Generally, herbal products are classified as medicinal products if they claim therapeutic or prophylactic indication, and are not considered as medicinal products when they do not make these claims. Products not classified as medicinal in most cases belong to the food or cosmetic areas, although they sometimes contain plants which have pharmacological properties. For example, senna pods (from *Cassia* plants, used as laxatives) (see General Remarks and monograph on *Rubia tinctorum*, *Morinda officinalis* and anthraquinones in this volume) can be marketed as food in Belgium. Specific categories of non-medicinal products exist in some Member States, such as the so-called 'therapeutic supplement products' in Austria. In Ireland, Spain and the United Kingdom, there exist preparations defined as medicinal products, which are under specific conditions exempt from licensing requirements.

3.2.3 *Combination products*

Herbal ingredients used in combination are widely used in Europe, and their assessment is often performed according to specific guidelines. Combinations of herbal and homeopathic ingredients exist in a few countries. Their assessment follows rather strict criteria, usually those of a 'full' application procedure. Combinations of herbal ingredients and vitamins are available in many countries.

3.2.4 *Documentation of quality, safety and efficacy*

A marketing authorization for a herbal medicinal product is, in principle, granted based on an extensive dossier in terms of proof of quality, safety and efficacy in all Member States, with the exception of Denmark and Finland, where it is possible to use only references to published data for herbal medicinal products. Luxembourg, in practice, only grants marketing authorization based on the assessment of other countries. In principle, according to Article 4.8 (a) (ii) of Council Directive 65/65/EEC (European Commission, 1965), the option of using reviews on published data is available in all Member States. However, this 'bibliographical' option is sometimes only available through assessment on a case-by-case basis or not used in practice. Austria permits this type of application for safety documentation only.

3.2.5 *ESCOP and WHO monographs*

European Scientific Cooperative on Phytotherapy (ESCOP) (see Awang, 1997) or WHO monographs may be used in many Member States as a summary of published data. Many regulatory authorities regard them as helpful documentation for clarifying efficacy and safety.

The European Commission (EC), the EMEA (European Agency for the Evaluation of Medicinal Products) Executive Director and the EMEA Management Board established the EMEA Ad Hoc Group in 1997. This Working Group is made up of representatives from the Member States (primarily health authorities) and representatives from the European Parliament, the EC and the European Pharmacopoeia. The Working Group has reviewed the criteria for the demonstration of quality, pre-clinical safety and clinical efficacy in marketing authorization applications for herbal medicinal products as set out in the Council Directives. The Working Group has proposed requirements for non-clinical testing of herbal drug preparations based on a draft EC Guideline for old substances with long market histories (EMEA, 2000). The Group has also discussed the appropriate role of scientific monographs prepared by the WHO and ESCOP.

3.2.6 *Simplified proof of efficacy*

Various traditional herbal medicinal products exist in many Member States in addition to fully licensed herbal medicinal products. For these products, national authorities usually verify the safety and ensure a sufficient level of quality. For proof of efficacy, the level of requirements is sometimes adjusted to take into account the long-term experience and is therefore simplified. For example, a specific simplified procedure exists in Austria, Belgium, France and Germany. Most other countries in the EU do not use this strategy.

3.2.7 *Further developed products*

For herbal medicinal products that have been proposed for non-traditional indications or are modified from their traditional form (e.g., highly processed or special extracts), a full licence is required in most cases, and efficacy has to be proven by clinical studies. In several countries, such products are not used.

3.2.8 *Individual supply*

Herbal medicinal products (like other medicinal products) are made up and/or supplied to individual patients following a one-to-one consultation between patient and practitioner. Some herbal medicinal products are made according to accepted formulae and are prepared by pharmacists. According to Article 2.4 of Council Directive 65/65/EEC (European Commission, 1965), a marketing authorization is not needed. A specific situation exists in the United Kingdom, where a practitioner, according to Section 12 of the Medicines Act 1968 (Griffin, 1998), may supply products to a customer without a licence.

3.2.9 *Products from foreign countries*

The quality of imported medicinal plants and their preparations is assessed differently in different Member States. In some cases, no specific regulations exist concerning the control of raw materials or crude drugs, particularly for products that enter the market as foodstuffs or other products that are not controlled in the same way as medicinal products. Finished products are often treated as new chemical entities with full proof of quality, safety and efficacy being required.

3.2.10 *Good manufacturing practices and quality control*

All Member States apply the manufacturing requirements of Council Directive 75/319/EEC (European Commission, 1975) to herbal medicinal products. Starting materials for herbal medicinal products are in principle controlled in accordance with the European Pharmacopoeia in all Member States. Good manufacturing practice inspections are carried out in nearly all Member States.

The European Pharmacopoeia was created in 1964; its efforts have resulted in the creation of 83 monographs on herbal drugs which are used either in their natural state after desiccation or concentration or for the isolation of natural active ingredients (Council of Europe, 1996).

3.2.11 *Post-marketing surveillance*

The adverse reaction reporting systems of the Member States also monitor herbal medicinal products if they are authorized medicinal products. This system has demonstrated its effectiveness in the case of several withdrawals of marketing authorizations for herbal medicinal products due to safety concern in connection with certain plants. Consumer reports could provide a picture of the spectrum of adverse reactions to herbal medicinal products and alert authorities to potential problems; the degree of acceptance of such reports varies between Member States.

3.2.12 *Advertising, distribution and retail sale*

All Member States have implemented Council Directive 92/28/EEC (European Commission, 1992a) on advertising in national law. This directive covers herbal products if they are authorized as medicinal products.

Wholesale marketing of all medicinal products as well as authorized herbal medicinal products is covered by Council Directive 92/25/EEC (European Commission, 1992b). The retail sale of herbal medicinal products is restricted to pharmacies in Belgium, France, Greece, Ireland, Italy, Luxembourg, Portugal and Spain. It is permitted in other outlets in the case of certain herbal medicinal products in Austria, Denmark, Finland, Germany, the Netherlands, Sweden and the United Kingdom. Distance selling and teleshopping are not permitted for herbal medicinal products in most countries.

3.2.13 *Differences between Member States*

Herbal medicinal products are regarded as medicinal products in most of the Member States and have, in theory, the option of obtaining marketing authorization in the same way as all other medicinal products. However, the legal systems of the Member States differ in the classification of herbal products, in the availability of an application process for a marketing authorization based on a full application, bibliographical application or simplified proof of efficacy, and in the permitted outlets for retail distribution. Member States have different traditions regarding the therapeutic use of medicinal plant preparations, which may make it more difficult for manufacturers of herbal medicinal products to apply for marketing authorization using the decentralized procedure.

3.3 **Individual countries** (Calixto, 2000)

3.3.1 *France*

The French Medicines Agency (Agence du Médicament) grants marketing authorizations based on abridged dossiers by making reference to traditional use. The shortened procedure requires limited or no pharmacological, toxicological and clinical tests and is detailed in the Agency Instructions No. 3. The list of drugs with accepted traditional uses was first published in 1985 by the Ministry of Health and has subsequently been revised several times (Table 5). Traditional use of approximately 200 herbal drugs or preparations derived from these drugs has been recognized for minor indications. Agency Instructions No. 3 includes rules for labelling and packaging of herbal medicinal products. If the drug is not specifically included in the list, there is no option to use an abridged procedure (AESGP, 1998). As of 1997, local medicinal plants were on the A list of the French Pharmacopoeia (Castot *et al.*, 1997) which groups the 454 herbs which benefit/risk ratio is considered as positive when traditionally used.

Table 5. Examples of plants and indications from the French Agency Instructions No. 3 (*Cahiers de l'Agence No. 3*) (Agence du Médicament)

Medicinal plant	Information for the medical profession	Information for the public
Valeriana officinalis	Traditionally used in the symptomatic treatment of neurotonic conditions of adults and children, notably in cases of mild sleeping disorders	Traditionally used to reduce nervousness in adults and children, notably in case of sleeping disorders
Matricaria chamomilla	Traditionally used topically as a soothing and antipruriginous application for dermatological ailments and as a protective treatment for cracks, grazes, chapped skin and insect bites.	Traditionally used topically as a soothing application and to calm the itching of skin ailments and in cases of cracks, grazes, chapped skin and insect bites.
	Traditionally used in the symptomatic treatment of digestive upsets such as epigastric distension, slow digestion, eructation and flatulence.	Traditionally used to promote digestion.
	Traditionally used to stimulate appetite.	Traditionally used to stimulate appetite.
	Traditionally used in cases of eye irritation or discomfort due to various causes (smoky atmospheres, sustained visual effort, swimming in the sea or swimming baths, etc.).	Traditionally used in cases of eye irritation or discomfort due to various causes (smoky atmospheres, sustained visual effort, swimming in the sea or swimming baths, etc.). (Precaution: use only for mild conditions. If the symptoms increase or persist for more than two days, consult a doctor).
	Traditionally used locally (mouth and throat washes, lozenges) as an analgesic in conditions of the oral cavity and/or larynx.	Traditionally used for the temporary relief of sore throat and/or transient hoarseness.
Cassia senna[a]	Short-term treatment of occasional constipation.	This medicinal product is a stimulant laxative; it stimulates bowel evacuation. It is intended for the short-term treatment of occasional constipation.
Hypericum perforatum	Traditionally used topically as a soothing and antipruriginous application for dermatological ailments and as a protective treatment for cracks, grazes, chapped skin and insect bites.	Traditionally used topically as a soothing application and to calm the itching of skin ailments and in cases of cracks, grazes, chapped skin and insect bites (Precaution: do not use before exposure to the sun).
	Traditionally used for sunburn, superficial burns or small area and nappy rash.	Traditionally used for sunburn, superficial burns or small area and nappy rash (Precaution: do not use before exposure to the sun).
	Traditionally used locally (mouth and throat washes, lozenges) as an analgesic in conditions of the oral cavity and/or larynx.	Traditionally used for the temporary relief of sore throat and/or transient hoarseness.

Table 5 (contd)

Medicinal plant	Information for the medical profession	Information for the public
Plantago major L.	Traditionally used topically as a soothing and antipruriginous application for dermatological ailments and as a protective treatment for cracks, grazes, chapped skin and insect bites.	Traditionally used topically as a soothing application and to calm the itching of skin ailments and in cases of cracks, grazes, chapped skin and insect bites (Precaution: do not use before exposure to the sun).
	Traditionally used in cases of eye irritation or discomfort due to various causes (smoky atmospheres, sustained visual effort, swimming in the sea or swimming baths, etc.).	Traditionally used in cases of eye irritation or discomfort due to various causes (smoky atmospheres, sustained visual effort, swimming in the sea or swimming baths, etc.). (Precaution: use only for mild conditions. If the symptoms increase or persist for more than two days, consult a doctor).

From AESGP (1998)
[a] See General Remarks

Castot *et al.* (1997) have reviewed the surveillance or pharmacovigilance of herbal medicines in France. Between 1 and 15 October 1996, the authors observed 15 publications or publicities in 23 magazines widely available in France; these publications/publicities offered for sale by mail a number of medicinal plants found or not found on the list of 34 'approved' plants. (These plants were listed in 1979 by the government in reason of lack of reported toxicity in traditional use or following complete bibliographic investigation.)

Between 1985 and 1995, the French national surveillance system registered 341 cases of undesirable effects possibly linked to herbal medicines; this figure represents only 0.35% of the total adverse effects from all drugs reported during the same period. The number of adverse effects from herbal medicines is almost certainly under-reported. The population concerned was largely female (73%) with a mean age of 50 years; reasons for taking herbal medicines were constipation, obesity and anxiety. Undesirable effects reported were quite diverse, including allergic and cutaneous responses, eczema, liver damage (linked to germander (*Teucrium chamaedrys*, tonic, diuretic)), digestive problems (linked to laxative plants), neurological effects such as vertigo (linked to plants classified as sedatives) and blood pressure fall and hypokalaemia (linked to plant laxatives containing anthraquinones (see monograph in this volume)). Outcome of these cases was generally favourable (Castot *et al.*, 1997).

3.3.2 *Germany* (see Kraft, 1999; Calixto, 2000)

Keller (1991) summarized the legal requirements for the use of phytopharmaceutical drugs in the Federal Republic of Germany. The legal status for herbal remedies was defined by the Medicines Act of 24 August 1976. For finished drugs a marketing authorization is obligatory. Herbal finished drugs have to comply with the same criteria for quality, safety and efficacy as all other finished drugs. Finished herbal drugs may be authorized for marketing in one of three ways:

(i) *Evaluation and validation of old medicines*. Finished drugs registered in 1978 possessed a provisional marketing authorization and could remain on the market until the end of April 1990. The medical evaluation of these drugs was mainly based on published data and was carried out by a special expert committee, the Commission E (Expert Commission for Herbal Remedies). The preparation of new monographs by Commission E ended in 1993 (Sandberg & Corrigan, 2001).

(ii) *Standardized marketing authorization*. Medicines that do not represent a direct or indirect risk for health can be exempted from the need for an individual marketing authorization by reference to a previously existing monograph.

(iii) *Individual application for marketing authorization*. In this procedure, complete documentation including the results of analytical tests, results of pharmacological and toxicological tests and results of clinical or other medical tests are required.

In addition, drugs sold outside pharmacies and only for traditional uses without clinical evidence for efficacy have to be labelled as 'traditionally used' (Table 6).

3.3.3 *United Kingdom*

A number of papers have discussed the situation regarding herbal medicines in the United Kingdom. De Smet (1995) recommended that herbal products be licensed as special products 'medicines'; he estimated that unlicensed preparations accounted for over 80% of herbal sales. Many medicine-like products on the British herbal market remain unregistered for two reasons: acceptable data on efficacy, safety and quality may not be available, and the licensing fee is high. Traditional experience with herbs can be a useful tool in detecting acute toxicity, but is less useful in detecting rare adverse reactions or those that develop after long-term exposure or after a latent period. Therefore, traditional experience needs to be supplemented with orthodox data from research and post-marketing surveillance. Such post-marketing surveillance is only partly helpful, as herbal suppliers and traditional practitioners are not obliged to report suspected adverse reactions and herbal products are of variable quality.

Table 6. Examples of plants and indications from the German 'traditional' list

Active ingredients	Dosage form	Indication 'Traditionally used for ...'
Ginseng root (liquid extract prepared with wine)	Liquid for oral administration	Improvement of the general condition
St John's wort (aqueous liquid extract)	Liquid for oral administration	Improvement of the condition in case of nervous stress
Garlic + mistle herb + hawthorn flowering tops	Sugar-coated tablets	Support of cardiovascular function
Garlic oil	Gastro-resistant capsules	For prevention of general atherosclerosis
Hamamelis leaf (aqueous liquid extract)	Cream	Support of skin function
Ginger + juniper berries	Tables	Support of digestive function
Onion (oily viscous extract)	Capsules	Prevention of general atherosclerosis
Melissa leaf	Sugar-coated tablets	Improvement of the condition in case of nervous stress, for support of stomach function
Dandelion root (aqueous solid extract)	Capsules	Support of the excretory function of the kidney

Adapted from AESGP (1998)

Shaw (1998) has discussed the safety aspects of herbal remedies in the United Kingdom. The legal status of herbal remedies/medicines in the United Kingdom can be broadly divided into three categories:

(i) Most herbal products are unlicensed and therefore no medicinal claims can be made. These are regarded as food supplements and come under food legislation (Ministry of Agriculture, Fisheries and Food (MAFF), 1998).

(ii) Licensed medicinal products require evidence of quality, safety and efficacy and are regulated by the Medicines Control Agency.

(iii) Herbal medicines supplied by a herbalist are exempt from licensing under the 1968 Medicines Act.

Since 1996, the Medicines Control Agency has included adverse reaction reports on unlicensed herbal remedies within its remit and now monitors all three categories (Griffin, 1998).

The House of Lords Science and Technology Committee in early 1999 reviewed a large amount of oral and written evidence from a wide variety of sources in order to scrutinize complementary and alternative medicine (CAM) including herbal medicines (Mills, 2001). The report noted that public satisfaction with CAM was high and that use of CAM was increasing. Evidence was required that CAM has an effect above and beyond placebo, and this information needed to be available to the public. The current lack of regulation of CAM did not protect the public interest adequately. Acupuncture and herbal medicine should be subject to statutory regulation under the Health Act 1999, as should possibly non-medical homeopathy. The regulatory status of herbal medicines was viewed as particularly unsatisfactory. The report recommended that training for CAM professionals should be standardized and independently accredited and, for many, should include basic biomedical science. Conventional health professionals should become more familiar with CAM and those working in the best-regulated CAM professions should work towards integration with conventional medicine.

3.3.4 *United States*

In the USA, the Food Drug and Cosmetics Act characterizes a product primarily on the basis of its intended use. For a botanical product, this intended use may be as a food (including a dietary supplement), a drug (including a biological drug), a medical device (e.g., gutta-percha) or a cosmetic as shown by, among other things, the products' accompanying labelling claims, advertising materials, and oral or written statements (21 Code of Federal Regulations (CFR) 201.128) (Food and Drug Administration (FDA), 2000).

For products classified as drugs, the FDA regulates them under the authority of the Food Drug and Cosmetics Act and its amendments. Under current regulations, if there is no marketing history in the USA for a botanical drug product, if available evidence of safety and effectiveness does not warrant inclusion of the product in an existing, approved category of OTC (over-the-counter) drugs, or if the proposed indication would not be appropriate for non-prescription use, the manufacturer must submit a new drug application to obtain FDA approval to market the product for the proposed use. If existing information on the safety and efficacy of a botanical drug product is insufficient to support a new drug application, new clinical studies will be needed to demonstrate safety and effectiveness.

Most botanical products in the USA are marketed as dietary supplements. Under the Dietary Supplement Health and Education Act of 1994 (DSHEA), an orally ingested product that meets the definition of a 'dietary supplement' under section 201(ff) of the Food Drug and Cosmetics Act may be lawfully marketed using a statement that (1) claims a benefit related to a classical nutrient deficiency disease (and discloses the prevalence of the disease in the USA); (2) describes how the product is intended to affect the structure or function of the human body; (3) characterizes the documented mechanism by which the product acts to maintain such structure or function; or (4) describes general well-being derived from consumption of the product (section 403 r (6)(A) of the Food Drug and Cosmetics Act, 21 U.S.C. 343 r(6)(A)). The term 'dietary supplement' is defined in section 201 (ff) of the Act and means a product (other than tobacco) intended to supplement the diet that contains one or more of certain dietary ingredients, such as a vitamin, a mineral, a herb or another botanical substance, an amino acid, a dietary substance for use by people to supplement the diet by increasing the total dietary intake, or a concentrate, metabolite, constituent, extract, or combination of the preceding ingredients (Chang, 1999). A dietary supplement is a product that is intended for ingestion in a form described in section 411C(1)(B)(i) of the Act (i.e., tablet, capsule powder, softgel, gelcap and liquid), which is not represented as conventional food, or as the sole item of a meal or of the diet, and which is labelled as a dietary supplement. It is the responsibility of the manufacturer to ensure that a dietary ingredient used in a dietary supplement is safe for its intended use.

The FDA has issued regulations defining the types of statement that can be made concerning the effect of a dietary supplement on the structure and function of the body. The regulations distinguish these statements from the types of statement that require prior approval as drug claims or prior authorization as health claims.

Safety monitoring of dietary supplements focuses on the post-marketing period. The FDA receives spontaneous reports of suspected adverse events through a variety of means, including through a programme called MEDWATCH, the FDA Medical Products Reporting Program (Goldman & Kennedy, 1998). The post-marketing surveillance system for foods and dietary supplements, called the Adverse Event Reporting Systems, is a passive system that relies on voluntary reporting by concerned parties, primarily health professionals and consumers (AESGP, 1998).

The DSHEA extended the definition of dietary supplements beyond vitamins and minerals and established a formal definition of a dietary supplement using new criteria. The Congressionally mandated Commission on Dietary Supplement Labels (CDSL) suggested that some botanicals may qualify as OTC products under existing statutes; these state that a product may avoid 'new drug' premarket approval requirements and may be eligible for marketing under an OTC drug monograph if the product is generally recognized as safe (GRAS) and effective under the conditions for use for which it is labelled and if the product has been used 'to a material extent and for a material time' under those conditions. The FDA's response to the Commission stated that it does not regard marketing experience outside the USA to meet conditions of historical use.

Angell and Kassirer (1998) stated that the primary factor that sets alternative medicine, including its most common form, herbal medicine, apart from conventional medicine is 'that it has not been scientifically tested and its advocates largely deny the need for such testing'. Angell and Kassirer defined 'testing' as the gathering of evidence of safety and efficacy, as required by the FDA. 'There cannot be two kinds of medicine — conventional and alternative. There is only medicine that has been adequately tested and medicine that has not, medicine that works and medicine that may or may not work. Once a treatment has been tested rigorously, it no longer matters whether it was considered alternative at the outset. If it is found to be reasonably safe and effective, it will be accepted. Alternative treatments should be subjected to scientific testing no less rigorous than that required for conventional treatments'.

3.3.5 *Canada*

The Canadian Food and Drug Act and findings of an Expert Advisory Committee on Herbs and Botanical Preparations were consulted by Kozyrskyj (1997) to provide an overview of the issues regarding regulation of herbal products in Canada. Case reports of herbal toxicity were identified to illustrate some of the hazards of herbal products, and references were provided to guide health professionals in searching the literature for clinical trials that have evaluated the efficacy of these drugs.

Herbal products not registered as drugs in Canada are sold as foods and are thus exempt from the drug review process that evaluates product efficacy and safety. An Expert Advisory Committee on Herbs and Botanical Preparations was formed in 1984 to advise the Health Protection Branch (HPB). HPB published lists of hazardous herbal products in 1987, 1989, 1992 and 1993. The last publication elicited a large response from consumers and the herbal industry. As of 1995, the list was still under review (Kozyrskyj, 1997).

The recently formed Office of Natural Health Products (currently the Natural Health Products Directorate) (Sibbald, 1999) is responsible for all regulatory functions including, but not limited to pre-market assessment for product labelling, licensing of manufacturers, post-approval monitoring and compliance and implementation of the recommendations of the standing House Health Committee.

In December 2000, the provincial government of British Columbia approved regulations that established traditional Chinese medicine as an alternative form of primary health care. The cost is not covered under Canadian medicare and practitioners face several practice restrictions. For example, 'no acupuncturist or herbalist may treat an active serious medical condition unless the client has consulted with a medical practitioner, naturopath or dentist or doctor of traditional Chinese medicine, as appropriate' (Johnson, 2001).

3.3.6 *Chile*

In 1992 the Unidad de Medicina Tradicional was established with the aims of incorporating traditional medicine with proven efficacy into health programmes and of contributing to the establishment of their practice. Herbal medicines are legally differentiated into: (a) drugs intended to cure, alleviate or prevent diseases; (b) food products for medicinal use and with therapeutic properties; and (c) food products for nutritional purposes (Calixto, 2000).

Herbal products with therapeutic indications and/or dosage recommendations are considered to be drugs. Distribution of these products is restricted to pharmacies. A registration for marketing authorization is needed for herbal products, homeopathic products, and other natural products. An application for such registration consists of the complete formula, the labelling, samples of the product, and a monograph which permits identification of the formula and characteristics of the product (Zhang, 1998).

3.3.7 *Japan* (Zhang, 1998; Eguchi *et al.*, 2000; Saito, 2000)

Japanese traditional medicine, as used in Japanese society for more than a thousand years, may be divided into folk medicine and Chinese medicine (or Kampo medicine). Kampo medicine is so popular that the per capita consumption of herbal medicine in Japan seems to be the highest in the world. One hundred and forty-six Kampo drugs are registered as drugs by the Ministry of Health and Welfare (MHW) and are included in coverage under the National Health Insurance. Acceptance of Kampo drugs took place without clinical validation studies. In 1989, about 80% of physicians reported prescribing Chinese medicine. Physicians generally recognize Chinese medicine as a complement to modern medicine; traditional drugs are viewed in Japanese society as safe.

Raw herbs which have long been used as folk medicine and which have also been used for a considerable period as components of an industrial product are each described in a corresponding monograph. These products are freely usable for the purposes indicated in the monograph. Local traditional usage is not sufficient for approval as a drug; the claims and rules of combinations of herbal ingredients are determined on the basis of the pharmacological actions of the ingredients. If a monograph is not available, the claims reported in the Japanese Pharmacopoeia are used as a guide.

In the evaluation of a Chinese medicine, importance is given to 'empirical facts or experience', such as reference data, clinical test reports, etc., rather than the pharmacological action of each ingredient. Safety and efficacy have been estimated based on general methods employed by modern medical science. In 1972 the MHW designated 210 formulae as OTC drugs; this selection was based primarily on the experience of doctors actually practising traditional Chinese medicine. In 1976, the MHW specified 146 formulations as 'National Health Insurance (NHI) applicable prescription drugs'. In the case of an application for approval of a prescription drug other than those previously listed, specified data on safety, stability, comparison with other drugs, clinical test results, etc. must be submitted.

New Kampo drugs are regulated in essentially the same way as 'western' drugs in Japan. The same data required for new 'western' drugs are required for new Kampo drugs, including data from three-phase clinical trials.

Since 1971, the MHW has been running a programme for re-evaluation of all drugs marketed before 1967; a new system to re-evaluate the efficacy and safety for all drugs every five years was launched in 1988.

An Advisory Committee for Kampo drugs was established in 1982 in close association with the MHW in order to improve quality control of Kampo drugs. Since the 1986 Good Manufacturing Practice Law, the standard applied to all pharmaceutical drugs has also applied to Kampo drugs. In addition, in 1985, guidelines for ethical extract products in oriental medicine formulations were developed.

The MHW has three major systems for collection of adverse reaction data. The first is a voluntary system involving 2915 monitoring hospitals. The second system — the Pharmacy Monitoring System — which includes 2733 pharmacies, collects data on cases of adverse reactions to OTC drugs. The third system is Adverse Reaction Reporting from Manufacturers. These cases are reported to the MHW by the responsible company, with information arising from medical conferences and from journals.

3.3.8 *Korea (Republic of)*

The Pharmaceutical Act of 1993 explicitly allowed pharmacists to prescribe and dispense herbal drugs (Cho, 2000).

3.3.9 *China*

Many herbal medicines have been used for hundreds of years and it is assumed in many cases that they must work. For example, about 7000 species of plants are used in China as herbal medicines, but only 230 of the most commonly used ones have been subject to in-depth pharmacological, analytical and clinical studies.

The 2000 edition of the Chinese pharmacopoeia included 784 items on traditional Chinese medicines and 509 on Chinese patent medicines. Herbal medicines in China are normally considered as medicinal products with special requirements for marketing. New drugs have to be investigated and approved according to the Drug Administration Law. New traditional Chinese medicines are classified under five categories based on the Amendment and Supplement Regulation of Approval of new traditional medicines:

Class 1

(1) Artificial alternatives of Chinese crude drugs.
(2) Newly discovered Chinese crude drugs and their preparations.
(3) Active constituents extracted from Chinese crude drugs and their preparations.
(4) Active constituents extracted from a composite formulation of traditional Chinese medicines.

Class 2

(1) Injection of traditional Chinese medicines
(2) Use of new medicinal parts of Chinese crude drugs and their preparations.
(3) Effective fractions extracted from Chinese crude drugs or natural drugs and their preparations.
(4) Chinese crude drugs artificially developed in an animal body and their preparations.
(5) Effective fractions extracted from a composite formulation.

Class 3

(1) New composite formulations of traditional Chinese medicines.
(2) Composite preparations of traditional Chinese medicines and chemical drugs with the main efficacy due to the traditional Chinese medicine.
(3) Domestically cultivated or bred crude drugs originally imported and commonly used in China, and their preparations.

Class 4

(1) Preparation with a change of dosage form or route of administration.

(2) Botanical crude drugs acclimatized from their origin, or crude drugs from a domesticated wild animal in China.

Class 5

Marketing drugs with new indications or syndromes.

In Hong Kong in 1989, the Government appointed a Working Party to review and make recommendations for the use and practice of traditional Chinese medicines. In 1995 the preparatory Committee on Chinese medicines was formed to manage the implementation of these recommendations: as a result 31 potent Chinese medicines that may potentially cause adverse effects have been identified. Proprietary preparations containing a combination of herbal ingredients and conventional drugs are regulated in the same manner as other conventional drugs.

The majority of suppliers are state-owned or state-connected. The extensive pharmacopoeia relating to traditional Chinese medicine allows the parallel manufacturing and sale of both pharmaceutical drugs and traditional herbal blends (Chan, 1997; Zhang, 1998).

3.3.10 *Saudi Arabia*

Registration of medicinal products by the Ministry of Health is obligatory, as is that of products, in addition to drugs, with medicinal claims or containing active ingredients having medicinal effects such as herbal preparations, health and supplementary food, medicated cosmetics, antiseptics or medical devices (Zhang, 1998).

3.3.11 *South Africa* (Zhang, 1998)

The trade in crude indigenous herbal products is completely unregulated. A large number of South Africans consult traditional healers, generally in addition to medical practitioners. There are about 200 000 traditional healers in the country.

Once a health-related claim is made for a finished herbal product, that product must go through a full drug evaluation in the Medicines Control Council (MCC) before marketing.

Specific regulations for registration and control of new 'traditional' herbal medicines do not exist. Old medicines, including such well known herbal medicines as senna or aloe, are already registered by the MCC, according to internationally accepted standards of efficacy and safety. Pharmaceutical standards need to be consistent with those of the United States Pharmacopeia or the British Pharmacopoeia.

3.3.12 *Australia and New Zealand* (Moulds & McNeil, 1988; Zhang, 1998)

The Therapeutic Goods Act 1989 sets out the legal requirements for the import, export, manufacture and supply of medicines in Australia. It details the requirements for listing or registering all therapeutic goods in the Australian Register of Therapeutic Goods (ARTG), as well as many other aspects of the law including advertising, labelling and product appearance. Australian manufacturers of therapeutic goods must be licensed and their manufacturing processes must comply with the principles of Good Manufacturing Practice (GMP). All medicines manufactured for supply in Australia must be listed or registered in the ARTG, unless they are specifically exempt or excluded. Listed medicines are considered to be of lower risk than registered medicines. Most complementary medicines (e.g., herbal, vitamin and mineral products) are examples of listed products. Medicines assessed as having a higher level of risk must be registered (not listed). Registered medicines include non-prescription (low-risk, OTC) medicines and prescription (high-risk) medicines. Complementary medicines (also known as 'traditional' or 'alternative' medicines) include vitamin, mineral, herbal, aromatherapy and homeopathic products. Complementary medicines may be either listed or registered, depending on their ingredients and the claims made. Most complementary medicines are listed in the ARTG and some are registered (Therapeutics Good Administration, 1999).

In New Zealand, supplements in the market place are largely manufactured in the USA. Regulations are not restrictive; there are no limits on ingredients or potencies and 'structure/function' claims are allowed.

4. References

AESGP (Association Européenne des Spécialités Pharmaceutiques Grand Public; The Association of the European Self-Medication Industry) (1998) *Herbal Medicinal Products in the European Union. Study Carried out on Behalf of the European Commission*, Brussels [http://pharmacos.eudra.org/F2/pharmacos/docs/doc99/Herbal%20Medecines%20EN.pdf]

Allison, D.B., Fontaine, K.R., Heshka, S., Mentore, J.L. & Heymsfield, S.B. (2001) Alternative treatments for weight loss: A critical review. *Crit. Rev. Food Sci. Nutr.*, **41**, 1–28

Angell, M. & Kassirer, J.P. (1998) Alternative medicine — The risks of untested and unregular remedies. *New Engl. J. Med.*, **339**, 839–841

Awang, D.V.C. (1997) Quality control and good manufacturing practices: Safety and efficacy of commercial herbals. *Food Drug Law Inst.*, **52**, 341–344

Berger, E. (2001) *The Canada Health Monitor Surveys of Health Issues in Canada, Survey 22*, Ottawa, Health Canada

Bhat, K.K.P. (1995) Medicinal plant information databases. In: *Non-Wood Forest Products. 11. Medicinal Plants for Conservation and Health Care*, Rome, Food and Agriculture Organization

Blumenthal, M., Busse, W.R., Goldberg, A., Gruenwald, J., Hall, T., Riggins, C.W. & Rister, R.S., eds (1998) *The Complete German Commission E Monographs: Therapeutic Guide to Herbal*

Medicines, Austin, TX/Boston, MA, American Botanical Council/Integrative Medicine Communications
Calixto, J.B. (2000) Efficacy, safety, quality control, marketing and regulatory guidelines for herbal medicines (phytotherapeutic agents). *Braz. J. med. biol. Res.*, **33**, 179–189
Castot, A., Djezzar, S., Deleau, N., Guillot, B. & Efthymiou, M.L. (1997) [Drug surveillance of herbal medicines] *Thérapie*, **52**, 97–103 (in French)
Chan, T.Y.K. (1997) Monitoring the safety of herbal medicines. *Drug Safety*, **17**, 209–215
Chang, J. (1999) Scientific evaluation of traditional Chinese medicine under DSHEA: A conundrum. *J. altern. complem. Med.*, **5**, 181–189
Cho, B.-H. (2000) The politics of herbal drugs in Korea. *Soc. Sci. Med.*, **51**, 505–509
Council of Europe (1996) *European Pharmacopoeia*, 3rd Ed., Strasbourg
De Smet, P.A. (1995) Should herbal medicine-like products be licensed as medicines? *Br. med. J.*, **310**, 1023–1024
Eguchi, K., Hyodo, I. & Saeki, H. (2000) Current status of cancer patients' perception of alternative medicine in Japan. A preliminary cross-sectional survey. *Sup. Care Cancer*, **8**, 28–32
EMEA (European Agency for the Evaluation of Medicinal Products) (2000) *Working Party on Herbal Medicinal Products: Position paper on the risks associated with the use of herbal products containing* Aristolochia species *(EMEA/HMPWP/23/00)*, London
ESCOP (European Scientific Cooperative on Phytotherapy) (1999) *ESCOP Monographs on the Medicinal Uses of Plant Drugs*, Exeter, UK
European Commission (1965) Council Directive 65/65/EEC of 26 January 1965 on the approximation of provisions laid down by Law, Regulation or Administrative Action relation to proprietary medicinal products. *Off. J.*, **P22**, 369–373
European Commission (1975) Second Council Directive 75/319/EEC of 20 May 1975 on the approximation of provisions laid down by Law, Regulation or Administrative Action relating to proprietary medicinal products. *Off. J.*, **L147**, 13–22
European Commission (1992a) Council Directive 92/28/EEC of 31 March 1992 on the advertising of medicinal products for human use. *Off. J.*, **L113**, 13–18
European Commission (1992b) Council Directive 92/25/EEC of 31 March 1992 on the wholesale distribution of medicinal products for human use. *Off. J.*, **L113**, 1–4
Food and Drug Administration (2000) *Guidance for Industry: Botanical Drug Products*, Washington DC, Center for Drug Evaluation and Research [http://www.fda.gov/cder/guidance/index.htm]
Food and Drug Administration (2002) *Good Manufacturing Practices (GMP)/Quality System (QS) Regulation* [http://www.fda.gov/cdrh/dsma/cgmphome.html]
Goldman, S.A. & Kennedy, D.L. (1998) FDA's Medical Products Reporting Program. A joint effort toward improved public health. *Postgr. Med.*, **103**, 13–16
Griffin, J.P. (1998) The evolution of human medicines control from a national to an international perspective. *Adverse Drug React. Toxicol. Rev.*, **17**, 19–50
Health Canada Online (2002a) Advisory: Health Canada Requests Recall of Certain Products containing Ephedra/ephedrine. Jan. 9, 2002 [http:www.hc.sc.gc.ca/english/protection/warnings/2002/2002_øie.htm]
Health Canada Online (2002b) Advisory: Health Canada is advising consumers not to use any products containing kava. Jan. 16, 2002 [http:www.hc.sc.gc.ca/english/protection/warnings/2002/2002_ø2e.htm]

Jellin, J.M. (2002) *Natural Medicines Comprehensive Database (Pharmacists Letter/Prescribers Letter)*, Stockton, CA, Therapeutic Research Faculty

Johnson, T. (2001) News. Chinese medicine now part of primary care scene in BC. *Can. med. Assoc. J.*, **164**, 1195

Keller, K. (1991) Legal requirements for the use of phytopharmaceutical drugs in the Federal Republic of Germany. *J. Ethnopharmacol.*, **32**, 225–229

Kozyrskyj, A. (1997) Herbal products in Canada. How safe are they ? *Can. Family Phys. Med. Famille Can.*, **43**, 697–702

Kraft, K. (1999) Herbal medicine products and drug law. *Forsch. Komplementärmed.*, **6**, 19–23

Li, L. (2000) [Opportunity and challenge of traditional Chinese medicine in face of the entrance to WTO (World Trade Organization)]. *Chin. Inform. trad. Chin. Med.*, **7**, 7–8 (in Chinese)

Mills, S.Y. (2001) The House of Lords report on complementary medicine. *Complem. Ther. Med.*, **9**, 34–39

Ministry of Agriculture, Fisheries and Food (MAFF) (1998) *Herb Legislation*, London, Herb Society

Morgan, K. (2002) Medicine of the Gods: Basic Principles of Ayurvedic Medicine [http://www.compulink.co.uk/~mandrake/ayurveda.htm]

Moulds, R.F.W. & McNeil, J.J. (1988) Herbal preparations — To regulate or not to regulate ? *Med. J. Austral.*, **149**, 572–574

NAPRALERT (2001) University of Illinois at Chicago, Program for Collaborative Research in the Pharmaceutical Sciences [http://pcog8.pmmp.inc.edu/mcp/mcp.html]

Non-prescription Drug Manufacturers Association of Canada (1998) *Health Vision 98*, Ottawa

Nutrition Business Journal (2000) *Global Nutrition Industry*, San Diego, CA

Nutrition Business Journal (2001) *US Nutrition Industry: Top 70 Supplements 1997–2000*, San Diego, CA

Saito, H. (2000) Regulation of herbal medicines in Japan. *Pharmacol. Regul.*, **41**, 515–519

Sandberg, F. & Corrigan, D. (2001) *Natural Remedies. Their Origins and Uses*, New York, Taylor & Francis

Schulz, V., Hänsel, R. & Tyler, V.E. (2001) *Rational Phytotherapy. A Physician's Guide to Herbal Medicine*, 4th Ed., Berlin, Springer-Verlag

Shaw, D. (1998) Risks or remedies ? Safety aspects of herbal remedies. *J. Roy. Soc. Med.*, **91**, 294–296

Sibbald, B. (1999) New federal office will spend millions to regulate herbal remedies, vitamins. *Can. med. Assoc. J.*, **160**, 1355–1357

Therapeutic Goods Administration (1999) *Medicines Regulation and the TGA* (December 1999), Woden, ACT, Australia

Tyler, V.E. (2000) Herbal medicine: From the past to the future. *Public Health Nutr.*, **3**, 447–452

WHO (1993) *Research Guidelines for Evaluating the Safety and Efficacy of Herbal Medicines*, Manila

WHO (1996) *Annex II. Guidelines for the Assessment of Herbal Medicines* (WHO Technical Report Series No. 863), Geneva

WHO (1999) *WHO Monographs on Selected Medicinal Plants*, Vol. 1, Geneva

Zhang, X. (1998) *Regulatory Situation of Herbal Medicines. A Worldwide Review* (WHO/trm/98.1), Geneva, World Health Organization

B. *ARISTOLOCHIA* SPECIES AND ARISTOLOCHIC ACIDS

1. Exposure Data

1.1 Origin, type and botanical data

Aristolochia species refers to several members of the genus (family *Aristolochiaceae*) (WHO, 1997) that are often found in traditional Chinese medicines, e.g., *Aristolochia debilis*, *A. contorta*, *A. manshuriensis* and *A. fangchi*, whose medicinal parts have distinct Chinese names. Details on these traditional drugs can be found in the *Pharmacopoeia of the People's Republic of China* (Commission of the Ministry of Public Health, 2000), except where noted. This Pharmacopoeia includes the following *Aristolochia* species:

Aristolochia species	Part used	Pin Yin Name
Aristolochia fangchi	Root	Guang Fang Ji
Aristolochia manshuriensis	Stem	Guan Mu Tong
Aristolochia contorta	Fruit	Ma Dou Ling
Aristolochia debilis	Fruit	Ma Dou Ling
Aristolochia contorta	Herb	Tian Xian Teng
Aristolochia debilis	Herb	Tian Xian Teng
Aristolochia debilis	Root	Qing Mu Xiang

In traditional Chinese medicine, *Aristolochia* species are also considered to be interchangeable with other commonly used herbal ingredients and substitution of one plant species for another is established practice. Herbal ingredients are traded using their common Chinese Pin Yin name and this can lead to confusion. For example, the name 'Fang Ji' can be used to describe the roots of *Aristolochia fangchi*, *Stephania tetrandra* or *Cocculus* species (EMEA, 2000).

Plant species supplied as 'Fang Ji'

Pin Yin name	Botanical name	Part used
Guang Fang Ji	*Aristolochia fangchi*	Root
Han Fang Ji	*Stephania tetrandra*	Root
Mu Fang Ji	*Cocculus trilobus*	Root
Mu Fang Ji	*Cocculus orbiculatus*	Root

Similarly, the name 'Mu Tong' is used to describe *Aristolochia manshuriensis*, and certain *Clematis* or *Akebia* species. There are some reports in Chinese literature where substitution can occur with 'Ma Dou Ling' (EMEA, 2000).

Plant species supplied as 'Mu Tong'

Pin Yin name	Botanical name	Part used
Guan Mu Tong	*Aristolochia manshuriensis*	Stem
Chuan Mu Tong	*Clematis armandii*	Stem
Chuan Mu Tong	*Clematis montana*	Stem
Bai Mu Tong	*Akebia quinata*	Stem
Bai Mu Tong	*Akebia trifoliata*	Stem

The Pin Yin name 'Mu Xiang' is applied to a number of species; there is no evidence of substitution between the species but the common names have potential for confusion in both Chinese and Japanese (EMEA, 2000).

Plant species supplied as 'Mu Xiang'

Pin Yin name	Botanical name	Japanese name
Qing Mu Xiang	*Aristolochia debilis*	Sei-Mokkou
Mu Xiang	*Aucklandii lappa*	
Guang Mu Xiang	*Saussurea lappa*	Mokkou
Tu Mu Xiang	*Inula helenium*	
	Inula racemosa	
Chuan Mu Xiang	*Vladimiria souliei*	Sen-Mokkou
	Vladimiria souliei var. cinerea	

1.1.1 Aristolochia contorta *Bunge and* Aristolochia debilis *Sieb. et Zucc.*

Aristolochia contorta (see Figure 1) is a perennial climbing herb. The stem is convoluted, smoothish, more than 2 metres long. Leaves are alternated, petioled, entire, triangular cordate-shaped, 3–13 cm long and 3–10 cm wide, with a soft (effeminate) stem (petiole), 1–7 cm long. Axillary racemes have 3–10 flowers clustered with a dark purple perianth which is 2–3 cm long and zygomorphic. Peduncles are 2 cm long, with a small ovate bract, 1.5 cm long and 1 cm wide, near the base. Flowers have an obliquely trumpet-shaped upper part with acuminate apex and a tubular middle part. The lower part is enclosing the style and globular, six stamens and six stigmas. Capsules, broadly obovate-shaped, burst into six valves when ripe. The plant grows in valleys along streams and in thickets (see WHO, 1997, 1998).

Figure 1. ***Aristolochia contorta*** **Bunge**

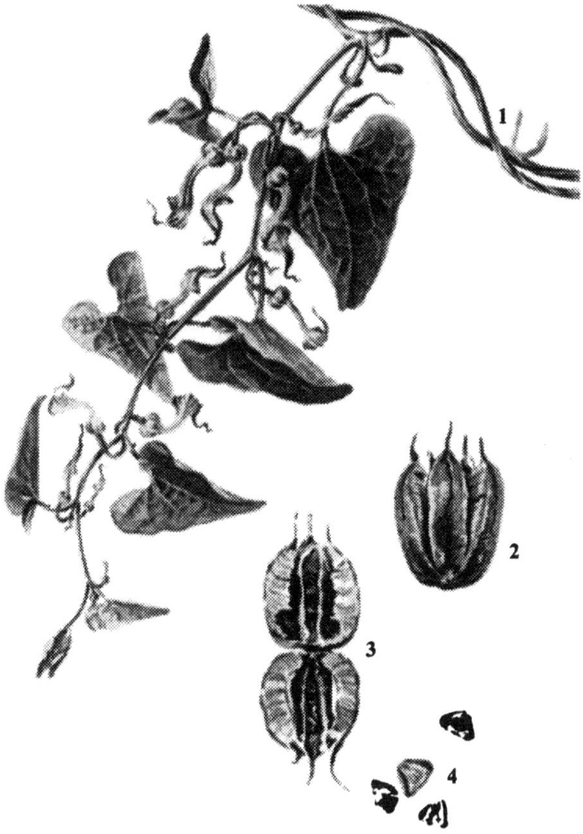

From Qian (1996)
1, flower twig; 2, fruit; 3, longitudinal section of fruit; 4, seed

Aristolochia debilis (see Figure 2) differs from *A. contorta* chiefly in the following: Leaves are 3–8 cm long and 2–4 cm wide, cordate at the base, with round auricles on both sides and a petiole 12 cm long. Capsules are subglobose or oblong.

The dried fruits of both *A. debilis* and *A. contorta* (Ma Dou Ling), also known as *Fructus Aristolochiae* in Latin and Dutchman's pipe fruit in English, are ovoid and 3–7 cm long and 2–4 cm in diameter. The outer surface is yellowish green, greyish green or brown, with 12 longitudinal ribs, from which extend numerous horizontal parallel veinlets. The apex is flattened and obtuse and the base has a slender fruit stalk. The pericarp, which is light and fragile, is easily divided into six valves; the fruit stalk is also divided into six splittings. The inner surface of the pericarp is smooth and lustrous, with dense transverse veins. The fruit is six-locular, with each locule containing many seeds, which are overlapped and arranged regularly. The seeds are flat and thin, obtuse triangular or fan-shaped, 6–10 mm long, 8–12 mm wide and winged all around, and pale brown. The fruit has a characteristic odour and a slightly bitter taste.

Figure 2. ***Aristolochia debilis*** **Siebold and Zuccarini**

From Qian (1996)
1, flower twig; 2, root; 3, fruit

Tian Xian Teng (*Herba Aristolochiae* in Latin and Dutchman's pipe vine in English) consists of stems of *A. debilis* and *A. contorta* that are slenderly cylindrical, slightly twisted, 1–3 mm in diameter, yellowish green or pale yellowish brown in colour, with longitudinal ridges and nodes and internodes varying in length. The texture is fragile; the stems are easily broken and when fractured exhibit several vascular bundles of variable size. The leaves are mostly crumpled and broken, but, when whole, are deltoid narrow ovate or deltoid broad ovate and cordate at the base, dark green or pale yellowish brown and basal leaves are clearly veined and slenderly petioled. It has a delicately aromatic odour and is weak to the taste.

Qing Mu Xiang (*Radix Aristolochiae* in Latin and slender Dutchman's pipe root in English) is the root of *A. debilis* and is cylindrical or compressed cylindrical, slightly tortuous, 3–15 cm long and 0.5–1.5 cm in diameter. It is yellowish brown or greyish brown in colour, rough and uneven, and exhibits longitudinal wrinkles and rootlet scars. The texture is fragile; the root is easily broken and when fractured shows an uneven, pale yellow bark and a wood with broad, whitish rays arranged radially and a distinct, yellowish brown cambium ring. Its odour is aromatic and characteristic and it has a bitter taste.

1.1.2 Aristolochia manshuriensis

The dried stem of *A. manshuriensis* (see Figure 3) is called Guan Mu Tong (*Caulis Aristolochiae manshuriensis* in Latin and commonly Manchurian Dutchman's pipe stem).

Figure 3. *Aristolochia manshuriensis*

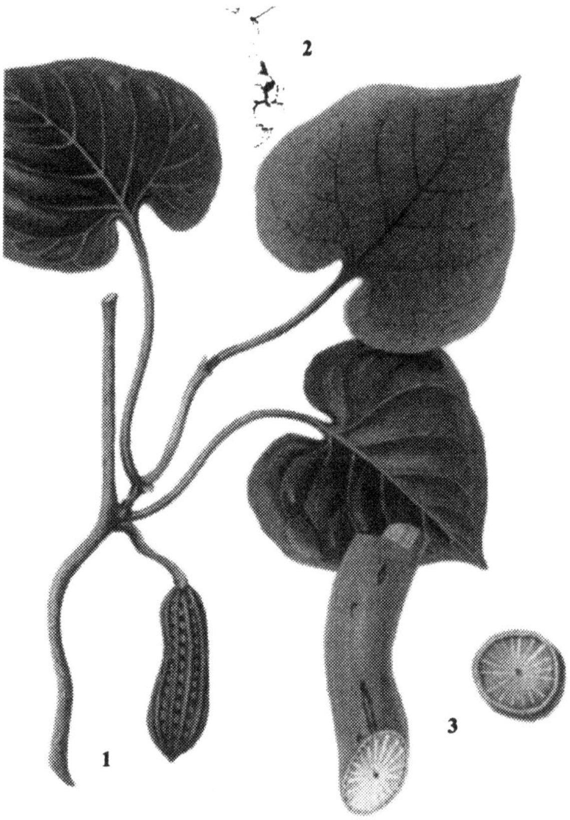

From Qian (1996)
1, fruit twig; 2, flower; 3, transverse section

The stem is twined and woody, grey and striated, from 6 to 14 m long. Its branches are dark purple, with whitish villi. Leaves are either petioled, entire or ovate-cordate and are 15–29 cm long and 13–28 cm wide with a whitish down on each side. Petioles are 6–8 cm long. Flowers are axillary, single and peduncled; the apex is disc-shaped, with a diameter of 4–6 cm or more, lobed, wide triangular and greenish outside; the lower part is tubular, 5–7 cm long, 1.5–2.5 cm in diameter and pink outside. Peduncles are drooping, with an ovate cordate, whitish, villous, sessile, greenish bract, 1 cm long. Six stamens adhibit to the outside of the stigma in pairs; they are inferior to the ovary, cylindrical, angular, whitish villi and 1–2 cm long. Capsules are cylindrical, dark brown, 9–10 cm long and 3–4 cm in diameter and burst into six valves when ripe. Seeds are numerous in each cell and cordate to triangular-shaped, with no winged margin.

The dried stem is long, cylindrical, slightly twisted, 1–2 m long and 1–6 cm in diameter. Externally, it is greyish yellow or brownish yellow, with shallow longitudinal grooves and has adhering remains of brown patches of coarse bark. Nodes are slightly swollen, with a branch scar. The stem is light and hard, not easily broken, but when fractured shows a narrow yellow or pale yellow bark, and broad wood having vessels arranged in many rings, with radial rays and indistinct pith. It has a slight odour and a bitter taste and gives off a smell like camphor when the remains of the coarse bark are rubbed.

1.1.3 Aristolochia fangchi

The dried root of *A. fangchi* (see Figure 4) is named *Guang Fang Ji* (*Radix Aristolochiae fangchi* in Latin and commonly southern fangchi root).

A. fangchi is a perennial climbing vine with crassitudinous roots. The stem is slender, greyish brown, with brownish villi and is 3–4 m long. The leaves are either petioled, entire, oblong or ovate-shaped, with grey whitish villi when young, and are 3–11 cm long and 2–6 cm wide. The petiole is 1–4 cm long. The perianth is canister-shaped, purple with yellowish spots and 2–3 cm long. Six stamens adhibit to the outside of the trilobed stigma. The capsules have numerous seeds. The plant grows in valleys along streams and in thickets and blooms from May to June.

The dried root is cylindrical or semi-cylindrical, slightly curved, 6–18 cm long and 1.5–4.5 cm in diameter. Externally it is greyish brown, rough and longitudinally wrinkled; the peeled stems are yellowish. The root is heavy and compact and is broken with difficulty; when fractured, the inside is starchy, exhibiting alternately greyish brown and whitish radial lines. It is odourless with a bitter taste.

1.1.4 Aristolochia clematitis

The medicinal parts of birthwort (*A. clematitis*) are the aerial portion (when in blossom) and the root. The plant has dirty yellow flowers, usually in axillary groups of seven. The perigone forms a straight tube, which is bulbous beneath and has a linguiform, oblong-ovate, obtuse border. There are six stamens, the style is upward-growing , and the stigma is six-lobed. The flower briefly traps insects that pollinate it. The fruit is a globose, pear-shaped capsule. The plant grows to a height of 30–100 cm. The stem is erect, simple, grooved and glabrous. The leaves are alternate, long-petioled, cordate-reniform, yellow-green, with prominent ribs. The plant has a fruit-like fragrance and is poiso-

Figure 4. ***Aristolochia fangchi***

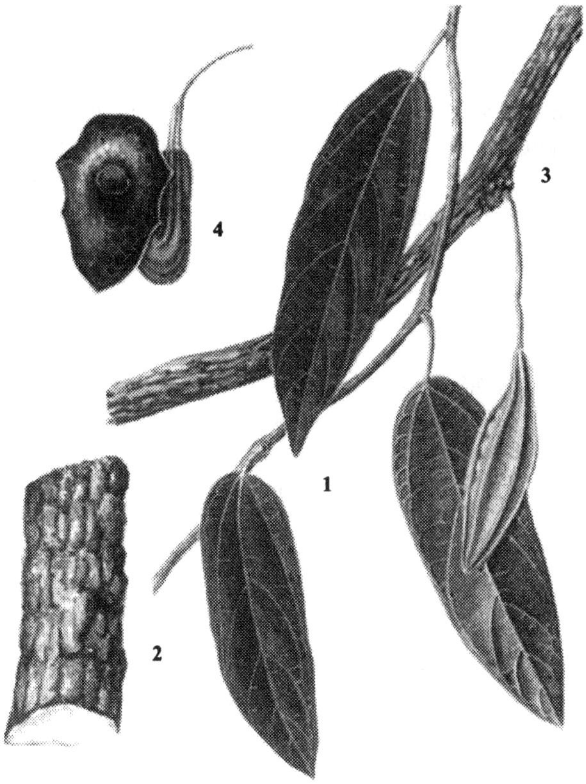

From Qian (1996)
1, twig leaf; 2, longitudinal section; 3, fruit twig; 4, flower

nous. The plant is indigenous to Mediterranean regions, Asia Minor and the Caucasus, but is found in numerous other regions (Medical Economics Co., 2000).

1.1.5 Aristolochia indica

Aristolochia indica (Indian birthwort) is a perennial climber with greenish white woody stems found throughout India in the plains and low hills. The leaves are glabrous and very variable, usually obovate-oblong to sub-pandurate, entire with somewhat undulate margins, somewhat cordate, acuminate. Flowers are few, in axillary racemes with a perianth up to 4 cm long having a glabrous pale-green inflated and lobed base narrowed into a cylindrical tube terminating in a horizontal funnel-shaped purple mouth and a lip clothed with purple-tinged hair. Capsules are oblong or globose-oblong, 3–5 cm long and the seeds are flat, ovate and winged (Anonymous, 1985).

1.2 Use

1.2.1 Aristolochia contorta *and* Aristolochia debilis

Several parts of *A. debilis* and *A. contorta* have been used for various therapeutic purposes in Chinese traditional medicine. The dried ripe fruits of both plants are used as a remedy for haemorrhoids, coughs and asthma. The dried stems or leaves are used for treatment of epigastric pain, arthralgia and oedema. The dried roots from *A. debilis* are used for treatment of dizziness, headache, abdominal pain, carbuncles, boils and snake and insect bites.

1.2.2 Aristolochia manshuriensis

There are no clinical reports concerning the use of dried stems of *A. manshuriensis* alone. It is usually used in complex prescriptions as an anti-inflammatory and diuretic for acute infections of the urinary system, and as emmenagogue and galactagogue for amenorrhoea and scanty lactation in traditional Chinese medicine.

1.2.3 Aristolochia fangchi

There are no clinical reports concerning the use of the stems of *A. fangchi* alone. It is usually used in complex prescriptions as a diuretic for oedema and for antipyretic and analgesic remedies in traditional Chinese medicine.

1.2.4 Aristolochia clematitis

Birthwort (*A. clematitis*) has been used to stimulate the immune system and in the treatment of allergically caused gastrointestinal and gall-bladder colic. The plant is used in a wide variety of ways in the folk medicine of nearly all European countries. In homeopathy, the drug is used for gynaecological disorders and climacteric symptoms, in addition to the treatment of wounds and ulcers. It is also used as a treatment after major surgery and in ear–nose–throat treatments (Medical Economics Co., 1998, 2000).

1.2.5 Aristolochia indica

The roots of Indian birthwort (*Aristolochia indica* L.) have been used in Indian folk medicine as an emmenagogue and an abortifacient (Che *et al.*, 1984).

1.3 Chemical constituents

[As most of the literature is in Chinese, the Working Group had difficulty in identifying some of the chemicals cited.]

1.3.1 Aristolochia contorta *and* Aristolochia debilis

Dried ripe fruits

Alkaloids: *A. debilis* contains aristolochic acids[1] I, II, IIIa and IVa, debilic acid, 7-hydroxyaristolochic acid I and 7-methoxyaristolochic acid I. *A. contorta* contains aristolochic acids I, IIIa and E, 7-methoxy-8-hydroxyaristolochic acid, methyl aristolochate, aristolic acid, aristolic acid methyl ester, aristolamide, aristolochic acid III methyl ester, aristolochic acid IV methyl ester, 6-methoxyaristolochic acid [methyl] ester, aristolochic acid BII methyl ester (3,4-dimethoxy-10-nitrophenanthrenic-1-acid methyl ester) and aristolophenanlactone I. Aristolactam (also known as aristololactam), magnoflorine and cyclanoline have also been detected in these plants (Zheng *et al.*, 1998; Commission of the Ministry of Public Health, 2000).

Terpenoids and steroids characterized in these fruits include aristolene, $\Delta^{1(10)}$-aristolene, $\Delta^{1(10)}$-aristolenone, debilone, $\Delta^{1(10),8}$-aristolodion-2-one, 9-aristolene, aristolone, Δ^{9}-aristolone, 3-oxoishwarane, β-sitosterol, stigmast-4-en-3-one, stigmast-4-en-3,6-dione and stigmastane-3,6-dione.

Other components include allantoin, flavones, coumarins, saccharides, gum, resinstanin, lignanoids and lipids (Zheng *et al.*, 1999).

Dried roots

Essential oil constituents found in *A. debilis* roots include 2-furaldehyde, camphene, 2-pentylfuran, 1,8-cineole, camphor, borneol, bornyl formate, bornyl acetate, α-copaene, β-elemene, *cis*-caryophyllene, 1,2-aristolene, α-gurjunene, β-gurjunene, δ-cadinene, tetradehydroaristolane, aristolenone, isoaristolenone, $\Delta^{1(10),8}$-aristol-2-one, 9-aristolene, Δ^{9}-aristolone and 3-oxoishwarane.

Alkaloids found in this product include aristolochic acids I, II, IIIa, IV, 7-hydroxyaristolochic acid I, 7-methylaristolochic acid I, aristolochic acid III methyl ester, methyl aristolochate, debilic acid, magnoflorine, aristolactam and cyclanoline.

Other components include allantoin (Zheng *et al.*, 1997, 1998).

1.3.2 Aristolochia manshuriensis

Components of this plant include aristolic acid II, aristolochic acids I, II, IV, IIIa and IVa, aristolactam IIIa, tannin, aristoloside, magnoflorine and β-sitosterol (Wang *et al.*, 2000).

---

[1] Aristolochic acids: I = A, II = B, IIIa = C, IVa = D

1.3.3 Aristolochia fangchi

Components of this plant include mufongchins A, B, C and D, *para*-coumaric acid, syringic acid, palmitic acid, aristolochic acids I, II and IIIa, allantoin, magnoflorine, aristolactam and β-sitosterol (Zheng *et al.*, 1999; Commission of the Ministry of Public Health, 2000).

1.3.4 Aristolochia clematitis

Components of this plant include aristolochic acids I and II, as well as the alkaloids magnoflorine and corytuberine.

1.3.5 Aristolochia indica

The essential oil of the aerial parts of *A. indica* is dominated by sesquiterpenes and monoterpenes such as β-caryophyllene, α-humulene, ishwarone, caryophyllene oxide I, ishwarol, ishwarane, aristolochene, linalool and α-terpinolene (Jirovetz *et al.*, 2000).

The roots of *A. indica* contain aristolindiquinone, aristololide, 2-hydroxy-1-methoxy-4*H*-dibenzo[de,g]quinoline-4,5-(6*H*)-dione, cepharadione, aristolactam IIa, β-sitosterol-β-D-glucoside, aristolactam glucoside I, stigmastenones II and III, methyl aristolate, ishwarol, ishwarane and aristolochene.

Other components found in *A. indica* include 12-nonacosenoic acid methyl ester, aristolic acid, (12*S*)-7,12-secoishwaran-12-ol, (+)-ledol, ishwarone, methyl aristolate, *para*-coumaric acid, 5βH,7β,10α-selina-4(14),11-diene, isoishwarane, aristolochic acids I, IVa, aristolochic acid IVa methyl ether lactam and aristolactam β-D-glucoside (Kupchan & Merianos, 1968; Ganguly *et al.*, 1969; Fuhrer *et al.*, 1970; Govindachari *et al.*, 1970; Govindachari & Parthasarathy, 1971; Teng & DeBardeleben, 1971; Govindachari *et al.*, 1973; Pakrashi *et al.*, 1977; Pakrashi & Chakrabarty, 1978a,b; Pakrashi & Pakrasi, 1978; Pakrashi & Shaha, 1978; Cory *et al.*, 1979; Pakrashi & Shaha, 1979a,b; Pakrashi *et al.*, 1980; Achari *et al.*, 1981, 1982, 1983; Che *et al.*, 1983, 1984; Achari *et al.*, 1985; Ganguly *et al.*, 1986; Mahesh & Bhaumik, 1987).

1.4 Active components

1.4.1 Aristolochia contorta *and* Aristolochia debilis

Hypotensive activity: magnoflorine, some aristolochic acid derivatives (Xu, 1957).
Analgesic effect: aristolochic acid

1.4.2 Aristolochia manshuriensis

Cardiotonic action: Calcium, tannin, dopamine (Zhou & Lue, 1958; Bulgakov *et al.*, 1996).

Antitumour action: Aristoloside, aristolic acid II, aristolochic acids I, II, IV, IIIa, IVa and aristolactam IIIa (Nagasawa *et al.*, 1997; Wang *et al.*, 2000).

1.4.3 Aristolochia fangchi

Hypotensive activity: magnoflorine (Zheng *et al.*, 1999).

1.5 Sales and consumption

About 320 tonnes of dried stems of *A. manshuriensis* were consumed in China in 1983 (Chinese Materia Medica, 1995).

1.6 Components with potential cancer hazard: aristolochic acids

For the purpose of this monograph, unless otherwise specified, the term 'aristolochic acids' refers to an extract of *Aristolochia* species comprising a mixture of aristolochic acid I and its demethoxylated derivative, aristolochic acid II. *Aristolochia* species also contain the related aristolactams, which are phenanthrene cyclic amides (EMEA, 2000). In some of the older literature, it is unclear whether individual compounds or mixtures are being discussed when referring to 'aristolochic acid'.

1.6.1 *Nomenclature*

Aristolochic acid I

Chem. Abstr. Serv. Reg. No.: 313-67-7
Deleted CAS Nos.: 12770-90-0; 61117-05-3
Chem. Abstr. Serv. Name: 8-Methoxy-6-nitrophenanthro[3,4-d]-1,3-dioxole-5-carboxylic acid
Synonyms and trade names: Aristinic acid; aristolochia yellow; aristolochic acid A; aristolochin; aristolochine; Descresept; isoaristolochic acid; 8-methoxy-3,4-methylenedioxy-10-nitrophenanthrene-1-carboxylic acid; 3,4-methylenedioxy-8-methoxy-10-nitro-1-phenanthrenecarboxylic acid; Tardolyt; TR 1736

Aristolochic acid II

Chem. Abstr. Serv. Reg. No.: 475-80-9
Deleted CAS No.: 79468-63-6
Chem. Abstr. Serv. Name: 6-Nitrophenanthro[3,4-d]-1,3-dioxole-5-carboxylic acid
Synonyms: Aristolochic acid B; 3,4-methylenedioxy-10-nitrophenanthrene-1-carboxylic acid

1.6.2 *Structural and molecular formulae and relative molecular mass*

Aristolochic acid I

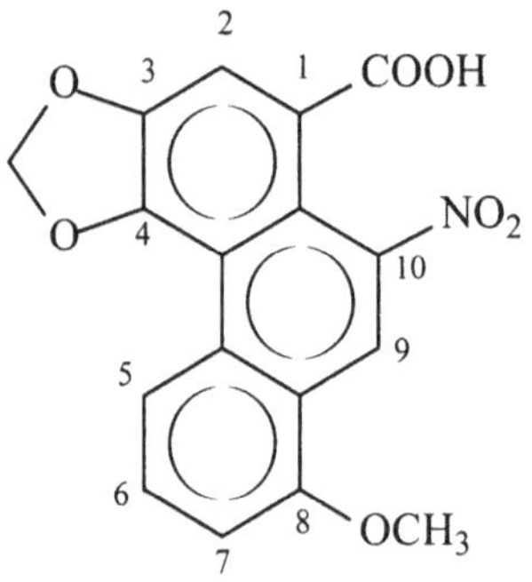

$C_{17}H_{11}NO_7$ Rel. mol. mass: 341.27

Aristolochic acid II

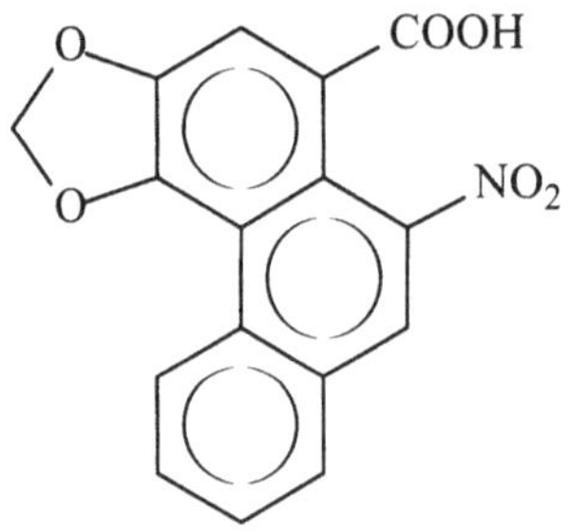

$C_{16}H_9NO_6$ Rel. mol. mass: 311.25

1.6.3 *Chemical and physical properties of the pure substance*

Aristolochic acids

(*a*) *Description*: Crystalline solid [aristolochic acid I] (Buckingham, 2001)
(*b*) *Melting-point*: 281–286 °C, decomposes [aristolochic acid I] (Buckingham, 2001)
(*c*) *Solubility*: Slightly soluble in water; soluble in acetic acid, acetone, aniline, alkalis, chloroform, diethyl ether and ethanol; practically insoluble in benzene and carbon disulfide (O'Neil, 2001)
(*d*) *Octanol/water partition coefficient (P)*: log P, 3.48 (Buckingham, 2001)

1.6.4 *Analysis*

A procedure based on an extraction method used in Germany for the determination of aristolochic acids in botanical products has been developed and applied to a variety of botanicals and botanical-containing dietary supplements. Aristolochic acids are extracted from the sample matrix with aqueous methanol/formic acid. The concentration of aristolochic acids in the extract is determined by gradient high-performance liquid chromatography (HPLC) with UV absorption detection at 390 nm and their identity is confirmed by liquid chromatography/mass spectrometry using either an ion-trap mass spectrometer or a triple quadrupole mass spectrometer. The quantitation limit is equivalent to 1.7 μg/g in solid samples and 0.14 μg/mL in liquid samples (Flurer *et al.*, 2000).

Lee *et al.* (2001) developed an HPLC procedure with a silica gel RP-18 reversed-phase column to determine aristolochic acids I and II in medicinal plants and slimming products. The recovery of these two compounds in medicinal plants and slimming products by extracting with methanol and purifying through a PHP-LH-20 (piperidino-hydroxypropyl Sephadex LH-20) column was better than 90%.

Targeted liquid chromatography/serial mass spectrometry (LC/MS/MS) analysis, using a quadrupole ion-trap mass spectrometer, permitted the detection of aristolochic acids I and II in crude 70% methanol extracts of multi-component herbal remedies without any clean-up or concentration stages. The best ionization characteristics were obtained using atmospheric pressure chemical ionization (APCI) and by including ammonium ions in the mobile phase. Limits of detection for aristolochic acids were influenced by the level of interference due to other components in the sample matrix. They were determined to be between 250 pg and 2.5 ng on-column within a matrix containing compounds extracted from 2 mg of herbal remedy (Kite *et al.*, 2002).

Ong and Woo (2001) developed a method for the analysis of aristolochic acids in medicinal plants or Chinese prepared medicines using capillary zone electrophoresis (CZE). The limits of detection for aristolochic acids I and II were 30 and 22.5 mg/kg, respectively. The proposed method using pressurized liquid extraction with CZE was used to determine the amount of aristolochic acids in medicinal plants or samples of Chinese prepared medicines with complex matrix and the results were compared with those from HPLC. Results obtained for aristolochic acids I and II in medicinal plants by CZE and HPLC are presented in Table 1.

Ong *et al.* (2000) compared extraction and analysis of aristolochic acids I and II in medicinal plants (*Radix aristolochiae*) using a pressurized liquid extraction method in a dynamic mode with ultrasonic and Soxhlet extraction. The effects of temperature, volume of solvent required and particle size were investigated. The pressurized liquid extraction method showed some advantages over ultrasonic and Soxhlet extraction methods.

Singh *et al.* (2001a,b) developed a reversed-phase HPLC method with photodiode array detection for quantitative detection of aristolochic acids in *Aristolochia* plant samples. The procedure involves extraction of aristolochic acids with methanol and

Table 1. Comparison of results obtained for aristolochic acids I and II in medicinal plants by capillary zone electrophoresis (CZE) and high-performance liquid chromatography (HPLC)[a]

	CZE (mg/kg)	HPLC (mg/kg)
Aristolochic acid I in *Radix aristolochiae fangchi*	541.3	479.8
Aristolochic acid II in *Radix aristolochiae fangchi*	314.6	57.0
Aristolochic acid I in *Radix aristolochiae fangchi* [b]	612.0	564.1
Aristolochic acid I in *Radix aristolochiae (qing mu xiang)*	654.9	567.6
Aristolochic acid II in *Radix aristolochiae (qing mu xiang)*	190.0	208.4

[a] From Ong & Woo (2001)
[b] Analysis of a different batch of *Radix aristolochiae fangchi*

chromatographic separation with a mobile phase of acetonitrile–water–trifluoroacetic acid–tetrahydrofuran (50:50:1:1). The average recovery of aristolochic acids was 97.8%; the minimum quantity detectable was 0.10 μg per injection with a 5 μL injection volume.

1.6.5 *Production*

Aristolochic acids are produced commercially only as a reference standard and as research chemicals (Sigma-Aldrich, 2002).

1.6.6 *Use*

The aristolochic acid occurring in *Aristolochia* species used in traditional herbal medicines has been reported to function as a phospholipase A_2 inhibitor and as an anti-neoplastic, antiseptic, anti-inflammatory and bactericidal agent (Buckingham, 2001).

1.6.7 *Occurrence*

Aristolochic acids are alkaloid components of a wide range of species of the family Aristolochiaceae (e.g., *Aristolochia*, *Asarum*). They are also found in several species of butterflies (e.g., *Atrophaneura*, *Battus, Pachliopta*, *Troides*) which feed on the *Aristolochia* plants (von Euw *et al.*, 1968; Urzúa & Priestap, 1985; Urzúa *et al.*, 1983, 1987;

Nishida & Fukami, 1989a,b; Nishida *et al.*, 1993; Sachdev-Gupta *et al.*, 1993; Fordyce, 2000; Klitzke & Brown, 2000; Sime *et al.*, 2000; Wu *et al.*, 2000).

Aristolochic acids I and II have been determined in several samples of medicinal plants and slimming products using HPLC. The major component was aristolochic acid I in *Aristolochia fangchi* and the level ranged from 437 to 668 ppm (mg/kg). Aristolochic acid II was the major component in *Aristolochia contorta*, at levels ranging from less than 1 to 115 ppm (mg/kg). Twelve out of 16 samples of slimming pills and powders contained aristolochic acids I and/or II. The major component in most slimming products was aristolochic acid II and the level ranged from less than 1 to 148 ppm (Lee *et al.*, 2001).

The amounts of aristolochic acids I and II in four groups of medicinal plants from the family Aristolochiaceae and some related plants were determined by HPLC. Aristolochic acids I and II were detected in all the plants from the genus *Aristolochia* (Aristolochiaceae) and at trace levels in some from the genus *Asarum* (Aristolochiaceae) (Hashimoto *et al.*, 1999). The levels of these compounds in several medicinal plants are presented in Table 2.

Table 2. Aristolochic acid content in individual samples of medicinal plants of the Aristolochiaceae family[a]

Botanical name	AA-I (mg/kg)	AA-II (mg/kg)
Aristolochia debilis	1010	180
	1080	120
	790	80
Aristolochia manshuriensis	3010	210
	1690	140
	8820	1000
Aristolochia fangchi	2220	220
	1430	60
	1030	40

[a] From Hashimoto *et al.* (1999)
AA-I, aristolochic acid I
AA-II, aristolochic acid II

1.6.8 *Regulations*

The Therapeutic Goods Administration (TGA) (2001a) of Australia has issued a Fact Sheet stating that all species of *Aristolochia* are prohibited for supply, sale or use in therapeutic goods in Australia. The Therapeutic Goods Administration (2001b) also issued a Practitioner Alert to communicate its concern about traditional Chinese medicine herbal products that are known to contain, or suspected to contain, *Aristolochia* species, which may contain aristolochic acids. The TGA has published three lists of botanicals or products at risk of containing aristolochic acids: botanicals known or suspected to contain aristolochic acid (Group A); botanicals which may be adulterated with aristolochic acid (Group B); and products which have Mu Tong and Fang Ji as declared ingredients (Group C).

The use of *Aristolochia* in unlicensed medicines was prohibited by the Medicines Control Agency (2001) in the UK in July 1999, and a further temporary prohibition covering certain herbal ingredients at risk of confusion with *Aristolochia* came into force in June 2000. The UK Committee on Safety of Medicines (2001) made these prohibitions permanent by issuing a Statutory Instrument to prohibit the sale, supply and importation of any medicinal product consisting of or containing certain plants belonging to a species of the genus *Aristolochia* or consisting of or containing Guan Mu Tong; or belonging to any of eight specifically listed plants (*Akebia quinata, Akebia trifoliata, Clematis armandii, Clematis montana, Cocculus laurifolius, Cocculus orbiculatus, Cocculus trilobus, Stephania tetrandra*); or consisting of or containing an extract from such a plant.

In addition, the European Agency for the Evaluation of Medicinal Products issued a position paper in October 2000 (EMEA, 2000), warning European Union Member States to 'take steps to ensure that the public is protected from exposure to aristolochic acids arising from the deliberate use of *Aristolochia* species or as a result of confusion with other botanical ingredients'.

The European Commission (EC) (2000) has prohibited 'aristolochic acid and its salts, as well as *Aristolochia* species', and their preparations in cosmetic products.

In 1999, Health Canada issued a warning not to use products containing *Aristolochia* due to potential risk of cancer, cell changes and kidney failure. Health Canada issued four additional warnings and advisories in 2001 advising not to use products labelled to contain *Aristolochia* (Health Canada, 1999, 2001a,b,c,d).

The Food and Drug Administration (2001a) of the USA has issued a Consumer Advisory to communicate its concern about the use and marketing of dietary supplements or other botanical-containing products that may contain aristolochic acids. It has also posted a listing of botanical ingredients of concern, including: botanicals known or suspected to contain aristolochic acids; botanicals which may be adulterated with aristolochic acids; products in which 'Mu Tong' and 'Fang Ji' are declared ingredients; and botanical products determined by FDA to contain aristolochic acids (Food and Drug Administration, 2001b).

2. Studies of Cancer in Humans

Aristolochia spp.

The possibility of adulteration of herbal products with *Aristolochia* species exists as a result of similarities in their common names, e.g. substitution of *Stephania tetrandra* (Han Fang Ji) with *Aristolochia* species (e.g. Guang Fang Ji).

In 1992, a cluster of patients with interstitial renal fibrosis rapidly progressing to end-stage renal disease after having followed a slimming regimen containing powdered extracts of Chinese herbs was recorded in Brussels, Belgium (Vanherweghem *et al.*, 1993; Cosyns *et al.*, 1994a; Depierreux *et al.*, 1994; Vanherweghem, 1998), followed by some reports from several other countries (see Section 4.2.1 for details). The herbal product was labelled as including *Stephania tetrandra*, but was later found to contain *Aristolochia fangchi*, which had been erroneously substituted for *Stephania tetrandra.*

Subsequent reports have shown the etiology of this disease to be related to exposure to aristolochic acids which are components of *Aristolochia* species (see Section 1.4.1(*c*)) (Vanhaelen *et al.*, 1994; Schmeiser *et al.*, 1996; Nortier *et al.*, 2000; Muniz Martinez *et al.*, 2002).

The renal disease associated with prolonged intake of some Chinese medicinal herbs is called by various names. The term 'Chinese herb nephropathy' has been widely used in scientific nephrology publications. However, this could be considered misleading in relation to the hundreds of Chinese medicinal herbs that are safely used throughout the world, including for renal diseases. Thus alternatives such as 'aristolochic acid-associated nephropathy' and 'Aristolochia nephropathy' are also used. However, throughout this monograph, the term 'Chinese herb nephropathy' is consistently used to refer specifically to the Aristolochia-associated disease.

2.1 Case reports

Mild-to-moderate atypia and atypical hyperplasia of the urothelium were detected in three women in Brussels, Belgium, who had undergone nephroureterectomies as part of a transplantation programme (Cosyns *et al.*, 1994a). All three cases of end-stage renal disease were attributed to the use of Chinese herbs containing aristolochic acids (Schmeiser *et al.*, 1996). One of these three women developed, 12 months after transplantation, at the age of 25 years, two papillary tumours of the posterior bladder wall, histologically classified as low-grade transitional-cell carcinomas, without invasion. Microscopic transitional-cell carcinomas of low-to-intermediate grade were also detected in the two ureters (right and remnant distal part of left ureter) and in the right pelvis (Cosyns *et al.*, 1994b).

A second report from Belgium described a 42-year-old woman presenting with end-stage renal disease and haematuria. Apart from the Chinese herbal product labelled as *Stephania tetrandra*, she had regularly taken paracetamol (1.2–2.4 g per day) for 27 years. In medical surveillance, no side-effects on renal function were recorded previous to the consumption of the herbal product. The haematuria was secondary to a transitional-cell carcinoma with moderate atypia of the right renal pelvis (Vanherweghem *et al.*, 1995).

In a report from Taiwan, China, of 12 cases with rapidly progressive interstitial fibrosis associated with Chinese herbal drugs, one case of a bladder carcinoma was observed (Yang *et al.*, 2000).

A bilateral multifocal transitional-cell urothelial carcinoma occurring six years after the onset of end-stage renal disease was described (Lord *et al.*, 1999, 2001) in one of two cases of Chinese herbal nephropathy in the United Kingdom related to aristolochic acid from *Aristolochia manshuriensis* contained in Mu Tong.

2.2 Prevalence of urothelial cancers among patients with Chinese herb nephropathy

Following the previously described case reports of rare urothelial tumours among some patients who had suffered end-stage renal disease after consumption of Chinese herbs, other patients were offered a bilateral removal of their native kidneys and ureters. High prevalence of urothelial cancers was documented in two series.

In the first series (Cosyns *et al.*, 1999), nephroureterectomies were performed in 10 renal-grafted Chinese herb nephropathy patients. The patients were all women and had a mean age of 40 years (range 27–59 years). In the pelviureteric urothelium, moderate atypia was observed in all samples. Multifocal high-grade carcinoma *in situ* was observed in four patients, in the renal pelvis (three patients), upper ureter (four patients), mid ureter (one patient) and lower ureter (three patients).

In the second series of 39 Chinese herb nephropathy patients with end-stage renal disease in Brussels (31 transplanted patients and eight dialysis patients), bilateral nephroureterectomy of the native kidneys and ureters was performed. Except for a 60-year-old man, all the patients were women (aged 54 ± 7 years). Aristolochic acid-specific DNA adducts were detected in tissue samples from kidneys. Among the 39 patients, 18 cases of urothelial carcinomas were found (prevalence, 46%; 95% confidence interval [CI], 29–62%). Except for one case of bladder cancer, all the carcinomas were located in the upper urinary tract and were almost equally distributed between the pelvis and the ureter. Mild to moderate dysplasia of the urothelium was found in 19 of the 21 patients without urothelial carcinoma. Cumulative doses of herbs labelled as *Stephania tetrandra* (which on analysis proved to contain various levels of *Aristolochia fangchi*) taken by the patients were significantly higher in the group of 18 Chinese herb nephropathy patients with urothelial cancer than in the group of 21 Chinese herb nephropathy patients without cancer (226 ± 23 g versus 167 ± 17 g; $p = 0.035$). Among the 24 patients with a cumulative dose of 200 g or less, eight cases of urothelial cancer were recorded and among the 15 patients who had ingested more than 200 g, 10 cases of urothelial cancer were observed ($p = 0.05$) (Nortier *et al.*, 2000).

3. Studies of Cancer in Experimental Animals

Aristolochic acids

3.1 Oral administration

Mouse: A group of 39 female NMRI mice [age not specified] was given daily doses of 5.0 mg/kg bw aristolochic acids (77.2% aristolochic acid I and 21.2% aristolochic acid II) by gavage for three weeks. A group of 11 vehicle controls was given solvent [unspecified]. The mice were kept for up to 56 weeks with interim sacrifice at 3, 9, 18, 26, 37 and 48 weeks. The remaining eight animals were killed at 56 weeks. At 18 and 26 weeks stages, low- to middle-grade papillomatosis was observed in the forestomach of all mice. Of the mice sacrificed at 37 and 48 weeks, 1/5 mice at each time point had squamous-cell carcinoma. Forestomach carcinoma was diagnosed in all of the eight mice killed at 56 weeks. Adenocarcinoma of the glandular stomach was observed in one mouse at 37 weeks. In addition, cystic papillary adenomas in the renal cortex (6/8 mice), malignant lymphomas (4/8 mice), alveologenic carcinomas (8/8 mice) and haemangiomas in the uterus (3/8 mice) were found at 56 weeks. No tumours were detected in 11 control animals at 56 weeks (Mengs, 1988).

Rat: Groups of 30 male and 30 female Wistar rats, 10 weeks of age, were given aristolochic acids (77.2% aristolochic acid I and 21.2% aristolochic acid II) as their sodium salts in distilled water at 10.0 or 1.0 mg/kg bw for three months and held for up to an additional six months or at 0.1 or 0 mg/kg bw for 3, 6 or 12 months and held for up to an additional four months. Forestomach carcinomas were observed in 13/18 males and 8/13 females given 10 mg/kg bw for three months and killed at six months. In addition, renal pelvis carcinomas (8/18) and urinary bladder carcinomas (3/18) were observed in the males given 10 mg/kg bw. Forestomach carcinomas were observed in 4/4 female rats treated with 10 mg/kg bw for three months and killed at nine months. In the groups dosed with 1.0 mg/kg bw for three months, forestomach carcinomas were observed in 3/11 males and 0/10 females after six months and 6/9 males and 2/11 females after nine months. In rats treated with 0.1 mg/kg bw for three months and killed at 12 months, 2/7 males and 0/6 females had forestomach carcinomas. In rats treated with 0.1 mg/kg bw for 12 months and killed at 16 months, 4/4 males and 1/5 females had forestomach tumours. Only one tumour (a spontaneous endometrial polyp) was observed in the controls (0/30 in males and 1/31 in females) (Mengs *et al.*, 1982).

Two groups of male Wistar rats, eight weeks old, were examined for the histogenesis of forestomach carcinoma caused by aristolochic acids [77.2% aristolochic acid I and 21.2% aristolochic acid II]. A group of 108 rats was given aristolochic acids daily by gavage at a dose of 10 mg/kg in distilled water for up to 90 days. A group of 37 controls received distilled water. The animals were sacrificed sequentially. Administration of the aristolochic acids caused extensive necrosis of the squamous epithelium in the forestomach, followed by regeneration and hyperplasia, papilloma formation and ultimately invasive squamous-cell carcinoma. No pathological changes were seen in the controls (Mengs, 1983).

Three groups of 20 male BD-6 rats weighing 140 g received twice-weekly doses of 10 mg/kg bw aristolochic acid [components not otherwise specified] by gavage for 12 weeks. One of these groups also received 150 mg/kg bw diallyl sulfide by gavage 4 h before aristolochic acid treatment, and another group received diallyl sulfide 24 h and 4 h before the treatment. A fourth group received diallyl sulfide only, four times per week, for 12 weeks. Treatment with diallyl sulfide 4 h before aristolochic acid treatment decreased the development of forestomach tumours that appeared within 6–9 months after the start of the experiment. The incidence of aristolochic acid-induced forestomach tumours (60%; 12/20 rats) was reduced to 10% (2/20 rats) by the prior 4 h-treatment with diallyl sulfide. The prior 4 and 24-h treatment with diallyl sulfide prevented the induction of squamous-cell carcinomas in the forestomach (aristolochic acid alone, 9/20; aristolochic acid with diallyl sulfide, 0/20), but did not prevent the induction of forestomach and urinary bladder papillomatosis [number of control animals and incidences of tumours not reported] (Hadjiolov *et al.*, 1993).

3.2 Intraperitoneal administration

Rabbit: Twelve female New Zealand white rabbits, 15 weeks of age, were given daily intraperitoneal injections of 0.1 mg/kg bw aristolochic acid [components not otherwise specified] in 25 mM NaOH on five days per week for 17–21 months. All 11 surviving rabbits developed fibrotic changes in the kidneys resembling Chinese herb nephropathy and two developed kidney tumours (renal-cell carcinoma or tubulopapillary adenoma). One rabbit had a transitional-cell carcinoma of the ureter as well as a peritoneal mesothelioma. Mild to moderate atypia of the epithelium of the collecting ducts and of the pelvis was present in 5/11 and 11/11 rabbits, respectively. These changes were not detected in 10 female control rabbits (Cosyns *et al.*, 2001).

3.3 Subcutaneous administration

Rat: In a study to model the renal fibrosis seen in Chinese herb nephropathy, 66 male Wistar rats, four weeks of age, were given a single intraperitoneal injection of furosemide and fed a low-salt normal-protein diet. This group was divided into three groups which received daily subcutaneous injections of either 1 mg/kg bw aristolochic acid [components not otherwise specified] (low-dose group; $n = 24$), 10 mg/kg bw aristolochic acid (high-dose group; $n = 24$) or vehicle (control group; $n = 18$) for 35 days. On days 10 and 35, six rats from each group were sacrificed for assessment of kidney function. Urothelial dysplasia was detected in two rats on day 10, one rat on day 35 and three of the remaining

11 rats on day 105 in the high-dose group. Urothelial dysplasia was also seen in a few rats of the low-dose group on day 10 or 105. Three of the high-dose group developed papillary urothelial carcinomas by day 105. Malignant fibrohistiocytic sarcomas were found on day 105 around the sites of subcutaneous injection in two of the six rats of the low-dose group and in seven of the 11 of the high-dose group. Marked interstitial fibrosis of the kidney was noted in the high-dose group on day 35. None was seen in the low-dose or controls groups (Debelle *et al.*, 2002).

4. Other Data Relevant to an Evaluation of Carcinogenicity and its Mechanisms

4.1 Absorption, distribution, metabolism and excretion

4.1.1 *Humans*

As part of a clinical study designed to investigate the effects of aristolochic acids on the phagocytic activity of granulocytes, six healthy volunteers were given a daily dose [presumably oral but not explicitly stated] of 0.9 mg of a mixture of aristolochic acids I and II [ratio not specified] for several days, and 24-h urine samples from day 3 of this trial were analysed for metabolites. The only metabolites detected were aristolactam I (metabolite of aristolochic acid I) and aristolactam II (metabolite of aristolochic acid II) (Figure 5). The percentage conversions to these two metabolites were not reported (Krumbiegel *et al.*, 1987). This contradicts an earlier report of oral absorption of aristolochic acid in humans resulting in the compound(s) being excreted unchanged in urine, bile, breast milk and cerebrospinal fluid (Schulz *et al.*, 1971).

4.1.2 *Experimental systems*

In an extensive study following oral administration of aristolochic acids I and II to male Wistar rats (pure compounds given; 3 mg), the following metabolites were detected in urine and faeces: from aristolochic acid I — aristolactam I, aristolactam Ia, aristolochic acid Ia, aristolic acid I and 3,4-methylenedioxy-8-hydroxy-1-phenanthrenecarboxylic acid; from aristolochic acid II — aristolactam II, aristolactam Ia and 3,4-methylenedioxy-1-phenanthrenecarboxylic acid (Krumbiegel *et al.*, 1987). The structures of these compounds (Figure 5) were determined by spectroscopic methods including low-resolution electron-impact mass spectrometry and low-field ^{1}H-NMR spectrometry (100 MHz) (Krumbiegel & Roth, 1987). A further metabolite of aristolochic acid II with a lactam moiety was not fully characterized. The principal metabolite of aristolochic acid I in rats was aristolactam Ia (46% of the dose converted to this was found in the urine and 37% in the faeces). In the urine, most of the aristolactam Ia was present as a conjugated form, which required alkaline treatment (3 M NaOH) for hydrolysis. The

Figure 5. Proposed metabolic transformations of aristolochic acids I and II in various species

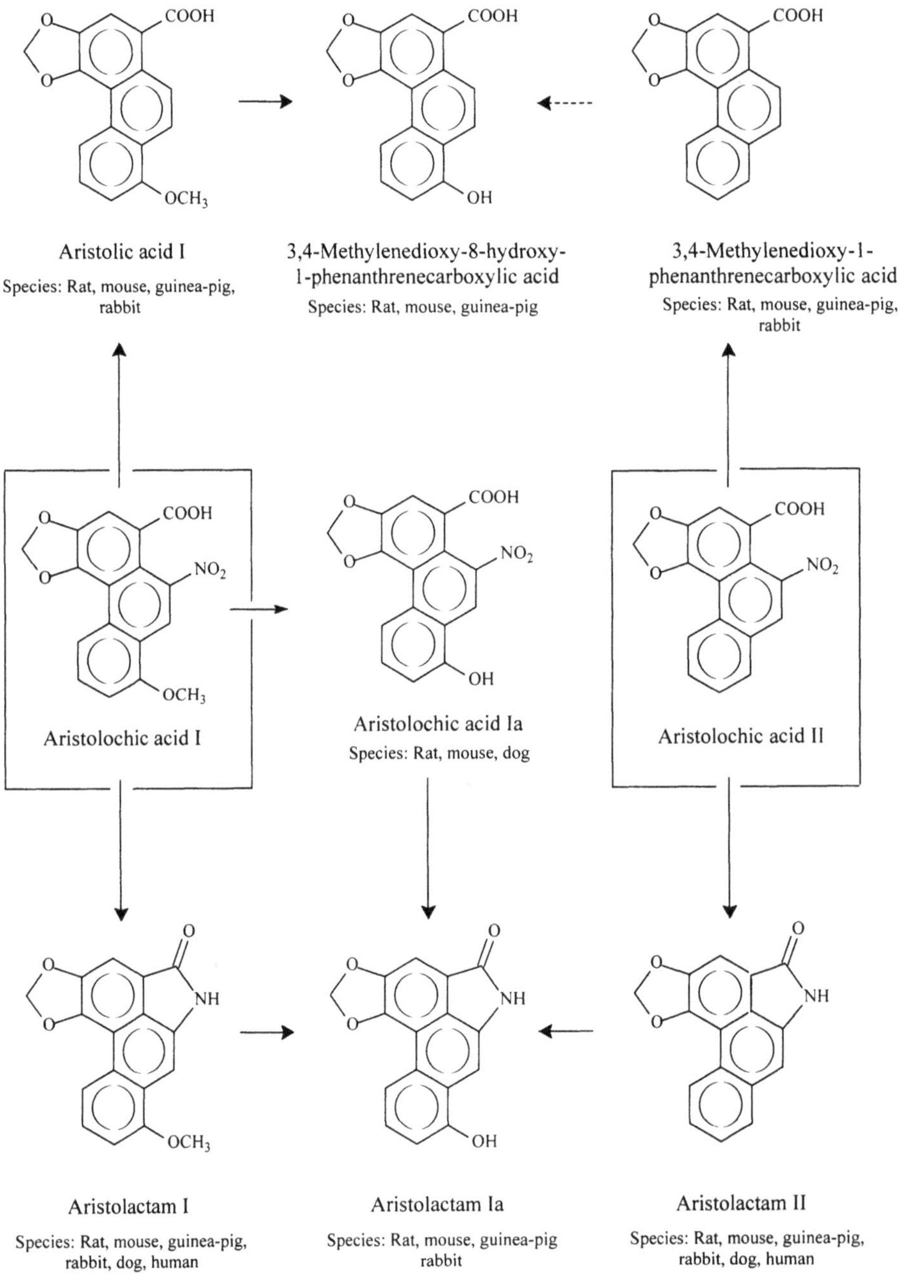

From Krumbiegel *et al.* (1987)
The metabolites were detected in the urine of the species indicated.

metabolites determined for aristolochic acid II in rats were all minor compounds, with the largest proportion accounted for by aristolactam II (4.6% in the urine and 8.9% in the faeces) (Krumbiegel *et al.*, 1987).

This study also examined the metabolism of aristolochic acids I and II by other laboratory animals. The amounts of aristolochic acids I and II administered were similar in proportion to the weight of each test animal. The mouse (NMRI, female) was the only species to exhibit the same metabolic profiles for these compounds as the rat. In general, dogs (beagle, male), rabbits (White Vienna, male) and guinea-pigs (Pirbright White, male) showed smaller numbers of metabolites than mice and rats (summarized in Figure 5) (Krumbiegel *et al.*, 1987).

4.2 Toxic effects

4.2.1 *Humans*

Aristolochic acid I [purity not specified] was given by infusion to 20 patients having various malignant tumours, at different dose schedules ranging from 0.1 mg/kg bw per day for five days to a single dose of 2 mg/kg bw. The compound was too toxic to the kidneys for further trial (Jackson *et al.*, 1964).

The so-called *Stephania tetrandra* powder, first introduced in a slimming regimen in a Belgian clinic in early 1990, was withdrawn from the Belgian market at the end of 1992. Altogether, 1500–2000 persons are thought to have followed the same regimen during the period 1990–92. Among them, about 100 patients with renal disease (70% of them being in end-stage renal disease that had to be treated by dialysis or kidney transplantation) had been recorded by 1998, or about 5% of the exposed population (Vanherweghem, 1998).

The pathology of the renal disease was characterized by extensive interstitial fibrosis with atrophy and loss of the tubules, while the glomeruli were relatively untouched (Vanherweghem *et al.*, 1993; Cosyns *et al.*, 1994a; Depierreux *et al.*, 1994). The typical slimming treatment followed by the patients included intradermal injection of artichoke extract (Chophytol S) and euphyllin as well as oral intake of a mixture of fenfluramine, diethylpropion, meprobamate, cascara powder, acetazolamide, belladonna extract, *Magnolia officinalis* and *Stephania tetrandra*. Replacement of *Stephania tetrandra* by the nephrotoxic *Aristolochia fangchi* was suspected because *Stephania tetrandra* (Han Fang Ji) belongs to the same family and the Chinese characters are identical to those for *Aristolochia fangchi* (Guang Fang Ji) (Vanherweghem *et al.*, 1993).

The inadvertent substitution of *Stephania* by *Aristolochia* was confirmed by phytochemical analysis of 12 different batches of herb powders delivered in Belgium under the name of *Stephania tetrandra*: only one batch contained tetrandrine and not aristolochic acids I and II, one contained both tetrandrine and aristolochic acids and 10 contained aristolochic acids only. The amount of aristolochic acids in the 12 batches varied from undetectable (< 0.02 mg/g) to 1.56 mg/g (mean ± SD, 0.66 ± 0.56 mg/g) (Vanhaelen *et al.*, 1994).

Further evidence supports the involvement of aristolochic acids I and II in the kidney disease: 7-(deoxyadenosin-N^6-yl)aristolactam I–DNA adduct (dA-AAI) was detected in renal tissue obtained from five patients with Chinese herb nephropathy, while none was found in renal tissue from six patients with other renal diseases (Schmeiser *et al.*, 1996; see Section 4.4.1). A larger series of kidney samples from 38 patients with Chinese herb nephropathy confirmed the presence of DNA adducts formed by aristolochic acid six years after their exposure to the so-called *Stephania tetrandra* powder (actually, *Aristolochia fangchi*). Such adducts were absent in kidney tissues obtained from eight patients with renal disease of other origin (Nortier *et al.*, 2000; see Section 4.4.1).

After the description of the initial cases (Vanherweghem *et al.*, 1993), similar cases of Chinese herb nephropathy were reported in many other countries: four cases in France secondary to the intake of slimming pills containing *Stephania tetrandra* which was, in fact, *Aristolochia fangchi* (Pourrat *et al.*, 1994; Stengel & Jones, 1998); one case in Spain after chronic intake of a tea made with a mixture of herbs containing *Aristolochia pistolochia*, a herb that was grown in the Catalonia region (Peña *et al.*, 1996); two cases in the United Kingdom after treatment of eczema with Mu Tong containing aristolochic acid (Lord *et al.*, 1999); 12 cases in Taiwan related to the use of various unidentified herbal medications for different purposes (Yang *et al.*, 2000); one case in the USA after intake of herbal medicine containing aristolochic acid for low back pain (Meyer *et al.*, 2000); and 12 cases in Japan, in five of which the presence of aristolochic acid was demonstrated in the herbal medicine; in the other cases, confusion of Mokutsu (*Akebia quinata*) with Kan-Mokutsu (*Aristolochia manshuriensis*) and Boui (*Sinomenium acutum*) with Kou-Boui (*Aristolochia fangchi*) or Kanchu-Boui (*Aristolochia heterophylla*) was suspected (Tanaka *et al.*, 2001). In Japan, the cases of Chinese herb nephropathy often presented with adult-onset Fanconi syndrome (Tanaka *et al.*, 2000a,b). A similar case was reported in Germany after intake of a purported *Akebia* preparation containing aristolochic acid (Krumme *et al.*, 2001).

4.2.2 *Experimental systems*

In a study of acute effects of a mixture of aristolochic acids I (77.2%) and II (21.2%), intragastric or intravenous administration at high doses to male and female mice (NMRI) and rats (Wistar) resulted in death from acute renal failure within 15 days. The oral LD_{50} ranged from 56 to 203 mg/kg bw and the intravenous LD_{50} from 38 to 83 mg/kg bw, depending on the species and sex. The predominant histological features were severe necrosis of the renal tubules, atrophy of the spleen and thymus, superficial ulceration of the forestomach by both routes, followed by hyperplasia and hyperkeratosis of the squamous epithelium (Mengs, 1987).

In a follow-up study, a no-effect level of 0.2 mg/kg bw — given daily by gavage for four weeks — was observed for aristolochic acids in male Wistar rats. Mild changes only were observed at 1.0 mg/kg bw, with clear toxic effects first seen at 5.0 mg/kg bw given daily for four weeks. Degenerative lesions in the kidneys, forestomach, urinary bladder and testes were observed at a dose of 25 mg/kg bw. Two of the rats in the 25-mg/kg bw group died due to tubular necrosis following renal failure. Renal lesions were also found within three days in female Wistar rats given single doses of 10, 50 or 100 mg/kg bw intragastrically (Mengs & Stotzem, 1992). There was a dose-dependent decrease in body weight, which was significant at doses above 10 mg/kg bw. At necropsy, a grey-brown discoloration of the kidneys was seen in the 100-mg/kg bw group and the relative kidney weights of this group were significantly greater than those of the controls. There was evidence of dose-dependent tubular epithelial necrosis of the renal tubules, with (in the 100-mg/kg bw group) degenerative changes predominantly localized in the *pars recta* of the proximal tubules and widespread necrosis affecting all of the nephrons. In the 50 mg/kg bw group, single cell necrosis predominated, while no necrotic lesions were observed in the 10-mg/kg bw group. There were specific increases in the urinary enzymes malate dehydrogenase, γ-glutamyltranspeptidase and *N*-acetyl-β-glucosaminidase in the 50- and 100-mg/kg bw groups. In these same two groups, urinary electrolyte determinations showed decreases in calcium and magnesium concentrations, although the levels of sodium, potassium and chloride remained unaffected. Urine testing showed significantly increased protein concentrations and some increases in haemoglobin in the 50- and 100-mg/kg bw groups, with glucose concentrations elevated significantly in all three dose groups. Urinary volume, specific gravity and pH were unaffected in all dosed animals (Mengs & Stotzem, 1993).

In a recent study on the effects of chronic administration of a commercially available mixture (97% purity) of aristolochic acids I (44%) and II (56%), female New Zealand white rabbits were given intraperitoneal injections of either the test mixture (0.1 mg/kg bw) or saline (control) for five days per week for 17–21 months. All dosed animals developed renal hypocellular interstitial fibrosis and urothelial atypia, whereas no significant pathological changes were seen in the control animals. Three animals developed tumours of the urinary tract. The treated group also showed impaired growth, increased serum creatinine, glucosuria, tubular proteinuria and anaemia (Cosyns *et al.*, 2001).

Groups of 5–6 male Fischer 344 rats, weighing 140–150 g, were given a single intraperitoneal injection of 10 mg/kg bw aristolochic acids (a commercially available mixture of aristolochic acids I and II) 18 h after a two-thirds partial hepatectomy. After a one-week recovery period, one group was kept on the basal diet while the second group was given a diet with 1% orotic acid to stimulate cell proliferation. In the second group, the percentage of rats with nodules and the number of nodules per rat were increased compared with the group given aristolochic acids alone, suggesting that aristolochic acids act as initiating agents (Rossiello *et al.*, 1993).

The aristolochic acids have a phagocytosis- and metabolism-activating effect. They are also thought to enhance the production of lymphokines. Activation of phagocytes has been demonstrated in tests in rabbits and guinea-pigs. In addition, in animal tests, aristolochic acids enhanced immune resistance to herpes simplex viruses in the eye (Medical Economics Co., 2000).

4.3 Reproductive and developmental effects

4.3.1 *Humans*

The roots of *Aristolochia indica L.* are reputedly used in Indian folk medicine as an abortifacient (references cited in Che *et al.*, 1984). However, no data concerning reproductive effects in humans were available to the Working Group.

4.3.2 *Experimental systems*

With a view to developing fertility-regulating agents from plants, various extracts of *A. indica* have been investigated for effects during pregnancy in mice, rats, hamsters and rabbits.

An ethanolic extract of *A. indica* roots decreased the number of pregnancies in both rats and hamsters when administered daily by gavage on days 1–10 and 1–6 post-coitum, respectively. The extract was fractionated and the various materials were tested. Aristolochic acid I was lethal to rats at a daily dose of 40 mg/kg bw and to hamsters at a daily dose of 25 mg/kg bw. A dose of 10 mg/kg bw per day was not lethal to rats and had no effect on the number of pregnancies, while a daily dose of 12.5 mg/kg bw caused some lethality in hamsters (4/10 died), and tended to decrease the number of pregnancies in survivors. However, this may have been a result of non-specific toxic effects of the test substance (Che *et al.*, 1984).

Aristolochic acid I from *A. mollisima* H. given orally to mice on gestation days 1–6 or 7–10 caused a marked decrease in the number of pregnancies at a daily dose of ≥ 3.7 mg/kg bw. In similarly dosed rats, no effect was found. Single injections given intra-amniotically to rats on one of gestation days 14–16 and to dogs on one of gestation days 30–45 led to fetal death and termination of pregnancy at doses per fetus of 50 or 100 μg (for rats) and 1–18 mg (dogs) (Wang & Zheng, 1984).

In female Swiss albino mice given a single oral dose of 100 mg/kg bw of a crude extract of *A. indica* on day 6 or 7 of gestation, the number of pregnancies was markedly reduced. Dosing with 50 mg/kg bw of various purified fractions also led to a marked effect. No toxic effects were reported at the doses used; however, no other data were given on, for example, body weight of the animals (Pakrashi *et al.*, 1976).

Aristolochic acid I metabolites

A number of studies have investigated *aristolic acid,* a metabolite of aristolochic acid I detected in rats, mice, guinea-pigs and rabbits, but not in humans (see Figure 5).

Oral administration of 90 mg/kg bw aristolic acid to Swiss albino mice on gestation day 6 resulted in termination of pregnancy and in-utero death (Pal *et al.*, 1982).

A single oral dose of 120 or 90 mg/kg bw aristolic acid given by gavage to Swiss albino mice on gestation day 1 or 6 caused marked reduction in uterine weight and increased concentrations of acid phosphatase and a decrease in alkaline phosphatase in the uterus. A subcutaneous dose of progesterone (1 mg per mouse) on gestation days 5–8 with aristolic acid on gestation day 6 did not prevent the effects. The authors concluded that aristolic acid did not appear to block hormone synthesis in the ovary (Pakrashi & Ganguly, 1982).

Aristolic acid disrupted nidation in Swiss albino mice when given at a dose of 150 mg/kg bw on gestation day 1. The treatment did not affect tubal transport of eggs, but affected implantation. In addition, it inhibited the increase in specific uterine alkaline phosphatase activity of the uterus, which in control mice was about three-fold through gestation days 4–6. Based on these results, it was inferred that aristolic acid interferes with estrogenic conditioning of the uterus (Ganguly *et al.*, 1986).

Studies of endocrine properties of aristolic acid revealed anti-estrogenic effects, as the compound inhibited estrogen-induced weight increase and epithelial growth of the uterus in immature female Swiss albino mice. Aristolic acid given on gestation day 1 caused total inhibition of implantation at 60 mg/kg bw and decreased implantation at dose levels down to 15 mg/kg bw (43% reduction) (Pakrashi & Chakrabarty, 1978a).

The methyl ester of aristolic acid caused a 100% abortifacient effect when a single oral dose of 60 mg/kg bw was administered to Swiss albino mice on gestation day 6 or 7. At a dose of 30 mg/kg bw, the effect was 40%. The 60-mg/kg bw dose caused 25% and 20% reduction of fertility when given on gestation day 10 or 12, respectively. No toxic effects were reported in the dams and no malformations were found in the offspring (Pakrashi & Shaha, 1978).

Aristolic acid administered orally to rabbits on gestation day 9 caused 65% fetal loss at a dose of 60 mg/kg bw and 80% fetal loss at 90 mg/kg bw (Pakrashi & Chakrabarty, 1978b).

In male mice, oral feeding of the water-soluble part of the chloroform extract of *A. indica* at a dose of 75 mg/kg bw caused a marked decrease in the weight of the testes (55%) and accessory genital organs. There were varying degrees of arrest of spermatogenesis and nuclear degeneration in various germinal cell types. The treatment also caused a decrease of approximately 30% in body weight. The effects on the male sex organs may have been a result of non-specific toxicity (Pakrashi & Pakrasi, 1977).

4.4 Genetic and related effects

4.4.1 *Humans*

(*a*) *DNA-adduct formation in patients with Chinese herb nephropathy* (see Table 3 for details of studies and references)

Aristolochic acid-specific DNA adducts have been detected by the ^{32}P-postlabelling method in the kidneys and ureters of patients with Chinese herb nephropathy (a total of 47 women and one man). The major DNA adduct, which co-chromatographed with 7-(deoxyadenosin-N^6-yl)aristolactam I (dA-AAI), was detected in all urothelial tissues analysed, whereas the two minor ones, chromatographically indistinguishable from 7-(deoxyguanosin-N^2-yl)aristolactam I (dG-AAI) and 7-(deoxyadenosin-N^6-yl)aristolactam II (dA-AAII), were found in most cases (see Figure 6). Total aristolochic acid-specific adduct levels in DNA obtained from whole organs or biopsies from Chinese herb nephropathy patients were in the range of 1.7–530 adducts per 10^9 normal nucleotides. All studies presented evidence that these patients had taken herbal preparations containing a natural mixture of aristolochic acids.

(*b*) *p53 overexpression in patients with Chinese herb nephropathy*

Overexpression of the p53 protein, a common finding in human tumours, was observed in carcinoma *in situ*, papillary transitional-cell carcinoma and urothelial atypia found in 10 Belgian patients with Chinese herb nephropathy (Cosyns *et al.*, 1999).

4.4.2 *Experimental systems*

(*a*) *DNA adduct formation by aristolochic acids in rats* in vivo (see Table 4 for details of studies and references)

Aristolochic acid–DNA adducts were formed *in vivo* in many organs of male rats given oral doses of aristolochic acid (natural mixture) or the pure major components aristolochic acid I or aristolochic acid II. Aristolochic acid–DNA adducts were also formed *in vivo* in the kidney (the only organ examined) of male and female rats given multiple oral doses of a slimming regimen of plant material that contained aristolochic acids. The results confirm that the three major DNA adducts formed *in vivo* co-chromatograph with 7-(deoxyadenosin-N^6-yl)aristolactam I (dA-AAI), 7-(deoxyguanosin-N^2-yl)-aristolactam I (dG-AAI) and 7-(deoxyadenosin-N^6-yl)aristolactam II (dA-AAII). Oral administration of a single dose of aristolochic acid I to rats led to formation of the dA-AAI adduct that persisted in DNA of several organs, consistent with the results obtained from studies in patients with Chinese herb nephropathy.

Table 3. DNA adduct formation in Chinese herb nephropathy (CHN) patients

Details of study	DNA binding	Reference
Detection of AA-specific DNA adducts in renal tissue from 5 female patients with CHN from Belgium by ^{32}P-postlabelling. DNA was obtained from cortical or cortico-medullary tissue. The patients had taken pills containing Chinese herbs for 13–23 months. The adduct chromatographically indistinguishable from dA-AAI was detectable up to 27 months after termination of the regimen.	dA-AAI: 0.7–5.3/10^7 nucleotides	Schmeiser *et al.* (1996)
Detection of AA-specific DNA adducts in the kidneys and one ureter from 6 female patients with CHN from Belgium by ^{32}P-postlabelling (five cases were from Schmeiser *et al.*, 1996). Three AA-specific adducts chromatographically indistinguishable from dA-AAI, dG-AAI and dA-AAII were detectable up to 44 months after termination of the regimen.	dA-AAI: 0.7–5.3/10^7 nucleotides dG-AAI: 0.02–0.12/10^7 nucleotides dA-AAII: 0.06–0.24/10^7 nucleotides	Bieler *et al.* (1997)
Detection of AA-specific DNA adducts in the kidneys of CHN patients (37 female and one male), and in 17 ureters from 11 of these patients from Belgium by ^{32}P-postlabelling. Among these patients were 18 cases of urothelial carcinoma. Data were related to cumulative doses of compounds in the weight-reducing pills on the basis of all prescriptions made during the period of exposure (1990–92). The cumulative dose of aristolochia was a significant risk factor for developing urothelial carcinoma. The kidneys and ureters from the CHN patients had the same pattern of adducts consisting of dA-AAI, dG-AAI and dA-AAII. The major adduct dA-AAI was detectable up to 89 months after discontinuation of use of the weight-reducing pills. No statistically significant difference was observed between mean levels of dA-AAI DNA adducts determined in renal tissue samples from patients who had developed urothelial carcinoma and those from tumour-free patients.	In the kidneys: dA-AAI: 1.2–165/10^9 nucleotides dG-AAI: 0.4–8.2/10^9 nucleotides dA-AAII: 0.6–6.8/10^9 nucleotides In the ureters: dA-AAI: 2.2–34/10^9 nucleotides	Nortier *et al.* (2000)
Detection of AA-specific DNA adducts in the kidneys from 2 new female CHN patients from the Belgian cohort by ^{32}P-postlabelling.	2.9–5.0/10^8 nucleotides	Arlt *et al.* (2001a)

Table 3 (contd)

Details of study	DNA binding	Reference
Detection of the dA-AAI DNA adduct in a renal biopsy from a female CHN patient outside the Belgian cohort by ^{32}P-postlabelling. The patient ingested a Chinese herbal preparation bought in Shanghai for the preceding 6 months (2 pills/day). The presence of AA in the pills was determined by HPLC analysis (0.3 mg AA/pill).	dA-AAI: $1.8/10^8$ nucleotides	Gillerot *et al.* (2001)
Detection of the dA-AAI DNA adduct in the kidney and ureter from a female CHN patient outside the Belgian cohort by ^{32}P-postlabelling. The patient took a herbal preparation containing aristolochic acid for 2 years prescribed for eczema. She developed invasive transitional cell carcinoma of the urinary tract.	In the kidney: dA-AAI: $3.8/10^9$ nucleotides In the ureter: dA-AAI: $40/10^9$ nucleotides	Lord *et al.* (2001)

AA, aristolochic acid; CHN, Chinese herb nephropathy; dA-AAI, 7-(deoxyadenosin-N^6-yl)aristolactam I; dA-AAII, 7-(deoxyadenosin-N^6-yl)aristolactam II; dG-AAI, 7-(deoxyguanosin-N^2-yl)aristolactam I; HPLC, high-performance liquid chromatography

Figure 6. Metabolic activation and DNA adduct formation of aristolochic acids I (R = OCH_3) and II (R–H)

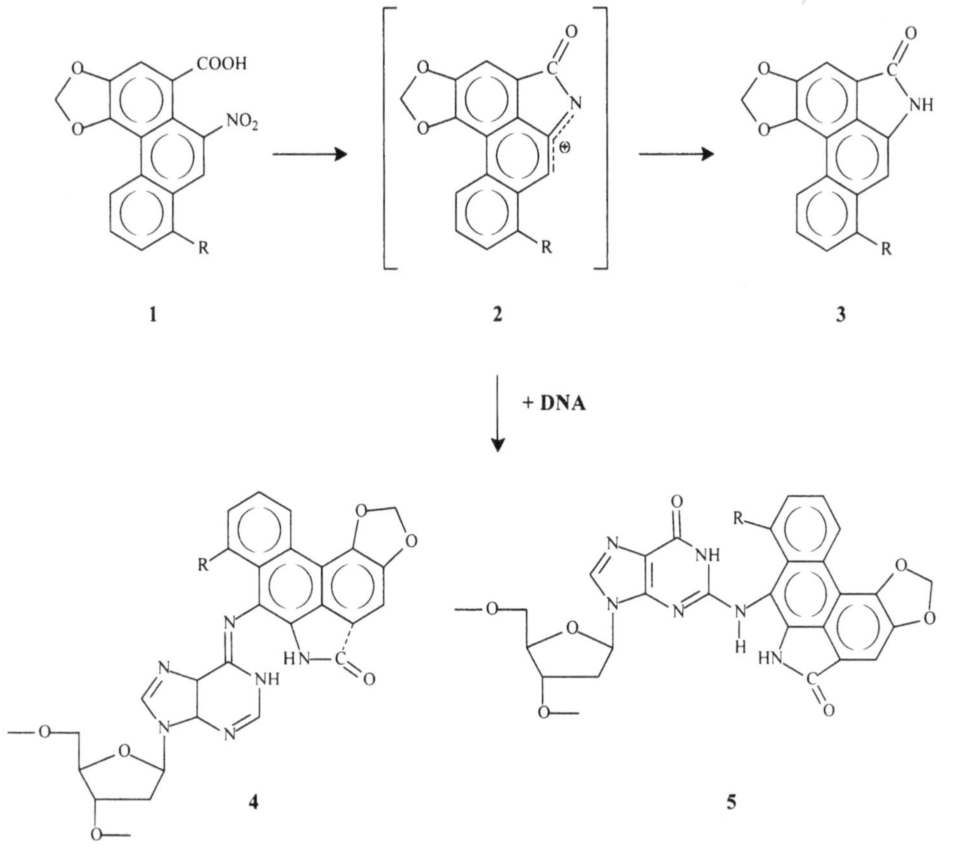

From Schmeiser *et al.* (1997)
(1) Aristolochic acid; (2) cyclic nitrenium ion of aristolochic acids I or II; (3) aristolactams; (4) 7-(deoxy-adenosin-N^6-yl)aristolactam I or II (dA-AAI or dA-AAII); (5) 7-(deoxyguanosin-N^2-yl)aristolactam I or II (dG-AAI or dG-AAII)

(*b*) *Mutations in proto-oncogenes in tumours induced by aristolochic acids in rodents* in vivo (see Table 5 for details of studies and references)

Activated c-Ha-*ras* proto-oncogenes were found in 7/7 ear-duct tumours (squamous-cell carcinoma), in 13/14 forestomach tumours (squamous-cell carcinoma) and in the transformant of the 14th forestomach tumour, and in one metastasis in the lung induced in rats by aristolochic acid I. The mutations were all A → T transversions at the second base of codon 61. In the same animal study, activated c-Ki-*ras* proto-oncogenes were found in 1/7 ear duct tumours and 1/8 tumours of the small intestine. In the one metas-

Table 4. DNA adduct formation with aristolochic acid (AA) in rats *in vivo*

Details of study	Compound, route and dose	DNA binding	Reference
Adduct formation in DNA of several organs of male Wistar rats *in vivo* given multiple doses of AAI or AAII (equimolar doses as sodium salts). Adduct formation by both compounds was detected by the standard ^{32}P-postlabelling method in forestomach, stomach, liver, kidney and lung. In addition, adduct formation by AAII was detected in bladder and brain.	AAI and AAII, gavage AAI: 11 mg/kg bw daily for 5 days AAII: 10 mg/kg bw daily for 5 days	 + [not quantified] + [not quantified]	Schmeiser *et al.* (1988)
Detection and quantitation of adducts in DNA of several organs of male Wistar rats *in vivo* given multiple doses of AAI or AAII (equimolar doses as sodium salts). AAI induced higher adduct levels than AAII in fore-stomach, glandular stomach, liver, kidney and urinary bladder analysed by the nuclease P1 enhancement of the ^{32}P-postlabelling method. Assignment of characterized *in vitro* nucleoside adducts of AAI to the biphosphate derivatives obtained by the ^{32}P-postlabelling procedure.	As above (Schmeiser *et al.*, 1988)	AAI: 17–330/10^8 nucleotides AAII: 24–80/10^8 nucleotides	Pfau *et al.* (1990a)
Adduct formation in DNA of forestomach and liver of male Wistar rats (APf SD strain) given multiple doses of AA. Adduct detection was by the nuclease P1 enhancement of the ^{32}P-postlabelling method. Butylated hydroxyanisole pretreatment enhanced adduct formation by AA.	AA mixture, gavage, 1 mg/kg bw daily for 5 days	Liver: 63/10^9 nucleotides forestomach: 77/10^9 nucleotides	Routledge *et al.* (1990)
Adduct formation in DNA of whole urinary bladder, urothelial cells and exfoliated cells in urine of male Wistar rats given daily doses of AAI. Animals were killed 3 months after the last dose. Adduct detection was by the nuclease P1 and the butanol extraction enhancement of the ^{32}P-postlabelling method. AA-specific adducts were detectable in whole urinary bladder, urothelial cells and exfoliated cells in urine pooled from several rats over a period of 2 weeks after the first dose. The major adduct spots were identified by co-chromatography.	AAI, gavage, 10 mg/kg bw daily, five times a week for 3 months	Total adduct levels (dG-AAI, dA-AAI) determined by the nuclease P1 version exfoliated cells: 5.8/10^9 nucleotides urinary bladder: 37/10^9 nucleotides urothelial cells: 126/10^9 nucleotides	Fernando *et al.* (1992)

Table 4 (contd)

Details of study	Compound, route and dose	DNA binding	Reference
Adduct formation in DNA of several organs of male Wistar rats given a single dose of AAI and sacrificed 1 day and 1, 2, 4, 16 and 36 weeks later. The nuclease P1 enhancement of the ^{32}P-postlabelling method was used to analyse adducts in forestomach, glandular stomach, liver, lung and urinary bladder. In the target organ (forestomach), dA-AAI and dG-AAI adducts were removed rapidly within the first 2 weeks; thereafter, extensive removal of the dG-AAI continued, whereas dA-AAI remained at constant levels from 4 to 36 weeks.	AAI, gavage, 5 mg/kg bw	Initial level in forestomach: dA-AAI: $30/10^8$ nucleotides, dG-AAI: $21/10^8$ nucleotides Level after 36 weeks in forestomach: dA-AAI: $2/10^8$ nucleotides, dG-AAI: $0.4/10^8$ nucleotides	Fernando *et al.* (1993)
Adduct formation in DNA of the forestomach of male BD-6 rats given multiple doses of AA mixture. The nuclease P1 enhancement of the ^{32}P-postlabelling method was used to analyse adducts in forestomach. Chronic diallyl sulfide co-administration decreased adduct levels in forestomach DNA.	AA mixture, gavage, 10 mg/kg bw twice a week for 12 weeks	Total level in forestomach: $87/10^8$ nucleotides	Hadjiolov *et al.* (1993)
Adduct formation in DNA of the forestomach of male Sprague Dawley rats given multiple doses of AAI or AAII or AA mixture. The AA mixture consisted of 65% AAI and 34% AAII. The nuclease P1 enhancement of the ^{32}P-postlabelling method was used to analyse adducts in forestomach. Adduct spots were identified by co-chromatography with in-vitro prepared standard compounds. Adduct formation with AAI was more efficient than with AAII.	AAI or AAII or AA mixture, gavage, 10 mg/kg bw twice a week for 2 weeks	Total level in forestomach (dA and dG adducts): AAI: $62/10^7$ nucleotides AAII: $2.5/10^7$ nucleotides AA: $3.2/10^7$ nucleotides	Stiborová *et al.* (1994)
Adduct formation in DNA of the kidney of male Wistar rats given a single dose of AAI and sacrificed 1 day and 1, 2, 4, 16 and 36 weeks later. The nuclease P1 enhancement of the ^{32}P-postlabelling method was used to analyse adducts. The dA-AAI adduct showed lifelong persistence.	AAI, gavage, 5 mg/kg bw	Initial level in kidney: dA-AAI: $6.5/10^8$ nucleotides, dG-AAI: $3.8/10^8$ nucleotides Level after 36 weeks in kidney: dA-AAI: $1.6/10^8$ nucleotides, dG-AAI: $0.5/10^8$ nucleotides	Bieler *et al.* (1997)

Table 4 (contd)

Details of study	Compound, route and dose	DNA binding	Reference
Adduct formation in DNA of the kidney of male and female Wistar rats given multiple doses of AA as plant material of the slimming regimen. Animals were sacrificed 11 months after last treatment. The nuclease P1 enhancement of the ^{32}P-postlabelling method was used to analyse adducts. AA-specific DNA adduct levels were higher in female than in male rats.	Slimming regimen containing AA mixture as plant material, gavage, 0.15 mg/kg bw per day, 5 times a week for 3 months	Total level in kidneys: 51–83/10^9 nucleotides	Arlt *et al.* (2001a)

AA, aristolochic acids (mixed); AAI and AAII, aristolochic acids I and II; dA-AAI, 7-(deoxyadenosin-N^6-yl)aristolactam I; dA-AAII, 7-(deoxyadenosin-N^6-yl)aristolactam II; dG-AAI, 7-(deoxyguanosin-N^2-yl)aristolactam I

Table 5. Mutations in oncogenes found in rodents treated *in vivo* with aristolochic acids (AA)

Species	Treatment	Incidence and type of tumours	Method of analysis	No. of mutated genes/no. of tumours analysed	Details of mutations		
					Gene, codon	Base change	Reference
Male Wistar rats	10 mg/kg bw AAI given daily by gavage to 8-week-old rats 5 times a week for 3 months. Rats were killed over a 15-week period after treatment	15/40 forestomach tumours (SCC), 7/40 ear duct tumours (SCC), 23/40 adeno-carcinomas or sarcomas of the small intestine, 2/40 metastases of SCC in lung and pancreas	DNA isolated from 5 excised forestomach tumours was transfected into NIH 3T3 cells which induced tumours in nude mice. c-Ha-*ras* fragments were amplified by PCR of DNA from nude mouse tumours and analysed by sequencing. DNA extracted from rat tumours was amplified by PCR for regions of c-Ha-*ras*, c-Ki-*ras* and c-N-*ras* gene and analysed by selective oligonucleotide hybridization with probes carrying different *ras* base-pair substitutions.	7/7 ear duct tumours	c-Ha-*ras*, 61 c-Ki-*ras*, 61	CAA → CTA (7/7) CAA → CAT (1/7)	Schmeiser *et al.* (1990)
				14/14 forestomach tumours	c-Ha-*ras*, 61	CAA → CTA (14/14)	
				1/8 tumours of the small intestine	c-Ki-*ras*, 61	CAA → CTA (1/8)	
				1/1 metastasis in the pancreas	c-N-*ras*, 61	CAA → CTA	
				1/1 metastasis in the lung	c-Ha-*ras*, 61	CAA → CTA	
Female NMRI mice	5 mg/kg bw AA mixture, by gavage, daily for 3 weeks (80% AAI, 20% AAII). Animals killed after 56 weeks	SCC of the forestomach, adenocarcinoma of the lung	DNA extracted from histologically normal and neoplastic tissue in paraffin sections, c-Ha-*ras* fragments around codon 61 amplified by PCR and analysed by oligo-nucleotide hybridization with probes carrying different c-Ha-*ras* base-pair substitutions	1/1 fore-stomach tumour 1/3 lung tumours	c-Ha-*ras*, 61	CAA → CTA (2/4)	Schmeiser *et al.* (1991)

Table 5 (contd)

Species	Treatment	Incidence and ype of tumours	Method of analysis	No. of mutated genes/no. of tumours analysed	Details of mutations		
					Gene, codon	Base change	Reference
Male Wistar rats	10 mg/kg bw AAI given daily by gavage to 8-week-old rats, 5 times a week for 3 months. Rats were killed over a 15-week period after treatment	SCC of the fore-stomach and pancreas	DNA extracted from histologically normal and neoplastic tissue in paraffin sections, c-Ha-*ras* fragments around codon 61 amplified by PCR and analysed by oligonucleotide hybridization with probes carrying different c-Ha-*ras* base-pair substitutions	2/2 fore-stomach tumours 0/1 pancreas tumour	c-Ha-*ras*, 61	CAA → CTA (2/3)˙	Schmeiser *et al.* (1991)

SCC, squamous-cell carcinoma; PCR, polymerase chain reaction; AA, aristolochic acid; AAI, aristolochic acid I; AAII, aristolochic acid II tasis in the pancreas, an activated c-N-*ras* proto-oncogene was detected. All mutations were A → T transversions at either the second or the third base of codon 61.

In mice, a mixture of aristolochic acids induced squamous-cell carcinoma in the forestomach and adenocarcinoma in the lung. In the tumours analysed, one forestomach squamous-cell carcinoma and 1/3 lung adenocarcinomas contained activated c-Ha-*ras* proto-oncogenes both mutated by A → T transversions at the second base of codon 61.

(*c*) *In-vitro studies* (see Tables 6–8 for details of studies and references)

After metabolic activation, aristolochic acid I and aristolochic acid II form adducts *in vitro* with calf thymus DNA, MCF-7 DNA, plasmids, polydeoxyribonucleotides, oligodeoxyribonucleotides, deoxyribonucleotide-3′-monophosphates (purines), deoxyadenosine and deoxyguanosine. In-vitro systems capable of activating aristolochic acids I and II to reactive species that may form adducts are S9 mix from Aroclor 1254- or β-naphthoflavone-pretreated rats, xanthine oxidase, peroxidases (horseradish peroxidase, lactoperoxidase, prostaglandin H synthase), zinc at pH 5.8 and microsomal preparations from various species other than the rat. Aristolochic acid-specific adducts were formed in calf thymus DNA after activation of aristolochic acids I and II with hepatic microsomes from humans, mini-pigs and rats, as well as with microsomes containing recombinant human CYP1A1 and CYP1A2. From studies with specific inducers and selective inhibitors, it can be concluded that most of the microsomal activation of aristolochic acids is due to CYP1A1 and CYP1A2.

Activated aristolochic acids I and II react with DNA to form three and two major adducts, respectively. These major adducts co-chromatograph with 7-(deoxyadenosin-N^6-yl)aristolactam I (dA-AAI), 7-(deoxyguanosin-N^2-yl)aristolactam I (dG-AAI), 7-(deoxyadenosin-N^6-yl)aristolactam II (dA-AAII) and 7-(deoxyguanosin-N^2-yl)aristolactam II (dG-AAII) (see Figure 6), indicating that aristolochic acid reacts preferentially with the exocyclic amino group of purine bases. On the basis of the adduct structures, it can be concluded that reduction of the nitro group is the main metabolic pathway for the activation of aristolochic acid.

The major metabolites, the aristolactams, form DNA adducts *in vitro* after activation by hepatic microsomes or horseradish peroxidase. Adducts with calf thymus DNA are also formed by aristolochic acids I and II *in vitro* in the presence of rat faecal bacteria. In explants of rat stomach tissue, both acids formed adducts in the DNA of the epithelial layer. DNA adducts have been detected in MCF-7 cells after exposure to aristolochic acid I and in opossum kidney cells after exposure to an aristolochic acid mixture.

After reaction of aristolochic acids with DNA, DNA synthesis by T7 DNA polymerase and human DNA polymerase α is mainly blocked at the nucleotide 3′ to the aristolochic acid-induced DNA adducts. This property has allowed the use of polymerase arrest assays that revealed binding of aristolochic acids I and II *in vitro* to the c-Ha-*ras* gene and the *TP53* gene.

Aristolochic acids I and II and the aristolochic acid mixture induced SOS repair and mutations in bacteria. In nitroreductase-deficient strains of *Salmonella typhimurium*,

Table 6. DNA adduct formation by aristolochic acids (AAs) *in vitro*

Test system	Assay	Dose	Result	Reference
Adduct formation by AAI in DNA (calf thymus) *in vitro* with S9 mix[a], aerobic	Standard ^{32}P-postlabelling	0.4 mM	4 adducts	Schmeiser *et al.* (1988)
Adduct formation by AAI in DNA (calf thymus) *in vitro* with S9 mix[a], anaerobic[b]			2 adducts	
Adduct formation by AAII in DNA (calf thymus) *in vitro* with S9 mix[a], aerobic			No adducts	
Adduct formation by AAII in DNA (calf thymus) *in vitro* with S9 mix[a], anaerobic[b]			2 adducts	
Adduct formation by AAI in DNA (calf thymus) *in vitro* with xanthine oxidase (0.1 μg/mL), anaerobic[b]			4 adducts	
Adduct formation by AAII in DNA (calf thymus) *in vitro* with xanthine oxidase (0.1 μg/mL), anaerobic[b]			2 adducts	
Adduct formation by AAI in polydG *in vitro* with S9 mix aerobic or anaerobic			2 adducts	
Adduct formation by AAI in DNA (calf thymus) *in vitro* with rat faecal bacteria, anaerobic[b]	Nuclease P1-enhanced ^{32}P-postlabelling	1 mmol/2 mL	Same pattern as *in vivo* (cf Table 4)	Pfau *et al.* (1990a)
Adduct formation by AAII in DNA (calf thymus) *in vitro* with rat faecal bacteria, anaerobic[b]			Same pattern as *in vivo* (cf Table 4)	
Adduct formation by AAI in DNA of forestomach and glandular stomach in explanted stomach of rats		30 μmol/kg bw daily for 5 days (oral)	4 adducts	
Adduct formation by AAII in DNA of forestomach and glandular stomach in explanted stomach of rats			2 adducts	
Adduct formation by AAI with deoxyguanosine and deoxyadenosine *in vitro* in the presence of xanthine oxidase, anaerobic	UV/vis and fluorescence spectroscopy. Adducts were spectroscopically identified as dG-AAI and dA-AAI.		Same adducts are formed in DNA	Pfau *et al.* (1990b)

Table 6 (contd)

Test system	Assay	Dose	Result	Reference
Adduct formation by AAII with deoxyadenosine *in vitro* in the presence of xanthine oxidase, anaerobic	UV/vis, fluorescence and NMR spectroscopy. Adduct was spectroscopically identified as dA-AAII.		Same adduct is formed in DNA *in vitro* and *in vivo*	Pfau *et al.* (1991)
Adduct formation by AAI with deoxyadenosine 3′-monophosphate or deoxyguanosine 3′-monophosphate *in vitro* in the presence of xanthine oxidase, anaerobic	Butanol extraction-enhanced ^{32}P-postlabelling	100 μM (with 0.3 μM substrate)	dA-AAI and dG-AAI	Stiborová *et al.* (1994)
Adduct formation by AAII with deoxyadenosine 3′-monophosphate or deoxyguanosine 3′-monophosphate *in vitro* in the presence of xanthine oxidase, anaerobic			dA-AAII and dG-AAII	
Adduct formation by AAI in 18-mer oligonucleotides containing either a single guanosine or adenosine in the presence of zinc at pH 5.8	Standard ^{32}P-postlabelling	2 mM (with 100 μM substrate)	dA-AAI and dG-AAI	Broschard *et al.* (1994)
Adduct formation by AAII in 18-mer oligonucleotides containing either a single guanosine or adenosine in the presence of zinc at pH 5.8			dA-AAII and dG-AAII Adenine adducts have greater miscoding potential than guanine adducts (see Table 7).	
Adduct formation by AAI in 30-mer oligonucleotides containing either a single guanosine or adenosine in the presence of zinc at pH 5.8	Standard ^{32}P-postlabelling	2 mM (with 100 μM substrate)	dA-AAI and dG-AAI	Broschard *et al.* (1995)
Adduct formation by AAII in 30-mer oligonucleotides containing either a single guanosine or adenosine in the presence of zinc at pH 5.8			dA-AAII and dG-AAII Adenine adducts formed by AAI and AAII and guanine adducts formed by AAI block DNA replication more efficiently than dG-AAII (see Table 7).	

Table 6 (contd)

Test system	Assay	Dose	Result	Reference
Adduct formation by aristolactam I in DNA (calf thymus) *in vitro* with microsomes[c]	Nuclease P1-enhanced ^{32}P-postlabelling	0.3 mM (with 1.3 mg DNA/mL)	Total adduct level: 1.3/10^7 nucleotides	Stiborová *et al.* (1995)
Adduct formation by aristolactam II in DNA (calf thymus) *in vitro* with microsomes[c]			Total adduct level: 0.5/10^7 nucleotides	
Adduct formation by aristolactam I in DNA (calf thymus) *in vitro* with horseradish peroxidase			Total adduct level: 93/10^7 nucleotides	
Adduct formation by aristolactam II in DNA (calf thymus) *in vitro* with horseradish peroxidase			Total adduct level: 19/10^7 nucleotides	
Adduct formation by AAI in DNA (calf thymus) *in vitro* with hepatic microsomes[c], aerobic	Nuclease P1-enhanced ^{32}P-postlabelling (adducts dG-AAI, dG-AAII, dA-AAI and dA-AAII identified by chromatography)	0.3 mM (with 1.3 mg DNA/mL)	Total adduct level: 14/10^7 nucleotides	Schmeiser *et al.* (1997)
Adduct formation by AAI in DNA (calf thymus) *in vitro* with hepatic microsomes [c], anaerobic[b]			Total adduct level: 34/10^7 nucleotides	
Adduct formation by AAII in DNA (calf thymus) *in vitro* with hepatic microsomes [c], aerobic			Total adduct level: 2.3/10^7 nucleotides	
Adduct formation by AAII in DNA (calf thymus) *in vitro* with hepatic microsomes [c], anaerobic[b]			Total adduct level: 7.1/10^7 nucleotides	
Adduct formation by AAI in DNA (calf thymus) *in vitro* with xanthine oxidase, aerobic			Total adduct level: 76/10^7 nucleotides	
Adduct formation by AAI in DNA (calf thymus) *in vitro* with xanthine oxidase, anaerobic[b]			Total adduct level: 145/10^7 nucleotides	
Adduct formation by AAII in DNA (calf thymus) *in vitro* with xanthine oxidase, aerobic			Total adduct level: 4.4/10^7 nucleotides	
Adduct formation by AAII in DNA (calf thymus) *in vitro* with xanthine oxidase, anaerobic[b]			Total adduct level: 6.1/10^7 nucleotides	
Adduct formation by AAI in DNA (calf thymus) *in vitro* with zinc pH 5.8, aerobic			Total adduct level: 22420/10^7 nucleotides	
Adduct formation by AAII in DNA (calf thymus) *in vitro* with zinc pH 5.8, aerobic			Total adduct level: 52700/10^7 nucleotides	
Adduct formation by AAI in DNA (calf thymus) *in vitro* with horseradish peroxidase, aerobic			Total adduct level: 74/10^7 nucleotides	
Adduct formation by AAI in DNA (calf thymus) *in vitro* with horseradish peroxidase, anaerobic[b]			Total adduct level: 100/10^7 nucleotides	

Table 6 (contd)

Test system	Assay	Dose	Result	Reference
Adduct formation by AAII in DNA (calf thymus) *in vitro* with horseradish peroxidase, aerobic		0.3 mM	Total adduct level: $4.1/10^7$ nucleotides	Schmeiser *et al.* (1997) (contd)
Adduct formation by AAII in DNA (calf thymus) *in vitro* with horseradish peroxidase, anaerobic[b]			Total adduct level: $19/10^7$ nucleotides	
Adduct formation by AAI in DNA (calf thymus) *in vitro* with lactoperoxidase, aerobic			Total adduct level: $10.2/10^7$ nucleotides	
Adduct formation by AAI in DNA (calf thymus) *in vitro* with lactoperoxidase, anaerobic			Total adduct level: $13.8/10^7$ nucleotides	
Adduct formation by AAII in DNA (calf thymus) *in vitro* with lactoperoxidase, aerobic			Total adduct level: $0.6/10^7$ nucleotides	
Adduct formation by AAII in DNA (calf thymus) *in vitro* with lactoperoxidase, anaerobic			Total adduct level: $3.6/10^7$ nucleotides	
Adduct formation by aristolactam I in DNA (calf thymus) *in vitro* with rat liver microsomes[c], aerobic	Nuclease P1-enhanced ^{32}P-postlabelling	0.3 mM (1.3 mg DNA/mL)	Total adduct level: $1.3/10^7$ nucleotides[d]	Stiborová *et al.* (1999)
Adduct formation by aristolactam I in DNA (calf thymus) *in vitro* with horseradish peroxidase, aerobic			Total adduct level: $93/10^7$ nucleotides	
Adduct formation by AAI activated by zinc in a plasmid containing exon 2 of the mouse c-Ha-*ras* gene	Standard ^{32}P-postlabelling	1.2 mM	Total adduct level: $5.2/10^3$ nucleotides	Arlt *et al.* (2000)
		0.12 mM	Total adduct level: $0.6/10^3$ nucleotides	
Adduct formation by AAII activated by zinc in a plasmid containing exon 2 of the mouse c-Ha-*ras* gene		1.2 mM	Total adduct level: $9.9/10^3$ nucleotides	
		0.12 mM	Total adduct level: $6.1/10^3$ nucleotides	
Adduct formation by AAI activated by zinc in polydeoxynucleotides poly(dA), poly(dG)-poly(dC) and poly (dC)		1.2 mM	dA-AAI, dG-AAI, dC-AAI	
Adduct formation by AAII activated by zinc in polydeoxynucleotides poly(dA), poly(dG)-poly(dC) and poly (dC)		1.2 mM	dA-AAI, dG-AAII, dC-AAII	

Table 6 (contd)

Test system	Assay	Dose	Result	Reference
Adduct formation of AAI in MCF-7 cells with different doses for 24 h	Nuclease P1-enhanced ^{32}P-postlabelling		Total adduct level:	Arlt *et al.* (2001b)
		10 μM	$1.2/10^7$ nucleotides	
		100 μM	$33/10^7$ nucleotides	
		200 μM	$52/10^7$ nucleotides	
Adduct formation of AAI in MCF-7 DNA *in vitro*, activated by zinc	Standard ^{32}P-postlabelling	0.24 mM	Total adduct level: $7.1/10^4$ nucleotides	
Adduct formation of AAII in MCF-7 DNA *in vitro*, activated by zinc		0.24 mM	Total adduct level: $55/10^4$ nucleotides	
Adduct formation of AAI in MCF-7 DNA *in vitro*, activated by xanthine oxidase		2.4 mM	Total adduct level: $1.3/10^4$ nucleotides	
Adduct formation of AAII in MCF-7 DNA *in vitro*, activated by xanthine oxidase		2.4 mM	Total adduct level: $6.3/10^4$ nucleotides	
Adduct formation of AAI in DNA (calf thymus) *in vitro*, activated by ram seminal vesicle microsomes, anaerobic	Nuclease P1-enhanced ^{32}P-postlabelling	0.5 mM	Total adduct level: $0.54/10^7$ nucleotides[e]	Stiborová *et al.* (2001a)
Adduct formation of AAII in DNA (calf thymus) *in vitro*, activated by ram seminal vesicle microsomes, anaerobic			Total adduct level: $0.3/10^7$ nucleotides[e]	
Adduct formation of AAI in DNA (calf thymus) *in vitro*, activated by pure prostaglandin H synthase (PHS-1), anaerobic			Total adduct level: $4.6/10^7$ nucleotides[e]	
Adduct formation of AAII in DNA (calf thymus) *in vitro*, activated by pure prostaglandin H synthase (PHS-1), anaerobic			Total adduct level: $1.9/10^7$ nucleotides[e]	
Adduct formation of AAI in DNA (calf thymus) *in vitro*, activated by human hepatic microsomes, anaerobic	Nuclease P1-enhanced ^{32}P-postlabelling (adducts identified by co-chromatography)	0.5 mM (4 mM DNA)	Total adduct level: $6.4/10^7$ nucleotides	Stiborová *et al.* (2001b)
Adduct formation of AAII in DNA (calf thymus) *in vitro*, activated by human microsomes, anaerobic			Total adduct level: $0.8/10^7$ nucleotides	
Adduct formation of AAI in DNA (calf thymus) *in vitro*, activated by hepatic microsomes from minipig, anaerobic			Total adduct level: $11/10^7$ nucleotides	
Adduct formation of AAII in DNA (calf thymus) *in vitro*, activated by hepatic microsomes from minipig, anaerobic			Total adduct level: $2.1/10^7$ nucleotides	
Adduct formation of AAI in DNA (calf thymus) *in vitro*, activated by hepatic microsomes from rats, anaerobic			Total adduct level: $5.7/10^7$ nucleotides	

Table 6 (contd)

Test system	Assay	Dose	Result	Reference
Adduct formation of AAII in DNA (calf thymus) *in vitro*, activated by hepatic microsomes from rats, anaerobic			Total adduct level: 0.7/10^7 nucleotides	Stiborová *et al.* (2001b) (contd)
Adduct formation of AAI in DNA (calf thymus) *in vitro* activated by recombinant human CYP1A1 in microsomes from a baculovirus insect cell expression system, anaerobic			Total adduct level: 22/10^7 nucleotides	
Adduct formation of AAII in DNA (calf thymus) *in vitro*, activated by recombinant human CYP1A1 in microsomes from a baculovirus insect cell expression system, anaerobic			Total adduct level: 2.5/10^7 nucleotides	
Adduct formation of AAI in DNA (calf thymus) *in vitro*, activated by recombinant human CYP1A2 in microsomes from a baculovirus insect cell expression system, anaerobic			Total adduct level: 19/10^7 nucleotides	
Adduct formation of AAII in DNA (calf thymus) *in vitro*, activated by recombinant human CYP1A2 in microsomes from a baculovirus insect cell expression system, anaerobic			Total adduct level: 2/10^7 nucleotides	
Adduct formation of AA mixture in DNA of opossum kidney cells after 24 h incubation	Nuclease P1-enhanced ^{32}P-postlabelling	10 µM 20 µM	Total adduct level: 47/10^7nucleotides 87/10^7nucleotides 15/10^7 nucleotides after 6 days of recovery	Lebeau *et al.* (2001)

AA, aristolochic acids (mixed); AAI and AAII, aristolochic acids I and II; dA-AAI, 7-(deoxyadenosin-N^6-yl)aristolactam I; dA-AAII, 7-(deoxyadenosin-N^6-yl)-aristolactam II; dG-AAI, 7-(deoxyguanosin-N^2-yl)aristolactam I; dG-AAII, 7-(deoxyguanosin-N^2-yl)aristolactam II; vis, visible

[a] S9 mix from Aroclor 1254-pretreated rats

[b] Anaerobic: reaction mixture purged with argon for 15 min before addition of substrates

[c] Microsomes from β-naphthoflavone-pretreated rats

[d] Adduct formed in incubations of aristolactam I with DNA and microsomes is found in the ureter of a Chinese herb nephropathy patient, shown by co-chromatography.

[e] Same adducts as found *in vivo* in the kidney of Chinese herb nephropathy patients, as shown by co-chromatography

Table 7. Polymerase action on aristolochic acid-induced adducts *in vitro*

Test system	Result	Reference
Site-specifically adducted oligonucleotides containing AA-DNA adducts were used as templates for T7 DNA polymerase in primer extension reactions. DNA sythesis products were analysed on polyacrylamide gels using 5′-^{32}P-labelled primers	Mainly block of DNA synthesis at the nucleotide 3′ to each adduct, but translesional synthesis was also observed. dA-AAI and dA-AAII allowed incorporation of dAMP and dTMP directly across equally well, deoxyguanosine adducts allowed preferential incorporation of dCMP	Broschard *et al.* (1994)
Site-specifically adducted oligonucleotides containing AA-DNA adducts were used as templates for human DNA polymerase α in primer extension reactions. DNA synthesis products were analysed on polyacrylamide gels using 5′-^{32}P-labelled primers	Mainly block of DNA synthesis at the nucleotide 3′ to adducts dA-AAI, dG-AAI and dA-AAII. Only dG-AAII allowed substantial translesional synthesis	Broschard *et al.* (1995)
Polymerase arrest assay was used to determine the distribution of DNA adducts formed by AAI and AAII *in vitro* in the mouse c-Ha-*ras* gene. Arrest spectra were obtained on sequencing gels using 5′-^{32}P-labelled primers and polymerase Sequenase	AAI-induced adducts showed preference for adenine residues. AAII-induced damage was found at guanine, adenine and cytosine residues. Adduct distribution was not random.	Arlt *et al.* (2000)
Polymerase arrest assay combined with a terminal transferase-dependent PCR (TD-PCR) was used to map the distribution of DNA adducts formed by AAI and AAII *in vitro* in the human *TP53* gene of MCF-7 DNA. Adducted DNA was used as template for TD-PCR.	AA-DNA binding spectrum in the *TP53* gene: CpG sites are not preferential targets for AAI or AAII.	Arlt *et al.* (2001b)

AA, aristolochic acids (mixed); AAI and AAII, aristolochic acids I and II; dA-AAI, 7-(deoxyadenosin-N^6-yl)aristolactam I; dA-AAII, 7-(deoxyadenosin-N^6-yl)aristolactam II; dG-AAI, 7-(deoxyguanosin-N^2-yl)aristolactam I; dG-AAII, 7-(deoxyguanosin-N^2-yl)aristolactam II; PCR, polymerase chain reaction; dAMP, deoxyadenosine 3′-monophosphate; dTMP, deoxythymidine 3′-monophosphate; dCMP, deoxycytidine 3′-monophosphate

Table 8. Genetic and related effects of aristolochic acid (AA)

Test system	Result[a]		Compound, dose (LED or HID)[b]	Reference
	Without exogenous metabolic system	With exogenous metabolic system		
Escherichia coli PQ37 SOS repair induction	+	+[c]	AA plant extract, 0.38 μg/assay	Kevekordes *et al.* (1999)
Escherichia coli PQ37 SOS repair induction	+	+[c]	AAI, 0.17 μg/assay	Kevekordes *et al.* (1999)
Escherichia coli PQ37 SOS repair induction	+	(+)[c]	AAII, 0.16 μg/assay	Kevekordes *et al.* (1999)
Salmonella typhimurium TM677, forward mutation, *hprt* locus *in vitro*	+	NT	AAI, 8.5	Pezzuto *et al.* (1988)
Salmonella typhimurium TA100, TA1537, reverse mutation	+	+	AA mixture, 50 μg/plate	Robisch *et al.* (1982)
Salmonella typhimurium TA1535, TA1538, TA98, reverse mutation	–	–	AA mixture, 200 μg/plate	Robisch *et al.* (1982)
Salmonella typhimurium TA100, TA1537, reverse mutation	+	+	AAI, 100 μg/plate	Schmeiser *et al.* (1984)
Salmonella typhimurium TA100, TA1537, reverse mutation	–	+	Aristolactam I and aristolactam II, 50 μg/plate	Schmeiser *et al.* (1986)
Salmonella typhimurium TA100, TA102, TA1537, reverse mutation	+	NT	AAI, 100 μg/plate	Pezzuto *et al.* (1988)
Salmonella typhimurium TA98, YG1020, YG1021, reverse mutation	(+)	NT	AAI, 170 μg/plate	Götzl & Schimmer (1993)
Salmonella typhimurium TA98, YG1020, YG1021, reverse mutation	(+)	NT	AAII, 78 μg/plate	Götzl & Schimmer (1993)
Salmonella typhimurium YG1024, reverse mutation	+	NT	AAI, 34 μg/plate	Götzl & Schimmer (1993)
Salmonella typhimurium YG1024, reverse mutation	+	NT	AAII, 31 μg/plate	Götzl & Schimmer (1993)
Salmonella typhimurium TA100NR, reverse mutation	–	–	AAI, 200 μg/plate	Schmeiser *et al.* (1984)
Salmonella typhimurium TA100NR, TA98NR, reverse mutation	–	NT	AAI, 200 μg/plate	Pezzuto *et al.* (1988)
Salmonella typhimurium TA100, YG1025, YG 1026, YG 1029, reverse mutation	+	NT	AAI and AAII, 34 μg/plate	Götzl & Schimmer (1993)

Table 8 (contd)

Test system	Result[a]		Compound, dose (LED or HID)[b]	Reference
	Without exogenous metabolic system	With exogenous metabolic system		
Drosophila melanogaster, somatic mutations, recombination, sex-linked recessive mutation	+	NT	AA mixture (65% AAI, 35% AAII), 17 μg/mL in feed	Frei *et al.* (1985)
DNA strand breaks, rat hepatocytes, *in vitro* (alkaline elution assay)	–	NT	AAI and AAII [dose not reported]	Pool *et al.* (1986)
Gene mutation, rat subcutaneous granuloma tissue, *Hprt* locus *in vitro*	+	NT	AAI and AAII, 20	Maier *et al.* (1987)
Gene mutation, Chinese hamster ovary cells, *Hprt* locus *in vitro*	+	NT	AAI, 18.2	Pezzuto *et al.* (1988)
Sister chromatid exchange, human lymphocytes *in vitro*	+	NT	AA mixture, 1.0	Abel & Schimmer (1983)
Micronucleus induction, human lymphocytes and hepatoma cells (Hep-G2) *in vitro*	+	+	AA mixture, 17	Kevekordes *et al.* (2001)
Chromosomal aberrations, human lymphocytes *in vitro*	+	NT	AA mixture, 1.0	Abel & Schimmer (1983)
Unscheduled DNA synthesis, male F344/Ducrj rat glandular stomach mucosa *in vitro*, after in-vivo treatment	–		AA mixture, 400 × 1 po	Furihata *et al.* (1984)
Unscheduled DNA synthesis, male PV6 rat glandular stomach mucosa *in vivo*	–		AA mixture, 300 × 1 po	Burlinson (1989)
Gene mutation, male SD rat subcutaneous granuloma tissue, *Hprt* locus *in vivo*	+		AA mixture, 40 μg in pouch	Maier *et al.* (1985)
Gene mutation, male SD rat subcutaneous granuloma tissue, *Hprt* locus *in vivo*	+		AAI, 80 μg; AAII, 32 μg in pouch	Maier *et al.* (1987)
Micronucleus induction, male and female NMRI mouse bone-marrow cells *in vivo*	+		AA mixture, 20 × 1 iv	Mengs & Klein (1988)

AA, aristolochic acids (mixed); AAI and AAII, aristolochic acids I and II

[a] +, positive; (+), weak positive; –, negative; NT, not tested

[b] LED, lowest effective dose; HID, highest ineffective dose; in-vitro tests, μg/mL; in-vivo tests, mg/kg bw/day; iv, intravenous; po, oral

[c] Activity is much less in the presence of metabolic activation.

aristolochic acids I and II were not mutagenic. Both aristolactams were mutagenic in *S. typhimurium* only with S9 mix. Aristolochic acids I and II did not induce DNA strand breaks in hepatocytes. In cultured human lymphocytes, the aristolochic acid mixture induced chromosomal aberrations, sister chromatid exchange and micronuclei and it induced somatic and sex-linked recessive lethal mutations in *Drosophila melanogaster*. It also induced micronuclei in human hepatoma cells *in vitro* and in mouse bone-marrow cells after administration *in vivo*, but it did not induce unscheduled DNA synthesis in rat stomach after in-vivo treatment. The aristolochic acid mixture as well as aristolochic acids I and II induced mutations at the *hprt* locus in *Salmonella typhimurium*, in the granuloma pouch assay, and in Chinese hamster cells.

4.5 Mechanistic considerations

Aristolochic acid must be activated in order to form DNA adducts *in vitro* and *in vivo*. From the structures of the major aristolochic acid–DNA adducts identified in various in-vitro systems and in animals and humans *in vivo*, it can be concluded that the major pathway of activation of aristolochic acid is nitroreduction (Figure 6): a cyclic *N*-acylnitrenium ion with a delocalized positive charge is the ultimate DNA-reactive species that binds preferentially to the exocyclic amino groups of purine nucleotides in DNA or is hydrolysed to the corresponding 7-hydroxyaristolactam. Therefore, the activation of aristolochic acid is a unique example of an intra-molecular acylation which leads to the DNA-reactive species. This view is supported by the results of the *Salmonella* mutagenicity assays showing that only the nitro group is important for the mutagenic activity of aristolochic acid and by the demonstration that the enzymatic activation of both aristolochic acids by xanthine oxidase, a mammalian nitroreductase, produced an adduct pattern identical to that seen after metabolism mediated by rat liver S9 mix (Schmeiser *et al.*, 1988). It was also demonstrated that both aristolochic acids could be activated by rat liver microsomes through simple nitroreduction (Schmeiser *et al.*, 1997). In a single study, this anaerobic hepatic microsomal activation of aristolochic acid could be attributed to CYP1A1 and CYP1A2 and — to a lesser extent — to NADPH:CYP reductase (CYPOR) using specific CYP/CYPOR inhibitors (Stiborová *et al.*, 2001a).

The adduct patterns in DNA from forestomach and kidney — target tissues of aristolochic acid-mediated carcinogenesis — and from non-target tissues, such as glandular stomach, liver and lung, were similar. This indicates that adduct formation is not sufficient to result in neoplasia.

In rodents, many chemical carcinogens activate the *ras* proto-oncogene by inducing a single point mutation, resulting in alteration of amino acid residue 12, 13 or 61. Similarly, aristolochic acid-initiated carcinogenesis in rodents is associated with a distinct molecular characteristic, the activation of *ras* by a specific AT → TA transversion mutation in codon 61 (CAA). This mutation occurs exclusively at the first adenine of codon 61 of Ha-*ras* (CAA → CTA) in all forestomach and ear-duct tumours of rats treated with aristolochic acid I (Schmeiser *et al.*, 1990) and was also found in tumours of the forestomach and lung of mice treated with a plant extract containing both aristolochic acids I and II (Schmeiser *et al.*, 1991). The selectivity of aristolochic acid I for mutations at adenine residues is consistent with the extensive formation of dA-AAI adducts in the target organ (Pfau *et al.*, 1990a; Stiborová *et al.*, 1994). Moreover, an apparently lifelong persistence of the dA-AAI adduct in rat forestomach DNA was observed, whereas the less abundant dG-AAI adduct was removed continuously from the same DNA over a 36-week period after treatment with a single dose of aristolochic acid I (Fernando *et al.*, 1993). Therefore, both the higher initial levels and the longer persistence of the dA-AAI adduct in urothelial tissue of patients with Chinese herb nephropathy probably contribute to the relative abundance of this adduct.

Investigations on the conversion of individual aristolochic acid–DNA adducts into mutations have shown that during in-vitro DNA synthesis, dAMP and dTMP were incorporated opposite the adenine adducts (dA-AAI, dA-AAII) equally well, whereas the guanine adducts (dG-AAI, dG-AAII) led to preferential incorporation of dCMP. The translesional by-pass of adenine adducts of aristolochic acid indicates a mutagenic potential resulting from dAMP incorporation by DNA polymerase, consistent with an AT → TA transversion as the mutagenic consequence. Therefore, the adenine adducts have a higher mutagenic potential than the guanine adducts, which may explain the apparent selectivity for mutations found at adenine residues in codon 61 of the *ras* genes in aristolochic acid-induced rodent tumours (Schmeiser *et al.*, 1990, 1991).

An adduct-specific polymerase arrest assay with a plasmid containing exon 2 of the mouse c-H-*ras* gene demonstrated that both adenines in codon 61 of this gene are aristolochic acid–DNA binding sites (Arlt *et al.*, 2000). Since *ras* genes are activated with high frequency by an AT → TA transversion mutation in codon 61 of *ras* DNA from aristolochic acid-induced tumours in animals (Schmeiser *et al.*, 1990), this suggests an important role of the dA-AAI adduct in Chinese herb nephropathy-related urothelial cancer in humans. Polymerase arrest spectra showed a preference for reaction with purine bases in human *TP53* for both aristolochic acids. The aristolochic acid–DNA binding spectrum in the *TP53* gene did not suggest the existence of any mutational hotspot in urothelial tumours of the current *TP53* mutation database. Thus, aristolochic acid is not a likely cause of urothelial tumours not associated with Chinese herb nephropathy (Arlt *et al.*, 2001b).

5. Summary of Data Reported and Evaluation

5.1 Exposure data

Several *Aristolochia* species (notably *A. contorta*, *A. debilis*, *A. fangchi* and *A. manshuriensis*) have been used in traditional Chinese medicine as anti-rheumatics, as diuretics and in the treatment of oedema. Aristolochic acids are nitrophenanthrene carboxylic acid derivatives that are constituents of these plant species.

5.2 Human carcinogenicity data

An outbreak of rapidly progressive renal fibrosis in Belgium involved at least 100 patients, mostly middle-aged women undergoing a weight-loss regimen that included use of a mixture of Chinese herbs containing *Aristolochia* species incorrectly labelled as *Stephania tetrandra.* Additional cases of rapidly progressive renal disease involving Chinese herbs have been reported from at least five other countries in Europe and Asia. This syndrome has been called 'Chinese herb nephropathy'.

Because of a few early cases of urothelial cancer among Belgian patients suffering from Chinese herb nephropathy, individuals with end-stage renal disease were offered prophylactic nephroureterectomy. This surgical procedure led to the identification of a high prevalence of pre-invasive and invasive neoplastic lesions of the renal pelvis, the ureter and the urinary bladder in patients with Chinese herb nephropathy. The number of malignancies detected (18 cancers in 39 women undergoing prophylactic nephroureterectomy) greatly exceeded the expected number of these uncommon tumours. There was a positive dose–response relationship between the consumption of the herbal mixture and the prevalence of the tumours. Some cases of clinically invasive disease have been described in the follow-up of end-stage Chinese herb nephropathy patients not undergoing prophylactic nephroureterectomy.

Subsequent phytochemical investigation led to the identification of aristolochic acids in the herbal mixture consumed by these patients. While there was batch-to-batch variation in the chemical composition of such mixtures employed in the weight-loss clinics in Belgium, specific aristolochic acid–DNA adducts were found in urothelial tissue specimens from all the urothelial cancer patients, providing conclusive evidence of exposure to plants of the genus *Aristolochia.*

One additional case of urothelial cancer following treatment for eczema with another herbal mixture that contained aristolochic acid has been reported.

5.3 Animal carcinogenicity data

Aristolochic acids, when tested for carcinogenicity by oral administration in mice and rats and by intraperitoneal injection in rabbits, induced forestomach carcinomas in mice and rats, and fibrotic changes in the kidney together with a low incidence of kidney tumours in rabbits.

Subcutaneous injection of aristolochic acids into rats induced a low incidence of urothelial carcinomas in the kidney and malignant fibrohistiocytic sarcomas at the injection site.

5.4 Other relevant data

Several structurally defined metabolites (mainly nitroreduction products) have been reported following oral administration of aristolochic acid I (five metabolites) and aristolochic acid II (three metabolites) to rats and mice. Fewer metabolites were observed in beagle dogs, rabbits, guinea-pigs and humans than in rats and mice.

The toxic effects of aristolochic acids I and II have been inferred from effects seen in patients suffering from kidney nephropathy as a result of consuming herbal mixtures containing *Aristolochia* species, which leads to rapidly progressive fibrosing interstitial nephritis. In experimental animals, high doses of aristolochic acids administered either orally or intravenously caused severe necrosis of the renal tubules, atrophy of the spleen and thymus, and ulceration of the forestomach, followed by hyperplasia and hyperkeratosis of the squamous epithelium.

Various constituents of *Aristolochia indica* including aristolochic acids and aristolic acid (a metabolite) caused termination of pregnancy in female mice, hamsters and rabbits, but not rats. The dose levels used, however, may also lead to general toxicity.

Aristolochic acids, when metabolically activated by nitroreduction, are consistently active in genotoxicity tests *in vivo* and *in vitro*. They form DNA adducts in rodent tissues and activate *ras* oncogenes through a specific transversion mutation in codon 61. Aristolochic acid-specific DNA adducts were identified in urothelial tissues of all patients with Chinese herb nephropathy.

5.5 Evaluation

There is *sufficient evidence* in humans for the carcinogenicity of herbal remedies containing plant species of the genus *Aristolochia*.

There are no data in experimental animals on the carcinogenicity of herbal remedies containing plant species of the genus *Aristolochia*.

There is *limited evidence* in humans for the carcinogenicity of naturally occurring mixtures of aristolochic acids.

There is *sufficient evidence* in experimental animals for the carcinogenicity of aristolochic acids.

Overall evaluation

Herbal remedies containing plant species of the genus *Aristolochia* are *carcinogenic to humans (Group 1)*.

Naturally occurring mixtures of aristolochic acids are *probably carcinogenic to humans (Group 2A)*.

6. References

Abel, G. & Schimmer, O. (1983) Induction of structural chromosome aberrations and sister chromatid exchanges in human lymphocytes *in vitro* by aristolochic acid. *Hum. Genet.*, **64**, 131–133

Achari, B., Chakrabarty, S. & Pakrashi, S.C. (1981) Studies on Indian medicinal plants. Part 63. An N-glycoside and steroids from *Aristolochia indica*. *Phytochemistry*, **20**, 1444–1445 [consulted as abstract: CA1982:35654]

Achari, B., Chakrabarty, S., Bandyopadhyay, S. & Pakrashi, S.C. (1982) Studies on Indian medicinal plants. Part 69. A new 4,5-dioxoaporphine and other constituents of *Aristolochia indica*. *Heterocycles*, **19**, 1203–1206 [consulted as abstract: CA1982:469282]

Achari, B., Bandyopadhyay, S., Saha, C.R. & Pakrashi, S.C. (1983) A phenanthroid lactone, steroid and lignans from *Aristolochia indica*. *Heterocycles*, **20**, 771–774 [consulted as abstract: CA1983:403041]

Achari, B., Bandyopadhyay, S., Basu, K. & Pakrashi, S.C. (1985) Studies on Indian medicinal plants. Part LXXIX. Synthesis proves the structure of aristolindiquinone. *Tetrahedron*, **41**, 107–110 [consulted as abstract: CA1985:406129]

Anonymous (1985) *The Wealth of India: A Dictionary of Indian Raw Materials and Industrial Products. Raw Materials*, Volume I: A (Revised), New Delhi, Publications and Information Directorate Council of Scientific and Industrial Research

Arlt, V.M., Wiessler, M. & Schmeiser, H.H. (2000) Using polymerase arrest to detect DNA binding specificity of aristolochic acid in the mouse H-*ras* gene. *Carcinogenesis*, **21**, 235–242

Arlt, V.M., Pfohl-Leszkowicz, A., Cosyns, J.-P. & Schmeiser, H.H. (2001a) Analyses of DNA adducts formed by ochratoxin A and aristolochic acid in patients with Chinese herbs nephropathy. *Mutat. Res.*, **494**, 143–150

Arlt, V.M., Schmeiser, H.H. & Pfeifer, G.P. (2001b) Sequence-specific detection of aristolochic acid-DNA adducts in the human p53 gene by terminal transferase-dependent PCR. *Carcinogenesis*, **22**, 133–140

Bieler, C.A., Stiborova, M., Wiessler, M., Cosyns, J.-P., van Ypersele de Strihou, C. & Schmeiser, H.H. (1997) ^{32}P-Post-labelling analysis of DNA adducts formed by aristolochic acid in tissues from patients with Chinese herbs nephropathy. *Carcinogenesis*, **18**, 1063–1067

Broschard, T.H., Wiessler, M., von der Lieth, C.-W. & Schmeiser, H.H. (1994) Translesional synthesis on DNA templates containing site-specifically placed deoxyadenosine and deoxyguanosine adducts formed by the plant carcinogen aristolochic acid. *Carcinogenesis*, **15**, 2331–2340

Broschard, T.H., Wiessler, M. & Schmeiser, H.H. (1995) Effect of site-specifically located aristolochic acid DNA adducts on in vitro DNA synthesis by human DNA polymerase α. *Cancer Lett.*, **98**, 47–56

Buckingham, J., ed. (2001) *Dictionary of Natural Products on CD-ROM*, Boca Raton, FL, CRC Press, Chapman & Hall/CRC

Bulgakov, V.P., Zhuravlev, Y.U.N. & Radchenko, S.V. (1996) Constituents of *Aristolochia manshuriensis* cell suspension culture possessing cardiotonic activity. *Fitoterapia*, **67**, 238–240

Burlinson, B. (1989) An *in vivo* unscheduled DNA synthesis (UDS) assay in the rat gastric mucosa: Preliminary development. *Carcinogenesis*, **10**, 1425–1428

Che, C.T., Cordell, G.A., Fong, H.H.S. & Evans, C.A. (1983) Studies on *Aristolochia*. Part 2. Aristolindiquinone — A new naphthoquinone from *Aristolochia indica* L. (Aristolochiaceae). *Tetrahedron Lett.*, **24**, 1333–1336 [consulted as abstract: CA1983:450253]

Che, C.-T., Ahmed, M.S., Kang, S.S., Waller, D.P., Bingel, A.S., Martin, A., Rajamahendran, P., Bunyapraphatsara, N., Lankin, D.C., Cordell, G.A., Soejarto, D.D., Wijesekera, R.O.B. & Fong, H.H.S. (1984) Studies on Aristolochia. III. Isolation and biological evaluation of constituents of *Aristolochia indica* roots for fertility-regulating activity. *J. nat. Prod.*, **47**, 331–341

Chinese Materia Medica (1995) *Safety, Efficacy and Modernization*, Beijing, China Academy of Trade

Commission of the Ministry of Public Health (2000) *Pharmacopoeia* (Part I), Beijing, Chemical Industry Press, pp. 31, 39, 41, 114, 154

Cory, R.M., Chan, D.M.T., McLaren, F.R., Rasmussen, M.H. & Renneboog, R.M. (1979) A short synthesis of ishwarone. *Tetrahedron Lett.*, **43**, 4133–4136 [consulted as abstract: CA1980:215559]

Cosyns, J.-P., Jadoul, M., Squifflet, J.P., de Plaen, J.F., Ferluga, D. & Van Ypersele de Strihou, C. (1994a) Chinese herbs nepropathy: A clue to Balkan endemic nephropathy? *Kidney int.*, **45**, 1680–1688

Cosyns, J.-P., Jadoul, M., Squifflet, J.P., Van Cangh, P.J. & Van Ypersele de Strihou, C. (1994b) Urothelial malignancy in nephropathy due to Chinese herbs (Letter to the Editor). *Lancet*, **344**, 188

Cosyns, J.-P., Jadoul, M., Squifflet, J.P., Wese, F.X. & Van Ypersele de Strihou, C. (1999) Urothelial lesions in Chinese herb nephropathy. *Am. J. Kidney Dis.*, **33**, 1011–1017

Cosyns, J.-P., Dehoux, J.-P., Guiot, Y., Goebbels, R.-M., Robert, A., Bernard, A.M. & Van Ypersele de Strihou, C. (2001) Chronic aristolochic acid toxicity in rabbits: A model of Chinese herbs nephropathy? *Kidney int.*, **59**, 2164–2173

Debelle, F.D., Nortier, J.L., De Prez, E.G., Garber, C.H., Vienne, A.R., Salmon, I.J., Deschodt-Lanckman, M.M. & Vanherweghem, J.L. (2002) Aristolochic acids induce chronic renal failure with interstitial fibrosis in salt-depleted rats. *J. Am. Soc. Nephrol.*, **13**, 431–436

Depierreux, M., Van Damme, B., Vanden Houte, K. & Vanherweghem, J.-L. (1994) Pathologic aspects of a newly described nephropathy related to the prolonged use of Chinese herbs. *Am. J. Kidney Dis.*, **24**, 172–180

EMEA (European Agency for the Evaluation of Medicinal Products) (2000) *Working Party on Herbal Medicinal Products: Position paper on the risks associated with the use of herbal products containing* Aristolochia species *(EMEA/HMPWP/23/00)*, London

European Commission (2000) Twenty-fifth Commission Directive 2000/11/EC of 10 March 2000 adapting to technical progress Annex II to Council Directive 76/768/EEC on the approximation of the laws of the Member States relating to cosmetic products. *Off. J.*, **L065**

von Euw, J., Reichstein, T. & Rothschild, M. (1968) Aristolochic acid-I in the swallowtail butterfly *Pachlioptera aristolochiae (Fabr.) (Papilionidae)*. *Israel J. Chem.*, **6**, 659-670

Fernando, R.C., Schmeiser, H.H., Nicklas, W. & Wiessler, M. (1992) Detection and quantitation of dG-AAI and dA-AAI adducts by ^{32}P-postlabelling methods in urothelium and exfoliated cells in urine of rats treated with aristolochic acid I. *Carcinogenesis*, **13**, 1835–1839

Fernando, R.C., Schmeiser, H.H., Scherf, H.R. & Wiessler, M. (1993) Formation and persistence of specific purine DNA adducts by ^{32}P-postlabelling in target and non-target organs of rats treated with aristolochic acid I. In: Phillips, D.H., Castegnaro, M. & Bartsch, H., eds, *Post-labelling Methods for Detection of DNA Adducts* (IARC Scientific Publications No. 124), IARC*Press*, Lyon, pp. 167–171

Flurer, R.A., Jones, M.B., Vela, N., Ciolino, L.A. & Wolnik, K.A. (2000) *Determination of Aristolochic Acid in Traditional Chinese Medicines and Dietary Supplements* (Laboratory Information Bulletin No. 4212), Cincinnati, OH, Forensic Chemistry Center, US Food and Drug Administration

Food and Drug Administration (2001a) *General Information on the Regulation of Dietary Supplements*, Washington DC, Center for Food Safety and Applied Nutrition, Office of Food Labeling

Food and Drug Administration (2001b) *Dietary Supplements: Aristolochic Acid* [http://vm.cfsan.fda.gov/~dms/ds-bot.html]

Fordyce, J.A. (2000) A model without a mimic: Aristolochic acids from the California pipevine swallowtail, *Battus philenor hirsuta*, and its host plant, *Aristolochia californica*. *J. chem. Ecol.*, **26**, 2567–2578

Frei, H., Würgler, F.E., Juon, H., Hall, C.B. & Graf, U. (1985) Aristolochic acid is mutagenic and recombinogenic in *Drosophila* genotoxicity tests. *Arch. Toxicol.*, **56**, 158–166

Fuhrer, H., Ganguly, A.K., Gopinath, K.W., Govindachari, T.R., Nagarajan, K., Pai, B.R. & Parthasarathy, P.C. (1970) Ishwarone. *Tetrahedron*, **26**, 2371–2390 [consulted as abstract: CA1970:435542]

Furihata, E., Yamawaki, Y., Jin, S.-S., Moriya, H., Kodama, K., Matsushima, T., Ishikawa, T., Takayama, S. & Nakadate, M. (1984) Induction of unscheduled DNA synthesis in rat stomach mucosa by glandular stomach carcinogens. *J. natl Cancer Inst.*, **72**, 1327–1333

Ganguly, A.K., Gopinath, K.W., Govindachari, T.R., Nagarajan, K., Pai, B.R. & Parthasarathy, P.C. (1969) Ishwarone, a tetracyclic sesquiterpene. *Tetrahedron Lett.*, **3**, 133–136 [consulted as abstract: CA1969:87973]

Ganguly, T., Pakrashi, A. & Pal, A.K. (1986) Disruption of pregnancy in mouse by aristolic acid: I. Plausible explanation in relation to early pregnancy events. *Contraception*, **34**, 625–637

Gillerot, G., Jadoul, M., Arlt, V.M., van Ypersele de Strihou, C., Schmeiser, H.H., But, P.P.H., Bieler, C.A. & Cosyns, J.-P. (2001) Aristolochic acid nephropathy in a Chinese patient: Time to abandon the term 'Chinese herbs nephropathy'? *Am. J. Kidney Dis.*, **38**, 1–5

Götzl, E. & Schimmer, O. (1993) Mutagenicity of aristolochic acids (I, II) and aristolic acid I in new YG strains in *Salmonella typhimurium* highly sensitive to certain mutagenic nitroarenes. *Mutagenesis*, **8**, 17–22

Govindachari, T.R. & Parthasarathy, P.C. (1971) Ishwarol, a new tetracyclic sesquiterpene alcohol from *Aristolochia indica*. *Indian J. Chem.*, **9**, 1310 [consulted as abstract: CA1972:138154]

Govindachari, T.R., Mohamed, P.A. & Parthasarathy, P. C. (1970) Ishwarane and aristolochene, two new sesquiterpene hydrocarbons from *Aristolochia indica*. *Tetrahedron*, **26**, 615–619 [consulted as abstract: CA1970:87185]

Govindachari, T.R., Parthasarathy, P.C., Desai, H.K. & Mohamed, P.A. (1973) 5βH,7β,10α-Selina-4(14),II-diene, a new sesquiterpene hydrocarbon from *Aristolochia indica*. *Indian J. Chem.*, **11**, 971–973 [consulted as abstract: CA1974:108696]

Hadjiolov, D., Fernando, R.C., Schmeiser, H.H., Wiessler, M., Hadjiolov, N. & Pirajnov, G. (1993) Effect of diallyl sulfide on aristolochic acid-induced forestomach carcinogenesis in rats. *Carcinogenesis*, **14**, 407–410

Hashimoto, K., Higuchi, M., Makino, B., Sakakibara, I., Kubo, M., Komatsu, Y., Maruno, M. & Okada, M. (1999) Quantitative analysis of aristolochic acids, toxic compounds, contained in some medicinal plants. *J. Ethnopharmacol.*, **64**, 185–189

Health Canada (1999) *Warning: Warning not to Use Products Containing* Aristolochia *due to Cancer, Cell Changes, and Kidney Failure* [No. 1999-129], Ottawa

Health Canada (2001a) *Advisory: Health Canada Advising not to Use Products Labeled to Contain* Aristolochia [No. 2001-91], Ottawa

Health Canada (2001b) *Warning: Health Canada is Warning Canadians not to Use the Pediatric Product Tao Chih Pien as it Contains Aristolochic Acid* [No. 2001-94], Ottawa

Health Canada (2001c) *Warning: Health Canada Market Survey Confirms Some Products Contain Aristolochic Acid* [No. 2001-100], Ottawa

Health Canada (2001d) *Advisory: Health Canada Advises Consumers about Additional Products that Could Contain Aristolochic Acid* [No. 2001-105], Ottawa

Jackson, L., Kofman, S., Weiss, A. & Brodovsky, H. (1964) Aristolochic acid (NSC-50413): Phase I clinical study. *Cancer Chemother. Rep.*, **42**, 35–37

Jirovetz, L., Buchbauer, G., Puschmann, C., Fleischhacker, W., Shafi, P.M. & Rosamma, M.K. (2000) Analysis of the essential oil of the aerial parts of the medicinal plant *Aristolochia indica* (Aristolochiaceae) from South-India. *Scientia Pharmaceutica*, **68**, 309–316 [consulted as abstract: 2000:CA759374]

Kevekordes, S., Mersch-Sundermann, V., Burghaus, C.M., Spielberger, J., Schmeiser, H.H., Arlt, V.M. & Dunkelberg, H. (1999) SOS induction of selected naturally occurring substances in *Escherichia coli* (SOS chromotest). *Mutat. Res.*, **445**, 81–91

Kevekordes, S., Spielberger, J., Burghaus, C.M., Birkenkamp, P., Zietz, B., Paufler, P., Diez, M., Bolten, C. & Dunkelberg, H. (2001) Micronucleus formation in human lymphocytes and in the metabolically competent human hepatoma cell line Hep-G2: results with 15 naturally occurring substances. *Anticancer Res.*, **21**, 461–469

Kite, G.C., Yule, M.A., Leon, C. & Simmonds, M.S.J. (2002) Detecting aristolochic acids in herbal remedies by liquid chromatography/serial mass spectrometry. *Rapid Commun. Mass Spectrom.*, **16**, 585–590

Klitzke, C.F. & Brown, K.S., Jr. (2000) The occurrence of aristolochic acids in neotropical troidine swallowtails (Lepidoptera: Papilionidae). *Chemoecology*, **10**, 99–102

Krumbiegel, G. & Roth, H.J. (1987) [Semisynthetic aristolochic acids I- and II-derivatives as reference compounds for metabolites in the rat.] *Arch. Pharmacol.*, **320**, 264–270 (in German)

Krumbiegel, G., Hallensleben, J., Mennicke, W.H., Rittmann, N. & Roth, H.J. (1987) Studies on the metabolism of aristolochic acids I and II. *Xenobiotica*, **17**, 981–991

Krumme, B., Endmeir, R., Vanhaelen, M. & Walb, D. (2001) Reversible Fanconi syndrome after ingestion of a Chinese herbal 'remedy' containing aristolochic acid. *Nephrol. Dial. Transplant.*, **16**, 400–402

Kupchan, S.M. & Merianos, J.J. (1968) Tumor inhibitors. XXXII. Isolation and structural elucidation of novel derivatives of aristolochic acid from *Aristolochia indica*. *J. org. Chem.*, **33**, 3735–3738 [consulted as abstract: CA1969:4529]

Lebeau, C., Arlt, V.M., Schmeiser, H.H., Boom, A., Verroust, P.J., Devuyst, O. & Beauwens, R. (2001) Aristolochic acid impedes endocytosis and induces DNA adducts in proximale tubule cells. *Kidney int.*, **60**, 1332–1342

Lee, T.-Y., Wu, M.-L., Deng, J.-F. & Hwang, D.-F. (2001) High-performance liquid chromatographic determination for aristolochic acid in medicinal plants and slimming products. *J. Chromatogr. B. Biomed. Sci. Appl.*, **766**, 169–174

Lord, G.M., Tagore, R., Cook, T., Gower, P. & Pusey, C.D. (1999) Nephropathy caused by Chinese herbs in the UK. *Lancet*, **354**, 481–482

Lord, G.M., Cook, T., Arlt, V.M., Schmeiser, H.H., Williams, H. & Pusey, C.D. (2001) Urothelial malignant disease and Chinese herbal nephropathy. *Lancet*, **358**, 1515–1516

Mahesh, V.K. & Bhaumik, H.L. (1987) Isolation of methyl ester of 12-nonacosenoic acid from *Aristolochia indica*. *Indian J. Chem., Sect. B*, **26B**, 86 [consulted as abstract: CA1987: 210985]

Maier, P., Schwalder, H.P., Weibel, B. & Zbinden, G. (1985) Aristolochic acid induces 6-thioguanine-resistant mutants in an extrahepatic tissue in rats after oral application. *Mutat. Res.*, **143**, 143–148

Maier, P., Schawalder, H. & Weibel, B. (1987) Low oxygen tension, as found in tissues *in vivo*, alters the mutagenic activity of aristolochic acids I and II in primary fibroblast-like rat cells *in vitro*. *Environ. mol. Mutag.*, **10**, 275–284

Medical Economics Co. (1998) *PDR for Herbal Medicines*, 1st Ed., Montvale, NJ, pp. 660–661, 1103

Medical Economics Co. (2000) *PDR for Herbal Medicines*, 2nd Ed., Montvale, NJ, pp. 80–81, 490

Medicines Control Agency's (MCA) (2001) *Licensing of Medicines: Policy on Herbal Medicines*, London [E-mail: info@mca.gov.uk]

Mengs, U. (1983) On the histopathogenesis of rat forestomach carcinoma caused by aristolochic acid. *Arch. Toxicol.*, **52**, 209–220

Mengs, U. (1987) Acute toxicity of aristolochic acid in rodents. *Arch. Toxicol.*, **59**, 328–331

Mengs, U. (1988) Tumour induction in mice following exposure to aristolochic acid. *Arch. Toxicol.*, **61**, 504–505

Mengs, U. & Klein, M. (1988) Genotoxic effects of aristolochic acid in the mouse micronucleus test. *Planta med.*, **54**, 502–503

Mengs, U. & Stotzem, C.D. (1992) Toxicity of aristolochic acid. A subacute study in male rats. *Med. Sci. Res.*, **20**, 223–224

Mengs, U. & Stotzem, C.D. (1993) Renal toxicity of aristolochic acid in rats as an example of nephrotoxicity testing in routine toxicology. *Arch. Toxicol.*, **67**, 307–311

Mengs, U., Lang, W. & Poch, J.-A. (1982) The carcinogenic action of aristolochic acid in rats. *Arch. Toxicol.*, **51**, 107–119

Meyer, M.M., Chen, T.-P. & Bennett, W.M. (2000) Chinese herb nephropathy. *BUMC (Baylor University Medical Center) Proc.*, **13**, 334–337

Muniz Martinez, M.C., Nortier, J.L., Vereerstraeten, P. & Vanherweghem, J.-L. (2002) Progression rate of Chinese-herb nephropathy: Impact of *Aristolochia fangchi* ingested dose. *Nephrol. Dial. Transplant*, **17**, 1–5

Nagasawa, H., Wu, G. & Inatomi, H. (1997) Effects of aristoloside, a component of Guan-mu-tong (*Caulis aristolochiae manshuriensis*), on normal and pre-neoplastic mammary gland growth in mice. *Anticancer Res.*, **17**, 237–240

Nishida, R. & Fukami, H. (1989a) Oviposition stimulants of an Aristolochiaceae-feeding swallowtail butterfly, *Atrophaneura alcinous*. *J. Chem. Ecol.*, **15**, 2565–2575

Nishida, R. & Fukami, H. (1989b) Ecological adaptation of an Aristolochiaceae-feeding swallowtail butterfly, *Atrophaneura alcinous*, to aristolochic acids. *J. Chem. Ecol.*, **15**, 2549–2563

Nishida, R., Weintraub, J.D., Feeny, P. & Fukami H. (1993) Aristolochic acids from *Thottea* spp. (Aristolochiaceae) and the osmeterial secretions of *Thottea*-feeding troidine swallowtail larvae (Papilionidae). *J. chem. Ecol.*, **19**, 1587–1594

Nortier, J.L., Martinez, M.C., Schmeiser, H.H., Arlt, V.M., Bieler, C.A., Petein, M., Depierreux, M.F., De Pauw, L., Abramowicz, D., Vereerstraeten, P. & Vanherweghem, J.-L. (2000) Urothelial carcinoma associated with the use of a Chinese herb (*Aristolochia fangchi*). *New Engl. J. Med.*, **342**, 1686–1692

O'Neil, M.J., ed. (2001) *The Merck Index*, 13th Ed., Whitehouse Station, NJ, Merck & Co., pp. 48, 134

Ong, E.S. & Woo, S.O. (2001) Determination of aristolochic acids in medicinal plants (Chinese) prepared medicine using capillary zone electrophoresis. *Electrophoresis*, **22**, 2236–2241

Ong, E.W., Woo, S.O. & Yong, Y.L. (2000) Pressurized liquid extraction of berberine and aristolochic acids in medicinal plants. *J. Chromatogr. A.*, **313**, 57–64

Pakrashi, A. & Chakrabarty, B. (1978a) Anti-oestrogenic & anti-implantation effect of aristolic acid from *Aristolochia indica* (Linn.). *Indian J. exp. Biol.*, **16**, 1283–1285

Pakrashi, A. & Chakrabarty, B (1978b) Antifertility effect of aristolic acid from *Aristolochia indica* (Linn) in female albino rabbits. *Experientia*, **34**, 1377

Pakrashi, A. & Ganguly, T. (1982) Changes in uterine phosphatase levels in mice treated with aristolic acid during early pregnancy. *Contraception*, **26**, 635–643

Pakrashi, A. & Pakrasi, P.L. (1977) Antispermatogenic effect of the extract of *Aristolochia indica* Linn on male mice. *Indian J. exp. Biol.*, **15**, 256–259

Pakrashi, A. & Pakrasi, P. (1978) Biological profile of *p*-coumaric acid isolated from *Aristolochia indica* Linn. *Indian J. exp. Biol.*, **16**, 1285–1287 [consulted as abstract: CA1979:115634]

Pakrashi, A. & Shaha, C. (1978) Effect of methyl ester of aristolic acid from *Aristolochia indica* Linn. on fertility of female mice. *Experientia*, **34**, 1192–1193

Pakrashi, A. & Shaha, C. (1979a) Short term toxicity study with methyl ester of aristolic acid from *Aristolochia indica* Linn. in mice. *Indian J. exp. Biol.*, **17**, 437–439 [consulted as abstract: CA1979:450514]

Pakrashi, A. & Shaha, C. (1979b) Effect of methyl aristolate from *A. indica* Linn. on implantation in mice. *IRCS Med. Sci.: Libr. Compend.*, **7**, 78 [consulted as abstract: CA1979:180684]

Pakrashi, A., Chakrabarty, B. & Dasgupta, A. (1976) Effect of the extracts from *Aristolochia indica* Linn. on interception in female mice. *Experientia*, **32**, 394–395

Pakrashi, S.C., Ghosh-Dastidar, P., Basu, S. & Achari, B. (1977) Studies on Indian medicinal plants. Part 46. New phenanthrene derivatives from *Aristolochia indica*. *Phytochemistry*, **16**, 1103–1104 [consulted as abstract: CA1977:514573]

Pakrashi, S.C., Dastidar, P.P.G., Chakrabarty, S. & Achari, B. (1980) (12S)-7,12-Secoishwaran-12-ol, a new type of sesquiterpene from *Aristolochia indica* Linn. *J. org. Chem.*, **45**, 4765–4767 [consulted as abstract: CA1981:4130]

Pal, A.K., Kabir, S.N. & Pakrashi, A. (1982) A probe into the possible mechanism underlying the interceptive action of aristolic acid. *Contraception*, **25**, 639–648

Peña, J.M., Borrás, M., Ramos, J. & Montoliu, J. (1996) Rapidly progressive interstitial renal fibrosis due to the chronic intake of a herb (*Aristolochia pistolochia*) infusion. *Nephrol. Dial. Transplant.*, **11**, 1359–1360

Pezzuto, J.M., Swanson, S.M., Mar, W., Che, C.-T., Cordell, G.A. & Fong, H.H.S. (1988) Evaluation of the mutagenic and cytostatic potential of aristolochic acid (3,4-methylenedioxy-8-

methoxy-10-nitrophenanthrene-1-carboxylic acid) and several of its derivatives. *Mutat. Res.*, **206**, 447–454

Pfau, W., Schmeiser, H.H. & Wiessler, M. (1990a) ^{32}P-Postlabelling analysis of the DNA adducts formed by aristolochic acids I and II. *Carcinogenesis*, **11**, 1627–1633

Pfau, W., Schmeiser, H.H. & Wiessler, M. (1990b) Aristolochic acid binds covalently to the exocyclic amino group of purine nucleotides in DNA. *Carcinogenesis*, **11**, 313–319

Pfau, W., Schmeiser, H.H. & Wiessler, M. (1991) *N*^{6}-Adenyl arylation of DNA by aristolochic acid II and a synthetic model for the putative proximate carcinogen. *Chem. Res. Toxicol.*, **4**, 581–586

Pool, B.L., Eisenbrand, G., Preussmann, R., Schlehofer, J.R., Schmezer, P., Weber, H. & Wiessler, M. (1986) Detection of mutations in bacteria and of DNA damage and amplified DNA sequences in mammalian cells as a systematic test strategy for elucidating biological activities of chemical carcinogens. *Food chem. Toxicol.*, **24**, 685–691

Pourrat, J., Montastruc, J.L., Lacombe, J.L., Cisterne, J.M., Rascol, O. & Dumazer, Ph. (1994) [Neuropathy associated with Chinese herbs – 2 cases (Letter).] *Presse méd.*, **23**, 1669 (in French)

Qian, X.-Z. (1996) *Colour Pictorial Handbook of Chinese Herbs*, Beijing, The Peoples Medical Publishing House

Robisch, G., Schimmer, O. & Göggelmann, W. (1982) Aristolochic acid as a direct mutagen in *Salmonella typhimurium*. *Mutat. Res.*, **105**, 201–204

Rossiello, M.R., Laconi, E., Rao, P.M., Rajalakshmi, S. & Sarma, D.S.R. (1993) Induction of hepatic nodules in the rat by aristolochic acid. *Cancer Lett.*, **71**, 83–87

Routledge, M.N., Orton, T.C., Lord, P.G. & Garner, R.C. (1990) Effect of butylated hydroxyanisole on the level of DNA adduction by aristolochic acid in the rat forestomach and liver. *Jpn. J. Cancer Res.*, **81**, 220–224

Sachdev-Gupta, K., Feeny, P.P. & Carter, M. (1993) Oviposition stimulants for the pipevine swallowtail butterfly, *Battus philenor* (Papilionidae), from an *Aristolochia* host plant: Synergism between inositols, aristolochic acids and a monogalactosyl diglyceride. *Chemoecology*, **4**, 19–28

Schmeiser, H.H., Pool, B.L. & Wiessler, M. (1984) Mutagenicity of the two main components of commercially available carcinogenic aristolochic acid in *Salmonella typhimurium*. *Cancer Lett.*, **23**, 97–101

Schmeiser, H.H., Pool, B.L. & Wiessler, M. (1986) Identification and mutagenicity of metabolites of aristolochic acid formed by rat liver. *Carcinogenesis*, **7**, 59–63

Schmeiser, H.H., Schoepe, K.-B. & Wiessler, M. (1988) DNA adduct formation of aristolochic acids I and II *in vitro* and *in vivo*. *Carcinogenesis*, **9**, 297–303

Schmeiser, H.H., Janssen, J.W., Lyons, J., Scherf, H.R., Pfau, W., Buchmann, A., Bartram, C.R. & Wiessler, M. (1990) Aristolochic acid activates *ras* genes in rat tumors at deoxyadenosine residues. *Cancer Res.*, **50**, 5464–5469

Schmeiser, H.H., Scherf, H.R. & Wiessler, M. (1991) Activating mutations at codon 61 of the c-Ha-*ras* gene in thin-tissue sections of tumors induced by aristolochic acid in rats and mice. *Cancer Lett.*, **59**, 139–143

Schmeiser, H.H., Bieler, C.A., Wiessler, M., van Ypersele de Strihou, C. & Cosyns, J.P. (1996) Detection of DNA adducts formed by aristolochic acid in renal tissue from patients with Chinese herbs nephropathy. *Cancer Res.*, **56**, 2025–2028

Schmeiser, H.H., Frei, E., Wiessler, M. & Stiborova, M. (1997) Comparison of DNA adduct formation by aristolochic acids in various in-vitro activation systems by ^{32}P-post-labelling: Evidence for reductive activation by peroxidases. *Carcinogenesis*, **18**, 1055–1062

Schulz, M., Weist, F. & Gemählich, M. (1971) [Thin layer chromatographic demonstration of aristolochic acids in various body fluids.] *Arzneim.-Forsch.*, **21**, 934–936 (in German)

Sigma-Aldrich (2002) *Biochemicals and Reagents for Life Science Research 2002–2003*, St. Louis, MO, p. 220

Sime, K.R., Feeny, P.P. & Haribal, M.M. (2000) Sequestration of aristolochic acids by the pipevine swallowtail, *Battus philenor* (L.): evidence and ecological implications. *Chemoecology*, **10**, 169–178

Singh, D.V., Singh, B.L., Verma, R.K., Gupta, M.M., Banerji, S. & Kumar, S. (2001a) Quantitation of aristolochic acid using high performance liquid chromatography with photodiode array detection. *J. Indian chem. Soc.*, **78**, 487–488

Singh, D.V., Singh, B.L., Verma, R.K., Gupta, M.M. & Kumar, S. (2001b) Reversed phase high performance liquid chromatographic analysis of aristolochic acid. *J. med. arom. Plant Sci.*, **22–23**, 29–31

Stengel, B. & Jones, E. (1998) [Terminal renal insufficiency associated with consumption of Chinese herbs in France.] *Néphrologie*, **19**, 15–20 (in French)

Stiborová, M., Fernando, R.C., Schmeiser, H.H., Frei, E., Pfau, W. & Wiessler, M. (1994) Characterization of DNA adducts formed by aristolochic acids in the target organ (forestomach) of rats by ^{32}P-postlabelling analysis using different chromatographic procedures. *Carcinogenesis*, **15**, 1187–1192

Stiborová, M., Frei, E., Schmeiser, H.H. & Wiessler, M. (1995) Cytochrome P-450 and peroxidase oxidize detoxication products of carcinogenic aristolochic acids (aristolactams) to reactive metabolites binding to DNA *in vitro*. *Collect. Czech. Chem. Commun.*, **60**, 2189–2199

Stiborová, M., Frei, E., Breuer, A., Bieler, C.A. & Schmeiser, H.H. (1999) Aristolactam I a metabolite of aristolochic acid I upon activation forms an adduct found in DNA of patients with Chinese herbs nephropathy. *Exp. Toxicol. Pathol.*, **51**, 421–427

Stiborová, M., Frei, E., Breuer, A., Wiessler, M. & Schmeiser, H.H. (2001a) Evidence for reductive activation of carcinogenic aristolochic acids by prostaglandin H synthase —^{32}P-postlabeling analysis of DNA adduct formation. *Mutat. Res.*, **493**, 149–160

Stiborová, M., Frei, E., Wiessler, E. & Schmeiser, H.H. (2001b) Human enzymes involved in the metabolic activation of carcinogenic aristolochic acids: Evidence for reductive activation by cytochromes P450 1A1 and 1A2. *Chem. Res. Toxicol.*, **14**, 1128–1137

Tanaka, A., Nishida, R., Maeda, K., Sugawara, A. & Kuwahara, T. (2000a) Chinese herb nephropathy in Japan presents adult onset Fanconi syndrome: Could different components of aristolochic acids cause a different type of Chinese herb nephropathy? *Clin. Nephrol.*, **53**, 301–306

Tanaka, A., Nishida, R., Yokoi, H. & Kuwahara, T. (2000b) The characteristic pattern of aminoaciduria in patients with aristolochic acid-induced Fanconi syndrome: Could iminoaciduria be the hallmark of this syndrome? *Clin. Nephrol.*, **54**, 198–202

Tanaka, A., Nishida, R., Yoshida, T., Koshikawa, M., Goto, M. & Kuwahara, T. (2001) Outbreak of Chinese herb nephropathy in Japan: Are there any differences from Belgium? *Intern. Med.*, **40**, 296–300

Teng, L.C. & DeBardeleben, J.F. (1971) Novel tetracyclic sesquiterpene from the oil of orejuela of *Cymbopetalum penduliforum*. *Experientia*, **27**, 14–15 [consulted as abstract: CA1971: 110470]

Therapeutic Goods Administration (2001a) *Aristolochia Fact Sheet — 25 May 2001* [http://www.health.gov.au/tga/docs/html/aristol.htm]

Therapeutic Goods Administration (2001b) *Practioner Alert*, Woden, Australia, Health and Aged Care

UK Committee on Safety of Medicines (2001) *The Medicines (Aristolochia and Mu Tong etc.) (Prohibition) Order 2001* (Statutory Instrument 2001 No. 1841), London, The Stationary Office [http://www.Scotland-legislation.hmso.gov.uk/si/si2001/20011841.htm]

Urzúa, A. & Priestap, H. (1985) Aristolochic acids from *Battus polydamas*. *Biochem. Syst. Ecol.*, **13**, 169-170

Urzúa, A., Salgado, G., Cassels, B.K. & Eckhardt, G. (1983) Aristolochic acids in *Aristolochia chilensis* and the Aristolochia-feeder *Battus archidamas* (Lepidoptera). *Collect. Czech. Chem. Commun.*, **48**, 1513–1519

Urzúa, A., Rodríguez, R. & Cassels, B. (1987) Fate of ingested aristolochic acids in *Battus archidamas*. *Biochem. Syst. Ecol.*, **15**, 687–689

Vanhaelen, M., Vanhaelen-Fastre, R., But, P. & Vanherweghem, J.-L. (1994) Identification of aristolochic acid in Chinese herbs (Letter to the Editor). *Lancet*, **343**, 174

Vanherweghem, J.-L. (1998) Misuse of herbal remedies: The case of an outbreak of terminal renal failure in Belgium (Chinese herbs nephropathy). *J. altern. complement. Med.*, **4**, 9–13

Vanherweghem, J.-L., Depierreux, M., Tielemans, C., Abramowicz, D., Dratwa, M., Jadoul, M., Richard, C., Vandervelde, D., Verbeelen, D., Vanhaelen-Fastre, R. & Vanhaelen, M. (1993) Rapidly progressive interstitial renal fibrosis in young women: Association with slimming regimen including Chinese herbs. *Lancet*, **341**, 387–391

Vanherweghem, J.-L., Tielemans, C., Simon, J. & Depierreux, M. (1995) Chinese herbs nephropathy and renal pelvic carcinoma. *Nephrol. Dial. Transplant.*, **10**, 270–273

Wang, W.-H. & Zheng, J.-H. (1984) The pregnancy terminating effect and toxicity of an active constituent of *Aristolochia mollissima* Hance, aristolochic acid A. *Yao-Xue-Xue-Bao*, **19**, 405–409

Wang, Y., Pan, J.X., Gao, J.J., Du, K. & Jia, Z.J. (2000) [The antitumor constituents from stems of *Aristolochia manshuriensis*.] *J. Beijing med. Univ.*, **32**, 18–21 (in Chinese)

Westendorf, J., Poginsky, B., Marquardt, H., Groth, G. & Marquardt, H. (1988) The genotoxicity of lucidin, a natural component of *Rubia tinctorum* L., and lucidinethylether, a component of ethanolic *Rubia* extracts. *Cell Biol. Toxicol.*, **4**, 225–239

WHO (1997) *Medicinal Plants in China. A Selection of 150 Commonly Used Species.* WHO Regional Publications, Western Pacific Series No. 2, Manila

WHO (1998) *Medicinal Plants in the Republic of Korea.* WHO Regional Publications, Western Pacific Series No. 21, Manila

Wu, T.S., Leu, Y.-L. & Chan, Y.-Y. (2000) Aristolochic acids as a defensive substance for the aristolochiaceous plant-feeding swallowtail butterfly, *Pachliopta aristolochiae interpositus*. *J. Chin. chem. Soc. (Taipei)*, **47**, 221–226

Xu, Z.F. (1957) [Studies on component of Chinese herb Tu Qing Mu Xiang.] *Acta pharm. sin.*, **5**, 235–247 (in Chinese)

Yang, C.-S., Lin, C.-H., Chang, S.-H. & Hsi, H.-C. (2000) Rapidly progressive fibrosing interstitial nephritis associated with Chinese herbal drugs. *Am. J. Kidney Dis.*, **35**, 313–318

Zheng, H.Z., Dong, Z.H. & Se, Q. (1997) *Current Studies and Application of Medica Materia*, 10th Ed., Zueyuan Press, pp. 2606–2613 (in Chinese)

Zheng, H.Z., Dong, Z.H. & Se, Q. (1998) [*Qing muxiang, Radix Aristolochiae (A. debilis)*. In: *Modern Study of Traditional Chinese Medicine*], Vol. III, Beijing, Xue Yuan Press, pp. 2606–2613 (in Chinese)

Zheng, H.Z., Dong, Z.H. & Se, Q. (1999) [*Guang fangji, Radix Aristolochiae Fangchi (A. Fangchi)*. In: [*Modern Study of Tradition of Chinese Medicine*], Vol. VI, Beijing, Xueyuan Press, pp. 5541–5545 (in Chinese)

Zhou, E.F. & Lue, F.H. (1958) [Cardiotonic action of *Caulis* aristolochiae *manshuriensis*.] *Acta pharm. sin.*, **6**, 341–346 (in Chinese)

C. RUBIA TINCTORUM, MORINDA OFFICINALIS AND ANTHRAQUINONES

1. Exposure Data

1.1 Origin, type and botanical data

1.1.1 Rubia tinctorum *L.* (*Rubiaceae*)

The medicinal part of *Rubia tinctorum* is the dried root (Blumenthal *et al.*, 1998) (common names: madder, dyer's madder (Felter & Lloyd, 2002)). The small yellowish-green flowers are in loose, leafy, long-peduncled terminal or axillary cymes. The margin of the calyx is indistinct, 4- to 5-sectioned and has a tip that is curved inward. There are five stamens and an inferior ovary. The fruit is a black, pea-sized glabrous, smooth drupe containing two seeds. The perennial plant grows to a height of 60 to 100 cm. The pencil-thick rhizome creeps widely underground. The stem is quadrangular with backward-turning prickles at the edges. The stems are at times so thin that they are more descendent than erect. The leaves are in whorls, in fours below, in sixes above. They are oblong to lanceolate with one rib and protrudingly reticulate beneath. The plant is indigenous to southern Europe, western Asia and North Africa, and is cultivated elsewhere (Medical Economics Co., 2000).

1.1.2 Morinda officinalis *How* (*Rubiaceae*) (Figure 1)

The medicinal root of *M. officinalis* is cylinder-shaped (with oblate circumference of the section) and slightly curved. The diameter of the root is usually 0.5–2 cm. The surface, with vertical wrinkles and transverse crackles, is yellowish-gray or dark gray. In some roots, the bark is thicker, violet or light violet in colour, and easy to separate from the xylem. The xylem, with a diameter of 1–5 mm and with slightly dentate margin, is hard and yellowish-brown or whitish-yellow in colour. Morinda root is odourless, sweet and slightly sour in taste. The fleshy root is used for medicinal purposes.

Figure 1. ***Morinda officinalis*** **How**

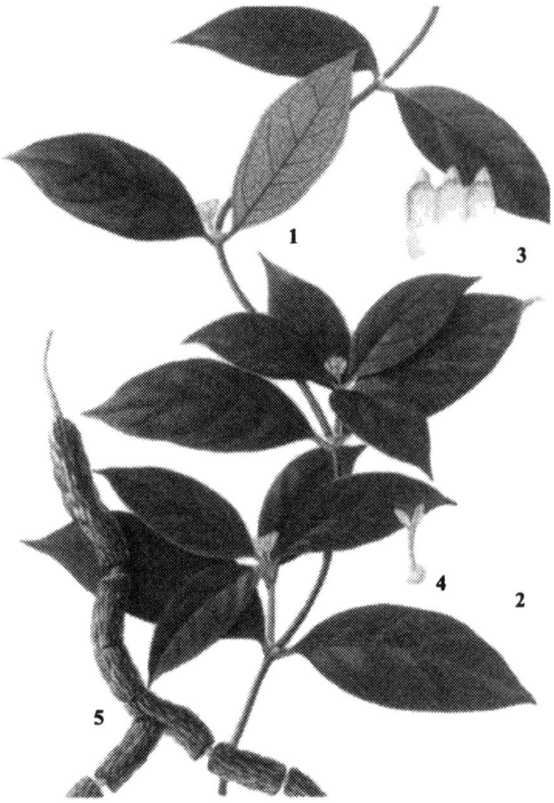

From Qian (1996)
1. flower twig; 2, flower; 3, stamen in dissected flower; 4. pistil; 5, root

1.2 Use

The roots of *R. tinctorum* contain a red colouring matter which is used for dyeing. Additionally, extracts from *R. tinctorum* are used for the treatment of kidney and bladder stones (Westendorf *et al.*, 1990; Blömeke *et al.*, 1992). Plants containing 1-hydroxy-anthraquinone have been widely used for pharmaceutical purposes such as treatment of kidney and bladder stones, as a laxative mixture, and as a mild sedative (Brown & Brown, 1976; Mori *et al.*, 1990; Wang *et al.*, 1992). Madder root has reportedly also been used medicinally for menstrual and urinary disorders (Medical Economics Co., 1998, 2000). The roots of *M. officinalis* have been used as a Chinese natural medicine for tonic and analgesic purposes.

Anthraquinone glycosides are the active principles of plant-derived laxatives such as senna, cascara, frangula, rhubarb and aloe. These five plant-derived laxative substances are not included in the present review, because there are no published reports on their potential carcinogenicity. 1,8-Dihydroxyanthraquinone, the aglycone moiety of the laxative ingredient of senna, was formerly marketed as a laxative under the trade name Dantron®, but human drug products containing Dantron® (see IARC, 1990) were withdrawn from commerce in the United States in 1987 after it was shown to cause intestinal tumours in experimental animals.

1.3 Chemical constituents

Anthracene derivatives are widely distributed in the plant kingdom. Especially in the dicotyledons, many families, such as the Hypericaceae (*Hypericum*), Polygonaceae (*Rheum*, *Rumex*, *Polygonum*), Rhamnaceae (*Rhamnus*) and Rubiaceae (*Rubia*, *Morinda*, *Galium*), are rich in anthracene derivatives. In the monocotyledons, only the family Liliaceae (*Aloe*) contains this class of chemicals. About 90% of these compounds occur as derivatives of 9,10-anthracenedione (anthraquinones) with several hydroxy and other functional groups, such as methyl, hydroxymethyl and carboxy groups. Hydroxyanthraquinones are the active principles of many phytotherapeutic drugs.

1.3.1 Rubia tinctorum

Compounds found in *Rubia tinctorum* include purpurin (oxyalizarin; 1,2,4-trihydroxyanthraquinone), mollugin (6-hydroxy-2,2-dimethyl-2*H*-naphtho[1,2-*b*]pyran-5-carboxylic acid, methyl ester), 1-hydroxy-2-methylanthraquinone, 2-ethoxymethylanthraquinone, rubiadin (1,3-dihydroxy-2-methylanthraquinone), 1,3-dihydroxyanthraquinone, 7-hydroxy-2-methylanthraquinone, lucidin (1,3-dihydroxy-2-hydroxymethylanthraquinone), 1-methoxymethylanthraquinone, 2,6-dihydroxyanthraquinone, lucidin-3-*O*-primeveroside [6-(β-D-xylosido)-D-glucoside], alizarin (1,2-dihydroxyanthraquinone), lucidin-*O*-ethyl ether, 1-hydroxy-2-hydroxymethylanthraquinone 3-glucoside, 2-hydroxymethylanthraquinone 3-glucoside, 3,8-dihydroxy-2-hydroxymethylanthraquinone 3-glucoside, ruberythric acid (alizarin primeveroside; alizarin glycoside), quinizarin and iridoid asperuloside (Schneider *et al.*, 1979; Kawasaki *et al.*, 1992; Derksen *et al.*, 1998; El-Emary & Backheet, 1998; Marczylo *et al.*, 2000).

A number of compounds have been characterized from the roots of *R. tinctorum* (the source of commercial madder colour) by various analytical methods. Among these compounds are alizarin, ruberythric acid, purpurin, lucidin, rubiadin, mollugin, 1-hydroxy-2-methylanthraquinone, tectoquinone (2-methylanthraquinone), nordamnacanthal (1,3-dihydroxy-2-anthraquinonecarboxaldehyde), 1-hydroxy-2-methoxyanthraquinone, 1,3-dihydroxy-2-ethoxymethylanthraquinone, scopoletin (7-hydroxy-6-methoxycoumarin) and the glucosides and/or the primeverosides of these compounds (Kawasaki *et al.*, 1992; Westendorf *et al.*, 1998; Medical Economics Co., 2000). The majority of the anthraquinones present in the plant itself or in plant extracts are glycosides (Blömeke *et al.*, 1992; Westendorf *et al.*, 1998).

1.3.2 Morinda officinalis

A number of compounds have been isolated from *M. officinalis*. Anthraquinones found in the plant include 1,6-dihydroxy-2,4-dimethoxyanthraquinone, 1,6-dihydroxy-2-methoxyanthraquinone, methylisoalizarin, methylisoalizarin-1-methyl ether, 1-hydroxy-2-methoxyanthraquinone, 1-hydroxy-2-methylanthraquinone, physcion, 1-hydroxy-anthraquinone, 2-methylanthraquinone, 2-hydroxy-3-hydroxymethylanthraquinone, rubiadin and rubiadin-1-methyl ether. Terpenoids found include asperuloside tetra-acetate, monotropein, morindolide and morofficinaloside. Glucosides found include nystose, 1F-fructofuranosylnystose, inulin-type hexasaccharide and heptasaccharide. β-Sitosterol, 24-ethylcholesterol, a ketone (officinalisin) and several amino acids have also been found (Li *et al.*, 1991; Yang *et al.*, 1992; Yoshikawa *et al.*, 1995; Zheng & Dong, 1997; Yao *et al.*, 1998).

1.4 Sales and consumption

No data were available to the Working Group.

1.5 Component(s) with potential cancer hazard (1-hydroxyanthraquinone; 1,3-dihydroxy-2-hydroxymethylanthraquinone (lucidin))

1.5.1 *Nomenclature*

1-Hydroxyanthraquinone

Chem. Abstr. Serv. Reg. No.: 129-43-1
Chem. Abstr. Serv. Name: 1-Hydroxy-9,10-anthracenedione
Synonyms: Erythroxyanthraquinone; α-hydroxyanthraquinone

1,3-Dihydroxy-2-hydroxymethylanthraquinone (lucidin)

Chem. Abstr. Serv. Reg. No.: 478-08-0
Chem. Abstr. Serv. Name: 1,3-Dihydroxy-2-(hydroxymethyl)-9,10-anthracenedione
Synonyms: Henine

1.5.2 *Structural and molecular formulae and relative molecular mass*

1-Hydroxyanthraquinone

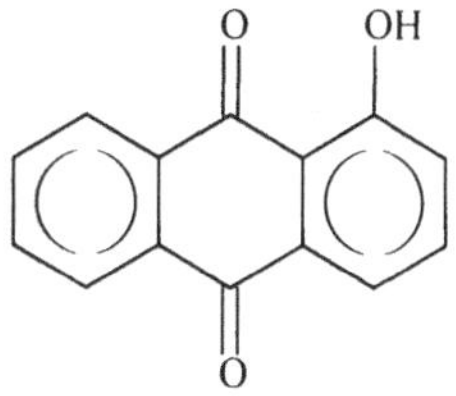

$C_{14}H_8O_3$ Relative molecular mass: 224.21

Lucidin

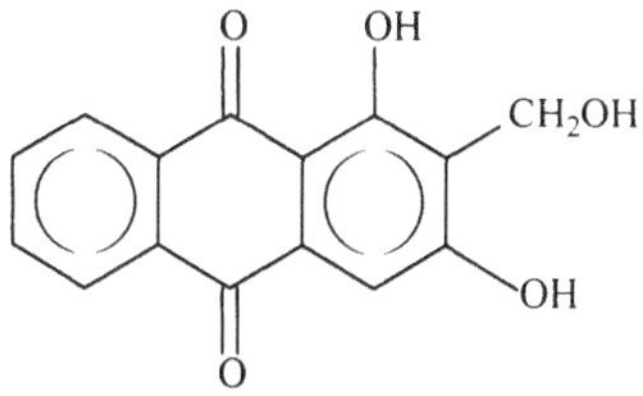

$C_{15}H_{10}O_5$ Relative molecular mass: 270.24

1.5.3 *Chemical and physical properties of the pure substance*

1-Hydroxyanthraquinone

(*a*) *Description*: Fine yellow crystals (Buckingham, 2001)
(*b*) *Melting-point*: 195–196 °C (Buckingham, 2001)

Lucidin

(*a*) *Description*: Yellow crystals (Buckingham, 2001)
(*b*) *Melting-point*: 330 °C (Buckingham, 2001)
(*c*) *Dissociation constant*: pK_a, 8.11 (20 °C, water) (Buckingham, 2001)

1.5.4 *Analysis*

A method using reversed-phase high-performance liquid chromatography (HPLC) has been described that allows the separation of 13 naturally occurring naphthoquinones and anthraquinones, including 1-hydroxyanthraquinone. The separation was achieved under isocratic and gradient conditions (Steinert *et al.*, 1996).

The following method has been used to determine hydroxyanthraquinones in experimental laboratory animal diet containing madder. Hydroxyanthraquinones were isolated by Soxhlet extraction with ethyl acetate, evaporation of the solvent, and dissolution in methanol. Separation of the hydroxyanthraquinones was accomplished by HPLC with spectrometric detection at 410 nm (Westendorf *et al.*, 1998).

Lucidin content in *R. tinctorum* crude drug powder has been determined using reversed-phase HPLC with methanol/acetic acid as the mobile phase and detection at 225 nm (Wang *et al.*, 1997).

An HPLC method with isocratic elution has been developed for the separation of anthraquinones with particular attention to the detection of lucidin in commercially available sources of *R. tinctorum* aglycones (Bosáková *et al.*, 2000).

A reversed-phase HPLC method has been developed for the simultaneous characterization of anthraquinone glycosides and aglycones in extracts of *R. tinctorum*. The anthraquinones, including lucidin, are separated on a reversed-phase column with a water–acetonitrile gradient as eluent and measured with ultraviolet detection at 250 nm (Krizsan *et al.*, 1996; Derksen *et al.*, 1998).

1.5.5 *Production*

1-Hydroxyanthraquinone has been synthesized by diazotization of 1-aminoanthraquinone and heating the diazonium salt with concentrated sulfuric acid. After dilution with water, the precipitated crude product was diluted with acetone and purified by preparative thin-layer chromatography (TLC) in toluene:ethyl formate:formic acid (75:24:1) (Blömeke *et al.*, 1992).

A general method for synthesis of anthraquinones, including 1-hydroxy-anthraquinone, has been developed. The anthraquinones were obtained under mild conditions from *ortho*-dicarboxylic acid chlorides and suitable aromatic substrates via a Friedel–Crafts process (Sartori *et al.*, 1990).

1-Hydroxyanthraquinone has also been prepared in 96.4% yield by reacting 1-nitroanthraquinone with sodium formate in dimethylformamide at 130 °C for 17 hours (Michalowicz, 1981).

Lucidin has been synthesized from nordamnacanthal (Prista *et al.*, 1965).

Available information indicates that 1-hydroxyanthraquinone is manufactured by one company each in China and Japan (Chemical Information Services, 2001).

1.5.6 *Use*

Anthraquinones are the largest group of naturally occurring quinones. Both natural and synthetic anthraquinones have been widely used as colourants in food, drugs, cosmetics, hair dyes and textiles (Brown & Brown, 1976; Mori *et al.*, 1990). 1-Hydroxyanthraquinone can be used as an intermediate in the production of dyes and drugs (Imaki & Fukumoto, 1988). The Working Group was not aware of any commercial use of purified lucidin.

1.5.7 *Occurrence*

1-Hydroxyanthraquinone has been isolated from the roots of *Rubia cordifolia*, *Morinda officinalis* and *Damnacanthus indicus*, from the heartwood of *Tabebuia avellanedae* and the herb *Cassia occidentalis* (Brown & Brown, 1976; Mori *et al*., 1990; Wang *et al*., 1992; Yang *et al*., 1992; Steinert *et al*., 1996; Buckingham, 2001). 1-Hydroxyanthraquinone has also been identified as a metabolite of alizarin primeveroside, found in *Rubia tinctorum*, when alizarin primeveroside was given orally to rats (Blömeke *et al*., 1992).

Lucidin has been identified in plants from *Rubia* (*R. tinctorum, R. iberica*), *Coprosma (C. lucida, C. rotundifolia, C. acerosa)*, *Morinda (M. citrifolia, M. umbellata)*, *Galium* (*G. fagetorum, G. mollugo, G. dasypodum*), *Hymenodictyon* (*H. excelsum*) and *Commitheca* (*C. liebrechtsiana*) species (Prista *et al*., 1965; Burnett & Thomson, 1968a,b; Zhural'ov & Borisov, 1970; Thomson & Brew, 1971; Murti *et al*., 1972; Leistner, 1975; Briggs *et al*., 1976; Hocquemiller *et al*., 1976; Bauch & Leistner, 1978; Inoue *et al*., 1981; Zhural'ov *et al*., 1987; Buckingham, 2001).

1.5.8 *Human exposure*

Thousands of patients in European countries have been treated chronically in the past, against kidney stones, with madder root preparations (*R. tinctorum*) at high doses. The daily amount of lucidin ingested by these patients was calculated to be 3–10 mg. Under certain circumstances, the daily lucidin intake may even reach several hundred milligrams (Westendorf *et al*., 1998).

2. Studies of Cancer in Humans

Laxatives based on naturally occurring anthraquinone derivatives

Herbs containing anthraquinone derivatives (rhubarb, senna, frangula, cascara, aloe) are used as laxatives. A meta-analysis of 11 studies dealing with colorectal cancers and laxative use showed that use of laxatives carried an increased risk (odds ratio, 1.5; 95% CI, 1.3–1.6) for colorectal cancer (Sonnenberg & Müller, 1993). A more recent and very large study among nurses in the USA found no association between colorectal cancer and laxative use (Dukas *et al*., 2000). The relevance of this evidence to the carcinogenicity of anthraquinone-containing herbs used as laxatives is unknown because it is uncertain whether the use of laxatives in general is an adequate proxy measure for the use of anthraquinones.

The following studies mentioned anthraquinones specifically in evaluating laxative use as a risk factor for cancer. Also reported below are studies that used melanosis coli as a marker of exposure, because this in turn may reflect consumption of anthraquinones (see Section 4.2).

2.1 Case–control studies

2.1.1 *Gastrointestinal cancer*

Boyd and Doll (1954) analysed data collected in a previous inquiry, which detailed histories of purgative use for 2249 patients. People whose history of taking purgatives extended over a continuous period of at least five years were considered as 'purgative users'. After excluding patients suffering from gastrointestinal diseases other than cancer, there remained 614 patients with gastrointestinal cancers (387 cancers of the large bowel and 227 cancers of the stomach) and 1313 control patients (647 cancers of the lung and 666 patients with non-cancer non-gastrointestinal diseases). A history of regular (at least once per week) use of purgatives was reported for 222/614 [36.2%] of the patients with gastrointestinal cancer and 343/1313 [26.1%] of the control patients. Use of cascara was reported by [6%] of cases and [5.4%] of controls. Use of senna was reported by [6.2%] of cases and [2.6%] of controls and when exposure was restricted to chronic users (regularly, at least once per week), the percentages were [5.2%] of cases and [1.7%] of controls ($p = 0.04$). The difference between cases and controls was similar for separate consideration of cancer of the large bowel.

Siegers *et al.* (1993) studied retrospectively 3049 patients and prospectively 1095 patients who underwent endoscopic examinations in Lübeck, Germany. In the retrospective study, melanosis coli was found in 3.1% of patients without other abnormalities, in [2.8%] of patients with colitis or diverticulosis, in 8.6% of patients with adenomas and in 3.9% of patients with carcinomas. In the prospective study, melanosis coli was found in 6.9% in patients without other abnormalities, in [4.5%] in patients with colitis or diverticulosis, in 9.8% of patients with adenomas and in 18.6% of patients with carcinomas. The relative risk for colorectal cancer among subjects with melanosis coli was 3.0 (95% CI, 1.2–4.9). Among 33 of the patients who had both melanosis coli and adenoma or carcinoma, all but two acknowledged abuse of anthranoid laxatives for 10–30 years. [The Working Group noted that detection of melanosis coli requires colonoscopy and that indication of colonoscopy may result in selection bias of the controls.]

Kune (1993) analysed laxative use in 685 colorectal cancer patients in Australia, diagnosed over a 12-month period (1980–81), compared with 723 age- and sex-matched community-based controls. Use of laxatives containing anthraquinones was reported by [13.9%] of the cases of colorectal cancers and [14.1%] of controls. Comparing cases and controls using laxatives containing anthraquinones with patients and controls using no laxatives, the relative risk for colorectal cancer related to the use of laxatives containing anthraquinones was 1.01 [confidence intervals were not reported].

In a retrospective analysis of 2229 consecutive patients having undergone a colonoscopy in Erlangen, Germany, between 1985 and 1992, the presence of colorectal cancer was not associated with melanosis coli or laxative use: colorectal cancer was present in 2.7% ($n = 60$) of all patients, 2.9 % ($n = 3$) of the 102 patients with melanosis coli (relative risk, 1.1; 95% CI, 0.35–3.4) and in 3.1% ($n = 9$) of the 286 patients with anthranoid laxative use (relative risk, 1.1; 95% CI, 0.48–2.3) (Nusko *et al.*, 1997). [The Working

Group noted some discrepancies with data from the same study reported by Nusko *et al.* (1993).]

Nusko *et al.* (2000) performed a prospective case–control study in Erlangen of the association between anthranoid laxative use and risk for development of colorectal adenomas or carcinomas. The study included a total of 202 patients with newly diagnosed colorectal carcinomas, 114 patients with adenomatous polyps and 238 patients (controls) with no colorectal neoplasms who had been referred for total colonoscopy between 1993 and 1996. For each subject, the use of anthranoid preparations was assessed by standardized non-blind interview after colonoscopy, and melanosis coli was studied by histopathological examination. Use of anthranoid laxatives was reported by 33/238 controls, 16/114 adenoma patients and 29/202 carcinoma patients. Anthranoid use did not confer any significantly elevated risk for development of colorectal adenomas (unadjusted odds ratio, 1.0; 95% CI, 0.5–1.9) or carcinomas (unadjusted odds radio, 1.0; 95% CI, 0.6–1.8). After adjustment for the risk factors age, sex and blood in the stools by logistic regression analysis, the odds ratio for adenomas was 0.84 (95% CI, 0.4–1.7) and that for carcinomas was 0.93 (95% CI, 0.5–1.7). Duration of anthranoid laxative use, when included in the logistic regression as a continuous variable, was not significantly associated with colorectal carcinoma ($p = 0.41$). Macroscopic and high-grade microscopic melanosis coli were not significantly associated with the development of adenomas or carcinomas.

2.1.2 *Urothelial cancer*

Bronder *et al.* (1999) reported on 766 cases of urothelial cancers (98% confirmed by histology) in Berlin, Germany, between 1990 and 1994. A control group (1:1) was obtained by sampling, from the West Berlin Population Registry, persons of German nationality who had lived in Germany for at least 20 years and matched with the patients for sex and age. Through a standardized questionnaire completed by 648 patients and 647 controls, social class was recorded as well as consumption of analgesics, laxatives and tobacco. After adjustment for tobacco use and social class, the risk of urothelial carcinoma was increased in laxative users. Use of contact laxatives was reported by 63 urothelial cancer patients versus 29 controls (odds ratio, 2.5; 95% CI, 1.5–4.2) and 13 renal pelvis and ureter cancer patients versus two controls (odds ratio, 9.3; 95% CI, 1.1–83.3). For different laxatives, the corresponding figures (urothelial cancer patients versus controls) were: chemical and anthranoid laxatives, five versus two (odds ratio, 2.7; 95% CI, 0.47–16); anthranoid laxatives alone, 37 versus 20 (odds ratio, 2.0; 95% CI, 1.1–3.7); aloe, 16 versus 11 (odds ratio, 1.6; 95% CI, 0.66–3.7); senna, 26 versus 13 (odds ratio, 2.4; 95% CI, 1.1–5.0); and rhubarb, eight versus four (odds ratio, 2.6; 95% CI, 0.68–9.6). [The Working Group noted that no results for laxatives adjusted for use of analgesics were presented.]

3. Studies of Cancer in Experimental Animals

3.1 1-Hydroxyanthraquinone

3.1.1 *Oral administration*

Rat: A group of 30 male ACI/N rats, 1.5 months of age, was fed 1% 1-hydroxyanthraquinone in CE-2 diet throughout the experimental period of 480 days. Thirty control rats were fed basal diet. Rats that survived more than 280 days developed various tumours of the intestine (25/29). These comprised caecal adenomas (10/29) or adenocarcinomas (5/29) and colonic adenomas (12/29) or adenocarcinomas (11/29). No such tumours were diagnosed in control rats. In addition, neoplastic nodules and hepatocellular adenomas (12/29) and forestomach papillomas and glandular stomach adenomas (5/29) were observed. No such tumours were observed in control animals (Mori *et al.*, 1990).

A group of 27 male ACI/N rats, six weeks of age, was fed 1.5% 1-hydroxyanthraquinone in the diet for 48 weeks. A second group (14 rats) was also given 16 mg/L indomethacin in the drinking-water for the experimental period. Fifteen control rats were fed basal diet. Rats fed with 1-hydroxyanthraquinone had incidences of 12/27 large intestinal neoplasms (adenomas and adenocarcinomas) and 14/27 forestomach tumours (papillomas). In rats fed 1-hydroxyanthraquinone and treated with indomethacin, these incidences were 0/14 and 2/14, respectively. Untreated animals and rats given indomethacin alone had no neoplasms in the large intestine and forestomach (Tanaka *et al.*, 1991).

3.1.2 *Administration with known carcinogens*

Rat: Two groups of male and female ACI/N rats, six weeks of age, were given weekly intraperitoneal injections of 25 mg/kg bw methylazoxymethanol acetate for two weeks. One group (21 males and 20 females) was subsequently fed 1% 1-hydroxyanthraquinone until the end of the experiment (44 weeks). The second group treated with methylazoxymethanol acetate (19 males and 19 females) was fed basal diet. A third group (17 males and 20 females) received 1-hydroxyanthraquinone alone and a control group (16 males and 22 females) was fed basal diet only. The incidences of caecal tumours (males, 15/17; females, 11/15) and colon tumours (males, 17/17; females, 14/15) in the group treated with methylazoxymethanol acetate and 1-hydroxyanthraquinone were significantly higher ($p < 0.05$) than those of the groups treated with methylazoxymethanol acetate alone (caecal tumours: males, 0/19; females, 3/15; colon tumours: males, 8/15; females, 11/15) or 1-hydroxyanthraquinone alone (caecal tumours: males, 1/17; females, 0/17; colon tumours: males, 0/17; females, 1/17). The incidence and multiplicity of caecum and colon tumours in the first group were significantly greater than the combined results for those of second and third groups, suggesting that 1-hydroxyanthraquinone acted synergistically with methylazoxymethanol acetate (Mori *et al.*, 1991).

3.2 1,3-Dihydroxy-2-hydroxymethylanthraquinone (lucidin)

No data on the carcinogenicity of lucidin in rodents were available to the Working Group, but this compound is an active principle of *Rubia tinctorum* used as a herbal medicine.

Rat: Groups of 15–20 male and female ACI rats, weighing 150–200 g, were fed a standard diet containing 0, 1 or 10% madder root (*Rubia tinctorum*) for 780 days. This root contained lucidin (0.34%) but also large amounts of alizarin (0.67%; 1,2-dihydroxyanthraquinone) and the primeverosides of both compounds. All surviving animals were killed and necropsied at 780 days. In the groups receiving 10% madder root diet, 2/16 males and 3/17 females developed hepatocellular adenomas, whereas none were observed in the controls or the 1% madder root group of either sex. In addition, renal tubule-cell adenomas were observed in 1/16 males and 2/16 females in the 10% madder root group and a renal tubule-cell carcinoma in 1/14 males in the 1% madder root group. No renal tumours were seen in the control groups (Westendorf *et al.*, 1998). [The Working Group noted the small number of animals used in this study and that the madder root contained large amounts of other compounds such as alizarin.]

4. Other Data Relevant to an Evaluation of Carcinogenicity and its Mechanisms

4.1 Absorption, distribution, metabolism and excretion

1-Hydroxyanthraquinone

(*a*) *Humans*

No data were available to the Working Group.

(*b*) *Experimental systems*

In an early study, 1-hydroxyanthraquinone was found to be absorbed when continuously administered to rats [strain not specified] by stomach tube (50 mg suspended in aqueous gum arabic), and its metabolites were identified by paper chromatography. Urine and faeces were collected during 48 h after treatment and extracted with diethyl ether. Of the 1-hydroxyanthraquinone originally present, 2.49% and 0.74% were converted into alizarin (1,2-dihydroxyanthraquinone), in urine and faeces, respectively. The alizarin was then excreted after sulfation and glucuronidation (Fujita *et al.*, 1961).

1,3-Dihydroxy-2-hydroxymethylanthraquinone (lucidin)

(*a*) *Humans*

No data were available to the Working Group.

(*b*) *Experimental systems*

No direct information was found on the absorption, distribution, metabolism or excretion of ludicin in experimental animal systems. However, a report has appeared on the oral administration of lucidin 3-*O*-primeveroside to rats, which resulted in the excretion in the urine of lucidin and (to a lesser extent) rubiadin (1,3-dihydroxy-2-methylanthraquinone) (Blömeke *et al.*, 1992).

4.2 Toxic effects

Chronic ingestion of laxatives of the anthracene group such as cascara, senna, frangula, aloe or rhubarb is considered to be an etiological factor in melanosis coli, a condition in which colonic macrophages accumulate a dark pigment with the staining characteristics of lipofuscin. In 14 patients submitted to repeated proctoscopies, Speare (1951) produced melanosis 11 times by prescribing cascara sagrada for 4–14 months and eliminated it nine times by withdrawing the laxative for 5–11 months. Steer and Colin-Jones (1975) studied histological patterns of rectal biopsies from seven patients with melanosis coli three months after they ceased to take anthraquinone-containing purgatives. A decrease was observed in the number of macrophages infiltrating the mucosa, as well as a reduction in the intensity of the acid phosphatase reaction.

The effect of anthraquinone glycosides on cell proliferation in the sigmoid colon of patients *in vivo* has been studied. Twenty-four hours before colonoscopy with a sigmoid biopsy, nine patients were given an oral dose of 1 mL/kg of a syrup containing 2.0 mg/mL of the anthraquinone glycosides sennosides A and B. Proliferative activity of epithelial cells was determined by incubating the biopsy specimens with the thymidine analogue 5-bromo-2′-deoxyuridine (BrdU) and then visualizing BrdU-labelled cells immunochemically. Proliferative activity was expressed as the labelling index, i.e. the number of labelled nuclei divided by the total number of nuclei multiplied by 100 (%). Data from these nine subjects were compared with those of 10 subjects who had a normal colonoscopy and 14 patients with sporadic colonic neoplasms who had not received anthraquinone laxatives before the colonoscopy. The labelling index (% ± SD) was significantly higher in the group given the sennosides (26.4 ± 6.1) than in normal patients (5.7 ± 1.8) or in patients with a colonic neoplasm (8.7 ± 2.6; $p < 0.005$). Furthermore, the reduction of the number of cells per crypt, observed as a reduced crypt height in the treated group, suggests that compensatory cell proliferation may occur as well (Kleibeuker *et al.*, 1995).

1-Hydroxyanthraquinone

(*a*) *Humans*

No data were available to the Working Group.

(*b*) *Experimental systems*

Little information is available on the toxic effects of 1-hydroxyanthraquinone in experimental animals, and no acute toxicity studies have been reported. A group of ACI/N rats fed 1.5% 1-hydroxyanthraquinone in the diet for 48 weeks showed inflammatory changes of various degrees (colitis, ulcerative colitis or melanosis coli) in the large bowel, but no clinical sign of toxicity (Tanaka *et al.*, 1995).

1-Hydroxyanthraquinone was fed in the diet (0.5%, 1%, 2%, 4%) to male Fischer 344 rats for seven days. Induction of cell proliferation in the intestines was analysed by BrdU labelling. At the high dose, 1-hydroxyanthraquinone induced cell proliferation in the caecum and the proximal colorectum, but there was little evidence of intestinal cytotoxicity (Toyoda *et al.*, 1994).

1-Hydroxyanthraquinone is a constituent of the heartwood of *Tabebuia avellanedae* (Burnett & Thompson, 1967; Steinert *et al.*, 1996) and the roots of *Morinda officinalis* and *Damnacanthus indicus* (Yang *et al.*, 1992). No reports documenting toxicity of any of these three plants appear to have been published.

1,3-Dihydroxy-2-hydroxymethylanthraquinone (lucidin)

(*a*) *Humans*

No data were available to the Working Group.

(*b*) *Experimental systems*

No information was available on general toxic effects of lucidin. However, this compound is a known constituent of the roots of *Rubia tinctorum* (madder) (Burnett & Thomson, 1968a), and one study has been published on the acute and subacute toxicity of madder root. A 14-day toxicity test was conducted on an aqueous extract of madder root administered by gavage to (C57BL/6 × C3H)F_1 mice. The maximum tolerated dose was between 3500 and 5000 mg/kg bw. A subacute toxicity test was then performed on 62 mice of each sex with madder root extract incorporated into their diet at 0, 0.3, 0.6, 1.25, 2.5 or 5% for 90 days. All mice tolerated these doses well, and none showed clinical signs of toxicity or adverse effects on body weight gain. Histopathological examination showed retention cysts of the kidneys and epidermal vaginal cysts in a few of the treated and control mice. It was concluded that dietary exposure to madder root at the doses tested had no significant acute or subacute toxic effects on mice (Ino *et al.*, 1995). [The Working Group noted that this study is more germane to the toxicity of any glycoside forms of lucidin present than to that of the aglycone itself.]

4.3 Reproductive and developmental effects

1-Hydroxyanthraquinone and 1,3-dihydroxy-2-hydroxymethylanthraquinone (lucidin)

(*a*) *Humans*

No data were available to the Working Group.

(*b*) *Experimental systems*

No data were available to the Working Group.

4.4 Genetic and related effects

(*a*) *Humans*

No data were available to the Working Group.

(*b*) *Experimental systems* (for references see Table 1)

1-Hydroxyanthraquinone

1-Hydroxyanthraquinone induced frameshift mutations in *Salmonella typhimurium* TA1537 in the absence of metabolic activation; addition of liver homogenate led to a reduction of the mutagenic activity. In strains TA98 and TA100, no mutagenic effects were seen. The induction of unscheduled DNA synthesis in rat hepatocytes *in vitro* was studied independently by two groups and a positive response was found in both studies.

A few studies were available on genotoxic effects of alizarin, a metabolite of 1-hydroxyanthraquinone. Alizarin was weakly mutagenic in *S. typhimurium* TA1537 in the presence of S9 and in rat hepatocyte DNA-repair assays, but it was consistently inactive in transformation experiments with C3H/M2 mouse fibroblasts and in *Hprt* mutation assays with Chinese hamster V79 cells (Westendorf *et al.*, 1990).

1,3-Dihydroxy-2-hydroxymethylanthraquinone (lucidin)

A positive response was obtained with lucidin in *Salmonella*/microsome reverse mutation assays in strains TA100, TA102, TA104, TA1537, TA1538 and TA98. The highest activity was seen in the frameshift strain TA1537 and in the base substitution strain TA100. The compound was mutagenic in these assays in the absence of metabolic activation and addition of liver homogenate had no major effect on the mutagenic potency. Lucidin also gave positive results in unscheduled DNA synthesis experiments with rat liver cells, it caused gene mutations in Chinese hamster V79 cells (without activation) and it transformed mouse fibroblasts. With polynucleotides *in vitro* the compound binds to DNA bases. The only data from an in-vivo study indicated that DNA adducts were present in various organs of mice fed lucidin (2 mg per day per animal for

Table 1. Genetic and related effects of hydroxyanthraquinones and their derivatives

Test system	Result[a]		Dose[b] (LED or HID)	Reference
	Without exogenous metabolic system	With exogenous metabolic system		
1-Hydroxyanthraquinone				
Salmonella typhimurium TA100, TA98 reverse mutation	–	–	100 μg/plate	Blömeke *et al.* (1992)
Salmonella typhimurium TA1537, reverse mutation	+	+[c]	30 μg/plate	Blömeke *et al.* (1992)
Unscheduled DNA synthesis, rat primary hepatocytes *in vitro*	+	NT	11.2	Kawai *et al.* (1986)
Unscheduled DNA synthesis, rat primary hepatocytes *in vitro*	+	NT	50	Blömeke *et al.* (1992)
1,3-Dihydroxy-2-hydroxymethylanthraquinone (lucidin)				
Salmonella typhimurium TA100, TA98, reverse mutation	+	+	10 μg/plate	Yasui & Takeda (1983)
Salmonella typhimurium TA100, reverse mutation	+	+	0.5 μg/plate	Westendorf *et al.* (1988)
Salmonella typhimurium TA104, TA98, reverse mutation	+	+	5 μg/plate	Westendorf *et al.* (1988)
Salmonella typhimurium TA100, reverse mutation	+	+	0.5 μg/plate	Poginsky (1989)
Salmonella typhimurium TA100, reverse mutation	+	+	5 μg/plate	Kawasaki *et al.* (1992)
Salmonella typhimurium TA102 reverse mutation	+	+	10 μg/plate	Westendorf *et al.* (1988, 1990)
Salmonella typhimurium TA1535 reverse mutation	–	?	10 μg/plate	Westendorf *et al.* (1988)
Salmonella typhimurium TA1537, reverse mutation	+	+	0.5 μg/plate	Westendorf *et al.* (1988, 1990)
Salmonella typhimurium TA1538 reverse mutation	+	+	10 μg/plate	Westendorf *et al.* (1988)
Salmonella typhimurium TA98 reverse mutation	+	+	5 μg/plate	Poginsky (1989)
Salmonella typhimurium TA98, reverse mutation	+	+	30 μg/plate	Kawasaki *et al.* (1992)
Drosophila melanogaster, somatic mutation, wing-spot test	–		1% in feed	Marec *et al.* (2001)
Unscheduled DNA synthesis, rat primary hepatocytes, *in vitro*	+	NT	6.3	Westendorf *et al.* (1988)
Unscheduled DNA synthesis, rat primary hepatocytes, *in vitro*	+	NT	12.5	Poginsky (1989)
Unscheduled DNA synthesis, rat primary hepatocytes, *in vitro*	+	NT	10	Blömeke *et al.* (1992)
Gene mutation, Chinese hamster lung V79 cells *Hprt* locus *in vitro*	+	NT	10	Westendorf *et al.* (1988)

Table 1 (contd)

Test system	Result[a]		Dose[b] (LED or HID)	Reference
	Without exogenous metabolic system	With exogenous metabolic system		
Gene mutation, Chinese hamster lung V79 cells *Hprt* locus *in vitro*	+	NT	10	Poginsky (1989)
Cell transformation, C3H/M2 mouse fibroblasts	+	NT	5	Westendorf *et al.* (1988)
DNA strand breaks, V79 Chinese hamster fibroblasts *in vitro* (alkaline elution)	+	NT	20	Westendorf *et al.* (1988)
DNA binding, primary rat hepatocytes *in vitro*, ^{32}P-postlabelling	+	NT	40	Poginsky *et al.* (1991)
Binding to adenine and guanine *in vitro*, HPLC, FAB-MS analysis	+	NT	0.2	Kawasaki *et al.* (1994)
Binding to poly [d(A-T)] and polydC*polydG *in vitro*, ^{32}P-postlabelling	+		200 000	Poginsky *et al.* (1991)
Binding to DNA from male Parkes mouse liver, kidney, duodenum and colon *in vivo*, ^{32}P-postlabelling	+		2 mg/day for 4 days; feed	Poginsky *et al.* (1991)

[a] +, positive; –, negative; NT, not tested; ?, equivocal; FAB, fast atom bombardment; MS, mass spectrometry
[b] LED, lowest effective dose; HID, highest ineffective dose; in-vitro tests, µg/mL; in-vivo tests, mg/kg bw/day
[c] Lower response in the presence of S9

four days). In hepatic tissue, 1.16 ± 0.2 adducts per 10^8 nucleotides were detected; slightly higher adduct levels were found in kidneys and duodenum, namely 2.19 ± 1.2 and 1.91 ± 0.6, respectively (Poginsky *et al.*, 1991). The only genotoxicity assay in which lucidin gave a negative response was the *Drosophila* wing-spot test.

4.5 Mechanistic considerations

1-Hydroxyanthraquinone

There is some evidence that 1-hydroxyanthraquinone is genotoxic. In rats, induction of cell proliferation *in vivo* was seen in the caecum which was paralleled by enhanced ornithine decarboxylase activity (Mori *et al.*, 1992). In a combination experiment with methylazoxymethanol acetate (given by single intraperitoneal injection) and 1-hydroxyanthraquinone (1% in diet; 40 weeks), the level of tumour necrosis factor α was increased when rats were exposed to the combination compared with the groups that received either compound alone. It was postulated that this might account for the synergistic carcinogenic effect of these compounds (see Section 3.2.2) (Yoshimi *et al.*, 1994). In studies of mutations in cancer-related genes from tumours of rats exposed to methylazoxymethanol acetate and 1-hydroxyanthraquinone, specific mutations were found in the gene encoding β-catenin but not in other genes such as *apc* (Suzui *et al.*, 1999).

1,3-Dihydroxy-2-hydroxymethylanthraquinone (lucidin)

Lucidin is genotoxic *in vitro* and *in vivo* (Westendorf *et al.*, 1988). Data on carcinogenic activity of this compound were not available, but it has been shown that madder root (*Rubia tinctorum*) — which contains lucidin — causes tumours in liver and kidneys of rats. The DNA-adduct pattern from colon, liver and kidneys of rats showed one adduct that was also found when deoxyguanosine-3′-monophosphate was incubated with lucidin *in vitro*. This supports the assumption that lucidin is involved in the carcinogenic effects of *Rubia tinctorum* (Westendorf *et al.*, 1998). It has been hypothesized that lucidin is activated by phase II enzymes through substitution of the hydroxymethyl group, followed by hydrolytic cleavage, leading to the formation of a DNA-reactive methylene metabolite or carbenium ion (Poginsky *et al.*, 1991).

5. Summary of Data Reported and Evaluation

5.1 Exposure data

Hydroxyanthraquinones are constituents of a number of plant species including *Rubia tinctorum* (madder root) and *Morinda officinalis* that are used in traditional herbal medicines. 1,3-Dihydroxy-2-hydroxymethylanthraquinone (lucidin) occurs in *R. tinctorum* in the form of glycoside conjugates, and 1-hydroxyanthraquinone similarly occurs in *M. officinalis*.

5.2 Human carcinogenicity data

Herbs containing anthraquinone derivatives

Herbs containing anthraquinone derivatives are used as laxatives and several studies have reported data relating indices of use of such laxatives to cancer outcomes. In an early British study, patients with gastrointestinal cancer reported higher past chronic use of senna than did patients with other diseases. Three German case–control studies on colorectal cancer using melanosis coli as an exposure indicator gave conflicting results. An Australian case–control study on colorectal cancer which assessed self-reported use of anthraquinone laxatives as an index of exposure found no excess risk. A study of urothelial cancer in Germany found elevated relative risks in relation to several types of laxative, including those containing anthraquinones. For all of these case–control studies, it is difficult to exclude bias and confounding from dietary habits, constipation or use of analgesics. Except for two studies (the Australian colorectal cancer and the German urothelial cancer studies), selection of controls is also a major concern.

5.3 Animal carcinogenicity data

1-Hydroxyanthraquinone was tested for carcinogenicity by oral administration in three studies in rats and induced adenocarcinomas of the large intestine in two studies.

No carcinogenicity tests have been carried out with 1,3-dihydroxy-2-hydroxymethylanthraquinone (lucidin) *per se*, although the herb madder root (*Rubia tinctorum*) (which contains this compound among others) was tested by oral administration in rats. Madder root caused an increase in hepatocellular adenomas and adenomas and carcinomas of the renal cortex in males and females in a single experiment.

5.4 Other relevant data

1-Hydroxyanthraquinone is metabolized by rats and excreted as alizarin (1,2-dihydroxyanthraquinone). When given orally to rats, it induced inflammatory changes in the colon. It is mutagenic in bacteria and causes unscheduled DNA synthesis in rat liver cells *in vitro*.

Lucidin primeveroside is hydrolysed in rats to the aglycones 1,3-dihydroxy-2-hydroxymethylanthraquinone (lucidin) and rubiadin, which are excreted in urine. Lucidin is mutagenic in bacteria, mutagenic and genotoxic in cultured mammalian cells and forms DNA adducts in mice.

5.5 Evaluation

There is *inadequate evidence* in humans for the carcinogenicity of laxatives containing anthraquinone derivatives.

There is *sufficient evidence* in experimental animals for the carcinogenicity of 1-hydroxyanthraquinone.

There is *limited evidence* in experimental animals for the carcinogenicity of madder root (*Rubia tinctorum*).

Overall evaluation

1-Hydroxyanthraquinone is *possibly carcinogenic to humans (Group 2B)*.

Madder root (*Rubia tinctorum*) is *not classifiable as to its carcinogenicity to humans (Group 3)*.

6. References

Bauch, H.J. & Leistner, E. (1978) Aromatic metabolites in cell suspension cultures of *Galium mollugo* L. *Planta med.*, **33**, 105–123

Blömeke, B., Poginsky, B., Schmutte, C., Marquardt, H. & Westendorf, J. (1992) Formation of genotoxic metabolites from anthraquinone glycosides present in *Rubia tinctorum* L. *Mutat. Res.*, **265**, 263–272

Blumenthal, M., Busse, W.R., Goldberg, A., Gruenwald, J., Hall, T., Riggins, C.W. & Rister, R.S., eds (1998) *The Complete German Commission E Monographs: Therapeutic Guide to Herbal Medicines*, Austin, TX/ Boston, MA, American Botanical Council/Integrative Medicine Communications

Bosáková, Z., Peršl, J. & Jegorov, A. (2000) Determination of lucidin in *Rubia tinctorum* aglycones by an HPLC method with isocratic elution. *J. high Resol. Chromatogr.*, **23**, 600–602

Boyd, J.T. & Doll, R. (1954) Gastro-intestinal cancer and the use of liquid paraffin. *Br. J. Cancer*, **8**, 231–237

Briggs, L.H., Beachen, J.F., Cambie, R.C., Dudman, N.P.B., Steggles, A.W. & Rutledge, P.S. (1976) Chemistry of *Coprosma* genus. Part XIV. Constituents of five New Zealand species. *J. Chem. Soc. Perkin Trans. I*, 1789–1792

Bronder, E., Klimpel, A., Helmert, U., Greiser, E., Molzahn, M. & Pommer, W. (1999) [Analgesics and laxatives as risk factors for cancer of the efferent urinary tract — Results of the Berlin Urothelial Carcinoma Study.] *Soz. Präventivmed.*, **44**, 117–125 (in German)

Brown, J.P. & Brown, R.J. (1976) Mutagenesis by 9,10-anthraquinone derivatives and related compounds in *Salmonella typhimurium*. *Mutat. Res.*, **40**, 203–224

Buckingham, J., ed. (2001) *Dictionary of Natural Products on CD-ROM*, Boca Raton, FL, Chapman & Hall/CRC Press

Burnett, A.R. & Thomson, R.H. (1967) Naturally occurring quinones. Part X. The quinonoid constituents of *Tabebuia avellanedae* (Bignoniaceae). *J. chem. Soc.(C)*, 2100–2104

Burnett, A.R. & Thomson, R.H. (1968a) Naturally occurring quinones. XV. Biogenesis of the anthraquinones in *Rubia tinctorum* (madder). *J. chem. Soc. C*, 2437–2441

Burnett, A.R. & Thomson, R.H. (1968b) Anthraquinones in *Morinda umbellata*. *Phytochemistry*, **7**, 1421–1422

Chemical Information Services (2001) *Directory of World Chemical Producers 2001.2 Edition*, Dallas, TX [CD-ROM]

Derksen, G.C.H., Van Beek, T.A., De Groot, A. & Capelle, A. (1998) High-performance liquid chromatographic method for the analysis of anthraquinone glycosides and aglycones in madder root (*Rubia tinctorum* L.). *J. Chromatogr. A.*, **816**, 277–281

Dukas, L., Willett, W.C., Colditz, G.A., Fuchs, C.S., Rosner, B. & Giovannucci, E.L. (2000) Prospective study of bowel movement, laxative use, and risk of colorectal cancer among women. *Am. J. Epidemiol.*, **151**, 958–964

El-Emary, N.A. & Backheet, E.Y. (1998) Three hydroxymethylanthraquinone glycosides from *Rubia tinctorum*. *Phytochemistry*, **49**, 277–279

Felter, H.W. & Lloyd, J.U. (2002) *King's American Dispensatory. Rubia-Madder* (http://ftp.oit.unc.edu/herbmed/eclectic/kings/rubia.html)

Fujita, M., Furuya, T. & Matsuo, M. (1961) Studies on the metabolism of naturally occurring anthraquinones. I. The metabolism of 1-hydroxyanthraquinone and 2-hydroxyanthraquinone. *Chem. pharm. Bull.*, **9**, 962–966

Hocquemiller, R., Fournet, A., Bouquet, A., Bruneton, J. & Cave, A. (1976) *Commitheca liebrechtsiana* (Rubiaceae). *Plant Med. Phytother.*, **10**, 248–250

IARC (1990) *IARC Monographs on the Evaluation of Carcinogenic Risk to Humans*, Vol. 50, *Pharmaceutical Drugs*, Lyon, IARC*Press*, pp. 265–275

Imaki, S. & Fukumoto, Y. (1988) [Process for the preparation of hydroxyanthraquinone derivatives as intermediates for dyes and drugs.] [Patent No. JP 63091347] *Jpn. Kokai Tokkyo Koho* (in Japanese)

Ino, N., Tanaka, T., Okumura, A., Morishita, Y., Makita, H., Kato, Y., Nakamura, M. & Mori, H. (1995) Active and subacute toxicity tests of madder root, natural colorant extracted from madder (*Rubia tinctorium*), in (C57BL/6 × C3H)F_1 mice. *Toxicol. ind. Health*, **11**, 449–458

Inoue, K., Nayeshiro, H., Inoue, H. & Zenk, M. (1981) Quinones and related compounds in higher plants. Part 12. Anthraquinones in cell suspension cultures of *Morinda citrifolia*. *Phytochemistry*, **20**, 1693–1700

Kawai, K., Mori, H., Sugie, S., Yoshimi, N., Inoue, T., Nakamaru, T., Nozawa, Y. & Matsushima, T. (1986) Genotoxicity in the hepatocyte/DNA repair test and toxicity to liver mitochondria of 1-hydroxyanthraquinone and several dihydroxyanthraquinones. *Cell Biol. Toxicol.*, **2**, 457–467

Kawasaki, Y., Goda, Y. & Yoshihira, K. (1992) The mutagenic constituents of *Rubia tinctorum*. *Chem. pharm. Bull.*, **40**, 1504–1509

Kawasaki, Y., Goda, Y., Noguchi, H. & Yamada, T. (1994) Identification of adducts formed by reaction of purine bases with a mutagenic anthraquinone, lucidin: mechanisms of mutagenicity by anthraquinones occurring in *Rubiaceae* plants. *Chem. pharm. Bull.*, **42**, 1971–1973

Kleibeuker, J.H., Cats, A., Zwart, N., Mulder, N.H., Hardonk, M.J. & de Vries, E.G.E. (1995) Excessively high cell proliferation in sigmoid colon after an oral purge with anthraquinone glycosides. *J. natl Cancer Inst.*, **87**, 452–455

Krizsan, K., Szokan, G., Toth, Z.A., Hollosy, F., Laszlo, M. & Khlafulla, A. (1996) HPLC analysis of anthraquinone derivatives in madder root (*Rubia tinctorum*) and its cell cultures. *J. liq. Chromatogr. relat. Technol.*, **19**, 2295–2314

Kune, G.A. (1993) Laxative use not a risk factor for colorectal cancer: Data from the Melbourne colorectal cancer study. *Z. Gastroenterol.*, **31**, 140–143

Leistner, E. (1975) [Isolation, identification, and biosynthesis of anthraquinones in cell suspension cultures of *Morinda citrifolia*.] *Planta med.*, **Suppl.**, 214–224 (in German)

Li, S., Ouyang, Q., Tan, X., Shi, S. & Yao, Z. (1991) [Chemical constitutents of *Morinda officinalis How*.] *Zhongguo Zhong Yao Za Zhi*, **16**, 675–675 (in Chinese)

Marczylo, T., Arimoto-Kobayashi, S. & Hayatsu, H. (2000) Protection against Trp-p-2 mutagenicity by purpurin: Mechanism of in vitro antimutagenesis. *Mutagenesis*, **15**, 223–228

Marec, F., Kollárová, I. & Jegorov, A. (2001) Mutagenicity of natural anthraquinones from *Rubia tinctorum* in the *Drosophila* wing spot test. *Planta med.*, **67**, 127–131

Medical Economics Co. (1998) *PDR for Herbal Medicines*, 1st Ed., Montvale, NJ, pp. 660–661, 1103

Medical Economics Co. (2000) *PDR for Herbal Medicines*, 2nd Ed., Montvale, NJ, pp. 80–81, 490

Michalowicz, W.A. (1981) *Preparation of Hydroxyanthraquinones* (US Patent No. 4,292,248), Charlotte, NC, American Color & Chemical Corporation

Mori, H., Yoshimi, N., Iwata, H., Mori, Y., Hara, A., Tanaka, T. & Kawai, K. (1990) Carcinogenicity of naturally occurring 1-hydroxyanthraquinone in rats: Induction of large bowel, liver and stomach neoplasms. *Carcinogenesis*, **11**, 799–802

Mori, Y., Yoshimi, N., Iwata, H., Tanaka, T. & Mori, H. (1991) The synergistic effect of 1-hydroxyanthraquinone on methylazoxymethanol acetate-induced carcinogenesis in rats. *Carcinogenesis*, **12**, 335–338

Mori, H., Mori, Y., Tanaka, T., Yoshimi, N., Sugie, S., Kawamori, T. & Narisawa, T. (1992) Cell kinetic analysis of the mucosal epithelium and assay of ornithine decarboxylase activity during the process of 1-hydroxyanthraquinone-induced large bowel carcinogenesis in rats. *Carcinogenesis*, **13**, 2217–2220

Murti, V.V.S., Seshadri, T.R. & Sivakumaran, S. (1972) Chemical components of *Rubia iberica*. *Indian J. Chem.*, **10**, 246–247

Nusko, G., Schneider, B., Müller, G., Kusche, J. & Hahn, E.G. (1993) Retrospective study on laxative use and melanosis coli as risk factor for colorectal neoplasma. *Pharmacology*, **47** (Suppl. 1), 234–241

Nusko, G., Schneider, B., Ernst, H., Wittekind, C. & Hahn, E.G. (1997) Melanosis coli — A harmless pigmentation or precancerous condition? *Z. Gastroenterol.*, **35**, 313–318

Nusko, G., Schneider, B., Schneider, I., Wittekind, C. & Hahn, E.G. (2000) Anthranoid laxative use is not a risk factor for colorectal neoplasia: Results of a prospective case control study. *Gut*, **46**, 651–655

Poginsky, B. (1989) [*Occurrence and Genotoxicity of Anthracene Derivatives in* Rubia tinctorum *(L)*] (PhD Thesis), Hamburg, University of Hamburg (in German)

Poginsky, B., Westendorf, J., Blömeke, B., Marquardt, H., Hewer, A., Grover, P.L. & Phillips, D.H. (1991) Evaluation of DNA-binding activity of hydroxyanthraquinones occurring in *Rubia tinctorum* L. *Carcinogenesis*, **12**, 1265–1271

Prista, L.N., Roque, A.S., Ferreira, M.A. & Alves, A.C. (1965) Synthesis of lucidin and lucidin 1-methyl ether starting from nordamnacanthal and damnacanthal. *Garcia Orta*, **13**, 45–50 [consulted as abstract CA66:46240]

Quian, X.-Z. (1996) *Colour Pictorial Handbook of Chinese Herbs*, Beijing, The Peoples Medical Publishing House

Sartori, G., Casnati, G., Bigi, F. & Foglio, F. (1990) A new methodological approach to anthraquinone and anthracyclidone synthesis. *Gazz. Chim. ital.*, **120**, 13–19

Schneider, H.J., Unger, G., Rossler, D., Bothor, C., Berg, W. & Ernst, G. (1979) [Effect of drugs used for the prevention of urinary calculi recurrence on the growth and metabolism of young experimental animals.] *Z. Urol. Nephrol.*, **72**, 237–247 (in German)

Siegers, C.P., Von Hertzberg-Lottin, E., Otte, M. & Schneider, B. (1993) Anthranoid laxative abuse — A risk for colorectal cancer? *Gut*, **34**, 1099–1101

Sonnenberg, A. & Müller, A.D. (1993) Constipation and cathartics as risk factors of colorectal cancer: A meta-analysis. *Pharmacology*, **47** (Suppl. 1), 224–233

Speare, G.S. (1951) Melanosis coli. Experimental observations on its production and elimination in twenty-three cases. *Am. J. Surgery*, **November**, 631–637

Steer, H.W. & Colin-Jones, D.G. (1975) Melanosis coli: Studies of the toxic effects of irritant purgatives. *J. Pathol.*, **115**, 199–205

Steinert, J., Khalaf, H. & Rimpler, M. (1996) High-performance liquid chromatographic separation of some naturally occurring naphthoquinones and anthraquinones. *J. Chromatogr.*, **A723**, 206–209

Suzui, M., Ushijima, T., Dashwood, R.H., Yoshimi, N., Sugimura, T., Mori, H. & Nagao, M. (1999) Frequent mutations of the rat β-catenin gene in colon cancers induced by methylazoxymethanol acetate plus 1-hydroxyanthraquinone. *Mol. Carcinog.*, **24**, 232–237

Tanaka, T., Kojima, T., Yoshimi, N., Sugie, S. & Mori, H. (1991) Inhibitory effect of the non-steroidal anti-inflammatory drug, indomethacin on the naturally occurring carcinogen, 1-hydroxyanthraquinone in male ACI/N rats. *Carcinogenesis*, **12**, 1949–1952

Tanaka, T., Suzui, M., Kojima, T., Okamoto, K., Wang, A. & Mori, H. (1995) Chemoprevention of the naturally occurring carcinogen 1-hydroxyanthraquinone-induced carcinogenesis by the non-steroidal anti-inflammatory drug indomethacin in rats. *Cancer Detect. Prev.*, **19**, 418–425

Thomson, R.H. & Brew, E.J.C. (1971) Naturally occurring quinones. XIX. Anthraquinones in *Hymenodictyon excelsum* and *Damnacanthus major*. *J. chem. Soc. C*, 2001–2007

Toyoda, K., Nishikawa, A., Furukawa, F., Kawanishi, T., Hayashi, Y., Takahashi, M. (1994) Cell proliferation induced by laxatives and related compounds in the rat intestine. *Cancer Lett.*, **83**, 43–49

Wang, S.X., Hua, H.M., Wu, L.J., Li, X. & Zhu, T.R. (1992) [Anthraquinones from the roots of *Rubia cordifolia* L.] *Yaoxue Xuebao*, **27**, 743–747 (in Chinese)

Wang, G.L., Tian, J.G. & Chen, D.C. (1997) [Study on chemical composition of *Rubia cordifolia* L. and *R. tinctorum* L. II. Quantitative determination of alizarin and lucidin by reversed-phase HPLC.] *Yaowu Fenxi Zazhi*, **17**, 219–221 [consulted as abstract: *Analytical Abstracts*, **60**, H199] (in Chinese)

Westendorf, J., Poginsky, B., Marquardt, H., Groth, G. & Marquardt, H. (1988) The genotoxicity of lucidin, a natural component of *Rubia tinctorum* L., and lucidinethylether, a component of ethanolic *Rubia* extracts. *Cell Biol. Toxicol.*, **4**, 225–239

Westendorf, J., Marquardt, H., Poginsky, B., Dominiak, M., Schmidt, J. & Marquardt, H. (1990) Genotoxicity of naturally occurring hydroxyanthraquinones. *Mutat. Res.*, **240**, 1–12

Westendorf, J., Pfau, W. & Schulte, A. (1998) Carcinogenicity and DNA adduct formation observed in ACI rats after long-term treatment with madder root, *Rubia tinctorum* L. *Carcinogenesis*, **19**, 2163–2168

Yang, Y.J., Shu, H.Y. & Min, Z.D. (1992) Anthraquinones isolated from *Morinda officinalis* and *Damnacanthus indicus*. *Yao Xue Xue Bao (Acta pharm. sin.)*, **27**, 358–364

Yao, Z.Q., Guo, Q. & Huang, H.H. (1998) [Officinalisin, a new compound isolated from root-bard of medicinal Indian mulberry (*Morinda officinalis*).] *Chinese trad. Herb. Drugs*, **29**, 217–219 (in Chinese)

Yasui, Y. & Takeda, N. (1983) Identification of a mutagenic substance in *Rubia tinctorum* L. (madder) root, as lucidin. *Mutat. Res.*, **121**, 185–190

Yoshikawa, M., Yanaguchi, S., Nishisaka, H., Yamahara, J. & Murakami, N. (1995) [Chemical constituents of Chinese natural medicine, *Morindae Radix*, the dried roots of *Morinda officinalis* How.: Structures of morindolide and morofficinaloside.] *Chem. pharm. Bull.*, **43**, 1462–1465 (in Japanese)

Yoshimi, N., Sato, S., Makita, H., Wang, A., Hirose, Y., Tanaka, T. & Mori, H. (1994) Expression of cytokines, TNF-α and IL-1α, in MAM acetate and 1-hydroxyanthraquinone-induced colon carcinogenesis of rats. *Carcinogenesis*, **15**, 783–785

Zheng, H.Z. & Dong, Z.H. (1997) [Radix Morindae officinalis]. In: [*Modern Study of Traditional Chinese Medicine*], Vol. II, Xue Yuan Press, 1234–1245 (in Chinese)

Zhuravl'ov, M.S. & Borisov, M.I. (1970) [Anthraquinones of *Galium dasypodum*]. *Farm. Zh. (Kiev)*, *25*, 76–79 (in Ukrainian)

Zhuravl'ov, M.S., Shtefan, L.M. & Luchkina, T.V. (1987) [Anthraquinones of *Galium fagetorum*. II.] *Khim. Prir. Soedin*, 908 [consulted as abstract: CA108:147204] (in Russian)

D. *SENECIO* SPECIES AND RIDDELLIINE

1. Exposure Data

1.1 Origin, type and botanical data (Molyneux *et al.*, 1991)

Senecio riddellii (Asteraceae) (Riddell groundsel) is a grey-white half-shrub, 30–90 cm tall with pinnatifid and relatively hairless leaves, revealing its bright green leaf colour. It has bright yellow flowers on the stems at about the same height above the ground. This gives the plant a flat-topped appearance when in bloom. It produces flowers in late summer to early autumn and dies back to ground level after the first frost. It grows in dry, sandy soils and its roots are long and about as thick as a lead pencil.

Senecio longilobus (also known as woolly groundsel and thread-leaf groundsel) is a shrubby, erect, branched, leafy plant, 30–60 cm tall. It has narrowly linear leaves which are thick, white, and occasionally pinnately lobed, up to 10 cm long. The composite yellow flower heads contain numerous clusters. In North America, this is a common plant with a range extending from Colorado to Utah, south to Texas and Mexico (Kingsbury, 1964).

1.2 Use

The 'bush tea' used in Jamaica to treat children for a cold an a herbal tea that is popular in the south-west USA, gordolobo yerba, may contain riddelliine (Stillman *et al.*, 1977a; Huxtable, 1980; National Toxicology Program, 2002).

1.3 Chemical constituents

The plants (ragworts) from which riddelliine and other pyrrolizidine alkaloids are isolated are found in the rangelands of the western USA. Cattle, horses and, less commonly, sheep that ingest these plants can succumb to their toxic effects (called pyrrolizidine alkaloidosis). Riddelliine residues have been found in meat, milk and honey. The plants may contaminate human food sources as intact plants, and their seeds may contaminate commercial grains such as wheat (Fu *et al.*, 2001; National Toxicology Program, 2002).

The main chemical constituents of importance in *S. riddellii* are pyrrolizidine alkaloids.

Four pyrrolizidine alkaloids have been identified as constituents of *S. longilobus*, namely riddelliine, retrorsine, senecionine and seneciphylline (Segall & Molyneux, 1978).

1.4 Sales and consumption

No information was available to the Working Group.

1.5 Component with potential cancer hazard (riddelliine)

Riddelliine was evaluated previously (IARC, 1976).

1.5.1 *Nomenclature*

Chem. Abstr. Serv. Reg. No.: 23246-96-0
Chem. Abstr. Serv. Name: 13,19-Didehydro-12,18-dihydroxysenecionan-11,16-dione
Synonym: Riddelline

1.5.2 *Structural and molecular formulae and relative molecular mass*

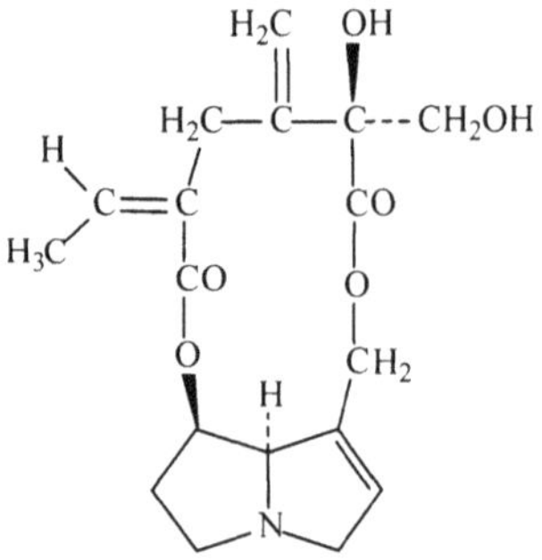

$C_{18}H_{23}NO_6$ Relative molecular mass: 349.38

1.5.3 *Chemical and physical properties of the pure substance*

(*a*) *Description*: Crystalline solid (National Toxicology Program, 2002)
(*b*) *Melting-point*: 197–198 °C (decomposes) (Buckingham, 2001)
(*c*) *Solubility*: Sparingly soluble in water; soluble in chloroform; slightly soluble in acetone and ethanol (National Toxicology Program, 2002)
(*d*) *Optical rotation*: $[\alpha]_D^{25}$ –109.5 (chloroform) (Buckingham, 2001)

1.5.4 *Analysis*

Underivatized pyrrolizidine alkaloids, including riddelliine, from natural sources (plants and insects) have been analysed by capillary gas chromatography–mass spectrometry (Witte *et al.*, 1992). Thin-layer chromatography (TLC) and nuclear magnetic resonance (NMR) spectroscopy have also been used to detect pyrrolizidine alkaloids in plant extracts (Molyneux *et al.*, 1979; Molyneux & Roitman, 1980).

Riddelliine, retrorsine, senkirkine, retronecine, integerrimine, seneciphylline and senecionine (0.007, 0.008, 0.012, 0.005, 0.008, 0.042 and 0.036%, respectively) were determined in methanol extracts from dry *Senecio vernalis* using reversed-phase high-performance liquid chromatography (HPLC) with spectrometric detection at 225 nm (Sener *et al.*, 1986).

Pyrrolic metabolites from pyrrolizidine alkaloids were detected in liver and dried blood samples from pigs fed varying amounts of riddelline using gas chromatography/tandem mass spectrometry (Schoch *et al.*, 2000).

1.5.5 *Production*

Riddelliine is produced commercially only as a reference standard and as a research chemical (National Toxicology Program, 2002).

1.5.6 *Use*

Riddelliine has no known commercial use.

1.5.7 *Occurrence*

Riddelliine is found in *S. riddellii*, *S. longiflorus*, *S. eremophilus*, *S. vernalis*, *S. cruentus*, *S. longilobus*, *S. aegyptus*, *S. desfontainei* (*S. coronopifolius*) and *S. jacobaea* (Klásek *et al.*, 1968; Segall & Molyneux, 1978; Asada *et al.*, 1982; Sener *et al.*, 1986; Mirsalis *et al.*, 1993; Fu *et al.*, 2001; Röder, 2002).

Structurally, riddelliine belongs to a class of toxic pyrrolizidine alkaloids that are esters of unsaturated basic alcohols (necines) and of a necic acid produced by plants growing in climates ranging from temperate to tropical. The pyrrolizidine alkaloid-producing plants are unrelated taxonomically. The alkaloids occur in different parts of the plants, with the highest content in the seeds and flowering tops. The quantity of the alkaloids varies, depending on the season, climate and soil constitution (Fu *et al.*, 2001; National Toxicology Program, 2002).

2. Studies of Cancer in Humans

Riddelliine

No report of cancer related to the intake of riddelliine or of *Senecio* spp. in humans was available to the Working Group.

3. Studies of Cancer in Experimental Animals

Riddelliine

Oral administration

Mouse: Groups of 50 male B6C3F$_1$ mice, 5–6 weeks of age, were administered 0, 0.1, 0.3, 1.0 or 3 mg/kg bw riddelliine by gavage in sodium phosphate buffer on five days per week for 105 weeks. Groups of 50 female B6C3F$_1$ mice, 5–6 weeks of age, were administered 0 or 3 mg/kg riddelliine by gavage in sodium phosphate buffer on five days per week for 105 weeks. Mean survival (days) among these groups was 696 (0.0 mg/kg), 705 (0.1 mg/kg), 716 (0.3 mg/kg), 701 (1 mg/kg) and 667 (3 mg/kg) for males and 670 (0.0 mg) and 678 (3 mg/kg) for females. The incidence of hepatic haemangiosarcomas in males that received 3 mg/kg was significantly greater than that of the controls (2/50, 1/50, 0/50, 2/50, 31/50 ($p < 0.001$) at doses of 0, 0.1, 0.3, 1.0 and 3 mg/kg, respectively). The incidence of hepatocellular tumours was negatively correlated with dose in male mice and was significantly decreased in females given the 3 mg/kg dose. The incidences of bronchiolo-alveolar adenoma (1/50 control versus 9/50 treated) and of adenoma and carcinoma combined (2/50 control versus 13/50 treated) were significantly increased in the highest-dose females compared with control females. The incidence of bronchiolo-alveolar tumours in this group exceeded the historical control range for this neoplasm (National Toxicology Program, 2002).

Rat: Groups of 50 female Fischer 344 rats, 5–6 weeks of age, were administered riddelliine in sodium phosphate buffer by gavage at doses of 0, 0.01, 0.03, 0.1, 0.3 and 1 mg/kg bw on five days per week for 105 weeks. Groups of 50 males were administered riddelliine similarly at doses of 0 or 1 mg/kg bw for only 72 weeks, due to mortality in high-dose males. All high-dose (1 mg/kg bw) females died before week 97. Survival of females in the other groups was not affected. In females, there was increased incidence of haemangiosarcoma at the highest dose (0/50, 0/50, 0/50, 0/50, 3/50 and 38/50 ($p < 0.001$) at doses of 0, 0.01, 0.03, 0.1, 0.3 and 1 mg/kg, respectively). Hepatocellular adenoma/carcinoma combined were also increased (1/50, 0/50, 0/50, 0/50, 2/50 and 8/50 ($p < 0.001$) at doses of 0, 0.01, 0.03, 0.1, 0.3 and 1 mg/kg, respectively). Mononuclear-cell leukaemia was found in 12/50, 8/50, 13/50, 18/50, 18/50 and 14/50 at doses of 0, 0.01, 0.03, 0.1, 0.3 and 1 mg/kg, respectively (trend test positive, $p = 0.009$). In treated males, 43/50 ($p < 0.001$) developed haemangiosarcomas of the liver, 4/50 ($p = 0.03$) developed hepatocellular adenomas and 9/50 versus 2/50 controls ($p = 0.004$) had mononuclear-cell leukaemia (National Toxicology Program, 2002).

Groups of 20 male and 20 female Fischer 344 rats, 6–8 weeks old, were given 0, 0.1, 0.33, 1.0, 3.3 or 10 mg/kg bw riddelliine by gavage in phosphate buffer five times per week. Ten rats per sex per dose were sacrificed after 13 weeks of treatment. Five rats per sex were killed after a seven-week recovery period and the remaining five rats per sex were killed after 14 weeks' recovery. No liver lesions were observed in control animals. Two of 10 female rats in the 10-mg/kg group examined at 13 weeks had hepatocellular adenomas. Three other females in this group had hepatocellular foci or focal nodular hyperplasia. One female in this group that died during the 14-week recovery period had multiple adenomas and multi-focal nodular hyperplasia and cholangiocellular hyperplasia (Chan *et al.*, 1994).

4. Other Data Relevant to an Evaluation of Carcinogenicity and its Mechanisms

4.1 Absorption, distribution, metabolism and excretion

4.1.1 *Humans*

No data were available to the Working Group.

4.1.2 *Experimental systems*

Several feeding experiments have indicated that riddelliine and its *N*-oxide are absorbed in the gastrointestinal tract of domestic farm animals. Calves treated for 20 days with either riddelliine *N*-oxide (40.5 mg/kg/day) or a mixture of riddelliine *N*-oxide (40.5 mg/kg/day) and riddelliine (4.5 mg/kg/day) showed 100% morbidity, implying absorption in each case (Molyneux *et al.*, 1991). Tissue samples from pigs fed various amounts (3, 10 or 15 mg/kg bw in gelatin capsules) of riddelliine for 40 days also showed absorption, as determined by liquid chromatography and tandem mass spectrometric (LC/MS) analysis of pyrrolic metabolites in blood and liver samples collected one day after the end of treatment (Schoch *et al.*, 2000).

Pyrrolizidine alkaloids exist in plants in two forms, the free-base alkaloids and their *N*-oxides, neither of which is toxic *per se*. In numerous experimental studies, especially in rats, free riddelliine base has been found to be absorbed and transported to the liver, where it is converted to hepatotoxic pyrroles by microsomal enzymes (Mattocks & White, 1971; Molyneux *et al.*, 1991). In liver microsomes of Fischer 344 rats pretreated with phenobarbital, the major metabolites of riddelliine obtained after aerobic incubation for 30 min were riddelliine *N*-oxide and dehydroretronecine, as determined by HPLC and LC/MS (Figure 1). The latter metabolite interacts with DNA to form various adducts (Yang *et al.*, 2001a).

Figure 1. Metabolism of riddelliine by rat liver microsomes

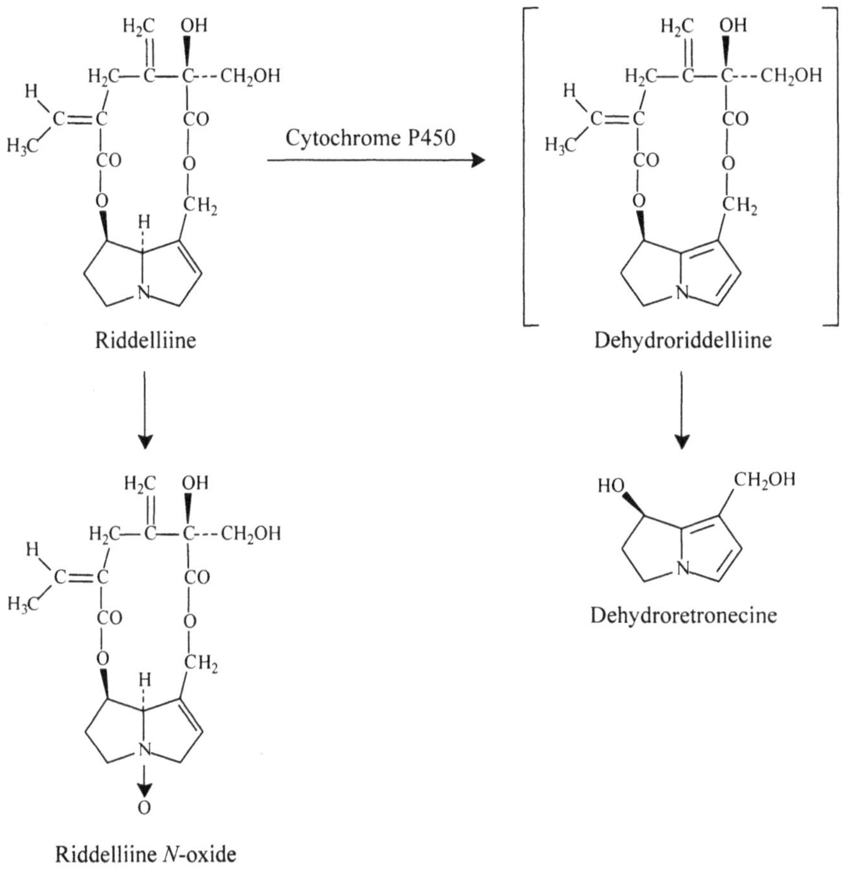

From Yang *et al.* (2001a)

4.2 Toxic effects

4.2.1 *Humans*

No data were available to the Working Group on the toxicity of riddelliine itself to humans.

Riddelliine is one constituent of the plant *Senecio longilobus*, which has been used as a folk remedy called gordolobo yerba by Mexican–Americans in the south-western USA (Segall & Molyneux, 1978). The consumption of gordolobo yerba has been linked with the incidence of acute hepatic veno-occlusive disease (Stillman *et al.*, 1977a,b). Two infants were diagnosed as exhibiting hepatic veno-occlusive disease due to pyrrolizidine alkaloid intoxication, as a result of ingesting *S. longilobus* as a herbal tea used as a cough medicine. One of these infants subsequently died (Stillman *et al.*, 1977a).

4.2.2 *Experimental systems*

Cheeke (1988) reviewed the general toxicity of pyrrolizidine alkaloids in *Senecio*, *Crotalaria* and other plant species in large animals, small herbivores and other laboratory animals. Cattle and horses are highly susceptible to pyrrolizidine poisoning, whereas sheep, goats, rabbits and guinea-pigs are much more resistant. A few studies have been published specifically on the toxic signs and symptoms of riddelliine. Thus, preflowering *Senecio riddellii* whole plants (Riddell's groundsel), a species that contains mainly (> 90%) riddelliine and its *N*-oxide as alkaloidal constituents, caused seneciosis when given to calves for 20 days either by gavage or in gelatin capsules at an equivalent dose of 15–20 mg/kg bw pyrrolizidine alkaloid per day. The main clinical signs were malaise, depression, erratic or unpredictable behaviour, aimless walking and ataxia. Diarrhoea with tecnesmus and rectal prolapse, and abdominal distension were frequently observed. Gross pathological findings included ascites, abomasal oedema, mesenteric lymph node enlargement and oedema, and oedema of the mesentery between loops of the ansa spiralis. Hepatobiliary lesions were present in all cattle necropsied. Portal biliary hyperplasia, formation of new bile ductules and periportal fibrosis were also seen. In more severe cases, vacuolated and enlarged hepatocytes were seen throughout the liver lobule. Central nervous system lesions were present in all cattle with seneciosis, such as spongy degeneration along the axonal tracts of the white matter. Other microscopic lesions included abomasal oedema, mucosal haemorrhage, lymph node oedema and occasional pulmonary haemorrhage (Johnson *et al.*, 1985).

In a follow-up study, under the experimental conditions described by Johnson *et al.* (1985), a group of calves fed 4.5 mg/kg bw riddelliine free base per day showed no sign of toxicosis or serum enzyme changes. However, two further groups fed pure riddelliine *N*-oxide (40.5 mg/kg/day) and a mixture of pure riddelliine (4.5 mg/kg/day) and pure riddelliine *N*-oxide (40.5 mg/kg/day) showed 100% morbidity, with the latter group showing fewer liver lesions. It was established from this study that the *N*-oxide of riddelliine alone is capable of inducing seneciosis in cattle (Molyneux *et al.*, 1991).

The toxicity of riddelliine was studied in male and female Fischer rats and in B6C3F$_1$ mice. The compound was given by gavage in 0.1 M phosphate buffer five times per week at daily doses of 0, 0.1, 0.33, 1, 3.3 and 10 mg/kg bw (rats) and of 0, 0.33, 1, 3.3, 10 and 25 mg/kg bw (mice). The animals were necropsied after 13 weeks of treatment or after an additional 7 or 14 weeks of recovery. Body weight gain was inversely related to dose in both rats and mice. The initial group sacrificed after 13 weeks showed dose-related hepatopathy and intravascular macrophage accumulation in rats (at doses ≥ 0.33 mg/kg bw) and hepatocytomegaly in mice (only at 25 mg/kg bw). Some of these lesions persisted throughout the 14-week recovery period, with hepatic foci and cellular alterations observed in male rats, and increasingly severe bile duct proliferation in female rats and in male and female mice (Chan *et al.*, 1994).

4.3 Reproductive and developmental effects

4.3.1 *Humans*

No data were available to the Working Group.

4.3.2 *Experimental systems*

The developmental toxicity of pyrrolizidine alkaloids has been evaluated (IARC, 1976; WHO, 1988). Several pyrrolizidine alkaloids, pyrrolizidine alkaloid derivatives and related compounds have been shown to produce teratogenic and fetotoxic effects in experimental animals (WHO, 1988). However, none of the studies reviewed was on riddelliine. According to Chan *et al.* (1994), pyrrolizidine alkaloids have been detected in milk of lactating rats and cows.

In 13-week gavage studies combined with mating trials, groups of male and female Fischer 344 rats and B6C3F$_1$ mice were administered riddelliine at doses of up to 10 and 25 mg/kg bw, respectively. Body weight gain was inversely related to dose in both rats and mice. In rats, decreased epididymal and testis weights were observed in males given 1.0 mg/kg bw, but spermatozoal measurements were not affected. At 10 mg/kg bw (a dose lethal to the males), riddelliine caused persistent estrus in females. In mating trials with 1.0 mg/kg bw as the highest dose, no effect was seen on fertility, weight gain of dams during gestation, litter size or percentage of live pups. At 14 and 21 days of age, female pups had lower body weight than control pups. In mice, females in the 25-mg/kg bw group showed marked prolongation of length of the estrus cycle. Despite this, dosed dams were able to conceive and continue with pregnancy. There was no effect on maternal body weight gain during pregnancy at the 25-mg/kg bw dose, but live litter size was reduced and pups of treated dams had lower body weight at birth and during the lactation period than control pups. There was no effect on litter size or pup body weight at 3.3 mg/kg bw (National Toxicology Program, 1993; Chan *et al.*, 1994).

4.4 Genetic and related effects

4.4.1 *Humans*

No data were available to the Working Group.

4.4.2 *Experimental systems* (see Table 1 for references)

Riddelliine was mutagenic in *Salmonella typhimurium* strain TA100 in the presence of a metabolic activation system, but not in strains TA1535, TA97 and TA98. It caused DNA–protein cross-links in bovine kidney CCL 22 cells *in vitro*. Sister chromatid exchange and chromosomal aberrations were induced in Chinese hamster ovary cells *in vitro*. Transformation of BALB/c3T3 cells was observed in a single study.

Table 1. Genetic and related effects of riddelliine and its metabolite dehydroxyretronecine

Test system	Result[a]		Dose[b] (LED or HID)	Reference
	Without exogenous metabolic system	With exogenous metabolic system		
Riddelliine				
Salmonella typhimurium TA1535, TA97, TA98, reverse mutation	–	–	5000 μg/plate	Zeiger *et al.* (1988)
Salmonella typhimurium TA100, reverse mutation	–	+	1000 μg/plate	Zeiger *et al.* (1988)
DNA-protein cross-links, bovine kidney (MDBK) CCL 22 cells *in vitro*	+	+	17.5	Hincks *et al.* (1991)
Sister chromatid exchange, Chinese hamster ovary cells *in vitro*	+	+	3	Galloway *et al.* (1987)
Chromosomal aberrations, Chinese hamster ovary cells *in vitro*	–	+	300	Galloway *et al.* (1987)
Cell transformation, BALB/c3T3 mouse cells, *in vitro*	+	NT	225	Matthews *et al.* (1993)
Unscheduled DNA synthesis, male and female Fischer 344 rat hepatocytes *in vivo*	+		50 po × 1	Mirsalis (1987)
Unscheduled DNA synthesis, male and female Fischer 344 rat hepatocytes *in vivo*	–		25 po 30 d[c]	Mirsalis *et al.* (1993)
Unscheduled DNA synthesis, male and female B6C3F$_1$ mouse hepatocytes *in vivo*	?[d]		25 po 30 d[c]	Mirsalis *et al.* (1993)
Unscheduled DNA synthesis, male and female B6C3F$_1$ mouse hepatocytes *in vivo*	+		25 po × 5	Chan *et al.* (1994)
Unscheduled DNA synthesis, male and female Fischer 344 rat hepatocytes *in vivo*	+		1 po × 5	Chan *et al.* (1994)
Micronucleus induction, male and female B6C3F$_1$ mouse bone marrow, *in vivo*	–		25 po 5 or 30 d[c]	Mirsalis *et al.* (1993)
Micronucleus induction, male B6C3F$_1$ mouse peripheral blood, *in vivo*	+		150 po × 1	Chan *et al.* (1994)
Micronucleus induction, male B6C3F$_1$ mouse bone marrow, *in vivo*	+		270 po × 1	Chan *et al.* (1994)

Table 1 (contd)

Test system	Result[a]		Dose[b] (LED or HID)	Reference
	Without exogenous metabolic system	With exogenous metabolic system		
Micronucleus induction, male Swiss mouse peripheral blood erythrocytes, *in vivo*	+		150 ip × 1	National Toxicology Program (2002)
Micronucleus induction, male Swiss mouse bone-marrow erythrocytes, *in vivo*	+		270 ip × 1	National Toxicology Program (2002)
Micronucleus induction, male and female B6C3F$_1$ mouse peripheral blood, *in vivo*	–		25 po 13 wk	Witt *et al.* (2000)
Micronucleus induction, male and female Fischer 344 rat bone marrow, *in vivo*	–		3.3 po 30 d[c]	Mirsalis *et al.* (1993)
DNA adducts, female Fischer 344 rat liver, *in vivo*, ^{32}P-postlabelling	+		0.01 po 3 mo[c]	Yang *et al.* (2001a)
Dehydroretronecine				
Salmonella typhimurium TA92, reverse mutation	+	NT	500 μg/plate	Ord *et al.* (1985)
DNA cross-links, pBR322 plasmid and M13 viral DNA	+	NT	29	Reed *et al.* (1988)
DNA–protein cross-links, bovine kidney (MDBK) CCL22 cells *in vitro*	+	NT	46	Kim *et al.* (1995)
Sister chromatid exchange, human lymphocytes *in vitro*	+	NT	0.12	Ord *et al.* (1985)
DNA adducts, calf thymus DNA *in vitro*	+	NT	3.9	Yang *et al.* (2001b)
Alkylation of N^2 of deoxyguanosine *in vitro*	+	NT	1530	Robertson (1982)

[a] +, positive; (+), weak positive; –, negative; NT, not tested; ?, inconclusive
[b] LED, lowest effective dose; HID, highest ineffective dose; in-vitro tests, μg/mL; in-vivo tests, mg/kg bw/day; d, day; po, oral; ip, intraperitoneal; mo, month; wk, week
[c] Treated 5 days/week
[d] Equivocal response in males; modest response in females at highest dose only

Riddelliine induced unscheduled DNA synthesis in rat and mouse hepatocytes *in vivo* in some studies, but not in others. It induced micronuclei in peripheral blood erythrocytes and bone marrow of mice after a single high dose (intraperitoneal or oral) but not after repeated treatment of rats or mice at low dose. Riddelliine did form DNA adducts in rat liver following oral exposure.

A major metabolite of riddelliine, dehydroretronecine, was mutagenic in *S. typhimurium* strain TA92 and induced sister chromatid exchange in human lymphocytes in the absence of exogenous metabolic activation. Dehydroretronecine also induced DNA–DNA cross-links in pBR322 plasmid and M13 viral DNA and DNA–protein cross-links in bovine kidney CCL 22 cells *in vitro*. Exposure of calf thymus DNA or deoxyguanosine to dehydroretronecine *in vitro* caused DNA-adduct formation and alkylation of the exocyclic amino group, respectively.

Eight different dehydroretronecine-derived DNA adducts were detected in all liver samples from female Fischer 344 rats fed riddelliine at five different doses (0.01, 0.033, 0.1, 0.33, 1.0 mg/kg bw per day). Two of these were characterized as dehydroretronecine-3*N*-deoxyguanosin-N^2-yl epimers (Figure 2) by a ^{32}P-postlabelling/HPLC technique, supported by spectroscopic and synthetic procedures (Yang *et al.*, 2001b). The other six adducts detected were not identified (Yang *et al.*, 2001a).

4.5 Mechanistic considerations

Orally administered riddelliine forms dehydroretronecine–DNA adducts and induces unscheduled DNA synthesis in rat liver, indicating that liver is a target for its genotoxic activity. Dehydroretronecine forms DNA adducts and cross-links *in vitro* without further metabolic activation. This metabolite is also mutagenic to *S. typhimurium*. This activity, as well as the hepatotoxicity with consequent hyperplasia, could be significant events in riddelliine-induced hepatocarcinogenicity. It is not known, however, whether the basic metabolic step occurs in humans.

5. Summary of Data Reported and Evaluation

5.1 Exposure data

Riddelliine is a pyrrolizidine alkaloid that is found in *Senecio riddellii* and other *Senecio* species, including *S. longilobus*, which is used as a herbal remedy in the southwestern USA.

5.2 Human carcinogenicity data

No data on the carcinogenicity of riddelliine to humans were available to the Working Group.

Figure 2. Proposed metabolic activation of riddelliine leading to dehydroretronecine–DNA adducts in female Fischer 344 rats fed riddelliine

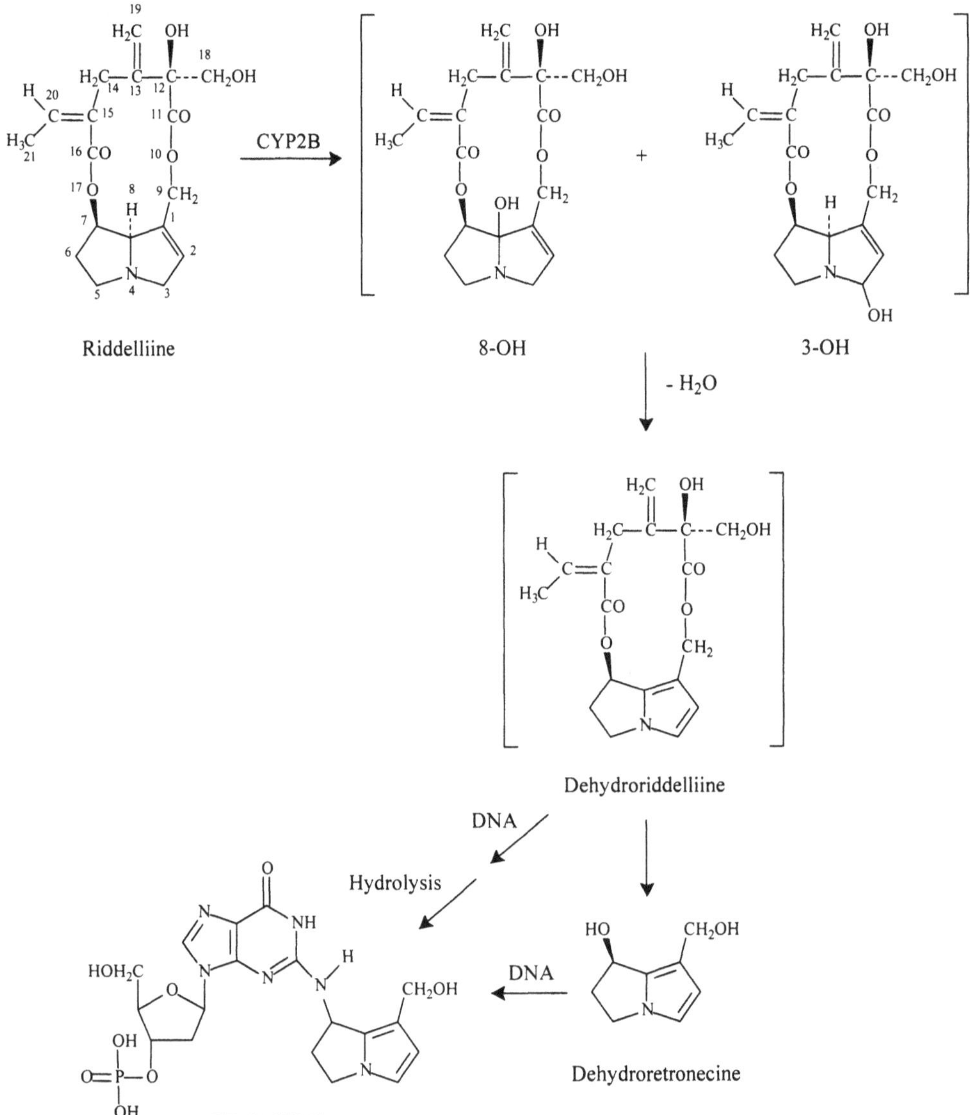

From Yang *et al.* (2001a)

5.3 Animal carcinogenicity data

In mice, oral administration of riddelliine induced hepatic haemangiosarcomas in males and bronchiolo-alveolar adenomas and carcinomas in females. In rats, oral administration of riddelliine increased the incidence of hepatic haemangiosarcomas, hepatocellular carcinomas and/or adenomas and mononuclear cell leukaemia in both males and females. In a short-term study, a few rats developed hepatocellular adenomas after 13 weeks of oral administration.

5.4 Other relevant data

Riddelliine and its *N*-oxide are absorbed from the gastrointestinal tract. Riddelliine is metabolized to riddelliine *N*-oxide and dehydroretronecine in rat liver microsomes.

A herbal preparation made from *Senecia longilobus*, of which riddelliine is a constituent, causes hepatic veno-occlusive disease in humans. Intoxication of calves with either *S. riddellii* or one of its two main pyrrolizidine alkaloid constituents, riddelliine *N*-oxide, led to the typical signs and symptoms of seneciosis. A similar toxicity profile was induced by riddelliine in rodents.

Riddelliine disturbs the estrus cycle in rodents. It causes developmental toxicity in the absence of marked toxicity in rodents.

DNA adducts of dehydroretronecine are found in rat liver following oral administration of riddelliine. Dehydroretronecine is genotoxic in a number of in-vitro systems. It induced sister chromatid exchange in human lymphocytes, DNA–protein cross-links in bovine kidney epithelial cells *in vitro* and gene mutations in bacteria.

5.5 Evaluation

There are no data on the carcinogenicity of riddelliine to humans.

There is *sufficient evidence* in experimental animals for the carcinogenicity of riddelliine.

Overall evaluation

Riddelliine is *possibly carcinogenic to humans (Group 2B)*.

6. References

Asada, Y., Furuya, T., Takeuchi, T. & Osawa, Y. (1982) Pyrrolizidine alkaloids from *Senecio cruentus*. *Planta med.*, **46**, 125–126

Buckingham, J., ed. (2001) *Dictionary of Natural Products on CD-ROM*, Boca Raton, FL, Chapman & Hall/CRC Press

Chan, P.C., Mahler, J., Bucher, J.R., Travlos, G.S. & Reid, J.B. (1994) Toxicity and carcinogenicity of riddelliine following 13 weeks of treatment to rats and mice. *Toxicon*, **32**, 891–908

Cheeke, P.R. (1988) Toxicity and metabolism of pyrrolizidine alkaloids. *J. anim. Sci.*, **66**, 2343–2350

Fu, P.P., Chou, M.W., Xia, Q., Yang, Y.C., Yan, J., Doerge, D.R. & Chan, P.C. (2001) Genotoxic pyrrolizidine alkaloids and pyrrolizidine alkaloid *N*-oxides — Mechanisms leading to DNA adduct formation and tumorigenicity. *Environ. Carcinog. Ecotoxicol. Rev.*, **C19**, 353–385

Galloway, S.M., Armstrong, M.J., Reuben, C., Colman, S., Brown, B., Cannon, C., Bloom, A.D., Nakamura, F., Ahmed, M., Duk, S., Rimpo, J., Margolin, B.H., Resnick, M.A., Anderson, B. & Zeiger, E. (1987) Chromosome aberrations and sister chromatid exchanges in Chinese hamster ovary cells: Evaluations of 108 chemicals. *Environ. mol. Mutag.*, **10** (Suppl. 10), 1–175

Hincks, J.R., Kim, H.-Y., Segall, H.J., Molyneux, R.J., Stermitz, F.R. & Coulombe, R.A., Jr (1991) DNA cross-linking in mammalian cells by pyrrolizidine alkaloids: Structure–activity relationships. *Toxicol. appl. Pharmacol.*, **111**, 90–98

Huxtable, R.J. (1980) Herbal teas and toxins: Novel aspects of pyrrolizidine poisoning in the United States. *Perspect. Biol. Med.*, **24**, 1–14

IARC (1976) *IARC Monographs on the Evaluation of the Carcinogenic Risk of Chemicals to Man*, Vol. 10, *Some Naturally Occurring Substances*, Lyon, IARC*Press*, pp. 313–317

Johnson, A.E., Molyneux, R.J. & Stuart, L.D. (1985) Toxicity of Riddell's groundsel (*Senecio riddellii*) to cattle. *Am. J. vet. Res.*, **46**, 577–582

Kim, H.-Y., Stermitz, F.R. & Coulombe, R.A., Jr (1995) Pyrrolizidine alkaloid-induced DNA-protein cross-links. *Carcinogenesis*, **16**, 2691–2697

Kingsbury, J.M. (1964) *Poisonous Plants of the United States and Canada*, Englewood Cliffs, NJ, Prentice-Hall, p. 427

Klásek, A., Svarovsky, V., Ahmed, S.S. & Santavy, F. (1968) Isolation of pyrrolizidine alkaloids from *Senecio aegyptus* and *S. desfontainei* (*S. coronopifolius*). *Collect. Czech. chem. Commun.*, **33**, 1738–1743

Matthews, E.J., Spalding, J.W. & Tennant, R.W. (1993) Transformation of BALB/c-3T3 cells: IV. Rank-ordered potency of 24 chemical responses detected in a sensitive new assay procedure. *Environ. Health Perspect.*, **101** (Suppl. 2), 319–345

Mattocks, A.R. & White, I.N.H. (1971) The conversion of pyrrolizidine alkaloids to *N*-oxides and to dihydropyrrolizidine derivatives by rat-liver microsomes *in vitro*. *Chem.-biol. Interact.*, **3**, 383–396

Mirsalis, J.C. (1987) In vivo measurement of unscheduled DNA synthesis and S-phase synthesis as an indicator of hepatocarcinogenesis in rodents. *Cell Biol. Toxicol.*, **3**, 165–173

Mirsalis, J.C., Steinmetz, K.L., Blazak, W.F. & Spalding, J.W. (1993) Evaluation of the potential of riddelliine to induce unscheduled DNA synthesis, S-phase synthesis, or micronuclei following in vivo treatment with multiple doses. *Environ. mol. Mutag.*, **21**, 265–271

Molyneux, R.J. & Roitman, J.N. (1980) Specific detection of pyrrolizidine alkaloids in thin-layer chromatograms. *J. Chromatogr.*, **195**, 412–415

Molyneux, R.J., Johnson, A.E., Roitman, J.N. & Benson, M.E. (1979) Chemistry of toxic range plants. Determination of pyrrolizidine alkaloid content and composition in *Senecio* species by nuclear magnetic resonance spectroscopy. *J. agric. Food Chem.*, **27**, 494–499

Molyneux, R.J., Johnson, A.E., Olsen, J.D. & Baker, D.C. (1991) Toxicity of pyrrolizidine alkaloids from Riddell groundsel (*Senecio riddellii*) to cattle. *Am. J. vet. Res.*, **52**, 146–151

National Toxicology Program (1993) *NTP Technical Report on Toxicity Studies of Riddelliine (CAS No. 23246-96-0) Administered by Gavage to F344/N Rats and B6C3F$_1$ Mice* (Toxicity Report Series No. 27; NIH Publication No. 94-3350). U.S. Department of Health and Human Services, Public Health Service, National Institutes of Health, Research Triangle Park, NC

National Toxicology Program (2002) *NTP Technical Report on the Toxicology and Carcinogenesis Studies of Riddelliine (CAS No. 23246-96-0) in F344/N Rats and B6C3F$_1$ Mice (Gavage Studies)* (NTP TR-508; NIH Publication No. 01-4442), Research Triangle Park, NC

Ord, M.J., Herbert, A. & Mattocks, A.R. (1985) The ability of bifunctional and monofunctional pyrrole compounds to induce sister-chromatid exchange (SCE) in human lymphocytes and mutations in *Salmonella typhimurium*. *Mutat. Res.*, **149**, 485–493

Reed, R.L., Ahern, K.G., Pearson, G.D. & Buhler, D.R. (1988) Crosslinking of DNA by dehydroretronecine, a metabolite of pyrrolizidine alkaloids. *Carcinogenesis*, **9**, 1355–1361

Robertson, K.A. (1982) Alkylation of N^2 in deoxyguanosine by dehydroretronecine, a carcinogenic metabolite of the pyrrolizidine alkaloid monocrotaline. *Cancer Res.*, **42**, 8–14

Röder, E. (2002) *Medicinal Plants in Europe Containing Pyrrolizidine Alcaloids in King's American Dispensatory* [http://ftp. ort.unc.edu/herbmed/PAs/PAs–plants.html]

Schoch, T.K., Gardner, D.R. & Stegelmeier, B.L. (2000) GC/MS/MS detection of pyrrolic metabolites in animals poisoned with the pyrrolizidine alkaloid riddelliine. *J. nat. Toxins*, **9**, 197–206

Segall, H.J. & Molyneux, R.J. (1978) Identification of pyrrolizidine alkaloids (*Senecio longilobus*). *Res. Commun. chem. Pathol. Pharmacol.*, **19**, 545–548

Sener, B., Temizer, H., Temizer, A. & Karakaya, A.E. (1986) High-performance liquid chromatographic determination of alkaloids in *Senecio vernalis*. *J. Pharm. belg.*, **41**, 115–117

Stillman, A.E., Huxtable, R.J., Fox, D., Hart, M. & Bergeson, P. (1977a) Pyrrolizidine (*Senecio*) poisoning in Arizona: Severe liver damage due to herbal teas. *Ariz. Med.*, **34**, 545–546

Stillman, A.E., Huxtable, R., Consroe, P., Kohnen, P. & Smith, S. (1977b) Hepatic veno-occlusive disease due to pyrrolizidine (*Senecio*) poisoning in Arizona. *Gastroenterology*, **73**, 349–352

WHO (1988) *Pyrrolizidine Alkaloids* (Environmental Health Criteria 80), Geneva, International Programme on Chemical Safety

WHO (1998) *Medicinal Plants in the Republic of Korea*. WHO Regional Publications, Western Pacific Series No. 21, Manila

Witt, K.L., Knapton, A., Wehr, C.M., Hook, G.J., Mirsalis, J., Shelby, M.D. & MacGregor, J.T. (2000) Micronucleated erythrocyte frequency in peripheral blood of B6C3F$_1$ mice from short-term, prechronic, and chronic studies of the NTP Carcinogenesis Bioassay Program. *Environ. mol. Mutag.*, **36**, 163–194

Witte, L., Rubiolo, P., Bicchi, C. & Hartmann, T. (1992) Comparative analysis of pyrrolizidine alkaloids from natural sources by gas chromatography-mass spectrometry. *Phytochemistry*, **32**, 187–196

Yang, Y.-C., Yan, J., Doerge, D.R., Chan, P.-C., Fu, P.P. & Chou, M.W. (2001a) Metabolic activation of the tumorigenic pyrrolizidine alkaloid, riddelliine, leading to DNA adduct formation *in vivo*. *Chem. Res. Toxicol.*, **14**, 101–109

Yang, Y.-C., Yan, J., Churchwell, M., Beger, R., Chan, P.-C., Doerge, D.R., Fu, P.P. & Chou, M.W. (2001b) Development of a ^{32}P-postlabeling/HPLC method for detection of dehydroretronecine-derived DNA adducts *in vivo* and *in vitro*. *Chem. Res. Toxicol.*, **14**, 91–100

Zeiger, E., Anderson, B., Haworth, S., Lawlor, T. & Mortelmans, K. (1988) *Salmonella* mutagenicity tests: IV. Results from the testing of 300 chemicals. *Environ. mol. Mutag.*, **11**, 1–158

SOME MYCOTOXINS

AFLATOXINS

These substances were considered by previous working groups, in December 1971 (IARC, 1972), October 1975 (IARC, 1976), March 1987 (IARC, 1987) and June 1992 (IARC, 1993). Since that time, new data have become available, and these have been incorporated into this updated monograph.

1. Exposure Data

1.1 Chemical and physical data

1.1.1 *Synonyms, structural and molecular data* (see Figure 1)

Aflatoxin B_1

Chem. Abstr. Services Reg. No.: 1162-65-8
Deleted CAS Nos: 13214-11-4; 11003-08-0; 27261-02-5
Chem. Abstr. Name: (6aR,9aS)-2,3,6a,9a-Tetrahydro-4-methoxycyclopenta[*c*]furo-(3′,2′:4,5)furo[2,3-*h*][*l*]benzopyran-1,11-dione (9CI)
Synonyms: 6-Methoxydifurocoumarone; 2,3,6aα,9aα-tetrahydro-4-methoxycyclopenta[*c*]furo[3′,2′:4,5]furo[2,3-*h*][*l*]benzopyran-1,11-dione; (6aR-*cis*)-2,3,6a,9a-tetrahydro-4-methoxycyclopenta[*c*]furo[3′,2′:4,5]furo[2,3-*h*][*l*]benzopyran-1,11-dione

Aflatoxin B_2

Chem. Abstr. Services Reg. No.: 7220-81-7
Chem. Abstr. Name: (6aR,9aS)-2,3,6a,8,9,9a-Hexahydro-4-methoxycyclopenta[*c*]-furo[3′,2′:4,5]furo[2,3-*h*][*l*]benzopyran-1,11-dione (9CI)
Synonyms: Dihydroaflatoxin B_1; 2,3,6aα,8,9,9aα-hexahydro-4-methoxycyclopenta-[*c*]furo[3′,2′:4,5]furo[2,3-*h*][*l*]benzopyran-1,11-dione; (6aR-*cis*)-2,3,6a,8,9,9a-hexahydro-4-methoxycyclopenta[*c*]furo[3′,2′:4,5]furo[2,3-*h*][*l*]benzopyran-1,11-dione

Figure 1. Structures of naturally occurring aflatoxins

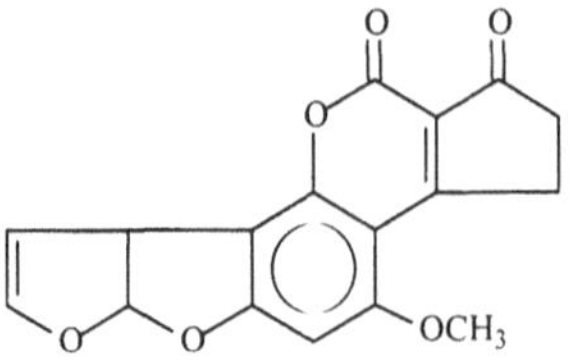

B_1: $C_{17}H_{12}O$ Mol. wt: 312.3

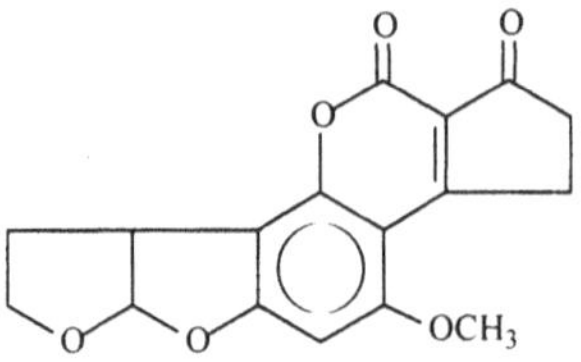

B_2: $C_{17}H_{14}O_6$ Mol. wt: 314.3

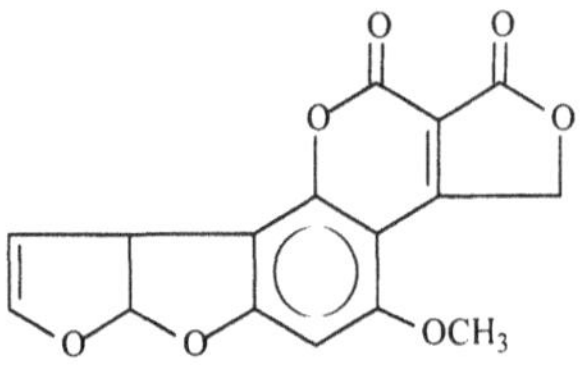

G_1: $C_{17}H_{12}O_7$ Mol. wt: 328.3

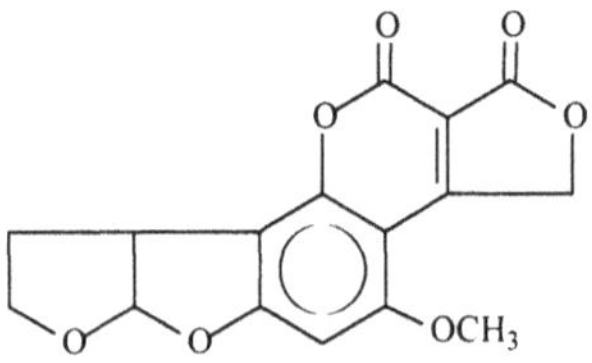

G_2: $C_{17}H_{14}O_7$ Mol. wt: 330.3

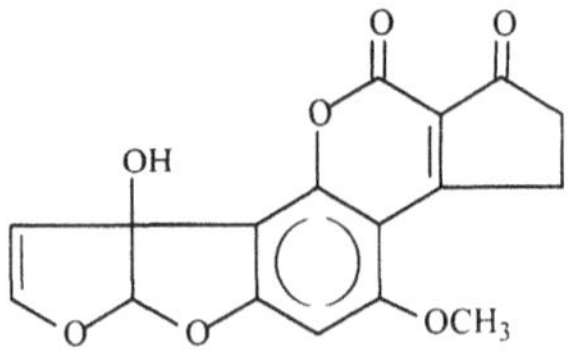

M_1: $C_{17}H_{12}O_7$ Mol. wt: 328.3

Aflatoxin G_1

Chem. Abstr. Services Reg. No.: 1165-39-5
Deleted CAS No.: 1385-95-1
Chem. Abstr. Name: (7aR,10aS)-3,4,7a,10a-Tetrahydro-5-methoxy-1*H*,12*H*-furo-[3′,2′:4,5]furo[2,3-*h*]pyrano[3,4-*c*][*l*]benzopyran-1,12-dione (9CI)
Synonym: 3,4,7aα,10aα-Tetrahydro-5-methoxy-1*H*,12*H*-furo[3′,2′:4,5]furo[2,3-*h*]-pyrano-[3,4-*c*][*l*]benzopyran-1,12-dione; (7aR-*cis*)-3,4,7a,10a-tetrahydro-5-methoxy-1*H*,12*H*-furo[3′,2′:4,5]furo[2,3-*h*]pyrano[3,4-*c*][*l*]benzopyran-1,12-dione

Aflatoxin G_2

Chem. Abstr. Services Reg. No.: 7241-98-7
Chem. Abstr. Name: (7aR,10aS)-3,4,7a,9,10,10a-Hexahydro-5-methoxy-1*H*,12*H*-furo[3′,2′:4,5]furo[2,3-*h*]pyrano[3,4-*c*][*l*]benzopyran-1,12-dione (9CI)
Synonyms: Dihydroaflatoxin G_1; 3,4,7aα,9,10,10aα-hexahydro-5-methoxy-1*H*,12*H*-furo[3′,2′:4,5]furo[2,3-*h*]pyrano[3,4-*c*][*l*]benzopyran-1,12-dione; (7aR-*cis*)-3,4,7a,9,10,10a-hexahydro-5-methoxy-1*H*,12*H*-furo[3′,2′:4,5]-furo[2,3-*h*]pyrano[3,4-*c*][*l*]-benzopyran-1,12-dione

Aflatoxin M_1

Chem. Abstr. Services Reg. No.: 6795-23-9
Chem. Abstr. Name: (6aR,9aR)-2,3,6a,9a-Tetrahydro-9a-hydroxy-4-methoxycyclopenta[*c*]furo[3′,2′:4,5]furo[2,3-*h*][*l*]benzopyran-1,11-dione (9CI)
Synonym: 4-Hydroxyaflatoxin B_1; (6aR-*cis*)-2,3,6a,9a-tetrahydro-9a-hydroxy-4-methoxycyclopenta[*c*]furo[3′,2′:4,5]furo[2,3-*h*][*l*]benzopyran-1,11-dione

1.1.2 *Chemical and physical properties of aflatoxins* (from Castegnaro *et al.*, 1980, 1991; O'Neil *et al.*, 2001, unless otherwise stated)

(*a*) *Description*: Colourless to pale-yellow crystals. Intensely fluorescent in ultraviolet light, emitting blue (aflatoxins B_1 and B_2) or green (aflatoxin G_1) and green–blue (aflatoxin G_2) fluorescence, from which the designations B and G were derived, or blue–violet fluorescence (aflatoxin M_1)
(*b*) *Melting-points*: see Table 1.
(*c*) *Absorption spectroscopy*: see Table 1.
(*d*) *Solubility*: Very slightly soluble in water (10–30 μg/mL); insoluble in non-polar solvents; freely soluble in moderately polar organic solvents (e.g., chloroform and methanol) and especially in dimethyl sulfoxide (Cole & Cox, 1981)
(*e*) *Stability*: Unstable to ultraviolet light in the presence of oxygen, to extremes of pH (< 3, > 10) and to oxidizing agents
(*f*) *Reactivity*: The lactone ring is susceptible to alkaline hydrolysis. Aflatoxins are also degraded by reaction with ammonia or sodium hypochlorite.

1.1.3 *Analysis*

Methods for determining aflatoxins in agricultural commodities and food products have been verified by AOAC International (IARC, 1993; AOAC International, 2000; Stroka *et al.*, 2001) and by various international committees (ISO, 1998; EN, 1999a,b; ISO, 2001), as shown in Table 2. The methods have greatly improved in recent years with the commercial availability of multifunctional columns and immunoaffinity

Table 1. Melting-points and ultraviolet absorption of aflatoxins

Aflatoxin	Melting-point (°C)	Ultraviolet absorption (ethanol)	
		λ_{max} (nm)	ε (L mol^{-1} cm^{-1})
B_1	268–269 (decomposition)	223	25 600
	(crystals from chloroform)	265	13 400
		362	21 800
B_2	286–289 (decomposition)	265	11 700
	(crystals from chloroform-pentane)	363	23 400
G_1	244–246 (decomposition)	243	11 500
	(crystals from chloroform-methane)	257	9 900
		264	10 000
		362	16 100
G_2	237–240 (decomposition)	265	9 700
	(crystals from ethyl acetate)	363	21 000
M_1	299 (decomposition)	226	23 100
	(crystals from methanol)	265	11 600
		357	19 000

From O'Neil *et al.* (2001)

columns, which are simple and rapid to use, and with reduction in the use of toxic solvents for extraction and clean-up.

Quality assurance for the analysis of aflatoxins B_1, B_2, G_1, G_2 and M_1 in foods is available for laboratories through the American Association of Cereal Chemists' Check Sample Program and the Analytical Proficiency Testing Programme administered in the USA and the United Kingdom, respectively.

As contamination may not occur in a homogeneous way throughout a sample of maize or peanuts[1], good sampling and sample preparation procedures must be used to obtain accurate quantitative results. Summaries of the procedures, variability and application of sampling plans for mycotoxins are included in Section 1.5 and in the European Commission directive 98/53/CE (European Commission, 1998a).

A number of approaches have been used to analyse aflatoxins and their metabolites in human tissues and body fluids. These include immunoaffinity purification, immunoassay (Wild *et al.*, 1987), high-performance liquid chromatography (HPLC) with fluorescence or ultraviolet detection and synchronous fluorescence spectroscopy (Groopman & Sabbioni, 1991). Molecular biomarkers, such as urinary markers, metabolites in milk and parent compounds in blood, are used for determining exposure to aflatoxins (Groopman, 1993).

[1] Maize (corn) and peanuts (groundnuts) will be used throughout this volume for corn and groundnuts.

Table 2. Analytical methods validated by AOAC International and the EU

Method no.	Aflatoxin	Food	Method[a]	Detection limit (µg/kg)
AOAC				
975.36	All[b]	Food and feeds (screening)	MC	5–15
979.18	All	Maize and peanuts (screening)	MC	10
990.31	All	Maize and peanuts (Aflatest screening)	IC	10
994.08	B_1, B_2, G_1, G_2	Maize, almond, Brazil nuts, peanuts, pistachio nuts (Mycosep)	MFC/HPLC	5
999.07	All, B_1	Peanut butter, pistachio paste, fig paste, paprika powder	IC/HPLC	NG
989.06	B_1	Cottonseed products and mixed feed (screening)	ELISA	15
990.32	B_1	Maize and roasted peanuts (screening)	ELISA	20
2000.16	B_1	Baby foods (infant formula)	IC/HPLC	0.1
990.34	B_1, B_2, G_1	Maize, cottonseed, peanuts, peanut butter (screening)	ELISA	20–30
991.45	B_1, B_2, G_1, G_2	Peanut butter	ELISA	9
993.16	B_1, B_2, G_1	Maize	ELISA	20
998.03	B_1, B_2, G_1, G_2	Peanuts	TLC	NG
968.22	B_1, B_2, G_1, G_2	Peanuts and peanut products	TLC	5
970.45	B_1, B_2, G_1, G_2	Peanuts and peanut products	TLC	10
971.23	B_1, B_2, G_1, G_2	Cocoa beans	TLC	10
971.24	B_1, B_2, G_1, G_2	Coconut, copra and copra meal	TLC	50
972.26	B_1, B_2, G_1, G_2	Maize	TLC	5
980.20	B_1, B_2, G_1, G_2	Cottonseed products and mixed fed (screening)	TLC, HPLC	10, 5
974.16	B_1, B_2, G_1, G_2	Pistachio nuts	TLC	15
972.27	B_1, B_2, G_1, G_2	Soya bean	TLC	10
990.33	B_1, B_2, G_1, G_2	Maize and peanut butter	HPLC	5
993.17	B_1, B_2, G_1, G_2	Maize and peanuts	TLC	1.5–10
991.31	B_1, B_2, G_1, G_2	Maize, peanuts, peanut butter (Aflatest)	IC/HPLC	10
970.46	B_1, B_2, G_1, G_2	Green coffee	TLC	25
978.15	B_1	Eggs	TLC	0.1
982.24	B_1 and M_1	Liver	TLC	0.1
974.17	M_1	Dairy products	TLC	0.1
980.21	M_1	Milk and cheese	TLC	0.1
986.16	M_1 and M_2	Fluid milk	HPLC	0.1

Table 2 (contd)

Method no.	Aflatoxin	Food	Method[a]	Detection limit (µg/kg)
EU				
NF EN 12955	All, B_1	Cereals, nuts and derived products	IC/HPLC	8 (all)
NF EN ISO 14501	M_1	Milk and milk powder	IC/HPLC	0.08 in powder 0.008 µg/L liquid
ISO 14718	B_1	Mixed feeding stuff	HPLC	1
ISO 6651	B_1	Animal feeding stuff	TLC/fluo-rescence	4

From IARC (1993); ISO (1998); EN (1999a,b); AOAC International (2000); ISO (2001); Stroka *et al.* (2001)
[a] MC, minicolumn; IC, immunoaffinity column, ELISA, enzyme-linked immunosorbent assay; TLC, thin-layer chromatography; HPLC, high-performance liquid chromatography; MFC, multifunctional column
[b] All, sum or total aflatoxins
AOAC, Association of Analytical Communities; EU, European Union; NG, not given

1.2 Sources, production and use

1.2.1 *Fungi producing aflatoxins*

Aflatoxins are produced by the common fungi *Aspergillus flavus* and the closely related species *A. parasiticus*. These are well defined species: *A. flavus* produces only B aflatoxins and sometimes the mycotoxin cyclopiazonic acid (CPA), while *A. parasiticus* produces both B and G aflatoxins, but never CPA (Schroeder & Boller, 1973; Dorner *et al.*, 1984; Klich & Pitt, 1988; Pitt, 1993) (see Annex).

This simple situation, of just two aflatoxigenic species, has been complicated by more recent taxonomic findings. Kurtzman *et al.* (1987) described *A. nomius*, a species closely related to *A. flavus* but which produces small bullet-shaped sclerotia, as distinct from the large spherical sclerotia produced by many *A. flavus* isolates. This species is also distinguished from *A. flavus* by the production of both B and G aflatoxins (Saito *et al.*, 1989; Pitt, 1993). A second new species, closely related to *A. nomius*, was described by Peterson *et al.* (2001) and named *A. bombycis*. These two species were distinguished from each other by differences in DNA, and also by differences in growth rates at 37 °C. Like *A. nomius*, *A. bombycis* produces both B and G aflatoxins.

The species *A. ochraceoroseus* described by Bartoli and Maggi (1978) was recently shown to be another aflatoxin producer. It also produces sterigmatocystin (Frisvad, 1997; Klich *et al.*, 2000). Saito and Tsuruta (1993) described *A. flavus* var. *parvisclerotigenus*, which produces small spherical sclerotia, but one isolate (NRRL 3251) (Agricultural Research Service Culture Collection) reported to be representative of the new variety was considered by both Christensen (1981) and Pitt (1993) to be a typical *A. flavus*. This same isolate was reported by Stubblefield *et al.* (1970) to produce B but not G aflatoxins, in line with those assessments. Moreover, Geiser *et al.* (2000) showed that the production of small versus large sclerotia does not have taxonomic significance within *A. flavus*.

Two aflatoxin-producing isolates from Japan, originally classified as aberrant *A. tamarii* (Goto *et al.*, 1996), were recently described as *A. pseudotamarii*. Like *A. flavus*, this species produces B aflatoxins and CPA, but differs from *A. flavus* by the production of orange-brown conidia (Ito *et al.*, 2001).

In studying population genetics of *A. flavus*, Geiser *et al.* (1998) showed that *A. flavus* from an Australian peanut field comprised two distinct subgroups, which they termed Group I and Group II, and suggested that Group II differed from Group I (*A. flavus sensu stricto*) sufficiently to be raised to species level. Further studies by Geiser *et al.* (2000) and independent observations have confirmed that *A. flavus* Group II comprises a distinct species, which will be described as '*Aspergillus australis*'. Unlike any other known species, *A. australis* produces both B and G aflatoxins and also CPA. It appears to occur almost exclusively in the southern hemisphere, where it has been found in Argentina, Australia, Indonesia and South Africa.

The current status of taxonomic information and mycotoxin production by species that produce aflatoxins is summarized in Table 3. This information, complex though it is, should not be allowed to obscure the importance of the older species. The evidence

Table 3. ***Aspergillus*** **species capable of producing aflatoxins**

Species	Mycotoxins produced			Major sources	Geographical distribution
	AFB	AFG	CPA		
A. flavus	+	–	–	All kinds of foods	Ubiquitous in warmer latitudes
A. parasiticus	+	+	–	Peanuts	Specific areas
A. nomius	+	+	–	Bees	USA, Thailand
A. pseudotamarii	+	–	+	Soil	Japan
A. bombycis	+	+	–	Silkworm frass	Japan, Indonesia
A. ochraceoroseus	+	–	–	Soil	Africa
A. australis	+	+	+	Soil, peanuts	Southern hemisphere

AFB, B aflatoxins; AFG, G aflatoxins; CPA, cyclopiazonic acid

indicates that *A. flavus* and *A. parasiticus* are responsible for the overwhelming proportion of aflatoxins found in foodstuffs throughout the world. Of the other species, only *A. australis*, which appears to be widespread in the southern hemisphere and is common in Australian peanut soils, may also be an important source of aflatoxins in a few countries.

1.2.2 *Production and reduction*

Apart from natural formation, aflatoxins are produced only in small quantities for research purposes, by *A. flavus* or *A. parasiticus* fermentations on solid substrates or media in the laboratory. Aflatoxins are extracted by solvents and purified by chromatography. Total annual production is less than 100 g (IARC, 1993).

Aflatoxins occurring naturally in foods and feeds may be reduced by a variety of procedures. Improved farm management practices, more rapid drying and controlled storage are now defined within GAP (Good Agricultural Practice) or HACCP (Hazard Analysis: Critical Control Point) (FAO, 1995). By segregation of contaminated lots after aflatoxin analyses and by sorting out contaminated nuts or grains by electronic sorters, contaminated lots of peanuts or maize can be cleaned up to produce food-grade products. Decontamination by ammoniation or other chemical procedures can be used for rendering highly contaminated commodities suitable as animal feeds. More detailed information on these topics is given in the Annex to this Monograph.

1.2.3 *Uses*

Aflatoxins are not used commercially, only for research.

1.3 Formation and occurrence

1.3.1 *Prevalence of toxigenic species in foods*

Because of the importance of aflatoxins, *A. flavus* has become the most widely reported foodborne fungus — even with the proviso that *A. parasiticus* is sometimes not differentiated from *A. flavus* in general mycological studies. *A. flavus* is especially abundant in the tropics. Levels of *A. flavus* in warm temperate climates such as in the USA and Australia are generally much lower, while the occurrence of *A. flavus* is uncommon in cool temperate climates except in foods and feeds imported from tropical countries (see Section 1.3.3).

The major hosts of *A. flavus* among food and feed commodities are peanuts, maize and cottonseed. In addition, various spices sometimes contain aflatoxins, while tree nuts are contaminated less frequently. Low levels may be found in a wide range of other foods (Pitt *et al.*, 1993, 1994; Pitt & Hocking, 1997)

It seems probable that although *A. parasiticus* has the same geographical range as *A. flavus*, it is less widely distributed. In particular, it has been found only rarely in south-east Asia. The food-related hosts of *A. parasiticus* are similar to those of *A. flavus*, except that *A. parasiticus* is very uncommon in maize (Pitt *et al.*, 1993, 1994).

1.3.2 *Factors affecting formation of aflatoxins in foods*

A fundamental distinction must be made between aflatoxin formation in crops before (or immediately after) harvest, and that occurring in stored commodities or foods. Peanuts, maize and cottonseed are associated with *A. flavus*, and in the case of peanuts, also with *A. parasiticus*, so that invasion of plants and developing seed or nut may occur before harvest. This close association results in the potential for high levels of aflatoxins in these commodities and is the reason for the continuing difficulty in eliminating aflatoxins from these products.

In contrast, *A. flavus* lacks this affinity for other crops, so it is not normally present at harvest. Prevention of the formation of aflatoxins therefore relies mainly on avoidance of contamination after harvest, using rapid drying and good storage practice (see Annex).

1.3.3 *Occurrence*

Aflatoxins have been found in a variety of agricultural commodities, but the most pronounced contamination has been encountered in maize, peanuts, cottonseed and tree nuts. Aflatoxins were first identified in 1961 in animal feed responsible for the deaths of 100 000 turkeys in the United Kingdom (Sargeant *et al.*, 1961). An extensive review of the levels of aflatoxins encountered in commodities in North America, South America, Europe, Asia and Africa was included in the previous IARC monograph (IARC, 1993).

A summary of data published since the previous monograph on the worldwide occurrence of aflatoxins is given in Table 4. From the point of view of dietary intake,

Table 4. Occurrence of aflatoxin B_1 in Latin America and Asia

Product	Region/Country	Detected/total no. of samples	Aflatoxin B_1 (μg/kg)	Compiled by the Working Group from the following references
	Latin America			
Maize	Argentina, Brazil, Costa Rica, Mexico, Venezuela	5086/15 555	0.2–560	Viquez *et al.* (1994); Torres Espinosa *et al.* (1995); Juan-López *et al.* (1995); Resnik *et al.* (1996); González *et al.* (1999) (none found in Argentina); Medina-Martínez & Martínez (2000); Ono *et al.* (2001); Vargas *et al.* (2001)
Maize foods	Brazil	30/322	2.80–1323[a]	Midio *et al.* (2001)
Peanuts and products	Brazil	41/80	Max. 1789	Freitas & Brigido (1998)
Soya bean	Argentina	9/94	< 1–11	Pinto *et al.* (1991)
Sorghum	Brazil	18/140	7–33 (mean)	da Silva *et al.* (2000)
Poultry feed	Argentina	41/300	17–197	Dalcero *et al.* (1997)
	Asia			
Maize and flour	China, India, Indonesia, Philippines, Thailand	1263/2541	0.11–4030	Yamashita *et al.* (1995); Yoshizawa *et al.* (1996); Zhang *et al.* (1996); Bhat *et al.* (1997); Ueno *et al.* (1997); Shetty & Bhat (1997); Ali *et al.* (1998); Vasanthi & Bhat (1998); Lipigorngoson *et al.* (1999); Li *et al.* (2001)
Maize products	Malaysia, Philippines	77/404	1–117	Ali *et al.* (1999); Arim (2000)
Maize feed	Viet Nam	27/32	8.6–96.0	Wang *et al.* (1995)
Peanuts	China, India, Japan, Thailand	1456/7796	0.2–833	Bhat *et al.* (1996); Zhang *et al.* (1996); Hirano *et al.* (1998); Lipigorngoson *et al.* (1999); Okano *et al.* (2002)
Peanut products, oil, butter	China, Malaysia, Philippines	235/594	1–244	Zhang *et al.* (1996); Ali *et al.* (1999); Arim (2000)

Table 4 (contd)

Product	Region/Country	Detected/total no. of samples	Aflatoxin B_1 (μg/kg)	Compiled by the Working Group from the following references
Peanut foods	India, Malaysia, Philippines	177/957	1–1500	Rati & Shantha (1994); Ali (2000); Arim (2000)
Nuts and products	Japan	23/673	0.3–128	Tabata *et al.* (1998)
Rice and wheat	China	0/92		Zhang *et al.* (1996)
Sorghum	India, Thailand	56/94	0.10–30.3	Shetty & Bhat (1997); Suprasert & Chulamorakot (1999)
Commercial foods	Japan, Malaysia	154/1053	0.1– > 50[a]	Taguchi *et al.* (1995); Tabata *et al.* (1998); Ali (2000)
Beer	Japan	13/116	0.0005–0.0831	Nakajima *et al.* (1999)

[a] Total aflatoxins

aflatoxins in foods used as staples such as maize assume considerable significance. Aflatoxins are a far greater problem in the tropics than in temperate zones of the world. However, because of the movement of agricultural commodities around the globe, no region of the world is free of aflatoxins.

With regard to aflatoxin contamination in foods imported into Japan, relatively low incidences and low levels of aflatoxins have been found in various commodities. Aflatoxin inspection of imported peanuts (1999–2000) indicated that 355 (6.9%) of 5108 samples were contaminated with aflatoxin B_1 at levels ranging from 0.2 to 760 μg/kg, and 145 samples (2.8%) contained over 10 μg/kg, the maximum permitted level in Japan (Okano *et al.*, 2002). In commercial nuts and nut products in markets, aflatoxin B_1 was found in 23 (3.4%) of 673 samples at levels of 0.3–128 μg/kg. Imported spices (white and red pepper, paprika and nutmeg) contained aflatoxin B_1 in 106 (19.4%) of 546 samples at levels of 0.2–27.7 μg/kg (Tabata *et al.*, 1998).

Information on the occurrence of aflatoxins in imported spices in the European Union (EU) is given in Table 5. Among the total of 3098 spice samples including nutmeg, pepper, chilli and paprika, 183 samples (5.9%) contained more than 10 μg/kg aflatoxins (European Commission, 1997).

Table 5. Aflatoxin B_1 in spices imported into the European Union

Product	Detected/ total samples	Aflatoxin B_1 (μg/kg)	
		> 2	> 10
Nutmeg	333/546	25%	8%
Pepper	282/828	7%	1%
Chilli and chilli powder	148/509	28%	9%
Paprika powder	195/1215	21%	7%
Total spices	958/3098	> 1 μg/kg	
	591/3098	> 2 μg/kg	
	183/3098	> 10 μg/kg	

From European Commission (1997)

In the United Kingdom, seven of 139 maize samples (5.0%) imported in 1998–99 contained total aflatoxins in the range of 4.9–29.1 μg/kg (3.7–16.4 μg/kg aflatoxin B_1) (MAFF, 1999).

The French Direction Générale de la Concurrence, de la Consommation et de la Répression des Fraudes (DGCCRF) surveyed 635 imported foods between 1992 and 1996, of which 227 (35.7%) had aflatoxin B_1 levels above 0.05 μg/kg. The highest levels were found in spices (up to 75 μg/kg) and dried fruits (up to 77 μg/kg) (Castegnaro & Pfohl-Leszkowicz, 1999).

Dietary intake of aflatoxin B_1 was monitored for one week in a number of households in a Chinese village. Aflatoxin B_1 was detected in 76.7% (23/30) of ground maize samples (range, 0.4–128.1 μg/kg), 66.7% (20/30) of cooking peanut oil samples (range, 0.1–52.5 μg/L) and 23.3% (7/30) of rice samples (range, 0.3–20 μg/kg) (Wang *et al.*, 2001).

(*a*) *Co-occurrence of aflatoxins and fumonisins*

Co-occurrence of aflatoxin B_1 and fumonisin B_1 in maize and sorghum from Latin America and Asia is shown in Table 6. Maize harvested in the tropical and subtropical areas of the world with hot and humid climates is the major commodity contaminated with the two mycotoxins.

Two studies were carried out on cross-contamination with aflatoxins and fumonisins in staple maize samples from two high-risk areas for human hepatocellular carcinoma in China; Haimen, Jiangsu Province, Shandong (Ueno *et al.*, 1997) and Chongzuo County, Guangxi in 1998 (Li *et al.*, 2001). Three-year (1993–95) surveys demonstrated that maize harvested in Haimen was highly contaminated with aflatoxins and fumonisins and that the levels of fumonisins were 10–50-fold higher than in a low-risk area (Ueno *et al.*, 1997). Staple maize samples from Guangxi were co-contaminated (14/20) with high levels of aflatoxin B_1 (11–2496 μg/kg) and fumonision B_1 (74–780 μg/kg), and the probable daily intake was estimated to be 3.68 μg/kg bw of aflatoxin B_1 and 3.02 μg/kg bw of fumonisin B_1 (Li *et al.*, 2001).

In India, rain-affected maize samples from rural households and retail shops had higher levels of contamination with fumonisins (250–6470 μg/kg) than normal samples (50–240 μg/kg) as well as with aflatoxin B_1 (250–25 600 versus 5–87 μg/kg), which co-occurred with fumonisins. The level of fumonisin B_1 was also higher in sorghum affected by rain (140–7800 μg/kg versus 70–360 μg/kg). No correlation was observed between levels of the two toxins in individual samples, indicating that the toxins are formed independently (Vasanthi & Bhat, 1998).

(*b*) *Occurrence of aflatoxin M_1*

Aflatoxin M_1 is a metabolite of aflatoxin B_1 that can occur in milk and milk products from animals consuming feed contaminated with B aflatoxins (Applebaum *et al.*, 1982). Data on occurrence of aflatoxin M_1 in milk were summarized earlier (IARC, 1993) and data reported subsequently are included in Table 7.

Galvano *et al.* (1996) reviewed the worldwide occurrence of aflatoxin M_1 in milk and milk products.

1.3.4 *Human biological fluids*

Covalent binding of aflatoxin to albumin in peripheral blood has been measured in a number of studies (Montesano *et al.*, 1997). The levels of these adducts are assumed to reflect exposure to aflatoxin over the previous 2–3 months, based on the half-life of

Table 6. Co-occurrence of aflatoxins and fumonisins in Asia and Latin America

Product	Region/Country	No. detected/total no. of samples			Range (μg/kg)		Compiled by the Working Group from the following references
		AFB_1	FB_1	$AFB_1 + FB_1$	AFB_1	FB_1	
	Asia						
Maize	China, India, Indonesia, Philippines, Thailand, Viet Nam	199/234	173/234	148/234	0.11–4030	10–18 800	Yamashita *et al.* (1995); Wang *et al.* (1995); Shetty & Bhat (1997); Ueno *et al.* (1997); Ali *et al.* (1998); Vasanthi & Bhat (1998); Li *et al.* (2001)
Maize flour	China	26/27	14/27	13/27	11–68	80–3190	Ueno *et al.* (1997)
Sorghum	India	2/44	9/44	2/44	0.18–30.3	150–500	Vasanthi & Bhat (1998)
	Latin America						
Maize	Brazil, Venezuela	88/251	233/251	88/251	0.2–129	25–15 050	Medina-Martínez & Martínez (2000); Vargas *et al.* (2001)
	Brazil (total aflatoxins and total fumonisins)	17/150	147/150	17/150	38–460	96–22 000	Ono *et al.* (2001)

AFB_1, aflatoxin B_1; FB_1, fumonisin B_1

Table 7. Occurrence of aflatoxin M_1 in milk

Country	No. positive/ no. of samples	Range of aflatoxin M_1 concentrations (μg/kg)	Reference
Brazil	4/52	0.05–0.37	de Sylos *et al.* (1996); JECFA (2001)
Cuba	22/85	> 0.5	Margolles *et al.* (1992); Galvano *et al.* (1996)
Cyprus	11/112	0.01–0.04	Ioannou-Kakouri *et al.* (1999)
France	5284/5489 200/5489 5/5489	< 0.05 0.05–0.5 > 0.5	Dragacci & Frémy (1993); Castegnaro & Pfohl-Leszkowicz (1999)
	0/562	–	Castegnaro & Pfohl-Leszkowicz (1999)
Greece	3/81	0.05–0.18	Markaki & Melissari (1997)
India	89/504	0.1–3.5	Rajan *et al.* (1995)
Italy	122/214	0.003–0.101	Bagni *et al.* (1993); Galvano *et al.* (1996)
Japan	0/37	–	Tabata *et al.* (1993); Galvano *et al.* (1996)
Korea (Republic of)	50/134	0.05–0.28	Kim *et al.* (2000); JECFA (2001)
Spain	29/155	0.015–0.04	Jalon *et al.* (1994); Galvano *et al.* (1996)
Thailand	58/310	0.5–6.6	Saitanu (1997); JECFA (2001)
Europe	314/7573	≤ 0.05	European Commission (1998a); JECFA (2001)

albumin. Experimental data have also shown that this biomarker reflects the formation of the reactive metabolite of aflatoxin B_1 and the level of DNA damage occurring in the livers of rats treated with aflatoxin B_1. Figure 2 shows data from a number of populations, with adduct levels expressed as picograms of aflatoxin B_1–lysine equivalents per milligram of serum albumin. Other measurements of aflatoxin–DNA and aflatoxin–protein adducts in humans are discussed in Sections 4.4 and 4.5.

Maxwell (1998) has discussed the presence of aflatoxins in human body fluids and tissues in relation to child health in the tropics. In Ghana, Kenya, Nigeria and Sierra

Figure 2. Level and prevalence of aflatoxin exposure

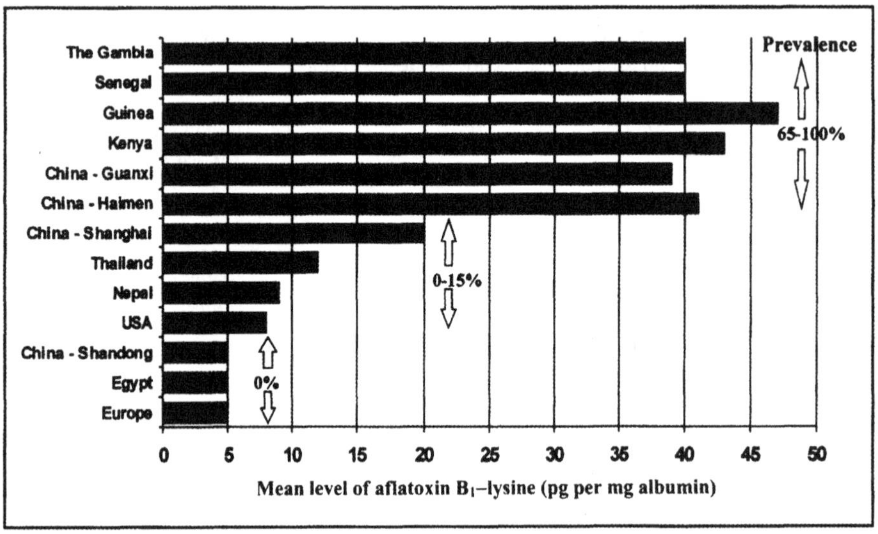

From Montesano *et al.* (1997)
Data are expressed as pg aflatoxin B_1–lysine equivalents/mg serum albumin and represent the mean levels in samples with levels above the detection limit of the enzyme-linked immunosorbent assay used (5 pg/mg). Shandong (China), Europe (France and Poland) and Egypt are represented at the detection limit, but no samples were above this level of adduct (0% prevalence). The number of sera analysed varies per country. Original data are from Wild *et al.* (1993a) and Yu (1995).

Leone, 25% of cord blood samples contained aflatoxins, primarily M_1 and M_2, as well as others in variable amounts (range: 1 ng aflatoxin M_1/L to 64 973 ng aflatoxin B_1/L).

Of 35 cord serum samples from Thailand, 48% contained aflatoxins at concentrations of 0.064–13.6 nmol/mL (mean, 3.1 nmol/mL). By comparison, only two of 35 maternal sera obtained immediately after birth contained aflatoxin (mean, 0.62 nmol/mL). These results show that transplacental transfer and concentration of aflatoxin by the feto-placental unit occur (Denning *et al.*, 1990).

Analyses of breast milk in Ghana, Nigeria, Sierra Leone and Sudan showed primarily aflatoxin M_1, aflatoxin M_2 and aflatoxicol. Aflatoxin exposure pre- or post-natally at levels ≥ 100 ng/L was very often associated with illness in the child (Maxwell, 1998).

Exposure of infants to aflatoxin M_1 from mothers' breast milk in the United Arab Emirates has been measured by Saad *et al.* (1995). Among 445 donors of breast milk, 99.5% of samples contained aflatoxin M_1 at concentrations ranging from 2–3 μg/L. The mothers were of a wide range of nationalities, ages and health status; no correlation was observed between these factors and aflatoxin M_1 content of the milk.

El-Nazami *et al.* (1995) measured levels of aflatoxin M_1 in breast milk in 73 women from Victoria, Australia and 11 women from Thailand. Aflatoxin M_1 was detected in 11 samples from Victoria (median concentration, 0.071 μg/L) and five samples from

Thailand (median concentration, 0.664 μg/L). Levels were significantly higher in the Thai samples.

1.3.5 *Occupational exposure to aflatoxins*

Kussak *et al.* (1995) demonstrated the presence of aflatoxins in airborne dust from feed factories.

During unloading of ships, aflatoxin B_1 has been found in bilge at levels as high as 300 ng/m^3 (Lafontaine *et al.*, 1994).

Autrup *et al.* (1993) assessed the exposure to aflatoxin B_1 of workers in animal feed processing plants in Denmark. The workers served as their own controls; blood samples were taken after their return from vacation and after four weeks of work. Binding of aflatoxin B_1 to serum albumin was measured. Seven of 45 samples were positive for aflatoxin B_1 with an average daily intake of 64 ng/kg bw aflatoxin B_1. The exposed workers had been unloading cargoes contaminated with aflatoxin B_1 or working at places where the dust contained detectable amounts of aflatoxin B_1. This level of exposure could partly explain the increased risk for liver cancer in workers in the animal feed processing industry.

Ghosh *et al.* (1997) assayed airborne aflatoxin in rice- and maize-processing plants in India using an indirect, competitive enzyme-linked immunosorbent assay. Levels of airborne aflatoxin were always higher in the respirable dust samples (< 7 μm) than in total dust samples. Concentrations of total airborne aflatoxin in the respirable dusts in the rice mill were 26 pg/m^3 and 19 pg/m^3 in the workplace and the storage area, respectively. Airborne aflatoxin was not detected in control sites of either of the grain-processing plants or in total dust samples obtained from the maize plant. At three sites in the maize-processing plant — the elevator (18 pg/m^3), the loading/unloading area (800 pg/m^3) and the oil mill (816 pg/m^3) — airborne aflatoxin was present only in the respirable dust samples.

In a study of factories in Thailand (Nuntharatanapong *et al.*, 2001), samples of airborne dust generated during handling of animal feed were analysed in order to assess worker exposure to aflatoxins. The average aflatoxin level in the control air samples was 0.99 ng/m^3. Higher levels of aflatoxins were found in the air samples taken by samplers carried by five workers adding hydrated sodium calcium aluminosilicate to animal feed (1.55 ng/m^3) and five workers adding glucomannan, a viscous polysaccharide, to animal feed (6.25 ng/m^3). The exposed workers had altered lactate dehydrogenase isoenzyme activity and tumour necrosis factor levels in plasma. These changes may be associated with inhalation of mycotoxins and other contaminants in foodstuffs.

1.4 International exposure estimates

1.4.1 *JECFA (1998)*

In 1997, the Joint Food and Agriculture Organization/World Health Organization (FAO/WHO) Expert Committee on Food Additives (JECFA) performed an exposure estimate for aflatoxins (JECFA, 1998). This report summarized the results of monitoring and available national estimates of intake of aflatoxins in order to provide a framework for estimating increments in intake of aflatoxins. Estimates were based on the results of available monitoring data. Total intake of aflatoxins based on the GEMS (Global Environment Monitoring System)/Food Regional Diets (WHO, 1998) was used to evaluate the impact of four different scenarios: no regulatory limit and limits set at 20, 15 and 10 μg/kg aflatoxins. The evaluation was carried out for total aflatoxins and aflatoxin B_1 in maize and peanuts. The data submitted were not considered to be representative because sampling largely focused on those lots that were most likely to contain the highest levels of aflatoxin. However, JECFA considered the analysis to provide useful qualitative comparisons between regulatory options.

JECFA received data for this analysis from at least one country on every continent. The submitters generally considered the data to be biased towards the upper end of intake. In some cases, JECFA required individual data points in order to generate distributions and to evaluate the impact of imposing upper limits on aflatoxin in foodstuffs. Hence, data reported by the USA, China and Europe were used because the raw data were available. The reader is referred to the original report for a more complete discussion of the data.

JECFA used three pieces of information to estimate the potential intakes due to aflatoxin in imported crops: (*a*) levels of aflatoxin in imported crops; (*b*) the amounts of each imported crop consumed; and (*c*) the effect of any subsequent processing on aflatoxin levels. It then applied methods for combining these three factors to estimate intake.

(*a*) *Aflatoxin levels in foods: general*

The 1995 compendium, *Worldwide Regulations for Mycotoxins* (FAO, 1997) summarized reports from 90 countries. The data submitted by 33 countries for aflatoxin B_1 and 48 countries for total aflatoxins (B_1, B_2, G_1 and G_2) were used to estimate median levels of 4 and 8 μg/kg, respectively, in foodstuffs. The range of levels reported for aflatoxin B_1 was from 0 to 30 μg/kg and for total aflatoxins from 0 to 50 μg/kg. Seventeen countries provided information on aflatoxin M_1 in milk, with a median of 0.05 μg/kg and a range of 0–1 μg/kg.

Participants in the European Union Scientific Cooperation (EU SCOOP) assessment of aflatoxin (SCOOP, 1996) reviewed data submitted by member countries and by Norway. JECFA concluded that the results were unlikely to be representative and should not be used to estimate total aflatoxin intake for individual countries or for Europe. However, some insights were gained. SCOOP concluded that aflatoxins were found in a broader range of foods than had been previously assumed, but that most samples did not contain any detectable aflatoxin. Sampling methods were very important in estimating aflatoxin levels accurately. In addition, different methods of collecting food consumption data may have made a difference in estimating aflatoxin intakes.

(*b*) *National estimates of aflatoxin intake (from JECFA, 1998)*

(i) *Australia*

From Australian market basket surveys, intake was estimated for average and extreme consumers. The average diet was estimated to give an intake of 0.15 ng aflatoxin per kg body weight per day and the upper 95th percentile diet approximately twice that level. Children's diets were estimated to give an intake up to approximately 0.45 ng/kg bw per day for the 95th percentile two-year-old (National Food Authority, 1992, reported by JECFA, 1998).

(ii) *China*

A series of intake and market basket studies have been conducted in China since 1980 to estimate the aflatoxin B_1 intakes, which were reported to range from 0 to 91 μg/kg bw per day (Chen, 1997).

(iii) *European Union*

Nine countries provided estimates of aflatoxin intake to the EU SCOOP project. None of these estimates was considered to be representative and all were viewed only as indicators of intake of aflatoxin. These estimates ranged from 2 to 77 ng per person per day for aflatoxin B_1 and from 0.4 to 6 ng per person per day for aflatoxin M_1. JECFA noted that these levels should not be used as estimates of intake either for a particular country or for Europe (JECFA, 1998).

(iv) *USA*

The US Food and Drug Administration (FDA) estimated intakes using data from the National Compliance Program for maize, peanut and milk products using Monte Carlo simulation procedures and data from the 1980s. Results differed only slightly from those of a repeat analysis in 1992 (Henry *et al.*, 1997). The lifetime intake of total aflatoxin was 18 ng per person per day for consumers only; intake for the 90th percentile individuals was 40 ng per person per day. For aflatoxin M_1, mean intake was 44 ng per person per day and for the 90th percentile individual 87 ng per person per day. Many assumptions were made in these estimates which would tend to bias the results upward.

(v) *Zimbabwe*

The theoretical maximum intake of aflatoxin M_1 for a child's diet containing 150 g maize with 5 μg/kg aflatoxin B_1 and 30 g peanuts with 10 μg/kg aflatoxin B_1 was estimated to be 1.05 μg per day (JECFA, 1998).

(c) *Impact of establishing maximum limits on estimate of intake*

Data from the EU, China and the USA were used to assess the potential impact of successfully eliminating aflatoxin levels above 20 μg/kg versus 15 μg/kg versus 10 μg/kg versus no limit for maize and peanuts (JECFA, 1998). The reader is referred to the full report for the tables, which more fully describe these four scenarios. JECFA emphasized that the aflatoxin levels presented in this report were not considered to be representative of the food supply in any country or of the commodities moving in international trade. The lack of representative data severely limited the ability to make quantitative estimates of aflatoxin intake; in general, the results were considered to be biased upwards. The data did provide, as JECFA stated, sufficient information to evaluate the likely impact of limiting aflatoxin levels in foodstuffs. Of the scenarios considered, the greatest relative impact on estimated average aflatoxin levels was achieved by limiting aflatoxin contamination to less than 20 μg/kg, i.e., eliminating all samples above 20 μg/kg from the food supply. Only small incremental reductions could be achieved by limiting aflatoxin levels to no more than 15 or 10 μg/kg.

1.4.2 *JECFA 2001*

In February 2001, at the request of the Codex Committee on Food Additives and Contaminants (Codex Alimentarius, 2000), JECFA (2001) evaluated the human health risks associated with consumption of milk contaminated with aflatoxin M_1 at two maximum regulatory levels of 0.05 μg/kg and 0.5 μg/kg. This project involved estimating exposure to aflatoxin M_1 in consumers from countries all over the world consuming milk and milk products.

Data on aflatoxin M_1 contamination of milk and/or milk products were submitted from Argentina, Canada, the Dubai Municipality of the United Arab Emirates, the EU, Indonesia, Korea (Republic of), Norway, the Philippines, Thailand and the USA. The majority of samples were submitted from the USA and the European Commission; very few were from south-east Asia and none from Africa. Of 6181 samples submitted from the USA and collected in south-western and southern states between 1998 and 2000, 1392 had aflatoxin M_1 levels between 0.05 and 0.5 μg/kg, and 113 had levels > 0.5 μg/kg. However, no samples were available from the north-eastern USA, where aflatoxin rarely occurs; most samples came from south-eastern parts where aflatoxin contamination often occurs in maize and peanuts consumed by dairy cows.

The data submitted by the European Commission represented 7573 samples from Austria, Belgium, Finland, France, Germany, Ireland, the Netherlands, Portugal, Sweden and the United Kingdom collected in 1999; 96% of the samples had aflatoxin M_1 levels below the limit of detection (which varied between countries: 0.001–0.03 μg/kg). The concentration in samples where aflatoxin M_1 was detected were ≤ 0.05 μg/kg (JECFA, 2001).

(*a*) *Estimates of aflatoxin M_1 intake using GEMS (Global Environment Monitoring System)/Food Regional Diets*

The GEMS/Food Regional Diets (WHO, 1998) are tables of dietary intakes of food commodities for five geographical areas. The major food class responsible for aflatoxin M_1 intake was identified as milk. The term 'milk' was assumed to include the mammalian milks (buffalo, camel, cattle, goat and sheep) listed in the GEMS/Food Regional Diets, but not to include cheese, butter or other dairy products derived from milk. In Table 8, exposure to aflatoxin M_1 from milk was calculated using three concentrations for the five geographical areas. [The use of dietary data to estimate mycotoxin intake can be misleading. Local climatic and other factors can greatly influence levels of mycotoxins in foods.] The first concentration was 0.05 μg/kg (the proposed maximum limit), the second was 0.5 μg/kg (the current maximum limit) and the third was the weighted mean of values for the geographical area. The use of 0.5 μg/kg as the level of aflatoxin M_1 in milk probably encompasses most of the milk samples and overestimates exposure. JECFA (2001)

Table 8. Estimated potential daily exposure to aflatoxin M_1 from all milks in five regional diets

Region/exposure	Aflatoxin M_1 in milk (μg/kg)	Aflatoxin M_1 intake (ng/person/day)
Europe/USA/Canada (0.294 kg milk/day)		
Proposed ML	0.05	14.7[a]
Current ML	0.5	147.0[a]
Weighted mean	0.023	6.8
Latin America (0.160 kg milk/day)		
Proposed ML	0.05	8.0[a]
Current ML	0.5	80.0[a]
Weighted mean	0.022	3.5
Far East (0.032 kg milk/day)		
Proposed ML	0.05	1.6[a]
Current ML	0.5	16.0[a]
Weighted mean	0.36	12
Middle East (0.116 kg milk/day)		
Proposed ML	0.05	5.8[a]
Current ML	0.5	58.0[a]
Weighted mean	0.005	0.6
Africa (0.042 kg milk/day)		
Proposed ML	0.05	2.9[a]
Current ML	0.5	20.9[a]
Weighted mean	0.002	0.1

From JECFA (2001)
ML, maximum level
[a] Calculated by the Working Group

used weighted means (including samples with zero values or values less than the limit of detection or quantification) to estimate dietary exposures for aflatoxin M_1. Because there were many non-detectable levels of aflatoxin M_1 in milk from the various studies and reports, the use of weighted means of all values could underestimate exposure for those individuals who are routinely exposed to higher levels of aflatoxin M_1 from milk.

(*b*) *Limitations of exposure estimates*

(i) The data submitted to FAO/WHO may not have been representative of countries or geographical areas, and not all member countries submitted data.
(ii) There were difficulties in attempting to compare and aggregate data on aflatoxin M_1 levels from different laboratories because the laboratories used different analytical methods. Also the data were presented in different ways (distributions, means of positive values, values less than a maximum limit).
(iii) The use of different analytical methods (thin-layer chromatography (TLC), HPLC) probably affected reported concentrations of aflatoxin M_1 in milk and therefore may affect intake estimates. Some methods, such as TLC, are more sensitive than others.
(iv) It was not possible to ascertain the effects of processing, season, climate or other environmental variables on the aflatoxin M_1 content of milk. These effects were often not addressed by the various studies and reports, and different descriptors were used for milk and other dairy products (e.g., 'raw' versus 'pasteurized') (JECFA, 2001).

1.4.3 *Exposure to aflatoxin M_1 in the French population*

Verger *et al.* (1999) have estimated exposures to aflatoxin M_1 in the French population (Table 9).

Table 9. Estimated average intake of aflatoxin M_1 in France

Type of product	Aflatoxin content (µg/kg)	ng/day per kg body weight				
		Mean	SD	95th percentile	Average % in the total intake	95th percentile/ mean
Milk and extra fresh milk[a]	0.014	[0.048]	0.107	0.261	59.2%	3.1
Cheeses[b]	0.093	0.058	0.050	0.143	40.8%	2.5
Total	–	0.142	0.122	0.362	100%	2.6

From Verger *et al.* (1999)
[a] Aflatoxin content of milk and extra fresh milk calculated from Direction Générale de l'Alimentation (DGAL), Paris (1995)
[b] Aflatoxin content of cheeses calculated from DGAL (1995) using a conversion coefficient from milk to cheese of 6.5

1.5 Regulations and guidelines

Efforts to reduce human and animal exposure to aflatoxins have resulted in the establishment of regulatory limits and monitoring programme worldwide. The rationale for the establishment of specific regulations varies widely; however, most regulations are based on some form of risk analysis including the availability of toxicological data, information on susceptible commodities, sampling and analytical capabilities, and the effect on the availability of an adequate food supply (Stoloff *et al.*, 1991). In 1995, among countries with more than five million inhabitants, 77 had known regulations for mycotoxins (all of which included aflatoxins) and 13 reported the absence of regulations. Data were not available for 40 countries (FAO, 1997). The regulation ranges for aflatoxin B_1 and total aflatoxins (B_1, B_2, G_1, G_2) were 'none detectable' to 30 or 50 μg/kg, respectively. Seventeen countries had regulations for aflatoxin M_1 in milk. The regulatory range for aflatoxin M_1 in milk was 'none detectable' to 1.0 μg/kg. New minimum EU regulations to which all EU countries must adhere were provided in 1998 (European Commission, 1998b). These regulations apply to all aflatoxins (B_1, B_2, G_1, G_2) in raw commodities and processed foods and to aflatoxin M_1 in milk. Regulations for other commodities include infant foods (European Commission, 2001) and selected spices (European Commission, 2002).

The Codex Alimentarius Commission (1999) is considering a recommendation to establish a limit for aflatoxins in foods of 15 μg/kg of total aflatoxins for all foods worldwide.

2. Studies of Cancer in Humans

Beginning in the 1960s and throughout the 1980s, a large number of ecological correlation studies were carried out to look for a possible correlation between dietary intake of aflatoxins and risk of primary liver cancer (IARC, 1993). Most of these studies were carried out in developing countries of sub-Saharan Africa or Asia, where liver cancer is common. With some notable exceptions, and despite the methodological limitations of these studies, they tended to show that areas with the highest presumed aflatoxin intake also had the highest liver cancer rates. However, the limitations of these studies, including questionable diagnosis and registration of liver cancer in the areas studied, questionable assessment of aflatoxin intake at the individual level, non-existent or questionable control for the effect of hepatitis virus and the usual problem of making inferences for individuals from observations on units at the ecological level, led to increasing recognition of the need for studies based on individuals as units of observation.

In the 1980s, some case–control studies were carried out in high-risk areas, generally based on reasonably reliable diagnostic criteria for liver cancer (IARC, 1993). The comparability of cases and controls was limited in some of these studies. Exposure to aflatoxins was sometimes assessed via dietary questionnaires and sometimes via biomarker measurements. As both of these were collected after disease onset, their relevance to past lifetime intake of aflatoxins was uncertain. Beginning in the mid 1980s, some prospective cohort studies were undertaken which avoided many of the methodological limitations of earlier studies. Among the major advantages of this new generation of studies were the following: new improved biomarkers of aflatoxin exposure, improved ability to measure hepatitis infection, better comparability of cases and controls within a well defined cohort, and control of the temporal sequence by measuring exposure before disease onset.

In 1992, an IARC Working Group described all relevant human studies that had been reported and concluded that there was *sufficient evidence* in humans for carcinogenicity of aflatoxin B_1 and of naturally-occurring mixtures of aflatoxins. The present monograph represents an update of evidence published since that evaluation was made, without describing the studies covered in the previous monograph, although brief summaries of the main studies are given in tabular format (Tables 10–12). These tables also provide summaries of the relevant studies that have been published since 1993 and which are described in the following sections. The outcome investigated in most studies was liver cancer. Different studies used different sources (e.g., death certificates, hospital registries, medical examinations) and different criteria (clinical, cytological) for definition of liver cancer. Different terms, such as liver cancer, primary liver cancer or hepatocellular carcinoma (HCC) were used. In the following descriptions, we have used the terminology used by the authors.

2.1 **Descriptive studies** (see Table 10)

Hatch *et al.* (1993) conducted a hybrid ecological cross-sectional study in eight areas of Taiwan (China), with a wide range of rates of mortality from primary hepatocellular carcinoma (HCC). In order to derive estimates of aflatoxin levels in the eight areas, they selected a representative sample of 250 adult residents in total (unequal numbers per area). Participants were interviewed and were asked to provide both morning urine and blood specimens. Serum was used for detecting hepatitis B surface antigen (HBsAg). Urine was used for detecting aflatoxins B_1 and G_1 and metabolites, including aflatoxins M_1 and P_1; the highest sensitivity was for aflatoxin B_1. Measured values ranged from 0.7 to 511.7 pg equivalents of aflatoxin B_1/mL of urine, with a mean of 41.3 pg/mL. Mean levels were similar in men and women, and in hepatitis B virus (HBV) carriers and HBV non-carriers. The primary analyses were carried out with individuals as the unit of analysis. In these analyses, the individual's measurements of aflatoxin B_1 equivalents and of HBsAg were used in conjunction with the HCC rate (sex-specific, age-adjusted) of the entire area in which the individual resided. There were 246 data points for these analyses (four individuals had missing blood specimens). Some bivariate correlation coefficients and some regression analyses in which aflatoxin levels were regressed on area HCC mortality, HBsAg, age and sex were calculated. In addition, in some analyses the data

Table 10. Summary of the principal ecological and cross-sectional studies on liver cancer and aflatoxins

Reference	Area	Units of observation/ number of units	Exposure measure(s)	Outcome measure(s)	Covariate	Results	Comments
Alpert *et al.* (1971)	Uganda	Main tribes and districts of Uganda; 7	Aflatoxin contamination of nearly 500 food samples taken from randomly selected native homes and markets; 1966–67	Hepatoma incidence identified from hospital records; 1963–66	Nil	The highest incidence of hepatoma occurred in areas with highest levels of aflatoxin contamination.	
Peers & Linsell (1973)	Kenya	Altitude areas of Murang'a district; 3	Aflatoxin extracted from food samples, repeated cluster sampling over 21 months	Incident hepatocellular cancers ascertained from local hospitals; 1967–70	Nil	Using 6 data points (3 areas, both sexes), correlation ($r = 0.87$) between aflatoxin intake and liver cancer	Questionable completeness of liver cancer registration. Small number of units of observation
Peers *et al.* (1976)	Swaziland	Altitude areas; 4	Aflatoxin from food and beer samples: every 2 months for 1 year, over 1000 samples analysed; 1972–73	PLC incidence rates, from national cancer registry; 1964–68	Nil	Correlation (males, $r = 0.99$; females, $r = 0.96$) between aflatoxin intake and PLC rates	Exposure post-dated cancer data
Wang *et al.* (1983)	China	29 provinces and municipalities; 552 cities	Grain oil contamination by aflatoxin B_1	PLC mortality	HBsAg, climate	Contamination by aflatoxin strongly correlated with liver cancer	Incomplete study description
Stoloff (1983)	USA	South-east, north and west regions of USA	Daily aflatoxin ingestion among males, based on historic food consumption surveys and historic estimates of aflatoxin contamination	PLC; 1968–71 and 1973–76	Nil	South-east had much higher aflatoxin ingestion and 10% higher PLC rates than 'north and west'	Considerable excess of PLC observed among Orientals and urban black males
Van Rensburg *et al.* (1985)	Southern Africa	7 districts of Mozambique; Transkei, South Africa	Mean aflatoxin contamination of food samples, over 2500 samples analysed; 1969–74	Mozambique: incidence rates of HCC; 1968–75 (variety of sources including local hospitals and South African mines); Transkei; 1965–69	Nil	Rank correlations between HCC and mean total aflatoxin 0.64 ($p < 0.05$) in men and 0.71 ($p < 0.01$) in women	

Table 10 (contd)

Reference	Area	Units of observation/ number of units	Exposure measure(s)	Outcome measure(s)	Covariate	Results	Comments
Autrup *et al.* (1987)	Kenya	Districts of Kenya; 9	Urinary 8,9-dihydro-8-(7-guanyl)-9-hydroxy-AFB_1 as ascertained in surveys at outpatient clinics in the 9 districts (total sample, 983); 1981–84	Primary hepatocellular carcinoma (PHC) incidence diagnosed at one large hospital in Nairobi; 1978–82	HBsAg and anti-HBc	*Spearman rank correlation* (*r*) with PHC rate Prevalence of AFB_1 0.75 HBV 0.19	Potential confounding by ethnicity. No interaction between AFB_1 and HBV
Peers *et al.* (1987)	Swaziland	Topographic (4) and administrative regions (10)	Aflatoxins measured in food samples from households and crop samples from fields; over 2500 samples analysed; 1982–83	Incidence rates of PLC; 1979–83	HBsAg and other markers of HBV infection in 3047 serum samples from the Swaziland blood bank	Significant correlation between estimated aflatoxin (and AFB_1) consumption and PLC; little effect of HBsAg on PLC	
Campbell *et al.* (1990)	China	48 widely scattered counties – out of a total of 2392 in China	Mean urinary aflatoxin metabolites, serum HBsAg and 3-day dietary intake, based on local sample surveys in 1983	PLC mortality rates; 1973–75	HBsAg (50 individuals per country), alcohol, some others	Urinary aflatoxin and PLC ($r = -0.17$)	Positive associations between liver cancer and HBsAg ($r = 0.45$), liquor intake ($r = 0.46$), dietary cadmium ($r = 0.40$), plasma cholesterol ($r = 0.42$). Exposure data post-dated cancer data
Van Rensburg *et al.* (1990)	South Africa	Districts of the Transkei; 4	AFB_1 contamination of local food samples, based on over 600 samples; 1976–77	PLC incidence in residents and in gold-miners	Nil	Rank order correlations between AFB_1 intake and PLC incidence in goldminers from the Transkei were significant at $p < 0.05$.	

Table 10 (contd)

Reference	Area	Units of observation/ number of units	Exposure measure(s)	Outcome measure(s)	Covariate	Results	Comments
Srivatanakul *et al.* (1991a)	Thailand	Selected areas of Thailand; 5	Surveys of local residents, aflatoxin measured in urine and in serum, 50–100 subjects per area	Incidence rates of HCC and cholangio-carcinoma 1980–82. Standardized propor-tionate incidence ratio (PIR)	Same 100–200 individuals per area as for aflatoxin, HBsAg, anti-HBs, anti-HBc and liver fluke (OV)	HCC / Cholangiocarcinoma Serum aflatoxin: –0.75 ($p = 0.14$) / –0.03 ($p = 0.96$) Urinary aflatoxin: –0.64 ($p = 0.25$) / 0.17 ($p = 0.78$) Anti-OV titre: –0.37 ($p = 0.54$) / 0.98 ($p = 0.004$) HBsAg: –0.45 ($p = 0.44$) / 0.27 ($p = 0.66$)	
Hatch *et al.* (1993)	Taiwan	Townships; 8	Mean urine levels of various aflatoxins. Measured on a total of 250 randomly selected subjects in 8 townships	HCC mortality rate in the area of the township	HBsAg; smoking status, alcohol consumption	Univariate correlation between mean urinary aflatoxin and area HCC rates: men, 0.83 ($p = 0.012$) women, 0.49 ($p = 0.22$)	Univariate correlations between area HCC rates and mean HBsAg were around 0.50. When individuals were used as units of observation, with the area mortality rate attributed to the individual, the correlations between urinary aflatoxins and HCC were considerably attenuated.
Omer *et al.* (1998)	Sudan	Two areas, one high- risk, one low-risk	Peanut butter samples collected in markets and analysed for AFB_1. Type of storage assessed	Fragmentary data indicate that risk of liver cancer is higher in one area than the other.	–	Aflatoxin consumption levels were higher in the presumed high-risk area than in the presumed low-risk area.	Only two areas compared. Unreliable measures of liver cancer incidence

AFB_1, aflatoxin B_1; anti-HBc, antibody to hepatitis B core antigen; HBsAg, hepatitis B surface antigen; HBV, hepatitis B virus; HCC, hepatocellular carcinoma; OV, *Opisthorchis viverrini*; PHC, primary hepatocellular carcinoma; PLC, primary liver cancer were collapsed into a conventional ecological study with eight units of observation and the mean levels of aflatoxin and HBsAg correlated with HCC rates. The univariate correlations between HCC and aflatoxins at the ecological level were 0.83 ($p = 0.01$) in men and 0.49 ($p = 0.22$) in women. The correlations were much lower, albeit statistically significant, when analysed at the individual level: 0.29 ($p = 0.002$) in men and 0.17 ($p = 0.047$) in women. In the multivariate regression analysis, HCC was significantly associated with aflatoxin levels, after adjusting for age, sex and HBsAg. Adjustments for smoking and alcohol, in a subset of 190 subjects with available interview data, and the inclusion of interaction terms did not materially affect the findings. Thus, the very different types of analysis all pointed to an association between urinary aflatoxins and HCC. [The Working Group noted that the calculation of the p value did not take into account the clustered sampling design.]

Omer *et al.* (1998) carried out a comparison of aflatoxin contamination of peanut products in two areas of Sudan. On the basis of 'clinical experience and Khartoum hospital records', the authors suspected that incidence of HCC was substantially higher in western Sudan than in central Sudan. The study was carried out in 1995 and involved selection of peanut butter samples from local markets using a staged sampling approach to identify markets in the two study areas. Samples were characterized as to how they had been stored and were analysed for aflatoxin B_1 by HPLC. Mean aflatoxin B_1 levels were much higher in 'high-risk' western Sudan (87.4 ± 197.3 μg/kg) than in central Sudan (8.5 ± 6.8 μg/kg). Also, dietary questionnaires among subjects recruited for a small case–control study indicated that residents in western Sudan consumed more peanut butter than residents in central Sudan. [The Working Group noted that only two areas were compared, and that there was no documented evidence of differences in the incidence rates of HCC between the two areas. A small case–control study was carried out as well, but the Working Group noted that this was probably a small feasibility study that was superseded by Omer *et al.* (2001).]

2.2 **Cohort studies** (see Table 11)

Qian *et al.* (1994) updated a cohort study previously described by Ross *et al.* (1992) of 18 244 male residents of Shanghai, China, 96% of whom were aged 45–64 years on entry to the study. The men were recruited by invitation from four geographically defined areas and responded to questionnaires administered by nurses, usually in their homes, on lifestyle (including smoking and alcohol consumption) and on food frequency. Blood and urine specimens were collected. The men were followed up by identification of death records in district vital statistics units and through linkage with the Shanghai Cancer Registry (estimated to be 85% complete). An attempt was also made to contact each cohort member annually. The cohort was established between January 1986 and September 1989 and was followed to 1 February 1992 for the current analysis, resulting in 69 393 person–years of follow-up. Of 364 cancer cases identified, 55 were diagnosed as primary liver cancers, nine of which were confirmed by biopsy. The reported diet

Table 11. Summary of the principal cohort and nested case–control studies on liver cancer and aflatoxins

Reference	Area	Study base	Type of analysis	Exposure measures	Outcome measures	Covariate	Results	Comments
Hayes *et al.* (1984)	Netherlands	71 male oil-press workers exposed to dust containing aflatoxin	Cohort – SMR comparisons with Dutch males	Worked > 2 years in exposed area during 1961–70	Cancer mortality	–	Cases / SMR / 95% CI Lung cancer: 7 / 2.5 / 1.0–5.0 Liver cancer: 0 / 0 / NA	
Yeh *et al.* (1985)	China	Selected villages in Fusui county, Guangxi	Cohort (ecological exposure)	Village mean for intake of aflatoxin based on food samples. ~10-fold difference between low and high contamination areas	Liver cancer mortality	HBsAg measured among cases	Aflatoxin contamination / HBsAg status / HCC deaths: No. / Rate per 1000 p–yr Heavy: Positive 13 / 649; Negative 2 / 99 Light: Positive 1 / 66; Negative 0 / 0	Incomplete study description [unit presumed to be: per 1000 p–yr]
Olsen *et al.* (1988)	Denmark	Male employees of 241 livestock feed companies, employed after 1964	Cohort SPIR	Longest-held job (> 1964) in one of 241 companies	Liver cancer incidence traced in Danish Cancer Registry; 1970–84		Longest employment / Cases / SPIR / 95% CI Ever: 6 / 1.4 / 0.57–2.9 ≥ 10 yrs before diagnosis: 7 / 2.5 / 1.1–4.9	Confounding by HBV and alcohol unlikely
Yeh *et al.* (1989)	China	Five communities of southern Guangxi Province, men enrolled in 1982–83; $n > 7917$	Cohort – PHC mortality rates in different categories of estimated mean AFB_1 consumption	Mean AFB_1 level of community of residence, as estimated from food samples collected from all over the region; 1978–84	Mortality from PHC based on follow-up 1984–86	HBsAg in cases and subcohort	Strong correlation between PHC mortality and estimated levels of AFB_1 in 4 communities; aflatoxin levels not available in one community (Pearson correlation coefficient, 1.00; $p = 0.004$)	Strong association observed between HBsAg and PHC mortality in cohort and nested case–control analysis (RR = 32); not seen in the ecological analysis ($r = 0.28$; $p = 0.65$)

Table 11 (contd)

Reference	Area	Study base	Type of analysis	Exposure measures	Outcome measures	Covariate	Results	Cases	RR/OR	95% CI	Comments
Qian *et al.* (1994); Ross *et al.* (1992)	Shanghai, China	Men, mainly 45–64 years old; *n* = 18 244. Resident in one of four areas. Recruited during 1986–89	Both cohort analysis and a nested case–control analysis with 50 cases and 267 matched controls	Detailed dietary history linked to measured levels in sample foods	Liver cancer mortality and incidence follow-up to 1992. Intensive tracing	Measured HBsAg, cigarette smoking	Aflatoxin	Cases	RR	95% CI	Cohort analysis
							Medium	25	1.6	0.8–3.1	
							Heavy	16	0.9	0.4–1.9	
								Cases	OR	95% CI	Nested case–control analysis
							Aflatoxin–guanine adducts	18	9.1	2.9–29	
				Measured urinary aflatoxin and aflatoxin–guanine adducts			Any bio-marker of aflatoxin	36	5.0	2.1–12	
							Joint exposure to any biomarker of aflatoxin and HBsAg	23	59	17–212	
Chen *et al.* (1996)	Taiwan, China	Penghu Islets. Over 6000 subjects enrolled. Possibly a subset of the cohort of Wang *et al.* (1996)	Nested case–control analysis based on 20 cases and 86 controls	Measured AFB_1–albumin adducts	HCC, ascertained by an active diagnostic procedure	Sociodemo-graphic cha-racteristics, HBsAg, anti-HCV, family history of HCC and liver cirrhosis		Cases	OR	95% CI	
							AFB_1–albumin adducts	13	5.5	1.2–25	
Wang *et al.* (1996a)	Taiwan, China	7 townships. Over 25 000 subjects enrolled in cohort	Nested case–control analysis based on 56 cases and 220 controls	Biomarker measurement of urinary aflatoxins and aflatoxin–albumin adducts	HCC, ascertained by a variety of tracing sources, 1991–95	Sociodemo-graphic characteristics and HBsAg		Cases	OR	95% CI	*Adjusted for HBsAg
							Aflatoxin–albumin adducts	31	1.6*	0.4–5.5	
							Urinary aflatoxin	26	3.8*	1.1–13	
							Urinary aflatoxin + HBsAg	22	112	14–905	

Table 11 (contd)

Reference	Area	Study base	Type of analysis	Exposure measures	Outcome measures	Covariate	Results	Cases	OR	95% CI	Comments
Sun *et al.* (2001)	Taiwan, China	Same as Wang *et al.* (1996)	Nested case–control analysis based on 79 cases and 149 controls, all HBsAg-positive	Aflatoxin–albumin adducts	Same as Wang *et al.* (1996), 1991–97	Same as Wang *et al.* (1996), anti-HCV, $GSTM_1$, $GSTT_1$	Aflatoxin–albumin adducts	47	2.0	1.1–3.7	Statistically significant interaction with $GSTT_1$ genotypes
Yu *et al.* (1997a)	Taiwan	Male patients from Government Employee Central Clinics and a Taipei hospital, aged 30–65 during enrolment, 1988–92; *n* = 4841 HBsAg-positive, *n* = 2501 HBsAg-negative	Nested case–control on 43 cases and 2 matched controls per case, one HBsAg positive and one negative	Baseline interviews and measurement of urinary aflatoxin and aflatoxin adducts	HCC	Sociodemographic, alcohol, smoking	AFM_1 AFP_1 AFB_1 AFB_1–*N*7-guanine adducts	23 18 17 6	6.0 2.0 2.0 2.8	1.2–29 0.5–8.0 0.7–5.8 0.6–13	All analyses restricted to HBsAg-positive subjects
Lu *et al.* (1998)	China	Seven townships in Qidong. Men aged 20–60 years during enrolment; follow-up: 1987–97	Nested case–control among HBsAg carriers; 30 cases and 5 controls per case (matched for age, place of residence)	AFB_1–albumin adducts	PLC		AFB_1-albumin adducts	23	3.5	[1.3–10]	
								Cases	RR	95% CI	
Sun *et al.* (1999)	China	Men in 2 townships screened and found positive for chronic HBV infection, recruited in 1987–98; *n* = 145	Cohort analysis	AFM_1 measured in 8 pooled urinary samples	HCC	Anti-HCV, family history of HCC, smoking, alcohol	AFM_1	17	4.5*	1.6–13	Increased risks for HCC among anti-HCV (RR, 6.0) and those with family history of HCC (RR, 4.7) *Adjusted for anti-HCV and HCC family history

AFB_1, aflatoxin B_1; AFM_1, aflatoxin M_1; AFP_1, aflatoxin P_1; CI, confidence interval; GST, glutathione *S*-transferase; HBsAg, hepatitis B surface antigen; HBV, hepatitis B virus; HCC, hepatocellular carcinoma; HCV, hepatitis C virus; NA, not applicable; OR, odds ratio; OV, *Opisthorchis viverrini*; PHC, primary hepatocellular carcinoma; p–yr, person–years; RR, relative risk; SMR, standardized mortality ratio; SPIR, standardized proportionate incidence ratio history based on a frequency checklist of 45 food items usually consumed as an adult was combined with a set of independently measured aflatoxin levels in various local foods to derive a quantitative measure of dietary aflatoxin exposure. In a cohort-type analysis, using the lowest tertile of aflatoxin exposure as reference, the middle tertile had a relative risk (adjusted for age and smoking) of 1.6 (95% confidence interval [CI], 0.8–3.1; 25 cases) and the highest tertile had an odds ratio (OR) of 0.9 (95% CI, 0.4–1.9; 16 cases).

To assess the risks in relation to biomarkers of aflatoxin exposure, a nested case–control study was conducted using 50 of the cases (Qian *et al.*, 1994). Controls were selected from among subjects who had no history of liver cancer on the date of cancer diagnosis of the index cases and were matched to cases in ratios ranging from 10:1 to 3:1, yielding a total of 267 controls. For each case and control, urine samples were analysed for aflatoxins B_1, P_1 and M_1 and for aflatoxin B_1–*N*7-guanine adducts, and among a subgroup of 28 cases and their matched controls for aflatoxins G_1 and Q_1. HBsAg was measured by radioimmunoassay. Thirty-two out of 50 cases and 31 out of 267 controls were HBsAg-positive. Each of the six biomarkers of aflatoxin exposure was more frequently present among cases than controls. For 36 of the 50 liver cancer cases and 109 of 267 controls, results were positive in at least one of the four assays analysed for the full set of cases and controls (adjusted relative risk, 5.0; 95% CI, 2.1–12). The highest risks were found among subjects with aflatoxin B_1–*N*7-guanine adducts. Compared with subjects who had no aflatoxin biomarkers and were HBsAg-negative, the interaction of the two factors was supra-multiplicative, with relative risks as follows: aflatoxin biomarker only, 3.4 (95% CI, 1.1–10); HBsAg only, 7.3 (95% CI, 2.2–24); both factors, 59 (95% CI, 17–212). [The Working Group noted inconsistencies between analyses based on dietary questionnaires and biomarkers.]

The Penghu Islets reportedly have the highest rates of HCC in Taiwan, China. Chen *et al.* (1996) enrolled 4691 men and 1796 women, aged 30–65 years, in a prospective cohort study. The subjects were selected from a housing register maintained by the local administration. Participants were interviewed on a variety of sociodemographic, dietary and medical history topics. Blood samples were collected and stored frozen. A two-stage screening process for HCC was undertaken which included serological markers and clinical assessments with ultrasonography. Subjects were further followed up with annual examinations. A total of 33 cases of HCC were diagnosed by December 1993, of whom two were negative for HBsAg. A total of 123 controls were selected from within the cohort among unaffected subjects, and matched with cases for age, sex, village and date of blood collection. Blood samples from cases and controls were analysed for HBsAg, for anti-hepatitis C virus (HCV) antibodies and aflatoxin B_1–albumin adducts, although samples for adduct analysis were usable for only 20 cases and 86 controls. Using logistic regression, with age and sex adjustment, and a detection limit for albumin adducts of 0.01 fmol/μg as the cut-off value, the OR for an association between presence of aflatoxin B_1–albumin adducts and HCC was 3.2 (13 cases; 95% CI, 1.1–8.9). When the statistical model also included several other covariates (HBsAg, anti-HCV, family history of liver cancer and cirrhosis), the odds ratio for aflatoxin B_1–albumin adducts rose to 5.5 (95% CI, 1.2–25). There was also an extremely high risk associated with positive HBsAg status (OR = 129; 95% CI, 25–659). The authors surmised that peanut contamination was a major source of aflatoxin in this population.

Wang *et al.* (1996a) carried out a cohort study in seven townships of Taiwan, China, including three on the Penghu Islets and four on Taiwan Island. Of the total population of 89 342 eligible subjects selected from local housing offices and mailed an invitation in a cancer screening project, 25 618 (29%) volunteered to participate. Among participants, 47% were men and enrolment occurred from July 1990 to June 1992. Participants were interviewed to elicit information on sociodemographic characteristics, alcohol and smoking habits, and medical history. Fasting blood and spot urine specimens were collected and stored frozen. Serum samples were assayed for HBsAg and α-fetoprotein, anti-HCV and various markers of liver function. Abdominal ultrasonography was carried out among a subgroup of high-risk persons from two Penghu Islets. All participants were recontacted by invitation to local research centres or by telephone interviews between 1992–94. Periodic searches for death certificates from local housing offices and in June 1995 through linkage with the national death and cancer registries were carried out. The overall follow-up rate was > 98%. Between February 1991 and June 1995, 56 HCC cases were identified in the cohort, of which 22 were histologically/cytologically confirmed. For each case, four controls were selected among cohort members who were free of liver cancer or cirrhosis at the time of case identification, and who were matched for age, sex, township and recruitment date. Altogether there were 56 HCC cases and 220 controls. Serum and urine specimens were available for analysis on subsets: serum for 52 cases and 168 controls, and urine for 38 cases and 137 controls. Urinary aflatoxin metabolites were determined using a monoclonal antibody with high affinity to aflatoxin B_1 and significant cross-reactivity to aflatoxins B_2, M_1, G_1 and P_1. Serum aflatoxin–albumin adducts were measured. Using conditional logistic regression, the OR for liver cancer corresponding to detectable levels of aflatoxin–albumin adducts was 4.6 (95% CI, 2.0–10) before adjustment for HBsAg and 1.6 (95% CI, 0.4–5.5) after adjustment. [The Working Group noted inconsistencies in numbers of available controls for serum aflatoxin–albumin adducts.] For high levels of urinary aflatoxin metabolites, the OR was 3.3 (95% CI, 1.4–7.7) before adjustment for HBsAg and 3.8 (95% CI, 1.1–13) after adjustment. While there was little or no effect of aflatoxin biomarkers on HCC among HBsAg-negative subjects, there were quite strong effects among HBsAg-positive subjects, especially in the analysis using aflatoxin metabolites as the exposure. [It seems that the present cohort and the cases identified in it include the cases studied by Chen *et al.* (1996) in the Penghu Islets. The low participation rate would not have affected the validity of the results unless individuals with preclinical liver cancer symptoms and high aflatoxin exposure were more likely to volunteer for participation in the study than others in the same population.]

Sun *et al.* (2001) reported the results from a nested case–control study of an extended follow-up of the cohort described by Wang *et al.* (1996a). Seventy-nine HBsAg-positive cases of HCC were identified between 1991 and 1997, and matched for age, gender, residence and date of recruitment to one or two randomly selected HBsAg-positive controls (total, 149). Blood samples were collected and analysed for HBV and HCV, for aflatoxin B_1–albumin adducts, and for glutathione *S*-transferase (GST) M1 and T1 genotypes. In a conditional logistic regression analysis, a significant relationship was observed between HCC risk and aflatoxin B_1–albumin adducts (OR = 2.0; 95% CI, 1.1–3.7). *GSTM1*- and *GSTT1*-null genotypes were associated with a decreased risk for HCC (OR = 0.4; 95% CI, 0.2–0.7 and OR = 0.5; 95% CI, 0.2–0.9). A statistically significant (p = 0.03) interaction was found between aflatoxin B_1–albumin adducts and *GSTT1* genotype, indicating a more pronounced risk among those who were *GSTT1*-null genotype (OR = 3.7; 95% CI, 1.5–9.3), and no risk among those who had the non-null genotype (OR = 0.9; 95% CI, 0.3–2.4).

Yu *et al.* (1997a) carried out a cohort study in Taiwan, China, to study the role of aflatoxin in the etiology of HCC. Between 1988 and 1992, a cohort of 4841 male asymptomatic HBsAg carriers and 2501 male non-carriers, aged 30–65 years, was recruited from the Government Employee Central Clinics and the Liver Unit of a hospital in Taipei. At entry into the study, each participant was interviewed to obtain information on demographic characteristics, habits of cigarette smoking and alcohol drinking, diet (including the frequency of consuming peanuts and fermented bean products, which are thought to be the major aflatoxin-contaminated foodstuffs in Taiwan), as well as personal and family history of major chronic diseases. Urine and blood samples from study subjects were stored frozen. All HBsAg carriers in this study had both ultrasonography and α-fetoprotein measurement every 6–12 months. Follow-up of HBsAg non-carriers was carried out by annual examination including a serum α-fetoprotein test. The response rate to the periodic follow-up examinations was approximately 72% for HBsAg carriers and 80% for HBsAg non-carriers. Information on HCC occurrence and vital status of study subjects who did not participate in the follow-up examinations was obtained from both computerized data files of the national death certification and the cancer registry. By 31 December 1994, 34 579 person-years of follow-up had been accumulated, an average of 4.7 years per person. Fifty HCC cases were identified during the follow-up period. All HCC cases were diagnosed on the basis of either pathological and cytological examinations or an elevated α-fetoprotein level combined with at least one positive image. To investigate the role of aflatoxin, a nested case–control comparison was carried out, in which two separate matched controls per case were selected, one who was HBsAg-positive and one who was HBsAg-negative. Levels of aflatoxin metabolites in urine were analysed by reverse-phase HPLC allowing measurement of aflatoxins M_1, P_1, B_1 and G_1 and aflatoxin B_1–*N*7-guanine. Most subjects were also tested for anti-HCV. After exclusion of subjects with missing specimens, analyses were available on 43 matched case–control sets. Among all HCC cases, only one occurred in the HBsAg-negative subcohort, and that one was positive for anti-HCV. All study subjects were positive for aflatoxin M_1, 81% for aflatoxin P_1, 43% for aflatoxin B_1–*N*7-guanine adducts, 28% for aflatoxin B_1 and 12% for aflatoxin G_1. There was a significant correlation (r = 0.35) between reported dietary intake of various foods thought to contain aflatoxins and levels of urinary aflatoxin M_1. No significant correlations with other aflatoxin metabolites were observed. The main analyses, using conditional logistic regression, were carried out among cases and controls who were HBsAg carriers. Four of the five aflatoxin markers, but not aflatoxin G_1, were associated with an elevated risk for HCC among subjects in the highest tertile of exposure, although only for aflatoxin M_1 was this significant. The OR in the highest tertile of aflatoxin M_1 exposure, after adjustment for education, ethnicity, alcohol and smoking, was 6.0 (23 cases; 95% CI, 1.2–29). When pairs of these aflatoxin biomarkers were examined, certain combinations were found to be associated with particularly high risk: thus, subjects with detectable aflatoxin B_1–*N*7-guanine and high levels of aflatoxin M_1 had an OR of 12 (16 cases; 95% CI, 1.2–117).

To investigate the role of HBV and aflatoxin in the etiology of liver cancer, Lu *et al.* (1998) carried out a nested case–control analysis within a cohort of 737 male HBsAg carriers and 699 HBsAg non-carriers in Qidong, China (follow-up was from 1987–97). Among the HBsAg carriers, 30 cases of liver cancer were matched for age and place of residence with 150 non-cases from the cohort. Levels of aflatoxin B_1–albumin adducts were significantly higher among cases than among controls, both in proportion and in concentration. The crude OR was 3.5 [95% CI, 1.3–10].

Sun *et al.* (1999) reported on the experience of a cohort of 145 men with chronic hepatitis B. These HBV-positive men had been detected in a prevalence survey carried out in 1981–82 in two townships in Qidong, China. They were recruited for the present follow-up study during 1987–98. At recruitment, they were interviewed and examined; eight urine samples were obtained at monthly intervals and blood was drawn periodically throughout the follow-up period. The urine samples for each individual were pooled and aflatoxin M_1 was measured in the pooled sample. No patients were lost to follow-up. The mean age of the cohort was 39 years in 1998. Over the period of follow-up, 22 of the 145 subjects were diagnosed with liver cancer, of whom 10 had pathological confirmation. Anti-HCV-positive subjects had an increased risk for HCC compared with subjects who were anti-HCV-negative and subjects with a family history of HCC had an increased risk compared with subjects who did not have a family history of HCC. The median concentration of aflatoxin M_1 in urine was 9.6 ng/L and the highest concentration was 243 ng/L. Using 3.6 ng/L as the cut-point in a Cox proportional hazard model, the relative risk for HCC among subjects with high aflatoxin M_1 compared with those having low aflatoxin M_1 was 3.3 (95% CI, 1.2–8.9). When anti-HCV status and family history of HCC were also included in the model, the relative risk for HCC associated with aflatoxin M_1 was 4.5 (95% CI, 1.6–13).

2.3 Case–control studies (see Table 12)

Olubuyide *et al.* (1993a,b) carried out a small case–control study in Nigeria to assess the role of HBV and aflatoxins in primary hepatocellular carcinoma. Cases were 22 patients at a university hospital in Ibadan in 1988. Controls were 22 patients from the gastroenterology ward of the same hospital with acid peptic disease unrelated to liver

Table 12. Summary of principal case–control studies on liver cancer and aflatoxins

Reference	Area	Study base	Cases	Controls	Exposure measures	Covariate	Results	Comments
Bulatao-Jayme *et al.* (1982)	Philippines	Three hospitals	PLC; $n = 90$	Patients with normal liver function; $n = 90$ matched by age and sex	Detailed dietary history, linked to measured levels in sample foods. Also biomarkers of aflatoxins	Alcohol and sociodemographic variables	Elevated risks with most aflatoxin-contaminated foods Medium–heavy exp., 15 cases, RR = 14 ($p < 0.05$) Heavy exp., 55 cases, RR = 17 ($p < 0.05$) Urinary aflatoxins B_1 and M_1 were significantly ($p < 0.05$) higher among cases than among controls.	
Lam *et al.* (1982)	Hong Kong	1 university hospital	PLC; $n = 107$ Chinese	Trauma patients, same hospital. $n = 107$ matched by age and sex	Dietary history, linked to earlier market survey of aflatoxin contamination	HBsAg, smoking, alcohol	No differences between cases and controls in reported consumption of different aflatoxin-contaminated foods	
Parkin *et al.* (1991)	Thailand	Three hospitals in Thailand	Cholangiocarcinoma; $n = 103$	Patients visiting clinics or admitted to same hospitals; matched by age, sex and residence to cases; $n = 103$	Dietary history. Aflatoxin–albumin adducts	HBsAg, anti-HBs, anti-HBc, anti-OV, smoking, alcohol	Cases / OR / 95% CI Consumption of presumed aflatoxin-contaminated food: NR / 1.4 / 0.8–2.7 Presence of aflatoxin–albumin adducts: 1 / 1.0 / 0.1–16.0	
Srivatanakul *et al.* (1991b)	Thailand	Three hospitals in Thailand	HCC; $n = 65$	Patients visiting clinics or admitted to same hospitals; matched by age, sex and residence to cases; $n = 65$	Dietary history. Aflatoxin–albumin adducts	HBsAg, anti-HBs, anti-HBc, anti-OV, anti-HCV, smoking, alcohol	Cases / OR / 95% CI Consumption of presumed aflatoxin-contaminated food: NR / 1.9 / Not significant Presence of aflatoxin–albumin adducts: 8 / 1.0 / 0.4–2.7	
Olubuyide *et al.* (1993a)	Nigeria	Hospital in Ibadan	Primary HCC diagnosed in 1988; $n = 22$	Matched patients from gastro-enterology ward; $n = 22$	Serum levels of aflatoxin	HBsAg measured but not included in analysis of aflatoxins	High aflatoxin levels were detected in 5 cases and 1 control ($p < 0.05$)	2 of these 5 cases were HBsAg-negative

Table 12 (contd)

Reference	Area	Study base	Cases	Controls	Exposure measures	Covariate	Results	Comments
Mandishona *et al.* (1998)	South Africa	Two hospitals in one province of South Africa	HCC; $n = 24$	Two control series: one hospital-based (trauma or infection patients), $n = 48$; family-based (including related and unrelated family members), $n = 75$	Measured AFB_1–albumin adducts	Several measured, but not used in analysis of aflatoxin	Median levels of AFB_1–albumin adducts were lower among cases than among both sets of controls	High risks of HCC found for HBsAg-positive subjects, alcohol, and iron overload. Questionable comparability of hospital control series, and possible over-matching with family control series
Omer *et al.* (2001)	Sudan	Residents of two regions of Sudan	Cases of HCC diagnosed in 5 out of 6 hospitals in Khartoum; $n = 150$	Community-based, selected from 'sugar shop' registries in same regions as cases; $n = 205$	Detailed diet history for peanut butter, and for storage of peanuts	HBsAg, anti-HCV, smoking alcohol, $GSTM_1$ genotype	Cases / OR / 95% CI Highest quartile of peanut butter intake*: 63 / 3.0 / 1.6–5.5 Humid storage: 99 / 1.6 / 1.1–2.5 Highest quartile of peanut butter intake + $GSTM_1$-null genotype**: NR / 17 / 2.7–105	Questionable comparability of cases and controls *Test for trend statistically significant **Test for trend among $GSTM_1$-null genotype subjects statistically significant

AFB_1, aflatoxin B_1; AFM_1, aflatoxin M_1; CI, confidence interval; GST, glutathione *S*-transferase; HBsAg, hepatitis B surface antigen; HBV, hepatitis B virus; HCC, hepatocellular carcinoma; HCV, hepatitis C virus; NR, not reported; OR, odds ratio; PLC, primary liver cancer; RR, relative risk diseases and matched to cases for sex and age. Blood samples were collected after subjects were on hospital diet for one week and were analysed for HBsAg and a number of aflatoxins (B_1, B_2, M_1, M_2, G_1, G_2) and aflatoxicol. HBsAg was detected in 16 cases and 8 controls. Elevated levels of aflatoxins were detected in five (23%) cases and one (5%) control, the difference being significant ($p < 0.05$). [The Working Group questioned the comparability of cases and controls.]

Mandishona *et al.* (1998) carried out a small case–control study in South Africa aiming primarily to determine the role of dietary iron overload in the etiology of HCC. They also collected information on exposure to aflatoxin B_1 and reported risks in relation to this. Cases were 24 consecutive patients with HCC in two hospitals of one province of South Africa. Two control series were assembled. A matched (sex, age, race) series of 48 (two controls per case) was selected from patients hospitalized mainly with trauma or infection. In addition, 75 relatives and family members of the cases constituted a second control series. Interviews were conducted and blood samples taken. Laboratory analyses yielded measures of serum aflatoxin B_1–albumin adducts, iron overload, HBsAg, anti-HCV and other biochemical parameters. The median level of aflatoxin B_1–albumin adducts (pg/mg) was lower among cases (7.3; range 2.4–91.2) than among hospital controls (21.7; range 0–45.6) and family controls (8.7; range 0.7–82.1). Several other parameters (HBsAg, serum ferritin) were higher among cases than controls. [The Working Group noted that the comparability of cases and controls was questionable for the purpose of investigating carcinogenicity of aflatoxin. The measure used to compare aflatoxin levels between groups — the median — may fail to reflect the numbers with high values in the different groups.]

Omer *et al.* (2001) conducted a case–control study in Sudan to assess the association between peanut butter intake as a source of aflatoxins and the *GSTM1* genotype in the etiology of HCC. Cases were 150 patients with HCC who were diagnosed in five out of six hospitals of Khartoum and whose place of residence was in either western Sudan, about 650 km from Khartoum or central Sudan, about 500 km from Khartoum. Controls were 205 residents of the two study areas, and selected by a two-stage process, the second stage of which involved random selection from local village 'sugar shops'. These lists are thought to be comprehensive. Data collection involved a questionnaire which included a particularly detailed history of peanut butter consumption and information on potential confounders. The peanut butter history was transformed into a quantitative cumulative index. Usable blood samples were analysed for HBsAg and anti-HCV (115 cases and 199 controls) and genotyped for *GSTM1* (110 cases and 189 controls). Cases consumed more peanut butter than controls. There was a clear dose–response relationship between average peanut butter consumption and risk for HCC. In the highest quartile of consumption, the OR ranged from 3.0 to 4.0 depending on the covariates included in the model, and all were statistically significant. The pattern of risk differed by region. Peanut butter consumption conferred no increased risk in central Sudan, but a very high risk in western Sudan (OR in the highest quartile, 8.7). Aflatoxin contamination of peanut butter was found to be a much greater problem in western Sudan than in central Sudan (Omer *et al.*, 1998). The authors also noted, however, that residents of the two regions are ethnically different, so effect modification by unmeasured genetic or environmental factors cannot be excluded. While *GSTM1* genotype was not a risk factor for HCC, it was a strong effect modifier. The excess risk due to peanut butter consumption was restricted to subjects with *GSTM1*-null genotype; the OR in the highest quartile of peanut butter exposure among *GSTM1*-null subjects was 17 (95% CI, 2.7–105).

2.4 Limitations of recent studies

While recent studies have incorporated many methodological improvements over studies described in the previous monograph on aflatoxins (IARC, 1993), there nevertheless remain certain problems that limit our ability to fully understand the role of aflatoxins in liver carcinogenesis.

Many recent studies have used HBsAg as the marker of exposure to HBV. However, among liver cancer cases that are negative for HBsAg, HBV DNA can be detected in 33% (serum) and 47% (liver) of the cases, notably those from areas of high viral prevalence. Similarly, HCV RNA can be found in 7% (serum) and 26% (liver) of anti-HCV-negative liver cancer cases (Bréchot *et al.*, 1998). Thus studies relying on HBsAg or anti-HCV measurements may underestimate viral exposure and this may affect an evaluation of interaction between hepatitis viruses and aflatoxin (Paterlini *et al.*, 1994; Kew *et al.*, 1997; Kazemi-Shirazi *et al.*, 2000).

Biomarkers of exposure to aflatoxin have been increasingly used to assess aflatoxin exposure. However, measurable urinary metabolites of aflatoxin or aflatoxin–albumin adducts in serum reflect only exposures in a recent period (days or weeks), and these may not be related to exposures during the etiologically relevant period (years earlier). Moreover, it is unclear whether the presence of liver disease before cancer modifies the levels of the markers found in serum or urine. In the presence of liver disease, comparisons of levels of the marker between liver cancer cases and controls may be biased. Follow-up studies of either general populations in areas of different aflatoxin exposure or of HBsAg carriers investigated with repeated measurements of aflatoxin biomarkers have not yet accumulated follow-up periods that are long enough to minimize the possibility that pre-existing liver disease led to bias in measurements of biomarker levels.

Dietary questionnaires and food measurement surveys at the population level, often used to estimate aflatoxin exposure, provide crude measurements and may fail to account for secular trends in exposure or individual variations in exposure. Similarly, mortality rates used in ecological studies to characterize regions at variable risk of liver cancer may suffer from misclassification of the diagnosis and reporting systems in some countries.

3. Studies of Cancer in Experimental Animals

Studies on the carcinogenicity of aflatoxins in experimental animals completed and reported up to 1993 have been previously evaluated (IARC, 1993). Oral administration of aflatoxin mixtures and aflatoxin B_1 in several strains of rats, hamsters, salmon, trout, ducks, tree shrews and monkeys induced benign and malignant hepatocellular and/or cholangiocellular tumours. Orally administered aflatoxin B_1 did not induce liver tumours in mice. Renal cell tumours were also found in rats following oral administration, while intraperitoneal administration to adult mice increased the incidence of lung adenomas. Intraperitoneal administration of aflatoxin B_1 to pregnant and lactating rats induced benign and malignant liver tumours in mothers and their progeny. Oral administration of aflatoxin B_2 to rats induced liver adenomas, while intraperitoneal administration induced a low incidence of hepatocellular carcinoma. Oral administration of aflatoxin G_1 induced hepatocellular adenomas and carcinomas and renal cell tumours in rats and liver-cell tumours in trout; however, the responses were less than with aflatoxin B_1 at the same dose level. A similar pattern was reported for aflatoxin M_1 and its metabolite aflatoxicol, while aflatoxin Q_1 induced a higher incidence of hepatocellular carcinoma in trout than aflatoxin B_1. Aflatoxin M_1 induced fewer hepatocellular carcinomas following oral administration to rats and trout than aflatoxin B_1 given at an equivalent dose by the same route. Aflatoxicol induced hepatocellular carcinomas in both species; the tumour incidences were lower than that in animals treated with aflatoxin B_1 at the same dose level.

The previous evaluations of aflatoxins (IARC, 1993) were that evidence for carcinogenicity in experimental animals was *sufficient* for aflatoxins B_1, G_1 and M_1, limited for aflatoxin B_2 and *inadequate* for aflatoxin G_2.

This monograph considers only relevant carcinogenicity studies published since 1993.

3.1 Intraperitoneal administration

3.1.1 *Transgenic mouse*

Groups of 11 wild-type $F_1 \times F_1$ (C57BL/6 × CBA) and 12 transgenic mice over-expressing porcine transforming growth factor β_1 (TGF-β_1) [sex unspecified], seven days of age, received aflatoxin B_1 as a single intraperitoneal dose of 6 μg/kg bw. No spontaneous tumours were detected after 12 months in either nine control wild-type or 19 control transgenic mice. Adenomas [assumed to be hepatocellular adenomas] (2/12 and 3/11) and hepatocellular carcinomas (1/12 and 0/11) were detected after 12 months in the liver of the aflatoxin-treated transgenic and wild-type mice, respectively (Schnur *et al.*, 1999). [The Working Group noted the limited reporting of this study.]

3.2 Oral administration of ammoniated forms

3.2.1 *Rat*

In a study to investigate the effect of ammoniation on the carcinogenicity of aflatoxin B_1-contaminated cakes, male and female Wistar WAG and Fischer 344 rats, 4–5 weeks of age, were fed: diet containing 30% peanut oil cake contaminated with 1000 ppb [μg/kg] aflatoxin B_1 and 170 ppb aflatoxin G_1; contaminated diet treated with pressurised ammonia gas (2 bar; 200 kPa) for 15 min at 95 °C (140 ppb aflatoxin B_1 and 20 ppb aflatoxin G_1); or contaminated diet treated with ammonia at a pressure of 3 bar (300 kPa) for 15 min at 95 °C (60 ppb aflatoxin B_1 and 10 ppb aflatoxin G_1). A control group received a diet containing 30% of uncontaminated peanut oil cakes (later determined to contain ~50 ppb aflatoxin B_1); this was reduced to 25% of diet after one month due to greater consumption than the other diets. Deaths occurred among rats receiving the contaminated, untreated diet (10% male Fischer 344, 25% female Fischer 344, 15% male Wistar and 25% female Wistar). Incidences of hepatic tumours at the 12-month termination are given in Table 13. Treatment with ammonia at 2 bar (200 kPa) greatly reduced the carcinogenic potential of the contaminated diet. Treatment with ammonia at 3 bar (300 kPa) eliminated induction of tumours occurring at 12 months (Frayssinet & Lafarge-Frayssinet, 1990).

Table 13. Effect of ammonia treatment on incidence of hepatic tumours in rats fed aflatoxin-contaminated diets for 12 months

Strain	Sex	Contaminated diet	Contaminated/ treated diet (200 kPa)	Contaminated/ treated diet (300 kPa)	Control
Fischer 344	Males	18/19	0/20	0/15	0/20
	Females	5/11	0/10	0/19	0/10
Wistar WAG	Males	17/17	2/20	0/31	0/20
	Females	9/11	1/10	0/20	0/10

From Frayssinet & Lafarge-Frayssinet (1990)

3.2.2 *Trout*

The effect of ammoniation on the hepatocarcinogenic potential of aflatoxin B_1 was investigated in Mount Shasta strain rainbow trout. Duplicate groups of 80 trout (average weight 63 g) were randomly distributed into ~380-L tanks and fed uncontaminated corn or corn contaminated with 180 μg/kg aflatoxins (B_1, 160 μg/kg; B_2, 10 μg/kg; G_1, 9 μg/kg; G_2, trace) treated by ammoniation or untreated, in the following diets for 12 months: uncontaminated untreated corn; uncontaminated treated corn; contaminated untreated corn; and contaminated treated corn. This corn was mixed with modified basal diet and controls received only the basal diet (Table 14). Ten fish were removed from each tank for histopathological evaluation at four and eight months and the remaining fish were held for 12 months. Liver nodules of doubtful classification and five randomly sampled livers at each sampling date were examined histopathologically. At eight months, fish fed contaminated untreated corn had a high incidence (19/20) of hepatoma (Brekke *et al.*, 1977). Ammoniation significantly reduced the carcinogenic potential at 12 months of aflatoxin B_1-contaminated corn (Table 14).

Table 14. Incidence of hepatoma in trout fed various diets for 12 months

Diet	Incidence of hepatoma
Basal diet	1/116
Uncontaminated corn untreated	2/115
Uncontaminated ammonia-treated corn	0/111
Contaminated corn untreated	109/112
Contaminated ammonia-treated corn	3/116

From Brekke *et al.* (1977)

In a more recent experiment, trout were given non-fat dried milk power prepared from the milk of cows that had received aflatoxin-contaminated diets. Thirty lactating cows (Holstein-Friesian) were fed a ration containing ammonia-treated (atmospheric pressure at ambient temperature; AP/AT) aflatoxin-contaminated whole cottonseed (aflatoxin B_1, 5200 μg/kg) for seven days followed by an untreated aflatoxin-containing seed for seven days. The final aflatoxin B_1 concentration in the cow ration was 780 μg/kg. In a second experiment, three lactating cows were fed for 10 days a ration containing ammonia-treated (high pressure at high temperature; HP/HT) aflatoxin-contaminated cottonseed (aflatoxin B_1, 1200 μg/kg), incorporated at 25% (w/w) of the total dry weight of the ration. Then, they were fed for 10 days a ration containing untreated aflatoxin-contaminated cottonseed (aflatoxin B_1, 1200 μg/kg), incorporated at 25% (aflatoxin B_1, 300 μg/kg, in final ration). Milk was collected daily from days 3 to 7 (for each period) when cows were fed the AP/AT material and from days 3 to 10 (for each period) when they were fed the HP/HT material. The milk was processed to prepare non-fat dried milk powder, which was fed as 25% of the diet to rainbow trout (*Oncorhynchus mykiss*) for 12 months. The aflatoxin M_1 levels in the milk powders from cows given untreated and treated seed were: AP/AT, 85 and < 0.05 μg/kg; and HP/HT, 32 and < 0.05 μg/kg, respectively. AP/AT treatment reduced the liver tumour incidence to 2.5% compared with 42% in the trout fed the milk from the cows that received the untreated cottonseed. In positive controls, feeding aflatoxin B_1 (4 μg/kg) continuously for 12 months resulted in a 34% tumour incidence, while feeding for two weeks a diet containing 20 μg/kg aflatoxin B_1 resulted in an incidence of 37% of liver tumours, and feeding of 80 μg/kg or 800 μg/kg aflatoxin M_1 resulted in tumour incidences of 5.7 and 50%, respectively, after nine months. In the separate HP/HT experiment, no tumours were observed in the livers of the trout fed diets containing milk from either the ammonia-treated or untreated source or the control diet containing 8 μg/kg aflatoxin M_1. In positive controls fed 64 μg/kg aflatoxin B_1 for two weeks, tumour incidence was 29% after 12 months. It was concluded that neither aflatoxin M_1 at 8 μg/kg nor any HP/HT-derived aflatoxin derivatives carried over into milk represented a detectable carcinogenic hazard to trout (Bailey *et al.*, 1994a).

3.3 Carcinogenicity of metabolites

Trout: Fry (*O. mykiss*, Shasta strain) were fed Oregon Test Diet (OTD) containing 0, 4, 8, 16, 32 or 64 ng aflatoxin B_1 or aflatoxicol per g dry weight of diet for two weeks. Each dietary group consisted of 400 treated fish or 200 control fish. The experiment was terminated after nine months. The incidence of hepatic tumours in the groups given the 4, 8, 16, 32 and 64 ng/g diet was 25/382 (7%), 98/387 (25%), 194/389 (50%), 287/389 (74%), 302/383 (80%) for aflatoxin B_1 and 57/200 (29%), 143/345 (41%), 183/386 (47%), 255/383 (66%) and 291/390 (75%) for aflatoxicol. No hepatic tumours (0/192) were seen in the controls. In the second protocol, quadruplicate groups of 120 eggs (21-day-old embryos) were exposed to various concentrations (0.01, 0.025, 0.05, 0.1, 0.25, 0.5 μg/mL) of aflatoxin B_1 or aflatoxicol for 1 h. At swimup (after hatching and yolk sac absorption), 360 healthy fry per treatment group (320 and 240 fish from the 0.5 ppm aflatoxin B_1- and aflatoxicol-treated embryos, respectively) were fed the OTD diet for 13 months. There was a dose-related incidence of hepatic tumours: 15/346 (4%), 59/348 (17%), 131/343 (38%), 191/343 (57%), 254/347 (73%), and 252/313 (80%) for the six aflatoxin concentrations and 28/347 (8%), 157/346 (45%), 245/353 (69%), 276/355 (78%), 275/338 (81%) and 148/220 (67%) for the six aflatoxicol concentrations. Aflatoxicol induced a slightly higher hepatic tumour response in fry and fish embryos than did aflatoxin B_1 (Bailey *et al.*, 1994b).

Groups of 120 rainbow trout fry, weighing 1.2 g, were treated with concentrations of aflatoxin B_1 and aflatoxicol ranging from 4 to 64 ng/g of diet and of aflatoxin M_1 and aflatoxicol M_1 ranging from 80 to 1280 ng/g of diet for two weeks. All the fry were then maintained on the control diet until termination at one year. The tumour responses relative to aflatoxin B_1 were: aflatoxin B_1, 1.00; aflatoxicol, 0.936; aflatoxin M_1, 0.086; and aflatoxicol M_1, 0.041. The authors also monitored DNA-adduct formation and concluded that the differences in tumour response were largely accounted for by differences in uptake and metabolism leading to DNA adduct formation, rather than differences in tumour-initiating potency per DNA adduct (Bailey *et al.*, 1998).

3.4 Administration with known carcinogens and other modifying factors

3.4.1 *Viruses*

Transgenic mouse: Hepatitis B virus-positive (HBV+) C57BL/6 mice were bred with *TP53*-null mice (*TP53*$^{-/-}$) to produce *TP53*$^{+/-}$, HBV+ mice. These mice and control litter mates (*TP53*$^{+/+}$, HBV+ and *TP53*$^{+/-}$, HBV–) were randomly divided into groups of 16–24 animals. Approximately half of the animals in each group were females. The experimental group received a single intraperitoneal injection of 10 mg/kg bw aflatoxin B_1 in tricaprylin, while the controls received tricaprylin alone at the age of one week. Surviving animals were sacrificed at 13 months and assessed for HBV positivity by HBsAg expression. The incidence of hepatocellular tumours of Beckers classification grade 2 or higher [adenomas and carcinomas] was 100% in males that were heterozygous for the *TP53* allele and that carried HBV and received aflatoxin B_1 and was 62.5% in males that were homozygous for the wild-type *TP53* allele and had both risk factors (HBV and aflatoxin B_1). The presence of HBV without aflatoxin in heterozygous animals was more potent (25%) than the presence of aflatoxin without HBV (14.2%). The wild-type *TP53* was capable of suppressing tumours when animals were exposed to either risk factor alone (0%). In a companion set of experiments reported later, the relative effect of a mutant allele for *TP53* at serine 246 (the mouse homologue to the human *TP53* 249^{ser} mutation) on the risk factors described above was examined. The 246^{ser} mutation, when present in heterozygous *TP53* male animals, led to the development of tumours in 25% of the animals even without virus or aflatoxin. In animals heterozygous for *TP53* with the 246^{ser} mutation, 71% of the animals receiving aflatoxin B_1 and not carrying HBV had tumours, whereas in the previous study, the heterozygous wild-type allele was associated with 14.2% tumours. Female mice in both studies had fewer tumours but exhibited similar patterns of response. The presence of the *TP53* 246^{ser} mutant not only enhanced the synergistic effect of HBsAg and aflatoxin B_1 but also increased tumorigenesis in aflatoxin B_1-treated mice not expressing HBsAg (Ghebranious & Sell, 1998a,b).

*Tree shrew (*Tupaia belangeri chinensis*)*: Male and female tree shrews, weighing 100–160 g [age unspecified], were divided into four groups. Normal and HBV (human)-infected animals were fed aflatoxin B_1 at 150 μg/kg bw per day in milk on six days per week for 105 weeks. Animals were held until 160 weeks. Hepatocellular carcinomas developed in 67% (14/21) of both male (5/10) and female (9/11) combined HBV-infected/aflatoxin B_1-treated animals and in 30% (3/10) animals treated with aflatoxin B_1 alone (male, 1/6; female, 2/4). The average time for development of hepatocellular carcinoma for males and females combined was significantly reduced ($p < 0.01$) with combined HBV-infection and aflatoxin B_1 treatment (120.3 ± 16.6 weeks) compared with aflatoxin B_1 treatment alone (153 ± 5.8 weeks) (Li *et al.*, 1999).

*Woodchuck (*Marmota monax*)*: Woodchucks were infected with the woodchuck hepatitis virus (WHV; closely related to HBV) at 2–7 days of age. From the age of 12 months, groups of six male and six female WHV-positive and six male and six female WHV-negative animals were fed 50 μg/kg bw aflatoxin B_1 (< 99% pure) in bananas, five days per week for four months followed by treatment with 20 μg/kg bw aflatoxin B_1, five days per week for life. Two untreated groups (WHV-positive and -negative) served as the controls. Woodchucks infected with WHV with or without aflatoxin B_1 treatment developed preneoplastic foci, hepatocellular adenomas and carcinomas between 6 and 26 months after commencing the treatment. Liver tumours were observed by ultrasound at 25 months in 5/9 animals infected with WHV and at 11 months in 1/11, at 19 months in 4/10 and at 25 months in 2/5 animals that received the combined WHV/aflatoxin treatment. No liver tumours were diagnosed in aflatoxin B_1-treated or untreated control animals. The combined treatment resulted in earlier tumour appearance than with WHV alone (Bannasch *et al.*, 1995).

4. Other Data Relevant to an Evaluation of Carcinogenicity and its Mechanisms

4.1 Absorption, distribution, metabolism and excretion

4.1.1 *Humans*

Rigorous quantitative comparisons of dietary intakes and aflatoxin metabolites in body fluids following absorption and distribution are lacking. As noted in the previous monograph (IARC, 1993), aflatoxin M_1 concentrations in urine and human milk have been correlated with dietary aflatoxin intake. However, studies of human exposure have yielded quantitatively very different correlations between aflatoxin concentrations in foods and either aflatoxin–protein or aflatoxin–DNA adducts in urine and sera (Hall & Wild, 1994). Hudson *et al.* (1992) very carefully measured aflatoxin intake based on plate foods in a village in The Gambia. They found intakes less than those estimated from aflatoxin–serum and urinary adduct levels in the same individuals. In humans, as with other species, the DNA binding and carcinogenicity of aflatoxin B_1 result from its conversion to the 8,9-epoxide by cytochrome P450 (CYP) enzymes (Essigman *et al.*, 1982). There is individual variability in the rate of activation of aflatoxin, including between children and adults, which may be material to the pharmacokinetics (Ramsdell & Eaton, 1990; Wild *et al.*, 1990). The pharmacokinetics of aflatoxins in humans are still not clearly known.

Factors that explain variation in response to aflatoxin between individual humans, animal species and strains include the proportion of aflatoxin metabolized to the 8,9-epoxide (mainly by CYP enzymes) relative to the other much less toxic metabolites and the prevalence of pathways forming non-toxic conjugates with reduced mutagenicity and cytotoxicity. Several excellent reviews have been published on the metabolism of aflatoxins since the last IARC monograph on this topic (Eaton & Gallagher 1994; McLean & Dutton 1995; Guengerich *et al.*, 1998).

The 8,9-epoxide of aflatoxin B_1 is short-lived but highly reactive and is the main mediator of cellular injury (McLean & Dutton, 1995). Formation of DNA adducts of aflatoxin B_1-epoxide is well characterized. The primary site of adduct formation is the *N*7 position of the guanine base (Guengerich *et al.*, 1998).

The metabolism of aflatoxins in humans has been extensively studied and the major CYP enzymes involved have been identified as CYP1A2 and CYP3A4 (Gallagher *et al.*, 1996; Ueng *et al.*, 1998). CYP3A4 mediates formation of the *exo*-epoxide and aflatoxin Q_1 while CYP1A2 can generate some *exo*-epoxide but also a high proportion of *endo*-epoxide and aflatoxin M_1. In-vitro evidence that both CYP3A4 and 1A2 are responsible for aflatoxin metabolism in humans has been substantiated by biomarker studies. Aflatoxins M_1 and Q_1, produced by CYP1A2 and 3A4, respectively, are present in the urine of individuals exposed to aflatoxin (Ross *et al.*, 1992; Qian *et al.*, 1994). The *N*7-guanine adducts of aflatoxin are generated primarily by the *exo*-8,9-epoxide, from which yields of adduct are > 98% (Guengerich *et al.*, 1998). The overall contribution of the above enzymes to aflatoxin B_1 metabolism *in vivo* will depend not only on their affinity but also on their expression in human liver, where CYP3A4 is predominant. CYP3A5, in contrast to CYP3A4, metabolizes aflatoxin B_1 mainly to the *exo*-8,9 epoxide but is about 100-fold less efficient in catalysing 3-hydroxylation of aflatoxin B_1 to yield the aflatoxin Q_1 metabolite (Wang *et al.*, 1998). Hepatic CYP3A5 expression differs markedly between individuals, with a proportion of the population, dependent on ethnic group, showing no expression; in particular, 40% of African Americans do not express this enzyme. Therefore, differences in expression of CYP3A5 could influence susceptibility to aflatoxins. Recently, polymorphisms have been identified in the promoter region of CYP3A5 leading to alternative splicing and truncated protein (Hustert *et al.*, 2001; Kuehl *et al.*, 2001). The role of these polymorphisms in susceptibility to aflatoxins is currently unknown.

CYP3A7 (also called P450 HFLa) is a major form of cytochrome P450 in human fetal liver, which has the capacity to activate aflatoxin B_1 to the 8,9-epoxide (Kitada *et al.*, 1989, 1990). This is consistent with the detection of aflatoxin–albumin adducts in the cord blood of newborns whose mothers were exposed to dietary aflatoxin in The Gambia (Wild *et al.*, 1991). Recombinant CYP3A7 conferred sensitivity to aflatoxin B_1 in transfected Chinese hamster lung cells (Kamataki *et al.*, 1995).

In humans, the reactive *exo*- and *endo*-epoxides of aflatoxin B_1 can be detoxified via a number of pathways. One route is glutathione *S*-transferase (GST)-mediated conjugation to reduced glutathione (GSH) to form aflatoxin B_1 *exo*- and *endo*-epoxide–GSH conjugates (Guengerich *et al.*, 1998). The *exo*- and *endo*-epoxides can also hydrolyse rapidly by a non-enzymatic process to an 8,9-dihydrodiol that in turn undergoes slow, base-catalysed ring opening to a dialdehyde phenolate ion (Johnson *et al.*, 1996; Johnson & Guengerich, 1997). The dialdehydes from aflatoxins B_1 and G_1 form Schiff bases with primary amine groups such as those in lysine, to yield protein adducts, for example with albumin (Sabbioni *et al.*, 1987; Sabbioni & Wild 1991). A further metabolic step involves aflatoxin B_1 aldehyde reductase (AFB_1-AR) which catalyses the NADPH-dependent reduction of the dialdehydic phenolate ion to a dialcohol; this enzyme has been characterized in both rats and humans (Hayes *et al.*, 1993; Ireland *et al.*, 1998; Knight *et al.*, 1999).

The role of epoxide hydrolase in hydrolysis of aflatoxin B_1 8,9-epoxide has been investigated (Guengerich *et al.*, 1996; Johnson *et al.*, 1997a,b) with respect to the observed association between epoxide hydrolase genotype and risk for hepatocellular carcinoma in aflatoxin-exposed populations (McGlynn *et al.*, 1995; Tiemersma *et al.*, 2001). If the enzyme is involved, its contribution may be limited, given the rapid non-enzymatic hydrolysis mentioned above (Guengerich *et al.*, 1998).

Oltipraz, an antischistosomal drug, acts as a potent inhibitor of aflatoxin-induced hepatocarcinogenesis in animal models. A total of 234 healthy adults from Qidong (China) were assigned to receive 125 mg oltipraz daily, 500 mg oltipraz weekly or a placebo. Urinary aflatoxin metabolites were quantified by sequential immunoaffinity chromatography and liquid chromatography coupled to mass spectrometry or fluorescence detection. One month of weekly administration of 500 mg oltipraz led to a 51% decrease in the amount of aflatoxin M_1 excreted in urine compared with administration of a placebo ($p = 0.030$), but it had no effect on concentrations of aflatoxin–mercapturic acid ($p = 0.871$). Daily intervention with 125 mg oltipraz led to a 2.6-fold increase in median aflatoxin–mercapturic acid excretion ($p = 0.017$) but had no effect on excreted aflatoxin M_1 levels ($p = 0.682$). It was concluded that the higher dose of oltipraz inhibited aflatoxin activation, whereas the lower dose increased GSH conjugation of aflatoxin 8,9-epoxide (Wang *et al.*, 1999a). Among other things, this clinical trial demonstrates that the results of studies conducted *in vitro* on the major pathways of aflatoxin processing (discussed below) are consistent with human data.

Kirby *et al.* (1993) examined liver tissues from 20 liver cancer patients from Thailand, an area where exposure to aflatoxin occurs. The activity of hepatic CYP isoenzymes and GST was examined and compared with the in-vitro metabolism of aflatoxin B_1. There was considerable inter-individual variation in activity of CYP enzymes, including CYP3A4 (57-fold), CYP2B6 (56-fold) and CYP2A6 (120-fold). In microsomal preparations from liver tumours, metabolism of aflatoxin B_1 to the 8,9-epoxide and aflatoxin Q_1 (the major metabolites) was related to the concentration of CYP3A3/4 and CYP2B6. There was significantly reduced activity of major CYP proteins in microsome preparations from liver tumours compared with those from adjacent non-tumour areas in the liver. The major classes of cytosolic GSTs (α, μ and π) were also analysed in normal and tumorous liver tissue. The activity of α and μ class proteins was decreased and π increased in the majority of tumour cytosols compared with normal liver. Cytosolic GST activity was significantly lower in liver tumours than in normal liver. There was no detectable conjugation of aflatoxin B_1 8,9-epoxide to GSH by microsomal preparations from either normal liver or liver tumour tissue.

Heinonen *et al.* (1996) studied aflatoxin B_1 metabolism in human liver slices from three donors by incubating the tissue with 0.5 μM [^{3}H]aflatoxin B_1 for 2 h. The rates of oxidative metabolism of aflatoxin B_1 to aflatoxins Q_1, P_1 and M_1 were similar to those observed in rat liver slices, albeit with significant interindividual variation. GSH-conjugate formation was not detected in the human liver samples.

It is probable that the apparent discrepancies between studies showing the elimination of mercapturic acids in the urine of aflatoxin-exposed individuals (Wang *et al.*, 1999a) and the apparent lack of formation of glutathione conjugates in cytosolic incubations with aflatoxin B_1 8,9-epoxide (Heinonen *et al.*, 1996) are due to differences in sensitivity of the analytical methods employed.

Rodent studies (see below) have demonstrated that viral damage to the liver affects the metabolism of aflatoxin. Kirby *et al.* (1996a) examined the expression of CYP enzymes in sections of normal human liver and in livers with hepatitis and cirrhosis. By use of immunohistochemical techniques, it was shown that in sections infected with hepatitis B virus (HBV) or hepatitis C virus (HCV), the concentration of CYP2A6 was increased in hepatocytes immediately adjacent to areas of fibrosis and inflammation. In the same tissues, CYP3A4 and CYP2B1 were somewhat increased and CYP1A2 was unaffected compared with normal liver. In HCV-infected liver, CYP2A6, CYP3A4 and CYP2B1 were increased in hepatocytes that had accumulated haemosiderin pigment.

4.1.2 *Experimental systems*

(*a*) *Human tissues*

Data published before 1993 were reviewed in *IARC Monographs* Volume 56 (IARC, 1993). The metabolism and major metabolites of aflatoxin B_1 are shown in Figures 3 and 4.

Gallagher *et al.* (1996) studied the kinetics of aflatoxin oxidation in human liver microsomes and in lymphoblastoid microsomes expressing human CYP3A4 and CYP1A2 cDNA: the K_m was 41 μM for CYP1A2 and 140–180 μM (average affinity for two binding sites) for CYP3A4. In the case of CYP3A4, the rate of product formation dropped as the substrate concentration was reduced. In contrast, CYP1A2 has a higher affinity for aflatoxin. In humans, at plausible serum aflatoxin concentrations, the rate of formation of aflatoxin 8,9-epoxide will be determined by both the lower K_m of CYP1A2 and the greater abundance of CYP3A4 in human liver.

The ability of the human lung to metabolize aflatoxin has been studied in the context of the risk of pulmonary carcinogenesis from handling crops contaminated by aflatoxins. Bioactivation of tritiated aflatoxin B_1 was demonstrated in fresh lung preparations from patients undergoing lobectomy for lung cancer. Lipoxygenase and prostaglandin H synthase activity was shown to be primarily responsible for aflatoxin activation, rather than the CYP enzymes, which display a low level of activity in this tissue (Donnelly *et al.*, 1996).

Neal *et al.* (1998) compared the metabolism of aflatoxins B_1 and M_1 *in vitro* using human liver microsomes. Indirect evidence was obtained for metabolism of aflatoxin M_1 to the 8,9-epoxide by trapping the reactive metabolite with Tris and GSH in the presence of mouse cytosolic fraction. Human liver cytosol did not appear to mediate GSH

Figure 3. Metabolic activation of aflatoxin B_1 to the 8,9-epoxide, leading to binding to glutathione, DNA and serum albumin

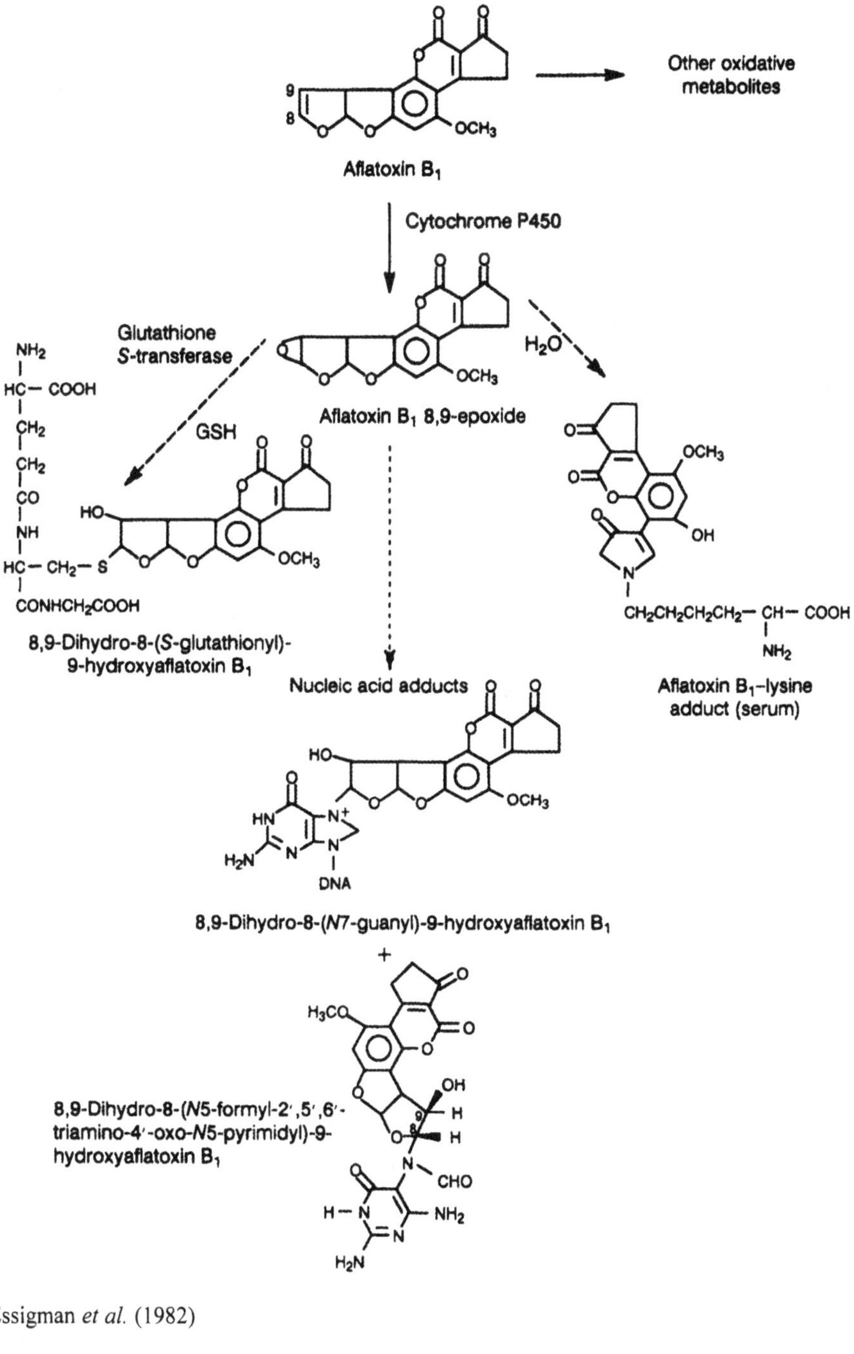

From Essigman *et al.* (1982)

Figure 4. Major metabolites of aflatoxin B_1

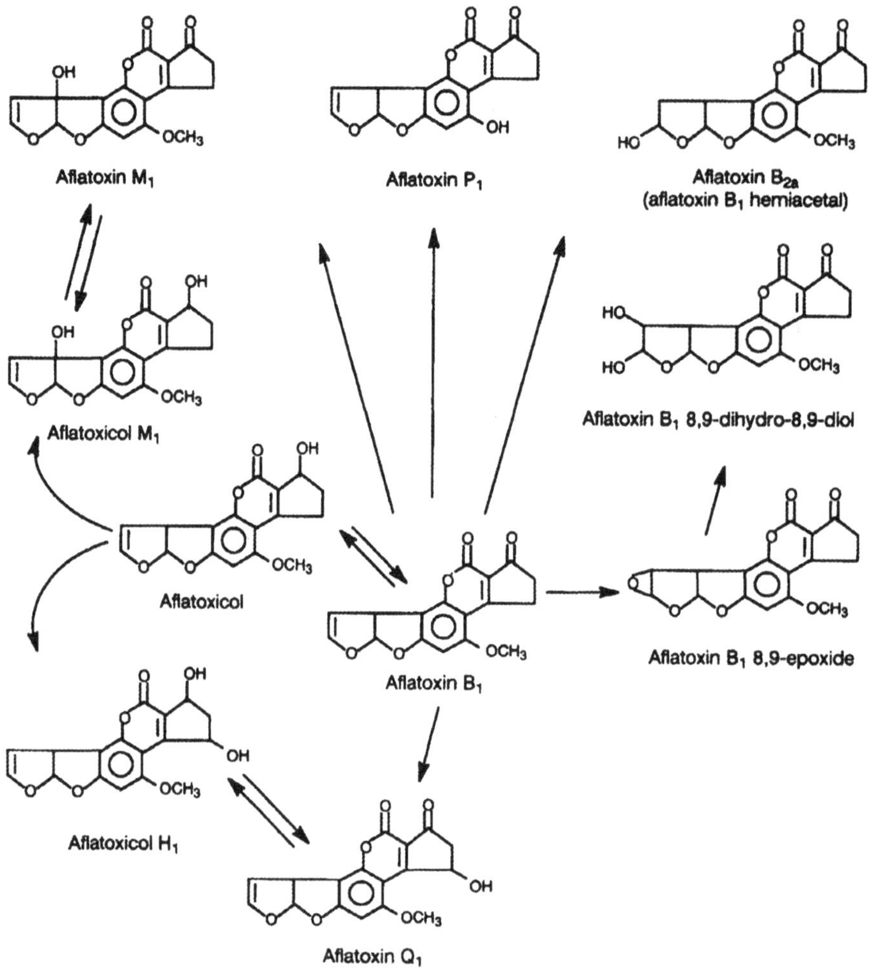

From Essigman *et al.* (1982)

conjugation of either microsomally activated aflatoxin B_1 or aflatoxin M_1. An interesting observation was the cytotoxicity of low doses of aflatoxin M_1 ($\geq$ 0.5 μg/mL) to lymphoblastoid cells in the absence of metabolic activation, which was not observed with aflatoxin B_1.

A non-tumorigenic SV40-immortalized human liver epithelial cell line expressing human CYP1A2 cDNA mediated the formation of both aflatoxin B_1– and aflatoxin M_1–DNA adducts, suggesting that aflatoxin B_1 hydroxylation to aflatoxin M_1 can subsequently lead to DNA damage (Macé *et al.*, 1997).

As mentioned above, human cytosolic fractions or liver slices *in vitro* show little detectable conjugation of aflatoxin B_1 8,9-epoxide. Purified recombinant human α-class

GSTs, namely hGSTA1-1 and hGSTA2-2, also lack significant activity (Raney *et al.*, 1992; Buetler *et al.*, 1996; Johnson *et al.*, 1997a,b). Some conjugating activity was expressed by human μ-class GSTs M1a-1a and M2-2, although mostly towards the *endo*-epoxide (Raney *et al.*, 1992).

Langouët *et al.* (1995) investigated metabolism of aflatoxin B_1 in primary human hepatocytes from eight human liver donors with or without pretreatment by oltipraz. Parenchymal cells obtained from the three GSTM1-positive livers metabolized aflatoxin B_1 to aflatoxin M_1 and to aflatoxin B_1–glutathione conjugates, but no such conjugates were formed in the cells lacking GSTM1. Although oltipraz treatment of the cells induced GSTs A2, A1 and M1, it resulted in decreased formation of aflatoxin M_1 and aflatoxin B_1 oxides due to inhibition of CYP1A2 and CYP3A4.

(*b*) *Experiments on animals and animal tissues*

Kirby *et al.* (1996b) demonstrated that viral infections causing liver injury alter the activity of aflatoxin-metabolizing enzymes in human liver. This has also been shown in transgenic mice that overproduce the HBV large envelope protein, which results in progressive liver cell injury, inflammation and regenerative hyperplasia. The activity of CYP2A5 and CYP3A and a GSTα isoenzyme was examined in these mice. Increased activity and altered distribution of CYP2A5 were shown to be associated with the development of liver injury. The amount of CYP3A was also increased, while GSTα enzyme concentrations were the same in transgenic mice and in otherwise isogenic, non-transgenic mice (Kirby *et al.*, 1994a).

Fetal rat liver contains a GST that forms a conjugate with aflatoxin B_1 8,9-epoxide, identified as a GSTα. By means of immunoblotting and enzyme assays it was shown that liver from adult female rats contains concentrations of one of the enzyme subunits (Yc_2) about 10-fold higher than those in liver from adult male rats. This may contribute to the relative insensitivity of female rats to aflatoxin B_1 (Hayes *et al.*, 1994).

In addition to CYP-mediated activation, aflatoxin B_1 8,9-epoxide can also be formed through metabolism by lipoxygenase and prostaglandin H synthase isolated from guinea-pig tissues and ram seminal vesicles, respectively (Battista & Marnett, 1985; Liu & Massey, 1992). In some organs, for example guinea-pig kidney, the contributions of CYP and prostaglandin H synthase to formation of the epoxide are similar (Liu *et al.*, 1990).

There are marked species differences in sensitivity to aflatoxin carcinogenesis (Gorelick, 1990; Eaton & Gallagher, 1994; Eaton & Groopman, 1994). For example, the adult mouse is almost completely refractory to tumour formation except under conditions of partial hepatectomy or liver injury through expression of HBV antigens. In contrast, the rat is extremely sensitive (see Section 3). A considerable part of this inter-species variation is now understood in terms of differences in activity of aflatoxin-metabolizing enzymes in the pathways described above. Microsomal preparations from mice actually exhibit higher specific activity for aflatoxin B_1 8,9-epoxide production than the rat (Ramsdell & Eaton, 1990). However, in the mouse, the resistance to aflatoxin carcinogenesis is largely if not exclusively explained by the constitutive hepatic expression of an α-class GST, mGSTA3-3, which has high affinity for aflatoxin B_1 8,9-epoxide (Buetler & Eaton, 1992; Hayes *et al.*, 1992). In contrast, rats do not constitutively express a GST isoform with high epoxide-conjugating activity but do express an inducible α-class GST (rGSTA5-5) with high activity. The induction of this enzyme plays a major role in the resistance of rats to aflatoxin B_1-induced hepatocarcinogenicity following treatment with enzyme inducers including oltipraz, ethoxyquin and butylated hydroxyanisole (Kensler *et al.*, 1986, 1987; Hayes *et al.*, 1991, 1994).

A cross-species study of rats (Fischer 344, Sprague-Dawley and Wistar), mice (C57BL), hamsters (Syrian golden) and guinea-pigs (Hartley) was conducted using doses of aflatoxin B_1 between 1 and 80 μg/kg bw per day for up to 14 days by gavage (Wild *et al.*, 1996). Aflatoxin–albumin adducts were measured at 1, 3, 7 and 14 days and hepatic aflatoxin B_1–DNA adducts were measured at the final time point. Both albumin and DNA adducts were formed in the order rat > guinea-pig > hamster > mouse, with similar ratios between the two biomarkers across species, suggesting that the albumin adducts reflected hepatic DNA damage. Calculations from human environmental exposure data and albumin adducts suggested that humans and rats — a sensitive species — have similar formation of albumin adducts for a given exposure to aflatoxin.

In rats, the μ-class enzymes rGSTM2-2 and rGSTM2-3 can conjugate both the *exo*- and *endo*-epoxide of aflatoxin but the latter is the preferred substrate (Raney *et al.*, 1992; Johnson *et al.*, 1997a). Wang *et al.* (2000) showed that the GST-conjugating ability of the non-human primate, *Macaca fascicularis* (mf), towards the 8,9-epoxide was partially due to a μ-class GST, mfaGSTM2-2, with 96% amino acid homology to the human hGSTM2. The enzyme mfaGSTM2-2 was predominantly active towards the *endo*-epoxide, whereas another enzyme, GSHA-GST, for which the encoding cDNA was not cloned, had activity towards the *exo*-epoxide of aflatoxin. However, the activity was about two orders of magnitude lower than that of the rodent α-class GSTs, mGSTA3-3 and rGSTA5-5.

In direct comparison, human and marmoset (*Callithrix jacchus*) hepatic microsomes had similar rates of oxidation of aflatoxin B_1 to the 8,9-epoxide to those of macaques (*Macaca nemestrina*), but GST activity towards the epoxide was below the detection limit in the former two species (Bammler *et al.*, 2000).

Stresser *et al.* (1994a) examined the influence of dietary indole-3-carbinol (found in cruciferous vegetables) on the relative levels of different CYP isozymes known to metabolize aflatoxin B_1 in male Fischer 344 rats. Diets containing 0.2% (w/w) indole-3-carbinol given for seven days were shown to increase the microsomal concentrations of CYP1A1, 1A2 and 3A1/2 (24-, 3.1- and 3.8-fold, respectively, compared with rats receiving a control diet), with a smaller effect on 2B1/2 (1.7-fold) and no effect on CYP2C11. The influence of dietary indole-3-carbinol on the aflatoxin B_1 glutathione detoxication pathway and aflatoxin B_1–DNA adduct formation was also investigated. After seven days of feeding a diet containing indole-3-carbinol (0.2% w/v), rats were administered [^{3}H]aflatoxin B_1 (0.5 mg/kg bw) by intraperitoneal injection and killed after 2 h. The diet with indole-3-carbinol inhibited the formation of aflatoxin B_1–DNA adducts in the liver by 68%, based on analysis of DNA-bound radioactivity (Stresser *et al.*, 1994b).

4.2 Toxic effects

4.2.1 *Humans*

Reports of toxic effects of aflatoxins in humans were reviewed in the previous IARC monograph (IARC, 1993).

There are data suggesting that children are more vulnerable than adults to acute hepatotoxicity resulting from ingestion of aflatoxin. In 1988, 13 Chinese children died of acute hepatic encephalopathy in Perak, Malaysia (Lye *et al.*, 1995). Common symptoms included vomiting, haematemesis and seizures; jaundice was detected in seven cases and all children had liver dysfuntion with elevated serum concentrations of hepatic enzymes (aspartate aminotransferase and alanine aminotransferase). The deaths occurred 1–7 days after hospital admission and were associated with consumption of Chinese rice noodles shortly before the outbreak. Aflatoxins were found in blood and organs from the children (Chao *et al.*, 1991). Pesticides, carbon tetrachloride and mushroom poisons were not found. The flour used to make the noodles was found to contain aflatoxin. Adults who presumably consumed the same contaminated food were not reported to have been affected (Lye *et al.*, 1995).

Children suffering from protein-energy malnutrition in developing countries may also be exposed to aflatoxin. In a study conducted in South Africa, aflatoxin concentrations in serum were higher in 74 children with protein-energy malnutrition than in 35 age-matched control children. The control group, however, had a higher concentration of aflatoxins in urine (Ramjee *et al.*, 1992). [Possible explanations for this result are that aflatoxin metabolism is affected in children with protein-energy malnutrition or that malnourished children are more highly exposed.] A second study compared children with protein-energy malnutrition with high ($n = 21$) and undetectable ($n = 15$) aflatoxin concentrations in serum and urine. The aflatoxin-positive group of children with protein-energy malnutrition showed a significantly lower haemoglobin level ($p = 0.02$), longer duration of oedema ($p = 0.05$), an increased number of infections ($p = 0.03$) and a longer duration of hospital stay ($p = 0.008$) than the aflatoxin-negative group (Adhikari *et al.*, 1994). This finding confirmed results of an earlier study which suggested that malarial infections were increased in children exposed to aflatoxin, as determined on the basis of the amounts of aflatoxin–albumin adducts (Allen *et al.*, 1992). However, a similar study from the Philippines gave inconclusive results (Denning *et al.*, 1995). [The Working Group noted that in these studies estimates of aflatoxin exposures were not available and that possible confounders were not considered.]

4.2.2 *Experimental systems*

No primary studies on the toxicity of aflatoxins were found other than those summarized in IARC (1993) and Eaton and Groopman (1994).

Experimental carcinogenicity studies with aflatoxin B_1 reported previously (IARC, 1993) described preneoplastic lesions of various types in addition to tumours of the liver (mainly hepatocellular carcinomas) in rodents (rats and Syrian hamsters) and non-human primates (rhesus, cynomolgus and African green monkeys). There have been numerous subsequent studies in aflatoxin B_1-exposed animals, especially rats, of GSTP-positive foci in the liver. Particular emphasis has been placed on modification by different co-exposures of the development and frequency of these foci. Often, there was an increase in the number of GSTP-positive foci by L-buthionine sulfoximine (which depletes reduced glutathione) and inhibition of their appearance or rate of development by phenobarbital, anti-oxidants and various sulfur compounds, including dithiolethiones (e.g., Bolton *et al.*, 1993; Gopalan *et al.*, 1993; Maxuitenko *et al.*, 1993; Primiano *et al.*, 1995; Hiruma *et al.*, 1996; Williams & Iatropoulos, 1996; Hiruma *et al.*, 1997; Soni *et al.*, 1997; Maxuitenko *et al.*, 1998).

(*a*) *Immunosuppression*

Studies on the immunosuppressive effects of aflatoxins published before 1993 were reviewed in the previous monograph (IARC, 1993).

Aflatoxins modulate the immune system in domestic and laboratory animals after dietary intake of up to several milligrams per kg feed (Hall & Wild, 1994; Bondy & Pestka, 2000). The major effects involve suppression of cell-mediated immunity, most notably impairment of delayed-type hypersensitivity, which has been a consistent observation at low dose levels in various species (Bondy & Pestka, 2000). Other notable effects include suppression of non-specific humoral substances, reduced antibody formation, suppression of allograft rejection, decreased phagocytic activity and decreased blastogenic response to mitogens (Pier & McLoughlin, 1985; Denning, 1987; WHO, 1990). Strong modification of cytokine secretion and interleukin gene expression has also been observed *in vitro* with mycotoxins, including aflatoxins (Han *et al.*, 1999; Moon *et al.*, 1999; Rossano *et al.*, 1999). The immune system of developing pigs was affected by maternal dietary exposure to aflatoxin B_1 or aflatoxin G_1 during gestation and lactation. Motility and chemotaxis of neutrophils were inhibited in piglets from aflatoxin-treated sows (Silvotti *et al.*, 1997). In a further study, thymic cortical lymphocytes were depleted and thymus weight was reduced in piglets from sows exposed to aflatoxin B_1 (800 ppb [μg/kg] in diet) from day 60 of gestation up to day 28 of lactation (Mocchegiani *et al.*, 1998).

The effects of aflatoxin B_1 on growing rats have been shown to be similar to those in adult animals. Weanling rats [strain unspecified] were given oral doses of 60, 300 or 600 μg/kg bw aflatoxin B_1 in corn oil every other day for four weeks. Aflatoxin B_1 selectively suppressed cell-mediated immunity, assessed by measuring the delayed-type hypersensitivity response, at the 300- and 600-μg/kg bw doses (Raisuddin *et al.*, 1993).

In order to determine the effect of aflatoxin B_1 on the activation of toxoplasmosis, CF1 mice were injected with the cyst-forming parasite *Toxoplasma gondii* one month before aflatoxin B_1 was given by gavage daily for 50 days at 100 μg/kg bw. Cysts developed in the brains of all mice, but the lesions were judged to be more severe in the aflatoxin B_1-treated animals (Venturini *et al.*, 1996).

Several studies have been reported on the effect of aflatoxin on isolated alveolar macrophages, but only few experiments in intact animals. One such study involved male Fischer 344 rats and female Swiss mice that were exposed to aflatoxin B_1 by either aerosol inhalation or intratracheal instillation. Nose-only inhalation exposure of rats to aflatoxin B_1 aerosols suppressed alveolar macrophage phagocytosis at an estimated dose of 16.8 μg/kg bw. The effect persisted for about two weeks. The effects after intratracheal exposures were similar but occurred at approximately 10-fold higher doses. Additionally, intratracheal administration of aflatoxin B_1 suppressed the release of tumour necrosis factor α (TNFα) and inhibited peritoneal macrophage phagocytosis (Jakab *et al.,* 1994).

The overall picture from studies of immunosuppressive effects of aflatoxins in animals is of increased susceptibility to bacterial and parasitic infections and an adverse effect on acquired immunity, as evidenced by experimental challenge with infectious agents after vaccination (reviewed by Denning, 1987). In contrast to the evidence of the immunosuppressive action of aflatoxins in animal studies, evidence in humans comes only from in-vitro experiments. Extremely low doses of aflatoxin B_1 (0.5–1.0 pg/mL) in cultures of human monocytes *in vitro* were shown to decrease phagocytosis and microbicidal activity against *Candida albicans* (Cusumano *et al.*, 1996). Concentrations as low as 0.05 pg/mL were shown to reduce the release of interleukins 1 and 6 and TNFα (Rossano *et al.*, 1999). Mycotoxin-induced immune disruption may influence susceptibility to childhood infections, but may also increase later susceptibility to hepatocellular carcinoma through the child's reduced immune response to hepatitis B virus (HBV) and risk of subsequent development of chronic HBV-carrier status (see Section 4.5.3).

4.3 Reproductive and developmental effects

Reproductive and developmental effects of aflatoxins were reviewed in the previous monograph (IARC, 1993). Aflatoxins cross the placental barrier, and there is some evidence that concentrations in cord blood are higher than those in maternal blood (Lamplugh *et al.*, 1988). Malformations and reduced fetal weight have been found in mice administered high doses (32–90 mg/kg bw) of aflatoxin intraperitoneally. No corresponding effect was seen after oral treatment (Tanimura *et al.*, 1982; Roll *et al.*, 1990). In rats, decreased fetal weight and behavioural changes, but not malformations, have been found at dose levels of around 2–7 mg/kg bw (Sharma & Sahai, 1987).

4.3.1 *Humans*

Several studies have reported high levels of free aflatoxins in maternal and umbilical cord blood in humans living in areas where consumption of large amounts of food highly contaminated with aflatoxins is suspected or has been demonstrated in previous studies. However, the chemical analysis in each study relied on a single method and the results were not confirmed by other means. A number of studies have reported effects in infants, but in most studies, various confounders were not controlled for and exposure levels were not investigated.

Aflatoxins have been reported to occur in up to 40% of samples of breast milk collected from women in tropical Africa (Hendrickse, 1997) (see also Section 1.3.3(*b*)).

Concentrations of aflatoxin M_1 were measured in breast milk of women from Victoria (Australia) and Thailand as a biomarker for exposure to aflatoxin B_1. Aflatoxin M_1 was detected in 11 of 73 samples from Victoria (median concentration, 0.071 ng/mL) and in five of 11 samples from Thailand (median concentration, 0.664 ng/mL) (El-Nezami *et al.*, 1995).

In a survey of the occurrence of aflatoxins in mothers' breast milk carried out in Abu Dhabi and involving 445 donors, 99.5% of samples contained concentrations of aflatoxin M_1 ranging from 2 pg/mL to 3 ng/mL (Saad *et al.*, 1995).

Maxwell (1998) reviewed the presence of aflatoxins in human body fluids and tissues in relation to child health in the tropics. In Ghana, Kenya, Nigeria and Sierra Leone, 25% of cord blood samples contained aflatoxins, primarily M_1 and M_2, in variable amounts (range for aflatoxin M_1: 7 ng/L–65 μg/L).

Of 35 cord serum samples from Thailand, 17 (48%) contained aflatoxin concentrations of 0.064–13.6 nmol/mL (mean, 3.1 nmol/mL). By comparison, only two (6%) of 35 maternal sera obtained immediately after birth of the child contained aflatoxin (mean, 0.62 nmol/mL). These results demonstrate transplacental transfer and indicate that aflatoxin is concentrated by the feto-placental unit (Denning *et al.*, 1990).

A study of 480 children (aged 1–5 years) in Benin and Togo examined aflatoxin exposure in relation to growth parameters. Mean concentrations of aflatoxin–albumin adducts in the blood were 2.5-fold higher in fully weaned children than in those who were still partially breast-fed. There was a strong negative correlation between aflatoxin–albumin adduct levels in the blood and both height-for-age (stunting) and weight-for-age (being underweight) compared with WHO reference population data after adjustment for age, sex, weaning status, socioeconomic status and geographical location. These data suggest that aflatoxin may inhibit growth in West African children (Gong *et al.*, 2002).

In a small study of the presence of aflatoxin in cord blood in Ibadan, Nigeria, a significant reduction in birth weight was found in jaundiced neonates, who had significantly higher serum aflatoxin concentrations compared with babies without jaundice (Abulu *et al.*, 1998).

In a study to investigate whether aflatoxins contribute to the occurrence of jaundice in Ibadan, blood samples were obtained from 327 jaundiced neonates and 60 non-jaundiced controls. Aflatoxins were detected in 24.7% of jaundiced neonates and in 16.6% of controls. Analysis of the data indicated that either glucose-6-phosphate dehydrogenase deficiency or serum aflatoxin are risk factors for neonatal jaundice; odds ratios were significantly increased: 3.0 (95% CI, 1.3–6.7) and 2.7 (95% CI, 1.2–6.1), respectively (Sodeinde *et al.*, 1995).

Aflatoxins were detected in 14 of 64 (37.8%) cord blood samples from jaundiced neonates compared with 9 of 60 (22.5%) samples from non-jaundiced control babies in another study in Nigeria, but the difference was not statistically significant (Ahmed *et al.*, 1995).

Aflatoxins were detected in 37% of cord blood samples in a study of 125 pregnancies in rural Kenya, with 53% of maternal blood samples being aflatoxin-positive. There was no correlation between aflatoxins in maternal and cord blood. A significantly lower mean birth weight of infants born to aflatoxin-positive mothers was recorded for female babies, but not for males (De Vries *et al.*, 1989).

In cord blood collected from 625 babies in Nigeria, aflatoxins were detected in 14.6% of the samples. There was no significant difference in birth weight between the groups positive or negative for aflatoxins (Maxwell *et al.*, 1994).

In a study of the presence of the imidazole ring-opened form of aflatoxin B_1–DNA adducts (see Figure 3) in placenta and cord blood, 69 of 120 (57.5%) placentas contained the adduct at 0.6–6.3 μmol/mol DNA and 5 of 56 (8.9%) cord blood samples contained the adduct at 1.4–2.7 μmol/mol DNA. The results indicate that transplacental transfer of aflatoxin B_1 and its metabolites to the progeny is possible (Hsieh & Hsieh, 1993).

A random sampling of semen from adult men, comprising 50 samples collected from infertile men and 50 samples from fertile men from the same community in Nigeria, revealed the presence of aflatoxin B_1 in 40% of samples from infertile men compared with 8% in fertile men. The mean concentration of aflatoxins in semen of the infertile men was significantly higher than that in semen of fertile men. Infertile men with aflatoxin in their semen showed a higher percentage of spermatozoal abnormalities (50%) than the fertile men (10–15%) (Ibeh *et al.*, 1994).

4.3.2 *Experimental systems*

(*a*) *Developmental toxicity studies*

Behavioural effects were observed in offspring born to Jcl:Wistar rats given subcutaneous injections of 0.3 mg/kg bw aflatoxin B_1 per day on gestation days 11–14 or 15–18. At birth, the number of live pups and their body weight were lower than those of controls. There were no effects on maternal body weight during gestation or lactation. The exposure produced a delay in early response development, impaired locomotor coordination and impaired learning ability. Exposure on days 11–14 of gestation appeared to produce more effects than later exposure (Kihara et al., 2000).

Aflatoxin B_1 produced embryonic mortality and decreased embryo weight and length when injected into embryonating chicken eggs. The number of abnormal embryos was not significantly increased (Edrington *et al.*, 1995).

(*b*) *Reproductive toxicity studies*

Effects suggesting severe impairment of fertility, i.e., reductions in ovarian and uterine size, increases in fetal resorption, disturbances of estrus cyclicity, inhibition of lordosis and reduction in conception rates and litter sizes, were observed in Druckrey rats exposed to 7.5 mg/kg bw aflatoxin B_1 per day for 14 days. An aflatoxin B_1 blood concentration of 86.2 [μg/L] ppb was found in the exposed animals (Ibeh & Saxena, 1997a).

Female Druckrey rats were given oral doses of 7.5 or 15 mg/kg bw aflatoxin B_1 daily for 21 days. Dose-dependent reductions were seen in the number of oocytes and large follicles. The blood hormone levels and sex organ weight were also disturbed (Ibeh & Saxena, 1997b).

Male mature rabbits were given oral doses of 15 or 30 μg/kg bw aflatoxin B_1 every other day for nine weeks followed by a nine-week recovery period. Body weight, relative testes weight, serum testosterone, ejaculate volume, sperm concentration and sperm motility were reduced and the number of abnormal sperm was increased in a dose-dependent manner. These effects continued during the recovery period. Simultaneous treatment with ascorbic acid (20 mg/kg bw) alleviated the effects of exposure to aflatoxin B_1 during the treatment and recovery period (Salem *et al.*, 2001).

The reproductive performance of female mink (Mustela vison) given a diet containing 5 or 10 ppb [μg/kg] total aflatoxins from naturally contaminated corn for 90 days was not impaired compared with a control group. Body weights of the kits were significantly decreased at the 10-ppb dose at birth and in both exposed groups at three weeks of age. Kit mortality was highest in the 10-ppb group and reached 33% by three weeks of age. In the 10-ppb dose group, analysis of milk samples showed very low concentrations of aflatoxin metabolites (Aulerich *et al.*, 1993).

In an experiment to determine the effects of aflatoxin B_1 (2–16 ppb [μg/L] in medium) on the in-vitro fertilizing ability of oocytes and epididymal sperm of albino rats, a significant reduction of the mean number of oocytes fertilized was observed, as well as a significant decrease in sperm motility (Ibeh *et al.*, 2000).

4.4 Genetic and related effects

4.4.1 *Humans*

(*a*) *General*

DNA and protein adducts of aflatoxin have been detected in many studies of human liver tissues and body fluids (IARC, 1993). Some studies related the level of adducts detected to polymorphisms in metabolizing enzymes, in order to investigate interindividual susceptibility to aflatoxin.

Wild *et al.* (1993b) measured serum aflatoxin–albumin adducts in 117 Gambian children in relation to *GSTM1* genotype and found no difference in adduct levels by genotype.

In a larger study of 357 adults in the same population, aflatoxin–albumin adduct levels were examined in relation to genetic polymorphisms in the *GSTM1*, *GSTT1*, *GSTP1* and epoxide hydrolase genes. Only the *GSTM1*-null genotype was associated with a modest increase in aflatoxin–albumin adduct levels and this effect was restricted to non-HBV-infected individuals. *CYP3A4* phenotype, as judged by urinary cortisol metabolite ratios, was also not associated with adduct level. The main factors affecting the level of aflatoxin–albumin adducts were place of residence (rural areas higher than urban areas) and season of blood sample collection (dry season higher than wet season) (Wild *et al.*, 2000). Kensler *et al.* (1998) also found no association between aflatoxin–albumin adducts and *GSTM1* genotype in 234 adults from Qidong County, China.

The role of polymorphisms in the DNA repair enzyme, XRCC1, in influencing the levels of aflatoxin B_1–DNA adducts in samples of placental DNA was studied in women at a Taiwanese maternity clinic. The presence of at least one allele of polymorphism, 399Gln, was associated with a two- to three-fold higher risk of having detectable aflatoxin B_1–DNA adducts compared with subjects homozygous for the 399Arg allele. However, when the association between polymorphism and tertiles of adduct level was examined, the 399Gln allele was associated with intermediate but not high adduct levels. The authors suggested that this may reflect saturation of repair pathways (Lunn *et al.*, 1999).

Studies of the types of genetic alteration associated with exposure to aflatoxin *in vivo* have been less extensive. In human subjects from Qidong County, China, aflatoxin exposure was determined as high or low (dichotomized around the population mean) by aflatoxin–albumin adduct level in serum and compared with the *HPRT* mutation frequency in lymphocytes. A raised *HPRT* mutant frequency was observed in subjects with high compared with low aflatoxin exposure (OR, 19; 95% CI, 2.0–183) (Wang *et al.*, 1999b).

The levels of chromosomal aberrations, micronuclei and sister chromatid exchange were studied in 35 Gambian adults, 32 of whom had measurable concentrations of aflatoxin–albumin adducts. There was no correlation within this group between the cytogenetic alterations and aflatoxin–albumin adducts in peripheral blood at the individual level. In a further study, blood samples of 29 individuals of the same Gambian group were tested for DNA damage in the single-cell gel electrophoresis (comet) assay but no correlation was observed with aflatoxin–albumin adducts or *GSTM1* genotype (Anderson *et al.*, 1999).

(*b*) *TP53 mutations in human hepatocellular carcinoma (HCC)*

Molecular analyses of human HCC have revealed a high prevalence of an AGG to AGT (arg to ser) transversion at codon 249 of the *TP53* tumour-suppressor gene (249^{ser} mutation) in tumours from areas of the world with reported high aflatoxin exposure (Montesano *et al.*, 1997). A large number of studies have been published since 1993 on aflatoxin exposure and *TP53* mutations; two recent meta-analyses examined the relationship between aflatoxin exposure, HBV infection and *TP53* mutations in 20 (Lasky & Magder, 1997) and in 48 published studies (Stern *et al.*, 2001). Table 15 summarizes the published data and the key findings are described below. [It is important to note that the specificity of *TP53* mutations in relation to aflatoxin exposure is associated only with G to T transversions at the third base of codon 249, whereas the meta-analysis by Stern *et al.* (2001) included a few G to T transversions in the second base of

Table 15. Analyses of *TP53* codon 249[ser] mutations in human hepatocellular carcinomas (HCC)

Region/country	No. of HCC analysed	No. with codon 249[ser] mutation	Reference
Africa			
Mozambique	15	8	Ozturk (1991)
South Africa — Transkei	12	1	Ozturk (1991)
Southern Africa	10	3	Bressac *et al.* (1991)
Senegal	15	10	Coursaget *et al.* (1993)
America			
USA			
Alaska	7	0	Buetow *et al.* (1992)
Alaskans	12	0	De Benedetti *et al.* (1995)
	12	0	Kazachkov *et al.* (1996)
	17	0	Wong *et al.* (2000)
Mexico	16	3	Soini *et al.* (1996)
Asia			
China			
Qidong	36	21	Scorsone *et al.* (1992)
	25	13	Fujimoto *et al.* (1994)
	20	9	Li *et al.* (1993)
Xian	45	1	Buetow *et al.* (1992)
Beijing	9	0	Fujimoto *et al.* (1994)
Tongan	21	7	Yang *et al.* (1997)
Jiang-su south	16	9	Shimizu *et al.* (1999)
Jiang-su north	15	1	Shimizu *et al.* (1999)
Shanghai	12	1	Buetow *et al.* (1992)
	18	1	Li *et al.* (1993)
	20	4	Wong *et al.* (2000)
Guanxi	50	18	Stern *et al.* (2001)
Hong Kong	26	1	Ng *et al.* (1994a,b)
	30	4	Wong *et al.* (2000)
India	21	2	Katiyar *et al.* (2000)
Indonesia	4	1	Oda *et al.* (1992)
Japan	128	1	Oda *et al.* (1992)
	10	0	Buetow *et al.* (1992)
	43	0	Murakami *et al.* (1991)
	60	0	Hayashi *et al.* (1993)
	52	0	Konishi *et al.* (1993)
	53	0	Nishida *et al.* (1993)
	20	0	Nose *et al.* (1993)
	34	3	Tanaka *et al.* (1993)
	41	0	Teramoto *et al.* (1994)
	41	0	Hsieh & Atkinson (1995)
	16	0	Wong *et al.* (2000)

Table 15 (contd)

Region/country	No. of HCC analysed	No. with codon 249ser mutation	Reference
Korea (Republic of)	6	0	Oda *et al.* (1992)
	35	0	Park *et al.* (2000)
Singapore (Chinese)	44	0	Shi *et al.* (1995)
Taiwan, China	2	0	Oda *et al.* (1992)
	12	0	Hosono *et al.* (1993)
Europe			
France	100	2	Laurent-Puig *et al.* (2001)
Germany	13	0	Kress *et al.* (1992)
	20	0	Kubicka *et al.* (1995)
Italy	20	0	Bourdon *et al.* (1995)
Spain	70	0	Boix-Ferrero *et al.* (1999)
United Kingdom	19	0	Challen *et al.* (1992)
	170	0	Vautier *et al.* (1999)

Adapted from Stern *et al.* (2001)

this codon. The authors of the meta-analyses pointed out that many studies, particularly the earlier ones, looked only for the presence or absence of the specific codon 249ser mutation and as a consequence may have overemphasized the importance of this particular mutation among the total of *TP53* mutations in HCC.]

The vast majority of these studies have not directly assessed either population exposure or individual exposure to aflatoxin in relation to *TP53* mutations. Instead, estimates were made from data on aflatoxin levels in food, frequency of consumption of those foods and extrapolation to expected aflatoxin levels based on climatic conditions likely to promote aflatoxin production. This limits the interpretation of these data. Montesano *et al.* (1997) attempted to use biomarker data on human exposure to provide information additional to the estimates based on geographical origin of the samples.

Fujimoto *et al.* (1994) studied 25 HCC tissue samples from Qidong County, China, an area of high aflatoxin exposure, and nine HCC samples from Beijing with lower aflatoxin exposure. Thirteen of 25 tumours (52%) from Qidong carried the 249ser mutation, while none of the nine from Beijing did. Shimizu *et al.* (1999) studied 31 HCC tissue samples from different parts of the Jiang-su province. In the northern region of this province, where aflatoxin exposure is expected to be lower, one of 15 (8%) tumours showed a 249ser mutation, while in the southern part, including Haimen City, nine of 16 (56%) HCC samples showed this mutation. Scorsone *et al.* (1992) found the 249ser mutation in 21 of 36 (58%) of HCC tumour tissue samples from Qidong. In contrast, of 26 HCC examined in Hong Kong, only one (4%) had the specific 249ser mutation, although another tumour had a G to T transversion at the second nucleotide (Ng *et al.*, 1994a,b). Wong *et al.* (2000) found the *TP53* 249ser mutation in 4/30 (13%) HCC samples from Hong Kong and in 4/20 (20%) from Shanghai. This mutation was not found in 16 samples from Japan and 17 from the USA, although there were other mutations in exon 7 in the Japanese samples.

In areas of expected low aflatoxin exposure (including Japan, Republic of Korea, Europe and North America), the prevalence of codon 249 mutations is extremely low (< 1%) and even those that do occur tend to be at the second nucleotide rather than the third. Oda *et al.* (1992) analysed 140 HCC tissue samples (128 Japanese, six Korean, four Indonesian and two Taiwanese); of these, only one Japanese and one Indonesian showed the specific 249^{ser} mutation. The limited information on residence and ethnicity in many studies has been commented upon (Laskey & Magder, 1997).

Hollstein *et al.* (1993) measured serum aflatoxin–albumin adducts and liver aflatoxin B_1–DNA adducts from 15 Thai patients, but only one had measurable concentrations of albumin adducts in serum. In none of the samples were aflatoxin B_1–DNA adducts found. Only one had the specific 249^{ser} mutation. In another study of 16 HCC cases from Mexico (Soini *et al.*, 1996), three tumours contained the 249^{ser} mutation; of these, sera were available for two patients and both contained aflatoxin–albumin adducts. Aflatoxin–albumin adducts were detected in sera from all of a further 14 patients without the mutation in the corresponding HCC.

Since chronic HBV infection is a strong and specific risk factor for HCC and aflatoxin exposure commonly co-occurs with viral infection, it is important to examine whether the 249^{ser} mutation is seen only in the presence of chronic HBV infection. Although it is clear from the studies summarized below of HCC in North America, Europe and Japan that HBV alone does not induce the 249^{ser} mutation, the high prevalence of HBV infection in aflatoxin-endemic areas has made it hard to establish whether both risk factors are required for the mutation to occur.

Lasky and Madger (1997) summarized thirteen studies that both ascertained HBV status and analysed *TP53* mutations. Data were available on 449 patients, of whom 201 were positive and 248 negative for HBV markers. The association between aflatoxin exposure and the 249^{ser} mutation was still observed when the analysis was restricted to HBV-positive patients in the groups with high and low aflatoxin exposure. However, the number of HBV-negative patients with high aflatoxin exposure was too small to allow a similar comparison in HBV-negative cases. Overall, it appears that the 249^{ser} mutation related to aflatoxin exposure is not explained by any confounding introduced by possible associations between aflatoxin and HBV.

It remains unclear at what stage in the natural history of HCC the *TP53* mutation occurs. Some information is available from the analysis of histologically normal liver samples from patients resident in areas reportedly differing in aflatoxin exposure level. Aguilar *et al.* (1993, 1994) examined non-tumorous liver tissue from small numbers of HCC patients from Qidong (China), Thailand and the USA and demonstrated the presence of *TP53* AGG to AGT mutations (in codon 249) at a higher frequency in samples from China than in those from Thailand or the USA. By use of an allele-specific polymerase chain reaction (PCR) assay, Kirby *et al.* (1996b) detected the *TP53* 249^{ser}

mutation in non-tumorous liver DNA from five of six HCC patients in Mozambique; none of seven samples from North American patients had a positive signal in the assay. These observations suggest that this type of mutation is found in histologically normal cells of patients with HCC.

The *TP53* 249^{ser} mutation has also been detected in blood samples from HCC patients, patients with cirrhosis and individuals without clinically diagnosed liver disease. Kirk *et al.* (2000) compared 53 HCC patients, 13 patients with cirrhosis and 53 control subjects in The Gambia with 60 non-African French patients, 50 of whom had HCC and 10 had cirrhosis. The 249^{ser} mutation was detected by restriction fragment length polymorphism (RFLP)-PCR in circulating DNA in plasma from 19 (36%) of the HCC patients, two (15%) cirrhosis patients and three (6%) of the African control subjects. The prevalence of the 249^{ser} mutation did not differ between HBsAg-positive and -negative individuals. The mutation was not detected in any of the French plasma samples. No tumour tissue was available from the Gambian patients and so the presence of the same mutations in the corresponding HCC could not be confirmed. The detection of the 249^{ser} mutation in circulating DNA in the plasma of non-cancer patients again could reflect either an early neoplastic event or exposure to aflatoxin.

Jackson *et al.* (2001) examined 20 paired plasma and HCC samples from patients from Qidong County, China, for the presence of the *TP53* 249^{ser} mutation analysed by short oligonucleotide mass spectrometry. Eleven tumours were positive for the mutation and the same mutation was detected in six of the paired plasma samples. An additional four plasma samples were positive for the mutation in the absence of a detectable mutation in the corresponding tumour. The authors suggested that this might be due to other non-sampled HCCs in those patients. In contrast to the findings of Aguilar *et al.* (1993), no 249^{ser} mutations were detected in DNA from normal tissue adjacent to the HCC.

(*c*) *Other genetic alterations in human HCC*

It would be unexpected if aflatoxin carcinogenesis were exclusively associated with a specific *TP53* mutation, given the multiple genetic alterations observed in human HCC. Consequently, several studies have tested the hypothesis that aflatoxin exposure is associated with other specific genetic alterations.

In the study by Fujimoto *et al.* (1994) described above, while the 249^{ser} mutation was more frequent in samples from Qidong than in those from Beijing, additional differences were found in the pattern of loss of heterozygosity (LOH). Specifically, in tumours from Qidong, four of 14 informative cases (28%) showed LOH on chromosome 4 (4p11-q21) and nine of 10 (90%) and 11 of 19 (58%) showed LOH on chromosome 16q22.1 and 16q22-24, respectively. In contrast, none of six informative cases from Beijing showed LOH at 16q22-24 and none of five at 4p11-q21.

Wong *et al.* (2000) studied 83 HCC samples from patients undergoing curative resection. Of these, 50 were from China (30 from Hong Kong, 20 from Shanghai), 16 from Japan and 17 from the USA. The Chinese subjects were all HBV-positive, the Japanese patients were HCV-positive and the patients from the USA were HBV-negative.

In eight subjects (four from Hong Kong and four from Shanghai), single-strand conformation polymorphism (SSCP) analysis and DNA sequencing of exons 5 to 9 of the *TP53* gene revealed the 249^{ser} mutation. However, the authors also performed comparative genomic hybridization. In HCC from Shanghai, there were significantly more alterations per sample in these HBV-related cases than in those from Hong Kong or in the HCCs from Japan and the USA; approximately double the number of alterations per sample was observed in Shanghai compared with Hong Kong. The most frequent changes responsible for this increase were deletions on chromosomes 4q, 8p and 16q and gain of 5p. The authors suggested that this might reflect broader genetic effects of aflatoxin than simply the 249^{ser} mutation in the *TP53* gene.

These studies show that, in addition to *TP53* mutations, geographical location may influence other genetic alterations in HCC, but the data are insufficient to ascribe any of these specifically to aflatoxin exposure.

4.4.2 *Experimental systems*

(*a*) *General*

Aflatoxin B_1 induces mutations in *Salmonella typhimurium* strains TA98 and TA100, and unscheduled DNA synthesis, chromosomal aberrations, sister chromatid exchange, micronucleus formation and cell transformation in various in-vivo and in-vitro mammalian systems (IARC, 1993; for references and details on results published since 1993, see Table 16).

Aflatoxin B_1 can induce mitotic recombination in addition to point mutations. This has been demonstrated in both yeast and mammalian cells. In human lymphoblastoid cells, aflatoxin B_1 treatment led to mitotic recombination and LOH. A reversion assay demonstrated aflatoxin B_1-induced intrachromosomal recombination in a mutant cell line derived from V79 cells harbouring an inactivating tandem duplication in the *Hprt* gene.

Aflatoxin B_1 also induced recombination in minisatellite sequences in yeast expressing recombinant human CYP1A2. In addition, liver tumours derived from HBV-transgenic mice treated with aflatoxin B_1 transplacentally contained rearrangements in minisatellite sequences, but no such alterations were observed in tumours from HBV-transgenic mice not exposed to aflatoxin B_1 (Kaplanski *et al.*, 1997). This suggests that aflatoxin can promote genetic instability in addition to point mutations. Mitotic recombination and genetic instability may therefore be alternative mechanisms by which aflatoxin contributes to genetic alterations such as LOH in HCC (see Section 4.4.1(*c*)).

As expected, aflatoxin B_1 is significantly more mutagenic following metabolic activation. The mutagenicity of aflatoxin B_1 in *Salmonella* tester strains TA98 and TA100 without S9 was approximately 1000 times lower than in the presence of S9.

Splenic lymphocytes were examined for mutant frequency at the *Hprt* locus in Fischer 344 rats exposed to aflatoxin B_1. *Hprt* mutants (frequency, $19.4–31.0 \times 10^{-6}$) were induced after a three-week exposure of male Fischer 344 rats to aflatoxin B_1 by repeated intragastric dosing to a total dose of 1500 μg/kg bw. In the same experiment,

Table 16. Genetic and related effects of aflatoxin B_1

Test system	Result[a]		Dose[b] (LED or HID)	Reference
	Without exogenous metabolic system	With exogenous metabolic system		
Salmonella typhimurium TA100, TA98, reverse mutation	NT	+	2.5 ng/tube	Loarca-Piña *et al.* (1996)
Salmonella typhimurium TA100, TA98, reverse mutation	+	NT	0.6 μg/tube	Loarca-Piña *et al.* (1998)
Salmonella typhimurium TA98, reverse mutation (S9 mix from HepG2 cells)	NT	+	2 μg/plate	Knasmüller *et al.* (1998)
Saccharomyces cerevisiae with recombinant CYP1A2 (human), mitotic recombination	+	NT	5	Kaplanski *et al.* (1998)
DNA damage (comet assay), human HepG2 cells *in vitro*	+		0.0025	Uhl *et al.* (2000)
Gene mutation, human hepatoma (HepG2) cells *in vitro*, *HPRT* locus	NT	+	0.5	Knasmüller *et al.* (1998)
Gene mutation, Chinese hamster ovary AS52 cells *in vitro*, *Gpt* locus	NT	+	0.16	Goeger *et al.* (1998)
Gene mutation, Chinese hamster ovary K_1BH_4 cells *in vitro*, *Hprt* locus	NT	+	0.31	Goeger *et al.* (1998)
Gene mutation, Chinese hamster ovary K_1BH_4 cells *in vitro*, *Hprt* locus	NT	+	0.31	Goeger *et al.* (1999)
Gene mutation, mouse lymphoma L5178Y cells *in vitro*, *tk* locus	NT	+[d]	0.005	Preisler *et al.* (2000)
Recombination, yeast *S. cerevisiae in vitro*[c]	+	NT	7.8	Sengstag *et al.* (1996)
Recombination, Chinese hamster V79 SP5 cells *in vitro*, reversion mutation assay	NT	+	0.19	Zhang & Jenssen (1994)
Micronucleus formation, Chinese hamster V97MZr2B1 cells *in vitro*[e]	+	NT	0.031	Reen *et al.* (1997)
Cell transformation, rat liver epithelial BL9 cells	–	+	7.5	Stanley *et al.* (1999)
Gene mutation, human lymphoblastoid cells (recombinant CYP1A1) *in vitro*, *HPRT* locus	+	NT	0.004	Cariello *et al.* (1994)
Recombination, human lymphoblastoid TK6 cells *in vitro*, *TK* locus	NT	+	0.016	Stettler & Sengstag (2001)
Sister chromatid exchange, human lymphocytes *in vitro*	+	+	9.4	Wilson *et al.* (1995)
Sister chromatid exchange, human leukocytes *in vitro*	NT	+[f]	0.31	Wilson *et al.* (1997)

Table 16 (contd)

Test system	Result[a]		Dose[b] (LED or HID)	Reference
	Without exogenous metabolic system	With exogenous metabolic system		
Gene mutation, male F344 rat splenic lymphocytes *in vivo*, *Hprt* locus	+		0.1 on 5 d/w for 3 w, po	Casciano *et al.* (1996)
Gene mutation, male F344 rat splenic lymphocytes *in vivo*, *Hprt* locus	+		0.01 ppm in diet (intermittent)[g]	Morris *et al.* (1999)
Gene mutation, Big Blue male C57BL/6 mice *in vivo*, *LacI* locus	–		2.5 × 1 ip	Dycaico *et al.* (1996)
Gene mutation, Big Blue male Fischer 344 rats *in vivo*, *LacI* locus	+		0.25 × 1 ip	Dycaico *et al.* (1996)
Micronucleus formation, male Swiss mouse bone marrow *in vivo*	–		1.0 ip × 1	Anwar *et al.* (1994)
Micronucleus formation, male Wistar rat bone marrow *in vivo*	+		0.1 ip × 1	Anwar *et al.* (1994)
Chromosomal aberrations, male Swiss mouse bone marrow *in vivo*	(+)		1.0 ip × 1	Anwar *et al.* (1994)
Chromosomal aberrations, male Wistar rat bone marrow *in vivo*	+		0.1 ip × 1	Anwar *et al.* (1994)
Gene mutation, intrasanguineous host-mediated assay (Wistar rat), *E. coli* K12, *LacI* locus	+		1 ip × 1	Prieto-Alamo *et al.* (1996)

[a] +, positive; (+), weak positive; –, negative; ?, inconclusive; NT, not tested
[b] LED, lowest effective dose; HID, highest ineffective dose; in-vitro tests, μg/mL; in-vivo tests, mg/kg bw/day; NG, not given; po, oral; im, intramuscular; ip, intraperitoneal
[c] Transfected with cDNA encoding human CYP1A1 or CYP1A2
[d] Majority of mutants due to mitotic recombination
[e] Transfected with cDNA encoding rat liver CYP2B1
[f] Activation by microsomes from mouse liver; those from rat or human liver were less active; presence of cytosol from mouse reduced activity, but rat or human cytosol did not.
[g] Animals received aflatoxin B_1 during alternating four-week periods (5–8, 13–16 and 21–24 weeks of age).

aflatoxin-treated rats on a calorically restricted diet showed much lower mutation frequencies (Casciano *et al.*, 1996).

In an intermittent feeding trial, Fischer 344 rats (aged four weeks at the start of the trial) were exposed either to (*a*) control diet, (*b*) various concentrations of aflatoxin B_1 (0.01, 0.10, 0.04, 0.4 or 1.6 ppm [mg/kg]) in the diet for three four-week periods separated by four-week periods of control diet or (*c*) a tumorigenic dose of aflatoxin B_1 (1.6 ppm) in the diet continuously for 20 weeks. A dose-dependent increase in the *Hprt* mutant frequency in splenic lymphocytes was observed after the second four-week feeding period. This effect was further enhanced after the third four-week feeding period, at which time a particularly strong response was observed with the 0.4-ppm dose (mean mutant frequency > 70×10^{-6}). These results may be explained by accumulation of DNA damage in splenic lymphocyte DNA (Morris *et al.*, 1999).

The species differences in susceptibility to the toxic and carcinogenic effects of aflatoxins (Section 4.1.2) are reflected in the differing activities of microsomal preparations to produce genetic damage following aflatoxin treatment. Human, mouse and rat liver preparations were used to activate aflatoxin B_1 and the induction of sister chromatid exchange in human mononuclear leukocytes was examined. In leukocytes treated with aflatoxin B_1 activated by human liver microsomes from six different donors, there was a 10-fold interindividual variation in the mean number (1.1–11.6) of sister chromatid exchanges per cell. The induction of sister chromatid exchange was correlated with CYP1A2 phenotype (using model substrates) in the same livers but not with *GSTM1* genotype or epoxide hydrolase phenotype. Mouse microsomes were more effective than rat or human at activating aflatoxin B_1 to induce sister chromatid exchange. The addition of mouse but not human or rat liver cytosol reduced aflatoxin B_1-induced genotoxicity (Wilson *et al.*, 1997).

Mutation assays of the xanthine-guanine phosphoribosyltransferase (*Gpt*) gene in Chinese hamster ovary AS52 cells and at the *Hprt* gene in Chinese hamster ovary K_1BH_4 cells were performed with aflatoxin B_1 metabolized by liver S9 either from chick embryos or rats; the effect of coumarin as a chemoprotectant was also examined. In the *Gpt* assay, 1 μM aflatoxin B_1 induced 25-fold more mutants when chick S9 was used than with rat S9. Coumarin (50 and 500 μM) decreased the mutant frequency by 52 and 88% with the chick embryo-activated aflatoxin B_1 but had no effect on the frequency following activation with rat S9. In K_1BH_4 cells, a dose of 1 μM aflatoxin B_1 induced approximately sixfold more *Hprt* mutants per 10^6 clonable cells when activated by chick embryo than by rat liver S9 (Goeger *et al.*, 1998).

Male rats and mice were treated with single doses of aflatoxin B_1 (0.01–1.0 mg/kg bw) and the frequency of chromosomal aberrations and micronuclei in bone marrow and the amount of aflatoxin B_1–albumin adducts in peripheral blood were measured. In rats, both chromosomal aberrations and micronuclei showed increased frequency at doses above 0.1 mg/kg, whereas in mice only a slight increase in chromosomal aberrations was seen with the highest dose (1.0 mg/kg) and no effect on micronuclei was detected. In rats, aflatoxin B_1–albumin adduct levels were correlated with chromosomal aberrations at the individual level (Anwar *et al.*, 1994).

Mutations *in vivo* were also studied in lambda/*lacI* (Big Blue®) transgenic C57BL/6 mice and Fischer 344 rats treated with aflatoxin B_1. Six mice were given a single intraperitoneal injection of 2.5 mg/kg bw aflatoxin B_1 but no mutations were detected in the liver 14 days after treatment. In contrast, in six rats treated with 0.25 mg/kg bw aflatoxin B_1, there was a nearly 20-fold increase in mutant frequency (mean, 49×10^{-5}) compared with controls. In mutated *lacI* DNA isolated from rats treated with aflatoxin B_1, the predominant mutations (78%) were GC to TA transversions, compared with 11% of spontaneous mutations of this type in control rats. Of the G to T transversions in the treated rats, 71% were at CpG sites. In particular, 5′-GC<u>G</u>G-3′ and 5′-CC<u>G</u>C-3′ sequences were hotspots (target G underlined) (Dycaico *et al.*, 1996).

Transgenic mice were developed carrying both the human *CYP3A7* gene, expressed in the small intestine but not kidney, and the *rsp*L gene from *Escherichia coli* as a target for mutations. Microsomal preparations from the small intestine of these transgenic mice had higher capacity to convert aflatoxin B_1 to a mutagen in the Ames test with *Salmonella typhimurium* TA98 strain than did kidney microsomes from the same mice or small intestine and kidney microsomes from non-transgenic mice. In addition, the target *rsp*L gene in mice carrying the *CYP3A7* transgene contained a significantly higher mutation frequency in the small intestine than kidney or than either organ in mice non-transgenic for *CYP3A7* (Yamada *et al.*, 1998).

A number of differences have been reported in metabolic activity of S9 fractions from organs other than liver. Ball *et al.* (1995) compared the ability of tracheal and lung S9 from rabbit (male, New Zealand white), hamster (male Syrian golden) and rat (male Sprague Dawley) to induce aflatoxin B_1 mutations in the TA98 strain of *Salmonella typhimurium*. Trachea from hamster and rabbit and lung from rabbit showed a positive response in the assay. In hamsters, trachea S9 was more efficient than lung S9 in producing aflatoxin B_1-induced mutations, while in rabbits the opposite was true.

(*b*) *TP53 mutations in animal tumours*

In order to test the plausibility of an association between aflatoxin exposure and *TP53* mutations, HCC or preneoplastic lesions from several species have been examined for mutations at the codon corresponding to codon 249 in humans, referred to as 'codon 249 equivalent' (Wild & Kleihues, 1996).

In contrast to the specific mutations found in human HCC from areas with high exposure to aflatoxin, no G to T transversions in 'codon 249 equivalent' have been identified in animal tumours. Two major limitations to making a valid comparison are the different DNA sequences in this region of the *TP53* gene across species and the relatively few animal tumours analysed. In ducks, for example, the third nucleotide is not a guanine but a cytosine and no mutations were observed at 'codon 249 equivalent' (Duflot *et al.*, 1994). In all the non-primates studied, codon sequences with any base following CG would lead to a silent mutation at the third nucleotide, each base change resulting in coding for arginine. In four rhesus monkeys (*Macaca mulatta*) and four cynomolgus monkeys (*M. fascicularis*), aflatoxin-induced HCC did not carry any mutation at codon 249, but only four HCC (two from one animal) were analysed (Fujimoto *et al.*, 1992).

Preneoplastic lesions have been examined to define the time point in the natural history of HCC when the *TP53* mutation occurs. Hulla *et al.* (1993) examined six hyperplastic nodules from rat liver following intraperitoneal treatment with 150 μg/kg aflatoxin B_1 for 10 days (2 × 5 days with a two-day break in between, followed by partial hepatectomy three weeks later and sacrifice after a further three weeks) and found no mutations at the codon 249 equivalent.

Female $AC3F_1$ (A/J × C3H/He/J) mice [numbers unspecified], 5–7 weeks of age, received intraperitoneal injections of aflatoxin B_1 three times per week for eight weeks (total dose, 150 mg/kg bw). Mice were killed between 6 and 14 months later. Of the 71 lung tumours examined, 79% showed positive nuclear p53 staining. SSCP analysis of microdissected tumour samples revealed mutations in different codons in exons 5, 6 and 7. Direct sequencing showed 26 mutations which included nine G:C to A:T transitions, 11 A:T to G:C transitions and five transversions (two G:C to T:A, two T:A to A:T and one A:T to C:G). The high mutation frequency and heterogeneous staining pattern suggested that *TP53* mutations occur relatively late in aflatoxin B_1-induced mouse lung tumorigenesis (Tam *et al.*, 1999).

Lee *et al.* (1998) treated 10 male Sprague-Dawley rats with 37.5 μg aflatoxin B_1 five times per week for eight weeks by gavage and a further 20 with the same treatment after partial hepatectomy. Of the latter group, 13 of 17 rats that survived to 60 weeks after treatment had either liver tumours (5), preneoplastic nodules (7) or both (1). Of the six surviving rats from the group that received aflatoxin B_1 alone, two had liver tumours, one had liver focal lesions and one had both. All rats were killed at 60 weeks and liver tumours and preneoplastic lesions were excised. A total of 19 abnormal liver specimens were obtained (10 liver focal lesions and nine liver tumours). The PCR SSCP method was used to screen the *TP53* gene; five rat livers (29%) exhibited abnormal conformational polymorphisms, all five being from the group receiving aflatoxin B_1 after partial hepatectomy. No *TP53* alterations were detected in four samples from the group that received aflatoxin B_1 alone. One liver tumour contained a silent mutation at codon 247 (CGG to CGT, Arg to Arg).

Aflatoxin B_1 activated with quail liver microsomes transformed BL9 rat epithelial cells *in vitro* and these transformed cells induced liver tumours in nude mice. However, the tumours did not contain mutations in codons 242–244 of the *TP53* gene (spanning the equivalent to codon 249 in human *TP53*) (Stanley *et al.*, 1999).

Tree shrews (*Tupaia belangeri chinensis*), which can be infected with human HBV, were used to study aflatoxin B_1–induced mutations in the presence or absence of viral infection. Park *et al.* (2000) studied eight tree shrews, four infected with HBV at 12 weeks of age and four not infected. Two of the uninfected and all four infected animals were treated with 400 μg/kg bw aflatoxin B_1 per day for six days. Five liver tumours were detected upon sacrifice at age 2–3 years, four from HBV- and aflatoxin-treated animals and one from one of the two animals receiving aflatoxin B_1 alone. No specific *TP53* mutations were observed in relation to the codon 249 hotspot.

TP53 mutations in extrahepatic tumours induced by aflatoxin have been rarely studied. Two cholangiocarcinomas, a spindle-cell carcinoma of the bile duct, a haemangioendothelial sarcoma of the liver and an osteogenic sarcoma of the tibia from rhesus and cynomolgus monkeys treated with aflatoxin B_1 were analysed (Fujimoto *et al.*, 1992), but no codon 249 mutations were found. Lung tumours induced in female AC3F$_1$ mice by 150 mg/kg bw aflatoxin B_1 (divided into 24 doses over eight weeks) were microdissected and *TP53* mutations were shown to be frequent events by SSCP and direct sequencing. Most mutations were base substitutions of different types and these were distributed across exons 5–7 of the gene (Tam *et al.*, 1999).

Lung cells isolated from AC3F$_1$ mice seven weeks after treatment with aflatoxin B_1 (2 × 50 mg/kg bw, given two weeks apart) were examined for point mutations in the Ki-*ras* gene. Ki-*ras* mutant alleles were detected in Clara cells but not in other enriched cell fractions. This result indicates the susceptibility of Clara cells to Ki-*ras* activation, an early event in aflatoxin B_1-induced mouse lung tumorigenesis (Donnelly & Massey, 1999).

(*c*) *Sequence-specific binding to DNA and induction of mutations*

Aflatoxin B_1 is metabolically activated to its 8,9-*exo*-epoxide, which reacts with DNA to form the 8,9-dihydro-8-(*N*7-guanyl)-9-hydroxy aflatoxin B_1 (AFB$_1$–*N*7-Gua) adduct (see Section 4.1.2). The positively charged imidazole ring of the guanine adduct promotes depurination and consequently, apurinic site formation. Under slightly alkaline conditions, the imidazole ring of AFB$_1$-*N*7-Gua is opened and forms a more stable and persistent ring-opened aflatoxin B_1–formamidopyrimidine adduct. Investigations have been conducted to establish which is the most likely precursor of the mutations induced by aflatoxin B_1.

The mutations induced by aflatoxin B_1 in a number of experimental systems are certainly consistent with the main carcinogen binding occurring at guanine in DNA, leading to G to T transversions (IARC, 1993). When a pS189 shuttle vector was aflatoxin B_1-modified and then replicated in human Ad293 cells, predominantly G to T transversions were detected (Trottier *et al.*, 1992). However, other types of mutation have also been observed with aflatoxin B_1. Levy *et al.* (1992) transfected an aflatoxin B_1-modified shuttle vector into DNA repair-deficient (XP) or -proficient human fibroblasts, and examined mutations in the *supF* marker gene. Higher mutation frequencies were observed in the DNA repair-deficient cells and the location of mutations was significantly affected by repair proficiency. The majority of mutations were at GC base pairs: 50–70% were G to T transversions, but G to C transversions and G to A transitions were also frequent. A polymerase stop assay was used to examine the location of aflatoxin B_1 binding within the shuttle vector, but no strong correlation was found between initial binding sites and subsequent hotspots for mutation. This suggests that the processing of the adducts, e.g., during DNA replication and repair, can influence not only the overall mutation frequency but also the distribution of mutations.

An intrasanguineous host-mediated assay was used to determine the pattern of mutagenesis induced by aflatoxin B_1 in the *lacI* gene of *E. coli* bacteria recovered from rat liver. Most of the 281 mutations analysed were base substitutions at GC base pairs; over half were GC to TA transversions, with other mutations evenly divided between GC to AT transitions and GC to CG transversions (Prieto-Alamo *et al.*, 1996).

In a human lymphoblastoid cell line (h1A2v2) expressing recombinant human CYP1A1 enzyme, aflatoxin B_1 (4 ng/mL; 25 h) produced a hotspot GC to TA transversion mutation at base pair 209 in exon 3 of the *HPRT* gene in 10–17% of all mutants (Cariello *et al.*, 1994). This hotspot occurred at a GG<u>G</u>GGG sequence (target base underlined).

Bailey *et al.* (1996) studied the induction of mutations with two of the principal forms of DNA damage induced by aflatoxin B_1, namely the AFB_1–*N*7-Gua adduct and the consequent apurinic sites, by site-directed mutagenesis. Single-stranded M13 bacteriophage DNA containing a unique AFB_1–*N*7-Gua adduct or an apurinic site was used to transform *E. coli.* The predominant mutations with AFB_1–*N*7-Gua were G to T transversions targeted to the site of the original adduct (approximately 74%), with lower frequencies of G to A transitions (13–18%) and G to C transversions (1–3%). Using *E. coli* strains differing in biochemical activity of UmuDC and MucAB — proteins involved in processing of apurinic sites by insertion of dAMP — the authors showed that the mutations observed with the AFB_1–*N*7-Gua were not predominantly a simple result of depurination of the initial adduct. A significant number of base substitutions were located at the base 5′ to the site of the original adduct, representing around 13% of the total mutations. This induction of mutations at the base adjacent to the original site of damage was not observed with apurinic sites as the mutagenic lesion. This was suggested to reflect interference with DNA replication following the intercalation of aflatoxin B_1 8,9-epoxide (Gopalakrishnan *et al.*, 1990).

Earlier studies suggested general sequence preferences for aflatoxin B_1 binding dependent on the target guanine being located in a sequence of guanines or with a 5′ cytosine (IARC, 1993). The base 3′ to the modified G appears less consistently predictive of reactivity. Results on sequence-specific binding have been reviewed (Smela *et al.*, 2001). In this context, the question has been raised as to whether cytosine methylation affects the binding of aflatoxin B_1. Ross *et al.* (1999) examined the effect of CpG methylation on binding of aflatoxin B_1 8,9-epoxide to an 11-mer oligodeoxynucleotide containing the sequence of codons 248 and 249 of the *TP53* gene (see Section 4.4.1(*b*)). Binding to methylated or unmethylated fragments of the human *HPRT* gene was also investigated. In neither instance did cytosine methylation affect aflatoxin B_1 binding to guanines within the sequence. In contrast, Chen *et al.* (1998) reported strongly enhanced binding of aflatoxin B_1 to methylated compared with non-methylated CpG sites in *TP53* mutational hotspots, including codon 248, using an UvrABC incision method. The difference between the two studies could reflect the different methods, sequence contexts of the target bases or the methods used to detect DNA adducts (Ross *et al.*, 1999).

4.5 Mechanistic considerations

4.5.1 *Specificity of 249^{ser} mutation in the TP53 gene*

A number of experimental approaches have been used to examine the plausibility of a causal link between aflatoxin-induced DNA damage and the common AGG to AGT transversion mutation at codon 249 of the *TP53* gene in human HCC from areas where aflatoxin exposure is high. Analysis of *TP53* mutations induced by aflatoxin B_1 reveals that G to T transversion is the most common base substitution (see Section 4.4.1(*b*)). A number of studies have examined the sequence specificity of the induction of either DNA damage or mutations in the *TP53* gene (Puisieux *et al.*, 1991; Aguilar *et al.*, 1993; Denissenko *et al.*, 1998, 1999). In these experimental systems, aflatoxin B_1 induces both damage and mutations at the third nucleotide of codon 249, with some evidence of preferential targeting of this latter site in comparison to the adjacent guanine (second nucleotide of codon 249) or guanines in surrounding codons. However, the degree of targeting to codon 249 does not appear to be sufficient by itself to explain the mutational specificity observed in human HCC. Aflatoxin B_1–DNA adducts at other sites within the *TP53* gene would also induce G to T transversions, with alterations of amino acids and associated changes in p53 protein function, but these mutations are far rarer than the codon 249^{ser} mutation in HCC from areas where aflatoxin exposure is high.

The possible effect of the codon 249^{ser} mutation on p53 protein function and any selective growth advantage conferred on hepatocytes carrying this mutation is also of importance. Overall, the codon 249^{ser} mutation in *TP53* appears insufficient to immortalize human hepatocyte cells in culture, but it does confer a growth advantage to previously immortalized cells (Ponchel *et al.*, 1994; Forrester *et al.*, 1995; Schleger *et al.*, 1999). These cell culture studies are consistent with a selective growth advantage resulting from the codon 249^{ser} mutation, but do not fully explain the high prevalence of this mutation in the *TP53* gene in human HCC. However, in-vitro studies generally do not address the role of the mutation in the intact tissue and the studies mentioned above did not assess the role of co-infection with HBV and its possible influence on both the induction and clonal selection of *TP53* codon 249^{ser} mutations. Sohn *et al.* (2000) transfected human liver epithelial cells with the *HBx* gene — encoding the HBVx protein — and these cells were more sensitive to the cytotoxic action of aflatoxin B_1 8,9-epoxide and to induction of apoptosis and mutations at codon 249, possibly as a result of altered excision repair of the aflatoxin B_1–DNA adduct (Hussain & Harris, 2000).

4.5.2 *Modulation of the effects of aflatoxin with chemopreventive agents*

The understanding of human metabolism of aflatoxin B_1 (see Section 4.1) and the extensive literature on chemoprevention of aflatoxin B_1-induced carcinogenesis in experimental animals have provided a rationale for chemoprevention studies in human populations (Kensler *et al.*, 1999). Notably, agents that induce hepatic GST and aflatoxin aldehyde reductase (AFAR) in rats give rise to decreased aflatoxin–DNA and –protein adduct formation and inhibition of aflatoxin-associated carcinogenicity (Roebuck *et al.*, 1991; Groopman *et al.*, 1992; Judah *et al.*, 1993; Kensler *et al.*, 1997; Groopman & Kensler, 1999). Consequently, a similar modulation of the balance between aflatoxin activation and detoxification in humans has been sought and two chemopreventive agents, oltipraz and chlorophyllin, have been evaluated in clinical trials in China.

The protective action of oltipraz is based on inhibition of the enzyme CYP1A2, resulting in reduced formation of the aflatoxin B_1 8,9-epoxide and aflatoxin M_1, and induction of GST enzymes, resulting in increased excretion of the 8,9-epoxide glutathione conjugate as a mercapturic acid (Morel *et al.*, 1993; Langouët *et al.*, 1995). In Chinese subjects exposed to aflatoxin through consumption of their regular diet, concurrent dietary intake of oltipraz was shown to modulate aflatoxin metabolism by increasing excretion of the mercapturic acid and decreasing urinary aflatoxin M_1 concentrations and blood aflatoxin–albumin levels (Jacobson *et al.*, 1997; Kensler *et al.*, 1998; Wang *et al.*, 1999a).

Chlorophyllin is an anti-mutagen in genotoxicity assays *in vitro* and *in vivo* (Dashwood *et al.*, 1998). Mechanistic studies of aflatoxin B_1-induced hepatocarcinogenesis in rainbow trout have revealed that chlorophyllin acts as an 'interceptor molecule' through the formation of tight molecular complexes with aflatoxin B_1 (Breinholt *et al.*, 1995). Consequently, it may diminish the bioavailability of aflatoxin B_1, leading to reduced DNA-adduct formation and tumour development (Breinholt *et al.*, 1999). Chlorophyllin has also been evaluated in a chemoprevention trial in China and consumption of 100 mg of this compound at each meal during four months led to an overall reduction of 55% ($p = 0.036$) in median urinary concentrations of AFB_1–*N*7-Gua compared with placebo controls (Egner *et al.*, 2001).

The above clinical trials confirm that aflatoxin metabolism occurring in people exposed to the toxin through the diet is consistent with the metabolic pathways deduced from *in vitro* and animal model studies. These trials also provide proof of principle that aflatoxin metabolism can be modulated *in vivo* to reduce genotoxic damage; this provides a basis for prevention strategies through dietary modulation.

4.5.3 *Interactions of hepatitis B virus and aflatoxins*

In countries with a high incidence of HCC, endemic infection with HBV is often associated with exposure to aflatoxins. Prospective cohort studies from Asia have observed a multiplicative increase in risk for HCC in individuals chronically infected with HBV and exposed to dietary aflatoxins (see Section 2). Experimental studies in

HBV-transgenic mice and woodchucks also suggest a synergism between the two risk factors in the induction of HCC (Sell *et al.*, 1991; Bannasch *et al.*, 1995). An understanding of the molecular mechanisms behind this interaction is relevant to public health measures aimed at reducing HCC incidence. These mechanisms are considered briefly below; for more detailed information, the reader is referred to a number of review articles (Harris & Sun, 1986; Wild *et al.*, 1993a,b; JECFA, 1998; Sylla *et al.*, 1999; Wild & Hall, 1999).

One possible mechanism of interaction between the virus and the chemical carcinogen is HBV infection altering the expression of aflatoxin-metabolizing enzymes. This has been addressed most extensively in HBV-transgenic mouse lineages carrying the gene for HBsAg, where induction of specific CYP isozymes, namely 1A and 2A5, is observed in association with expression of the *HBsAg* transgene (Chemin *et al.*, 1996; Kirby *et al.*, 1994a), but only in lineages in which transgene expression was associated with induction of liver injury (Chomarat *et al.*, 1998; Chemin *et al.*, 1999). Similar induction of CYP enzymes is observed with liver injury associated with bacterial and parasitic infections (Kirby *et al.*, 1994b; Chomarat *et al.*, 1997), suggesting a general mechanism involving liver injury *per se* rather than a specific effect of HBV on enzyme expression. The modifying effects of HBV-related transgene expression are not limited to CYP enzymes but also include effects on GST enzymes (Chemin *et al.*, 1999). Kirby *et al.* (1996a) showed increased CYP2A6 and CYP3A4 activity in human liver in relation to hepatitis infection. Human liver specimens with evidence of HBV infection had significantly lower total GST activity than non-infected livers (Zhou *et al.*, 1997). In HBV-transfected HepG2 human hepatoma cells, expression of GST α-class enzymes was significantly decreased (Jaitovitch-Groisman *et al.*, 2000).

Some studies have examined aflatoxin metabolism in HBV-infected individuals exposed to the toxin through the diet. Cortisol metabolism was measured as a marker of CYP3A4 activity in relation to aflatoxin–albumin adducts in 357 Gambian adults. No association was observed between CYP3A4 activity and either HBV infection status or adduct levels. The level of aflatoxin–albumin adduct was not related to the HBV status of the individual, including HBV DNA and HBe antigen, which are markers of active viral replication. In addition, there was no correlation between adducts and serum transaminases, markers of liver injury (Wild *et al.*, 2000). In a study of nine adult Gambian HBV carriers and 11 non-carriers with measured dietary intakes of aflatoxin, no differences in levels of urinary aflatoxin–DNA adducts were observed between the two groups (Groopman *et al.*, 1992). These data suggest that the aflatoxin–albumin and urinary aflatoxin–DNA adducts are not influenced by HBV infection in adults. In contrast, in HBV-infected Gambian children there was a higher level of aflatoxin–albumin adducts than in non-infected children, an observation consistent with altered aflatoxin metabolism (Allen *et al.*, 1992; Turner *et al.*, 2000). Similar observations of higher aflatoxin–albumin adduct levels in HBsAg carriers have been reported in a study of 200 Chinese adolescents (Chen *et al.*, 2001).

Thus overall, there is potential for HBV infection to modulate aflatoxin metabolism but the effects are likely to be complex, involving the possibility of both altered activation and detoxification.

An alternative mechanism of interaction between HBV and aflatoxin is that carcinogen exposure may modulate the course of viral infection and replication. In ducklings infected with duck hepatitis virus and treated with aflatoxin B_1, there was a significant increase in markers of viral replication in liver and serum (Barraud *et al.*, 1999, 2000), supporting the hypothesis that aflatoxins can enhance hepadnaviral gene expression. Human hepatoma HepG2 cells transfected with HBV and treated with aflatoxin B_1 also showed an increase in the concentration of transcription factors that may influence HBV expression (Banerjee *et al.*, 2000).

A further hypothesis for the probable interaction of the two risk factors is that the DNA adducts formed by aflatoxin are more likely to be fixed as mutations in the presence of the increased cell proliferation induced by chronic HBV infection.

In summary, plausible mechanisms of interaction between aflatoxins and HBV exist, but to date no conclusion can be drawn as to the most relevant mechanism in terms of the synergistic effects observed in epidemiological studies.

5. Summary of Data Reported

5.1 Exposure data

Aflatoxins are a family of fungal toxins produced mainly by two *Aspergillus* species which are especially abundant in areas of the world with hot, humid climates. *Aspergillus flavus*, which is ubiquitous, produces B aflatoxins. *A. parasiticus*, which produces both B and G aflatoxins, has more limited distribution. Major crops in which aflatoxins are produced are peanuts, maize and cottonseed, crops with which *A. flavus* has a close association. Human exposure to aflatoxins at levels of nanograms to micrograms per day occurs mainly through consumption of maize and peanuts, which are dietary staples in some tropical countries. Maize is also frequently contaminated with fumonisins. Aflatoxin M_1 is a metabolite of aflatoxin B_1 in humans and animals. Human exposure to aflatoxin M_1 at levels of nanograms per day occurs mainly through consumption of aflatoxin-contaminated milk, including mothers' milk. Measurement of biomarkers is being used increasingly to confirm and quantify exposure to aflatoxins.

5.2 Human carcinogenicity data

Studies evaluated in Volume 56 of the *IARC Monographs* led to the classification of naturally occurring aflatoxins as *carcinogenic to humans (Group 1)*. Recent studies have incorporated improvements in study design, study size and accuracy of measurement of markers of exposure to aflatoxin and hepatitis viruses.

In a large cohort study in Shanghai, China, risk for hepatocellular carcinoma was elevated among people with aflatoxin metabolites in urine, after adjustment for cigarette smoking and hepatitis B surface antigen positivity. No association was observed between dietary aflatoxin levels, as ascertained by a diet frequency questionnaire, and risk for hepatocellular carcinoma.

There were four reports from cohort studies in Taiwan, China, although three of them partly overlapped. Selected subjects in the three overlapping studies were enrolled, were interviewed, had biological specimens taken, and were followed up intensively for liver cancer. In nested case–control studies, including some prevalent cases, subjects with exposure to aflatoxin, as assessed by biomarker measurements, had elevated risks for liver cancer, after adjustment for hepatitis B surface antigen positivity. The effect due to aflatoxin exposure was especially high among those who were positive for hepatitis B surface antigen, but there were few liver cancer cases negative for hepatitis B surface antigen. The other Taiwan study was carried out in a large cohort of chronic carriers of hepatitis B virus. These subjects were interviewed at baseline, had biological specimens taken, and were followed up intensively for liver cancer. Several aflatoxin metabolites and albumin adducts were measured in a nested case–control series. Subjects with quantified levels of most of the biomarkers of exposure to aflatoxin showed elevated risk for liver cancer.

In two studies in Qidong, China, cohorts of hepatitis B carriers were tested for biomarkers of aflatoxin and followed up for liver cancer. In both studies, subjects with aflatoxin biomarkers had excess risks for liver cancer.

In a Sudanese case–control study of liver cancer, a relationship was found between reported ingestion of peanut butter and liver cancer in a region with high aflatoxin contamination of peanuts, but no such relationship in a region with low contamination of peanuts.

In a hybrid ecological cross-sectional study in Taiwan, China, a number of subjects were selected from eight regions; for each subject several biomarkers of aflatoxin and hepatitis B viral infection were assessed in relation to the liver cancer rates in the region of residence. There were correlations between aflatoxin metabolites and liver cancer rates after adjustment for hepatitis B status.

The overall body of evidence supports a role of aflatoxins in liver cancer etiology, notably among subjects who are carriers of hepatitis B surface antigen. Nevertheless, the interpretation of human studies is hampered by the difficulties in properly assessing an individual's lifetime exposure to aflatoxins and the difficulties in disentangling the effects of aflatoxins from those of hepatitis infections. Novel biomarkers, some still under development and validation, should bring greater clarity to the issue.

5.3 Animal carcinogenicity data

Extensive experimental studies on the carcinogenicity of aflatoxins led to a previous *IARC Monographs* evaluation of the evidence as follows: *sufficient evidence* for carcinogenicity of naturally occurring mixtures of aflatoxins and of aflatoxins B_1, G_1 and M_1, *limited evidence* for aflatoxin B_2 and *inadequate evidence* for aflatoxin G_2. The principal tumours induced were liver tumours.

Carcinogenicity studies in experimental animals since 1993 were limited to a few experiments in rats, trout, mice, tree shrews and woodchucks. Under certain conditions, including increased pressure, decontamination of feed containing aflatoxins by ammoniation almost completely eliminated the induction of hepatic tumours in rats. Studies in trout showed that ammoniation of aflatoxin-contaminated maize significantly reduced the incidence of liver tumours. In trout fed non-fat dried milk from cows fed ammoniated or non-ammoniated aflatoxin-contaminated whole cottonseed, ammoniation almost eliminated the liver tumour response. Less hepatic tumours were induced in trout after exposure to aflatoxin M_1 than with aflatoxin B_1. One aflatoxin metabolite, aflatoxicol, elicited a slightly higher hepatic tumour response in fry and fish embryos than aflatoxin B_1.

A study in transgenic mice overexpressing transforming growth factor β showed no increased susceptibility to induction of hepatocellular adenomas and carcinomas after intraperitoneal administration of aflatoxin B_1. In another study, induction of hepatocellular tumours by aflatoxin B_1 was significantly enhanced in transgenic mice heterozygous for the *TP53* gene and expressing hepatitis B surface antigen. The tumour response for aflatoxin B_1 was reduced in the absence of either one of these risk factors. The presence of the *TP53* 246^{ser} mutant not only enhanced the synergistic effect of hepatitis B virus and aflatoxin B_1 but also increased tumorigenesis due to aflatoxin B_1 in the absence of hepatitis B virus.

In tree shrews, the incidence of hepatocellular carcinomas was significantly increased and the time of occurrence was shortened in animals treated with aflatoxin B_1 and infected with (human) hepatitis B virus compared with aflatoxin B_1-treated animals. Woodchucks infected with woodchuck hepatitis virus were more sensitive to the carcinogenic effects of aflatoxin B_1 than uninfected woodchucks. The combined woodchuck hepatitis virus/aflatoxin B_1 treatment not only reduced the time of appearance but also resulted in a higher incidence of liver tumours.

In conclusion, recent studies continue to confirm the carcinogenicity of aflatoxins in experimental animals.

5.4 Other relevant data

Metabolism of aflatoxin B_1 in humans has been well characterized, with activation to aflatoxin B_1 8,9-*exo*-epoxide resulting in DNA adduct formation. CYP1A2, 3A4, 3A5, 3A7 and GSTM1 enzymes among others mediate metabolism in humans. The expression of these enzymes can be modulated with chemopreventive agents, resulting in inhibition of DNA-adduct formation and hepatocarcinogenesis in rats. Oltipraz is a chemopreventive agent that increases glutathione conjugation and inhibits some cytochrome P450 enzymes. Results from clinical trials in China using oltipraz are consistent with experimental data in showing that, following dietary exposure to aflatoxins, modulation of the metabolism of aflatoxins can lead to reduced levels of DNA adducts.

Aflatoxin B_1 is immunosuppressive in animals, with particularly strong effects on cell-mediated immunity. Exposure to aflatoxin results in increased susceptibility to bacterial and parasitic infections. Human monocytes treated with aflatoxin B_1 had impaired phagocytic and microbicidal activity and decreases in specific cytokine secretion. Studies have linked human exposure to aflatoxins to increased prevalence of infection.

Aflatoxins cross the human placenta. Aflatoxin exposure has been associated with growth impairment in young children. Malformations and reduced fetal weight have been seen after mice were treated intraperitoneally with high doses of aflatoxin. In rats, decreased pup weight and behavioural changes have been found at low doses. Effects suggesting impairment of fertility have been reported in female and male rats and in male rabbits.

Aflatoxin B_1 is genotoxic in prokaryotic and eukaryotic systems *in vitro*, including human cells, and *in vivo* in humans and in a variety of animal species. It forms DNA and albumin adducts and induces gene mutations and chromosomal alterations including micronuclei, sister chromatid exchange and mitotic recombination.

In geographical correlation studies, exposure to aflatoxin is associated with a specific G to T transversion in codon 249 of the *TP53* gene in human hepatocellular carcinoma. This alteration is consistent with the formation of the major aflatoxin B_1–*N*7-guanine adduct and the observation that G to T mutations are predominant in cell and animal model systems. The high prevalence of the codon 249 mutation in human hepatocellular carcinoma, however, is not fully explained in experimental studies either by the sequence-specific binding and mutation induced by aflatoxin B_1 or by altered function of the p53 protein in studies of hepatocyte growth and transformation.

Current knowledge of the molecular mechanisms contributes to the understanding of the nature of the interaction between hepatitis B virus and aflatoxins in determining risk for hepatocellular carcinoma. Infection with hepatitis B virus may increase aflatoxin metabolism; in hepatitis B virus-transgenic mice, liver injury is associated with increased expression of cytochrome P450 (CYP) enzymes. Glutathione *S*-transferase activity is also reduced in human liver in the presence of hepatitis B virus infection. Other molecular mechanisms are, however, also likely to be relevant to aflatoxin-induced carcinogenesis.

On the basis of the data described above, the existing Group 1 evaluation of naturally occurring aflatoxins was reaffirmed.

5.5 Further research needs

Some research areas are identified here for the purpose of assisting in any future update by an IARC Monographs Working Group. It is not implied that these areas listed override the importance of other research areas or needs, nor should this be construed as endorsement for any specific studies planned or in progress.

- Development and use of molecular dosimetry and/or biomarkers to identify high-risk groups, e.g., in view of children's susceptibility
- Further study of interaction between aflatoxin B_1 and hepatitis B virus and the role of both factors in hepatocarcinogenesis
- Use of biomarker measurements in assessing the association between aflatoxins and hepatocellular carcinoma in epidemiological studies
- Pharmacokinetic studies of ingested aflatoxins in humans (with and without liver disease)
- New epidemiological studies of, for example, liver cancer among populations vaccinated against hepatitis B, populations with exposure to aflatoxins and (limited) exposure to hepatitis B virus, as in Latin America, and joint effects of aflatoxins and hepatitis C virus on liver cancer.

6. References

Abulu, E.O., Uriah, N., Aigbefo, H.S., Oboh, P.A. & Agbonlahor, D.E. (1998) Preliminary investigation on aflatoxin in cord blood of jaundiced neonates. *West Afr. J. Med.*, **17**, 184–187

Adhikari, M., Ramjee, G. & Berjak, P. (1994) Aflatoxin, kwashiorkor and morbidity. *Natural Toxins*, **2**, 1–3

Aguilar, F., Hussain, S.P. & Cerutti, P. (1993) Aflatoxin B_1 induces the transversion of G $\rightarrow$ T in codon 249 of the p53 tumor suppressor gene in human hepatocytes. *Proc. natl Acad. Sci. USA*, **90**, 8586–8590

Aguilar, F., Harris, C.C., Sun, T., Hollstein, M. & Cerutti, P. (1994) Geographic variation of *p53* mutational profile in nonmalignant human liver. *Science*, **264**, 1317–1319

Ahmed, H., Hendrickse, R.G., Maxwell, S.M. & Yakubu, A.M. (1995) Neonatal jaundice with reference to aflatoxins, an aetiological study in Zaria: Northern Nigeria. *Ann. trop. Paediatr.*, **15**, 11–20

Ali, N. (2000) Aflatoxins in Malaysian food. *Mycotoxins*, **50**, 31–35

Ali, N., Sardjono, Yamashita, A. & Yoshizawa, T. (1998) Natural co-occurrence of aflatoxins and *Fusarium* mycotoxins (fumonisins, deoxynivalenol, nivalenol and zearalenone) in corn from Indonesia. *Food Addit. Contam.*, **15**, 377–384

Ali, N., Hashim, N.H. & Yoshizawa, T. (1999) Evaluation and application of a simple and rapid method for the analysis of aflatoxins in commercial foods from Malaysia and the Philippines. *Food Addit. Contam.*, **16**, 273–280

Allen, S.J., Wild, C.P., Wheeler, J.G., Riley, E.M., Montesano, R., Bennett, S., Whittle, H.C., Hall, A.J. & Greenwood, B.M. (1992) Aflatoxin exposure, malaria and hepatitis B infection in rural Gambian children. *Trans. R. Soc. trop. Med. Hyg.*, **86**, 426–430

Alpert, M.E., Hutt, M.S., Wogan, G.N. & Davidson, C.S. (1971) Association between aflatoxin content of food and hepatoma frequency in Uganda. *Cancer*, **28**, 253–260

Anderson, D., Yu, T.-W., Hambly, R.J., Mendy, M. & Wild, C.P. (1999) Aflatoxin exposure and DNA damage in the comet assay in individuals from the Gambia, West Africa. *Teratog. Carcinog. Mutag.*, **19**, 147–155

Anwar, W.A., Khalil, M.M. & Wild, C.P. (1994) Micronuclei, chromosomal aberrations and aflatoxin-albumin adducts in experimental animals after exposure to aflatoxin B_1. *Mutat. Res.*, **322**, 61–67

AOAC International (2000) *AOAC Official Methods 991.31, 991.45, 993.16, 999.17, 994.08, 998.03, 999.07*, Gaithersburg, MD

Applebaum, R.S., Brackett, R.E., Wiseman, D.W. & Marth, E.H. (1982) Aflatoxin: Toxicity to dairy cattle and occurrence in milk and milk products. A review. *J. Food Prot.*, **45**, 752–777

Arim, R.H. (2000) Recent status of mycotoxin research in the Philippines. *Mycotoxins*, **50**, 23–26

Aulerich, R.J., Bursian, S.J. & Watson, G.L. (1993) Effects of sublethal concentrations of aflatoxins on the reproductive performance of mink. *Bull. environ. Contam. Toxicol.*, **50**, 750–756

Autrup, H., Seremet, T., Wakhisi, J. & Wasunna, A. (1987) Aflatoxin exposure measured by urinary excretion of aflatoxin B1-guanine adduct and hepatitis B virus infection in areas with different liver cancer incidence in Kenya. *Cancer Res.*, **47**, 3430–3433

Autrup, J.L., Schmidt, J., Seremet, T. & Autrup, H. (1993) Exposure to aflatoxin B in animal feed production plant workers. *Environ. Health Perspect.*, **99**, 195–197

Bagni, A., Castagnetti, G.B., Chiavarti, C., Ferri, G., Losi, G. & Montanari, G. (1993) [Investigation of the presence of aflatoxins M_1 and M_2 in cows' milk from sampling in the Province of Reggio Emilia.] *Ind. Latte*, **4**, 55–66 (in Italian)

Bailey, G.S., Price, R.L., Park, D.L. & Hendricks, J.D. (1994a) Effect of ammoniation of aflatoxin B_1-contaminated cottonseed feedstock on the aflatoxin M_1 content of cows' milk and hepatocarcinogenicity in the trout bioassay. *Food chem. Toxicol.*, **32**, 707–715

Bailey, G.S., Loveland, P.M., Pereira, C., Pierce, D., Hendricks, J.D. & Groopman, J.D. (1994b) Quantitative carcinogenesis and dosimetry in rainbow trout for aflatoxin B_1 and aflatoxicol, two aflatoxins that form the same DNA adduct. *Mutat. Res.*, **313**, 25–38

Bailey, E.A., Iyer, R.S., Stone, M.P., Harris, T.M. & Essigmann, J.M. (1996) Mutational properties of the primary aflatoxin B_1–DNA adduct. *Proc. natl Acad. Sci. USA*, **93**, 1535–1539

Bailey, G.S., Dashwood, R., Loveland, P.M., Pereira, C. & Hendricks, J.D. (1998) Molecular dosimetry in fish: Quantitative target organ DNA adduction and hepatocarcinogenicity for four aflatoxins by two exposure routes in rainbow trout. *Mutat. Res.*, **399**, 233–244

Ball, R.W., Huie, J.M. & Coulombe, R.A., Jr (1995) Comparative activation of aflatoxin B_1 by mammalian pulmonary tissues. *Toxicol. Lett.*, **75**, 119–125

Bammler, T.K., Slone, D.H. & Eaton, D.L. (2000) Effects of dietary oltipraz and ethoxyquin on aflatoxin B_1 biotransformation in non-human primates. *Toxicol. Sci.*, **54**, 30–41

Banerjee, R., Caruccio, L., Zhang, Y.J., McKercher, S. & Santella, R.M. (2000) Effects of carcinogen-induced transcription factors on the activation of hepatitis B virus expression in human hepatoblastoma HepG2 cells and its implication on hepatocellular carcinomas. *Hepatology*, **32**, 367–374

Bannasch, P., Khoshkhou, N.I., Hacker, H.J., Radaeva, S., Mrozek, M., Zillmann, U., Kopp-Schneider, A., Haberkorn, U., Elgas, M., Tolle, T., Roggendorf, M. & Toshkov, I. (1995) Synergistic hepatocarcinogenic effect on hepadnaviral infection and dietary aflatoxin B_1 in woodchucks. *Cancer Res.*, **55**, 3318–3330

Barraud, L., Guerret, S., Chevallier, M., Borel, C., Jamard, C., Trepo, C., Wild, C.P. & Cova, L. (1999) Enhanced duck hepatitis B virus gene expression following aflatoxin B_1 exposure. *Hepatology*, **29**, 1317–1323

Barraud, L., Douki, T., Guerret, S., Chevallier, M., Jamard, C., Trepo, C., Wild, C.P., Cadet, J. & Cova, L. (2001) The role of duck hepatitis B virus and aflatoxin B_1 in the induction of oxidative stress in the liver. *Cancer Detect. Prev.*, **25**, 192–201

Bartoli, A. & Maggi, O. (1978) Four new species of *Aspergillus* from Ivory Coast soil. *Trans. Br. mycol. Soc.*, **71**, 393–394

Battista, J.R. & Marnett, L.J. (1985) Prostaglandin H synthase-dependent epoxidation of aflatoxin B_1. *Carcinogenesis*, **6**, 1227–1229

Bhat, R.V., Vasanthi, S., Rao, B.S., Rao, R.N., Rao, V.S., Nagaraja, K.V., Bai, R.G., Prasad, C.A.K., Vanchinathan, S., Roy, R., Saha, S., Mukherjee, A., Ghosh, P.K., Toteja, G.S. & Saxena, B.N. (1996) Aflatoxin B_1 contamination in groundnut samples collected from different geographical regions of India: A multicentre study. *Food Addit. Contam.*, **13**, 325–331

Bhat, R.V., Vasanthi, S., Rao, R.S., Rao, R.N., Rao, V.S., Nagaraja, K.V., Bai, R.G., Prasad, C.A.K., Vanchinathan, S., Roy, R., Saha, S., Mukherjee, A., Ghosh, P.K., Toteja, G.S. & Saxena, B.N. (1997) Aflatoxin B_1 contamination in maize samples collected from different geographical regions of India — A multicentre study. *Food Addit. Contam.*, **14**, 151–156

Boix-Ferrero, J., Pellin, A., Blesa, R., Adrados, M. & Llombart-Bosch, A. (1999) Absence of p53 gene mutations in hepatocarcinomas from a Mediterranean area of Spain. A study of 129 archival tumour samples. *Virchows Arch.*, **434**, 497–501

Bolton, M.G., Munoz, A., Jacobson, L.P., Groopman, J.D., Maxuitenko, Y.Y., Roebuck, B.D. & Kensler, T.W. (1993) Transient intervention with oltipraz protects against aflatoxin-induced hepatic tumorigenesis. *Cancer Res.*, **53**, 3499–3504

Bondy, G.S. & Pestka, J.J. (2000) Immunomodulation by fungal toxins. *J. Toxicol. environ. Health (part B)*, **3**, 109–143

Bourdon, J.C., D'Errico, A., Paterlini, P., Grigioni, W., May, E. & Debuire, B. (1995) p53 Protein accumulation in European hepatocellular carcinoma is not always dependent on p53 gene mutation. *Gastroenterology*, **108**, 1176–1182

Breinholt, V., Schimerlik, M., Dashwood, R. & Bailey, G. (1995) Mechanisms of chlorophyllin anticarcinogenesis against aflatoxin B_1: Complex formation with the carcinogen. *Chem. Res. Toxicol.*, **8**, 506–514

Breinholt, V., Arbogast, D., Loveland, P., Pereira, C., Dashwood, R., Hendricks, J. & Bailey, G. (1999) Chlorophyllin chemoprevention in trout initiated by aflatoxin B_1 bath treatment: An evaluation of reduced bioavailability vs. target organ protective mechanisms. *Toxicol. appl. Pharmacol.*, **158**, 141–151

Brekke, O.L., Sinnhuber, R.O., Peplinski, A.J., Wales, J.H., Putnam, G.B., Lee, D.J. & Ciegler, A. (1977) Aflatoxin in corn: Ammonia inactivation and bioassay with rainbow trout. *Appl. environ. Microbiol.*, **34**, 34–37

Bressac, B., Kew, M., Wands, J. & Ozturk, M. (1991) Selective G to T mutations of p53 gene in hepatocellular carcinoma from southern Africa. *Nature*, **350**, 429–431

Buetler, T.M. & Eaton, D.L. (1992) Complementary DNA cloning, messenger RNA expression, and induction of alpha-class glutathione S-transferases in mouse tissues. *Cancer Res.*, **52**, 314–318

Buetler, T.M., Bammler, T.K., Hayes, J.D. & Eaton, D.L. (1996) Oltipraz-mediated changes in aflatoxin B_1 biotransformation in rat liver: Implications for human chemointervention. *Cancer Res.*, **56**, 2306–2313

Buetow, K.H., Sheffield, V.C., Zhu, M., Zhou, T., Shen, F.-M., Hino, O., Smith, M., McMahon, B.J., Lanier, A.P., London, W.T., Redeker, A.G. & Govindarajan, S. (1992) Low frequency of *p53* mutations observed in a diverse collection of primary hepatocellular carcinomas. *Proc. natl Acad. Sci. USA*, **89**, 9622–9626

Bulatao-Jayme, J., Almero, E.M., Castro, M.C., Jardeleza, M.T. & Salamat, L.A. (1982) A case–control dietary study of primary liver cancer risk from aflatoxin exposure. *Int. J. Epidemiol.*, **11**, 112–119

Campbell, T.C., Chen, J.S., Liu, C.B., Li, J.Y. & Parpia, B. (1990) Nonassociation of aflatoxin with primary liver cancer in a cross-sectional ecological survey in the People's Republic of China. *Cancer Res.*, **50**, 6882–6893

Cariello, N.F., Cui, L. & Skopek, T.R. (1994) *In vitro* mutational spectrum of aflatoxin B_1 in the human hypoxanthine guanine phosphoribosyltransferase gene. *Cancer Res.*, **54**, 4436–4441

Casciano, D.A., Chou, M., Lyn-Cook, L.E. & Aidoo, A. (1996) Calorie restriction modulates chemically induced in vivo somatic mutation frequency. *Environ. mol. Mutag.*, **27**, 162–164

Castegnaro, M. & Pfohl-Leszkowicz, A. (1999) Aflatoxins. In: *Les Mycotoxines dans l'Alimentation: Evaluation et Gestion du Risque [Mycotoxins in Food: Risk Evaluation and Management]*, Paris, Technique et Documentation, pp. 199–247

Castegnaro, M., Hunt, D.C., Sansone, E.B., Schuller, P.L., Siriwardana, M.G., Telling, G.M., van Egmond, H.P. & Walker, E.A., eds (1980) *Laboratory Decontamination and Destruction of Aflatoxins B_1, B_2, G_1, G_2 in Laboratory Wastes* (IARC Scientific Publications No. 37), Lyon, IARC*Press*

Castegnaro, M., Pleština, R., Dirheimer, G., Chernozemsky, I.N. & Bartsch, H., eds (1991) *Mycotoxins, Endemic Nephropathy and Urinary Tract Tumours* (IARC Scientific Publications No. 115), Lyon, IARC*Press*

Challen, C., Lunec, J., Warren, W., Collier, J. & Bassendine, M.F. (1992) Analysis of the *p53* tumor-suppressor gene in hepatocellular carcinomas from Britain. *Hepatology*, **16**, 1362–1366

Chao, T.C., Maxwell, S.M. & Wong, S.Y. (1991) An outbreak of aflatoxicosis and boric acid poisoning in Malaysia: A clinicopathological study. *J. Pathol.*, **164**, 225–233

Chemin, I., Takahashi, S., Belloc, C., Lang, M.A., Ando, K., Guidotti, L.G., Chisari, F.V. & Wild, C.P. (1996) Differential induction of carcinogen metabolizing enzymes in a transgenic mouse model of fulminant hepatitis. *Hepatology*, **24**, 649–656

Chemin, I., Ohgaki, H., Chisari, F.V. & Wild, C.P. (1999) Altered expression of hepatic carcinogen metabolizing enzymes with liver injury in HBV transgenic mouse lineages expressing various amounts of hepatitis B surface antigen. *Liver*, **19**, 81–87

Chen, J. (1997) *Dietary Aflatoxin Intake Levels in China: Data Compilation* (information submitted to WHO)

Chen, C.J., Wang, L.Y., Lu, S.N., Wu, M.H., You, S.L., Zhang, Y.J., Wang, L.W. & Santella, R.M. (1996) Elevated aflatoxin exposure and increased risk of hepatocellular carcinoma. *Hepatology*, **24**, 38–42

Chen, J.X., Zheng, Y., West, M. & Tang, M.-S. (1998) Carcinogens preferentially bind at methylated CpG in the p53 mutational hot spots. *Cancer Res.*, **58**, 2070–2075

Chen, S.-Y., Chen, C.-J., Chou, S.-R., Hsieh, L.-L., Wang, L.-Y., Tsai, W.-Y., Ahsan, H. & Santella, R.M. (2001) Association of aflatoxin B_1-albumin adduct levels with hepatitis B surface antigen status among adolescents in Taiwan. *Cancer Epidemiol. Biomarkers Prev.*, **10**, 1223–1226

Chomarat, P., Sipowicz, M.A., Diwan, B.A., Fornwald, L.W., Awasthi, Y.C., Anver, M.R., Rice, J.M., Anderson, L.M. & Wild, C.P. (1997) Distinct time courses of increase in cytochromes P450 1A2, 2A5 and glutathione S-transferases during the progressive hepatitis associated with *Helicobacter hepaticus*. *Carcinogenesis*, **18**, 2179–2190

Chomarat, P., Rice, J.M., Slagle, B.L. & Wild, C.P. (1998) Hepatitis B virus-induced liver injury and altered expression of carcinogen metabolising enzymes: The role of the HBx protein. *Toxicol. Lett.*, **102–103**, 595–601

Christensen, M. (1981) A synoptic key and evaluation of species in the *Aspergillus flavus* group. *Mycologia*, **73**, 1056–1084

Codex Alimentarius Commission (1999) *Codex Alimentarius, Twenty-third Session, Rome, Italy, 28 June–3 July 1999, Report of the Thirtieth Session of the Codex Committee on Food Additives and Contaminants, The Hague, The Netherlands, 9–13 March 1998, ALINORM 99/12*, Rome, FAO

Codex Alimentarius (2000) *Codex Committee on Food Additives and Contaminants. CL 1999/13 GEN-CX 0016 FAC-Agenda item 16a. ALINORM 99/37, paras 103–105*. Draft maximum level for aflatoxin M_1 in milk, 20–24 March, Rome, FAO

Cole, R.J. & Cox, R.H. (1981) *Handbook of Toxic Fungal Metabolites*, New York, Academic Press, pp. 1–66

Coursaget, P., Depril, N., Chabaud, M., Nandi, R., Mayelo, V., LeCann, P. & Yvonnet, B. (1993) High prevalence of mutations at codon 249 of the p53 gene in hepatocellular carcinomas from Senegal. *Br. J. Cancer*, **67**, 1395–1397

Cusumano, V., Rossano, F., Merendino, R.A., Arena, A., Costa, G.B., Mancuso, G., Baroni, A. & Losi, E. (1996) Immunobiological activities of mould products: Functional impairment of human monocytes exposed to aflatoxin B_1. *Res. Microbiol.*, **147**, 385–391

Dalcero, A., Magnoli, C., Chiacchiera, S., Palacios, G. & Reynoso, M. (1997) Mycoflora and incidence of aflatoxin B_1, zearalenone and deoxynivalenol in poultry feeds in Argentina. *Mycopathologia*, **137**, 179–184

Dashwood, R., Negishi, T., Hayatsu, H., Breinholt, V., Hendricks, J. & Bailey, G. (1998) Chemopreventive properties of chlorophylls towards aflatoxin B_1: A review of the antimutagenicity and anticarcinogenicity data in rainbow trout. *Mutat. Res.*, **399**, 245–253

da Silva, J.B., Pozzi, C.R., Mallozzi, M.A.B., Ortega, E.M. & Correa, B. (2000) Mycoflora and occurrence of aflatoxin B_1 and fumonisin B_1 during storage of Brazilian sorghum. *J. agric. Food Chem.*, **48**, 4352–4356

De Benedetti, V.M., Welsh, J.A., Trivers, G.E., Harpster, A., Parkinson, A.J., Lanier, A.P., McMahon, B.J. & Bennett, W.P. (1995) *p53* is not mutated in hepatocellular carcinomas from Alaska natives. *Cancer Epidemiol. Biomarkers Prev.*, **4**, 79–82

Denissenko, M.F., Koudriakova, T.B., Smith, L., O'Connor, T.R., Riggs, A.D. & Pfeifer, G.P. (1998) The *p53* codon 249 mutational hotspot in hepatocellular carcinoma is not related to selective formation or persistence of aflatoxin B_1 adducts. *Oncogene*, **17**, 3007–3014

Denissenko, M.F., Cahill, J., Koudriakova, T.B., Gerber, N. & Pfeifer, G.P. (1999) Quantitation and mapping of aflatoxin B_1-induced DNA damage in genomic DNA using aflatoxin B_1-8,9-epoxide and microsomal activation systems. *Mutat. Res.*, **425**, 205–211

Denning, D.W. (1987) Aflatoxin and human disease. *Adv. Drug React. Ac. Pois. Rev.*, **6**, 175–209

Denning, D.W., Allen, R., Wilkinson, A.P. & Morgan, M.R. (1990) Transplacental transfer of aflatoxin in humans. *Carcinogenesis*, **11**, 1033–1035

Denning, D.W., Quiepo, S.C., Altman, D.G., Makarananda, K., Neal, G.E., Camallere, E.L., Morgan, M.R.A. & Tupasi, T.E. (1995) Aflatoxin and outcome from acute lower respiratory infection in children in The Philippines. *Ann. trop. Paediatr.*, **15**, 209–216

De Vries, H.R., Maxwell, S.M. & Hendrickse, R.G. (1989) Foetal and neonatal exposure to aflatoxins. *Acta paediatr. scand.*, **78**, 373–378

Donnelly, P.J. & Massey, T.E. (1999) Ki-*ras* activation in lung cells isolated from AC3F1 (A/J x C3H/HeJ) mice after treatment with aflatoxin B_1. *Mol. Carcinog.*, **26**, 62–67

Donnelly, P.J., Stewart, R.K., Ali, S.L., Conlan, A.A., Reid, K.R., Petsikas, D. & Massey, T.E. (1996) Biotransformation of aflatoxin B_1 in human lung. *Carcinogenesis*, **17**, 2487–2494

Dorner, J.W., Cole, R.J. & Diener, U.L. (1984) The relationship of *Aspergillus flavus* and *Aspergillus parasiticus* with reference to production of aflatoxins and cyclopiazonic acid. *Mycopathologia*, **87**, 13–15

Dragacci, S. & Frémy, J.M. (1993) [Milk contamination by aflatoxin M_1: Results of 15 years of surveillance]. *Sci. Alim.*, **13**, 711–722 (in French)

Duflot, A., Hollstein, M., Mehrotra, R., Trepo, C., Montesano, R. & Cova, L. (1994) Absence of *p53* mutation at codon 249 in duck hepatocellular carcinomas from the high incidence area of Qidong (China). *Carcinogenesis*, **15**, 1353–1357

Dycaico, M.J., Stuart, G.R., Tobal, G.M., de Boer, J.G., Glickman, B.W. & Provost, G.S. (1996) Species-specific differences in hepatic mutant frequency and mutational spectrum among lambda/*lacI* transgenic rats and mice following exposure to aflatoxin B_1. *Carcinogenesis*, **17**, 2347–2356

Eaton, D.L. & Gallagher, E.P. (1994) Mechanisms of aflatoxin carcinogenesis. *Annu. Rev. Pharmacol. Toxicol.*, **34**, 135–172

Eaton, D.L. & Groopman, J.D., eds (1994) *The Toxicology of Aflatoxins: Human Health, Veterinary, and Agricultural Significance*, San Diego, CA, Academic Press

Edrington, T.S., Harvey, R.B. & Kubena, L.F. (1995) Toxic effects of aflatoxin B_1 and ochratoxin A, alone and in combination, on chicken embryos. *Bull. environ. Contam. Toxicol.*, **54**, 331–336

Egner, P.A., Wang, J.-B., Zhu, Y.-R., Zhang, B.-C., Wu, Y., Zhang, Q.-N., Qian, G.-S., Kuang, S.-Y., Gange, S.J., Jacobson, L.P., Helzlsouer, K.J., Bailey, G.S., Groopman, J.D. & Kensler, T.W. (2001) Chlorophyllin intervention reduces aflatoxin-DNA adducts in individuals at high risk for liver cancer. *Proc. natl Acad. Sci. USA*, **98**, 14601–14606

El-Nezami, H.S., Nicoletti, G., Neal, G.E., Donohue, D.C. & Ahokas, J.T. (1995) Aflatoxin M_1 in human breast milk samples from Victoria, Australia and Thailand. *Food chem. Toxicol.*, **33**, 173–179

EN (European Norm) (1999a) *Milk and Milk Powder — Determination of Aflatoxin M_1 Content — Cleanup by Immunoaffinity Chromatography and Determination by High-performance Liquid Chromatography* (EN ISO 14504), Brussels, European Committee for Standardization

EN (European Norm) (1999b) *Foodstuffs — Determination of Aflatoxin B_1 and the Sum of Aflatoxins B_1, B_2, G_1 and G_2 in Cereals, Cell-fruits and Derived Products — High-performance Liquid Chromatographic Method with Post-column Derivatization and Immunoaffinity Column Clean up* (EN 12955), Brussels, European Committee for Standardization

Essigmann, J.M., Croy, R.G., Bennett, R.A. & Wogan, G.N. (1982) Metabolic activation of aflatoxin B_1: Patterns of DNA adduct formation, removal, and excretion in relation to carcinogenesis. *Drug Metab. Rev.*, **13**, 581–602

European Commission (1997) *Coordinated Programme for the Official Control of Foodstuff*, Luxembourg

European Commission (1998a) *Directive No. 98/53/CE de la Commission sur les Modes d'Echantillonnage et les Méthodes d'Analyse pour Contrôle des Aflatoxines, 16 juillet 1998*, Luxembourg

European Commission (1998b) *Règlement (CE) No. 1525/98 de la Commission sur les Teneurs Maximales en Aflatoxines*, Luxembourg

European Commission (2001) Commission Regulation (EC) No. 466/2001 of 8 March 2001 setting maximum levels for certain contaminants in foodstuffs. *Off. J. Europ. Comm., L 77/1*

European Commission (2002) Commission Regulation (EC) No. 472/2002 of 12 March 2002 amending Regulation (EC) No. 466/2001 setting Maximum levels for certain contaminants in foodstuffs. *Off. J. Europ. Comm.*, *L75*, 18–20

FAO (1995) Guidelines for the application of the hazard analysis critical control point (HACCP) system (CAC/GL 18-1993). In: *Codex Alimentarius*, Vol. 1B, *General Requirements (Food Hygiene)*, pp. 21–30, Rome, FAO/WHO

FAO (1997) *Worldwide Regulations for Mycotoxins 1995: A Compendium* (FAO Food and Nutrition Paper 64), Rome

Forrester, K., Lupold, S.E., Ott, V.L., Chay, C.H., Band, V., Wang, X.W. & Harris, C.C. (1995) Effects of *p53* mutants on wild-type *p53*-mediated transactivation are cell type dependent. *Oncogene*, **10**, 2103–2111

Foster, P.L., Eisenstadt, E. & Miller, J.H. (1983) Base substitution mutations induced by metabolically activated aflatoxin B_1. *Proc. natl Acad. Sci. USA*, **80**, 2695–2698

Frayssinet, C. & Lafarge-Frayssinet, C. (1990) Effect of ammoniation on the carcinogenicity of aflatoxin-contaminated groundnut oil cakes: Long-term feeding study in the rat. *Food Addit. Contam.*, **7**, 63–68

Freitas, V.P.S. & Brigido, B.M. (1998) Occurrence of aflatoxins B_1, B_2, G_1, and G_2 in peanuts and their products marketed in the region of Campinas, Brazil in 1995 and 1996. *Food Addit. Contam.*, **15**, 807–811

Frisvad, J.C. (1997) New producers of aflatoxin. In: *Third International Workshop on* Penicillium *and* Aspergillus. *May 26–29, 1997, Baarn, The Netherlands*, Centraalbureau voor Schimmelcultures, International Commission of Penicillium and Aspergillus (ICPA), Paris, International Union of Microbiological Sciences (IUMS)

Fujimoto, Y., Hampton, L.L., Luo, L.-D., Wirth, P.J. & Thorgeirsson, S.S. (1992) Low frequency of *p53* gene mutation in tumors induced by aflatoxin B_1 in nonhuman primates. *Cancer Res.*, **52**, 1044–1046

Fujimoto, Y., Hampton, L.L., Wirth, P.J., Wang, N.J., Xie, J.P. & Thorgeirsson, S.S. (1994) Alterations of tumor suppressor genes and allelic losses in human hepatocellular carcinomas in China. *Cancer Res.*, **54**, 281–285

Gallagher, E.P., Kunze, K.L., Stapleton, P.L. & Eaton, D.L. (1996) The kinetics of aflatoxin B_1 oxidation by human cDNA-expressed and human liver microsomal cytochromes P450 1A2 and 3A4. *Toxicol. appl. Pharmacol.*, **141**, 595–606

Galvano, F., Galofaro, V. & Galvano, G. (1996) Occurrence and stability of aflatoxin M_1 in milk and milk products: A worldwide review. *J. Food Protect.*, **59**, 1079–1090

Geiser, D.M., Pitt, J.I. & Taylor, J.W. (1998) Cryptic speciation and recombination in the aflatoxin-producing fungus *Aspergillus flavus*. *Proc. natl Acad. Sci. USA*, **95**, 388–393

Geiser, D.M., Dorner, J.W., Horn, B.W. & Taylor, J.W. (2000) The phylogenetics of mycotoxin and sclerotium production in *Aspergillus flavus* and *Aspergillus oryzae*. *Fungal Genet. Biol.*, **31**, 169–179

Ghebranious, N. & Sell, S. (1998a) Hepatitis B injury, male gender, aflatoxin, and p53 expression each contribute to hepatocarcinogenesis in transgenic mice. *Hepatology*, **27**, 383–391

Ghebranious, N. & Sell, S. (1998b) The mouse equivalent of the human p53ser246 mutation p53ser246 enhances aflatoxin hepatocarcinogenesis in hepatitis B surface antigen transgenic and p53 heterozygous null mice. *Hepatology*, **27**, 967–973

Ghosh, S.K., Desai, M.R., Pandya, G.L. & Venkaiah, K. (1997) Airborne aflatoxin in the grain processing industries in India. *Am. ind. Hyg. Assoc. J.*, **58**, 583–586

Goeger, D.E., Anderson, K.E. & Hsie, A.W. (1998) Coumarin chemoprotection against aflatoxin B_1-induced gene mutation in a mammalian cell system: A species difference in mutagen activation and protection with chick embryo and rat liver S9. *Environ. mol. Mutag.*, **32**, 64–74

Goeger, D.E., Hsie, A.W. & Anderson, K.E. (1999) Co-mutagenicity of coumarin (1,2-benzopyrone) with aflatoxin B_1 and human liver S9 in mammalian cells. *Food chem. Toxicol.*, **37**, 581–589

Gong, Y.Y., Cardwell, K., Hounsa, A., Egal, S., Turner, P.C., Hall, A.J. & Wild, C.P. (2002) Dietary aflatoxin exposure and impaired growth in young children from Benin and Togo: Cross-sectional study. *Br. med. J.*, **325**, 20–21

González, H.H.L., Martínez, E.J., Pacin, A.M., Resnik, S.L. & Sydenham, E.W. (1999) Natural co-occurrence of fumonisins, deoxynivalenol, zearalenone and aflatoxins in field trial corn in Argentina. *Food Addit. Contam.*, **16**, 565–569

Gopalakrishnan, S., Harris, T.M. & Stone, M.P. (1990) Intercalation of aflatoxin B_1 in two oligodeoxynucleotide adducts: Comparative ^{1}H NMR analysis of d(ATC^{AFB}GAT).d(ATCGAT) and d(AT^{AFB}GCAT)$_2$. *Biochemistry*, **29**, 10438–10448

Gopalan, P., Tsuji, K., Lehmann, K., Kimura, M., Shinozuka, H., Sato, K. & Lotlikar, P.D. (1993) Modulation of aflatoxin B_1-induced glutathione S-transferase placental form positive hepatic foci by pretreatment of rats with phenobarbital and buthionine sulfoximine. *Carcinogenesis*, **14**, 1469–1470

Gorelick, N.J. (1990) Risk assessment for aflatoxin: I. Metabolism of aflatoxin B_1 by different species. *Risk Anal.*, **10**, 539–559

Goto, T., Wicklow, D.T. & Ito, Y. (1996) Aflatoxin and cyclopiazonic acid production by a sclerotium-producing *Aspergillus tamarii* strain. *Appl. environ. Microbiol.*, **62**, 4036–4038

Groopman, J.D. (1993) Molecular dosimetry methods for assessing human aflatoxin exposures. In: Eaton, D.L. & Groopman, J.D., eds, *The Toxicology of Aflatoxins: Human Health, Veterinary and Agricultural Significance*, New York, Academic Press, pp. 259–279

Groopman, J.D. & Kensler, T.W. (1999) The light at the end of the tunnel for chemical-specific biomarkers: Daylight or headlight? *Carcinogenesis*, **20**, 1–11

Groopman, J.D. & Sabbioni, G. (1991) Detection of aflatoxin and its metabolites in human biological fluids. In: Bray, G.A. & Ryan, D.H., eds, *Mycotoxins, Cancer and Health* (Pennington Center Nutrition Series, Vol. 1), Baton Rouge, LA, Louisiana State University Press, pp. 18–31

Groopman, J.D., Hall, A.J., Whittle, H., Hudson, G.J., Wogan, G.N., Montesano, R. & Wild, C.P. (1992) Molecular dosimetry of aflatoxin-*N*7-guanine in human urine obtained in the Gambia, West Africa. *Cancer Epidemiol. Biomarkers. Prev.*, **1**, 221–227

Guengerich, F.P., Johnson, W.W., Ueng, Y.F., Yamazaki, H. & Shimada, T. (1996) Involvement of cytochrome P450, glutathione S-transferase, and epoxide hydrolase in the metabolism of aflatoxin B_1 and relevance to risk of human liver cancer. *Environ. Health Perspect.*, **104** (Suppl. 3), 557–562

Guengerich, F.P., Johnson, W.W., Shimada, T., Ueng, Y.-F., Yamazaki, H. & Langouët, S. (1998) Activation and detoxication of aflatoxin B_1. *Mutat. Res.*, **402**, 121–128

Hall, A.J. & Wild, C.P. (1994) Epidemiology of aflatoxin-related disease. In: Eaton, D.L. & Groopman, J.D., eds, *The Toxicology of Aflatoxins: Human Health, Veterinary and Agricultural Significance*, New York, Academic Press, pp. 233–258

Han, S.H., Jeon, Y.J., Yea, S.S. & Yang, K.-H. (1999) Suppression of the interleukin-2 gene expression by aflatoxin B_1 is mediated through the down-regulation of the NF-AT and AP-1 transcription factors. *Toxicol. Lett.*, **108**, 1–10

Harris, C.C. & Sun, T.-T. (1986) Interactive effects of chemical carcinogens and hepatitis B virus in the pathogenesis of hepatocellular carcinoma. *Cancer Surv.*, **5**, 765–780

Hatch, M.C., Chen, C.J., Levin, B., Ji, B.T., Yang, G.Y., Hsu, S.W., Wang, L.W., Hsieh, L.L. & Santella, R.M. (1993) Urinary aflatoxin levels, hepatitis-B virus infection and hepatocellular carcinoma in Taiwan. *Int. J. Cancer*, **54**, 931–934

Hayashi, H., Sugio, K., Matsumata, T., Adachi, E., Urata, K., Tanaka, S. & Sugimachi, K. (1993) The mutation of codon 249 in the *p53* gene is not specific in Japanese hepatocellular carcinoma. *Liver*, **13**, 279–281

Hayes, R.B., van Nieuwenhuize, J.P., Raatgever, J.W. & ten Kate, F.J. (1984) Aflatoxin exposures in the industrial setting: An epidemiological study of mortality. *Food chem. Toxicol.*, **22**, 39–43

Hayes, J.D., Judah, D.J., McLellan, L.I., Kerr, L.A., Peacock, S.D. & Neal, G.E. (1991) Ethoxyquin-induced resistance to aflatoxin B_1 in the rat is associated with the expression of a novel alpha-class glutathione S-transferase subunit, Yc2, which possesses high catalytic activity for aflatoxin B_1-8,9-epoxide. *Biochem. J.*, **279**, 385–398

Hayes, J.D., Judah, D.J., Neal, G.E. & Nguyen, T. (1992) Molecular cloning and heterologous expression of a cDNA encoding a mouse glutathione S-transferase Yc subunit possessing high catalytic activity for aflatoxin B_1-8,9-epoxide. *Biochem. J.*, **285**, 173–180

Hayes, J.D., Judah, D.J. & Neal, G.E. (1993) Resistance to aflatoxin B_1 is associated with the expression of a novel aldo-keto reductase which has catalytic activity towards a cytotoxic aldehyde-containing metabolite of the toxin. *Cancer Res.*, **53**, 3887–3894

Hayes, J.D., Nguyen, T., Judah, D.J., Petersson, D.G., & Neal, G.E. (1994) Cloning of CDNAS from fetal rat liver encoding glutathione S-transferase Yc polypeptides. *J. biol. Chem.*, **269**, 20707–20717

Heinonen, J.T., Fisher, R., Brendel, K. & Eaton, D.L. (1996) Determination of aflatoxin B_1 biotransformation and binding to hepatic macromolecules in human precision liver slices. *Toxicol. appl. Pharmacol.*, **136**, 1–7

Hendrickse, R.G. (1997) Of sick turkeys, kwashiorkor, malaria, perinatal mortality, heroin addicts and food poisoning: Research on the influence of aflatoxins on child health in the tropics. *Ann. trop. Med. Parasitol.*, **91**, 787–793

Henry, S. H., DiNovi, M.J., Bowers, J.C. & Bolger, P.M. (1997) Risk assessment for aflatoxin in corn and peanuts in the United States. *Fund. appl. Toxicol.* (Suppl. (The Toxicologist)), **36**, 172–173

Hirano, S., Okawara, N. & Ono, M. (1998) Relationship between the incidence of the aflatoxin detection in imported peanuts and the climate condition of its producing area. *J. Food Hyg. Soc. Jap.*, **39**, 440–443

Hiruma, S., Qin, G.Z., Gopalan-Kriczky, P., Shinozuka, H., Sato, K. & Lotlikar, P.D. (1996) Effect of cell proliferation on initiation of aflatoxin B_1-induced enzyme altered hepatic foci in rats and hamsters. *Carcinogenesis*, **17**, 2495–2499

Hiruma, S., Kimura, M., Lehmann, K., Gopalan-Kriczky, P., Qin, G.Z., Shinozuka, H., Sato, K. & Lotlikar, P.D. (1997) Potentiation of aflatoxin B_1-induced hepatocarcinogenesis in the rat by pretreatment with buthionine sulfoximine. *Cancer Lett.*, **113**, 103–109

Hollstein, M.C., Wild, C.P., Bleicher, F., Chutimataewin, S., Harris, C.C., Srivatanakul, P. & Montesano, R. (1993) *p53* Mutations and aflatoxin B_1 exposure in hepatocellular carcinoma patients from Thailand. *Int. J. Cancer*, **53**, 51–55

Hosono, S., Chou, M.J., Lee, C.S. & Shih, C. (1993) Infrequent mutation of *p53* gene in hepatitis B virus positive primary hepatocellular carcinomas. *Oncogene*, **8**, 491–496

Hsieh, D.-P. & Atkinson, D.N. (1995) Recent aflatoxin exposure and mutation at codon 249 of the human *p53* gene: Lack of association. *Food Addit. Contam.*, **12**, 421–424

Hsieh, L.-L. & Hsieh, T.-T. (1993) Detection of aflatoxin B_1–DNA adducts in human placenta and cord blood. *Cancer Res.*, **53**, 1278–1280

Hudson, G.J., Wild, C.P., Zarba, A. & Groopman, J.D. (1992) Aflatoxins isolated by immunoaffinity chromatography from foods consumed in The Gambia, West Africa. *Natural Toxins*, **1**, 100–105

Hulla, J.E., Chen, Z.Y. & Eaton, D.L. (1993) Aflatoxin B_1-induced rat hepatic hyperplastic nodules do not exhibit a site-specific mutation within the *p53* gene. *Cancer Res.*, **53**, 9–11

Hussain, S.P. & Harris, C.C. (2000) Molecular epidemiology and carcinogenesis: Endogenous and exogenous carcinogens. *Mutat. Res.*, **462**, 311–322

Hustert, E., Haberl, M., Burk, O., Wolbold, R., He, Y.-Q., Klein, K., Nuessler, A.C., Neuhaus, P., Klattig, J., Eiselt, R., Koch, I., Zibat, A., Brockmöller, J., Halpert, J.R., Zanger, U.M. & Wojnowski, L. (2001) The genetic determinants of the CYP3A5 polymorphism. *Pharmacogenetics*, **11**, 773–779

IARC (1972) *IARC Monographs on the Evaluation of the Carcinogenic Risk of Chemicals to Man*, Vol. 1, *Some Inorganic Substances, Chlorinated Hydrocarbons, Aromatic Amines,* N-*Nitroso Compounds, and Natural Products*, Lyon, IARC*Press*, pp. 145–156

IARC (1976) *IARC Monographs on the Evaluation of the Carcinogenic Risk of Chemicals to Man*, Vol. 10, *Some Naturally Occurring Substances*, Lyon, IARC*Press*, pp. 51–72

IARC (1987) *IARC Monographs on the Evaluation of Carcinogenic Risks to Humans*, Suppl. 7, *Overall Evaluations of Carcinogenicity: An Uptading of* IARC Monographs *Volumes 1 to 42*, Lyon, IARC*Press*, pp. 83–87

IARC (1993) *IARC Monographs on the Evaluation of Carcinogenic Risks to Humans*, Vol. 56, *Some Naturally Occurring Substances: Food Items and Constituents, Heterocyclic Aromatic Amines and Mycotoxins*, Lyon, IARC*Press*, pp. 245–395

Ibeh, I.N. & Saxena, D.K. (1997a) Aflatoxin B_1 and reproduction. I. Reproductive performance in female rats. *Afr. J. reprod. Health*, **1**, 79–84

Ibeh, I.N. & Saxena, D.K. (1997b) Aflatoxin B_1 and reproduction. II. Gametoxicity in female rats. *Afr. J. reprod. Health*, **1**, 85–89

Ibeh, I.N., Uraih, N. & Ogonar, J.I. (1994) Dietary exposure to aflatoxin in human male infertility in Benin City, Nigeria. *Int. J. Fertil.*, **39**, 208–214

Ibeh, I.N., Saxena, D.K. & Uraih, N. (2000) Toxicity of aflatoxin, effects on spermatozoa, oocytes, and in vitro fertilization. *J. environ. Pathol. Toxicol. Oncol.*, **19**, 357–361

Ioannou-Kakouri, E., Aletrari, M., Christou, E., Hadjioannou-Ralli, A. & Akkelidou, D. (1999) Surveillance and control of aflatoxins B_1, B_2, G_1, G_2, and M_1 in foodstuffs in the Republic of Cyprus: 1992–1996. *J. Assoc. off. anal. Chem.*, **82**, 883–892

Ireland, L.S., Harrison, D.J., Neal, G.E. & Hayes, J.D. (1998) Molecular cloning, expression and catalytic activity of a human AKR7 member of the aldo-keto reductase superfamily: Evidence that the major 2-carboxybenzaldehyde reductase from human liver is a homologue of rat aflatoxin B_1-aldehyde reductase. *Biochem. J.*, **332**, 21–34

ISO (1998) *Animal Feeding Stuffs — Determination of Aflatoxin B_1 Content of Mixed Feeding Stuffs — Method using High-performance Liquid Chromatography* (ISO 14718), Geneva, International Organization for Standardization

ISO (2001) *Animal Feeding Stuffs — Semi-quantitative Determination of Aflatoxin B_1 — Thin-layer Chromatographic Methods* (ISO 6651), Geneva, International Organization for Standardization

Ito, Y., Peterson, S.W., Wicklow, D.T. & Goto, T. (2001) *Aspergillus pseudotamarii*, a new aflatoxin producing species in *Aspergillus* section *Flavi*. *Mycol. Res.*, **105**, 233–239

Jackson, P.E., Qian, G.-S., Friesen, M.D., Zhu, Y.-R., Lu, P., Wang, J.-B., Wu, Y., Kensler, T.W., Vogelstein, B. & Groopman, J.D. (2001) Specific *p53* mutations detected in plasma and tumors of hepatocellular carcinoma patients by electrospray ionization mass spectrometry. *Cancer Res.*, **61**, 33–35

Jacobson, L.P., Zhang, B.-C., Zhu, Y.-R., Wang, J.-B., Wu, Y., Zhang, Q.-N., Yu, L.-Y., Qian, G.-S., Kuang, S.-Y., Li, Y.-F., Fang, X., Zarba, A., Chen, B., Enger, C., Davidson, N.E., Gorman, M.B., Gordon, G.B., Prochaska, H.J., Egner, P.A., Groopman, J.D., Muñoz, A., Helzlsouer, K.J. & Kensler, T.W. (1997) Oltipraz chemoprevention trial in Qidong, People's Republic of China: Study design and clinical outcomes. *Cancer Epidemiol. Biomarkers Prev.*, **6**, 257–265

Jaitovitch-Groisman, I., Fotouhi-Ardakani, N., Schecter, R.L., Woo, A., Alaoui-Jamali, M.A. & Batist, G. (2000) Modulation of glutathione S-transferase alpha by hepatitis B virus and the chemopreventive drug oltipraz. *J. biol. Chem.*, **275**, 33395–33403

Jakab, G.J., Hmieleski, R.R., Zarba, A., Hemenway, D.R. & Groopman, J.D. (1994) Respiratory aflatoxicosis: Suppression of pulmonary and systemic host defenses in rats and mice. *Toxicol. appl. Pharmacol.*, **125**, 198–205

Jalon, M., Urieta, B.P. & Macho, M.L. (1994) [Surveillance of contamination with aflatoxin M_1 of milk and milk products in the Autonomous Community of Vasca.] *Alimentaria*, **3**, 25–29 (in Spanish)

JECFA (1998) *Safety Evaluation of Certain Food Additives and Contaminants: Aflatoxins (WHO Food Additives Series 40), 49th Meeting of the Joint FAO/WHO Expert Committee on Food Additives (JECFA)*, Geneva, International Programme on Chemical Safety, World Health Organization

JECFA (2001) *Safety Evaluation of Certain Mycotoxins in Food (WHO Food Additives Series No. 47), 56th Meeting of the Joint FAO/WHO Expert Committee on Food Additives (JECFA)*, Geneva, International Programme on Chemical Safety, World Health Organization

Johnson, W.W. & Guengerich, F.P. (1997) Reaction of aflatoxin B_1 *exo*-8,9-epoxide with DNA: Kinetic analysis of covalent binding and DNA-induced hydrolysis. *Proc. natl Acad. Sci. USA*, **94**, 6121–6125

Johnson, W.W., Harris, T.M. & Guengerich, F.P. (1996) Kinetics and mechanism of hydrolysis of aflatoxin B_1 *exo*-8,9-oxide and rearrangement of the dihydrodiol. *J. Am. chem. Soc.*, **118**, 8213–8220

Johnson, W.W., Ueng, Y.-F., Widersten, M., Mannervik, B., Hayes, J.D., Sherratt, P.J., Ketterer, B. & Guengerich, F.P. (1997a) Conjugation of highly reactive aflatoxin B_1 *exo*-8,9-epoxide catalyzed by rat and human glutathione transferases: Estimation of kinetic parameters. *Biochemistry*, **36**, 3056–3060

Johnson, W.W., Yamasaki, H., Shimada, T., Ueng, Y.F. & Guengerich, F.P. (1997b) Aflatoxin B_1 8,9-epoxide hydrolysis in the presence of rat and human epoxide hydrolase. *Chem. Res. Toxicol.*, **10**, 672–676

Juan-López, M., Carvajal, M. & Ituarte, B. (1995) Supervising programme of aflatoxins in Mexican corn. *Food Addit. Contam.*, **12**, 297–312

Judah, D.J., Hayes, J.D., Yang, J.-C., Lian, L.-Y., Roberts, G.C., Farmer, P.B., Lamb, J.H. & Neal, G.E. (1993) A novel aldehyde reductase with activity towards a metabolite of aflatoxin B_1 is expressed in rat liver during carcinogenesis and following the administration of an anti-oxidant. *Biochem. J.*, **292**, 13–18

Kamataki, T., Hashimoto, H., Shimoji, M., Itoh, S., Nakayama, K., Hattori, K., Yokoi, T., Katsuki, M. & Aizawa, S. (1995) Expression of CYP3A7, a human fetus-specific cytochrome P450, in cultured cells and in the hepatocytes of p53-knockout mice. *Toxicol. Lett.*, **82–83**, 879–882

Kaplanski, C., Chisari, F.V. & Wild, C.P. (1997) Minisatellite rearrangements are increased in liver tumours induced by transplacental aflatoxin B_1 treatment of hepatitis B virus transgenic mice, but not in spontaneously arising tumours. *Carcinogenesis*, **18**, 633–639

Kaplanski, C., Wild, C.P. & Sengstag, C. (1998) Rearrangements in minisatellite sequences induced by aflatoxin B_1 in a metabolically competent strain of *Saccharomyces cerevisiae*. *Carcinogenesis*, **19**, 1673–1678

Katiyar, S., Dash, B.C., Thakur, V., Guptan, R.C., Sarin, S.K. & Das, B.C. (2000) *p53* Tumor suppressor gene mutations in hepatocellular carcinoma patients in India. *Cancer*, **88**, 1565–1573

Kazachkov, Y., Khaoustov, V., Yoffe, B., Solomon, H., Klintmalm, G.B. & Tabor, E. (1996) *p53* Abnormalities in hepatocellular carcinoma from United States patients: Analysis of all 11 exons. *Carcinogenesis*, **17**, 2207–2212

Kazemi-Shirazi, L., Peterman, D. & Müller, C. (2000) Hepatitis B virus DNA in sera and liver tissue of HBsAg negative patients with chronic hepatitis C. *J. Hepatol.*, **33**, 785–790

Kensler, T.W., Egner, P.A., Davidson, N.E. Roebuck, B.D., Pikul, A. & Groopman, J.D. (1986) Modulation of aflatoxin metabolism, aflatoxin-N7-guanine formation, and hepatic tumorigenesis in rats fed ethoxyquin: Role of induction of glutathione S-transferases. *Cancer Res.*, **46**, 3924–3931

Kensler, T.W., Egner, P.A., Dolan, P.M., Groopman, J.D. & Roebuck, B.D. (1987) Mechanism of protection against aflatoxin tumorigenicity in rats fed 5-(2-pyrazinyl)-4-methyl-1,2-dithiol-3-thione (oltipraz) and related 1,2-dithiol-3-thiones and 1,2-dithiol-3-ones. *Cancer Res.*, **47**, 4271–4277

Kensler, T.W., Gange, S.J., Egner, P.A., Dolan, P.M., Muñoz, A., Groopman, J.D., Rogers, A.E. & Roebuck, B.D. (1997) Predictive value of molecular dosimetry: Individual *versus* group effects of oltipraz on aflatoxin-albumin adducts and risk of liver cancer. *Cancer Epidemiol. Biomarkers Prev.*, **6**, 603–610

Kensler, T.W., He, X., Otieno, M., Egner, P.A., Jacobson, L.P., Chen, B., Wang, J.-S., Zhu, Y.-R., Zhang, B.-C., Wang, J.-B., Wu, Y., Zhang, Q.-N., Qian, G.-S., Kuang, S.-Y., Fang, X., Li, Y.-F., Yu, L.-Y., Prochaska, H.J., Davidson, N.E., Gordon, G.B., Gorman, M.B., Zarba, A., Enger, C., Muñoz, A., Helzlsouer, K.J. & Groopman, J.D. (1998) Oltipraz chemoprevention trial in Qidong, People's Republic of China: Modulation of serum aflatoxin albumin adduct biomarkers. *Cancer Epidemiol. Biomarkers Prev.*, **7**, 127–134

Kensler, T.W., Groopman, J.D., Sutter, T.R., Curphey, T.J. & Roebuck, B.D. (1999) Development of cancer chemopreventive agents: Oltipraz as a paradigm. *Chem. Res. Toxicol.*, **12**, 113–126

Kew, M.C., Yu, M.C., Kedda, M.A., Coppin, A., Sarkin, A. & Hodkinson, J. (1997) The relative roles of hepatitis B and C viruses in the etiology of hepatocellular carcinoma in southern African blacks. *Gastroenterology*, **112**, 184–187

Kihara, T., Matsuo, T., Sakamoto, M., Yasuda, Y., Yamamoto, Y. & Tanimura, T. (2000) Effects of prenatal aflatoxin B_1 exposure on behaviors of rat offspring. *Toxicol. Sci.*, **53**, 392–399

Kim, E.K., Shon, D.H., Ryu, D., Park, J.W., Hwang, H.J. & Kim, Y.B. (2000) Occurrence of aflatoxin M_1 in Korean dairy products determined by ELISA and HPLC. *Food Addit. Contam.*, **17**, 59–64

Kirby, G.M., Wolf, C.R., Neal, G.E., Judah, D.J., Henderson, C.J., Srivatanakul, P. & Wild, C.P. (1993) *In vitro* metabolism of aflatoxin B_1 by normal and tumorous liver tissue from Thailand. *Carcinogenesis*, **14**, 2613–2620

Kirby, G.M., Chemin, I., Montesano, R., Chisari, F.V., Lang, M.A. & Wild, C.P. (1994a) Induction of specific cytochrome P450s involved in aflatoxin B_1 metabolism in hepatitis B virus transgenic mice. *Mol. Carcinog.*, **11**, 74–80

Kirby, G.M., Pelkonen, P., Vatanasapt, V., Camus, A.-M., Wild, C.P. & Lang, M.A. (1994b) Association of liver fluke (*Opisthorchis viverrini*) infestation with increased expression of cytochrome P450 and carcinogen metabolism in male hamster liver. *Mol. Carcinog.*, **11**, 81–89

Kirby, G.M., Batist, G., Alpert, L., Lamoureux, E., Cameron, R.G., Alaoui-Jamali, M.A. (1996a) Overexpression of cytochrome P-450 isoforms involved in aflatoxin B_1 bioactivation in human liver with cirrhosis and hepatitis. *Toxicol. Pathol.*, **24**, 458–467

Kirby, G.M., Batist, G., Fotouhi-Ardakani, N., Nakazawa, H., Yamasaki, H., Kew, M., Cameron, R.G. & Alaoui-Jamali, M.A. (1996b) Allele-specific PCR analysis of *p53* codon 249 AGT transversion in liver tissues from patients with viral hepatitis. *Int. J. Cancer*, **68**, 21–25

Kirk, G.D., Camus-Randon, A.-M., Mendy, M., Goedert, J.J., Merle, P., Trepo, C., Bréchot, C., Hainaut, P. & Montesano, R. (2000) Ser-249 *p53* mutations in plasma DNA of patients with hepatocellular carcinoma from The Gambia. *J. natl Cancer Inst.*, **92**, 148–153

Kitada, M., Taneda, M., Ohi, H., Komori, M., Itahashi, K., Nagao, M. & Kamataki, T. (1989) Mutagenic activation of aflatoxin B_1 by P-450 HFLa in human fetal livers. *Mutat. Res.*, **227**, 53–58

Kitada, M., Taneda, M., Ohta, K., Nagashima, K., Itahashi, K. & Kamataki, T. (1990) Metabolic activation of aflatoxin B_1 and 2-amino-3-methylimidazo[4,5-*f*]-quinoline by human adult and fetal livers. *Cancer Res.*, **50**, 2641–2645

Klich, M.A. & Pitt, J.I. (1988) Differentiation of *Aspergillus flavus* from *A. parasiticus* and other closely related species. *Trans. Br. Mycol. Soc.*, **91**, 99–108

Klich, M.A., Mullaney, E.J., Daly, C.B. & Cary, J.W. (2000) Molecular and physiological aspects of aflatoxin and sterigmatocystin biosynthesis by *Aspergillus tamarii* and *A. ochraceoroseus*. *Appl. microbiol. Biotechnol.*, **53**, 605–609

Knasmüller, S., Parzefall, W., Sanyal, R., Ecker, S., Schwab, C., Uhl, M., Mersch-Sundermann, V., Williamson, G., Hietsch, G., Langer, T., Darroudi, F. & Natarajan, A.T. (1998) Use of metabolically competent human hepatoma cells for the detection of mutagens and anti-mutagens. *Mutat. Res.*, **402**, 185–202

Knight, L.P., Primiano, T., Groopman, J.D., Kensler, T.W. & Sutter, T.R. (1999) cDNA cloning, expression and activity of a second human aflatoxin B_1-metabolizing member of the aldo-keto reductase superfamily, AKR7A3. *Carcinogenesis*, **20**, 1215–1223

Konishi, M., Kikuchi-Yanoshita, R., Tanaka, K., Sato, C., Tsuruta, K., Maeda, Y., Koike, M., Tanaka, S., Nakamura, Y., Hattori, N. & Miyaki, M. (1993) Genetic changes and histo-pathological grades in human hepatocellular carcinomas. *Jpn. J. Cancer Res.*, **84**, 893–899

Kress, S., Jahn, U.R., Buchmann, A., Bannasch, P. & Schwarz, M. (1992) *p53* Mutations in human hepatocellular carcinomas from Germany. *Cancer Res.*, **52**, 3220–3223

Kubicka, S., Trautwein, C., Schrem, H., Tillmann, H. & Manns, M. (1995) Low incidence of *p53* mutations in European hepatocellular carcinomas with heterogeneous mutation as a rare event. *J. Hepatol.*, **23**, 412–419

Kuehl, P., Zhang, J., Lin, Y., Lamba, J., Assem, M., Schuetz, J., Watkins, P.B., Daly, A., Wrighton, S.A., Hall, S.D., Maurel, P., Relling, M., Brimer, C., Yasuda, K., Venkataramanan, R., Strom, S., Thummel, K., Boguski, M.S. & Schuetz, E. (2001) Sequence diversity in *CYP3A* promoters and characterization of the genetic basis of polymorphic CYP3A5 expression. *Nature Genet.*, **27**, 383–391

Kurtzman, C.P., Horn, B.W. & Hesseltine, C.W. (1987) *Aspergillus nomius*, a new aflatoxin-producing species related to *Aspergillus flavus* and *Aspergillus tamarii*. *Antonie van Leeuwenhoek*, **53**, 147–158

Kussak, A., Andersson, B. & Andersson, K. (1995) Determination of aflatoxins in airborne dust from feed factories by automated immunoaffinity column clean up and liquid chromato-graphy. *J. Chromatogr.*, **708**, 55–60

Lafontaine, M., Delsaut, P., Morelle, Y. & Taiclet, A. (1994) [Aflatoxins: Sampling and analysis in animal feed production plant.] *Cahiers Notes documentaires*, **156**, 297–305 (in French)

Lam, K.C., Yu, M.C., Leung, J.W. & Henderson, B.E. (1982) Hepatitis B virus and cigarette smoking: Risk factors for hepatocellular carcinoma in Hong Kong. *Cancer Res.*, **42**, 5246–5248

Lamplugh, S.M., Hendrickse, R.G., Apeagyei, F. & Mwanmut, D.D. (1988) Aflatoxins in breast milk, neonatal cord blood, and serum of pregnant women. *Br. med. J.*, **296**, 968

Langouët, S., Coles, B., Morel, F., Becquemont, L., Beaune, P., Guengerich, F.P., Ketterer, B. & Guillouzo, A. (1995) Inhibition of CYP1A2 and CYP3A4 by oltipraz results in reduction of aflatoxin B_1 metabolism in human hepatocytes in primary culture. *Cancer Res.*, **55**, 5574–5579

Lasky, T. & Magder, L. (1997) Hepatocellular carcinoma *p53* G > T transversions at codon 249: The fingerprint of aflatoxin exposure? *Environ. Health Perspect.*, **105**, 392–397

Laurent-Puig, P., Legoix, P., Bluteau, O., Belghiti, J., Franco, D., Binot, F., Monges, G., Thomas, G., Bioulac-Sage, P. & Zucman-Rossi, J. (2001) Genetic alterations associated with hepatocellular carcinomas define distinct pathways of hepatocarcinogenesis. *Gastroenterology*, **120**, 1763–1773

Lee, C.-C., Liu, J.-Y., Lin, J.-K., Chu, J.-S. & Shew, J.-Y. (1998) *p53* Point mutation enhanced by hepatic regeneration in aflatoxin B_1-induced rat liver tumors and preneoplastic lesions. *Cancer Lett.*, **125**, 1–7

Levy, D.D., Groopman, J.D., Lim, S.E., Seidman, M.M. & Kraemer, K.H. (1992) Sequence specificity of aflatoxin B_1-induced mutations in a plasmid replicated in xeroderma pigmentosum and DNA repair proficient human cells. *Cancer Res.*, **52**, 5668–5673

Li, D., Cao, Y., He, L., Wang, N.J. & Gu, J.R. (1993) Aberrations of *p53* gene in human hepatocellular carcinoma from China. *Carcinogenesis*, **14**, 169–173

Li, Y., Su, J.J., Qin, L.L., Yang, C., Ban, K.C. & Yan, R.Q. (1999) Synergistic effect of hepatitis B virus and aflatoxin B_1 in hepatocarcinogenesis in tree shrews. *Ann. Acad. Med. Singapore*, **28**, 67–71

Li, F.-Q., Yoshizawa, T., Kawamura, O., Luo, X.-Y. & Li, Y.-W. (2001) Aflatoxins and fumonisins in corn from the high-incidence area for human hepatocellular carcinoma in Guangxi, China. *J. agric. Food Chem.*, **49**, 4122–4126

Lipigorngoson, S., Limtrakul, P.-N., Khangtragool, W. & Suttajit, M. (1999) Quantitation of aflatoxin B_1 in corn seeds and ground peanut by ELISA method using in-house monoclonal antibody preparation. *Mycotoxins*, **Suppl. 99**, 197–200

Liu, L. & Massey, T.E. (1992) Bioactivation of aflatoxin B_1 by lipoxygenases, prostaglandin H synthase and cytochrome P450 monooxygenase in guinea-pig tissues. *Carcinogenesis*, **13**, 533–539

Liu, L., Daniels, J.M., Stewart, R.K. & Massey, T.E. (1990) In vitro prostaglandin H synthase- and monooxygenase-mediated binding of aflatoxin B_1 to DNA in guinea-pig tissue microsomes. *Carcinogenesis*, **11**, 1915–1919

Loarca-Piña, G., Kuzmicky, P.A., González de Mejía, E., Kado, N.Y. & Hsieh, D.P. (1996) Antimutagenicity of ellagic acid against aflatoxin B_1 in the *Salmonella* microsuspension assay. *Mutat. Res.*, **360**, 15–21

Loarca-Piña, G., Kuzmicky, P.A., González de Mejía, E. & Kado, N.Y. (1998) Inhibitory effects of ellagic acid on the direct-acting mutagenicity of aflatoxin B_1 in the *Salmonella* microsuspension assay. *Mutat. Res.*, **398**, 183–187

Lu, P., Kuang, S., Wang, J., Fang, X., Zhang, Q.N., Wu, Y., Lu, Z.H. & Qian, G.L. (1998) [Hepatitis B virus infection and aflatoxin exposure in the development of primary liver cancer.] *Zhonghua Yi Xue Za Zhi (Nat. Med. J. China)*, **78**, 340–342 (in Chinese)

Lunn, R.M., Langlois, R.G., Hsieh, L.L., Thompson, C.L. & Bell, D.A. (1999) XRCC1 polymorphisms: Effects on aflatoxin B_1–DNA adducts and glycophorin A variant frequency. *Cancer Res.*, **59**, 2557–2561

Lye, M.S., Ghazali, A.A., Mohan, J., Alwin, N. & Nair, R.C. (1995) An outbreak of acute hepatic encephalopathy due to severe aflatoxicosis in Malaysia. *Am. J. trop. Med. Hyg.*, **53**, 68–72

Macé, K., Aguilar, F., Wang, J.-S., Vautravers, P., Gómez-Lechón, M., Gonzalez, F.J., Groopman, J., Harris, C.C. & Pfeifer, A.M.A. (1997) Aflatoxin B_1-induced DNA adduct formation and *p53* mutations in CYP450-expressing human liver cell lines. *Carcinogenesis*, **18**, 1291–1297

MAFF (Ministry of Agriculture, Fisheries and Food) (1999) MAFF — UK — *Survey of Aflatoxins, Ochratoxin A, Fumonisins and Zearalenone in Raw Maize* (Food Surveillance Information Sheet No. 192), Edinburgh, Scottish Executive Department of Health

Mandishona, E., MacPhail, A.P., Gordeuk, V.R., Kedda, M.A., Paterson, A.C., Rouault, T.A. & Kew, M.C. (1998) Dietary iron overload as a risk factor for hepatocellular carcinoma in black Africans. *Hepatology*, **27**, 1563–1566

Margolles, E., Escobar, A. & Acosta, A. (1992) Aflatoxin B_1 residuality determination directly in milk by ELISA. *Rev. Salud Anim.*, **12**, 35–38

Markaki, P. & Melissari, E. (1997) Occurrence of aflatoxin M_1 in commercial pasteurized milk determined with ELISA and HPLC. *Food Addit. Contam.*, **14**, 451–456

Maxuitenko, Y.Y., MacMillan, D.L., Kensler, T.W. & Roebuck, B.D. (1993) Evaluation of the post-initiation effects of oltipraz on aflatoxin B_1-induced preneoplastic foci in a rat model of hepatic tumorigenesis. *Carcinogenesis*, **14**, 2423–2425

Maxuitenko, Y.Y., Libby, A.H., Joyner, H.H., Curphey, T.J., MacMillan, D.L., Kensler, T.W. & Roebuck, B.D. (1998) Identification of dithiolethiones with better chemopreventive properties than oltipraz. *Carcinogenesis*, **19**, 1609–1615

Maxwell, S.M. (1998) Investigations into the presence of aflatoxins in human body fluids and tissues in relation to child health in the tropics. *Ann. Trop. Paediatr.*, **18** (Suppl.), 41–46

Maxwell, S.M., Familusi, J.B., Sodeinde, O., Chan, M.C.K. & Hendrickse, R.G. (1994) Detection of naphthols and aflatoxins in Nigerian cord blood. *Ann. trop. Paediatr.*, **14**, 3–5

McGlynn, K.A., Rosvold, E.A., Lustbader, E.D., Hu, Y., Clapper, M.L., Zhou, T., Wild, C.P., Xia, X.-L., Baffoe-Bonnie, A., Ofori-Adjei, D., Chen, G.-C., London, W.T., Shen, F.-M. & Buetow, K.H. (1995) Susceptibility to hepatocellular carcinoma is associated with genetic variation in the enzymatic detoxification of aflatoxin B_1. *Proc. natl Acad. Sci. USA*, **92**, 2384–2387

McLean, M. & Dutton, M.F. (1995) Cellular interactions and metabolism of aflatoxin: An update. *Pharmacol. Ther.*, **65**, 163–192

Medina-Martínez, M.S. & Martínez, A.J. (2000) Mold occurrence and aflatoxin B_1 and fumonisin B_1 determination in corn samples in Venezuela. *J. agric. Food Chem.*, **48**, 2833–2836

Midio, A.F., Campos, R.R. & Sabino, M. (2001) Occurrence of aflatoxins B_1, B_2, G_1 and G_2 in cooked food components of whole meals marketed in fast food outlets of the city of São Paulo, SP, Brazil. *Food Addit. Contam.*, **18**, 445–448

Mocchegiani, E., Corradi, A., Santarelli, L., Tibaldi, A., DeAngelis, E., Borghetti, P., Bonomi, A., Fabris, N. & Cabassi, E. (1998) Zinc, thymic endocrine activity and mitogen responsiveness (PHA) in piglets exposed to maternal aflatoxicosis B_1 and G_1. *Vet. Immunol. Immunopathol.*, **62**, 245–260

Montesano, R., Hainaut, P. & Wild, C.P. (1997) Hepatocellular carcinoma: From gene to public health. *J. natl Cancer Inst.*, **89**, 1844–1851

Moon, E.-Y., Rhee, D.-K. & Pyo, S. (1999) In vitro suppressive effect of aflatoxin B_1 on murine peritoneal macrophage functions. *Toxicology*, **133**, 171–179

Morel, F., Fardel, O., Meyer, D.J., Langouët, S., Gilmore, K.S., Meunier, B., Tu, C.-P.D., Kensler, T.W., Ketterer, B. & Guillouzo, A. (1993) Preferential increase of glutathione *S*-transferase class α transcripts in cultured human hepatocytes by phenobarbital, 3- methylcholanthrene, and dithiolethiones. *Cancer Res.*, **53**, 231–234

Morris, S.M., Aidoo, A., Chen, J.J., Chou, M.W. & Casciano, D.A. (1999) Aflatoxin B_1-induced *Hprt* mutations in splenic lymphocytes of Fischer 344 rats. Results of an intermittent feeding trial. *Mutat. Res.*, **423**, 33–38

Murakami, Y., Hayashi, K., Hirohashi, S. & Sekiya, T. (1991) Aberrations of the tumor suppressor *p53* and retinoblastoma genes in human hepatocellular carcinomas. *Cancer Res.*, **51**, 5520–5525

Nakajima, M., Tsubouchi, H. & Miyabe, M. (1999) A survey of ochratoxin A and aflatoxins in domestic and imported beers in Japan by immunoaffinity and liquid chromatography. *J. Assoc. off. anal. Chem. int.*, **82**, 897–902

National Food Authority (1992) *Australia Market Basket Survey* (information submitted to WHO by Australia)

Neal, G.E., Eaton, D.L., Judah, D.J. & Verma, A. (1998) Metabolism and toxicity of aflatoxins M_1 and B_1 in human-derived *in vitro* systems. *Toxicol. appl. Pharmacol.*, **151**, 152–158

Ng, I.O.L., Srivastava, G., Chung, L.P., Tsang, S.W.Y. & Ng, M.M.T. (1994a) Overexpression and point mutations of *p53* tumor suppressor gene in hepatocellular carcinomas in Hong Kong Chinese people. *Cancer*, **74**, 30–37

Ng, I.O.L., Chung, L.P., Tsang, S.W.Y., Lam, C.L., Lai, E.C.S., Fan, S.T. & Ng, M. (1994b) *p53* Gene mutation spectrum in hepatocellular carcinomas in Hong Kong Chinese. *Oncogene*, **9**, 985–990

Nishida, N., Fukuda, Y., Kokuryu, H., Toguchida, J., Yandell, D.W., Ikenega, M., Imura, H. & Ishizaki, K. (1993) Role and mutational heterogeneity of the *p53* gene in hepatocellular carcinoma. *Cancer Res.*, **53**, 368–372

Nose, H., Imazeki, F., Ohto, M. & Omata, M. (1993) *p53* Gene mutations and 17p allelic deletions in hepatocellular carcinoma from Japan. *Cancer*, **72**, 355–360

Nuntharatanapong, N., Suramana, T., Chaemthanorn, S., Zapuang, R., Ritta, E., Semathong, S.,. Chuamorn, S., Niyomwan, V., Dusitsin, N., Lohinavy, O. & Sinhaseni, P. (2001) Increase in tumour necrosis factor-alpha and a change in the lactate dehydrogenase isoenzyme pattern in plasma of workers exposed to aflatoxin-contaminated feeds. *Arh. Hig. Rada Toksikol.*, **52**, 291–298

Oda, T., Tsuda, H., Scarpa, A., Sakamoto, M. & Hirohashi, S. (1992) *p53* Gene mutation spectrum in hepatocellular carcinoma. *Cancer Res.*, **52**, 6358–6364

Okano, K., Tomita, T. & Chonan, M. (2002) Aflatoxins inspection in groundnuts imported into Japan in 1994–2000. *Mycotoxins*, **52** (in press)

Olsen, J.H., Dragsted, L. & Autrup, H. (1988) Cancer risk and occupational exposure to aflatoxins in Denmark. *Br. J. Cancer*, **58**, 392–396

Olubuyide, I.O., Maxwell, S.M., Hood, H., Neal, G.E. & Hendrickse, R.G. (1993a) HBsAg, aflatoxins and primary hepatocellular carcinoma. *Afr. J. Med. med. Sci.*, **22**, 89–91

Olubuyide, I.O., Maxwell, S.M., Akinyinka, O.O., Hart, C.A., Neal, G.E. & Hendrickse, R.G. (1993b) HBsAg and aflatoxins in sera of rural (Igbo-Ora) and urban (Ibadan) populations in Nigeria. *Afr. J. Med. med. Sci.*, **22**, 77–80

Omer, R.E., Bakker, M.I., van't Veer, P., Hoogenboom, R.L.A.P., Polman, T.H.G., Alink, G.M., Idris, M.O., Kadaru, A.M.Y. & Kok, F.J. (1998) Aflatoxin and liver cancer in Sudan. *Nutr. Cancer*, **32**, 174–180

Omer, R.E., Verhoef, L., Van't Veer, P., Idris, M.O., Kadaru, A.M.Y., Kampman, E., Bunschoten, A. & Kok, F.J. (2001) Peanut butter intake, GSTM1 genotype and hepatocellular carcinoma: A case–control study in Sudan. *Cancer Causes Control*, **12**, 23–32

O'Neil, M.J., Smith, A. & Heckelman, P.E. (2001) *The Merck Index, 13th Ed.*, Whitehouse Station, NJ, Merck & Co., pp. 34–35

Ono, E.Y.S., Ono, M.A., Funo, F.Y., Medina, A.E., Oliveira, T.C.R.M., Kawamura, O., Ueno, Y. & Hirooka, E.Y. (2001) Evaluation of fumonisin-aflatoxin co-occurrence in Brazilian corn hybrids by ELISA. *Food Addit. Contam.*, **18**, 719–729

Ozturk, M. (1991) *p53* Mutation in hepatocellular carcinoma after aflatoxin exposure. *Lancet*, **338**, 1356–1359

Park, U.S., Su, J.J., Ban, K.C., Qin, L., Lee, E.H. & Lee, Y.I. (2000) Mutations in the p53 tumor suppressor gene in tree shrew hepatocellular carcinoma associated with hepatitis B virus infection and intake of aflatoxin B_1. *Gene*, **251**, 73–80

Parkin, D.M., Srivatanakul, P., Khlat, M., Chenvidhya, D., Chotiwan, P., Insiripong, S., L'Abbe, K.A. & Wild, C.P. (1991) Liver cancer in Thailand. I. A case–control study of cholangiocarcinoma. *Int. J. Cancer*, **48**, 323–328

Paterlini, P. & Bréchot, C. (1994) Hepatitis B virus and primary liver cancer in hepatitis B surface antigen-positive and negative patients. In: Bréchot, C., ed., *Primary Liver Cancer: Etiological and Progression Factors*, Paris, CRC Press

Peers, F.G. & Linsell, C.A. (1973) Dietary aflatoxins and liver cancer — A population based study in Kenya. *Br. J. Cancer*, **27**, 473–484

Peers, F.G., Gilman, G.A. & Linsell, C.A. (1976) Dietary aflatoxins and human liver cancer. A study in Swaziland. *Int. J. Cancer*, **17**, 167–176

Peers, F., Bosch, X., Kaldor, J., Linsell, A. & Pluijmen, M. (1987) Aflatoxin exposure, hepatitis B virus infection and liver cancer in Swaziland. *Int. J. Cancer*, **39**, 545–553

Peterson, S.W., Ito, Y., Horn, B.W. & Goto, T. (2001) *Aspergillus bombycis*, a new aflatoxigenic species and genetic variation in its sibling species, *A. nomius*. *Mycologia*, **93**, 689–703

Pier, A.C. & McLoughlin, M.E. (1985) Mycotoxic suppression of immunity. In: Lacey, J., ed., *Trichothecenes and Other Mycotoxins*, New York, John Wiley & Sons, pp. 507–519

Pinto, V.E.F., Vaamonde, G., Brizzio, S.B. & Apro, N. (1991) Aflatoxin production in soybean varieties grown in Argentina. *J. Food Prot.*, **54**, 542–545

Pitt, J.I. (1993) Corrections to species names in physiological studies on *Aspergillus flavus* and *Aspergillus parasiticus*. *J. Food Prot.*, **56**, 265–269

Pitt, J.I. & Hocking, A.D. (1997) *Fungi and Food Spoilage*, 2nd Ed, Gaithersburg, MD, Aspen Publishers

Pitt, J.I., Hocking, A.D., Bhudhasamai, K., Miscamble, B.F., Wheeler, K.A. & Tanboon-Ek, P. (1993) The normal mycoflora of commodities from Thailand. 1. Nuts and oilseeds. *Int. J. Food Microbiol.*, **20**, 211–226

Pitt, J.I., Hocking, A.D., Bhudhasamai, K., Miscamble, B.F., Wheeler, K.A. & Tanboon-Ek, P. (1994) The normal mycoflora of commodities from Thailand. 2. Beans, rice, small grains and other commodities. *Int. J. Food Microbiol.*, **23**, 35–53

Ponchel, F., Puisieux, A., Tabone, E., Michot, J.P., Froschl, G., Morel, A.P., Frébourg, T., Fontanière, B., Oberhammer, F. & Ozturk, M. (1994) Hepatocarcinoma-specific mutant p53-249ser induces mitotic activity but has no effect on transforming growth factor β1-mediated apoptosis. *Cancer Res.*, **54**, 2064–2068

Preisler, V., Caspary, W.J., Hoppe, F., Hagen, R. & Stopper, H. (2000) Aflatoxin B_1-induced mitotic recombination in L5178Y mouse lymphoma cells. *Mutagenesis*, **15**, 91–97

Prieto-Alamo, M.-J., Jurado, J., Abril, N., Díaz-Pohl, C., Bolcsfoldi, G. & Pueyo, C. (1996) Mutational specificity of aflatoxin B_1. Comparison of *in vivo* host-mediated assay with *in vitro* S9 metabolic activation. *Carcinogenesis*, **17**, 1997–2002

Primiano, T., Egner, P.A., Sutter, T.R., Kelloff, G.J., Roebuck, B.D. & Kensler, T.W. (1995) Intermittent dosing with oltipraz: Relationship between chemoprevention of aflatoxin-induced tumorigenesis and induction of glutathione *S*-transferases. *Cancer Res.*, **55**, 4319–4324

Puisieux, A., Lim, S., Groopman, J. & Ozturk, M. (1991) Selective targeting of *p53* gene mutational hotspots in human cancers by etiologically defined carcinogens. *Cancer Res.*, **51**, 6185–6189

Qian, G.S., Ross, R.K., Yu, M.C., Yuan, J.M., Gao, Y.T., Henderson, B.E., Wogan, G.N. & Groopman, J.D. (1994) A follow-up study of urinary markers of aflatoxin exposure and liver cancer risk in Shanghai, People's Republic of China. *Cancer Epidemiol. Biomarkers Prev.*, **3**, 3–10

Raisuddin, S., Singh, K.P., Zaidi, S.A.I., Paul, B.N. & Ray, P.K. (1993) Immunosuppressive effects of aflatoxin in growing rats. *Mycopathologia*, **124**, 189–194

Rajan, A., Ismail, P.K. & Radhakrishnan, V. (1995) Survey of milk samples for aflatoxin M_1 in Thrissur, Kerala. *Indian J. Dairy Sci.*, **48**, 302–305

Ramjee, G., Berjak, P., Adhikari, M. & Dutton, M.F. (1992) Aflatoxins and kwashiorkor in Durban, South Africa. *Ann. Trop. Paediatr.*, **12**, 241–247

Ramsdell, H.S. & Eaton, D.L. (1990) Species susceptibility to aflatoxin B_1 carcinogenesis: Comparative kinetics of microsomal biotransformation. *Cancer Res.*, **50**, 615–620

Raney, K.D., Meyer, D.J., Ketterer, B., Harris, T.M. & Guengerich, F.P. (1992) Glutathione conjugation of aflatoxin B_1 *exo*- and *endo*-epoxides by rat and human glutathione *S*-transferases. *Chem. Res. Toxicol.*, **5**, 470–478

Rati, E.R. & Shantha, T. (1994) Incidence of aflatoxin in groundnut-based snack products. *J. Food Sci. Technol.*, **31**, 327–329

Reen, R.K., Wiebel, F.J. & Singh, J. (1997) Piperine inhibits aflatoxin B_1-induced cytotoxicity and genotoxicity in V79 Chinese hamster cells genetically engineered to express rat cytochrome P4502B1. *J. Ethnopharmacol.*, **58**, 165–173

Resnik, S., Neira, S., Pacin, A., Martinez, E., Apro, N. & Latreite, S. (1996) A survey of the natural occurrence of aflatoxins and zearalenone in Argentine field maize: 1983–1994. *Food Addit. Contam.*, **13**, 115–120

Roebuck, B.D., Liu, Y.-L., Rogers, A.E., Groopman, J.D. & Kensler, T.W. (1991) Protection against aflatoxin B_1-induced hepatocarcinogenesis in F344 rats by 5-(2-pyrazinyl)-4-methyl-1,2-dithiole-3-thione (oltipraz): Predictive role for short-term molecular dosimetry. *Cancer Res.*, **51**, 5501–5506

Roll, R., Matthiaschk, G. & Korte, A. (1990) Embryotoxicity and mutagenicity of mycotoxins. *J. environ. Pathol. toxicol. Oncol.*, **10**, 1–7

Ross, R.K., Yuan, J.-M., Yu, M.C., Wogan, G.N., Qian, G.-S., Tu, J.-T., Groopman, J.D., Gao, Y.-T. & Henderson, B.E. (1992) Urinary aflatoxin biomarkers and risk of hepatocellular carcinoma. *Lancet*, **339**, 943–946

Ross, M.K., Mathison, B.H., Said, B. & Shank, R.C. (1999) 5-Methylcytosine in CpG sites and the reactivity of nearest neighboring guanines toward the carcinogen aflatoxin B_1-8,9-epoxide. *Biochem. biophys. Res. Commun.*, **254**, 114–119

Rossano, F., Ortega De Luna, L., Buommino, E., Cusumano, V., Losi, E. & Catania, M.R. (1999) Secondary metabolites of *Aspergillus* exert immunobiological effects on human monocytes. *Res. Microbiol.*, **150**, 13–19

Saad, A.M., Abdelgadir, A.M. & Moss, M.O. (1995) Exposure of infants to aflatoxin M_1 from mothers' breast milk in Abu Dhabi, U.A.E. *Food Addit. Contam.*, **12**, 255–261

Sabbioni, G. & Wild, C.P. (1991) Identification of an aflatoxin G_1-serum albumin adduct and its relevance to the measurement of human exposure to aflatoxins. *Carcinogenesis*, **12**, 97–103

Sabbioni, G., Skipper, P.L., Buchi, G. & Tannenbaum, S.R. (1987) Isolation and characterization of the major serum albumin adduct formed by aflatoxin B_1 *in vivo* in rats. *Carcinogenesis*, **8**, 819–824

Saitanu, K. (1997) Incidence of aflatoxin M_1 in Thai milk products. *J. Food Prot.*, **60**, 1010–1012

Saito, M. & Tsuruta, O. (1993) A new variety of *Aspergillus flavus* from tropical soil in Thailand and its aflatoxin productivity. *Proc. Jpn. Assoc. Mycotoxicol.*, **37**, 31–36

Saito, M., Tsuruta, O., Siriacha, P. & Manabe, M. (1989) Atypical strains of *Aspergillus flavus* isolated in maize fields. *Jpn. Agric. Res. Q.*, **23**, 151–154

Salem, M.H., Kamel, K.I., Yousef, M.I., Hassan, G.A. & EL-Nouty, F.D. (2001) Protective role of ascorbic acid to enhance semen quality of rabbits treated with sublethal doses of aflatoxin B_1. *Toxicology*, **162**, 209–218

Sargeant, K., O'Kelly, J., Carnaghan, R.B.A. & Allcroft, R. (1961) The assay of a toxic principle in certain groundnut meals. *Vet. Rec.*, **73**, 1219–1222

Schleger, C., Becker, R., Oesch, F. & Steinberg, P. (1999) The human *p53* gene mutated at position 249 *per se* is not sufficient to immortalize human liver cells. *Hepatology*, **29**, 834–838

Schnur, J., Nagy, P., Sebestyen, A., Schaff, Z. & Thorgeirsson, S.S. (1999) Chemical hepatocarcinogenesis in transgenic mice overexpressing mature TGF beta-1 in liver. *Eur. J. Cancer*, **35**, 1842–1845

Schroeder, H.W. & Boller, R.A. (1973) Aflatoxin production of species and strains of the *Aspergillus flavus* group isolated from field crops. *Appl. Microbiol.*, **25**, 885–889

SCOOP (1996) *Scientific Co-operation on Questions Relating to Food: Working Document in Support of a SCF Risk Assessment of Aflatoxin: Task 3.2.1 (SCOOP/CNTM/1)*, Task Coordinateur, UK

Scorsone, K.A., Zhou, Y.-Z., Butel, J.S. & Slagle, B.L. (1992) *p53* Mutations cluster at codon 249 in hepatitis B virus-positive hepatocellular carcinomas from China. *Cancer Res.*, **52**, 1635–1638

Sell, S., Hunt, J.M., Dunsford, H.A. & Chisari, F.V. (1991) Synergy between hepatitis B virus expression and chemical hepatocarcinogens in transgenic mice. *Cancer Res.*, **51**, 1278–1285

Sengstag, C., Weibel, B. & Fasullo, M. (1996) Genotoxicity of aflatoxin B_1 — Evidence for a recombination-mediated mechanism in *Saccharomyces cerevisiae*. *Cancer Res.*, **56**, 5457–5465

Sharma, A. & Sahai, R. (1987) Teratological effects of aflatoxin on rats (*Rattus norvegicus*). *Indian J. Anim. Res.*, **21**, 35–40

Shetty, P.H. & Bhat, R.V. (1997) Natural occurrence of fumonisin B_1 and its co-occurrence with aflatoxin B_1 in Indian sorghum, maize, and poultry feeds. *J. agric. Food Chem.*, **45**, 2170–2173

Shi, C.Y., Phang, T.W., Lin, Y., Wee, A., Li, B., Lee, H.P. & Ong, C.N. (1995) Codon 249 mutation of the *p53* gene is a rare event in hepatocellular carcinomas from ethnic Chinese in Singapore. *Br. J. Cancer*, **72**, 146–149

Shimizu, Y., Zhu, J.-J., Han, F., Ishikawa, T. & Oda, H. (1999) Different frequencies of *p53* codon-249 hot-spot mutations in hepatocellular carcinomas in Jiang-su province of China. *Int. J. Cancer*, **82**, 187–190

Silvotti, L., Petterino, C., Bonomi, A. & Cabassi, E. (1997) Immunotoxicological effects on piglets of feeding sows diets containing aflatoxins. *Vet. Rec.*, **141**, 469–472

Smela, M.E., Currier, S.S., Bailey, E.A. & Essigmann, J.M. (2001) The chemistry and biology of aflatoxin B_1: From mutational spectrometry to carcinogenesis. *Carcinogenesis*, **22**, 535–545

Sodeinde, O., Chan, M.C.K., Maxwell, S.M., Familusi, J.B. & Hendrickse, R.G. (1995) Neonatal jaundice, aflatoxins and naphthols: Report of a study in Ibadan, Nigeria. *Ann. trop. Paediatr.*, **15**, 107–113

Sohn, S., Jaitovitch-Groisman, I., Benlimame, N., Galipeau, J., Batist, G. & Alaoui-Jamali, M.A. (2000) Retroviral expression of the hepatitis B virus x gene promotes liver cell susceptibility to carcinogen-induced site specific mutagenesis. *Mutat. Res.*, **460**, 17–28

Soini, Y., Chia, S.C., Bennett, W.P., Groopman, J.D., Wang, J.-S., DeBenedetti, V.M.G., Cawley, H., Welsh, J.A., Hansen, C., Bergasa, N.V., Jones, E.A., DiBisceglie, A.M., Trivers, G.E., Sandoval, C.A., Calderon, I.E., Munoz Espinosa, L.E. & Harris, C.C. (1996) An aflatoxin-associated mutational hotspot at codon 249 in the *p53* tumor suppressor gene occurs in hepatocellular carcinomas from Mexico. *Carcinogenesis*, **17**, 1007–1012

Soni, K.B., Lahiri, M., Chackradeo, P., Bhide, S.V. & Kuttan, R. (1997) Protective effect of food additives on aflatoxin-induced mutagenicity and hepatocarcinogenicity. *Cancer Lett.*, **115**, 129–133

Srivatanakul, P., Parkin, D.M., Jiang, Y.Z., Khlat, M., Kao–Ian, U.T., Sontipong, S. & Wild, C. (1991a) The role of infection by Opisthorchis viverrini, hepatitis B virus, and aflatoxin exposure in the etiology of liver cancer in Thailand. A correlation study. *Cancer*, **68**, 2411–2417

Srivatanakul, P., Parkin, D.M., Khlat, M., Chenvidhya, D., Chotiwan, P., Insiripong, S., L'Abbe, K.A. & Wild, C.P. (1991b) Liver cancer in Thailand. II. A case–control study of hepatocellular carcinoma. *Int. J. Cancer*, **48**, 329–332

Stanley, L.A., Mandel, H.G., Riley, J., Sinha, S., Higginson, F.M., Judah, D.J. & Neal, G.E. (1999) Mutations associated with in vivo aflatoxin B_1-induced carcinogenesis need not be present in the in vitro transformations by this toxin. *Cancer Lett.*, **137**, 173–181

Stern, M.C., Umbach, D.M., Yu, M.C., London, S.J., Zhang, Z.-Q. & Taylor, J.A. (2001) Hepatitis B, aflatoxin B_1, and *p53* codon 249 mutation in hepatocellular carcinomas from Guangxi, People's Republic of China, and a meta-analysis of existing studies. *Cancer Epidemiol. Biomarkers Prev.*, **10**, 617–625

Stettler, P.M. & Sengstag, C. (2001) Liver carcinogen aflatoxin B_1 as an inducer of mitotic recombination in a human cell line. *Mol. Carcinog.*, **31**, 125–138

Stoloff, L. (1983) Aflatoxin as a cause of primary liver-cell cancer in the United States: A probability study. *Nutr. Cancer*, **5**, 165–186

Stoloff, L., van Egmond, H.P. & Park, D.L. (1991) Rationales for the establishment of limits and regulations for mycotoxins. *Food Addit. Contam.*, **8**, 213–222

Stresser, D.M., Bailey, G.S. & Williams, D.E. (1994a) Indole-3-carbinol and β-naphthoflavone induction of aflatoxin B_1 metabolism and cytochromes P-450 associated with bioactivation and detoxication of aflatoxin B_1 in the rat. *Drug Metab. Dispos.*, **22**, 383–391

Stresser, D.M., Williams, D.E., McLellan, L.I., Harris, T.M. & Bailey, G.S. (1994b) Indole-3-carbinol induces a rat liver glutathione transferase subunit (Yc2) with high activity toward aflatoxin B_1 *exo*-epoxide. *Drug Metab. Dispos.*, **22**, 392–399

Stroka, J., Anklam, E., Joerissen, U. & Gilbert, J. (2001) Determination of aflatoxin B_1 in baby food (infant formula) by immunoaffinity column cleanup liquid chromatography with post-column bromination: Collaborative study. *J. Assoc. off. anal. Chem. int.*, **84**, 1116–1123

Stubblefield, R.D., Shannon, G.M. & Shotwell, O.L. (1970) Aflatoxins M_1 and M_2: Preparation and purification. *J. Am. Oil chem. Soc.*, **47**, 389–390

Sun, Z., Lu, P., Gail, M.H., Pee, D., Zhang, Q., Ming, L., Wang, J., Wu, Y., Liu, G., Wu, Y. & Zhu, Y. (1999) Increased risk of hepatocellular carcinoma in male hepatitis B surface antigen carriers with chronic hepatitis who have detectable urinary aflatoxin metabolite M1. *Hepatology*, **30**, 379–383

Sun, C.A., Wang, L.Y., Chen, C.J., Lu, S.N., You, S.L., Wang, L.W., Wang, Q., Wu, D.M. & Santella, R.M. (2001) Genetic polymorphisms of glutathione *S*-transferases M1 and T1 associated with susceptibility to aflatoxin-related hepatocarcinogenesis among chronic hepatitis B carriers: A nested case–control study in Taiwan. *Carcinogenesis*, **22**, 1289–1294

Suprasert, D. & Chulamorakot, T. (1999) Mycotoxin contamination in detected sorghum in Thailand. *Food (Inst. Food Res. Products Develop., Kasartsart Univ.)*, **29**, 187–192

Sylla, A., Diallo, M.S., Castegnaro, J. & Wild, C.P. (1999) Interactions between hepatitis B virus infection and exposure to aflatoxins in the development of hepatocellular carcinoma: A molecular epidemiological approach. *Mutat. Res.*, **428**, 187–196

de Sylos, C.M., Rodriguez-Amaya, D.B. & Carvalho, P.R. (1996) Occurrence of aflatoxin M_1 in milk and dairy products commercialized in Campinas, Brazil. *Food Addit. Contam.*, **13**, 169–172

Tabata, S., Kamimura, H., Ibe, A., Hashimoto, H., Iida, M., Tamura, Y. & Nishima, T. (1993) Aflatoxin contamination in foods and foodstuffs in Tokyo: 1986–1990. *J. Assoc. off. anal. Chem. int.*, **76**, 32–35

Tabata, S., Ibe, A., Ozawa, H., Kamimura, H. & Yasuda, K. (1998) Aflatoxin contamination in foods and foodstuffs in Tokyo: 1991–1996. *J. Food Hyg. Soc. Jpn*, **39**, 444–447

Taguchi, S., Fukushima, S., Sumimoto, T., Yoshida, S. & Nishimune, T. (1995) Aflatoxins in foods collected in Osaka, Japan, from 1988 to 1992. *J. Assoc. off. anal. Chem. int.*, **78**, 325–327

Tam, A.S., Foley, J.F., Devereux, T.R., Maronpot, R.R. & Massey, T.E. (1999) High frequency and heterogeneous distribution of *p53* mutations in aflatoxin B_1-induced mouse lung tumors. *Cancer Res.*, **59**, 3634–3640

Tanaka, S., Toh, Y., Adachi, E., Matsumata, T., Mori, R. & Sugimachi, K. (1993) Tumor progression in hepatocellular carcinoma may be mediated by *p53* mutation. *Cancer Res.*, **53**, 2884–2887

Tanimura, T., Kihara, T. & Yamamoto, Y. (1982) Teratogenicity of aflatoxin B_1 in the mouse (in Japanese). *Kankyo Kagaku Kenkyusho Kenkyu Hokoku (Kinki Daigaku)*, **10**, 247–256

Teramoto, T., Satonaka, K., Kitazawa, S., Fujimori, T., Hayashi, K. & Maeda, S. (1994) *p53* Gene abnormalities are closely related to hepatoviral infections and occur at a late stage of hepatocarcinogenesis. *Cancer Res.*, **54**, 231–235

Tiemersma, E.W., Omer, R.E., Bunschoten, A., van't Veer, P., Kok, F.J., Idris, M.O., Kadaru, A.M.Y., Fedail, S.S. & Kampman, E. (2001) Role of genetic polymorphism of glutathione-*S*-transferase T1 and microsomal epoxide hydrolase in aflatoxin-associated hepatocellular carcinoma. *Cancer Epidemiol. Biomarkers Prev.*, **10**, 785–791

Torres Espinosa, E., Asakr, K.A., Torres, L.R.N., Olvera, R.M. & Anna, J.P.C.S. (1995) Quantification of aflatoxins in corn distributed in the city of Monterrey, Mexico. *Food Addit. Contam.*, **12**, 383–386

Trottier, Y., Waithe, W.I. & Anderson, A. (1992) Kinds of mutations induced by aflatoxin B_1 in a shuttle vector replicating in human cells transiently expressing cytochrome P4501A2 cDNA. *Mol. Carcinog.*, **6**, 140–147

Turner, P.C., Mendy, M., Whittle, H., Fortuin, M., Hall, A.J. & Wild, C.P. (2000) Hepatitis B infection and aflatoxin biomarker levels in Gambian children. *Trop. Med. int. Health*, **5**, 837–841

Ueng, Y.-F., Shimada, T., Yamazaki, H. & Guengerich, F.P. (1998) Aflatoxin B_1 oxidation by human cytochrome P450s. *J. Toxicol. Sci.*, **23** (Suppl. II), 132–135

Ueno, Y., Iijima, K., Wang, S.D., Dugiura, Y., Sekijima, M., Tanaka, T., Chen, C. & Yu, S.Z. (1997) Fumonisins as a possible contributory risk factor for primary liver cancer: A 3-year study of corn harvested in Haimen, China, by HPLC and ELISA. *Food chem. Toxicol.*, **35**, 1143–1150

Uhl, M., Helma, C. & Knasmüller, S. (2000) Evaluation of the single cell gel electrophoresis assay with human hepatoma (Hep G2) cells. *Mutat. Res.*, **468**, 213–225

Van Rensburg, S.J., Cook-Mozaffari, P., Van Schalkwyk, D.J., Van der Watt, J.J., Vincent, T.J. & Purchase, I.F. (1985) Hepatocellular carcinoma and dietary aflatoxin in Mozambique and Transkei. *Br. J. Cancer*, **51**, 713–726

Van Rensburg, S.J., Van Schalkwyk, G.C. & Van Schalkwyk, D.J. (1990) Primary liver cancer and aflatoxin intake in Transkei. *J. environ. Pathol. Toxicol. Oncol.*, **10**, 11–16

Vargas, E.A., Preis, R.A., Castro, L. & Silva C.M.G. (2001) Co-occurrence of aflatoxin B_1, B_2, G_1, G_2, zearalenone and fumonisin B_1 in Brazilian corn. *Food Addit. Contam.*, **18**, 981–986

Vasanthi, S. & Bhat, R.V. (1998) Mycotoxins in foods — Occurrence, health and economic significance and food control measures. *Indian J. med. Res.*, **108**, 212–224

Vautier, G., Bomford, A.B., Portmann, B.C., Metivier, E., Williams, R. & Ryder, S.D. (1999) *p53* Mutations in British patients with hepatocellular carcinoma: Clustering in genetic hemochromatosis. *Gastroenterology*, **117**, 154–160

Venturini, M. C., Quiroga, M.C., Risso, M.A., Di Lorenzo, C., Omata, Y., Venturini, L. & Godoy, H. (1996) Mycotoxin T-2 and aflatoxin B_1 as immunosuppressors in mice chronically infected with *Toxoplasma gondii*. *J. comp. Pathol.*, **115**, 229–237

Verger, P.K., Vlatier, J.-L. & Dufour, A. (1999) [Estimation of theoretic intake of aflatoxins and ochratoxin.] In: Pfohl-Leszkowicz, A., ed., *Les Mycotoxines dans l'Alimentation: Evaluation et Gestion du Risque* [Mycotoxins in Food: Evaluation and Management of Risk], Paris, Technique et Documentation, pp. 371–384 (in French)

Viquez, O.M., Castell-Perez, M.E., Shelby, R.A. & Brown, G. (1994) Aflatoxin contamination in corn samples due to environmental conditions, aflatoxin-producing strains, and nutrients in grain grown in Costa Rica. *J. agric. Food Chem.*, **42**, 2551–2555

Wang, Y.B., Lan, L.Z., Ye, B.F., Xu, Y.C., Liu, Y.Y. & Li, W.G. (1983) Relation between geographical distribution of liver cancer and climate–aflatoxin B1 in China. *Sci. Sin. [B]*, **26**, 1166–1175

Wang, D.S., Liang, Y.X., Chau, N.T., Dien, L.D., Tanaka, T. & Ueno, Y. (1995) Natural co-occurrence of *Fusarium* toxins and aflatoxin B_1 in corn for feed in North Vietnam. *Nat. Toxins*, **3**, 445–449

Wang, L.Y., Hatch, M., Chen, C.J., Levin, B., You, S.L., Lu, S.N., Wu, M.H., Wu, W.P., Wang, L.W., Wang, Q., Huang, G.T., Yang, P.M., Lee, H.S. & Santella, R.M. (1996a) Aflatoxin exposure and risk of hepatocellular carcinoma in Taiwan. *Int. J. Cancer*, **67**, 620–625

Wang, J.-S., Qian, G.-S., Zarba, A., He, X., Zhu, Y.-R., Zhang, B.-C., Jacobson, L., Gange, S.J., Munoz, A., Kensler, T.W. & (1996b) Temporal patterns of aflatoxin-albumin adducts in hepatitis B surface antigen-positive and antigen-negative residents of Daxin, Qidong County, People's Republic of China. *Cancer Epidemiol. Biomarkers Prev.*, **5**, 253–261

Wang, H., Dick, R., Yin, H., Licad-Coles, E., Kroetz, D.L., Szklarz, G., Harlow, G., Halpert, J.R. & Correia, M.A. (1998) Structure-function relationships of human liver cytochromes P450 3A: Aflatoxin B_1 metabolism as a probe. *Biochemistry*, **37**, 12536–12545

Wang, J.-S., Shen, X., He, X., Zhu, Y.-R., Zhang, B.-C., Wang, J.-B., Qian, G.-S., Kuang, S.-Y., Zarba, A., Egner, P.A., Jacobson, L.P., Muñoz, A., Helzlsouer, K.J., Groopman, J.D. & Kensler, T.W. (1999a) Protective alterations in phase 1 and 2 metabolism of aflatoxin B_1 by oltipraz in residents of Qidong, People's Republic of China. *J. natl Cancer Inst.*, **91**, 347–354

Wang, S.S., O'Neill, J.P., Qian, G.-S., Zhu, Y.-R., Wang, J.-B., Armenian, H., Zarba, A., Wang, J.-S., Kensler, T.W., Cariello, N.F., Groopman, J.D. & Swenberg, J.A. (1999b) Elevated *HPRT* mutation frequencies in aflatoxin-exposed residents of Daxin, Qidong County, People's Republic of China. *Carcinogenesis*, **20**, 2181–2184

Wang, C., Bammler, T.K., Guo, Y., Kelly, E.J. & Eaton, D.L. (2000) *Mu*-Class GSTs are reponsible for aflatoxin B_1-8,9-epoxide-conjugating activity in the nonhuman primate *Macaca fascicularis* liver. *Toxicol. Sci.*, **56**, 26–36

Wang, J.S., Huang, T., Su, J., Liang, F., Wei, Z., Liang, Y., Luo, H., Kuang, S.Y., Qian, G.S., Sun, G., He, X., Kensler, T.W. & Groopman, J.D. (2001) Hepatocellular carcinoma and aflatoxin exposure in Zhuqing Village, Fusui County, People's Republic of China. *Cancer Epidemiol. Biomarkers Prev.*, **10**, 143–146

Whitaker, T.B., Hagler, W.M., Jr, Giesbrecht, F.G., Dorner, J.W., Dowell, F.E. & Cole, R.J. (1998) Estimating aflatoxin in farmers' stock peanut lots by measuring aflatoxin in various peanut-grade components. *J. Assoc. off. anal. Chem. Int.*, **81**, 61–67

WHO (1998) *GEMS/Food Regional Diets. Regional Per Capita Consumption of Raw and Semi-processed Agricultural Commodities* (WHO/FSF/FOS/98.3), Geneva

Wild, C.P. & Hall, A.J. (1999) Hepatitis B virus and liver cancer: Unanswered questions. *Cancer Surv.*, **33**, 35–54

Wild, C.P. & Hall, A.J. (2000) Primary prevention of hepatocellular carcinoma in developing countries. *Mutat. Res.*, **462**, 381–393

Wild, C.P. & Kleihues, P. (1996) Etiology of cancer in humans and animals. *Exp. Toxicol. Pathol.*, **48**, 95–100

Wild, C.P., Pionneau, F.A., Montesano, R., Mutiro, C.F. & Chetsanga, C.J. (1987) Aflatoxin detected in human breast milk by immunoassay. *Int. J. Cancer*, **40**, 328-333

Wild, C.P., Jiang, Y.Z., Allen, S.J., Jansen, L.A., Hall, A.J. & Montesano, R. (1990) Aflatoxin–albumin adducts in human sera from different regions of the world. *Carcinogenesis*, **11**, 2271–2274

Wild, C.P., Rasheed, F.N., Jawla, M.F.B., Hall, A.J., Jansen, L.A.M. & Montesano, R. (1991) In-utero exposure to aflatoxin in West Africa (Letter to the Editor). *Lancet*, **337**, 1602

Wild, C.P., Jansen, L.A., Cova, L. & Montesano, R. (1993a) Molecular dosimetry of aflatoxin exposure: Contribution to understanding the multifactorial etiopathogenesis of primary hepatocellular carcinoma with particular reference to hepatitis B virus. *Environ. Health Perspect.*, **99**, 115–122

Wild, C.P., Fortuin, M., Donato, F., Whittle, H.C., Hall, A.J., Wolf, C.R. & Montesano, R. (1993b) Aflatoxin, liver enzymes, and hepatitis B virus infection in Gambian children. *Cancer Epidemiol. Biomarkers Prev.*, **2**, 555–561

Wild, C.P., Hasegawa, R., Barraud, L., Chutimataewin, S., Chapot, B., Ito, N. & Montesano, R. (1996) Aflatoxin-albumin adducts: A basis for comparative carcinogenesis between animals and humans. *Cancer Epidemiol. Biomarkers Prev.*, **5**, 179–189

Wild, C.P., Yin, F., Turner, P.C., Chemin, I., Chapot, B., Mendy, M., Whittle, H., Kirk, G.D. & Hall, A.J. (2000) Environmental and genetic determinants of aflatoxin-albumin adducts in The Gambia. *Int. J. Cancer*, **86**, 1–7

Williams, G.M. & Iatropoulos, M.J. (1996) Inhibition of the hepatocarcinogenicity of aflatoxin B_1 in rats by low levels of the phenolic antioxidants butylated hydroxyanisole and butylated hydroxytoluene. *Cancer Lett.*, **104**, 49–53

Wilson, A.S., Tingle, M.D., Kelly, M.D. & Park, B.K. (1995) Evaluation of the generation of genotoxic and cytotoxic metabolites of benzo[*a*]pyrene, aflatoxin B_1, naphthalene and tamoxifen using human liver microsomes and human lymphocytes. *Human exper. Toxicol.*, **14**, 507–515

Wilson, A.S., Williams, D.P., Davis, C.D., Tingle, M.D. & Park, B.K. (1997) Bioactivation and inactivation of aflatoxin B_1 by human, mouse and rat liver preparations: Effect on SCE in human mononuclear leucocytes. *Mutat. Res.*, **373**, 257–264

Wong, N., Lai, P., Pang, E., Fung, L.-F., Sheng, Z., Wong, V., Wang, W., Hayashi, Y., Perlman, E., Yuna, S., Lau, J.W.-Y. & Johnson, P.J. (2000) Genomic aberrations in human hepatocellular carcinomas of differing etiologies. *Clin. Cancer Res.*, **6**, 4000–4009

Yamada, A., Fujita, K., Yokoi, T., Muto, S., Suzuki, A., Gondo, Y., Katsuki, M. & Kamataki, T. (1998) In vivo detection of mutations induced by aflatoxin B_1 using human CYP3A7/HITEC hybrid mice. *Biochem. Biophys. Res. Commun.*, **250**, 150–153

Yamashita, A., Yoshizawa, T., Aiura, Y., Sanchez, P.C., Dizon, E.I., Arim, R.H. & Sardjono (1995) *Fusarium* mycotoxins (fumonisins, nivalenol, and zearalenone) and aflatoxins in corn from Southeast Asia. *Biosci. Biotech. Biochem.*, **59**, 1804–1807

Yang, M., Zhou, H., Kong, R.Y., Fong, W.F., Ren, L.Q., Liao, X.H., Wang, Y., Zhuang, W. & Yang, S. (1997) Mutations at codon 249 of *p53* gene in human hepatocellular carcinomas from Tongan, China. *Mutat. Res.*, **381**, 25–29

Yeh, F.S., Mo, C.C. & Yen, R.C. (1985) Risk factors for hepatocellular carcinoma in Guangxi, People's Republic of China. *Natl Cancer Inst. Monogr.*, **69**, 47–48

Yeh, F.S., Yu, M.C., Mo, C.C., Luo, S., Tong, M.J. & Henderson, B.E. (1989) Hepatitis B virus, aflatoxins, and hepatocellular carcinoma in southern Guangxi, China. *Cancer Res.*, **49**, 2506–2509

Yoshizawa, T., Yamashita, A. & Chokethaworn, N. (1996) Occurrence of fumonisins and aflatoxins in corn from Thailand. *Food Addit. Contam.*, **13**, 163–168

Yu, S.Z. (1995) Primary prevention of hepatocellular carcinoma. *J. Gastroenterol. Hepatol.*, **10**, 674–682

Yu, M.W., Lien, J.P., Chiu, Y.H., Santella, R.M., Liaw, Y.F. & Chen, C.J. (1997a) Effect of aflatoxin metabolism and DNA adduct formation on hepatocellular carcinoma among chronic hepatitis B carriers in Taiwan. *J. Hepatol.*, **27**, 320–330

Yu, M.W., Chiang, Y.C., Lien, J.P. & Chen, C.J. (1997b) Plasma antioxidant vitamins, chronic hepatitis B virus infection and urinary aflatoxin B_1-DNA adducts in healthy males. *Carcinogenesis*, **18**, 1189–1194

Zhang, L.-H. & Jenssen, D. (1994) Studies on intrachromosomal recombination in SP5/V79 Chinese hamster cells upon exposure to different agents related to carcinogenesis. *Carcinogenesis*, **15**, 2303–2310

Zhang, D., Zhang, J., Liu, C. & Luo, X.Y. (1996) [Survey on natural occurrence of AFB1 in cereals and oils harvested in 1992 collected from some of provinces in China]. *Chinese J. Food Hyg.*, **8**, 35–36 (in Chinese)

Zhou, T., Evans, A.A., London, W.T., Xia, X., Zou, H., Shen, F. & Clapper, M.L. (1997) Glutathione *S*-transferase expression in hepatitis B virus-associated human hepatocellular carcinogenesis. *Cancer Res.*, **57**, 2749–2753

ANNEX. AFLATOXINS IN FOODS AND FEEDS: FUNGAL SOURCES, FORMATION AND STRATEGIES FOR REDUCTION

1. Major Mycotoxins and Crops

Five types of mycotoxins of agricultural importance occur in staple crops (Miller, 1995): aflatoxins, fumonisins, ochratoxin A, specific trichothecenes (deoxynivalenol and nivalenol) and zearalenone. These mycotoxins can cause various forms of poisoning in animals and in humans, and some are carcinogenic (Table 1). Aflatoxins are produced by *Aspergillus* species in nuts and oilseeds, particularly maize, peanuts (groundnuts) and cottonseed, especially in tropical and subtropical climates. Fumonisins are produced by *Fusarium verticillioides* (formerly known as *F. moniliforme*) and the closely related *F. proliferatum*, in maize and sorghum. Ochratoxin A occurs in cereals as a result of growth of *Penicillium verrucosum* and in other crops, especially grape products (grape juice, wines and dried vine fruit), coffee and long-stored commodities as a result of growth of several *Aspergillus* species. Deoxynivalenol, nivalenol and zearalenone are formed as a result of growth of *F. graminearum* and *F. culmorum* in maize, wheat, barley and other small grains (JECFA, 2001). Ergot, the toxic product of the fungus *Claviceps purpurea* which grows on rye and to a lesser extent on other grains, was historically a significant source of epidemic poisoning in Europe, but due to effective inspection of grain it is rarely a public health problem today in Europe and North America.

Some crops are infected by only one toxigenic fungus: aflatoxins in peanuts and cottonseed, and ochratoxin A in susceptible crops are usually found by themselves. In small grains, trichothecenes and zearalenone usually occur together as the result of infection by one or more *Fusarium* species. Of greater importance, fumonisins and aflatoxins, and to a lesser extent trichothecenes and zearalenone, frequently occur simultaneously in maize. Any conclusions regarding reduction strategies should take this into account.

This Annex deals only with aflatoxins, but many of the points made have relevance to the other toxins.

Table 1. IARC evaluations

Previous evaluation[a]				This volume			
Agent	Degree of evidence of carcinogenicity		Overall evaluation of carcinogenicity to humans	Agent	Degree of evidence of carcinogenicity		Overall evaluation of carcinogenicity to humans
	Human	Animal			Human	Animal	
Aflatoxins, naturally occurring mixtures of	S	S	1	Naturally occurring aflatoxins			1 (reaffirmed)
Aflatoxin B_1	S	S					
Aflatoxin B_2		L					
Aflatoxin G_1		S					
Aflatoxin G_2		I					
Aflatoxin M_1	I	S	2B				
Toxins derived from *Fusarium moniliforme* (now called *F. verticillioides*)	I	S	2B				
Fumonisin B_1		L		Fumonisin B_1	I	S	2B
Fumonisin B_2		I					
Fusarin C		L					
Ochratoxin A	I	S	2B				
Trichothecenes							
Toxins derived from *Fusarium graminearum* and *F. culmorum*	I		3				
Zearalenone		L					
Deoxynivalenol		I					
Nivalenol		I					

S, sufficient evidence of carcinogenicity; L, limited evidence of carcinogenicity; I, inadequate evidence of carcinogenicity; group 1, carcinogenic to humans; group 2B, possibly carcinogenic to humans; group 3, not classifiable as to its carcinogenicity to humans

[a] IARC (1993) *IARC Monographs on the Evaluation of Carcinogenic Risks to Humans*, Vol. 56, *Some Naturally Occurring Substances: Food Items and Constituents, Heterocyclic Aromatic Amines and Mycotoxins*, Lyon, IARC*Press*

2. Aflatoxins

2.1 Introduction

Aflatoxins frequently contaminate certain types of foods and feeds in warm and tropical regions. Limiting the formation of aflatoxins in such commodities cannot usually be achieved by any single technique. To be effective, approaches to this problem require an overall strategy, which involves a knowledge of crops likely to be affected, the time when infection by the fungi is likely to occur, and careful management of crops both before and after harvest. Once aflatoxins have been formed, reduction again relies on a range of strategies, which if well managed can result in the removal of the major part of aflatoxin contamination. Contaminated crops can be used for animal feed after chemical treatment. This Annex provides an overview of the crops and conditions that favour aflatoxin contamination, and the various management strategies available to limit formation or reduce levels of aflatoxins in commodities.

2.2 Fungi producing aflatoxins

Aflatoxins in food and feed crops are almost entirely produced by the common fungi *Aspergillus flavus* and the closely related species *A. parasiticus*. *A. flavus* produces only B aflatoxins while *A. parasiticus* produces both B and G aflatoxins (Schroeder & Boller, 1973; Dorner *et al.*, 1984; Klich & Pitt, 1988; Pitt, 1993). Several other *Aspergillus* species are now known to produce aflatoxins but they are of little practical importance in foods.

2.3 Occurrence of toxigenic species in foods

2.3.1 *Aspergillus flavus*

A. flavus is the most widely reported food-borne fungus outside north-temperate areas. It is especially abundant in the tropics: *A. flavus* was isolated from 97% of nearly 500 peanut samples examined from south-east Asian sources over the years 1989–91, with an average infection rate of more than 40% of all surface-disinfected kernels examined from Thailand and the Philippines, and more than 60% of those from Indonesia. For maize, the figures were 89% of 380 samples, at an average of 38% of all grains infected from Indonesia and the Philippines, and 17% of those from Thailand (Pitt & Hocking, 1997).

Levels in food commodities from more temperate climates, such as Australia or the USA, are much lower. *A. flavus* was present in only 1.2% of surface-disinfected maize kernels and in less than 0.1% of all oats and wheat kernels examined from a large number of samples in the USA (Sauer *et al.*, 1984).

The major food and feed commodities where *A. flavus* is found are peanuts (McDonald, 1970; Pitt *et al.*, 1993, 1998), maize (Diener *et al.*, 1983; Pitt *et al.*, 1993, 1998) and cottonseed (Simpson & Batra, 1984). Spices of many kinds frequently contain *A. flavus* (ICMSF, 1998). From time to time, *A. flavus* occurs in most types of tree nuts, including pistachios, pecans, hazelnuts and walnuts, copra and kola nuts (Pitt & Hocking, 1997). Aflatoxins are sometimes produced in these commodities (Pohland & Wood, 1987). Low levels of *A. flavus* in small grain cereals and pulses, and many other kinds of foods, e.g. soybean, have been reported, but the possibility of significant aflatoxin accumulation is much lower (Pitt *et al.*, 1994; Pitt & Hocking, 1997; Pitt *et al.*, 1998).

2.3.2 *Aspergillus parasiticus*

A. parasiticus seems to be less widely distributed than *A. flavus*. During a major study, more than 30 000 *A. flavus* cultures from south-east Asian foods were isolated and identified, but not more than 20 isolates of *A. parasiticus* were found. Although *A. parasiticus* is certainly widely distributed in soils and foodstuffs in the USA, Latin America, South Africa, India and Australia, it is essentially unknown in south-east Asia. Like *A. flavus*, it is a tropical and subtropical species, less prevalent in warm temperate zones, and rare in the cool temperate regions of the world. The most important food source is peanuts, in which *A. parasiticus* is endemic. Other types of nuts may be infected, including hazelnuts and walnuts, pistachios and pecans. *A. parasiticus* is much less common than *A. flavus* on grains, and perhaps does not invade maize at all. A variety of other minor sources have been reported (Pitt & Hocking, 1997).

2.4 Formation of aflatoxins in foods

A fundamental distinction must be made between aflatoxins formed in crops before or immediately after harvest, and those occurring in stored commodities or food products. In subtropical and tropical areas, certain crop plants, notably peanuts, maize and cottonseed are associated with *A. flavus*, or in the case of peanuts, also for *A. parasiticus*, so that invasion of plants and developing seeds or nuts may occur before harvest. This is the cause of the frequent occurrence of high levels of aflatoxins in these crops, and is the reason for the difficulties still being experienced in eliminating aflatoxins from these commodities. In contrast, *A. flavus* is less common in other plants, seeds or nuts before harvest. In consequence, aflatoxins are not normally a problem with other crops at harvest and their elimination relies on preventing post-harvest contamination, by rapid drying and good storage practice (Pitt, 1989; Chatterjee *et al.*, 1990; Miller, 1995). Therefore, if infection of peanuts and maize by *A. flavus* could be controlled before harvest, excessive aflatoxin production would not normally occur in storage, even under somewhat unsatisfactory conditions. In temperate maize production, *A. flavus* contamination is associated mainly with insect damage during drought conditions (Miller, 1995).

2.5 Formation of aflatoxins in susceptible crops

2.5.1 *Peanuts*

Peanuts are susceptible to infection by both *A. flavus* and *A. parasiticus* (Hesseltine *et al.*, 1970; Diener *et al.*, 1987; Pitt & Hocking, 1997). The primary source of these fungi is soil, where high numbers may build up because some peanuts are not harvested, but remain in the ground and act as a nutrient source (Griffin & Garren, 1976a). Uncultivated soils contain very low amounts of *A. flavus*, but soils in peanut fields usually contain 100–5000 propagules (spores) per gram (Pitt, 1989). Under drought stress conditions, this number may rise to 10^4 or 10^5/g (Horn *et al.*, 1995). Large numbers of *A. flavus* spores are also airborne over susceptible crops (Holtmeyer & Wallin, 1981).

Direct entry to developing peanuts through the shell by *A. flavus* in the soil appears to be the main route of nut infection (Diener *et al.*, 1987). Infection can also occur through the pegs and flowers (Wells & Kreutzer, 1972; Griffin & Garren, 1976b; Pitt, 1989). *A. flavus* sometimes grows within peanut plants themselves. Growth in plant tissue is not pathogenic, but commensal: neither the seed pod (Lindsey, 1970) nor the plant (Pitt, 1989; Pitt *et al.*, 1991) shows any visible sign of colonization by the fungus.

A variety of factors influence invasion of developing peanuts by *A. flavus*. Infection occurs before harvest only if substantial numbers of spores or other propagules (thousands per gram of soil) exist in the soil. Other important factors are drought stress (Sanders *et al.*, 1981) and soil temperatures around 30 °C (Blankenship *et al.*, 1984; Sanders *et al.*, 1984; Cole *et al.*, 1985; Cole, 1989; Dorner *et al.*, 1989) during the last 30–50 days before harvest (Sanders *et al.*, 1985) .

2.5.2 *Maize*

So far as is known, maize is infected only by *A. flavus* (Lillehoj *et al.*, 1980; Angle *et al.*, 1982; Horn *et al.*, 1995). In temperate areas, the most important route for entry of *A. flavus* to maize is through insect damage (Lillehoj *et al.*, 1982; Bilgrami *et al.*, 1992; Miller, 1995). Invasion via the silks (the styles of the female maize flower) is also possible (Marsh & Payne, 1984; Diener *et al.*, 1987). High-temperature (32–38 °C) stress increases infection (Jones *et al.*, 1980). The critical time for high temperatures to favour infection is between 16 and 24 days after inoculation at silking (Payne, 1983). The time of infection is also important: inoculation two to three weeks after silk emergence produced much higher rates of infection than inoculation one or five weeks after silking (Jones *et al.*, 1980).

2.5.3 *Cottonseed*

A. flavus is also a commensal in the cotton plant (Klich *et al.*, 1984). Infection occurs through the nectaries (natural openings in the cotton stem), which are important in pollination (Klich & Chmielewski, 1985), or through cotyledonary leaf scars (Klich *et al.*, 1984). Upward movement occurs in the stem towards the boll, but not downwards from boll to stem (Klich *et al.*, 1986). Insect damage is also a potential cause of infection (Lee *et al.*, 1987), but insects are often well controlled in cotton crops. As in peanuts and maize, temperature appears to be a major environmental factor in pre-harvest infection of cottonseed (Marsh *et al.*, 1973; Simpson & Batra, 1984). Daily minimum temperatures above 24 °C, in combination with precipitation exceeding 2–3 cm, appear to lead to extensive aflatoxin formation (Diener *et al.*, 1987).

3. Management of Aflatoxin Contamination

A variety of approaches exist to limit aflatoxin production in crops. Limiting aflatoxins in crops before and immediately after harvest involves strategies aimed at reducing drought stress, by irrigation and weed control, by control of insect damage and by the use of fungicides. Improvements in harvesting procedures, better drying, and sorting of defective grains or nuts are all beneficial. The principles of safe storage are well known, and cannot be overemphasized. Physical methods to reduce aflatoxins in crops are widely practised, especially in developed countries. The technology involved in these approaches to limiting and reducing aflatoxins are relatively simple and inexpensive, and can be practised even by small-scale farmers. For heavily contaminated commodities, the use of chemical treatments can effectively reduce aflatoxins, but then use of the resulting materials is limited to animal feed. These procedures, systems and approaches are outlined below.

3.1 Intervention strategies

Interventions to reduce aflatoxin-related exposures can be considered in terms of those which are applicable at the individual level or those applicable at the community level (Figure 1).

3.1.1 *Individual level*

Dietary changes to avoid foods contaminated with aflatoxin are rarely an option in countries of high exposure where staple foods are contaminated (e.g. maize or peanuts). However, efforts to improve crop and dietary diversity must be made. Sorting procedures can be of value, but require education at the consumer level. Chemoprevention aims to diminish the toxicological effects of aflatoxins once exposure has occurred. Clinical

Figure 1. Intervention strategies

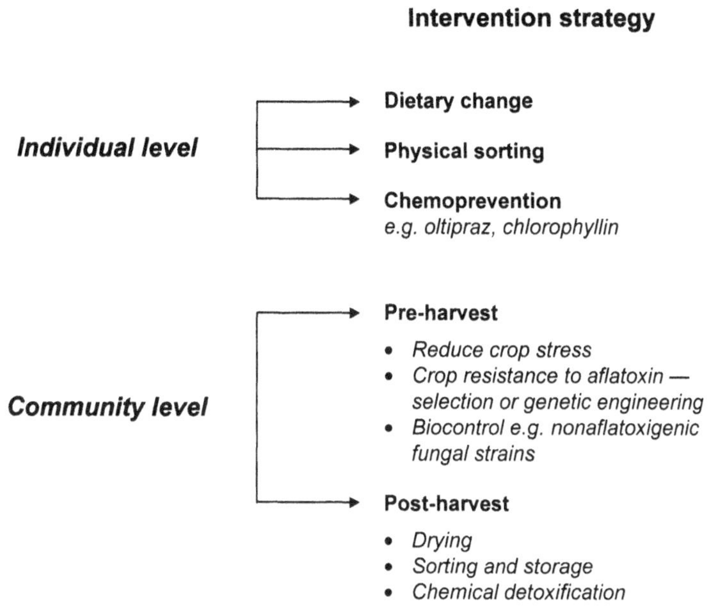

trials of two chemopreventive agents, oltipraz and chlorophyllin, have demonstrated that aflatoxin metabolism can be modified *in vivo* and the levels of aflatoxin bound to DNA and proteins can be reduced as a result. This approach may be valuable in individuals at particularly high risk of exposure. Work on these chemical agents also provides a valuable scientific basis for the exploration of dietary constituents consumed by populations exposed to aflatoxins, which may modify the toxicity of the aflatoxins in a similar way.

3.1.2 *Community level*

(*a*) *Limiting aflatoxin formation in susceptible crops before and after harvest*

In peanuts, it appears likely that infection by *A. flavus* while nuts are still in the ground is a prerequisite for high levels of aflatoxins to be formed after harvest (Pitt, 1989). In the absence of high pre-harvest infection levels, and with rapid and effective drying, peanuts can be produced free of any appreciable level of aflatoxin. The major causes of pre-harvest infection are high numbers of propagules in the soil, and drought stress during the days leading up to harvest (see Section 2.5.1).

Partial control of propagule numbers can be achieved by crop rotation: in particular, numbers of *A. flavus* in soil decrease under small grain cultivation or pasture. Irrigation, which eliminates drought stress, is regarded as the most effective method for reducing aflatoxin formation in peanuts (Pettit *et al.*, 1971; Cole *et al.*, 1982). However, peanuts throughout the world are recognized as a drought-resistant crop and are mostly grown under dry culture, with irrigation reserved for more moisture-sensitive crops such as rice or vegetables. In many areas where peanuts are grown, irrigation is not an option. Under these circumstances, reduction in drought stress by good agricultural practices can be a beneficial approach. For example, weed control and wider spacing between peanut rows can both assist in reducing drought stress (Rachaputi, 1999). Rapid drying of peanuts using mechanical dryers as soon as possible after pulling has a major effect in reducing the levels of aflatoxins in peanuts.

For maize also, irrigation and improved farm management practices have a beneficial effect on aflatoxin formation (Payne *et al.*, 1986). Resistant breeding stocks have been identified (Widstrom *et al.*, 1987; Campbell & White, 1995) and resistant maize genotypes, dependent on kernel pericarp wax (Guo *et al.*, 1995; Brown *et al.*, 1999) or kernel proteins (Huang *et al.*, 1997) have been developed recently. As maize is usually dried in the field, rapid drying techniques are not commonly practised, but should be in tropical countries.

Cottonseed is a by-product of cotton production, so field drying is normal. Breeding of cotton without nectaries has been proposed as one means of limiting *A. flavus* access to cotton bolls (Klich & Chmielewski, 1985).

Genetic engineering may offer novel ways of limiting pre-harvest contamination by mycotoxins, provided attention is paid to questions of importance specifically to developing countries (Wambugu, 1999). Genetic approaches to aflatoxin control include engineering of genes in *Aspergillus* species to influence the ability of the fungus to colonize the host plant. An alternative approach is to select or engineer varieties of cereal grains and oilseeds resistant to fungal infection or aflatoxin biosynthesis by the fungus once infection occurs.

(*b*) *Control of aflatoxin formation in other crops before and after harvest*

Entry of *A. flavus* into pistachio nuts depends on the time of splitting of hulls. Nuts in which hull splitting occurs early are much more susceptible to *A. flavus* invasion on the tree (Doster & Michailides, 1995). It is known that some cultivars are more prone to early splitting than others, and this is especially important where nuts are harvested from the ground, after contact with the soil.

In tree nut crops, various techniques, including timing of harvesting, are used to keep aflatoxin formation to a minimum.

Figs are sometimes infected by *A. flavus*, both because of their unique structure developed for insect fertilization and also because figs are harvested from the ground in some countries. Immature figs are not colonized by *A. flavus*, but once they are ripe infection occurs readily and fungal growth continues during drying (Buchanan *et al.*, 1975; Le Bars, 1990). The proportion of figs infected is only about 1% (Steiner *et al.*, 1988).

(*c*) *Control of aflatoxins in dried food commodities by physical means*

A range of physical factors control fungal growth: temperature, water activity (a_w), pH, gas atmospheres and oxygen concentration (Pitt & Hocking, 1997) together with the use of insecticides or preservatives in some cases. The pH of any commodity cannot be altered (Wheeler *et al.*, 1991), while temperature is usually not controllable in bulk storage. Gas atmospheres are increasingly used in developed countries to limit insect growth, and, if well maintained, can also limit fungal growth (Hocking, 1990). Storage of grain in a phosphine atmosphere used to control insects at a water activity of 0.80 or 0.86 reduced growth of *Aspergillus flavus*, but had little effect on the survival of spores (Hocking & Banks, 1991). In practice, in most storage systems, a_w is the principal variable that can be modified for preserving commodities.

The basic advice for handling any grain, nut, other bulk food commodity or feed after harvesting is to dry it rapidly and completely and to keep it dry. Food commodities are perishable, and must be kept free of insect infestations or water ingress or heating and cooling gradients which will cause migration of moisture. Full description of the methods for successful storage is beyond the scope of this document. Many good texts on grain storage exist: those by Champ and Highley (1988), Champ *et al.* (1990) and Highley *et al.* (1994) are recommended.

The prime consideration for storage of grains and nuts is to maintain the moisture content below that which permits fungal growth of any sort over a normal storage life, about one year, i.e. at a water activity below 0.65. This corresponds to different moisture contents for different commodities: 8% for peanuts and other nuts, 12% for grains and 22% for raisins, which contain a higher level of soluble carbohydrate (Iglesias & Chirife, 1982).

The limits for growth of *A. flavus* are now reasonably well defined: *A. flavus* is able to grow between 10–12 °C and 43–48 °C, with an optimum near 33 °C (Pitt & Hocking, 1997); the minimum a_w for growth is near 0.82 at 25 °C, 0.81 at 30 °C and 0.80 at 37 °C (Pitt & Miscamble, 1995). Growth can occur over the pH range 2–11 at least, at 25–37 °C, with optimal growth over a broad range from pH 3.4 to 10 (Wheeler *et al.*, 1991). *A. parasiticus* is very similar physiologically to *A. flavus* (Pitt & Hocking, 1997). Data from Pitt and Miscamble (1995) were used to provide a predictive model for *A. flavus* growth in relation to water activity and temperature (Gibson *et al.*, 1994).

Aflatoxin production has been reported to occur at water activity as low as 0.82 but is very slow below about 0.90 and optimal above 0.99, i.e. near the water activity of fresh grains or nuts (ICMSF, 1996; Gqaleni *et al.*, 1997).

It is evident that reduction of water activity of fresh commodities to below 0.80 will positively prevent aflatoxin production. However, it must be kept in mind that holding commodities above a water activity of 0.65 renders them susceptible to the growth of fungi. As such fungi grow, they release water by metabolism and produce heating, both of which in due course lead to conditions conducive to aflatoxin production. To prevent aflatoxin production in stored commodities, bulk foods or feeds, water activity must be maintained below 0.70 (Pitt & Hocking, 1997).

(*d*) *Reduction of aflatoxins in stored commodities by physical means*

For some crops, notably maize and figs, it is possible to sort grains or fruit using ultraviolet light, under which aflatoxins (and perhaps some other compounds) produce bright greenish yellow fluorescence. This test (qualitative but not quantitative) is best carried out on cracked maize grains (Shotwell *et al.*, 1972; Shotwell, 1983), but can be used on whole dried figs (Steiner *et al.*, 1988; Le Bars, 1990). Sorting out contaminated fruit, together with toxin analysis, has been effective in controlling aflatoxins in figs (Sharman *et al.*, 1991). Sieving of contaminated maize has been shown to reduce aflatoxin and the co-occurring fumonisin (Murphy *et al.*, 1993).

During wet milling of maize, aflatoxin is segregated primarily in the steep water (40%) and fibre (38%) and germ (6%), with less in the gluten (15%) and starch (1%) (Bennett & Anderson, 1978; Wood, 1982; Njapau *et al.*, 1998). In dry milling of maize, artificially contaminated rice and durum wheat, less than 10% of the aflatoxin in the original material remained in the prime products (grits and low-fat flour) (Schroder *et al.*, 1968; Scott, 1984).

After peanuts are shelled, several physical procedures such as colour sorting, density flotation, blanching and roasting are routinely used by processors to reduce aflatoxin levels by 99% (Park, 1993a; López-García *et al.*, 1999). The colour sorting process was developed originally to reject commercially unacceptable discoloured nuts, regardless of cause, but as fungal growth is a prime cause of discolouration, the process is also an effective non-destructive means of removing most nuts containing aflatoxins. In crops under severe drought stress, peanuts begin to dry in the ground, and under these conditions luxuriant growth of *A. flavus* can occur, with high aflatoxin production. In this case, blanching to remove skins and roasting to increase discolouration permits effective colour sorting to be carried out. Roasted peanuts must be sold under inert gas atmospheres to suppress development of rancidity (Read, 1989).

Colour sorting of other commodities is not easy. No effective non-chemical testing technique exists for cottonseed or pistachios and, as with other commodities, non-destructive chemical assays are not available.

The extent to which aflatoxins are destroyed during heating is largely dependent on the process used. Less than 25% of the aflatoxin content of a commodity is destroyed by boiling water (Christensen *et al.*, 1977; Njapau *et al.*, 1998), extrusion (Cazzaniga *et al.*, 2001) and autoclaving (Stoloff *et al.*, 1978). However, dry roasting of peanuts can reduce aflatoxin levels by up to 80% (Conway *et al.*, 1978; Njapau *et al.*, 1998). Heating of peanut oil at 250 °C for 3.5 h reduced aflatoxin by 99% (Peers & Linsell, 1975).

Heating at neutral pH at 125 °C had little effect on fumonisin, usually present as a co-contaminant in maize, but heating above 150 °C causes significant reduction in levels of fumonisin in processed maize products (Dupuy *et al.*, 1993; Jackson *et al.*, 1996a,b).

3.2 Chemical methods

The alkali process usually practised to produce refined table oil completely removes aflatoxin (ICMSF, 1996).

The use of chemicals to inactivate, bind or remove aflatoxins has been studied extensively. Any such procedure must effectively inactivate or remove the toxin, while maintaining the nutritional and technological properties of the product and without generating toxic reaction products (López-García & Park, 1998). Food safety demonstration studies must be conducted to ensure compliance with regulatory requirements. To date, these chemical methods have been approved only for the reduction of aflatoxins in animal feed commodities. Reacting the toxin chemically with another compound intentionally introduced in the vicinity of the toxin molecule holds the greatest promise for rapid and effective removal or inactivation of aflatoxin. Among such techniques are the use of chemosorbents and ammoniation. Other than the demonstrated reduction in bioavailability of aflatoxin as a result of hydrated sodium calcium aluminosilicate binding (Phillips *et al.*, 1988), ammoniation is the only chemical inactivation process that has been shown to destroy aflatoxin efficiently in cottonseed and cottonseed meal, peanuts and peanut meal and maize (Park *et al.*, 1988; Park & Price, 2001).

3.2.1 *Aflatoxin-binding agents*

Adsorption of aflatoxin using activated carbons, clays and aluminosilicates has been demonstrated in a number of studies. Bentonite clays (Masimango *et al.*, 1979) and activated charcoal (Decker, 1980), both used in oil purification, can adsorb up to 92% of aflatoxin present. A phyllosilicate clay currently used as an anti-caking agent has been shown to bind aflatoxin tightly and diminish markedly its uptake into the circulatory system, preventing aflatoxicoses and reducing levels of aflatoxin M_1 in milk (Phillips *et al.*, 1988). The specificity of the clays and their potential for binding nutrients in addition to aflatoxin remains a concern. Further research is required to determine whether these materials can be used in human foods.

3.2.2 *The ammoniation process*

The ammoniation process has been used to reduce aflatoxin levels in feed ration components in order to prevent the presence of aflatoxin in tissues and animal products such as milk. In the USA, the States of Arizona, California and Texas permit the ammoniation of cottonseed products. Texas has, in addition, approved the ammoniation procedure for aflatoxin-contaminated corn, but the treated corn may only be used in finishing beef cattle diets. Ammoniation is used in Brazil, France, Mexico, Senegal, Sudan and some states of the USA to lower aflatoxin contamination levels in animal feeds. The two procedures in widespread use are: (*a*) a high-pressure and high-temperature process (HP/HT) used at treatment plants, and (*b*) an atmospheric-pressure and ambient-temperature procedure (AP/AT) that can be used on the farm (Table 2). The HP/HT process

Table 2. Parameters and applications of ammonia procedures for aflatoxin decontamination

	Process	
	High pressure/high temperature	Atmospheric pressure/ ambient temperature
Ammonia level (%)	0.2–2	1–5
Pressure (psi)	35–50	Atmospheric
Temperature (°C)	80–120	Ambient
Duration	20–60 min	14–21 days
Moisture (%)	12–16	12–16
Commodities	Whole cottonseed, corn, cottonseed meal and peanuts	Whole cottonseed, corn
Application	Feed mill	Farm

Adapted from Park (1993b)
1 psi = 6.9 kPa

involves spraying the contaminated product with anhydrous ammonia (or introduction of ammonia gas) and water in a contained vessel. The treatment conditions, i.e. amount of ammonia (0.5–2%), moisture (12–16%), pressure (35–55 psi [240–380 kPa]), time (20–60 min) and temperature (80–120 °C) vary according to the initial levels of aflatoxin in the product (Park & Price, 2001).

The AP/AT process also uses anhydrous ammonia and water, but the commodity is packed in a plastic silage-type bag or more recently simply covered with a tarpaulin and sealed. The sealed container is then held for 14–42 days, depending on the initial aflatoxin levels and the ambient temperature (25–40 °C); higher aflatoxin levels and a lower ambient temperature require a longer holding time. Similarly, the amount of ammonia (1–5%) can be varied according to the initial level of aflatoxin present and the moisture content of the material to be treated. Completion of treatment has to be predetermined in the HP/HT process, whereas with the AP/AT technique, the bag is probed and tested periodically until the results show that aflatoxin levels are equal to or below 20 ppb (Park & Price, 2001).

The addition of a formaldehyde anti-caking agent can improve the process (Prevot, 1986) and leads to no changes in milk production and composition (Calet, 1984). However, the addition of formaldehyde is not recommended in view of potential human exposure (see IARC, 1995).

(*a*) *Aflatoxin/ammonia chemistry*

Sequential fractionation of meals spiked with uniformly ^{14}C-labelled aflatoxin B_1 (Park *et al.*, 1984) has allowed the partial isolation (Figure 2) and identification

Figure 2. Scheme for isolation and approximate concentrations of aflatoxin–ammonia reaction products in cottonseed and corn meals

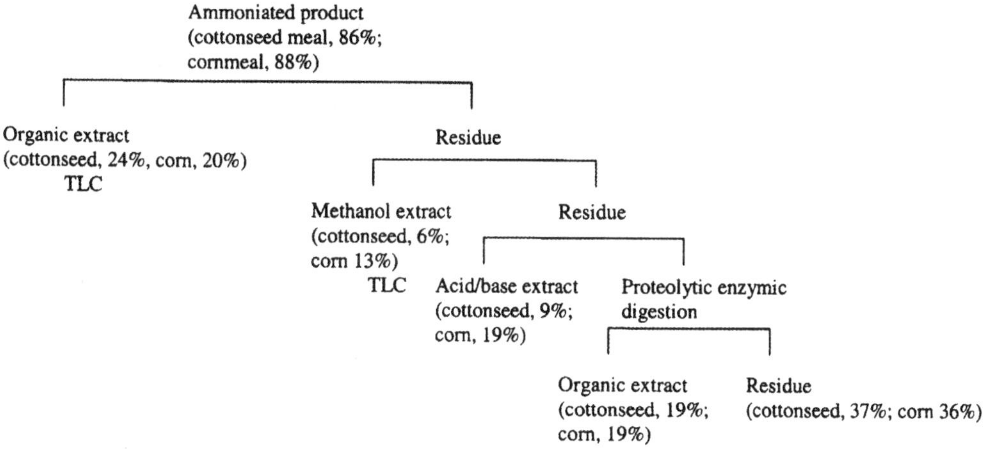

From Park & Price (2001)
TLC, thin-layer chromatography

(Figure 3) of some of the decomposition products formed as a result of the aflatoxin–ammonia reaction. Hydrolytic scission of the aflatoxin lactone ring — the first step in the reaction — readily occurs under basic conditions, but is reversible when the ammoniation process is carried out under less drastic conditions. Under HP/HT and well controlled AP/AT conditions, the reaction proceeds to low-molecular-weight compounds, among them aflatoxin D_1 (molecular weight 286) and others of molecular weight 256 and 236 (Park *et al.*, 1988).

(*b*) *Efficacy of aflatoxin decontamination using ammonia*

Studies with peanuts, cottonseed and corn have demonstrated the effectiveness of the ammoniation process (Dollear *et al.*, 1968; Brekke *et al.*, 1977; Park *et al.*, 1984; Martinez *et al.*, 1994; Neal *et al.*, 2001). Other studies are summarized in Table 3.

(*c*) *Safety of ammoniated commodities*

In vitro, sub-chronic and chronic studies have shown no mutagenic or tumorigenic lesions or toxic effects related to the HP/HT ammoniation procedure. Metabolism studies suggest poor absorption of decontamination reaction products compared with aflatoxin B_1. Livestock feeding studies with feedlot beef, dairy cattle, poultry, turkeys and ducklings have shown that the toxic effects observed following exposure to aflatoxin contamination are absent after ammoniation. These studies have been reviewed (Park *et al.*, 1988; Park & Price, 2001).

Figure 3. Proposed formation of aflatoxin-related reaction products following exposure to ammonia

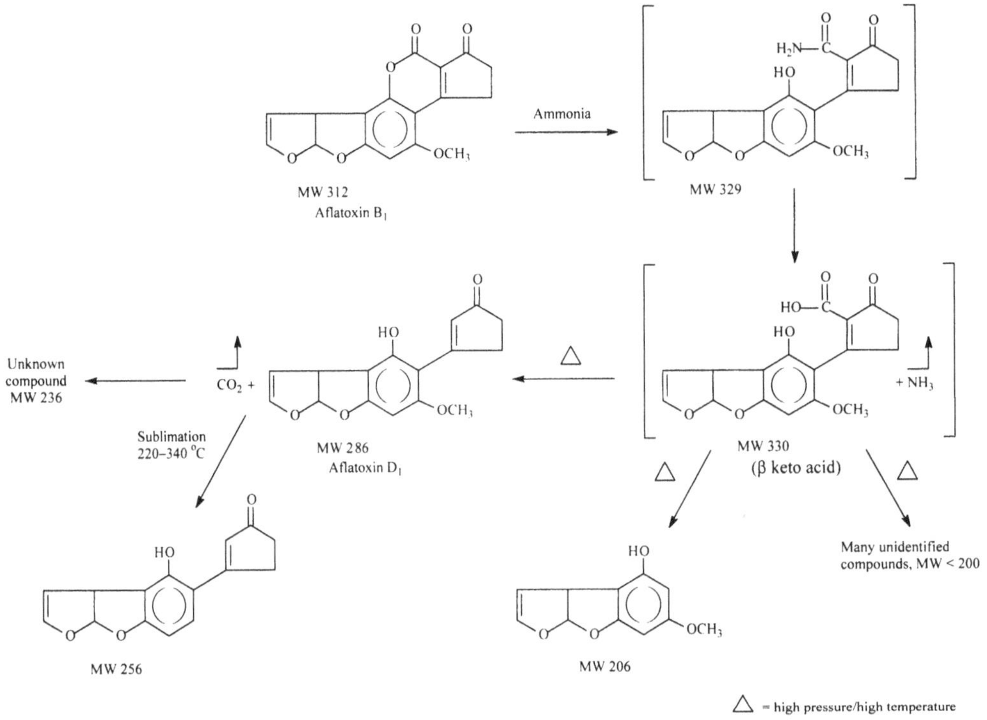

From Park *et al.* (1988)
Relative abundance and formation of reaction products are dependent on the conditions of ammoniation used.

Lactating mammals that ingest aflatoxin B_1 deposit the 4-hydroxylated metabolite, aflatoxin M_1, in their milk. Since infants and children have potentially greater vulnerability and sensitivity than adults, monitoring levels of aflatoxin M_1 in milk is important. Among human foods of animal origin, the rate of feed-to-tissue transfer of aflatoxin is highest for milk (Tables 4 and 5). Therefore, aflatoxins or their metabolites are not found in edible tissues except milk from animals fed aflatoxin-contaminated feed. The conversion rate of aflatoxin B_1 to aflatoxin M_1 in milk has been shown to vary between 1.1 and 14.7% for dairy rations containing between 20 and 800 ppb aflatoxin B_1 (Price *et al.*, 1982; Fremy & Quillardet, 1985; Bailey *et al.*, 1994). It is higher than 1% when the amount of ingested aflatoxin B_1 is low. The primary purpose of the ammoniation procedure was to reduce aflatoxin M_1 residues in milk, and numerous studies have demonstrated its efficacy in eliminating aflatoxin residues in milk (Price *et al.*, 1982; Fremy *et al.*, 1987; Bailey *et al.*, 1994).

Overall, decontamination reaction products in the feed matrix are usually derived from < 1% of the original aflatoxin content and large portions of these products are

Table 3. Studies on reduction of aflatoxin content by ammoniation of commodities

Commodity	Investigators	Initial AFB_1 content (ppb)	Process	Process parameters	Final AFB_1 content (ppb)	Reduction (%)
Corn	Hughes *et al.* (1979)	754 total [603 AFB_1]	AP/AT	1% NH_3, 18% m, 12–13 days	3.5	> 99
	Weng *et al.* (1994)	7500	HP/HT	2% NH_3, 16% m, 55 psi, 40–45 °C, 60 min	517	93
	Weng *et al.* (1994)	7500	HP/HT	2% NH_3, 16% m, 17 psi, 121°C, 60 min	31	> 99
Cottonseed	Jorgensen & Price (1981)	800 total	AP/AT	2% NH_3, 7.5% m, 21 °C, 15 days	< 20	99
	Bailey *et al.* (1994)	5200	AP/AT	1.5 % NH_3 17% m, 42 days	< 10	> 99
	Bailey *et al.* (1994)	1200	HP/HT	4% NH_3, 14% m, 40 psi, 100 °C, 30 min	ND (≤ 0.1)	~> 99
Peanut meal	Gardner *et al.* (1971)	121 total	HP/HT	NH_3 concentration not specified, 15% m, 30 psig, 93 °C, 15 min	ND	~> 99

AFB_1, aflatoxin B_1; m, moisture; AP/AT, atmospheric pressure/ambient temperature; HP/HT, high pressure/high temperature; ND, not detected; psig, pound per inch^2 gauge (indicating the pressure above atmospheric pressure); 1 psi = 6.9 kPa

Table 4. Relation of aflatoxin levels in feed to aflatoxin residue levels in edible tissues

Animal	Tissue	Aflatoxin	Feed:tissue ratio[a]
Beef cattle	Liver	B_1	14 000
Dairy cattle	Milk	M_1	75
		Aflatoxicol[b]	195 000
Swine	Liver	B_1	800
Layers	Eggs	B_1	2200
Broilers	Liver	B_1	1200

From Park & Pohland (1986); Park & Stoloff (1989)
[a] Level of aflatoxin B_1 in the feed divided by the level of the specified aflatoxin in the specified tissue
[b] A metabolite of aflatoxin B_1

Table 5. Aflatoxin B_1 levels in feed components required to yield 0.1 ng/g residue levels of aflatoxins in edible tissues

Species	Contamination level of aflatoxin B_1 (ng/g) in rations			
	Corn	Peanut meal	Cottonseed meal	Cottonseed
Beef cattle	1800	1400	12 725	14 000
Dairy cattle	14	54	54	38
Swine	105	730	1600	–
Layer	325	1835	2445	–
Broiler	180	925	1200	–

From Park & Pohland (1986)

strongly bound to feed components so that they are biologically unavailable to the animals (Park *et al.*, 1984) or are eliminated by excretion (Hoogenboom *et al.*, 2001).

4. Outstanding Health Questions in Aflatoxin Management

Increased liver cancer incidence associated with aflatoxins occurs in areas of the world with chronic high levels of toxins, frequently exceeding the regulatory levels under consideration by the Codex Alimentarius Commission (2001) by large amounts, sometimes factors of 10 or 100, and endemic infection with hepatitis B or C viruses (HBV or HCV). Some basis exists for quantifying the effects of aflatoxin exposure on liver cancer risk, and for the greater impact of aflatoxins in areas of high HBV or HCV incidence (JECFA, 2001; see Section 5). However, uncertainty remains regarding (1) the health effects of occasional high exposures occurring due to unusual weather patterns, in comparison with those due to normal chronic exposure; (2) the effect of combinations of mycotoxins, e.g. aflatoxins and fumonisins, on cancer risk; and (3) the broader health effects of aflatoxin exposure. Under the latter consideration, the most important aspects are (1) the greater sensitivity of children to acute toxicity of aflatoxins as compared with adults; (2) the in-utero effects of aflatoxins, known to cross the placenta; (3) the immunosuppressive effects of aflatoxins, which may influence susceptibility to infectious disease; (4) the suppressive effects of aflatoxins on growth; and (5) the interactions between HCV, HBV and aflatoxins in liver cancer development in various populations in the world.

The areas mentioned are important ones for future research, to provide information of direct relevance to public health decisions regarding aflatoxin exposure.

5. Previous Recommendations

Although some countries with fully developed market economies have enforced regulations on aflatoxins for some 30 years, it is only recently that a broader consensus has been developed on the impact of aflatoxins and other agriculturally important toxins on human populations. The health effects include those resulting from exposure in foods, from reductions in the quality of the crops affected and from reduced animal production. The health consequences of lowered income are also pertinent in some communities. This consensus was reached when the Joint Food and Agriculture Organization/World Health Organization (FAO/WHO) Expert Committee on Food Additives (JECFA) established an hypothetical standard for aflatoxins (JECFA, 1998), provisional maximum tolerable daily intakes (PMTDIs) for deoxynivalenol and fumonisins and a provisional tolerable weekly intake (PTWI) for ochratoxin A (JECFA, 2001). For aflatoxin and fumonisin, exposure in excess of the PMTDI occurs in much of Africa as well as parts of Asia and South America.

JECFA attempted to separate the portion of risk of liver cancer attributable to aflatoxin from the risk attributable to other factors (JECFA, 1998, 2001). JECFA modelled dose–response curves from laboratory animal studies as well as human epidemiological studies of aflatoxin carcinogenicity. The risk of liver cancer from aflatoxin consumption was significantly higher (perhaps 30-fold) in individuals who were chronically infected with hepatitis B virus (hepatitis B surface antigen (HBsAg)-positive) than in those who were not (HBsAg-negative).

It was concluded that populations such as those in western Europe and the USA with a low prevalence of HBsAg-positive individuals or populations with a low mean intake of aflatoxins are unlikely to achieve a decrease in liver cancer cases from more stringent aflatoxin standards (such as lowering the regulatory hypothetical standard for aflatoxin B_1 in peanuts from 20 μg/kg to 10 μg/kg) (JECFA, 2001).

Populations with a high prevalence of HBsAg-positive persons and high aflatoxin intake would benefit from reductions of aflatoxin intake, and such a reduction in aflatoxin intake could be achieved without loss of valuable food sources and risk of malnutrition or starvation (JECFA, 1998).

The FAO/WHO/UNEP (United Nations Environment Programme) (1999) conference, representing 38 countries and 10 international organizations, made a series of recommendations reflecting a new perspective on the health and economic impact of mycotoxins in staple foods. It was recommended that:

- Surveillance should be targeted to staple foods.
- HACCP (hazard analysis: critical control point) principles can be used to highlight the roles that fungal ecology, crop physiology and agronomic practices play in mycotoxin contamination prevention and control.
- The stakeholders in the production chain, particularly farmers, should be made aware of the importance of measures to reduce mycotoxin contamination.
- Before recommending the introduction of crops or new genotypes into new environments, consideration should be given to the potential for increased [toxigenic] fungal infections.
- Training programmes for the development of practical control and management strategies should be [developed and] conducted in developing countries in order to set up strong mycotoxin management programmes.

6. Trade in Crops

Mycotoxin contamination in crops is a major determinant of trade for all economies. Typically, the higher-quality grain or nuts are traded and the poorer-quality product remains with the farmer for consumption at the village level or for use in animal production. Major risks exist in this situation, especially where unseasonable rains or other climate variations result in excessive mycotoxin contamination of crops, where consumer standards for urban customers increase, or where commodities intended for international trade are upgraded by sampling or sorting.

Where industrialized domestic animal production takes place, there are a number of strategies to make use of crops with undesirable mycotoxin concentrations. These include dilution with other feed ingredients, addition of absorbents for aflatoxins, additions of sugar and, where permitted, ammoniation. In developing countries, however, diverting grains contaminated with aflatoxins into rural animal protein production may result in a substantial reduction in feed conversion as well as increased herd or flock mortality and morbidity. The economic impact of such reduced feed conversion can eliminate the expected increased return on investments in animal protein production.

Furthermore, the reduced animal production increases protein energy malnourishment in vulnerable groups, e.g. children and pregnant women.

7. Conclusions

This Annex has dealt in general terms with a number of very complex issues. Many areas have been highlighted which are in need of much future research to provide information that is directly relevant not only to public health, but also to availability of a wholesome food and feed supply worldwide. The following general conclusions and recommendations emerged from the deliberations of the Working Group.

1. Limiting aflatoxin occurrence in crops before harvest can be achieved by limiting drought and high temperature stress, controlling weeds, reducing insect damage, using effective harvesting techniques and reducing *Aspergillus* spore numbers in soil by crop rotation.
2. Genetic engineering may offer new ways of limiting the pre-harvest contamination of some crops by aflatoxins.
3. Aflatoxins can be controlled in susceptible crops after harvest by controlling factors which affect fungal growth, e.g. water activity, temperature, gas atmospheres, and the use of insecticides or preservatives. The prime consideration for storage of grains and nuts is to maintain the water activity (by control of moisture content) below the limit for fungal growth.
4. Aflatoxin levels can be reduced in stored commodities by physical means, such as colour sorting, density flotation, blanching and roasting.
5. Where approved, aflatoxin levels in commodities destined for animal feeds can be reduced by chemical processes. Such processes include agents which bind aflatoxins, such as adsorbent clays, and the ammoniation process. The main use for ammoniation is in elimination of aflatoxin from feed for dairy cows.
6. Increased liver cancer incidence associated with aflatoxin exposure occurs in areas of the world where chronic high levels of aflatoxins (often many times higher than regulatory limits) and endemic infection with HBV or HCV occur together. Populations with a low prevalence of HBV chronic carriers and a low mean aflatoxin intake are unlikely to achieve a decrease in liver cancer cases by introducing lower aflatoxin limits. In contrast, in populations with a high prevalence of HBV chronic carriers and high exposure to aflatoxin, measures to reduce aflatoxin exposure would be desirable and beneficial.

8. References

Angle, J.S., Dunn, K.A. & Wagner, G.H. (1982) Effect of cultural practices on the soil population of *Aspergillus flavus* and *Aspergillus parasiticus*. *Soil Sci. Soc. Am. J.*, **46**, 301–304

Bailey, G.S., Price, R.L., Park, D.L. & Hendricks, J.D. (1994) Effect of ammoniation of aflatoxin B_1-contaminated cottonseed feedstock on the aflatoxin M_1 content of cows' milk and hepatocarcinogenicity in the trout bioassay. *Food chem. Toxicol.*, **32**, 701–715

Bennett, G.A. & Anderson, R.A. (1978) Distribution of aflatoxin and/or zearalenone in wet-milled corn products: A review. *J. agric. Food Chem.*, **26**, 1055–1060

Bilgrami, K.S., Ranjan, K.S. & Sinha, A.K. (1992) Impact of crop damage on occurrence of *Aspergillus flavus* and aflatoxin in rainy-season maize (*Zea mays*). *Indian J. agric. Sci.*, **62**, 704–709

Blankenship, P.D., Cole, R.J., Sanders, T.H. & Hill, R.A. (1984) Effect of geocarposphere temperature on pre-harvest colonization of drought-stressed peanuts by *Aspergillus flavus* and subsequent aflatoxin contamination. *Mycopathologia*, **85**, 69–74

Brekke, O.L., Peplinski, A.J. & Lancaster, E.B. (1977) Aflatoxin inactivation in corn by aqua ammonia. *Trans. Am. Soc. agric. Eng.*, **20**, 1160–1166

Brown, R.L., Chen, Z.-Y., Cleveland, T.E. & Russin, J.S. (1999) Advances in the development of host resistance in corn to aflatoxin contamination by *Aspergillus flavus*. *Phytopathology*, **89**, 113–117

Buchanan, J.R., Sommer, N.F. & Fortlage, R.J. (1975) *Aspergillus flavus* infection and aflatoxin production in fig fruits. *Appl. Microbiol.*, **30**, 238–241

Calet, C. (1984) Consequences of ingesting aflatoxin-contaminated feeds. Efficacy of methods used for their destruction. *Rev. Aliment. Anim.*, **375**, 39–41; 43–45

Campbell, K.W. & White, D.G. (1995) Evaluation of corn genotypes for resistance to *Aspergillus* ear rot, kernel infection, and aflatoxin production. *Plant Dis.*, **79**, 1039–1045

Cazzaniga, D., Basílico, J.C., González, R.J., Torres, R.L. & de Greef, D.M. (2001) Mycotoxins inactivation by extrusion cooking of corn flour. *Lett. appl. Microbiol.*, **33**, 144–147

Champ, B.R. & Highley, E., eds (1988) *Bulk Handling and Storage of Grain in the Humid Tropics. Proceedings of an International Workshop held at Kuala Lumpur, Malaysia, 6–9 October 1987* (ACIAR Proceedings No. 22), Canberra, ACT, Australian Centre for International Agricultural Research

Champ, B.R., Highley. E. & Banks, H.J., eds (1990) *Fumigation and Controlled Atmosphere Storage of Grain. Proceedings of an International Conference held at Singapore, 14-18 February 1989* (ACIAR Proceedings No. 25), Canberra, ACT, Australian Centre for International Agricultural Research

Chatterjee, D., Chattopadhyay, B.K. & Mukherje, S.K. (1990) Storage deterioration of maize having pre-harvest infection with *Aspergillus flavus*. *Lett. appl. Microbiol.*, **11**, 11–14

Christensen, C.M., Mirocha, C.J. & Meronuck, R.A. (1977) *Mold, Mycotoxins and Mycotoxicoses* (Agricultural Experiment Station Report 142), St Paul, University of Minnesota

Codex Alimentarus Commission (2001) *Codex Alimentarius, 24th Session*, Geneva, WHO

Cole, R.J. (1989) Preharvest aflatoxin in peanuts. *Int. Biodet.*, **25**, 253–257

Cole, R.J., Hill, R.A., Blankenship, P.D., Sanders, T.H. & Garren, K.H. (1982) Influence of irrigation and drought stress on invasion by *Aspergillus flavus* of corn kernels and peanut pods. *Dev. ind. Microbiol.*, **23**, 229–236

Cole, R.J., Sanders, T.H., Hill, R.A. & Blankenship, P.D. (1985) Mean geocarposphere temperatures that induce preharvest aflatoxin contamination of peanuts under drought stress. *Mycopathologia*, **91**, 41–46

Conway, H.F., Anderson, R.A. & Bagley, E.B. (1978) Detoxification of aflatoxin contaminated corn by roasting. *Cereal Chem.*, **55**, 115–117

Decker, W.J. (1980) Activated charcoal absorbs aflatoxin B_1. *Vet. hum. Toxicol.*, **22**, 388–389

Diener, U.L., Asquith, R.L. & Dickens, J.W., eds (1983) *Aflatoxin and* Aspergillus flavus *in Corn* (Southern Cooperative Series Bulletin 279), Opalika, AL, Craftmaster Printers

Diener, U.L., Cole, R.J., Sanders, T.H., Payne, G.A., Lee, L.S. & Klich, M.A. (1987) Epidemiology of aflatoxin formation by *Aspergillus flavus*. *Ann. Rev. Phytopathol.*, **25**, 249–270

Dollear, F.G., Mann, G.E., Codifer, L.P., Jr, Gardner, H.K., Jr, Koltun, S.P. & Vix, H.L.E. (1968) Elimination of aflatoxins from peanut meal. *J. Am. Oil Chem. Soc.*, **45**, 862–865

Dorner, J.W., Cole, R.J. & Diener, U.L. (1984) The relationship of *Aspergillus flavus* and *Aspergillus parasiticus* with reference to production of aflatoxins and cyclopiazonic acid. *Mycopathologia*, **87**, 13–15

Dorner, J.W., Cole, R.J., Sanders, T.H. & Blankenship, P.D. (1989) Interrelationships of kernel water activity, soil temperature, maturity and phytoalexin production in pre-harvest aflatoxin contamination of drought-stressed plants. *Mycopathologia*, **105**, 117–128

Doster, M.A. & Michailides, T.J. (1995) The relationship between date of hull splitting and decay of pistachio nuts by *Aspergillus* species. *Plant Dis.*, **79**, 766–769

Dupuy, J., Le Bars, P., Boudra, H. & Le Bars, J. (1993) Thermostability of fumonisin B_1, a mycotoxin from *Fusarium moniliforme*, in corn. *Appl. environ. Microbiol.*, **59**, 2864–2867

FAO/WHO/UNEP (1999) *Third Joint FAO/WHO/UNEP International Conference on Mycotoxins, Tunis, Tunisia, 3–6 March 1999* (MYC-CONF/99/8a), Geneva, WHO

Fremy, J.M. & Quillardet, P. (1985) The 'carry-over' of aflatoxin into milk of cows fed ammoniated rations: Use of an HPLC method and a genotoxicity test for determining milk safety. *Food Addit. Contam.*, **2**, 201–207

Fremy, J.M., Gautier, J.P., Herry, M.P., Terrier, C. & Calet, C. (1987) Effects of ammoniation on the 'carry-over' of aflatoxin into bovine milk. *Food Addit. Contam.*, **5**, 39–44

Gardner, H.K., Jr, Koltun, S.P., Dollear, F.G. & Rayner, E.T. (1971) Inactivation of aflatoxins in peanut and cottonseed meals by ammoniation. *J. Am. Oil Chem. Soc.*, **48**, 70–73

Gibson, A.M., Baranyi, J., Pitt, J.I., Eyles, M.J. & Roberts, T.A. (1994) Predicting fungal growth: The effect of water activity on *Aspergillus flavus* and related species. *Int. J. Food Microbiol.*, **23**, 419–431

Gqaleni, N., Smith, J.E., Lacey, J. & Gettinby, G. (1997) Effects of temperature, water activity, and incubation time on production of aflatoxins and cyclopiazonic acid by an isolate of *Aspergillus flavus* in surface agar culture. *Appl. environ. Microbiol.*, **63**, 1048–1053

Griffin, G.J. & Garren, K.H. (1976a) Colonization of rye green manure and peanut fruit debris by *Aspergillus flavus* and *Aspergillus niger*-group in field soils. *Appl. environ. Microbiol.*, **32**, 28–32

Griffin, G.J. & Garren, K.H. (1976b) Colonization of aerial peanut pegs by *Aspergillus flavus* and *A. niger*-group fungi under field conditions. *Phytopathology*, **66**, 1161–1162

Guo, B.Z., Russin, J.S., Cleveland, T.E., Brown, R.L. & Widstrom, N.W. (1995) Wax and cutin layers in maize kernels associated with resistance to aflatoxin production by *Aspergillus flavus*. *J. Food. Prot.*, **58**, 296–300

Hesseltine, C.W., Shotwell, O.L., Smith, M., Ellis, J.J., Vandegraft, E. & Shannon, G. (1970) Production of various aflatoxins by strains of the *Aspergillus flavus* series. In: Herzberg, M., ed., *Proceedings of the First US-Japan Conference on Toxic Micro-organisms*, Washington, DC, US Department of the Interior, pp. 202–210

Highley, E., Wright, E.J., Banks, H.J. & Champ, B.R., eds (1994) *Stored Product Protection. Proceedings of the 6th International Working Conference on Stored-Product Protection*, 17–23 April 1994, Canberra, Australia (Wallingford UK, CAB International), Cambridge, University Press

Hocking, A.D. (1990) Responses of fungi to modified atmospheres. In: Champ, B.R., Highley, E. & Banks, H.J., eds, *Fumigation and Controlled Atmosphere Storage of Grain, Proceedings of an International Conference Held at Singapore, 14–18 February 1989* (ACIAR Proceedings No 25), Canberra, Australian Centre for International Agricultural Research, pp. 70–82

Hocking, A.D. & Banks, H.J. (1991) Effects of phosphine fumigation on survival and growth of storage fungi in wheat. *J. stored Prod. Res.*, **27**, 115–120

Holtmeyer, M.G. & Wallin, J.R. (1981) Incidence and distribution of airborne spores of *Aspergillus flavus* in Missouri. *Plant Dis.*, **65**, 58–60

Hoogenboom, L.A.P., Tulliez, J., Gautier, J.-P., Coker, R.D., Melcion, J.-P., Nagler, M.J., Polman, T.H.G. & Delort-Laval, J. (2001) Absorption, distribution and excretion of aflatoxin-derived ammoniation products in lactating cows. *Food Addit. Contam.*, **18**, 47–58

Horn, B.W., Greene, R.L. & Dorner, J.W. (1995) Effect of corn and peanut cultivation on soil populations of *Aspergillus flavus* and *A. parasiticus* in southwestern Georgia. *Appl. environ. Microbiol.*, **61**, 2472–2475

Huang, Z., White, D.G. & Payne, G.A. (1997) Corn seed proteins inhibitory to *Aspergillus flavus* and aflatoxin biosynthesis. *Phytopathology*, **87**, 622–627

Hughes, B.L., Barnett, B.D., Jones, J.E., Dick, J.W. & Norred, W.P. (1979) Safety of feeding aflatoxin-inactivated corn to white leghorn layer-breeders. *Poult. Sci.*, **58**, 1202–1209

IARC (1995) *IARC Monographs on the Evaluation of Carcinogenic Risks to Humans*, Vol. 62, *Wood Dust and Formaldehyde*, Lyon, IARC*Press*, pp. 217–362

ICMSF (International Commission on Microbiological Specifications for Foods) (1996) Toxigenic fungi: *Aspergillus*. In: *Microorganisms in Foods. 5. Microbiological Specifications of Food Pathogens*, London, Blackie Academic and Professional, pp. 347–381

ICMSF (International Commission on Microbiological Specifications for Foods) (1998) Spices, dry soups and oriental flavourings. In: *Microorganisms in Foods. 6. Microbial Ecology of Food Commodities*, London, Blackie Academic and Professional, pp. 274–312

Iglesias, H.A. & Chirife, J. (1982) *Handbook of Food Isotherms: Water Sorption Parameters for Food and Food Components*, New York, Academic Press

Jackson, L.S., Hlywka, J.J., Senthil, K.R. & Bullerman, L.B. (1996a) Effects of thermal processing on the stability of fumonisins. In: Jackson, L.S., de Vries, J.W. & Bullerman, L.B., eds, *Fumonisins in Food*, New York, Plenum Press, pp. 345–353

Jackson, L.S., Hlywka, J.J., Senthil, K.R., Bullerman, L.B. & Musser, S.M. (1996b) Effects of time, temperature, and pH on the stability of fumonisin B_1 in an aqueous model system. *J. agric. Food Chem.*, **44**, 906–912

JECFA (1998) *Safety Evaluation of Certain Food Additives and Contaminants (49th Meeting of the Joint FAO/WHO Expert Committee on Food Additives (WHO Food Additives Series 40)*, Geneva, International Programme on Chemical Safety, WHO

JECFA (2001) *Safety Evaluation of Certain Mycotoxins in Food, 56th Meeting of the Joint FAO/WHO Expert Committee on Food Additives (WHO Food Additives Series 47)*, Geneva, International Programme on Chemical Safety, WHO

Jones, R.K., Duncan, H.E., Payne, G.A. & Leonard, K.J. (1980) Factors influencing infection by *Aspergillus flavus* in silk-inoculated corn. *Plant Dis.*, **64**, 859–863

Jorgensen, K.V. & R.L. Price (1981) Atmospheric pressure-ambient temperature reduction of aflatoxin B_1 in ammoniated cottonseed. *J. agric. Food Chem.*, **29**, 555–558

Klich, M.A. & Chmielewski, M.A. (1985) Nectaries as entry sites for *Aspergillus flavus* in developing cotton bolls. *Appl. environ. Microbiol.*, **50**, 602–604

Klich, M.A. & Pitt, J.I. (1988) Differentiation of *Aspergillus flavus* from *A. parasiticus* and other closely related species. *Trans. Br. mycol. Soc.*, **91**, 99–108

Klich, M.A., Thomas, S.H. & Mellon, J.E. (1984) Field studies on the mode of entry of *Aspergillus flavus* into cotton seeds. *Mycologia*, **76**, 665–669

Klich, M.A., Lee, L.S. & Huizar, H.E. (1986) Occurrence of *Aspergillus flavus* in vegetative tissue of cotton plants and its relation to seed infection. *Mycopathologia*, **95**, 171–174

Le Bars, J. (1990) Contribution to a practical strategy for preventing aflatoxin contamination of dried figs. *Microbiol. Aliment. Nutr.*, **8**, 265–270

Lee, L.S., Lacey, P.E. & Goynes, W.R. (1987) Aflatoxin in Arizona cottonseed: A model study of insect-vectored entry of cotton bolls by *Aspergillus flavus*. *Plant Dis.*, **71**, 997–1001

Lillehoj, E.B., McMillian, W.W., Guthrie, W.D. & Barry, D. (1980) Aflatoxin-producing fungi in preharvest corn: Inoculum source in insects and soils. *J. environ. Qual.*, **9**, 691–694

Lillehoj, E.B., Kwolek, W.F., Guthrie, W.D., Barry, D., McMillian, W.W. & Widstrom, N.W. (1982) Aflatoxin accumulation in preharvest maize kernels: Interaction of three fungal species, European corn borer and two hybrids. *Plant Soil*, **65**, 95–102

Lindsey, D.L. (1970) Effect of *Aspergillus flavus* on peanuts grown under gnotobiotic conditions. *Phytopathology*, **60**, 208–211

López-García, R.L. & Park, D.L. (1998) Management of mycotoxin hazards through post-harvest procedures. In: Bhatnagar, D. & Sinha, K.K., eds, *Mycotoxins in Agriculture and Food Safety*, New York, Marcell Dekker, pp. 407–433

López-García, R.L., Park, D.L. & Gutierrez de Zubiaurre, M.B. (1999) [Procedures to reduce the presence of mycotoxins in foodstuffs.] In: Pfohl-Leszkowicz, ed., *Les Mycotoxines dans l'Alimentation. Evaluation et Gestion du Risque*, Paris, Editions TEC & TOC, pp. 387–408 (in French)

Masimango, N., Remacle, J. & Ramaut, J. (1979) [Elimination of aflatoxin B_1 by absorbent clays from contaminated substrates.] *Ann. Nutr. Alim.*, **33**, 137–147 (in French)

Marsh, S.F. & Payne, G.A. (1984) Preharvest infection of corn silks and kernels by *Aspergillus flavus*. *Phytopathology*, **74**, 1284–1289

Marsh, P.B., Simpson, M.E., Craig, G.O., Donoso, J. & Ramsey, H.H., Jr (1973) Occurrence of aflatoxins in cotton seeds at harvest in relation to location of growth and field temperatures. *J. environ. Qual.*, **2**, 276–281

Martinez, A.J., Weng, C.Y. & Park, D.L. (1994) Distribution of ammonia/aflatoxin reaction products in corn following exposure to ammonia decontamination procedure. *Food Addit. Contam.*, **11**, 659–667

Miller, J.D. (1995) Fungi and mycotoxins in grain: Implications for stored product research. *J. stored Prod. Res.*, **31**, 1–16

Murphy, P.A., Rice, L.G. & Ross, P.F. (1993) Fumonisin B_1, B_2, and B_3 content of Iowa, Wisconsin, and Illinois corn and corn screenings. *J. agric. Food Chem.*, **41**, 263–266

Neal, G.E., Judah, D.J., Carthew, P., Verma, A., Latour, I., Weir, L., Coker, R.D., Nagler, M.J. & Hoogenboom, L.A.P. (2001) Differences detected *in vivo* between samples of aflatoxin-contaminated peanut meal, following decontamination by two ammonia-based processes. *Food Addit. Contam.*, **18**, 137–149

Njapau, H., Muzungaile, E.M. & Changa, R.C. (1998) The effect of village processing techniques on the content of aflatoxins in corn and peanuts in Zambia. *J. Sci. Food Agric.*, **76**, 450–456

Park, D.L. (1993a) Controlling aflatoxin in food and feed. *Food Technol.*, **47**, 92–96

Park, D.L. (1993b) Perspectives on mycotoxin decontamination procedures. *Food Add. Contam.*, **10**, 49–60

Park, D.L. & Pohland, A.E. (1986) A rationale for the control of aflatoxin in animal feeds. In: Steyn, P.S. & Vleggaar, R., eds, *Mycotoxins and Phycotoxins*, Amsterdam, Elsevier, pp. 473–482

Park, D.L. & Price, W.D. (2001) Reduction of aflatoxin hazards using ammoniation. *Rev. environ. Contam. Toxicol.*, **171**, 139–175

Park, D.L. & Stoloff, L. (1989) Aflatoxin control — How a regulatory agency managed risk from an unavoidable natural toxicant in food and feed. *Regul. Toxicol. Pharmacol.*, **9**, 109–130

Park, D.L., Lee, L. & Koltun, S.A. (1984) Distribution of ammonia-related aflatoxin reaction products in cottonseed meal. *J. Am. Oil Chem. Soc.*, **61**, 1071–1074

Park, D.L., Lee, L.S., Price, R.L. & Pohland, A.E. (1988) Review of the decontamination of aflatoxins by ammoniation: Current status and regulation. *J. Assoc. off. anal. Chem.*, **71**, 685–703

Payne, G.A. (1983) Nature of field infection of corn by *Aspergillus flavus*. In: Diener, U.L., Asquith, R.L. & Dickens, J.W., eds, *Aflatoxin and* Aspergillus flavus *in Corn* (Southern Cooperative Series Bulletin 279), Opelika, AL, Craftmaster Printers, pp. 16–19

Payne, G.A., Cassel, D.K. & Adkins, C.R. (1986) Reduction in aflatoxin contamination in corn by irrigation and tillage. *Phytopathology*, **76**, 679–684

Peers, F.G. & Linsell, C.A. (1975) Aflatoxin contamination and its heat stability in Indian cooking oils. *Trop. Sci.*, **17**, 229–232

Pettit, R.E., Taber, R.A., Schroeder, H.W. & Harrison, A.L. (1971) Influence of fungicides and irrigation practice on aflatoxin in peanuts before digging. *Appl. Microbiol.*, **22**, 629–634

Phillips, T.D., Kubena, L.F., Harvey, R.B., Taylor, D.R. & Heidelbaugh, N.D. (1988) Hydrated sodium calcium aluminosilicates: A high affinity sorbent for aflatoxin. *Poult. Sci.*, **67**, 243–247

Pitt, J.I. (1989) Field studies on *Aspergillus flavus* and aflatoxins in Australian groundnuts. In: *Aflatoxin Contamination of Groundnut: Proceedings of the International Workshop, 6–9 October, 1987, ICRISAT Center, India*, Patancheru, ICRISAT (International Crops Research Institute for the Semi-Arid Tropics), pp. 223–235

Pitt, J.I. (1993) Corrections to species names in physiological studies on *Aspergillus flavus* and *Aspergillus parasiticus*. *J. Food Prot.*, **56**, 265–269

Pitt, J.I. & Hocking, A.D. (1997) *Fungi and Food Spoilage*, 2nd Ed., Cambridge, University Press

Pitt, J.I. & Miscamble, B.F. (1995) Water relations of *Aspergillus flavus* and closely related species. *J. Food Prot.*, **58**, 86–90

Pitt, J.I., Dyer, S.K. & McCammon, S. (1991) Systemic invasion of developing peanut plants by *Aspergillus flavus*. *Lett. appl. Microbiol.*, **13**, 16–20

Pitt, J.I., Hocking, A.D., Bhudhasamai, K., Miscamble, B.F., Wheeler, K.A. & Tanboon-Ek, P. (1993) The normal mycoflora of commodities from Thailand. 1. Nuts and oilseeds. *Int. J. Food Microbiol.*, **20**, 211–226

Pitt, J.I., Hocking, A.D., Bhudhasamai, K., Miscamble, B.F., Wheeler, K.A. & Tanboon-Ek, P. (1994) The normal mycoflora of commodities from Thailand. 2. Beans, rice, small grains and other commodities. *Int. J. Food Microbiol.*, **23**, 35–53

Pitt, J.I., Hocking, A.D., Miscamble, B.F., Dharmaputra, O.S., Kuswanto, K.R., Rahayu, E.S. & Sardjono, T. (1998) The mycoflora of food commodities from Indonesia. *J. Food Mycol.*, **1**, 41–60

Pohland, A.E. & Wood, G.E. (1987) Occurrence of mycotoxins in food. In: Krogh, P., ed., *Mycotoxins in Food*, London, Academic Press, pp. 35– 64

Prevot, A. (1986) Commercial detoxification of aflatoxin-contaminated peanut meal. *Bioact. Mol.*, **1**, 341–351

Price, R.L., Lough, O.G. & Brown, W.H. (1982) Ammoniation of whole cottonseed at atmospheric pressure and ambient temperature to reduce aflatoxin M_1 in milk. *J. Food Prot.*, **45**, 341–344

Rachaputi, N. (1999) Pre-harvest management trials. In: *Proceedings of the Aflatoxin Research Update Workshop, Kingaroy, Queensland, 25th August, 1999*, Kingaroy, Qld, Queensland Department of Primary Industries

Read, M. (1989) Removal of aflatoxin contamination from the Australian groundnut crop. In: *Aflatoxin Contamination of Groundnut: Proceedings of the International Workshop, 6-9 October, 1987, ICRISAT Center, India*, Patancheru, ICRISAT (International Crops Institute for the Semi-Arid Tropics), pp. 133–140

Sanders, T.H., Hill, R.A., Cole, R.H. & Blankenship, P.D. (1981) Effect of drought on occurrence of *Aspergillus flavus* in maturing peanuts. *J. Am. Oil Chem. Soc.*, **58**, 966A–970A

Sanders, T.H., Blankenship, P.D., Cole, R.J. & Hill, R.A. (1984) Effect of soil temperature and drought on peanut pod and stem temperatures relative to *Aspergillus flavus* invasion and aflatoxin contamination. *Mycopathologia*, **86**, 51–54

Sanders, T.H., Cole, R.J., Blankenship, P.D. & Hill, R.A. (1985) Relation of environmental stress duration to *Aspergillus flavus* invasion and aflatoxin production in preharvest peanuts. *Peanut Sci.*, **12**, 90–93

Sauer, D.B., Storey, C.L. & Walker, D.E. (1984) Fungal populations in US farm-stored grain and their relationship to moisture, storage time, regions and insect infestation. *Phytopathology*, **74**, 1050–1053

Schroeder, H.W. & Boller, R.A. (1973) Aflatoxin production of species and strains of the *Aspergillus flavus* group isolated from field crops. *Appl. Microbiol.*, **25**, 885–889

Schroder, H.W., Boller, R.A. & Hein, H., Jr (1968) Reduction in aflatoxin contamination of rice by milling procedures. *Cereal Chem.*, **45**, 574–580

Scott, P.M. (1984) Effects of food processing on mycotoxins. *J. Food Prot.*, **47**, 489–499

Sharman, M., Patey, A.L., Bloomfield, D.A. & Gilbert, J. (1991) Surveillance and control of aflatoxin contamination of dried figs and fig paste imported into the United Kingdom. *Food Addit. Contam.*, **8**, 299–304

Shotwell, O.L. (1983) Aflatoxin detection and determination in corn. In: Diener, U.L., Asquith, R.L. & Dickens, J.W., eds, *Aflatoxin and* Aspergillus flavus *in Corn* (Southern Cooperative Series Bulletin 279), Opelika, AL, Craftmaster Printers, pp. 38–45

Shotwell, O.L., Goulden, M.L. & Hesseltine, C.W. (1972) Aflatoxin contamination: Association with foreign material and characteristic fluorescence in damaged corn kernels. *Cereal Chem.*, **49**, 458–465

Simpson, M.E. & Batra, L.R. (1984) Ecological relations in respect to a boll rot of cotton caused by *Aspergillus flavus*. In: Kurata, H. & Ueno, Y., eds, *Toxigenic Fungi — Their Toxins and Health Hazard (Developments in Food Science 7)*, Amsterdam, Elsevier, pp. 24–32

Steiner, W.E., Rieker, R.H. & Battaglia, R. (1988) Aflatoxin contamination in dried figs: Distribution and association with fluorescence. *J. agric. Food Chem.*, **36**, 88–91

Stoloff, L., Trucksess, M., Anderson, P.W. & Glabe, E.F. (1978) Determination of the potential for mycotoxin contamination of pasta products. *J. Food Sci.*, **3**, 228–230

Wambugu, F. (1999) Why Africa needs agricultural biotech. *Nature*, **400**, 15–16

Wells, T.R. & Kreutzer, W.A. (1972) Aerial invasion of peanut flower tissues by *Aspergillus flavus* under gnotobiotic conditions (Abstract). *Phytopathology*, **62**, 797

Weng, C.Y., Martinez, A.J. & Park, D.L. (1994) Efficacy and permanency of ammonia treatment in reducing aflatoxin levels in corn. *Food Addit. Contam.*, **11**, 649–658

Wheeler, K.A., Hurdman, B.F. & Pitt, J.I. (1991) Influence of pH on the growth of some toxigenic species of *Aspergillus, Penicillium* and *Fusarium*. *Int. J. Food Microbiol.*, **12**, 141–150

Widstrom, N.W., McMillian, W.W. & Wilson, D. (1987) Segregation for resistance to aflatoxin contamination among seeds on an ear of hybrid maize. *Crop Sci.*, **27**, 961–963

Wood, G.M. (1982) Effects of processing on mycotoxins in maize. *Chem. Ind.*, **18 Dec.**, 972–974

FUMONISIN B_1

This substance was considered by a previous working group in June 1992 (IARC, 1993). Since that time, new data have become available, and these have been incorporated into the monograph and taken into consideration in the present evaluation.

1. Exposure Data

1.1 Chemical and physical data

1.1.1 *Nomenclature*

Chem. Abstr. Serv. Reg. No.: 116355-83-0
Chem. Abstr. Serv. Name: 1,2,3-Propanetricarboxylic acid, 1,1′-[1-(12-amino-4,9,11-trihydroxy-2-methyltridecyl)-2-(1-methylpentyl)-1,2-ethanediyl] ester
Synonyms: FB_1; macrofusine

1.1.2 *Structural and molecular formulae and relative molecular mass*

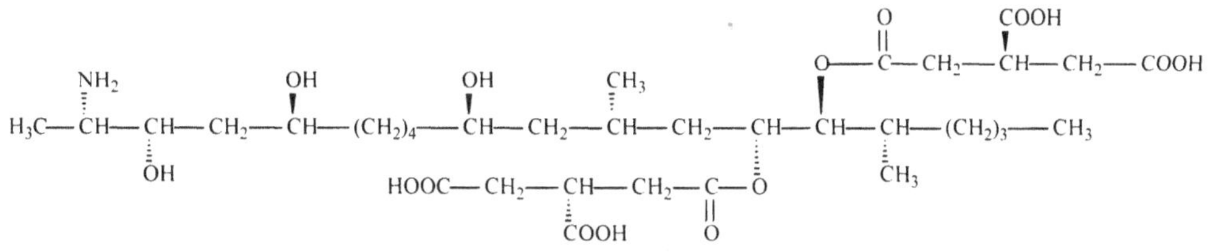

$C_{34}H_{59}NO_{15}$ Relative molecular mass: 721

1.1.3 *Chemical and physical properties of the pure substance*

From WHO (2000) unless otherwise noted

(*a*) *Description*: White hygroscopic powder
(*b*) *Melting-point*: Not known (has not been crystallized)
(*c*) *Spectroscopy*: Mass spectral and nuclear magnetic resonance spectroscopy data have been reported (Bezuidenhout *et al.*, 1988; Laurent *et al.*, 1989a; Plattner *et al.*, 1990; Savard & Blackwell, 1994)

(*d*) *Solubility*: Soluble in water to at least to 20 g/L (National Toxicology Program, 2000); soluble in methanol, acetonitrile–water

(*e*) *Octanol/water partition coefficient* (log P): 1.84 (Norred *et al.*, 1997)

(*f*) *Stability*: Stable in acetonitrile–water (1:1) at 25 °C; unstable in methanol at 25 °C, forming monomethyl or dimethyl esters (Gelderblom *et al.*, 1992a; Visconti *et al.*, 1994); stable in methanol at –18 °C (Visconti *et al.*, 1994); stable at 78 °C in buffer solutions at pH between 4.8 and 9 (Howard *et al.*, 1998)

1.1.4 *Analysis*

Methods for the analysis of fumonisins have been extensively reviewed (WHO, 2000). Six general analytical methods have been reported: thin-layer chromatographic (TLC), liquid chromatographic (LC), mass spectrometric (MS), post-hydrolysis gas chromatographic, immunochemical and electrophoretic methods (Sydenham & Shephard, 1996; Shephard, 1998).

An LC method for the determination of fumonisins B_1 and B_2 in maize and corn (maize) flakes was collaboratively studied. The method involves double extraction with acetonitrile–methanol–water (25:25:50), clean-up through an immunoaffinity column, and LC determination of the fumonisins after derivatization with *ortho*-phthaldialdehyde. This method has been proposed as the AOAC Official Method 2001.14, First Action (Visconti *et al.*, 2001).

The majority of studies have been performed using LC analysis of a fluorescent derivative (WHO, 2000). There are no validated biomarkers for human exposure to fumonisin B_1.

Many studies, both *in vivo* and *in vitro*, have demonstrated a correlation between disruption of sphingolipid metabolism — as measured by an increase in free sphinganine — and exposure to fumonisin B_1 (WHO, 2000, 2002).

Sphingolipids are a highly diverse class of lipids found in all eukaryotic cells. Their biological functions are equally diverse: the compounds serve as structural components required for maintenance of membrane integrity, as receptors for vitamins and toxins, as sites for cell–cell recognition and cell–cell and cell–substrate adhesion, as modulators of receptor function and as lipid second messengers in signalling pathways responsible for cell growth, differentiation and death (Merrill *et al.*, 1997).

Ceramide synthase is a key enzyme in the biosynthesis of sphingolipids. Alterations in the ratio free sphinganine/free sphingosine, a consequence of ceramide synthase inhibition, are now used as a biomarker for exposure to fumonisins in domestic animals and humans. The mechanistic aspects of the effects of fumonisins on sphingolipid metabolism are discussed more fully in Section 4.5.1.

Methods have been reported for extraction of fumonisin B_1 from human urine (Shetty & Bhat, 1998), plasma and urine of rats (Shephard *et al.*, 1992a, 1995a), bile of rats and vervet monkeys (Shephard *et al.*, 1994a, 1995b), faeces of vervet monkeys (Shephard *et al.*, 1994b), liver, kidney and muscle of beef cattle (Smith & Thakur, 1996) and milk (Maragos & Richard, 1994; Scott *et al.*, 1994; Prelusky *et al.*, 1996a).

1.2 Formation

Fumonisin B_1 was isolated in 1988 by Gelderblom *et al.* (1988). It was chemically characterized by Bezuidenhout *et al.* (1988), and shortly thereafter as 'macrofusine' by Laurent *et al.* (1989a), from cultures of *Fusarium verticillioides* (Sacc.) Nirenberg (formerly known as *Fusarium moniliforme* Sheldon) (Marasas *et al.*, 1979) as well as *Gibberella fujikuroi* (Leslie *et al.*, 1996). The absolute stereochemical configuration of fumonisin B_1 (see section 1.1.2) was determined by ApSimon (2001).

Fumonisin B_1 is produced by isolates of *F. verticillioides, F. proliferatum, F. anthophilum, F. beomiforme, F. dlamini, F. globosum, F. napiforme, F. nygamai, F. oxysporum, F. polyphialidicum, F. subglutinans* and *F. thapsinum* (*Gibberella thapsina*) isolated from Africa, the Americas, Oceania (Australia), Asia and Europe (Gelderblom *et al.*, 1988; Ross *et al.*, 1990; Nelson *et al.*, 1991; Thiel *et al.*, 1991a; Chelkowski & Lew, 1992; Leslie *et al.*, 1992; Nelson *et al.*, 1992; Miller *et al.*, 1993; Rapior *et al.*, 1993; Desjardins *et al.*, 1994; Visconti & Doko, 1994; Abbas & Ocamb, 1995; Abbas *et al.*, 1995; Logrieco *et al.*, 1995; Miller *et al.*, 1995; Leslie *et al.*, 1996; Klittich *et al.*, 1997; Musser & Plattner, 1997; Sydenham *et al.*, 1997). *Alternaria alternata* f. sp. *lycopersici* has also been shown to synthesize B fumonisins (Abbas & Riley, 1996). Fumonisins can be produced by culturing strains of the *Fusarium* species that produce these toxins on sterilized maize (Cawood *et al.*, 1991) and yields of up to 17.9 g/kg (dry weight) have been obtained with *F. verticillioides* strain MRC 826 (Alberts *et al.*, 1990). Yields of 500–700 mg/L for fumonisin B_1 plus fumonisin B_2 have been obtained in liquid fermentations and high recoveries of the toxins are possible (Miller *et al.*, 1994). The predominant toxin produced is fumonisin B_1. Fumonisin B_1 frequently occurs together with fumonisin B_2, which may be present at levels of 15–35% of fumonisin B_1 (IARC, 1993; Diaz & Boermans, 1994; Visconti & Doko, 1994).

F. verticillioides and *F. proliferatum* are among the most common fungi associated with maize. These fungi can be recovered from most maize kernels including those that appear healthy (Bacon & Williamson, 1992; Pitt *et al.*, 1993; Sanchis *et al.*, 1995). The level of formation of fumonisins in maize in the field is positively correlated with the occurrence of these two fungal species, which are predominant during the late maturity stage (Chulze *et al.*, 1996). These species can cause *Fusarium* kernel rot of maize, which is one of the most important ear diseases in hot maize-growing areas (King & Scott, 1981; Ochor *et al.*, 1987; De León & Pandey, 1989) and is associated with warm, dry ears and/or insect damage (Shurtleff, 1980).

1.3 Use

Fumonisin B_1 is not used commercially.

1.4 Occurrence

Fumonisins have been found worldwide, primarily in maize. More than 10 compounds have been isolated and characterized. Of these, fumonisins B_1, B_2 and B_3 are the major fumonisins produced. The most prevalent in contaminated maize is fumonisin B_1, which is believed to be the most toxic (Thiel *et al.*, 1992; Musser & Plattner, 1997; Food and Drug Administration, 2001a,b). A selection of data on the occurrence of fumonisin B_1 in maize and food products is given in Table 1.

(*a*) *Formation in raw maize*

The concentrations of fumonisins in raw maize are influenced by environmental factors such as temperature, humidity, drought stress and rainfall during pre-harvest and harvest periods. For example, high concentrations of fumonisins are associated with hot and dry weather, followed by periods of high humidity (Shelby *et al.*, 1994a,b). Magan *et al.* (1997) have studied the effects of temperature and water activity (a_w) on the growth of *F. moniliforme* and *F. proliferatum*. Growth increases with a_w (between 0.92 to 0.98) and is maximum at 30 °C for *F. moniliforme* and at 35 °C for *F. proliferatum*.

High concentrations of fumonisins may also occur in raw maize that has been damaged by insects (Bacon & Nelson, 1994; Miller, 2000). However, maize hybrids genetically engineered to carry genes from the bacterium *Bacillus thuringiensis* (Bt maize) that produce proteins that are toxic to insects, specifically the European maize borer, have been found to be less susceptible to *Fusarium* infection and contain lower concentrations of fumonisins than the non-hybrid maize in field studies (Munkvold *et al.*, 1997, 1999).

(*b*) *Occurrence in processed maize products*

One of the major factors that determine the concentration of fumonisins in processed maize products is whether a dry- or wet-milling process is used. The whole maize kernel consists of the following major constituents: (i) starch, which is the most abundant constituent from which maize starches and maize sweeteners are produced; (ii) germ, which is located at the bottom of the centre of the kernel from which maize oil is produced; (iii) gluten, which contains the majority of the protein found in maize kernel; and (iv) hull (pericarp), which is the outer coat of the kernel from which maize bran is produced.

Dry milling of whole maize kernel generally results in the production of fractions called bran, flaking grits, grits, meal and flour. Because fumonisins are concentrated in the germ and the hull of the whole maize kernel, dry milling results in fractions with different concentrations of fumonisins. For example, dry-milled fractions (except for the bran fraction) obtained from degermed maize kernels contain lower concentrations of fumonisins than dry-milled fractions obtained from non-degermed or partially degermed maize. Industry information indicates that dry milling results in fumonisin-containing fractions in the following order of descending fumonisin concentrations: bran, flour,

Table 1. Worldwide occurrence of fumonisin B_1 in maize-based products[a]

Product	Region/Country	Detected/ total samples	Fumonisin B_1 (mg/kg)
	North America		
Maize	Canada, USA	324/729	0.08–37.9
Maize flour, grits	Canada, USA	73/87	0.05–6.32
Miscellaneous maize foods[b]	USA	66/162	0.004–1.21
Maize feed	USA	586/684	0.1–330
	Latin America		
Maize	Argentina, Brazil, Uruguay	126/138	0.17–27.05
Maize flour, alkali-treated kernels, polenta	Peru, Uruguay, Venezuela	5/17	0.07–0.66
Miscellaneous maize foods[b]	Uruguay, Texas–Mexico border	63/77	0.15–0.31
Maize feed	Brazil, Uruguay	33/34	0.2–38.5
	Europe		
Maize	Austria, Croatia, Germany, Hungary, Italy, Poland, Portugal, Romania, Spain, United Kingdom	248/714	0.007–250
Maize flour, maize grits, polenta, semolina	Austria, Bulgaria, Czech Republic, France, Germany, Italy, Netherlands, Spain, Switzerland, United Kingdom	181/258	0.008–16
Miscellaneous maize foods[b]	Czech Republic, France, Germany, Italy, Netherlands, Spain, Sweden, Switzerland, United Kingdom	167/437	0.008–6.10
Imported maize, grits and flour	Germany, Netherlands, Switzerland	143/165	0.01–3.35
Maize feed	France, Italy, Spain, Switzerland, United Kingdom	271/344	0.02–70
	Africa		
Maize	Benin, Kenya, Malawi, Mozambique, South Africa, Tanzania, Uganda, Zambia, Zimbabwe	199/260	0.02–117.5
Maize flour, grits	Botswana, Egypt, Kenya, South Africa, Zambia, Zimbabwe	73/90	0.05–3.63
Miscellaneous maize foods[b]	Botswana, South Africa	8/17	0.03–0.35
Maize feed	South Africa	16/16	0.47–8.85

Table 1 (contd)

Product	Region/Country	Detected/ total samples	Fumonisin B_1 (mg/kg)
	Asia		
Maize	China, Indonesia, Iran[c], Nepal, Philippines, Thailand, Viet Nam	380/633	0.01–155
Maize flour, grits, gluten	China, India, Japan, Thailand, Viet Nam	44/53	0.06–2.60
Miscellaneous maize foods[b]	Japan, Taiwan	52/199	0.07–2.39
Maize feed	Korea (Republic of), Thailand	10/34	0.05–1.59
	Oceania		
Maize	Australia	67/70	0.3–40.6
Maize flour	New Zealand	0/12	—

[a] Adapted from WHO (2000) and Plattner *et al.* (1990); Shephard *et al.* (1990); Sydenham *et al.* (1990a,b); Wilson *et al.* (1990); Lew *et al.* (1991); Ross *et al.* (1991a,b); Sydenham *et al.* (1991); Thiel *et al.* (1991b); Bane *et al.* (1992); Colvin & Harrison (1992); Minervini *et al.* (1992); Osweiler *et al.* (1992); Park *et al.* (1992); Pittet *et al.* (1992); Rheeder *et al.* (1992); Stack & Eppley (1992); Sydenham *et al.* (1992); Caramelli *et al.* (1993); Chamberlain *et al.* (1993); Holcomb *et al.* (1993); Hopmans & Murphy (1993); Murphy *et al.* (1993); Price *et al.* (1993); Scudamore & Chan (1993); Sydenham *et al.* (1993a,b); Ueno *et al.* (1993); Wang *et al.* (1993); Chu & Li (1994); Doko & Visconti (1994); Doko *et al.* (1994); Kang *et al.* (1994); Lee *et al.* (1994); Pestka *et al.* (1994); Sanchis *et al.* (1994); Shelby *et al.* (1994a); Sydenham (1994); Usleber *et al.* (1994a,b); Viljoen *et al.* (1994); Yoshizawa *et al.* (1994); Zoller *et al.* (1994); Bottalico *et al.* (1995); Doko *et al.* (1995); Miller *et al.* (1995); Pascale *et al.* (1995); Trucksess *et al.* (1995); Visconti *et al.* (1995); Wang *et al.* (1995); Yamashita *et al.* (1995); Bryden *et al.* (1996); Burdaspal & Legarda (1996); Doko *et al.* (1996); Dragoni *et al.* (1996); Hirooka *et al.* (1996); Meister *et al.* (1996); Ramirez *et al.* (1996); Castella *et al.* (1997); Gao & Yoshizawa (1997); Patel *et al.* (1997); Piñeiro *et al.* (1997); Rumbeiha & Oehme (1997); Tseng & Liu (1997); Ueno *et al.* (1997); Ali *et al.* (1998); Fazekas *et al.* (1998); de Nijs *et al.* (1998a,b); Ostrý & Ruprich (1998); Scudamore *et al.* (1998); Stack (1998)

[b] Includes maize snacks, canned maize, frozen maize, extruded maize, bread, maize-extruded bread, biscuits, cereals, chips, flakes, pastes, starch, sweet maize, infant foods, gruel, purée, noodles, popcorn, porridge, tortillas, tortilla chips, masas, popped maize, soup, taco and tostada

[c] From Shephard *et al.* (2000)

meal, grits and flaking grits. Consequently, maize products such as corn bread, maize grits and maize muffins made from the grits and flour fractions may contain low concentrations of fumonisins. Ready-to-eat breakfast cereals made from flaking grits, such as corn (maize) flakes and puffed type cereals, contain very low concentrations (from non-detectable to 10 ppb) of fumonisins (Stack & Eppley, 1992).

Wet milling of whole maize generally results in the production of fumonisin-containing fractions in the following order of descending fumonisin concentrations: gluten, fibre, germ and starch. No fumonisins have been detected in the starch fraction obtained from wet milling of fumonisin-contaminated maize. The starch fraction is further processed for production of high-fructose maize syrups and other maize sweeteners (JECFA, 2001). Therefore, products of these types do not contain any detectable concentration of fumonisins. Maize oil, extracted from maize germ and refined, does not contain any detectable fumonisins (Patel *et al.*, 1997). The gluten and fibre fractions from the wet-milling process do contain fumonisins; these fractions are used to produce animal feed, such as maize gluten meal and maize gluten feed (JECFA, 2001).

Another process to which whole maize may be subjected is nixtamalization, which consists of boiling the raw maize kernels in aqueous calcium hydroxide solution (lye), cooling and washing to remove the pericarp and excess calcium hydroxide. The washed kernels are then ground to produce the 'masa', from which maize chips and tortillas are made. This process has been shown to reduce concentrations of fumonisins in raw maize kernels (Dombrink-Kurtzman & Dvorak, 1999). However, the reaction product, hydrolysed fumonisin B_1 (HFB_1) is highly hepatotoxic and nephrotoxic (Voss *et al.*, 1996a, 1998, 2001a).

Available data indicate the presence of low concentrations (4–82 ppb (μg/kg)) of fumonisins in sweet maize (Trucksess *et al.*, 1995). Fumonisins can be present in beer but at low concentrations (4.8–85.5 ppb (μg/L)) (from maize-based brewing adjuncts) (Torres *et al.*, 1998; Hlywka & Bullerman, 1999), but distilled spirits made from maize do not contain fumonisins (Bennett & Richard, 1996). The fermentation process does not destroy fumonisins and 85% of the toxin may be recovered in fermented products. Products from ethanol fermentations generally used as animal feeds may be detrimental if consumed by pigs or horses (Bennett & Richard, 1996).

Broken kernels of maize which have been screened from bulk lots of maize before any milling process contain higher concentrations of fumonisins than whole kernels, and are often used in animal feeds. Higher fumonisin concentrations are found in maize screenings. Fumonisin-contaminated maize at concentrations of 330 and 160 mg/kg has caused porcine pulmonary oedema and equine leukoencephalomalacia, respectively (Ross *et al.*, 1991a,b). However, fumonisin residues in milk (Maragos & Richard, 1994; Scott *et al.*, 1994; Becker *et al.*, 1995; Richard *et al.*, 1996), eggs (Vudathala *et al.*, 1994) and meat (Prelusky *et al.*, 1994, 1996a,b; Smith & Thakur, 1996) have been either undetectable or were detected at extremely low concentrations.

(*c*) *Formation in commodities other than maize*

Fumonisins have also been reported in other food products, notably sorghum in Botswana, Brazil, India, South Africa and Thailand (Bhat *et al.*, 1997; Siame *et al.*, 1998; Vasanthi & Bhat, 1998; Suprasert & Chulamorakot, 1999; da Silva *et al.*, 2000; Gamanya & Sibanda, 2001). In these countries, about 40% of the samples screened contained low concentrations of fumonisin B_1 (0.11–0.55 mg/kg). However, higher amounts (up to 7.8 mg/kg) were observed in rain-damaged sorghum, still lower than the concentrations reported for maize (up to 65 mg/kg) (Vasanthi & Bhat, 1998). Other commodities in which fumonisin B_1 has been detected include millets, rice, wheat, barley, cereal-based food products, soybean and pastes and animal feeds (Nelson *et al.*, 1992; Castella *et al.*, 1997; Abbas *et al.*, 1998; Siame *et al.*, 1998; Hlywka & Bullerman, 1999; Chulamorakot & Suprasert, 2000; Gamanya & Sibanda, 2001; JECFA, 2001).

Fumonisin B_1 has been found in black tea (80–280 mg/kg) as well as in some medicinal plants, such as leaves of the orange tree (350–700 mg/kg) and leaves and flowers of the linden tree (20–200 mg/kg) (Martins *et al.*, 2001).

Twenty-five asparagus plants affected by crown rot were analysed for *Fusarium* infestation and fumonisin contamination. *F. proliferatum* was found in all plants. Fumonisin B_1 was detected in crowns and stems at concentrations of 7.4 and 0.83 mg/kg dry weight, respectively (Logrieco *et al.*, 1998).

Co-occurrence of fumonisins and aflatoxins in maize is reviewed in the monograph on aflatoxins in this volume.

1.5 Human exposure to fumonisins

A number of estimates of human exposure to fumonisins have been made. In a preliminary estimate for the Food and Drug Administration (FDA) in the USA, exposure to fumonisins for consumers of maize in the USA was estimated as 0.08 μg/kg bw per day (Humphreys *et al.*, 1997; WHO, 2000). In Canada for the period 1991 to early 1995, Kuiper-Goodman *et al.* (1996) estimated human exposure to be 0.017–0.089 μg/kg bw per day. In Switzerland, mean daily intake of fumonisins has been estimated to be 0.03 μg/kg bw (Zoller *et al.*, 1994).

As a conservative estimate, de Nijs *et al.* (1998c) found that in the Netherlands 97% of individuals with gluten intolerance had a daily exposure of at least 1 μg fumonisin B_1 and 37% of at least 100 μg; in the general population it was estimated that 49% and 1%, respectively, were exposed to these levels of fumonisin B_1.

Human exposures in the Transkei, South Africa, were earlier estimated to be 14 and 440 μg/kg bw fumonisin B_1 per day for good quality and mouldy maize, respectively (Thiel *et al.*, 1992). More recent estimates of probable daily intake by South Africans vary from 1.2 to 355 μg/kg bw per day in a rural population in Transkei consuming home-grown mouldy maize (Marasas, 1997).

Since fumonisin B_1 is present in the spores and mycelia of *F. verticillioides* (Tejada-Simon *et al.*, 1995), occupational inhalation exposure could be a problem, but data are lacking on airborne levels of fumonisins during the harvesting, processing and handling of fumonisin-contaminated maize.

In a study conducted in China, urine was collected from volunteers before and after consumption of a fumonisin B_1-contaminated diet for one month. The ratio of free sphinganine to free sphingosine (Sa/So) was increased threefold in the urine of the men, but was unchanged in that of the women. This increase was also apparent when the data were pooled for men and women and grouped into those individuals who had estimated intakes of fumonisin B_1 greater than or less than 110 μg/kg bw per day (Qiu & Liu, 2001).

This result is similar to that reported for swine, where the sphinganine/sphingosine ratio became significantly different from controls at a fumonisin B_1 intake of 500 μg/kg bw per day (Rotter *et al.*, 1996).

1.6 Regulations and guidelines

The Joint FAO/WHO Expert Committee on Food Additives (JECFA) has recommended a provisional maximum tolerable daily intake (PMTDI) of 2 μg/kg bw for fumonisins B_1, B_2 and B_3, alone or in combination (JECFA, 2001).

An official tolerance value for dry maize products (1 mg/kg fumonisin B_1 plus fumonisin B_2) has been issued in Switzerland (Canet, 1999).

The recommended maximum levels for fumonisins in human foods and in animal feeds in the USA that the FDA considers achievable with the use of good agricultural and good manufacturing practices are presented in Tables 2 and 3. Human exposure to fumonisins should not exceed levels achievable with the use of such practices (Food and Drug Administration, 2001c).

2. Studies of cancer in humans

Studies on the relationship between *Fusarium verticillioides* (formerly known as *F. moniliforme*) toxins (of which fumonisin B_1 and fumonisin B_2 are the major toxic secondary metabolites) and oesophageal cancer in areas of South Africa and China were summarized in Volume 56 of the *IARC Monographs* (IARC, 1993). The evidence in humans was judged to be '*inadequate*' at that time.

The only subsequent study that investigated the relationship between fumonisins and cancer was carried out in the People's Republic of China. Yoshizawa and Gao (1999) collected 76 corn samples from the homes of oesophageal cancer patients, selected at random in Linxian, China (a high-risk area for oesophageal cancer) as well as 55 samples from homes of peasant families with no oesophageal cancer patient in Shangqiu (a low-

Table 2. Maximum levels of fumonisins in human foods and in animal feeds in the USA[a]

Product	Total fumonisins ($B_1 + B_2 + B_3$) ppm (mg/kg)
Human foods	
Degermed dry milled corn products (e.g., flaking grits, corn grits, corn meal, corn flour with fat content of < 2.25%, dry weight basis)	2
Whole or partially degermed dry milled corn products (e.g., flaking grits, corn grits, corn meal, corn flour with fat content of ≥ 2.25%, dry weight basis)	4
Dry milled corn bran	4
Cleaned corn intended for masa production	4
Cleaned corn intended for popcorn	3
Animal feeds	
Corn and corn by-products intended for:	
Equids and rabbits	5[b]
Swine and catfish	20[c]
Breeding ruminants, breeding poultry and breeding mink[d]	30[c]
Ruminants ≥ 3 months old being raised for slaughter and mink being raised for pelt production	60[c]
Poultry being raised for slaughter	100[c]
All other species or classes of livestock and pet animals	10[c]

[a] From Food and Drug Administration (2001a)
[b] No more than 20% of diet on a dry weight basis
[c] No more than 50% of diet on a dry weight basis
[d] Includes lactating dairy cattle and hens laying eggs for human consumption

risk comparison area). Homegrown samples of corn intended for human consumption were collected in 1989, 1995 and 1997 (Yoshizawa *et al.*, 1994; Yoshizawa & Gao, 1999). Fumonisins B_1, B_2 and B_3 were analysed by high-performance liquid chromatography (HPLC). Mean concentrations of fumonisin B_1 in the high- and low-risk areas, respectively, were 872 and 890 ng/g in 1989, 2730 and 2702 ng/g in 1995, and 2028 and 2082 ng/g in 1997. Maximum concentrations were 2960, 21 000 and 8290 ng/g in the high-risk area and 1730, 8470 and 5330 ng/g in the low-risk area, respectively, in the three years studied. There was no significant difference in any of the measured fumonisin levels between the two areas ($p > 0.05$). The percentages of samples with detectable

Table 3. Levels of total fumonisins ($B_1 + B_2 + B_3$) in corn, corn by-products and the total ration for various animal species recommended in the USA[a]

Animal or class	Recommended maximum level of total fumonisins in corn and corn by-products (ppm; mg/kg)	Feed factor[b]	Recommended maximum level of total fumonisins in the total ration (ppm)
Horse[c]	5	0.2	1
Rabbit	5	0.2	1
Catfish	20	0.5	10
Swine	20	0.5	10
Ruminants[d]	60	0.5	30
Mink[e]	60	0.5	30
Poultry[f]	100	0.5	50
Ruminant, poultry and mink breeding stock[g]	30	0.5	15
All others[h]	10	0.5	5

[a] From Food and Drug Administration (2001c)
[b] Fraction of corn or corn by-product mixed into the total ration
[c] Includes asses, zebras and onagers
[d] Cattle, sheep, goats and other ruminants that are ≥ 3 months old and fed for slaughter
[e] Fed for pelt production
[f] Turkeys, chickens, ducklings and other poultry fed for slaughter
[g] Includes laying hens, roosters, lactating dairy cows and bulls
[h] Includes dogs and cats

fumonisins in the high-risk area were 48% in 1989, 79% in 1995 and 73% in 1997; the corresponding figures in the low-risk area were 25% in 1981, 50% in 1995 and 47% in 1997. Based on local dietary habits, the estimated daily intake of fumonisin B_1 was 1.6–1.9 times higher in the high-risk than in the low-risk area for oesophageal cancer. The authors noted that aflatoxin B_1 was detected at very low levels in corn samples from both of the areas. [The Working Group noted that cancer families in high-risk areas were compared with non-cancer families in low-risk areas.]

3. Studies of Cancer in Experimental Animals

Toxins derived from *Fusarium moniliforme* were considered by a previous Working Group in 1992 (IARC, 1993). Since that time, new data have become available and these have been incorporated into the monograph and taken into consideration in the present evaluation.

3.1 Oral administration of fumonisin mixtures

3.1.1 *Studies using naturally contaminated maize (fumonisins including other* Fusarium *mycotoxins)*

Rat: A group of 12 male Fischer 344 rats (average body weight, 125 g) was fed a maize diet, naturally contaminated with *Fusarium verticillioides* (*F. moniliforme*) over a period of 4–6 months. The maize sample was obtained from feed being fed to horses during an outbreak of equine leukoencephalomalacia (ELEM). A group of 12 control rats was fed a commercial rodent chow (Purina 5001) (Wilson *et al.*, 1985). No aflatoxins were detected in the maize diet (detection limit, < 0.9 μg/kg), which was deficient in many nutrients including choline and methionine. Maize samples contained moniliformin (2.82 mg/kg) and fusarin C (0.39 mg/kg), but no trichothecene or aflatoxin (Thiel *et al.*, 1986). Retrospective mycological and chemical analyses indicated the presence of *F. verticillioides* and *Aspergillus flavus* as major fungal contaminants and a total fumonisin B (fumonisin B_1 and B_2) concentration of 33.1 mg/kg, while only trace amounts of aflatoxin B_1 and B_2 ranging between 0.05 and 0.1 μg/kg were detected (JECFA, 2001). A mean fumonisin B intake of between 1.6 and 2.0 mg/kg bw per day was estimated based on an apparent feed intake of 50–60 g/kg bw per day (Wilson *et al.*, 1985; JECFA, 2001). One of the treated rats died after 77 days, three were killed on days 123, 137 and 145, and the remaining eight animals were killed on day 176: all animals showed multiple hepatic nodules, large areas of adenofibrosis and cholangiocarcinomas. The controls had no liver lesions.

3.1.2 *Studies using fungal culture material*

Rat: A group of 31 female Wistar rats [age unspecified] was fed a diet containing maize bread inoculated with *F. verticillioides*. After 554–701 days of feeding, four papillomas and two carcinomas had developed in the forestomach. No epithelial lesion of the forestomach was seen in a control group of 10 female rats fed conventional maize bread, not inoculated with the mould, for 330–700 days (Li *et al.*, 1982). [The Working Group noted the inadequate reporting of the study.]

Groups of 20 male BD IX rats (weighing 80–100 g) were fed commercial rat feed containing *F. verticillioides* MRC 826 mouldy meal (freeze-dried or oven-dried) at levels of either 8% for up to 57 or 75 days or 4% for 286 days followed by 2% until 763 days (Marasas *et al.*, 1984). The estimated intakes of fumonisin B were 138 (8% in diet), 69 (4%) and 32 (2%) mg/kg bw per day for the respective diets, with an average feed intake of 32 g feed/kg bw per day (JECFA, 2001). All rats given 8% diet between 57 and 75 days had severe liver damage. Among rats fed the 4%/2% freeze-dried diet regimen and surviving beyond 450 days, 12/14 had hepatocellular carcinomas and 10/14 had hepatic ductular carcinoma. In the rats fed the 4%/2% oven-dried material, 12/16 had hepatocellular carcinoma and 9/16 had hepatic ductular carcinoma. In both treatment groups with freeze-dried or oven-dried diet, three rats had pulmonary metastases. No liver tumours were seen in rats fed control diet (Marasas *et al.*, 1984).

Groups of 30 male BD IX rats, weighing approximately 110 g, were fed for 27 months freeze-dried culture material of *F. verticillioides* MRC 826 at levels from 0.25% to 0.75% in either a semi-synthetic diet (marginally deficient in certain vitamins and minerals) or a semi-synthetic diet containing 5% culture material of *F. verticillioides* MRC 1069 (to contain 18.2 mg/kg fusarin C). Thirty control rats received commercial maize (5%) in the diet. Fumonisin B intake of between 4 and 13 mg/kg bw fumonisin B_1 in the rats fed MRC 826 was estimated assuming consumption of 32 g/kg bw feed per day. In animals fed the MRC 826 diet and necropsied between 23 and 27 months, neoplastic nodules (21/21), hepatocellular carcinoma with lung metastases (2/21), adenofibrosis (19/21) and cholangiocarcinoma (8/21) were observed. In the group fed the MRC 1069 diet, there were 1/22 neoplastic nodule and 1/22 adenofibrosis. No such lesions were seen in the control group. Forestomach papillomas were observed in 13/21 MRC 826-treated, 3/22 MRC 1069-treated and 5/22 control animals, respectively, and forestomach carcinomas were observed in 4/21 MRC 826-treated animals versus none in the other groups. Basal-cell hyperplasia of the oesophageal epithelium was observed in 12/21 rats fed the MRC 826 diet (Jaskiewicz *et al.*, 1987).

3.2 Oral administration of purified fumonisin B_1

3.2.1 *Mouse*

Groups of 48 male and female B6C3F$_1$ mice, four weeks of age, were fed fumonisin B_1 (> 96% pure, ammonium salt) at concentrations of 0, 5, 15, 80 or 150 mg/kg of diet (males) and 0, 5, 15, 50 or 80 mg/kg of diet (females) in NIH 46 diet over a period of two years (equivalent to average daily doses of ~0.6, 1.7, 9.5 and 17 mg/kg bw for males or 0.7, 2.1, 7.0 and 12.5 mg/kg bw for females). Survival of the female mice was significantly reduced in the group treated with 80 mg/kg fumonisin B_1 (60%), while that in treated male mice was not significantly different from the controls. The low incidence of spontaneous liver tumours in control female (11%) and male (26%) mice compared with historical controls was ascribed to feed restriction, as previously suggested by Haseman *et al.* (1998). After two years, the incidences of hepatocellular adenomas in female mice were 5/47, 3/48, 1/48, 16/47 [p = 0.0047] and 31/45 [p = 0.001] and those of hepatocellular carcinomas were 0/47, 0/48, 15/48 [p = 0.0007], 10/47 [p = 0.0007] and 9/45 for the groups treated with 0, 5, 15, 50 and 80 mg/kg of diet, respectively. The incidences of hepatocellular adenomas and carcinomas in treated males were not significantly increased compared with control males (National Toxicology Program, 2000). [No analyses to determine the presence of *Helicobacter hepaticus* were reported.]

3.2.2 *Rat*

Two groups of 25 male BD IX rats, weighing between 70 and 80 g, were fed a semi-purified diet (intentionally marginally deficient in minerals and vitamins) in the absence or presence of 50 ppm (mg/kg diet) fumonisin B_1 (90% pure) for 26 months (Gelderblom *et al.*, 1991). An average daily fumonisin B_1 intake of 1.6 mg/kg bw was calculated on the basis of a mean average feed intake of 32 g/kg bw per day (JECFA, 2001). Groups of five treated and five control rats were killed at 6 and 12 months. At 6 months, regenerative nodules were observed in all animals and cholangiofibrosis was observed in all but one; at 12 months, regenerative nodules and cholangiofibrosis were observed in all animals. In all 15 rats that died or were killed between 18 and 26 months, hepatocyte nodules, cholangiofibrosis and cirrhosis were observed. Hepatocellular carcinomas, two of which metastasized — one to the heart and lungs and one to the kidneys — were also observed in 10/15 animals (Gelderblom *et al.*, 1991).

Four groups of 20 male BD IX rats, weighing approximately 100 g, were fed a semi-purified diet (intentionally marginally deficient in minerals and vitamins) containing 0, 1, 10 or 25 mg/kg of diet fumonisin B_1 (purity, 92–95%) over a period of two years. The mean intakes of fumonisin B_1 (mg/kg bw) were 0.005, 0.03, 0.3 and 0.8 for control, low-, mid- and high-dose groups, respectively. The survival rates at two years were 16/20, 14/20, 18/20 and 17/20 in the four groups, respectively. There was a significant ($p < 0.05$) increase in the incidence of portal fibrosis (5/17), ground glass foci (7/17) and hepatocyte nodules (9/17) in the liver of the rats fed fumonisin B_1 at 25 mg/kg of diet. One rat had a large focal area of adenofibrosis. Some of these hepatic changes were detected to a smaller extent in rats treated with fumonisin B_1 at 1 and 10 mg/kg of diet. No such lesions were observed in the livers of control rats (Gelderblom *et al.*, 2001a).

Groups of 40–48 male and 40–48 [40 for the 15-mg/kg group] female Fischer 344 rats, eight weeks of age, were fed fumonisin B_1 (> 96% pure, ammonium salt) at concentrations of 0, 5, 15, 50 or 100 mg/kg of diet for females and 0, 5, 15, 50 or 150 mg/kg of diet for males in a powdered NIH 36 diet that was available *ad libitum* over a period of two years (equivalent to average daily doses of 0, 0.25, 0.8, 2.5 and 7.5 mg/kg bw for females and 0, 0.3, 0.9, 3.0 and 6.0 mg/kg bw for males). Survival rates were similar in the treated and control rats. In males at two years, the incidences of renal tubule adenomas were 0/38, 0/40, 0/48, 2/48 and 5/48, those of carcinomas were 0/48, 0/40, 0/48, 7/48 and 10/48 and those of adenomas and carcinomas combined were 0/48, 0/48, 0/48, 9/48 [$p = 0.001$] and 14/48 [$p = 0.0001$] for the controls and increasing doses, respectively. The occurrence of renal tumours in males was accompanied by an increased incidence of renal tubule epithelial cell hyperplasia at two years (2/48, 1/40, 4/48, 14/48 and 8/48 of the male rats receiving fumonisin B_1 at 0, 5, 15, 50 and 150 mg/kg of diet, respectively). In female rats, there were no significant fumonisin B_1-dependent changes in the incidence of tumours. One renal adenoma was detected in a female rat fed fumonisin B_1 at 50 mg/kg of diet, and one renal tubule carcinoma was detected in a female rat fed 100 mg/kg (National Toxicology Program, 2000).

A re-evaluation of the renal pathology of the National Toxicology Program (2000) study characterized toxic lesions as cytotoxic/regenerative (graded from 0 to 4) and atypical tubule hyperplasia (Hard *et al.*, 2001). There was a progressive increase in the grade of severity of the former lesion in the rats fed 15, 50 and 150 mg fumonisin B_1 per kg of diet. Atypical hyperplasia was observed in 4/48 and 9/48 rats fed the 50- and 150-mg/kg diets. Adenomas (4/48) and carcinomas (6/48) were observed in the kidneys of the rats fed 50 mg/kg and also in the group fed 150 mg/kg (8/48 adenomas, 10/48 carcinomas). Two of 8 and 5/8 carcinomas in the 50- and 150-mg/kg treatment groups, respectively, metastasized to the lung. Only one of the 18 carcinomas displayed the conventional reasonably differentiated phenotype. Cellular pleomorphisms were noticed in 3/8 and 1/10 carcinomas in rats fed 50 and 150 mg/kg, respectively. Among the carcinomas observed in these studies, 61% were an anaplastic variant. [The Working Group noted that this re-evaluation did not affect the conclusions of the National Toxicology Program (2000) study.]

3.3 Administration with known carcinogens and other modifying factors

3.3.1 *Mixtures of fumonisins*

Rat: Groups of six male Fischer rats, 10 days of age, were given an intraperitoneal injection of *N*-nitrosodiethylamine (NDEA) (15 mg/kg bw) and fed ground maize containing culture material of *F. proliferatum* (50 mg/kg fumonisin B_1 in the diet) or nixtamalized [calcium hydroxide-treated] corn culture material (8–11 mg/kg hydrolysed fumonisin B_1 (HFB_1)), in the absence or presence of nutrient supplementation for a period of 30 days, at which time animals were killed. No aflatoxins were detected in the maize, while nutritional modulation stimulated the toxic effects in the rats treated with the nixtamalized and untreated maize cultures in rats. Hepatocellular adenomas developed in 83% and 14% of the rats on diets containing the untreated and nixtamalized maize cultures, respectively. Cholangiomas were induced in 33% of the animals in both groups (with or without nixtamalization) (Hendrich *et al.*, 1993). [The Working Group noted the small number of animals.]

3.3.2 *Purified fumonisins*

Mouse: Three groups of 15 female SENCAR mice, seven weeks of age, were treated with a single application of 390 nmol 7,12-dimethylbenz[*a*]anthracene (DMBA) on their shaven backs. After one week, fumonisin B_1 was applied at doses of 0, 1.7 or 17 nmol twice a week and continued for 20 weeks. The highest dose of fumonisin B_1 induced skin tumours in all the mice with an average of 3.6 tumours per animal. In a similar experiment, groups of 10 female SENCAR mice were given a single intraperitoneal injection of 1.8 mg NDEA per mouse followed by fumonisin B_1 treatment (0.0025 % in the drinking-water) for 20 weeks. Lung tumours were found in 90% of the mice treated with NDEA and fumonisin B_1, while none were observed in the mice treated with NDEA alone (Nishino *et al.*, 2000).

Rat: In an initiation/promotion study in male BD IV rats, the promotional activity of fumonisin B_1 was tested with *N*-nitrosomethylbenzylamine as initiator. Fumonisin B_1 did not show any activity as a tumour promoter in the oesophagus over a 48-week period (Wild *et al.*, 1997).

Rainbow trout: Groups of 150 three-month-old rainbow trout fry were fed diet containing 0, 3, 23 or 104 mg/kg fumonisin B_1 (> 90% pure) for 34 weeks. No liver tumours were seen when the fish were killed at 60 weeks. Groups of 150 three-month-old rainbow trout fry pretreated with 100 mg/kg aflatoxin B_1 were fed diets containing 0, 3, 23 or 104 mg/kg fumonisin B_1 for 42 weeks. At 60 weeks, promotion of liver tumours was seen at 23 mg/kg (61% of fish had gross or confirmed tumours) and 104 mg/kg (74% of fish had gross or confirmed tumours). Groups of three-month-old rainbow trout fry pretreated with *N*-methyl-*N'*-nitro-*N*-nitrosoguanidine (MNNG; 35 mg/kg) were fed diets containing 0, 3, 23 or 104 mg/kg fumonisin B_1. At 60 weeks, promotion of liver tumours was seen in fish given 104 mg/kg fumonisin B_1 (55%) compared with 33% of fish treated with MNNG only (Carlson et al., 2001).

4. Other Data Relevant to an Evaluation of Carcinogenicity and its Mechanisms

4.1 Absorption, distribution, metabolism and excretion

4.1.1 *Humans*

No studies of the absorption, distribution, metabolism and excretion of fumonisin B_1 in humans have been reported. Chelule *et al.* (2001) measured fumonisin B_1 in staple maize and in faeces in rural and urban populations in KwaZulu Natal, South Africa. Faecal concentrations were of the same order of magnitude as those in the maize consumed.

4.1.2 *Experimental systems*

Studies have been conducted with fumonisins B_1 and B_2 biosynthesized using deuterated or ^{14}C-labelled methionine, resulting in fumonisins labelled at C-12 and C-16 or C-21 and C-22 (the methyl groups and adjacent carbons; see section 1.1.2), respectively (Plattner & Shackelford, 1992; Alberts *et al.*, 1993). 1,2-[^{14}C]Acetate has also been used, resulting in fumonisin labelled uniformly along the backbone with some label in the two tricarboxylic acid side-chains (Blackwell *et al.*, 1994). In these studies, the two forms of [^{14}C]fumonisin B_1 had specific activities of 36 and 650 μCi/mmol, respectively, and radiochemical purities of > 95%.

The kinetics of absorption of fumonisin B_1 and of fumonisin B_2 in rats are similar, involving rapid distribution and elimination (Shephard *et al.*, 1995c). In vervet monkeys (*Cercopithecus aethiops*), the bioavailability of fumonisin B_2 may be less than that of fumonisin B_1 and proportionally less fumonisin B_2 is excreted in bile (Shephard & Snijman, 1999).

The quantity of fumonisin B_1 detected in plasma after oral administration to pigs, laying hens, vervet monkeys, dairy cows and rats was very low. In rats (BD IX, Fischer 344, Sprague-Dawley, Wistar) given [^{14}C]fumonisin B_1 orally, accumulation of radioactivity in tissues was also very low. This demonstrates that absorption is very poor (< 4% of dose) (Shephard *et al.*, 1992b,c; Norred *et al.*, 1993; Shephard *et al.*, 1994c). Fumonisins are also poorly absorbed (2–< 6% of dose) in vervet monkeys, dairy cows and pigs (Prelusky *et al.*, 1994; Shephard *et al.*, 1994a,b; Prelusky *et al.*, 1996b). In orally dosed laying hens and dairy cows, systemic absorption based on plasma levels and accumulation of radioactivity in tissues was estimated to be < 1% of the dose (Scott *et al.*, 1994; Vudathala *et al.*, 1994; Prelusky *et al.*, 1995).

In rats and pigs given [^{14}C]fumonisin B_1 via the diet or by gavage, ^{14}C was distributed to various tissues, with the liver and kidney containing the highest concentration of radiolabel (Norred *et al.*, 1993; Prelusky *et al.*, 1994, 1996b). In chickens given a single oral dose of [^{14}C]fumonisin B_1, trace amounts of radioactivity were recovered in tissues, but no residues were detectable in eggs laid during the 24-h period after dosing (Vudathala *et al.*, 1994). No fumonisin B_1 or aminopentol hydrolysis products were recovered in milk from cows that had received an oral dose of fumonisin B_1 (Scott *et al.*, 1994). In pregnant rats dosed intravenously with [^{14}C]fumonisin B_1, approximately 14% and 4% of the dose was recovered in liver and kidney, respectively, after 1 h. In contrast, the uteri contained 0.24–0.44%, individual placentae contained 0–0.04% and total fetal recovery of radioactivity was ≤ 0.015% of the dose per dam (Voss *et al.*, 1996b).

When [^{14}C]fumonisin B_1 was administered by intraperitoneal or intravenous injection to rats (BD IX, Sprague-Dawley, Wistar), initial elimination (subsequent to the distribution phase) was rapid (half-life, approximately 10–20 min) with little evidence of metabolism (Shephard *et al.*, 1992b; Norred *et al.*, 1993; Shephard *et al.*, 1994c). In rats, the elimination kinetics based on intraperitoneal or intravenous dosing of fumonisin B_1 are consistent with a one- (Shephard *et al.*, 1992b) or two-compartment model (Norred *et al.*, 1993). However, one study using Wistar rats dosed orally with fumonisin B_1 indicated that the kinetics were probably best described by a three-compartment model (Martinez-Larranaga *et al.*, 1999), as was the case in swine (see below).

In vervet monkeys, as in rats, the radioactivity was widely distributed and rapidly eliminated (mean half-life, 40 min) after intravenous injection of [^{14}C]fumonisin B_1 (Shephard *et al.*, 1994a). The elimination kinetics after oral dosing in non-human primates have not been determined; however, peak plasma levels of fumonisin B_1 and B_2 occurred between one and several hours after a gavage dose of 7.5 mg/kg bw in vervet monkeys and the plasma concentrations ranged from 25–40 ng/mL for fumonisin B_2 to nearly 210 ng/mL for fumonisin B_1 (Shephard *et al.*, 1995b; Shephard & Snijman, 1999).

In pigs, clearance of [^{14}C]fumonisin B_1 from blood after an intravenous injection was best described by a three-compartment model (half-lives, 2.2, 10.5 and 182 min, respectively, averaged over five animals). Cannulation of the bile duct (which prevents enterohepatic circulation) resulted in much more rapid clearance, which was best described by a two-compartment model. A similar effect of bile removal was observed whether the dosing was intravenous or intragastric. The elimination half-life in pigs dosed intragastrically without bile removal was 96 min (averaged over four animals). The studies with pigs clearly show the importance of enterohepatic circulation of fumonisin B_1 in pigs. As with rats, over 90% of radioactivity was recovered in the faeces, with less than 1% recovered in urine after an oral dose of [^{14}C]fumonisin B_1 (Prelusky *et al.*, 1994).

After intraperitoneal injection in rats, fumonisin B_1 was excreted unchanged in bile (Shephard *et al.*, 1994c). In vervet monkeys after intravenous injection, there was evidence of metabolism to partially hydrolysed fumonisin B_1 and to a much lesser extent the fully hydrolysed aminopentol backbone in faeces. In urine, 96% of the radioactivity was recovered as fumonisin B_1 (Shephard *et al.*, 1994a). In further experiments, it was shown that metabolism was likely to be mediated by the bacteria in the gut, since partially hydrolysed and fully hydrolysed fumonisin B_1 were recovered in faeces but not bile of vervet monkeys (Shephard *et al.*, 1995b).

In-vitro studies using primary rat hepatocytes with microsomal preparations (Cawood *et al.*, 1994) and with a renal epithelial cell line (Enongene *et al.*, 2002a,b) indicated that there was no metabolism of fumonisin B_1 in these systems.

In rats given three oral doses of [^{14}C]fumonisin B_1 at 24-h intervals, the specific radioactivity in liver and kidney increased with each successive dose and remained unchanged for at least 72 h after the last dose (Norred *et al.*, 1993). In pigs, it was estimated that exposure to dietary fumonisin B_1 at 2–3 mg/kg feed would require a withdrawal period of at least two weeks for the [^{14}C]radiolabel to be eliminated from liver and kidney (Figure 1; Prelusky *et al.*, 1996b). Fumonisins B_1, B_2 and B_3 and aminopolyol hydrolysis products were detected in the hair of vervet monkeys exposed to fumonisin B_1 in the feed and of Fischer rats after oral exposure to culture material of *F. verticillioides* containing fumonisins (Sewram *et al.*, 2001).

Fumonisins do not appear to be metabolized in animal systems *in vitro* or *in vivo*, apart from some evidence for removal of the tricarboxylic acid side-chains. This is thought to be effected by the microbial flora of the gut.

4.1.3 *Comparison of humans and animals*

Several experiments have indicated that the rate of elimination of fumonisin B_1 is a function of body weight. In mice, elimination is very rapid, whereas fumonisin B_1 is predicted to be retained much longer in humans (Figure 2; Delongchamp & Young, 2001).

Figure 1. Percentage cumulative recovery of fumonisin B_1 in urine and faeces of swine fed [^{14}C]fumonisin B_1 (uniformly labelled) at 3 mg/kg diet from days 1 to 12, then 2 mg/kg diet from days 13 to 24 and clean feed from days 25 to 33

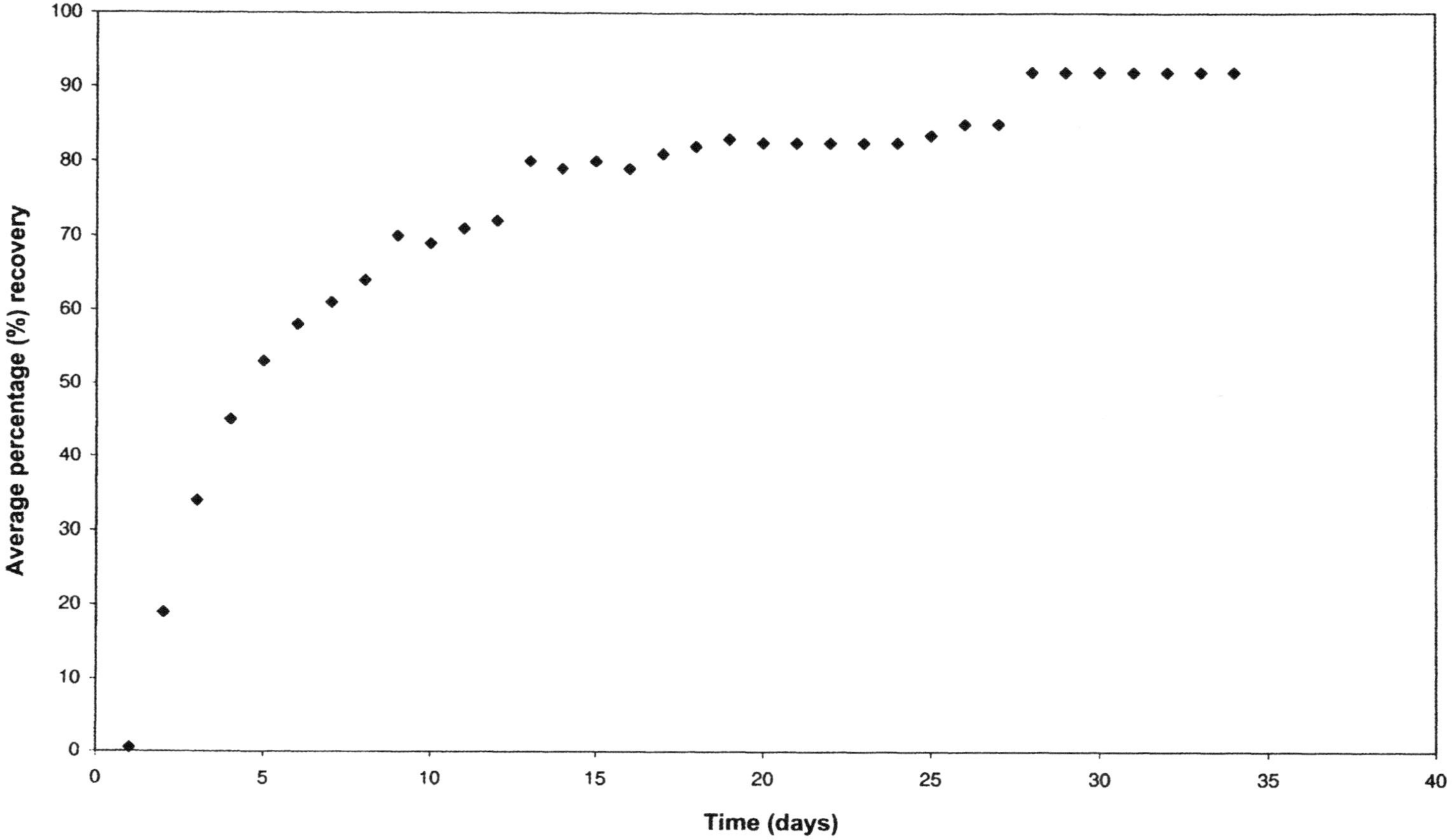

Adapted from Prelusky *et al*. (1996b)

Figure 2. Allometric relationship between body weight and fumonisin B_1 half-life for elimination

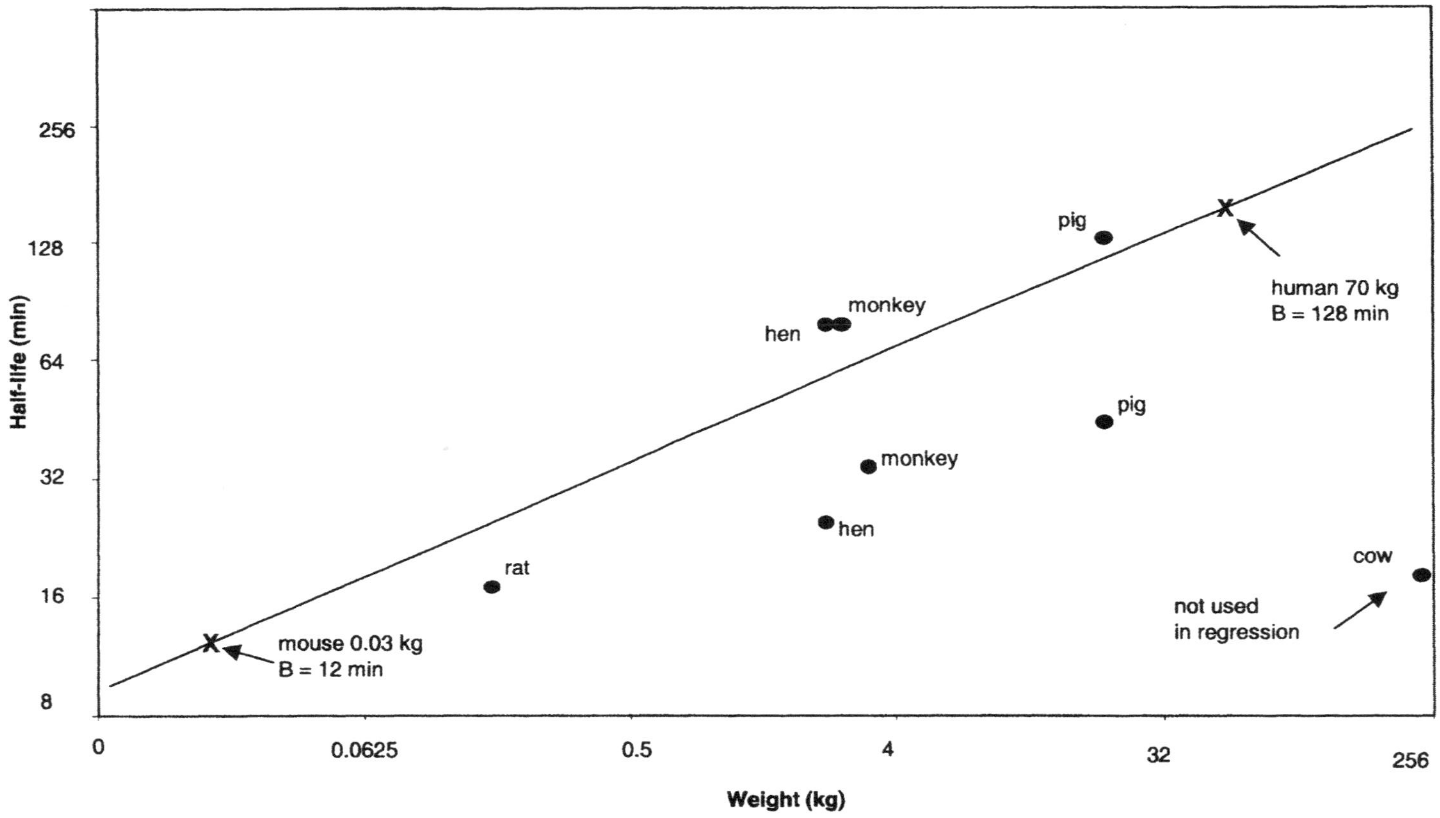

Adapted from Delongchamp & Young (2001)
The data for mouse and human are extrapolated from the linear regression model.

4.2 Toxic effects

4.2.1 *Humans*

As noted above, Chelule *et al.* (2001) reported that fumonisin levels in faeces were of the same order of magnitude as those in the maize consumed. In a study conducted in China, urine was collected from volunteers before and after consumption of a fumonisin B_1-contaminated diet for one month. The ratio of free sphinganine to free sphingosine (Sa/So) was increased threefold in the urine of the men, but was unchanged in that of the women. This increase was also apparent when the data were pooled for men and women and grouped into those individuals who had estimated intakes of fumonisin B_1 greater than or less than 110 μg/kg bw per day (Qiu & Liu, 2001). For a detailed discussion of the effects of fumonisin B_1 on sphingolipid metabolism, see section 4.5.1.

One report from India described gastric and other symptoms possibly associated with high exposures to fumonisins from consumption of rain-damaged mouldy sorghum or maize (Bhat *et al.*, 1997).

4.2.2 *Experimental systems*

Studies on culture material were reviewed in IARC (1993), WHO (2000) and in the background papers for the 56th Meeting of the Joint FAO/WHO Expert Committee on Food Additives (JECFA, 2001). Because, for most species, there are now adequate studies using fumonisin B_1 with reported purities of 96–98%, these will be emphasized here (Bondy *et al.*, 1996; Rotter *et al.*, 1996; National Toxicology Program, 2000).

The single-dose LD_{50} of fumonisin B_1 is unknown. Single gavage doses of 50, 100 and 200 mg/kg bw fumonisin B_1 significantly inhibited hepatocyte proliferation when given to male Fischer rats six hours after partial hepatectomy (Gelderblom *et al.*, 1994). In male Sprague-Dawley rats, intravenous injection of fumonisin B_1 at 1.25 mg/kg bw resulted in histological changes in the outer medulla of the kidney, with an increased number of mitotic figures and apoptosis followed by severe nephrosis (Lim *et al.*, 1996).

Equine leukoencephalomalacia (ELEM) syndrome is caused by ingestion of fumonisin B_1 in contaminated feed and is characterized by the presence of liquefactive necrotic lesions in the white matter of the cerebrum (Marasas *et al.*, 1988). The first symptoms are lethargy, head pressing and inability to eat or drink, followed by convulsions and death after several days. In addition to the brain lesions, histopathological abnormalities in liver and kidney have been reported in horses orally dosed with pure fumonisins (Kellerman *et al.*, 1990). Marasas *et al.* (1988) reported that high doses of fumonisin B_1 induced fatal hepatotoxicity with mild brain lesions, while low doses caused mild hepatotoxicity and severe brain lesions. Fatal liver disease in the absence of any brain lesions was induced in a mare by intravenous injection of large doses of fumonisin B_1, while gastric intubation of the mycotoxin had no effect (Laurent *et al.*, 1989b). However, signs of neurotoxicosis and liver lesions in the absence of elevated clinical chemistry parameters, and ELEM concurrent with significant liver disease have been observed in horses and ponies after ingestion of feeds naturally contaminated with fumonisins at low concentrations (Wilson *et al.*, 1992; Ross *et al.*, 1993). The development of brain lesions in the absence of major liver lesions does not preclude a contribution of biochemical dysfunction in non-brain tissue to the development of brain lesions. Length of exposure, level of contamination, individual animal differences, previous exposure or pre-existing liver impairment may all contribute to the appearance of the clinical disease (Ross *et al.*, 1993).

The lowest dietary dose observed to induce ELEM was 22 mg/kg fumonisin B_1 in a diet formulated with naturally contaminated maize screenings: one pony died of ELEM after consumption of contaminated diet for 235 days, of which the final 55 days' diet contained 22 ppm fumonisin B_1 (Wilson *et al.*, 1992). Analysis of feeds from confirmed cases of ELEM indicated that consumption of feed with a fumonisin B_1 concentration greater than 10 mg/kg diet is associated with increased risk of development of ELEM, whereas a concentration less than 6 mg/kg is not (Ross *et al.*, 1994). The minimum toxic dose of pure fumonisins is unknown.

In swine, fumonisin B_1 causes damage to the liver, lungs and cardiovascular and immune systems. Liver lesions have been induced with fumonisin-contaminated maize screenings at 1.1 mg/kg per day (fumonisins B_1 and B_2; 17 mg/kg fumonisin B_1 and 6 mg/kg fumonisin B_2 in the diet). Intravenous exposures resulted in changes similar to those recorded in rodents including necrosis and cell proliferation (Motelin *et al.*, 1994; Haschek *et al.*, 2001). When pure fumonisin B_1 was fed to Yorkshire swine at dietary levels of 0.1, 1 or 10 mg/kg (0.005, 0.052 or 0.496 mg/kg bw), apart from reduced organ weights (pancreas, adrenals), no histopathological signs of organ damage were observed. There were changes in sphingolipid ratios in lung, liver and kidney at the highest dose, as well as increased serum cholesterol (Rotter *et al.*, 1996). Fumonisin B_1 given to young adult swine at several doses up to 1 mg/kg in the diet resulted in changes in serum cholesterol and in altered carcass fat distribution at 0.05 mg/kg bw (Rotter *et al.*, 1997).

Lung oedema occurs in pigs following very high fumonisin B_1 exposure ($\geq$ 100 ppm in diet, or $\geq$ 16 mg/kg bw per day). Clinical signs of lung oedema typically occur 2–7 days after exposure, and usually include dyspnoea, weakness, cyanosis and death (Osweiler *et al.*, 1992; Haschek *et al.*, 2001). At necropsy, the animals exhibit varying degrees of interstitial and interlobular oedema, with pulmonary oedema and hydrothorax, with varying amounts of clear yellow fluid accumulating in the pleural cavity (Colvin & Harrison, 1992; Colvin *et al.*, 1993). Fumonisin B_1 is believed to be a negative osmotropic agent causing decreased cardiac contractility. It has been hypothesized that the cardiovascular alterations are a consequence of sphingosine-induced inhibition of L-type calcium channels. Pulmonary oedema results from left-sided heart failure (Smith *et al.*, 1996, 2000; Haschek *et al.*, 2001). Porcine pulmonary oedema was produced within 3–4 days after pigs started consuming a diet of culture material that provided 20 mg/kg bw fumonisin B_1 per day (Smith *et al.*, 1999). There are no published studies on pulmonary oedema induced by oral exposure to pure fumonisin B_1.

Because poultry are very resistant to fumonisins, toxicity studies all involve culture material obtained from fermentation of autoclaved corn inoculated with cultures of a fumonisin-producing fungus (WHO, 2000). Approximately 200 broiler chicks were fed *F. verticillioides* M-1325 culture material from hatching to 21 days. The concentrations of fumonisin B_1 were 75, 150, 225, 300, 375, 450 and 525 mg/kg feed. The total dietary fumonisin levels (B1 + B2 + B3) were reported as 89, 190, 283, 389, 481, 592 and 681 mg/kg feed. Broilers fed diets containing 89 and 190 ppm of the mixture showed no significant difference from controls in feed intake, weight gain, feed conversion, organ weights, haematology or clinical chemistry. Compared with controls, all chicks fed diets containing fumonisins had significantly increased ($p < 0.05$) serum sphinganine:sphingosine ratios. Histopathological examination of an extensive array of tissues showed no lesions or incidental findings in any treatment group. Isolated foci of hepatic necrosis with mild heterophil and macrophage infiltration, moderate diffuse hepatocellular hyperplasia, mild biliary hyperplasia and moderate to severe periportal granulocytic cell proliferation were noted only in broilers fed at least 283 mg/kg total fumonisins (Weibking *et al.*, 1993).

In an attempt to identify food-borne carcinogens contributing to the high incidence of oesophageal cancer in the Transkei region of the Eastern Cape Province, South Africa, various food mixtures, collected from households of local inhabitants, were fed to BD IX rats for 829–980 days. Diets consisted of maize, beans and a salt mixture with or without the addition of edible wild vegetables called *imifino* (Purchase *et al.*, 1975). Based on reported fumonisin levels in maize from the Transkei during six different seasons and an apparent mean dietary intake of the BD IX rats of 32 g/kg bw feed per day, a total fumonisin intake of 0.65–1.4 mg/kg per day was estimated (Rheeder *et al.*, 1992). Advanced toxic lesions in the liver included extensive bile duct proliferation, hepatocellular degeneration and development of hyperplastic liver nodules (Purchase *et al.*, 1975).

In male BD IX rats fed 1 g/kg fumonisin B_1 in the diet for 33 days, major changes seen in the liver included bile duct proliferation, fibrosis and hepatocyte nodules. Changes in kidney were mild and included fatty changes and scant necrosis in the proximal convoluted tubule. Gavage dosing with 238 mg/kg bw fumonisin B_1 per day caused the death of three of four rats and produced lesions in liver with minor changes in kidney. Severe disseminated acute myocardial necrosis and severe pulmonary oedema were observed in two rats. At lower dosages but with longer exposures (9–12 days), pathological changes were observed in liver. Early signs of bile duct proliferation and fibrosis, radiating from the portal areas, were noticed and the nuclei of a few hepatocytes were enlarged (Gelderblom *et al.*, 1988).

Additional feeding studies were performed using concentrations of 25–750 mg fumonisin B_1 per kg diet. Histopathological lesions in the liver (hepatocyte nodules) after 21 days were observed in rats that had received the high doses (750, 500 and 250 mg/kg diet), with severity decreasing with decreasing dose. A few necrotic cells were detected in the liver of rats treated with 50 mg fumonisin B_1 per kg diet, whereas no lesions were observed at the 25-mg/kg dietary level (Gelderblom *et al.*, 1994, 1996a).

Hepatocyte injury was investigated in male Fischer 344 rats fed a diet containing fumonisin B_1 at 250 mg/kg for five weeks. Fumonisin B_1 induced hepatocyte necrosis and apoptosis mainly in zone 3 of the liver lobule. Hepatocyte injury and death were reflected by desmin-positive hepatic stellate cell proliferation and marked fibrosis, with changes in architecture and formation of regenerative nodules. Oval cell proliferation was noted from week 2 and occurred in parallel with continuing hepatocyte mitotic activity. Nodules developed and, at later time points, oval cells were noted inside some of the nodules (Lemmer *et al.*, 1999a).

Male and female Fischer 344 rats were fed doses of approximately 12, 20, 28 or 56 mg/kg bw fumonisin B_1 per day for 28 days (National Toxicology Program, 2000). Body weight in both males and females was decreased at doses ≥ 20 mg/kg bw per day. The kidney was more sensitive to fumonisin B_1-induced changes in males than in females, but the liver was more affected in females than in males. The earliest cellular response in both liver and kidney was increased apoptosis accompanied by increased cell proliferation. Structural degeneration as a result of apoptosis was noted in both liver and kidney. In females, the lowest effective dose for bile duct hyperplasia and decreased liver weight was 56 mg/kg bw per day and that for liver degeneration and increased hepatocellular mitosis was 28 mg/kg bw per day. The lowest effective dose for increased hepatocellular apoptosis was 20 mg/kg bw per day. Decreased kidney weight, increased structural degeneration and increased renal tubule epithelial cell apoptosis were seen even at 12 mg/kg bw per day in males but only at ≥ 20 mg/kg per day in females (Howard *et al.*, 2001).

Male and female Sprague-Dawley rats were fed 15, 50 or 150 mg fumonisin B_1 per kg of diet over a period of four weeks. The estimated daily intake of fumonisin B_1 was 1.4, 4.4 and 13.6 mg/kg bw for males and 1.4, 4.1 and 13.0 mg/kg bw for females. In liver, mild histopathological changes were observed by light microscopy only in rats fed the high dose. Nephrotoxic changes were found in the proximal convoluted tubules in males fed diets containing ≥ 15 mg fumonisin B_1 per kg and in females at diets containing ≥ 50 mg/kg. Serum levels of enzymes, cholesterol and triglycerides were increased at dietary fumonisin B_1 concentrations of 150 mg/kg (Voss *et al.*, 1993, 1995a).

Male and female Sprague-Dawley rats were given oral doses of 1, 5, 15, 35 or 75 mg/kg bw fumonisin B_1 daily for 11 days. Histopathological changes in the kidneys were similar to those seen in other studies, males being more sensitive, with a lowest effective dose of 1 mg/kg bw per day versus 5 mg/kg bw for females. Hepatotoxicity was associated with reduced liver weight, as well as increased vacuolization of adrenal cortex cells, which occurred in female and male rats treated at doses ≥ 15 mg/kg bw per day. Elevated cholesterol concentrations in serum were observed in female rats at doses ≥ 5 mg/kg bw per day, but only at the highest dose (75 mg/kg bw per day) in males. Serum glucose was significantly reduced and alanine transaminase, aspartate transaminase and creatinine were significantly elevated at the highest dose in males, and in the two highest-dose groups in females. Single-cell necrosis and mitosis were seen at doses of 15 to 75 mg/kg bw per day in both males and females. Mild lymphocytosis in the thymic cortex of the fumonisin B_1-treated rats was evident at ≥ 5 mg/kg bw per day in males and at 75 mg/kg bw per day in females (Bondy *et al.*, 1996, 1998).

Male RIVM rats were treated with fumonisin B_1 at 0.19, 0.75 or 3 mg/kg bw per day by gavage for 28 days. The treatment had no effect on body weight, but kidney weight was significantly reduced in the highest-dose group. Increased apoptosis in the medulla of the kidney and renal tubule cell death were seen in the mid- and high-dose groups but not at the 0.19-mg/kg bw dose. There was no histological indication of liver toxicity (de Nijs, 1998).

Male BALB/c mice were given five subcutaneous injections of fumonisin B_1 in sterile water over a period of five days at doses of 0.25, 0.75, 2.25 or 6.75 mg/kg bw per day. Apoptosis was detected in the liver at doses above 0.75 mg/kg bw and in the kidneys at all doses. The relative kidney weights (% of bw) were decreased at all dose levels except 0.75 mg/kg bw per day, while no effect was observed with respect to relative liver weights (Sharma *et al.*, 1997; Tsunoda *et al.*, 1998).

In adult male and female B6C3F$_1$ mice treated with daily doses of 1, 5, 15, 35 or 75 mg/kg bw fumonisin B_1 by gavage during 14 days, hepatotoxicity was observed in both sexes, but kidney toxicity was seen only in females. Females were more sensitive than males to the effects in liver and kidney. Single-cell necrosis was detected in the liver at doses ≥ 35 mg and ≥ 15 mg/kg bw per day in males and females, respectively. Hepatocyte mitosis was elevated in males at 75 mg/kg bw per day and in females at ≥ 5 mg/kg bw per day. Mild single-cell necrosis in the kidney was detected in the cortical and medullary tubules only in female mice at 15–75 mg/kg bw per day. Males (≥ 35 mg/kg bw per day) and females (≥ 15 mg/kg bw per day) exhibited moderate diffuse vacuolization of adrenal cortical cell cytoplasm. Mild thymic cortical lymphocytolysis was noticed in a few female mice that received ≥ 35 mg/kg bw per day (Bondy *et al.*, 1997).

Male and female B6C3F$_1$ mice were fed 99, 163, 234 or 484 mg fumonisin B_1 per kg of diet over a period of 28 days (National Toxicology Program, 2000). The average daily intake of fumonisin B_1 was slightly higher in females than in males. Males developed liver lesions at 484 mg/kg in the diet, while such changes were seen in females at all dose levels. The lowest effective doses with respect to liver pathology were 93 mg/kg bw per day in males and 24 mg/kg bw per day in females.

Fumonisin B_1 at 1, 3, 9, 27 or 81 mg/kg of diet was fed to male and female B6C3F$_1$ mice over a period of 90 days. The mean daily intake of fumonisin B_1 was 0.3, 0.8, 2.4, 7.4 or 23 mg/kg bw for males and 0.3, 1, 3, 9.7 or 29 mg/kg bw for females. Serum levels of cholesterol, alanine transaminase, aspartate transaminase, alkaline phosphatase, lactate dehydrogenase and total bilirubin were significantly increased in the high-dose female mice, while no effect was reported in male mice. The clinical findings paralleled histological observations in the liver of the female mice, which were mainly restricted to the centrilobular zone. No lesions were reported in the kidneys of the mice (Voss *et al.*, 1995b).

4.2.3 *Related studies*

Alkaline hydrolysis of fumonisins B_1 and B_2 removes the carboxylic acid side-chains producing hydrolysed fumonisin B_1 (HFB_1) and hydrolysed fumonisin B_2 (HFB_2). HFB_1 and HFB_2 are major breakdown products in nixtamalized corn. Feeding nixtamalized *F. verticillioides* corn culture material containing 58 mg/kg HFB_1 to rats during four weeks caused lesions in the liver and kidney that were indistinguishable from those caused by feeding culture material that was not nixtamalized and contained predominantly fumonisin B_1 (71 mg/kg). Liver lesions included apoptosis, sloughing of epithelial cells into the limina and an increased nucleus-to-cytoplasm ratio. However, the extent and severity of the liver lesions, the decrease in weight gain and the elevation of free sphingoid bases were less in animals that received the nixtamalized culture material than in rats that received non-treated material, even though the molar concentration of HFB_1 (58 μg/g [143 nmol/g]) was greater than that of fumonisin B_1 (71 μg/g [98.5 nmol/g]) in the culture material diets that had not been nixtamalized (Voss *et al.*, 1996c).

Male Fischer 344 rats (8–10 animals per group) were treated by gavage with 1.4, 4.2, 14.3, 21.0 or 35.0 mg/kg bw fumonisin B_1 (92–95% pure) per day for 14 days. After 14 days, degenerative changes in the liver were seen in the two high-dose groups and included apoptosis, mild proliferation of oval cells and increased mitotic figures. One week after the start of the fumonisin B_1 treatment, separate groups were treated either intravenously with 100 μmol/kg bw lead nitrate, by partial hepatectomy or with a single gavage dose of 2 mL/kg bw carbon tetrachloride to stimulate cell proliferation. Three weeks after the fumonisin B_1 treatment, rats were subjected to 2-acetylaminofluorene (2-AAF)/partial hepatectomy or 2-AAF/carbon tetrachloride promotion treatmentsduring four days and the incidence of placental glutathione *S*-transferase (GSTP)-positive lesions was monitored two weeks later. In groups receiving partial hepatectomy or carbon tetrachloride during the initiation phase followed by 2-AAF/carbon tetrachloride or 2-AAF/partial hepatectomy, respectively, enhanced induction of GSTP-positive lesions was observed in the high-dose groups. This effect was not seen in the group treated with the mitogen lead nitrate followed by 2-AAF/partial hepatectomy (Gelderblom *et al.*, 2001b).

A total of 38 male Fischer 344 rats were divided into four groups and fed 250 mg fumonisin B_1 per kg diet (92–95% pure; fumonisin intake, 16.4 mg/kg bw per day) for five weeks in the absence or presence of 1–2% dietary iron in a modified American Institute of Nutrition (AIN) 76 diet. One group received dietary iron but no fumonisin B_1 and one group received the control diet. The dietary iron treatment included one week at 2%, one week on control diet followed by two weeks at 1% to avoid excessive toxic effects. Two animals in each treatment group and one control rat were killed at three and four weeks, and the remaining rats (six per group) were sacrificed after five weeks. Hepatocyte necrosis, mitosis and apoptosis and GSTP-positive hepatic lesions were noted in the fumonisin B_1-treated group (5.34 ± 1.42 lesions/cm^2), while the fumonisin

B_1/iron-treated group showed fewer GSTP-positive lesions (1.50 ± 0.52 lesions/cm^2). The concentration of alanine transaminase in serum was increased, reflecting hepatotoxicity, in both the fumonisin B_1- and fumonisin B_1/iron-treated groups. Body weight gain was decreased in the fumonisin B_1-, iron- and fumonisin B_1/iron-treated groups, while relative liver weights were decreased only in the fumonisin B_1-treated rats. Lipid peroxidation in the liver was increased in fumonisin B_1/iron- and iron-treated rats (Lemmer *et al.*, 1999b).

Male Fischer 344 rats (5–8 per group) were fed modified AIN-76 diets containing 250 mg fumonisin B_1 per kg diet for three weeks. Other groups received 17 μg/kg bw aflatoxin B_1 per day by gavage for 14 days (total dose, 240 μg/kg bw) or a single intraperitoneal injection of 200 mg/kg bw *N*-nitrosodiethylamine (NDEA). The three groups were compared using the resistant hepatocyte model (Semple-Roberts *et al.*, 1987) which consisted of treatment with 20 mg/kg bw 2-AAF by gavage on each of three consecutive days followed by partial hepatectomy on day 4. The induction of GSTP-positive lesions was monitored three weeks after the latter treatment. GSTP-positive lesions were increased by treatment with both fumonisin B_1 and aflatoxin B_1 in combination with the 2-AAF/partial hepatectomy promoting stimulus, but to a much lesser extent than by treatment with NDEA (ratio 1:3:10 for fumonisin B_1, aflatoxin B_1 and NDEA). In a second set of experiments, the separate and combined effects of aflatoxin B_1 and fumonisin B_1 on the induction of GSTP-positive lesions were determined in the absence of the 2-AAF/partial hepatectomy promoting treatment. When rats were treated sequentially with aflatoxin B_1 followed three weeks later by fumonisin B_1, a synergistic interaction was found based on increased numbers and size of the GSTP-positive lesions in the liver (total of 72, 5 and 1.6 lesions consisting of more than 5 cells/cm^2 for aflatoxin B_1 + fumonisin B_1, aflatoxin B_1 and fumonisin B_1, respectively (Gelderblom *et al.*, 2002).

4.3 Reproductive and developmental effects

4.3.1 *Humans*

A specific role for fumonisins in the development of neural tube defects was suggested after the appearance of a cluster of such defects in Texas associated with consumption of corn from the heavily fumonisin-contaminated 1989 corn crop (Hendricks, 1999). More recent studies have shown that fumonisin B_1 inhibits folate metabolism in cultured cells (Stevens & Tang, 1997). The relationship between folate deficiency and neural tube defects is well established, but there are no specific studies to confirm the association with exposure to fumonisins.

4.3.2 *Experimental systems*

(*a*) *Developmental and reproductive toxicity studies*

Pregnant CD CRL rats were given oral doses of 0, 1.875, 3.75, 7.5 or 15 mg/kg bw fumonisin B_1 per day on gestation days 3–16. Feed consumption and body weight gain were significantly decreased at the 15-mg/kg bw dose. Fetal body weights at day 17 were similar in control and treated groups, but in day-20 fetuses, female weight and crown–rump length were significantly decreased at the highest dose. In day-17 animals, dose-related increases in sphinganine/sphingosine ratios were seen in maternal livers, kidneys and serum. Sphinganine/sphingosine ratios in maternal brains were not affected, nor were those of fetal kidneys, livers or brains (Collins *et al.*, 1998a). In a similar study using dose levels of 0, 6.25, 12.5, 25 or 50 mg/kg bw fumonisin B_1 per day, maternal toxicity and fetal toxicity were seen at the 50-mg/kg bw dose. The effects on the fetuses included increased numbers of late deaths, decreased body weight and crown–rump length and increased incidence of hydrocephalus and skeletal anomalies. Dose-related increases in sphinganine/sphingosine ratios were seen in maternal livers, kidneys, serum and brain, but not in fetal livers, kidneys or brain (Collins *et al.*, 1998b). The data from these two studies suggest either that fumonisin B_1 does not cross the placenta, the observed fetal toxicity being a secondary consequence of maternal toxicity, or that a potential direct effect of fumonisin B_1 on fetal development is not related to changes in sphinganine/sphingosine ratios in the fetuses.

Groups of pregnant Fischer 344 rats were dosed by gavage daily on gestation days 8 to 12 with 30 or 60 mg/kg bw purified fumonisin B_1 or with a fat-soluble extract of *F. proliferatum*/corn culture that would provide a dose of approximately 60 mg fumonisin B_1 per kg body weight. Lower fetal litter weight and delayed ossification were observed in the rats given 60 mg/kg bw fumonisin B_1, but not in rats given 30 mg/kg bw fumonisin B_1 or the fat-soluble extract (Lebepe-Mazur *et al.*, 1995).

The neurobehavioural and developmental effects of fumonisin B_1 were studied in Sprague-Dawley rats treated by gavage on gestation days 13–20 with 0, 0.8 or 1.6 mg/kg bw fumonisin B_1 obtained from culture material or 0, 1.6 or 9.6 mg/kg bw purified fumonisin B_1. There was no effect on reproductive outcomes or offspring body weight through adulthood in either experiment. Some effects on acoustic startle response and play behaviour were found in male but not in female offspring prenatally treated with any dose of purified fumonisin B_1. Fumonisin B_1 treatment had no effect on complex maze performance or open field and running wheel activity (Ferguson *et al.*, 1997).

Pregnant Charles River CD-1 mice were treated orally with a semipurified extract of *F. verticillioides* culture providing 0, 12.5, 25, 50 or 100 mg/kg bw fumonisin B_1 daily on gestation days 7–15. Maternal mortality was observed at doses of 50 and 100 mg/kg bw. Signs of liver damage and decreased maternal body weight gain were observed at ≥ 25 mg/kg bw. The percentage of implants resorbed was increased at all doses in a dose-dependent manner. The number of live fetuses per litter and the mean fetal body weight were decreased and the incidence of ossification deficits, short and wavy ribs and hydrocephalus was increased at the 50- and 100-mg/kg bw doses (Gross *et al.*, 1994).

Pregnant Charles River CD-1 mice were administered 0 to 100 mg/kg bw pure fumonisin B_1 by gavage on gestation days 7–15. Doses ≥ 25 mg/kg bw induced maternal liver lesions and a dose-dependent increase in the incidence and severity of hydrocephalus in the fetuses. Reduced fetal body weight was found at ≥ 50 mg/kg bw, while increased frequency of resorptions and decreased litter size were present only at 100 mg/kg bw. Doses ≥ 25 mg/kg bw increased the sphinganine/sphingosine ratios in maternal but not fetal livers. The effects of fumonisin B_1 on the fetuses and the alteration of the sphinganine/sphingosine ratio in maternal but not fetal liver suggest that the effects of fumonisin B_1 on the fetuses are not mediated by changes in sphinganine/sphingosine ratios in the fetuses. The association with effects on the maternal liver may indicate that developmental effects are mediated by maternal hepatotoxicity (Reddy *et al.*, 1996).

Six groups of Syrian hamsters were dosed with 0–18 mg/kg bw purified fumonisin B_1 by gavage daily on days 8–12 of gestation and killed on day 15. The treatment caused fetal death, decreased fetal body weight and skeletal variations consistent with delayed development in a dose-dependent manner, without causing maternal toxicity (Penner *et al.*, 1998).

In timed-bred Syrian hamsters dosed daily with 0–12 mg/kg bw fumonisin B_1 by gavage on gestation days 8–10 or 12, reduced maternal weight gain was observed at doses ≥ 8 mg/kg bw. Maternal aspartate transaminase and total bilirubin, used as indices of maternal hepatotoxicity, showed no significant difference between groups. At doses higher than 2 mg/kg bw fumonisin B_1, there was an increased incidence of prenatal loss (death and resorptions). At 12 mg/kg bw, all litters were affected and 100% of the fetuses were dead and resorbing (Floss *et al.*, 1994a).

A significant increase in litters with fetal deaths occurred in Syrian hamsters given 18 mg/kg bw purified fumonisin B_1 or culture-extracted fumonisins (18 mg fumonisin B_1 plus 4.5 mg fumonisin B_2) by gavage on gestation days 8 and 9. There were no clinical signs of maternal intoxication (Floss *et al.*, 1994b).

New Zealand White rabbits were dosed by gavage on gestation days 3–19 with purified fumonisin B_1 at 0.1, 0.5 or 1.0 mg/kg bw. Maternal lethality occurred at the 0.5- and 1.0-mg/kg bw doses (10–20%), but there was no difference in maternal weight gain during pregnancy. Fetal weight and liver and kidney weights were decreased at 0.5 and 1.0 mg/kg bw. Increased sphinganine/sphingosine ratios were found in maternal serum, liver and kidney, but there was no significant effect of fumonisin B_1 on the sphinganine/sphingosine ratio in fetal brain, liver or kidney (LaBorde *et al.*, 1997).

Diet formulated with culture material of *F. verticillioides* strain MRC 826 to provide 0, 1, 10 or 55 mg fumonisin B_1 per kg diet was fed to male and female rats beginning 9 and 2 weeks before mating, respectively, and continuing throughout the mating, gestational and lactational phases of the study. Nephropathy was found in males at dietary doses of ≥ 10 mg/kg and in females fed 55 mg/kg diet. No significant reproductive effects were found in males or dams and fetuses examined on gestation day 15, or dams and litters on postnatal day 21. Litter weight gain in the 10- and 55-mg/kg groups was slightly decreased; however, gross litter weight and physical development of offspring were not affected. Increased sphinganine/sphingosine ratios were found in the livers of dams from the high-dose group on gestation day 15. However, sphinganine/sphingosine ratios in abdominal slices containing liver and kidney of fetuses from the control and high-dose groups did not differ. In an additional experiment, two dams were given an intravenous injection of 101 μg [^{14}C]fumonisin B_1 on gestation day 15. After 1 h, about 98% of the dose had disappeared from the maternal blood, but only negligible amounts of radioactivity were found in the fetuses (Voss et al., 1996b).

(*b*) *Mechanistically oriented developmental toxicity studies*

Doses of 0.8 or 8 mg/kg bw of fumonisin B_1 were given subcutaneously to male Sprague Dawley rats on postnatal day 12. Brain tissue and blood were collected at ten time points up to 24 h after fumonisin B_1 administration. The sphinganine levels in brain and plasma showed dose-dependent increases; the brain sphinganine level during the 24 h was much higher than plasma sphinganine, with an area under the concentration–time curve (AUC) ratio of 40:1. In addition, fumonisin B_1 was found in the brain tissue after the higher dose. These data indicate that alterations of the brain sphinganine levels are the result of a direct action of fumonisin B_1 on the brain rather than transport of peripheral sphinganine to the brain (Kwon *et al.*, 1997a).

Subcutaneous dosing of Sprague-Dawley rats with 0.4 or 0.8 mg/kg bw fumonisin B_1 from postnatal day 3 to day 12 resulted in reduced body weight gain and decreased survival. Both sphinganine concentration and sphinganine/sphingosine ratios in the brain were increased at the higher dose. To investigate the effects of limited nutrition on sphinganine levels and myelinogenesis, rats were given 0.8 mg/kg bw fumonisin B_1 or subjected to limited nutrition (temporary removal from dam in the postnatal period) and compared with a saline control group. Sphinganine levels were increased in rats treated with 0.8 mg/kg fumonisin B_1, but not in those given limited nutrition. Myelin deposition was decreased in both the nutritionally limited and the fumonisin B_1-exposed rats. These data indicate that sphingolipid metabolism in developing rats is vulnerable to fumonisin B_1, while hypomyelination associated with fumonisin B_1 may be mediated by limited nutrition (Kwon *et al.*, 1997b).

Concentration- and time-dependent increases in sphinganine/sphingosine ratios were found in developing chick embryos after injection of 72 or 360 μg of fumonisin B_1 per egg. A close correlation was observed between disruption of sphingolipid metabolism and tissue lesions detectable by light microscopy (Zacharias *et al.*, 1996).

4.4 Genetic and related effects

4.4.1 *Humans*

No data were available to the Working Group.

4.4.2 *Experimental systems* (see Table 4 for references)

Fumonisin B_1 was not mutagenic in *Salmonella typhimurium*/microsome assays with strains TA100, TA98 or TA97 or in the SOS repair test with *Escherichia coli*, whereas a positive result was reported from a Mutatox® assay (luminescence induction) in the absence of metabolic activation. The compound did not induce unscheduled DNA synthesis in liver cells of rats *in vitro* or *in vivo* and no evidence for DNA-adduct formation with oligonucleotides *in vitro* was found; however, positive results were obtained in chromosomal aberration assays and in the micronucleus test with rat hepatocytes. Furthermore, evidence for induction of DNA damage by fumonisin B_1 was found with C6 rat brain glioma cells and human fibroblasts *in vitro*, and in spleen and liver cells isolated from fumonisin B_1-exposed rats. The in-vivo effect could be reversed with α-tocopherol and selenium (Atroshi *et al.*, 1999). Positive results were obtained in micronucleus assays *in vitro* with human-derived hepatoma (HepG2) cells but not with rat hepatocytes. In bone marrow of mice, an increase in formation of micronuclei was found after intraperitoneal injection of fumonisin B_1, whereas in a transformation study with a mouse embryo cell line, no response was observed.

4.5 Mechanistic considerations

There are no published data demonstrating that fumonisins form DNA adducts (WHO, 2002). Early studies indicated that fumonisin B_1 gave negative results in bacterial mutation assays and in the unscheduled DNA synthesis assay using primary rat hepatocytes (IARC, 1993). More recent studies with rat hepatocytes *in vitro* and *in vivo* using the Comet assay (DNA migration) have shown that fumonisin B_1 induces DNA damage in rodent- and human-derived cells (Atroshi *et al.*, 1999; Erlich *et al.*, 2002; Galvano *et al.*, 2002) and also chromosomal aberrations or micronucleus formation in human hepatoma cells (Erlich *et al.*, 2002) and primary rat hepatocytes (Knasmüller *et al.*, 1997). In some studies, addition of antioxidants reduced the amount of DNA damage, leading to the conclusion that oxidative stress is the cause of the DNA damage (Atroshi *et al.*, 1999; Mobio *et al.*, 2000b).

Numerous studies since the previous evaluation of fumonisins (IARC, 1993) have demonstrated that fumonisins alter signalling pathways that control cell behaviour. Thorough reviews of the biochemical and cellular mechanisms implicated in fumonisin B_1 toxicity and carcinogenicity are available (WHO, 2000; Allaben *et al.*, 2001; WHO, 2002) and form the basis for much of what follows.

Two biochemical modes of action proposed to explain fumonisin-induced diseases in animals, including cancer, invoke disruption of lipid metabolism as the initial phase. Both hypothesized mechanisms are supported by data on carcinogenicity in animal models (Gelderblom *et al.*, 20001a,b,c; Merrill *et al.*, 2001; Riley *et al.*, 2001; WHO, 2002) and are similar in many respects.

Table 4. Genetic and related effects of fumonisin B_1

Test system	Result[a]		Dose (LED or HID)[b]	Reference
	Without exogenous metabolic system	With exogenous metabolic system		
SOS repair, *Escherichia coli* PQ37	–	–	500 μg/plate	Knasmüller *et al.* (1997)
Escherichia coli rec strains, differential toxicity	–	–	500 μg/plate	Knasmüller *et al.* (1997)
Salmonella typhimurium TA100, TA102, TA97a, TA98, reverse mutation	–	–	5000 μg/plate[c]	Gelderblom & Snyman (1991)
Salmonella typhimurium TA100, reverse mutation	–	–	100 μg/plate	Park *et al.* (1992)
Salmonella typhimurium TA100, TA98, reverse mutation	–	–	500 μg/plate	Knasmüller *et al.* (1997)
Salmonella typhimurium TA100, TA102, TA98, reverse mutation	–	–	114 μg/plate	Aranda *et al.* (2000)
Salmonella typhimurium TA100, TA102, TA1535, TA1537, TA98, reverse mutation	NT	–[d]	200 μg/plate	Ehrlich *et al.* (2002)
Luminescence induction, *Vibrio fischeri*, mutation *in vitro*	+	NT	5	Sun and Stahr (1993)
DNA strand breaks (DNA-unwinding method), rat liver cells *in vitro*	+	NT	29	Sahu *et al.* (1998)
DNA strand breaks (Comet assay), C6 rat brain glioma cells *in vitro*	+	NT	2.2	Mobio *et al.* (2000a)
DNA adduct formation, oligonucleotides *in vitro*[e]	–	NT	360	Pocsfalvi *et al.* (2000)
Unscheduled DNA synthesis, rat primary hepatocytes, *in vitro*	–	NT	58	Gelderblom *et al.* (1992b)
Unscheduled DNA synthesis, rat primary hepatocytes, *in vitro*	–	NT	180	Norred *et al.* (1992)
DNA hypermethylation, C6 rat brain glioma cells, *in vitro*	+	NT	6.5	Mobio *et al.* (2000b)
Micronucleus formation, rat hepatocytes *in vitro*	–	NT	100	Knasmüller *et al.* (1997)
Chromosomal aberrations, rat hepatocytes *in vitro*	+	NT	1	Knasmüller *et al.* (1997)
Cell transformation, BALB/3T3 A31-1-1 mouse embryo cells	–	NT	1000	Sheu *et al.* (1996)
DNA damage (Comet assay), human hepatoma (HepG2) cells *in vitro*	+	NT	25	Ehrlich *et al.* (2002)
DNA damage (Comet assay), human fibroblasts *in vitro*	+	NT	7.2	Galvano *et al.* (2002)
Micronucleus formation, human hepatoma (HepG2) cells *in vitro*	+	NT	25	Ehrlich *et al.* (2002)

Table 4 (contd)

Test system	Result[a]		Dose (LED or HID)[b]	Reference
	Without exogenous metabolic system	With exogenous metabolic system		
DNA fragmentation, male Sprague-Dawley rat liver and spleen *in vivo*	+		1.55 × 1 iv	Atroshi *et al.* (1999)
Unscheduled DNA synthesis, male Fischer 344 rat hepatocytes *in vivo*	–		100 × 1 po	Gelderblom *et al.* (1992b)
Micronucleus formation, male CF1 mouse bone-marrow cells *in vivo*	+		25 × 1 ip	Aranda et al. (2000)

[a] +, positive; –, negative; NT, not tested
[b] LED, lowest effective dose; HID, highest ineffective dose; in-vitro tests, μg/mL; in-vivo tests, mg/kg bw/day; iv, intravenous; po, oral; ip, intraperitoneal
[c] A dose of 10 mg/plate was inactive in the pre-incubation assay and toxic in the plate-incorporation assay.
[d] Metabolic activation with S9 from human hepatoma (HepG2) cells
[e] Analysed by HPLC-mass spectrometry

The first proposed lipid-based mechanism involves inhibition of ceramide synthase (Wang *et al.*, 1991), a key enzyme in the biosynthesis of sphingolipids. In line with findings in human cell lines (Biswal *et al.*, 2000; Charles *et al.*, 2001), human primary cell cultures (Tolleson *et al.*, 1999), non-human primates (Van der Westhuizen *et al.*, 2001) and all other animals tested (reviewed in WHO, 2000, 2002), human exposure to fumonisins is also associated with evidence of disruption of sphingolipid metabolism (Qiu & Liu, 2001). Alterations in the free sphinganine/free sphingosine ratio, a consequence of ceramide synthase inhibition, are now used as a biomarker for exposure to fumonisins in domestic animals (Riley *et al.*, 1994a,b) and humans (Van der Westhuizen *et al.*, 1999; Qiu & Liu, 2001; Ribar *et al.*, 2001). Turner *et al.* (1999) reviewed potential problems of using sphingoid base ratios as a functional biomarker for exposure to fumonisin B_1 in humans.

The second biochemical mechanism proposes changes in polyunsaturated fatty acids and phospholipid pools (Gelderblom *et al.*, 1996b). This mechanism is supported by data from studies with rat liver (reviewed in WHO, 2002) and human cell lines (Pinelli *et al.*, 1999; Seegers *et al.*, 2000).

The cellular consequences of both biochemical modes of action provide support for a non-genotoxic mechanism of carcinogenicity. It is proposed that alterations in cell growth, death and differentiation due to disruption of lipid-mediated signalling and regulatory pathways lead to an imbalance between the rates of apoptosis and proliferation and that this imbalance is a critical determinant in the process of hepato- and nephrotoxicity and tumorigenesis in animal models (reviewed in WHO, 2002).

4.5.1 *Interference with sphingolipid metabolism*

(*a*) *Sphingolipid chemistry and function*

Sphingolipids are a highly diverse class of lipids found in all eukaryotic cells. The biological functions are equally diverse: the compounds serve as structural components required for maintenance of membrane integrity, as receptors for vitamins and toxins, as sites for cell–cell recognition and cell–cell and cell–substrate adhesion, as modulators of receptor function and as lipid second messengers in signalling pathways responsible for cell growth, differentiation and death (Merrill *et al.*, 1997).

(*b*) *Inhibition of ceramide synthase*

In every cell line and animal, plant or fungus in which it has been tested, fumonisin B_1 inhibits the coenzyme A (CoA)-dependent acylation of sphinganine and sphingosine via interaction with the enzyme sphinganine/sphingosine *N*-acyltransferase (ceramide synthase). This enzyme recognizes both the amino group (sphingoid-binding domain) and the tricarboxylic acid side-chains (fatty acyl-CoA domain) of fumonisin B_1 (Merrill *et al.*, 2001) (Figure 3).

Figure 3. Proposed model illustrating how the tricarboxylic acid groups and free amino group of fumonisin B_1 mimic the fatty acyl–coenzyme A (CoA) and free sphinganine substrates, respectively, in the active site of ceramide synthase. For fumonisin B_1, the interaction is primarily electrostatic; for the normal substrates, there are also hydrophobic interactions involving partitioning into the lipid bilayer

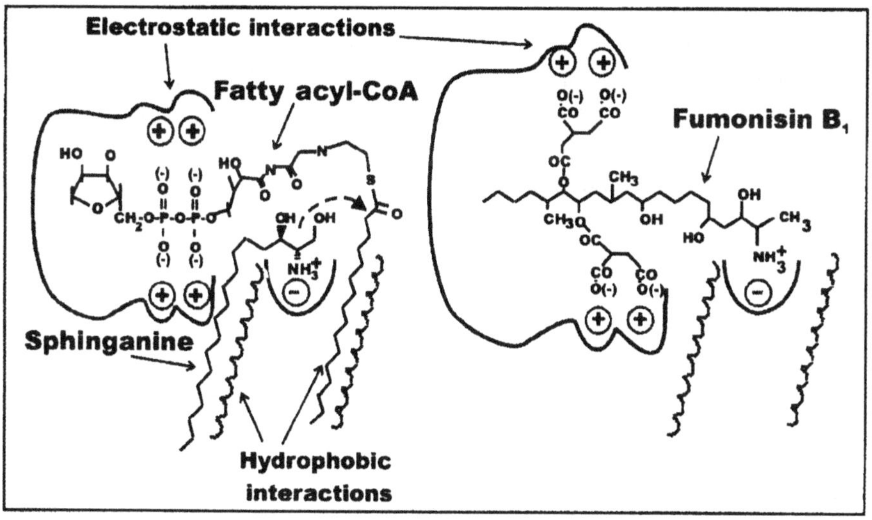

Modified from Merrill *et al.* (1996, 2001)

(*c*) *Sphingoid base accumulation*

When ceramide synthase is completely inhibited, either *in vitro* or *in vivo*, the intracellular sphinganine and sometimes sphingosine concentration increases rapidly. *In vivo* there is a close relationship between the amount of sphinganine accumulated and the expression of fumonisin toxicity in liver and kidney (Riley *et al.*, 1994a,b; Tsunoda *et al.*, 1998; Riley *et al.*, 2001; Voss *et al.*, 2001). Accumulated free sphingoid bases can persist in tissues (especially kidney) much longer than fumonisin B_1 (most recently shown by Enongene *et al.*, 2000; Garren *et al.*, 2001; Enongene *et al.*, 2002a,b). In urine from rats fed fumonisin B_1, nearly all the free sphinganine is recovered in dead cells. A sub-threshold dose in rats or mice can prolong the elevation of free sphinganine in urine or kidney caused by a higher dose (Wang *et al.*, 1999; Enongene *et al.*, 2002a,b). Fumonisin B_1-induced elevation of free sphingoid base levels and toxicity are both reversible, although elimination of free sphinganine from the liver is more rapid than from the kidney (Enongene *et al.*, 2000; Garren *et al.*, 2001; Enongene *et al.*, 2002a,b).

In ponies given fumonisin B_1-contaminated feed, changes in the sphinganine/sphingosine ratio in serum were seen before hepatic enzymes were notably elevated (Wang *et al.*, 1992; Riley *et al.*, 1997).

(*d*) *Sphingoid base metabolite, fatty acid and glycerophospholipid imbalances*

Inhibition of ceramide synthase by fumonisin B_1 can result in the redirection of substrates and metabolites to other pathways. For example, when sphinganine accumulates, it is metabolized to sphinganine 1-phosphate. The breakdown of sphinganine 1-phosphate results in the production of a fatty acid aldehyde and ethanolamine phosphate. Both products are redirected to other biosynthetic pathways, in particular increased biosynthesis of phosphatidylethanolamine (Badiani *et al.*, 1996). Disrupted sphingolipid metabolism leads to imbalances in phosphoglycerolipid, fatty acid metabolism and cholesterol metabolism via free sphingoid base- and sphingoid base 1-phosphate-induced alterations in phosphatidic acid phosphatase and monoacylglycerol acyltransferase. Thus, fumonisin B_1 inhibition of ceramide synthase can cause a wide spectrum of changes in lipid metabolism and associated lipid-dependent signalling pathways (reviewed in Merrill *et al.*, 2001).

(*e*) *Disruption of sphingolipid metabolism and in-vivo toxicity*

Disruption of sphingolipid metabolism, as shown by statistically significant increases in free sphinganine concentration, usually occurs at or below doses of fumonisin that cause liver or kidney lesions in short-term studies with rats, rabbits, mice, pigs, horses and many other species of animals and plants (reviewed in WHO, 2002). In some studies, significant increases in free sphingoid bases occur at doses that are higher than for other markers of hepatic effects (Liu *et al.*, 2001). Nevertheless, many studies show a close correlation between elevation of free sphinganine levels and increased apoptosis in liver and kidney (Riley *et al.*, 2001). For example, fumonisin B_1 induced an increase in the sphinganine/sphingosine ratio in kidney tissue and urine, which correlated with increased incidence of non-neoplastic and neoplastic kidney lesions in a long-term feeding study with Fischer 344/N Nctr rats (National Toxicology Program, 2000; Howard *et al.*, 2001). However, in livers of female B6C3F_1/Nctr mice, elevation of free sphinganine and the sphinganine/sphingosine ratio were significantly increased only after 3 and 9 weeks at 50 and 80 mg fumonisin B_1 per kg of diet, doses that also induced liver adenoma and carcinoma (National Toxicology Program, 2000).

Fumonisin B_1-induced hepatotoxicity in both female and male Sprague-Dawley rats *in vivo* was associated with free sphinganine concentrations in liver tissue of approximately 20 nmol/g fresh tissue. At the non-hepatotoxic dietary concentration of 50 ppm [mg/kg] fumonisin B_1, the free sphinganine levels were 12 and 4 nmol/g tissue in females and males, respectively. In contrast, nephrotoxic concentrations of fumonisin B_1 in the diet (50 ppm for females, 15 ppm for males) were associated with free sphinganine levels of 146 and 129 nmol/g tissue, respectively. For the female rats, this was a 10-fold increase of free sphinganine over the level measured at the non-nephrotoxic concentration of 15 ppm fumonisin B_1 in the diet (Voss *et al.*, 1996a). This is similar to the renal free sphinganine concentrations (100–134 nmol/g fresh tissue) associated with significantly increased nephropathy and the hepatic free sphinganine concentrations (5–15 nmol/g fresh tissue) associated with significantly increased hepatopathy in male BALB/c mice (Sharma *et al.*, 1997; Tsunoda *et al.*, 1998). Fumonisin B_1 is not a complete carcinogen in the rainbow trout model; however, a close correlation was reported between the elevated level of free sphinganine in liver and fumonisin B_1-mediated promotion of aflatoxin B_1-induced hepatocarcinogenicity (Carlson *et al.*, 2001).

(*f*) *Free sphingoid bases as functional biomarkers in humans*

Several studies have examined the use of the elevation of free sphinganine in human urine or blood as an indicator of exposure to fumonisin B_1, with mixed success. For example, van der Westhuizen *et al.* (1999) found no relationship between urine or serum sphingoid base levels and dietary fumonisin B_1 intake. However, in this study, the levels of fumonisin B_1 in the diet were low. A recent study conducted in China found that the free sphinganine to free sphingosine ratio was significantly greater in urine collected in households with an estimated fumonisin B_1 intake above 110 μg/kg bw/day (Qiu & Liu, 2001). A study conducted in the endemic nephropathy area of Croatia found statistically significant differences (compared with control groups) in free sphingoid bases in serum and urine of individuals living in the region but not affected by endemic nephropathy. However, in this study dietary exposure to fumonisin B_1 was not established, although exposure to a mycotoxin or environmental factor that impaired sphingolipid metabolism was suggested (Ribar *et al.*, 2001).

(*g*) *Sphingolipid metabolites and apoptosis*

Numerous studies using cultured cells have demonstrated sphingolipid-dependent mechanisms for inducing apoptosis. For example, accumulation of excess ceramide, glucosylceramide (Korkotian *et al.*, 1999) or sphingoid bases, or depletion of ceramide or more complex sphingolipids have all been shown to induce apoptotic or oncotic cell death (WHO, 2000; Merrill *et al.*, 2001; Riley *et al.*, 2001; WHO, 2002). Conversely, the balance between sphingosine 1-phosphate and ceramide is critical for signalling proliferation and cell survival (Spiegel, 1999). It can be expected that there will also be a diversity of alterations in cellular regulation resulting from imbalances in sphingolipid metabolite and product pools resulting from inhibition by fumonisin B_1 of ceramide synthase (Figure 4). This is best demonstrated by the numerous recent studies with fumonisin B_1 identifying cell processes that are ceramide-mediated; for example, the ability of fumonisin B_1 to protect oxidant-damaged cells from apoptosis and to alter the proliferative response (WHO, 2000; Riley *et al.*, 2001; WHO, 2002). In addition to the many studies cited previously, experiments with fumonisin B_1 have revealed new roles for de-novo ceramide production in inhibition of apoptosis and other ceramide-mediated processes (for example, Biswal *et al.*, 2000; Blázquez *et al.*, 2000; Chi *et al.*, 2000; Herget *et al.*, 2000; Kawatani *et al.*, 2000; Kirkham *et al.*, 2000; Lee *et al.*, 2000; Charles *et al.*, 2001; Dyntar *et al.*, 2001; Iacobini *et al.*, 2001; Kroesen et al., 2001; Maedler *et al.*, 2001; Wu *et al.*, 2001; Zhong *et al.*, 2001). In many of the examples cited above, short-term treatment with fumonisin B_1 protected against ceramide-mediated cell death. In contrast, prolonged exposure to fumonisin B_1 in vivo and in vitro is toxic to cells and induces apoptosis (WHO, 2000, 2002). Perhaps the best evidence for a cause-and-effect relationship between disruption of sphingolipid metabolism and the toxic effects of fumonisin B_1 has come from studies conducted *in vitro* using inhibitors of serine palmitoyltransferase to prevent sphinganine accumulation and reverse the increased apoptosis and altered cell growth induced by fumonisin B_1 treatment (WHO, 2000; Kim *et al.*, 2001; Riley *et al.*, 2001; Yu *et al.*, 2001; He *et al.*, 2002; WHO, 2002).

Figure 4. Pathways of sphingolipid biosynthesis and turnover in a mammalian cell. In boxes are the known biological activities affected by fumonisin B_1 (FB_1) inhibition of ceramide synthase and associated with changes in the biosynthesis of various sphingolipid intermediates and products

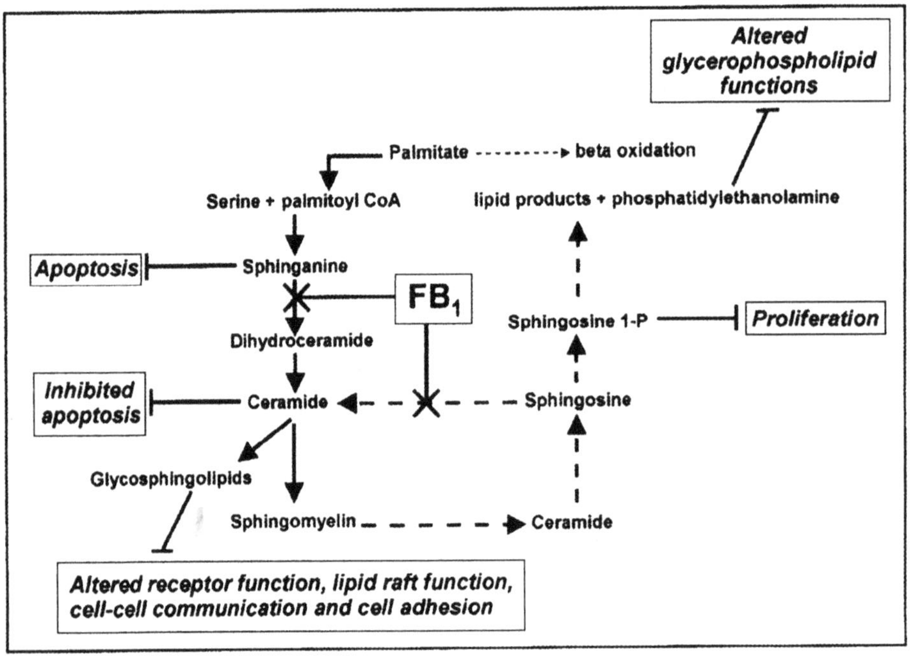

For additional details, see Merrill *et al.* (2001); Riley *et al.* (2001)
Sphingosine 1-P, sphingosine 1-phosphate

(*h*) *Depletion of complex sphingolipids*

Depletion of more complex sphingolipids also plays a role in the abnormal behaviour, altered morphology and altered proliferation of fumonisin-treated cells (WHO, 2000; Merrill *et al.*, 2001; Riley *et al.*, 2001; WHO, 2002), and this has been proposed as a mechanism in fumonisin B_1-induced nephrotoxicity in male rats (Hard *et al.*, 2001) through the disruption of cell–cell interactions. Numerous studies have demonstrated the ability of fumonisin to alter the function of specific glycosphingolipids and lipid rafts (membrane associations of sphingolipids, ceramide-anchored proteins and other lipids). Examples of these functions are inhibition of folate transport, bacterial toxin binding and transport (e.g., *Shigella* and cholera toxin), cell–cell and cell–substratum contact and cell–cell communication (WHO, 2000; Riley *et al.*, 2001; Merrill *et al.*, 2001; WHO, 2002).

(*i*) *Increased dihydroceramide* in vivo

In fumonisin-treated animals (pigs, horses, mice), there is an increased amount of complex sphingolipids containing sphinganine as the long-chain sphingoid-base backbone (see for example, Riley *et al.*, 1993). The ceramide generated from these complex sphingolipids is dihydroceramide, which is inactive in ceramide signalling and does not induce death of oxidant-damaged hepatocytes (Arora *et al.*, 1997). Dihydroceramide is also enriched in mouse hepatoma-22 cells, in which sphinganine comprised 37% of the ceramides as compared with 5% in normal rat liver (Rylova *et al.*, 1999).

(*j*) *Hypothesized cellular mechanism*

In cultured cells, the balance between the intracellular concentration of sphingolipid effectors that protect cells from apoptosis (decreased ceramide, increased sphingosine 1-phosphate) and the effectors that induce apoptosis (increased ceramide, increased free sphingoid bases, increased fatty acids) determines the observed cellular response (reviewed in Merrill *et al.*, 2001; Riley *et al.*, 2001). Cells sensitive to the proliferative effect of decreased ceramide and increased sphingosine 1-phosphate will be selected to survive and proliferate. Conversely, when the increase in free sphingoid bases exceeds the ability of a cell to convert sphinganine/sphingosine to dihydroceramide/ceramide or their sphingoid base 1-phosphate, free sphingoid bases will accumulate to toxic levels. Cells that are sensitive to sphingoid base-induced growth arrest will cease growing and insensitive cells will survive. Thus, the kinetics of fumonisin B_1 elimination (rapid), the affinity of fumonisin B_1 for ceramide synthase (competitive and reversible) and the kinetics of fumonisin-induced sphinganine elevation will influence the time course, amplitude and frequency of variations in the concentration of intracellular ceramide, sphingoid base-1 phosphates and free sphinganine in tissues of animals consuming fumonisins (Enongene *et al.*, 2002a,b). This is important, because the balance between the rates of apoptosis and cell proliferation is a critical determinant in the process of hepato- and nephrotoxicity and tumorigenesis in animal models (Dragan *et al.*, 2001; Howard *et al.*, 2001; Voss *et al.*, 2001). At the cellular level, it is hypothesized that apoptotic necrosis should be considered to be similar to oncotic necrosis (as defined in Levin *et al.*, 1999), in that both will lead to a regenerative process involving sustained cell proliferation (Dragan *et al.*, 2001; Hard *et al.*, 2001). Numerous endogenous processes can cause DNA damage that, if unrepaired, can give rise to a mutation in the DNA. Increased cell proliferation may thus involve replication of mutated DNA, resulting in an increased risk for cancer (Dragan *et al.*, 2001).

4.5.2 *Interference with fatty acid and glycerophospholipid metabolism*

(*a*) *Importance of fatty acids*

Essential fatty acids are major constituents of all cell membrane glycerophospholipids, sphingolipids and triglycerides. In addition to their important role as structural components of all cell membranes, essential fatty acids are precursors of many bioactive lipids known to regulate cell growth, differentiation and cell death.

(*b*) *Interference with fatty acid metabolism*

In rat liver and primary hepatocytes exposed to fumonisin B_1, changes in the phospholipid profile and fatty acid composition of phospholipids indicate that fumonisin B_1 interferes with fatty acid metabolism (Gelderblom *et al.*, 1996b). The following summary is taken from the review by Gelderblom *et al.* (2001a) and the WHO monograph (WHO, 2002).

(*c*) *Altered lipid metabolism in rat hepatocytes* in vitro

Gelderblom *et al.* (1996b) showed that, in fumonisin B_1-treated rat hepatocytes, the pattern of changes in specific polyunsaturated fatty acids suggested disruption of the Δ6 desaturase and cyclo-oxygenase metabolic pathways (Figure 5). These changes were considered to be important in the fumonisin B_1-induced toxicity observed in primary hepatocytes (Gelderblom *et al.*, 2001a; WHO, 2002).

(*d*) *Altered lipid metabolism in rat liver* in vivo

In-vivo studies have confirmed that fumonisin B_1 disrupts fatty acid and phospholipid biosynthesis, but the pattern of changes is different from that observed *in vitro* (Gelderblom *et al.*, 1997). Major changes are associated with both the phosphatidylethanolamine and the phosphatidylcholine phospholipid fractions, while cholesterol levels are increased in both the serum and liver (Gelderblom *et al.*, 2001a; WHO, 2002). A characteristic fatty acid pattern (Figure 6) is seen in the liver of rats exposed to dietary fumonisin B_1 levels associated with the development of preneoplastic lesions and in liver of rats fed fumonisin B_1 after treatment with cancer initiators.

(*e*) *Altered signalling for cell survival*

At the fumonisin B_1 doses that have been shown to alter fatty acid and glycerophospholipid profiles in rat liver, there are numerous changes in expression of proteins known to be involved in the regulation of cell growth, apoptosis and cell differentiation (WHO, 2002). For example, expression of hepatocyte growth factor (HGF), transforming growth factor α (TGF α), TGF β_1 and the c-*myc* oncogene were all increased during short-term feeding of fumonisin B_1. Overexpression of TGF β_1 could play a role in the increased apoptosis, while the increased expression of the proto-oncogene c-*myc* could contribute to the enhanced cell proliferation that is required for the tumour progression

Figure 5. A model for the proposed interference by fumonisin B_1 (FB_1) with delta 6 ($\Delta 6$) desaturase activity and consequent effects on the fatty acid composition of phosphatidylcholine (PC) and phosphatidylethanolamine (PE) and changes in the cyclo-oxygenase metabolic pathway

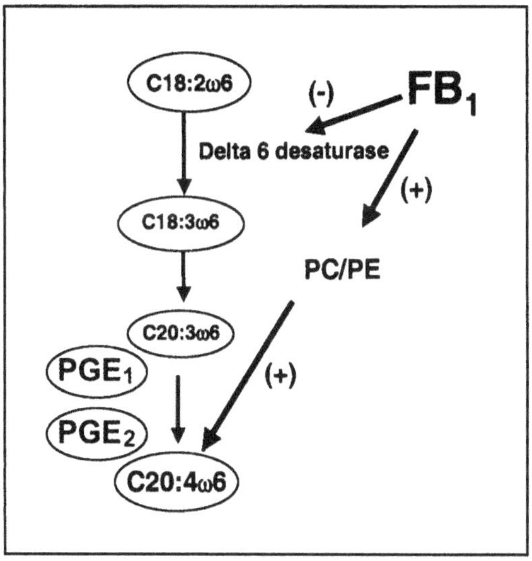

For additional details, see Gelderblom *et al.* (2001a).
PGE_1 and PGE_2, prostaglandin E_1 and E_2

observed in the liver of rats and mice exposed to hepatotoxic levels of fumonisin B_1 (Figure 6). Increased expression of c-*myc* and TGF β_1 may also play a role in the promotion of liver tumours by fumonisin B_1 (Lemmer *et al.*, 1999a).

(f) Altered cell cycle progression

Fumonisin B_1 disruption of sphingolipid metabolism and altered membrane phospholipids in the liver of BD IX rats have been suggested to cause the changes seen in several proteins (e.g., cyclin D1, retinoblastoma protein) that regulate cell cycle progression. Accumulation of cyclin D1 was due to post-translational stabilization of the protein (Ramljak *et al.*, 2000).

Fumonisin B_1-induced alterations in cellular glycerophospholipid content and the sphingomyelin cycle have been proposed to interact so as to modify a variety of cellular processes, resulting in the increased apoptosis and altered hepatocyte proliferation that are seen in liver of rats fed toxic doses of fumonisin B_1 (Figure 6). The balance between lipid mediators generated via the cyclo-oxygenase-2 and ceramide cycle could regulate processes related to cell proliferation and apoptosis. As summarized in Figure 6, fumonisin B_1-induced changes in ceramide, prostaglandins and other lipid mediators could alter the growth and survival of normal hepatocytes. Overexpression of TGF β_1 and

Figure 6. Proposed biochemical effects and cellular responses associated with fumonisin B_1 (FB_1)-induced alterations in delta 6 (Δ6) desaturase, sphingomyelinase (SMase), and ceramide synthase activity and biosynthesis of phosphotidylcholine (PC) and phosphatidylethanolamine (PE)

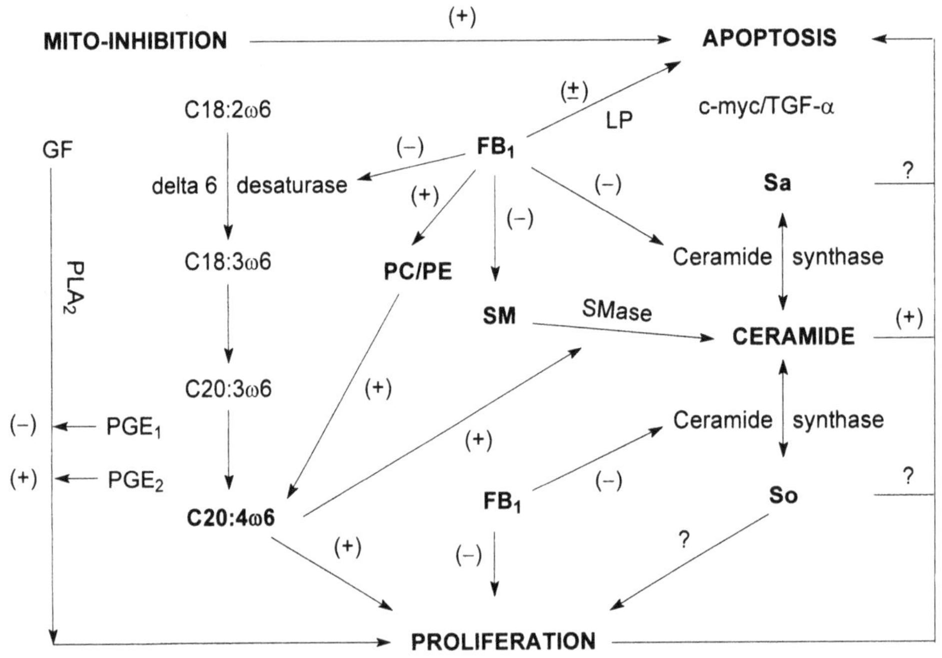

Abbreviations: PLA_2, phospholipase A_2; LP, products of lipid peroxidation; GF, growth factors; PGE_1 and PGE_2, prostaglandin E_1 and E_2; Sa, free sphinganine; So, free sphingosine; +, stimulatory; –, inhibitory or decreased; ?, response may be either increased or decreased (see Section 4.5.1); SM, sphingomyelin
Modified from Gelderblom *et al.* (2001a)

c-*myc* and oxidative damage could further enhance apoptosis and alter cell growth in affected hepatocytes (Gelderblom *et al.*, 2001a; WHO, 2002).

(*g*) *Hypothesized cellular mechanism*

Fumonisin B_1 has been shown to alter cell proliferation both *in vitro* and *in vivo*. The fumonisin B_1-induced effects on membrane lipids and the resultant effects on signalling pathways that involve lipid mediators could create an environment in which the growth of normal cells is impaired. Differential inhibition of cell proliferation is a possible mechanism by which hepatocytes resistant to fumonisin B_1-induced inhibition of cell growth are selectively stimulated, while growth of normal hepatocytes is inhibited. This selective inhibition of normal cell growth could increase the chances of survival of DNA-damaged hepatocytes, resulting in an increased likelihood of cancer development (Gelderblom *et al.*, 2001a; WHO, 2002).

Three lines of evidence support the hypothesis that fumonisin B_1-induced alterations in lipid metabolism contribute to the establishment of a growth differential in rat liver that could influence the process of neoplastic development. First, fumonisin B_1 induces an increase in phosphatidylethanolamine and arachidonic acid (C20 : 4 ω6) (Gelderblom *et al.*, 2001a), lipid mediators that are known to regulate many processes related to cell growth, such as proliferation and apoptosis (Khan *et al.*, 1995; Gelderblom *et al.*, 1999; Pinelli *et al.*, 1999; Seegers *et al.*, 2000; Abel *et al.*, 2001). Second, the decrease in the concentration of long-chain polyunsaturated fatty acids in hepatocytes exposed to fumonisin B_1 will produce a more rigid membrane structure, resulting in increased resistance to lipid peroxidation. Third, lipid metabolites, and in particular glycerophospholipids, are important components of many cellular signalling systems that control the balance between cell growth and cell death. Thus, changes in these lipid pools will alter response to growth factors and other mediators of cell survival.

Fumonisin B_1-induced disruption of lipid metabolism and the consequent induction of oxidative damage and lipid peroxidation (Abel & Gelderblom, 1998; WHO, 2002) could be important events leading to DNA damage, whereas changes in the balance of the different cell regulatory molecules such as those seen in livers of rats fed fumonisin B_1 are likely to be involved in the induction of a growth differential that selectively stimulates the survival of damaged hepatocytes and the development of cancer in rats.

4.5.3 *Other biochemical mechanisms*

Several in-vitro studies with fumonisins have found changes in cellular regulation and cell functions that have been attributed to processes other than lipid metabolism (WHO, 2000, 2002). Many of these effects could be relevant to the organ toxicity of fumonisins. Examples are the direct or indirect effects on protein kinase C (Huang *et al.*, 1995; Yeung et al., 1996), activity of extracellular regulated kinases (Pinelli *et al.*, 1999), altered DNA methylation and increased lipid peroxidation (Mobio *et al.*, 2000a,b) and alterations in the tumour necrosis factor α (TNFα) signalling pathways (He *et al.*, 2001; Jones *et al.*, 2001; Sharma *et al.*, 2001).

5. Summary of Data Reported and Evaluation

5.1 Exposure data

Fumonisin B_1 is the most prevalent member of a family of toxins produced by several species of *Fusarium* moulds which occur mainly in maize. Fumonisin B_1 contamination of maize has been reported worldwide at mg/kg levels. Human exposure occurs at levels of micrograms to milligrams per day and is greatest in regions where maize products are the dietary staple.

5.2 Human carcinogenicity data

No new studies on the human carcinogenicity of fumonisins were available to the Working Group.

5.3 Animal carcinogenicity data

Fumonisin B_1 has been tested for carcinogenicity by oral administration in one study in mice, one study in male rats and one study in male and female rats. In female mice, it caused an increase in hepatocellular adenomas and carcinomas. In one study in male rats, it caused an increase in cholangiocarcinomas and hepatocellular carcinomas. In the other rat study, it induced renal tubule carcinomas in male rats, over half of which were classified as a rare highly malignant variant.

Fumonisin B_1 has also been shown to promote tumours in mouse skin and trout livers when 7,12-dimethylbenz[*a*]anthracene and aflatoxin B_1, respectively, were used as tumour initiators.

5.4 Other relevant data

Fumonisins are poorly absorbed, rapidly excreted and not metabolized in animal systems. The half-life for elimination in animal species is directly related to the average body weight of the species, suggesting that the half-life in humans will be longer than those determined experimentally in rats and other animals.

Fumonisin B_1 is hepatotoxic and nephrotoxic in all animal species tested. The earliest histological change to appear in either the liver or kidney of fumonisin-treated animals is increased apoptosis followed by regenerative cell proliferation. While the acute toxicity of fumonisin is low, it is the known cause of two diseases which occur in domestic animals with rapid onset: equine leukoencephalomalacia and porcine pulmonary oedema syndrome. Both of these diseases involve disturbed sphingolipid metabolism and cardiovascular dysfunction.

Fumonisin B_1 causes developmental toxicity in several animal species. In rats, mice and rabbits, developmental effects occurred at dose levels associated with disruption of sphingolipid metabolism and maternal toxicity in liver and kidney.

Postnatal dosing causes decreased survival of rat pups and results indicate that sphingolipid metabolism is vulnerable after birth.

Fumonisin B_1 is inactive in bacterial mutation assays and in the unscheduled DNA synthesis assay with rat hepatocytes, but induces DNA damage, such as micronuclei, *in vitro* and *in vivo*. In some studies, addition of antioxidants reduced the DNA-damaging effects of fumonisin B_1, suggesting that the effects may be due to oxidative stress.

Disruption of various aspects of lipid metabolism, membrane structure and signal transduction pathways mediated by lipid second messengers appears to be an important aspect of all the various proposed mechanisms of action of fumonisin B_1, including its mechanism of carcinogenicity.

Fumonisin B_1-induced disruption of sphingolipid, phospholipid and fatty acid metabolism is observed both *in vitro* and *in vivo* in all animal models and in a single human study. Disruption of sphingolipid metabolism by fumonisin B_1 in animal and human systems *in vitro* causes cell death and regenerative cell proliferation mediated through sphingolipid signalling pathways. The kinetics of the increases and decreases in the various bioactive sphingolipid pools in liver, kidney, lung and heart are correlated with the observed toxicity.

5.5 Evaluation

There is *inadequate evidence* in humans for the carcinogenicity of fumonisins.

There is *sufficient evidence* in experimental animals for the carcinogenicity of fumonisin B_1.

Overall evaluation

Fumonisin B_1 is *possibly carcinogenic to humans (Group 2B)*.

6. References

Abbas, H.K. & Ocamb, C.M. (1995) First report of production of fumonisin B_1 by *Fusarium polyphialidicum* collected from seeds of *Pinus strobus* (Abstract). *Plant Dis.*, **79**, 642

Abbas, H.K. & Riley, R.T. (1996) The presence and phytotoxicity of fumonisins and AAL-toxin in *Alternaria alternata*. *Toxicon*, **34**, 133–136

Abbas, H.K., Ocamb, C.M., Xie, W., Mirocha, C.J. & Shier, W.T. (1995) First report of fumonisin B_1, B_2, and B_3 production by *Fusarium oxysporum* var. *redolens* (Abstract). *Plant Dis.*, **79**, 968

Abbas, H.K., Cartwright, R.D., Shier, W.T., Abouzied, M.M., Bird, C.B., Rice, L.G., Ross, P.F., Sciumbato, G.L. & Meredith, F.I. (1998) Natural occurrence of fumonisins in rice with Fusarium sheath rot disease. *Plant Dis.*, **82**, 22–25

Abel, S. & Gelderblom, W.C.A. (1998) Oxidative damage and fumonisin B_1-induced toxicity in primary rat hepatocytes and rat liver *in vivo*. *Toxicology*, **131**, 121–131

Abel, S., Smuts, C.M., de Villiers, C. & Gelderblom, W.C.A. (2001) Changes in essential fatty acid patterns associated with normal liver regeneration and the progression of hepatocyte nodules in rat hepatocarcinogenesis. *Carcinogenesis*, **22**, 795–804

Alberts, J.F., Gelderblom, W.C.A., Thiel, P.G., Marasas, W.F.O., Van Schalkwyk, D.J. & Behrend, Y. (1990) Effects of temperature and incubation period on production of fumonisin B_1 by *Fusarium moniliforme*. *Appl. environ. Microbiol.*, **56**, 1729–1733

Alberts, J.F., Gelderblom, W.C.A., Vleggaar, R., Marasas, W.F.O., & Rheeder, J.P. (1993) Production of [^{14}C]fumonisin B_1 by *Fusarium moniliforme* MRC 826 in corn cultures. *Appl. environ. Microbiol.*, **59**, 2673–2677

Ali, N.S., Yamashita, A. & Yoshizawa, T. (1998) Natural co-occurrence of aflatoxins and *Fusarium* mycotoxins (fumonisins, deoxynivalenol, nivalenol and zearalenone) in corn from Indonesia. *Food Addit. Contam.*, **15**, 377–384

Allaben, W., Bucher, J.R. & Howard, P.C., eds (2001) International conference on the toxicology of fumonisin. *Environ. Health Perspect.*, **109**, 237–356

ApSimon, J.W. (2001) Structure, synthesis, and biosynthesis of fumonisin B_1 and related compounds. *Environ. Health Perspect.*, **109** (Suppl. 2), 245–249

Aranda, M., Pérez-Alzola, L.P., Ellahueñe, M.F. & Sepúlveda, C. (2000) Assessment of *in vitro* mutagenicity in *Salmonella* and *in vivo* genotoxicity in mice of the mycotoxin fumonisin B_1. *Mutagenesis*, **15**, 469–471

Arora, A.S., Jones, B.J., Patel, T.C., Bronk, S.F. & Gores, G.J. (1997) Ceramide induces hepatocyte cell death through disruption of mitochondrial function in the rat. *Hepatology*, **25**, 958–963

Atroshi, F., Rizzo, A., Biese, I., Veijalainen, P., Saloniemi. H., Sankari, S. & Andersson, K. (1999) Fumonisin B_1-induced DNA damage in rat liver and spleen: Effects of pre-treatment with coenzyme Q_{10}, L-carnitine, α-tocopherol and selenium. *Pharmacol. Res.*, **40**, 459–467

Bacon, C.W. & Nelson, P.E. (1994) Fumonisin production in corn by toxigenic strains of *Fusarium moniliforme* and *Fusarium proliferatum*. *J. Food Prot.*, **57**, 514–521

Bacon, C.W. & Williamson, J.W. (1992) Interactions of *Fusarium moniliforme*, its metabolites and bacteria with corn. *Mycopathologia*, **117**, 65–71

Badiani, K., Byers, D.M., Cook, H.W. & Ridgway, N.D. (1996) Effect of fumonisin B_1 on phosphatidylethanolamine biosynthesis in Chinese hamster ovary cells. *Biochim. biophys. Acta*, **1304**, 190–196

Bane, D.P., Neumann, E.J., Hall, W.F., Harlin, K.S. & Slife, R.L.N. (1992) Relationship between fumonisin contamination of feed and mystery swine disease. *Mycopathologia*, **117**, 121–124

Becker, B.A., Pace, L., Rottinghaus, G.E., Shelby, R., Misfeldt, M. & Ross, P.F. (1995) Effects of feeding fumonisin B_1 in lactating sows and their suckling pigs. *Am. J. vet. Res.*, **56**, 1253–1258

Bennett, G.A. & Richard, J.L. (1996) Influence of processing on *Fusarium* mycotoxins in contaminated grains. *Food Technol.*, **50**, 235–238

Bezuidenhout, S.C., Gelderblom, W.C.A., Gorst-Allman, C.P., Horak, R.M., Marasas, W.F.O., Spiteller, G. & Vleggaar, R. (1988) Structure elucidation of the fumonisins, mycotoxins from *Fusarium moniliforme*. *J. chem. Soc. chem. Commun.*, **11**, 743–745

Bhat, R.V., Shetty, P.H., Amruth, R.P. & Sudershan, R.V. (1997) A foodborne disease outbreak due to the consumption of moldy sorghum and maize containing fumonisin mycotoxins. *J. Toxicol. clin. Toxicol.*, **35**, 249–255

Biswal, S.S., Datta, K., Acquaah-Mensah, G.K. & Kehrer, J.P. (2000) Changes in ceramide and sphingomyelin following fludarabine treatment of human chronic B-cell leukemia cells. *Toxicology*, **154**, 45–53

Blackwell, B.A., Miller, J.D., & Savard, M.E. (1994) Production of carbon 14-labeled fumonisin in liquid culture. *J. Am. off. anal. Chem. Int.*, **77**, 506–511

Blázquez, C., Galve-Roperh, I. & Guzmán, M. (2000) *De novo*-synthesized ceramide signals apoptosis in astrocytes via extracellular signal-regulated kinases. *FASEB J.*, **14**, 2315–2322

Bondy, G., Barker, M., Mueller, R., Fernie, S., Miller, J.D., Armstrong C., Hierlihy, S.L., Rowsell, P. & Suzuki, C. (1996) Fumonisins B_1 toxicity in male Sprague-Dawley rats. In: Jackson, L.S., De Vries, J.W. & Bullerman, L.B., eds, *Fumonisins in Food* (Advances in Experimental Medicine and Biology, Vol. 392), New York, Plenum Press, pp. 251–264

Bondy, G.S., Suzuki, C.A.M., Fernie, S.M., Armstrong, C.L., Hierlihy, S.L., Savard, M.E. & Barker, M.G. (1997) Toxicity of fumonisin B_1 to B6C3F_1 mice: A 14-day gavage study. *Food chem. Toxicol.*, **35**, 981–989

Bondy, G.S., Suzuki, C.A.M., Mueller, R.W., Fernie, S M., Armstrong C.L., Hierlihy, S.L., Savard, M.E. & Barker, M.G. (1998) Gavage administration of the fungal toxin fumonisin B_1 to female Sprague-Dawley rats. *J. Toxicol. environ. Health*, **53**, 135–151

Bottalico, A., Logrieco, A., Ritieni, A., Moretti, A., Randazzo, G. & Corda, P. (1995) Beauvericin and fumonisin B_1 in preharvest *Fusarium moniliforme* maize ear rot in Sardinia. *Food Addit. Contam.*, **12**, 599–607

Bryden, W.L., Ravindran, G., Amba, M.T., Gill, R.J. & Burgess, L.W. (1996) Mycotoxin contamination of maize grown in Australia, the Philippines and Vietnam. In: Miraglia, M., Brera, C. & Onori, R., eds, *Ninth International IUPAC Symposium on Mycotoxins and Phycotoxins, Rome, 27-31 May 1996* (Abstract book), Rome, Istituto Superiore di Sanità, p. 41

Bullerman, L.B. & Tsai, W.-Y.J. (1994) Incidence and levels of *Fusarium moniliforme*, *Fusarium proliferatum*, and fumonisins in corn and corn-based foods and feeds. *J. Food Prot.*, **57**, 541–546

Burdaspal, P.A. & Legarda, T.M. (1996) Occurrence of fumonisins in corn and processed corn-based commodities for human consumption in Spain. In: Miraglia, M., Brera, C. & Onori, R., eds, *Ninth International IUPAC Symposium on Mycotoxins and Phycotoxins, Rome, 27–31 May 1996* (Abstract book), Rome, Istituto Superiore di Sanità, p. 167

Canet, C. (1999) [World regulation]. In: Pfohl-Leszkowicz, ed., *Les Mycotoxines dans l'Alimentation: Evaluation et Gestion du Risque* [Mycotoxins in Food: Evaluation and Management of Risk], Paris, Technique et Documentation, pp. 409–428

Caramelli, M., Dondo, A., Cantini Cortellazzi, G., Visconti, A., Minervini, F., Doko, M.B. & Guarda, F. (1993) [Equine leukoencephalomalacia from fumonisin: First case in Italy.] *Ippologia*, **4**, 49–56 (in Italian)

Carlson, D.B., Williams, D.E., Spitsbergen, J.M., Ross, P.F., Bacon, C.W., Meredith, F.I. & Riley, R.T. (2001) Fumonisin B_1 promotes aflatoxin B_1 and *N*-methyl-*N'*-nitro-nitrosoguanidine-initiated liver tumors in rainbow trout. *Toxicol. appl. Pharmacol.*, **172**, 29–36

Castella, G., Bragulat, M.R. & Cabañes, F.J. (1997) Occurrence of *Fusarium* species and fumonisins in some animal feeds and raw materials. *Cereal Res. Commun.*, **25**, 355–356

Cawood, M.E., Gelderblom, W.C.A., Vleggaar, R., Behrend, Y., Thiel, P.G. & Marasas, W.F.O. (1991) Isolation of the fumonisin mycotoxins: A quantitative approach. *J. agric. Food Chem.*, **39**, 1958–1962

Cawood, M.E., Gelderblom, W.C.A., Alberts, J.F. & Snyman, S.D. (1994) Interaction of ^{14}C-labelled fumonisin B mycotoxins with primary rat hepatocyte cultures. *Food chem. Toxicol.*, **32**, 627–632

Chamberlain, W.J., Voss, K.A. & Norred, W.P. (1993) Analysis of commercial laboratory rat rations for fumonisin B_1, a mycotoxin produced on corn by *Fusarium moniliforme*. *Contemp. Top.*, **32**, 26–28

Charles, A.G., Han, T.-Y., Lui, Y.Y., Hansen, N., Giuliano, A.E., & Cabot, M.C. (2001) Taxol-induced ceramide generation and apoptosis in human breast cancer cells. *Cancer Chemother. Pharmacol.*, **47**, 444–450

Chelkowski, J. & Lew, H. (1992) *Fusarium* species of Liseola section — Occurrence in cereals and ability to produce fumonisins. *Microbiol. Alim. Nutr.*, **10**, 49-53

Chelule P.K., Gqaleni, N., Dutton, M.F. & Chuturgoon, A.A. (2001) Exposure of rural and urban populations in KwaZulu Natal, South Africa, to fumonisin B_1 in maize. *Environ. Health Perspect.*, **109**, 253–256

Chi, M.M.-Y., Schlein, A.L. & Moley, K.H. (2000) High insulin-like growth factor 1 (IGF-1) and insulin concentrations trigger apoptosis in the mouse blastocyst via down-regulation of the IGF-1 receptor. *Endocrinology*, **141**, 4784–4792

Chu, F.S. & Li, G.Y. (1994) Simultaneous occurrence of fumonisin B_1 and other mycotoxins in moldy corn collected from the People's Republic of China in regions with high incidences of esophageal cancer. *Appl. environ. Microbiol.*, **60**, 847–852

Chulamorakot, T. & Suprasert, D. (2000) Contamination of mycotoxin in cereal grains composition of R.C. product. *Food (Inst. Food Res. Product Develop., Kasetsart Univ.)*, **30**, 107–114

Chulze, S.N., Ramirez, M.L., Farnochi, M.C., Pascale, M., Visconti, A. & March, G. (1996) *Fusarium* and fumonisin occurrence in Argentinian corn at different ear maturity stages. *J. agric. Food Chem.*, **44**, 2797–2801

Collins, T.F.X., Shackelford, M.E., Sprando, R.L., Black, T.N., Laborde, J.B., Hansen, D.K., Eppley, R.M., Trucksess, M.W., Howard, P.C., Bryant, M.A., Ruggles, D.I., Olejnik, N. & Rorie, J.I. (1998a) Effects of fumonisin B1 in pregnant rats. *Food chem. Toxicol.*, **36**, 397–408

Collins, T.F.X., Sprando, R.L., Black, T.N., Shackelford, M.E., Laborde, J.B., Hansen, D.K., Eppley, R.M., Trucksess, M.W., Howard, P.C., Bryant, M.A., Ruggles, D.I., Olejnik, N. & Rorie, J.I. (1998b) Effects of fumonisin B1 in pregnant rats. Part 2. *Food. chem. Toxicol.*, **36**, 673–685

Colvin, B.M. & Harrison, L.R. (1992) Fumonisin-induced pulmonary edema and hydrothorax in swine. *Mycopathologia*, **117**, 79–82

Colvin, B.M., Cooley, A.J. & Beaver, R.W. (1993) Fumonisin toxicosis in swine: Clinical and pathologic findings. *J. vet. diagn. Invest.*, **5**, 232–241

da Silva, J.B., Pozzi, C.R., Mallozzi, M.A.B., Ortega, E.M. & Corrêa, B. (2000) Mycoflora and occurrence of aflatoxin B_1 and fumonisin B_1 during storage of Brazilian sorghum. *J. agric. Food Chem.*, **48**, 4352–4356

De León, C. & Pandey, S. (1989) Improvement of resistance to ear and stalk rots and agronomic traits in tropical maize gene pools. *Crop Sci.*, **29**, 12–17

Delongchamp, R.R. & Young, J.F. (2001) Tissue sphinganine as a biomarker of fumonisin-induced apoptosis. *Food Add. Contam.*, **18**, 255–261

Desjardins, A.E., Plattner, R.D. & Nelson, P.E. (1994) Fumonisin production and other traits of *Fusarium moniliforme* strains from maize in northeast Mexico. *Appl. environ. Microbiol.*, **60**, 1695–1697

Diaz, G.J. & Boermans, H.J. (1994) Fumonisin toxicosis in domestic animals: A review. *Vet. hum. Toxicol.*, **36**, 548–555

Doko, M.B. & Visconti, A. (1994) Occurrence of fumonisins B_1 and B_2 in corn and corn-based human foodstuffs in Italy. *Food Addit. Contam.*, **11**, 433–439

Doko, M.B., Rapior, S. & Visconti, A. (1994) Screening for fumonisins B_1 and B_2 in corn and corn-based foods and feeds from France. In: *Abstracts of the 7th International Congress of the IUMS Mycology Division, Prague, Czech Republic, 3-8 July 1994*, p. 468

Doko, M.B., Rapior, S., Visconti, A. & Schjøth, J.E. (1995) Incidence and levels of fumonisin contamination in maize genotypes grown in Europe and Africa. *J. agric. Food Chem.*, **43**, 429–434

Doko, M.B., Canet, C., Brown, N., Sydenham, E.W., Mpuchane, S. & Siame, B.A. (1996) Natural co-occurrence of fumonisins and zearalenone in cereals and cereal-based foods from Eastern and Southern Africa. *J. agric. Food Chem.*, **44**, 3240–3243

Dombrink-Kurtzman, M.A. & Dvorak, T.J. (1999) Fumonisin content in masa and tortilla from Mexico. *J. agric. Food Chem.*, **47**, 622–627

Dragan, Y.P., Bidlack, W.R., Cohen, S.M., Goldsworthy, T.L., Hard, G.C., Howard, P.C., Riley, R.T. & Voss, K.A. (2001) Implications of apoptosis for toxicity, carcinogenicity and risk assessment: Fumonisin B_1 as an example. *Toxicol. Sci.*, **61**, 6–17

Dragoni, I., Pascale, M., Piantanida, L., Tirilly, Y. & Visconti, A. (1996) [Presence of fumonisin in foodstuff destined for feeding to pigs in Brittany (France).] *Microbiol. Alim. Nutr.*, **14**, 97–103 (in Italian)

Dyntar, D., Eppenberger-Eberhardt, M., Maedler, K., Pruschy, M., Eppenberger, H.M., Spinas, G.A. & Donath, M.Y. (2001) Glucose and palmitic acid induce degeneration of myofibrils and modulate apoptosis in rat adult cardiomyocytes. *Diabetes*, **50**, 2105–2113

Ehrlich, V., Darroudi, F., Uhl, M., Steinkellner, H., Zsivkovits, M. & Knasmueller, S. (2002) Fumonisin B_1 is genotoxic in human derived hepatoma (HepG2) cells. *Mutagenesis*, **17**, 257–260

Enongene, E.N., Sharma, R.P., Bhandari, N., Voss, K.A. & Riley, R.T. (2000) Disruption of sphingolipid metabolism in small intestines, liver and kidney of mice dosed subcutaneously with fumonisin B_1. *Food chem. Toxicol.*, **38**, 793–799

Enongene, E.N., Sharma, R.P., Neetesh, B., Meredith, F.I., Voss, K.A., & Riley, R.T. (2002a) Time and dose-related changes in mouse sphingolidid metabolism following gavage fumonisin B1 administration. *Toxicol. Sci.*, **67**, 173–181

Enongene, E.N., Sharma, R.P., Bhandari, N., Miller, J.D., Meredith, F.I., Voss, K.A. & Riley, R.T. (2002b) Persistence and reversibility of the elevation in free sphingoid bases induced by fumonisin inhibition of ceramide synthase. *Toxicol. Sci.*, **67**, 173–181

Fazekas, B., Bajmócy, E., Glávits, R., Fenyvesi, A. & Tanyi, J. (1998) Fumonisin B_1 contamination of maize and experimental acute fumonisin toxicosis in pigs. *J. vet. Med.*, **B45**, 171–181

Ferguson, S.A., St. Omer, V.E.V., Kwon, O.S., Holson, R.R., Houston, R.J., Rottinghaus, G.E. & Slikker, W., Jr (1997) Prenatal fumonisin (FB_1) treatment in rats results in minimal maternal or offspring toxicity. *Neurotoxicology*, **18**, 561–569

Floss, J.L., Casteel, S.W., Johnson, G.C., Rottinghaus, G.E. & Krause, G.F. (1994a) Developmental toxicity in hamsters of an aqueous extract of *Fusarium moniliforme* culture material containing known quantities of fumonisin B_1. *Vet. hum. Toxicol.*, **36**, 5–10

Floss, J.L., Casteel, S.W., Johnson, G.C., Rottinghaus, G.E. & Krause, G.F. (1994b) Development toxicity of fumonisin in Syrian hamsters. *Mycopathologia*, **128**, 33–38

Food and Drug Administration (2001a) *Guidance for Industry: Fumonisin Levels in Human Foods and Animal Feeds. Final Guidance*, Center for Food Safety and Applied Nutrition, Center for Veterinary Medicine (http://www.cfsan.fda.gov/~dms/fumongu2.html)

Food and Drug Administration (2001b) *Background Paper in Support of Fumonisin Levels in Corn and Corn Products Intended for Human Consumption*, Center for Food Safety and Applied Nutrition, Center for Veterinary Medicine (http://www.cfsan.fda.gov/~dms/fumonbg4.html)

Food and Drug Administration (2001c) *Background Paper in Support of Fumonisin Levels in Animal Feed: Executive Summary of this Scientific Support Document*, Center for Food Safety and Applied Nutrition, Center for Veterinary Medicine (http://www.cfsan.fda.gov/~dms/fumonbg3.html)

Galvano, F., Russo, A., Cardile, V., Galvano, G., Valnella, A. & Renis, M. (2002) DNA damage in human fibroblasts exposed to fumonisin B_1. *Food chem. Toxicol.*, **40**, 25–31

Gamanya, R. & Sibanda, L. (2001) Survey of *Fusarium moniliforme* (*F. verticillioides*) and production of fumonisin B_1 in cereal grains and oilseeds in Zimbabwe. *Int. J. Food Microbiol.*, **30**, 145–149

Gao, H.P. & Yoshizawa, T. (1997) Further study on *Fusarium* mycotoxins in corn and wheat from a high-risk area for human esophageal cancer in China. *Mycotoxins*, **45**, 51–55

Garren, L., Galendo, D., Wild, C.P. & Castegnaro, M. (2001) The induction and persistence of altered sphingolipid biosynthesis in rats treated with fumonisin B_1. *Food Add. Contam.*, **18**, 850–856

Gelderblom, W.C.A. & Snyman, S.D. (1991) Mutagenicity of potentially carcinogenic mycotoxins produced by *Fusarium moniliforme*. *Mycotoxin Res.*, **7**, 46–52

Gelderblom, W.C.A., Jaskiewicz, K., Marasas, W.F.O., Thiel, P.G., Horak, R.M., Vleggaar, R. & Kriek, N.P.J. (1988) Fumonisins — Novel mycotoxins with cancer-promoting activity produced by *Fusarium moniliforme*. *Appl. environ. Microbiol.*, **54**, 1806–1811

Gelderblom, W.C.A., Kriek, N.P.J., Marasas, W.F.O. & Thiel, P.G. (1991) Toxicity and carcinogenicity of the *Fusarium moniliforme* metabolite, fumonisin B_1, in rats. *Carcinogenesis*, **12**, 1247–1251

Gelderblom, W.C.A., Marasas, W.F.O., Vleggaar, R., Thiel, P.G. & Cawood, M.E. (1992a) Fumonisins: Isolation, chemical characterization and biological effects. *Mycopathologia*, **117**, 11–16

Gelderblom, W.C.A., Semple, E., Marasas, W.F.O. & Farber, E. (1992b) The cancer-initiating potential of the fumonisin B mycotoxins. *Carcinogenesis*, **13**, 433–437

Gelderblom, W.C.A., Cawood, M.E., Snyman, S.D. & Marasas, W.F.O. (1994) Fumonisin B_1 dosimetry in relation to cancer initiation in rat liver. *Carcinogenesis*, **15**, 209–214

Gelderblom, W.C.A., Snyman, S.D., Lebepe-Mazur, S., van der Westhuizen, L., Kriek, N.P.J. & Marasas, W.F.O. (1996a) The cancer-promoting potential of fumonisin B_1 in rat liver using diethylnitrosamine as a cancer initiator. *Cancer Lett.*, **109**, 101–108

Gelderblom, W.C.A., Smuts, C.M., Abel, S., Snyman, S.D., Cawood, M.E. & Van der Westhuizen, L. & Swanevelder, S. (1996b) Effect of fumonisin B_1 on protein and lipid synthesis in primary rat hepatocytes. *Food chem. Toxicol.*, **34**, 361–369

Gelderblom, W.C.A., Smuts, C.M., Abel, S., Snyman, S.D., Van der Westhuizen L., Huber, W.W. & Swanevelder, S. (1997) The effect of fumonisin B_1 on the levels and fatty acid composition of selected lipids in rat liver *in vivo*. *Food chem. Toxicol.*, **35**, 647–656

Gelderblom, W.C.A., Abel, S., Smuts, C.M., Swanevelder, S. & Snyman, S.D. (1999) Regulation of fatty acid biosynthesis as a possible mechanism for the mitoinhibitory effect of fumonisin B_1 in primary rat hepatocytes. *Prost. Leuk. essen. Fatty Acids*, **61**, 225–234

Gelderblom, W.C.A., Abel, S., Smuts, C.M., Marnewick, J., Marasas, W.F.O., Lemmer, E.R. & Ramljak, D. (2001a) Fumonisin-induced hepatocarcinogenesis: Mechanisms related to cancer initiation and promotion. *Environ. Health Perspect.*, **109** (Suppl. 2), 291–300

Gelderblom, W.C.A., Galendo, D., Abel, S., Swanevelder, S., Marasas, W.F.O. & Wild, C.P. (2001b) Cancer initiation by fumonisin B_1 in rat liver — Role of cell proliferation. *Cancer Lett.*, **169**, 127–137

Gelderblom, W.C.A., Lebepe-Mazur, S., Snijman, P.W., Abel, S., Swanevelder, S., Kriek, N.P.J. & Marasas, W.F.O. (2001c) Toxicological effects in rats chronically fed low dietary levels of fumonisin B_1. *Toxicology*, **161**, 39–51

Gelderblom, W.C.A., Marasas, W.F.O., Lebepe-Mazur, S., Swanevelder, S., Vessey, C.J. & de la M. Hall, P. (2002) Interaction of fumonisin B_1 and aflatoxin B_1 in a short-term carcinogenesis model in rat liver. *Toxicology*, **171**, 161–173

Gross, S.M., Reddy, R.V., Rottinghaus, G.E., Johnson, G. & Reddy, C.S. (1994) Developmental effects of fumonisin B_1-containing *Fusarium moniliforme* culture extract in CD1 mice. *Mycopathologia*, **128**, 111–118

Hard, G.C., Howard, P.C., Kovatch, R.M. & Bucci, T.J. (2001) Rat kidney pathology induced by chronic exposure to fumonisin B_1 includes rare variants of renal tubule tumor. *Toxicol. Pathol.*, **29**, 379–386

Haschek, W.M., Gumprecht, L.A., Smith, G., Tumbleson, M.E. & Constable, P.D. (2001) Fumonisin toxicosis in swine: An overview of porcine pulmonary edema and current perspectives. *Environ. Health Perspect.*, **109** (Suppl. 2), 251–257

Haseman, J.K., Hailey, J.R. & Morris, R.W. (1998) Spontaneous neoplasm incidences in Fischer 344 rats and B6C3F$_1$ mice in two-year carcinogenicity studies: A national toxicology program update. *Toxicol. Pathol.*, **26**, 428–441

He, Q., Riley, R.T. & Sharma, R.P. (2001) Fumonisin-induced tumor necrosis factor-α expression in a porcine kidney cell line is independent of sphingoid base accumulation induced by ceramide synthase inhibition. *Toxicol. appl. Pharmacol.*, **174**, 69–77

He, Q., Riley, R.T. & Sharma, R.P. (2002) Pharmacological antagonism of fumonisin B_1 cytotoxicity in porcine renal epithelial cells (LLC-PK$_1$): A model for reducing fumonisin-induced nephrotoxicity *in vivo*. *Pharmacol. Toxicol.*, **90**, 268–277

Hendrich, S., Miller, K.A., Wilson, T.M. & Murphy, P.A. (1993) Toxicity of *Fusarium proliferatum*-fermented nixtamalized corn-based diets fed to rats: Effect of nutritional status. *J. agric. Food Chem.*, **41**, 1649–1654

Hendricks, K. (1999) Fumonisins and neural tube defects in South Texas. *Epidemiology*, **10**, 198–200

Herget, T., Esdar, C., Oehrlein, S.A., Heinrich, M., Schütze, S., Maelicke, A. & Van Echten-Deckert, G. (2000) Production of ceramides causes apoptosis during early neural differentiation *in vitro*. *J. biol. Chem.*, **275**, 30344–30354

Hirooka, E.Y., Yamaguchi, M.M., Aoyama, S., Sugiura, Y. & Ueno, Y. (1996) The natural occurrence of fumonisins in Brazilian corn kernels. *Food Addit. Contam.*, **13**, 173–183

Hlywka, J.J. & Bullerman, L.B. (1999) Occurrence of fumonisin B_1 and B_2 in beer. *Food Addit. Contam.*, **16**, 319–324

Holcomb, M., Thompson, H.C., Jr & Hankins, L.J. (1993) Analysis of fumonisin B_1 in rodent feed by gradient elution HPLC using precolumn derivatization with FMOC [9-fluorenylenethyl chloroformate] and fluorescence detection. *J. agric. Food Chem.*, **41**, 764–767

Hopmans, E.C. & Murphy, P.A. (1993) Detection of fumonisins B_1, B_2, and B_3 and hydrolyzed fumonisin B_1 in corn-containing foods. *J. agric. Food Chem.*, **41**, 1655–1658

Howard, P.C., Churchwell, M.I., Couch, L.H., Marques, M.M. & Doerge, D.R. (1998) Formation of *N*-(carboxymethyl)fumonisin B_1, following the reaction of fumonisin B_1 with reducing sugars. *J. agric. Food Chem.*, **46**, 3546–3557

Howard, P.C., Warbritton, A., Voss, K.A., Lorentzen, R.J., Thurman, J.D., Kovach, R.M. & Bucci, T.J. (2001) Compensatory regeneration as a mechanism for renal tubule carcinogenesis of fumonisin B_1 in the F344/N/Nctr BR rat. *Environ. Health Perspect.*, **109** (Suppl. 2), 309–314

Huang, C., Dickman, M., Henderson, G. & Jones, C. (1995) Repression of protein kinase C and stimulation of cyclic AMP response elements by fumonisin, a fungal encoded toxin which is a carcinogen. *Cancer Res.*, **55**, 1655–1659

Humphreys, S.H., Carrington, C. & Bolger, P.M. (1997) Fumonisin risk scenarios (Abstract No. 867). *Toxicologist*, **36**, 170

Iacobini, M., Menichelli, A., Palumbo, G., Multari, G., Werner, B. & Principe, D. del (2001) Involvement of oxygen radicals in cytarabine-induced apoptosis in human polymorphonuclear cells. *Biochem. Pharmacol.*, **61**, 1033-1040

IARC (1993) *IARC Monographs on the Evaluation of Carcinogenic Risks to Humans*, Vol. 56, *Some Naturally Occurring Substances: Food Items and Constituents, Heterocyclic Aromatic Amines and Mycotoxins*, Lyon, IARC*Press*, pp. 445–466

Jaskiewicz, K., Van Rensburg, S.J., Marasas, W.F. & Gelderblom, W.C. (1987) Carcinogenicity of *Fusarium moniliforme* culture material in rats. *J. natl Cancer Inst.*, **78**, 321–325

JECFA (2001) *Safety Evaluation of Certain Mycotoxins in Food (WHO Food Additives Series No. 47), 56th Meeting of the Joint FAO/WHO Expert Committee on Food Additives (JECFA)*, Geneva, International Programme on Chemical Safety, World Health Organization

Jones, C., Ciacci-Zanella, J.R., Zhang, Y., Henderson, G. & Dickman, M. (2001) Analysis of fumonisin B_1-induced apoptosis. *Environ. Health Perspect.*, **109** (Suppl. 2), 315–320

Kang, H.-J., Kim, J.-C., Seo, J.-A., Lee, Y.-W. & Son, D.-H. (1994) Contamination of *Fusarium* mycotoxins in corn samples imported from China. *Agric. Chem. Biotechnol.*, **37**, 385–391

Kawatani, M., Simizu, S., Osada, H., Takada, M., Arber, N., & Imoto, M. (2000) Involvement of protein kinase C-regulated ceramide generation in inostamycin-induced apoptosis. *Exp. Cell Res.*, **259**, 389–397

Kellerman, T.S., Marasas, W.F.O., Thiel, P.G., Gelderblom, W.C.A., Cawood, M. & Coetzer, J.A.W. (1990) Leukoencephalomalacia in two horses induced by oral dosing of fumonisin B_1. *Onderstepoort J. vet. Res.*, **57**, 269–275

Khan, W.A., Blobe, G.C. & Hannun, Y.A. (1995) Arachidonic acid and free fatty acids as second messengers and the role of protein kinase C. *Cell Sign.*, **7**, 171–184

Kim, M.S., Lee, D.-T., Wang, T. & Schroeder, J.J. (2001) Fumonisin B_1 induces apoptosis in LLC-PK_1 renal epithelial cells via a sphinganine- and calmodulin-dependent pathway. *Toxicol. appl. Pharmacol.*, **176**, 118–126

King, S.B. & Scott, G.E. (1981) Genotypic differences in maize to kernel infection by *Fusarium moniliforme*. *Phytopathology*, **71**, 1245–1247

Kirkham, P.A., Takamatsu, H.-H., Lam, E.W.F. & Parkhouse, R.M.E. (2000) Ligation of the WC1 receptor induces γ δ T cell growth arrest through fumonisin B_1-sensitive increases in cellular ceramide. *J. Immunol.*, **165**, 3564–3570

Klittich, C.J.R., Leslie, J.F., Nelson, P.E. & Marasas, W.F.O. (1997) *Fusarium thapsinum* (*Gibberella thapsina*): A new species in section *Liseola* from sorghum. *Mycologia*, **89**, 643–652

Knasmüller, S., Bresgen, N., Kassie, F., Mersch-Sundermann, V., Gelderblom, W., Zöhrer, E & Eckl, P.M. (1997) Genotoxic effects of three *Fusarium* mycotoxins, fumonisin B_1, moniliformin and vomitoxin in bacteria and in primary cultures of rat hepatocytes. *Mutat. Res.*, **391**, 39–48

Korkotian, E., Schwarz, A., Pelled, D., Schwarzmann, G., Segal, M. & Futerman, A.H. (1999) Elevation of intracellular glucosylceramide levels results in an increase in endoplasmic reticulum density and in functional calcium stores in cultured neurons. *J. biol. Chem.*, **274**, 21673–21678

Kroesen, B.-J., Pettus, B., Luberto, C., Busman, M., Sietsma, H., de Leij, L. & Hannun, Y.A. (2001) Induction of apoptosis through B-cell receptor cross-linking occurs via *de novo* generated C16-ceramide and involves mitochondria. *J. biol. Chem.*, **276**, 13606–13614

Kuiper-Goodman, T., Scott, P.M., McEwen, N.P., Lombaert, G.A. & Ng, W. (1996) Approaches to the risk assessment of fumonisins in corn-based foods in Canada. *Adv. exp. Med. Biol.*, **392**, 369–393

Kwon, O.-S., Sandberg, J.A. & Slikker, W., Jr (1997a) Effects of fumonisin B_1 treatment on blood-brain barrier transfer in developing rats. *Neurotoxicol. Teratol.*, **19**, 151–155

Kwon, O.-S., Schmued, L.C. & Slikker, W., Jr (1997b) Fumonisin B_1 in developing rats alters brain sphinganine levels and myelination. *Neurotoxicology*, **18**, 571–579

LaBorde, J.B., Terry, K.K., Howard, P.C., Chen, J.J., Collins, T.F.X., Shackelford, M.E. & Hansen, D.K. (1997) Lack of embryotoxicity of fumonisin B_1 in New Zealand white rabbits. *Fundam. appl.Toxicol.*, **40**, 120–128

Laurent, D., Platzer, N., Kohler, F., Sauviat, M.P. & Pellegrin, F. (1989a) [Macrofusin and micromonilin: Two new mycotoxins isolated from corn infested with *Fusarium moniliforme* Sheld.] *Microbiol. Aliment. Nutr.*, **7**, 9–16 (in French)

Laurent, D., Pellegrin, F., Kohler, F., Costa, R., Thevenon, J., Lambert, C. & Huerre, M. (1989b) [Fumonisin B1 in the pathogenesis of equine leukoencephalomalacia.] *Microbiol. Alim. Nutr.*, 7, 285–291 (in French)

Le Bars, J., Le Bars, P., Dupuy, J., Boudra, H. & Cassini, R. (1994) Biotic and abiotic factors in fumonisin B_1 production and stability. *J. Assoc. off. anal. Chem. int.*, **77**, 517–521

Lebepe-Mazur, S., Bal, H., Hopmans, E., Murphy, P. & Hendrich, S. (1995) Fumonisin B_1 is fetotoxic in rats. *Vet. hum. Toxicol.*, **37**, 126–130

Lee, U.-S., Lee, M.-Y., Shin, K.-S., Min, Y.-S., Cho, C.-M. & Ueno, Y. (1994) Production of fumonisins B_1 and B_2 by *Fusarium moniliforme* isolated from Korean corn kernels for feed. *Mycotoxin Res.*, **10**, 67–72

Lee, J.Y., Hannun, Y.A. & Obeid, L.M. (2000) Functional dichotomy of protein kinase C (PKC) in tumor necrosis factor-α (TNF-α) signal transduction in L929 cells. Translocation and inactivation of PKC by TNF-α. *J. biol. Chem.*, **275**, 29290–29298

Lemmer, E.R., Hall, P. De la M., Omori, N., Omori, M., Shephard, E.G., Gelderblom, W.C.A., Cruse, J.P., Barnard, R.A., Marasas, W.F.O., Kirsch, R.E. & Thorgeirsson, S.S. (1999a) Histopathology and gene expression changes in rat liver during feeding of fumonisin B_1, a carcinogenic mycotoxin produced by *Fusarium moniliforme*. *Carcinogenesis*, **20**, 817–824

Lemmer, E.R., Gelderblom, W.C.A., Shephard, E.G., Abel, S., Seymour, B.L., Cruse, J.P., Kirsch, R.E., Marasas, W.F.O. & Hall, P. de la M. (1999b) The effects of dietary iron overload on fumonisin B_1-induced cancer induction in rat liver. *Cancer Lett.*, **146**, 207–215

Leslie, J.F., Plattner, R.D., Desjardins, A.E. & Klittich, C.J.R. (1992) Fumonisin B_1 production by strains from different mating populations of *Gibberella fujikuroi* (*Fusarium* section *Liseola*). *Phytopathology*, **82**, 341–345

Leslie, J.F., Marasas, W.F.O., Shephard, G.S., Sydenham, E.W., Stockenström, S. & Thiel, P.G. (1996) Duckling toxicity and the production of fumonisin and moniliformin by isolates in the A and F mating populations of *Gibberella fujikuroi* (*Fusarium moniliforme*). *Appl. environ. Microbiol.*, **62**, 1182–1187

Levin, S., Bucci, T.J., Cohen, S.M., Fix, A.S., Hardistry, J.F., LeGrand, E.K., Maronpot, R.R. & Trump, B.F. (1999) The nomenclature of cell death: Recommendations of an ad hoc committee of the Society of Toxicological Pathologists. *Toxicol. Pathol.*, **27**, 484–490

Lew, H., Adler, A. & Edinger, W. (1991) Moniliformin and the European corn borer. *Mycotoxin Res.*, **7**, 71–76

Li, M., Tian, G., Lu, S., Kuo, S., Ji, C. & Wang, Y. (1982) Forestomach carcinoma induced in rats by cornbread inoculated with *Fusarium moniliforme*. *Chinese J. Oncol.*, **4**, 241–244

Lim, C.W., Parker, H.M., Vesonder, R.F. & Haschek, W.M. (1996) Intravenous fumonisin B_1 induces cell proliferation and apoptosis in the rat. *Natural Toxins*, **4**, 34–41

Liu, H., Lu, Y., Haynes, J.S., Cunnick, J.E., Murphy, P. & Hendrich, S. (2001) Reaction of fumonisin with glucose prevents promotion of hepatocarcinogenesis in female F344/N rats while maintaining normal hepatic sphinganine/sphingosine ratios. *J. agric. Food Chem.*, **49**, 4113–4121

Logrieco, A., Moretti, A., Ritieni, A., Bottalico, A. & Corda, P. (1995) Occurrence and toxigenicity of *Fusarium proliferatum* from preharvest maize ear rot, and associated mycotoxins, in Italy. *Plant Dis.*, **79**, 727–731

Logrieco, A., Doko, B., Moretti, A., Frisullo, S. & Visconti, A. (1998) Occurrence of fumonisins B_1 and B_2 in *Fusarium proliferatum* infected asparagus plants. *J. agric. Food Chem.*, **46**, 5201–5204

Maedler, K., Spinas, G.A., Dyntar, D., Moritz, W., Kaiser, N. & Donath, M.Y. (2001) Distinct effects of saturated and monounsaturated fatty acids on β-cell turnover and function. *Diabetes*, **50**, 69–76

Magan, N., Marin, S., Ramos, A.J. & Sanchis, V. (1997) The impact of ecological factors on germination, growth, fumonisin production of *F. moniliforme* and *F. proliferatum* and their interactions with other common maize funci. *Cereal Res. Commun.*, **25**, 643–645

Maragos, C.M. & Richard, J.L. (1994) Quantitation and stability of fumonisins B_1 and B_2 in milk. *J. Assoc. off. anal. Chem. int.*, **77**, 1162–1167

Marasas, W.F.O. (1997) Risk assessment of fumonisins produced by *Fusarium moniliforme* in corn. *Cereal Res. Commun.*, **25**, 399–406

Marasas, W.F.O., van Rensburg, S.J. & Mirocha, C.J. (1979) Incidence of *Fusarium* species and the mycotoxins, deoxynivalenol and zearalenone, in corn produced in esophageal cancer areas in Transkei. *J. agric. Food Chem.*, **27**, 1108–1112

Marasas, W.F.O., Kriek, N.P.J., Fincham, J.E. & Van Rensburg, S.J. (1984) Primary liver cancer and oesophageal basal cell hyperplasia in rats caused by *Fusarium moniliforme*. *Int. J. Cancer*, **34**, 383–387

Marasas, W.F.O., Kellerman, T.S., Gelderblom, W.C.A., Coetzer, J.A.W., Thiel, P.G. & Van der Lugt, J.J. (1988) Leukoencephalomalacia in a horse induced by fumonisin B_1 isolated from *Fusarium moniliforme*. *Onderstepoort J. vet. Res.*, **55**, 197–203

Martinez-Larrabbaga, M.R., Anadon, A., Diaz M.J., Fernandez-Cruz, M.L., Martinex, M.W., Frejo, M.T., Martinez, M, Fernandez, R., Anton R.M., Morales, M.E. & Tafur, M. (1999) Toxicokinics and oral bioavailability of fumonisin B1. *Vet. Hum. Toxicol.*, **41**, 357–362

Martins, M.L., Martins, H.M. & Bernardo, F. (2001) Fumonisins B_1 and B_2 in black tea and medicinal plants. *J. Food Prot.*, **64**, 1268–1270

Meister, U., Symmank, H. & Dahlke, H. (1996) [Investigation and evaluation of the contamination of native and imported cereals with fumonisins.] *Z. Lebensm. Unters. Forsch.*, **203**, 528–533 (in German)

Merrill, A.H., Wang, E., Vales, T.R., Smith, E.R., Schroeder, J.J., Menaldino, D.S., Alexander, C., Crane, H.M., Xia, J., Liotta, D.C., Meredith, F.I. & Riley, R.T. (1996) Fumonisin toxicity and sphingolipid biosynthesis. *Adv. exp. Med. Biol.*, **392**, 297–306

Merrill, A.H., Jr, Schmelz, E.-M., Dillehay, D.L., Spiegel, S., Shayman, J.A., Schroeder, J.J., Riley, R.T., Voss, K.A. & Wang, E. (1997) Sphingolipids — The enigmatic lipid class: Biochemistry, physiology, and pathophysiology. *Toxicol. appl. Pharmacol.*, **42**, 208–225

Merrill, A.H., Jr, Sullards, M.C., Wang, E., Voss, K.A. & Riley, R.T. (2001) Sphingolipid metabolism: Roles in signal transduction and disruption by fumonisins. *Environ. Health Perspect.*, **109** (Suppl. 2), 283–289

Miller, J.D. (2000) Factors that affect the occurrence of fumonisin. *Environ. Health Perspect.*, **109**, 321–324

Miller, J.D., Savard, M.E., Sibilia, A., Rapior, S., Hocking, A.D. & Pitt, J.I. (1993) Production of fumonisins and fusarins by *Fusarium moniliforme* from southeast Asia. *Mycologia*, **85**, 385–391

Miller, J.D., Savard, M.E. & Rapior, S. (1994) Production and purification of fumonisins from a stirred jar fermenter. *Natural Toxins*, **2**, 354–359

Miller, J.D., Savard, M.E., Schaafsma, A.W., Seifert, K.A. & Reid, L.M. (1995) Mycotoxin production by *Fusarium moniliforme* and *Fusarium proliferatum* from Ontario and occurrence of fumonisin in the 1993 corn crop. *Can. J. Plant Pathol.*, **17**, 233–239

Minervini, F., Bottalico, C., Pestka, J. & Visconti, A. (1992) On the occurrence of fumonisins in feeds in Italy. In: *Proceedings of the 46th National Congress of the Italian Society of Veterinary Science, Venice, Italy, 30 September–3 October 1992*, pp. 1365–1368 (in Italian)

Mobio, T.A., Baudrimont, I., Sanni, A., Shier, T.W., Saboureau, D., Dano, S.D., Ueno, Y., Steyn, P.S. & Creppy, E.E. (2000a) Prevention by vitamin E of DNA fragmentation and apoptosis induced by fumonisin B_1 in C6 glioma cells. *Arch. Toxicol.*, **74**, 112–119

Mobio, T.A., Anane, R., Baudrimont, I., Carratú, M.-R., Shier, T.W., Dano, S.D., Ueno, Y. & Creppy, E.E. (2000b) Epigenetic properties of fumonisin B_1: Cell cycle arrest and DNA base modification in C6 glioma cells. *Toxicol. appl. Pharmacol.*, **164**, 91–96

Motelin, G.K., Haschek, W.M., Ness, D.K., Hall, W.F., Harlin, K.S., Schaeffer, D.J. & Beasley, V.R. (1994) Temporal and dose-response features in swine fed corn screenings contaminated with fumonisin mycotoxins. *Mycopathologia*, **126**, 27–40

Munkvold, G.P., Hellmich, R.L. & Showers, W.B. (1997) Reduced *Fusarium* ear rot and symptomless infection in kernels of maize genetically engineered for European corn borer resistance. *Phytopathology*, **87**, 1071–1077

Munkvold, G.P., Hellmich, R.L. & Rice, L.G. (1999) Comparison of fumonisin concentrations in kernels of transgenic Bt maize hybrids and nontransgenic hybrids. *Plant Dis.*, **83**, 130–138

Murphy, P.A., Rice, L.G. & Ross, P.F. (1993) Fumonisins B_1, B_2, and B_3 content of Iowa, Wisconsin, and Illinois corn and corn screenings. *J. agric. Food Chem.*, **41**, 263–266

Musser, S.M. & Plattner, R.D. (1997) Fumonisin composition in cultures of *Fusarium moniliforme, Fusarium proliferatum,* and *Fusarium nygamai. J. agric. Food Chem.*, **45**, 1169–1173

National Toxicology Program (NTP) (2000) *NTP Technical Report on the Toxicology and Carcinogenesis Studies of Fumonisin B_1 (CAS No 116355-83-0) in F344/N Rats and B6C3F$_1$ Mice (Feed Studies)* (TR 496; NIH Publication No 99-3955), Research Triangle Park, NC

Nelson, P.E., Plattner, R.D., Shackelford, D.D. & Desjardins, A.E. (1991) Production of fumonisins by *Fusarium moniliforme* strains from various substrates and geographic areas. *Appl. environ. Microbiol.*, **57**, 2410–2412

Nelson, P.E., Plattner, R.D., Shackelford, D.D. & Desjardins, A.E. (1992) Fumonisin B_1 production by *Fusarium* species other than *F. moniliforme* in section *Liseola* and by some related species. *Appl. environ. Microbiol.*, **58**, 984–989

de Nijs M. (1998) *Public Health Aspects of* Fusarium *Mycotoxins in Food in The Netherlands — A Risk Assessment*, Wageningen, Agricultural University (Thesis)

de Nijs, M., Sizoo, E.A., Rombouts, F.M., Notermans, S.H.W. & van Egmond, H.P. (1998a) Fumonisin B_1 in maize for food production imported in The Netherlands. *Food Addit. Contam.*, **15**, 389–392

de Nijs, M., Sizoo, E.A., Vermunt, A.E.M., Notermans, S.H.W. & van Egmond, H.P. (1998b) The occurrence of fumonisin B_1 in maize-containing foods in The Netherlands. *Food Addit. Contam.*, **15**, 385–388

de Nijs, M., van Egmond, H.P., Nauta, M., Rombouts, F.M. & Notermans, S.H.W. (1998c) Assessment of human exposure to fumonisin B_1. *J. Food Prot.*, **61**, 879–884

Nishino, H., Tokuda, H., Vesonder, R.F. & Nagao, M. (2000) Protein phosphatases as a target of fumonisin B_1. *Mycotoxins*, **50**, 61–64

Norred, W.P., Plattner, R.D., Vesonder, R.F., Bacon, C.W. & Voss, K.A. (1992) Effects of selected secondary metabolites of *Fusarium moniliforme* on unscheduled synthesis of DNA by rat primary hepatocytes. *Food chem. Toxicol.*, **30**, 233–237

Norred, W.P., Plattner, R.D. & Chamberlain, W.J. (1993) Distribution and excretion of [^{14}C]fumonisin B_1 in male Sprague-Dawley rats. *Natural Toxins*, **1**, 341–346

Norred, W.P., Plattner, R.D., Dombrink-Kurtzman, M.A., Meredith, F.I. & Riley, R.T. (1997) Mycotoxin-induced elevation of free sphingoid bases in precision-cut rat liver slices: Specificity of the response and structure–activity relationships. *Toxicol. appl. Pharmacol.*, **147**, 63–70

Ochor, T.E., Trevathan, L.E. & King, S.B. (1987) Relationship of harvest date and host genotype to infection of maize kernels by *Fusarium moniliforme*. *Plant Dis.*, **71**, 311-313

Ostrý, V. & Ruprich, J. (1998) Determination of the mycotoxin fumonisins in gluten-free diet (corn-based commodities) in the Czech Republic. *Cent. Eur. J. public Health*, **6**, 57–60

Osweiler, G.D., Ross, P.F., Wilson, T.M., Nelson, P.E., Witte, S.T., Carson, T.L., Rice, L.G. & Nelson, H.A. (1992) Characterization of an epizootic of pulmonary edema in swine associated with fumonisin in corn screenings. *J. vet. diagn. Invest.*, **4**, 53–59

Park, D.L., Rua, S.M., Jr, Mirocha, C.J., Abd-Alla, E.A.M. & Weng, C.Y. (1992) Mutagenic potentials of fumonisin contaminated corn following ammonia decontamination procedure. *Mycopathologia*, **117**, 105–108

Pascale, M., Doko, M.B. & Visconti, A. (1995) [Determination of fumonisins in polenta by high performance liquid chromatography.] In: *Proceedings of the 2nd National Congress on Food Chemistry, Giardini-Naxos, 24-27 May 1995*, Messina, La Grafica Editoriale, pp. 1067–1071 (in Italian)

Patel, S., Hazel, C.M., Winterton, A.G.M. & Gleadle, A.E. (1997) Surveillance of fumonisins in UK maize-based foods and other cereals. *Food Addit. Contam.*, **14**, 187–191

Penner, J.D., Casteel, S.W., Pittman, L., Jr, Rottinghaus, G.E. & Wyatt, R.D. (1998) Developmental toxicity of purified fumonisin B_1 in pregnant Syrian hamsters. *J. appl. Toxicol.*, **18**, 197–203

Pestka, J.J., Azcona-Olivera, J.I., Plattner, R.D., Minervini, F., Doko, M.B. & Visconti, A. (1994) Comparative assessment of fumonisin in grain-based foods by ELISA, GC-MS and HPLC. *J. Food Prot.*, **57**, 169–172

Piñeiro, M.S., Silva, G.E., Scott, P.M., Lawrence, G.A. & Stack, M.E. (1997) Fumonisin levels in Uruguayan corn products. *J. Assoc. off. anal. Chem. int.*, **80**, 825–828

Pinelli, E., Poux, N., Garren, L., Pipy, B., Castegnaro, M., Miller, D.J. & Pfohl-Leszkowicz (1999) Activation of mitogen-activated protein kinase by fumonisin B_1 stimulates $cPLA_2$ phosphorylation, the arachidonic acid cascade and cAMP production. *Carcinogenesis*, **20**, 1683–1688

Pitt, J.I., Hocking, A.D., Bhudhasamai, K., Miscamble, B.F., Wheeler, K.A. & Tanboon-Ek, P. (1993) The normal mycoflora of commodities from Thailand: 1. Nuts and oilseeds. *Int. J. Food Microbiol.*, **20**, 211–226

Pittet, A., Parisod, V. & Schellenberg, M. (1992) Occurrence of fumonisins B_1 and B_2 in corn-based products from the Swiss market. *J. agric. Food Chem.*, **40**, 1352–1354

Plattner, R.D. & Shackelford, D.D. (1992) Biosynthesis of labeled fumonisins in liquid Cultures of *Fusarium moniliforme*. *Mycopathologia*, **117**, 17–22

Plattner, R.D., Norred, W.P., Bacon, C.W., Voss, K.W., Peterson, R., Shackelford, D.D. & Weisleder, D. (1990) A method of detection of fumonisins in corn samples associated with field cases of equine leukoencephalomalacia. *Mycologia*, **82**, 698–702

Pocsfalvi, G., Ritieni, A., Randazzo, H., Dobó, A. & Malorni, A. (2000) Interaction of *Fusarium* mycotoxins, fusaproliferin and fumonisin B_1, with DNA studied by electrospray ionization mass spectrometry. *J. agric. Food Chem.*, **48**, 5795–5801

Prelusky, D.B., Trenholm, H.L. & Savard, M.E. (1994) Pharmacokinetic fate of ^{14}C-labelled fumonisin B_1 in swine. *Natural Toxins*, **2**, 73–80

Prelusky, D.B., Savard, M.E. & Trenholm, H.L.(1995) Pilot study on the plasma pharmacokinetics of fumonisin B_1 in cows following a single dose by oral gavage or intravenous administration. *Natural Toxins*, **3**, 389–394

Prelusky, D.B., Trenholm, H.L., Rotter, B.A., Miller, J.D., Savard, M.E., Yeung J.M. & Scott, P.M. (1996a) Biological fate of fumonisin B_1 in food-producing animals. *Adv. exp. Med. Biol.*, **392**, 265–278

Prelusky, D.B., Miller, J.D. & Trenholm, H.L. (1996b) Disposition of ^{14}C-derived residues in tissues of pigs fed radiolabelled fumonisin B_1. *Food Addit. Contam.*, **13**, 155–162

Price, W.D., Lovell, R.A. & McChesney, D.G. (1993) Naturally occurring toxins in feedstuffs: Center for Veterinary Medicine perspective. *J. Anim. Sci.*, **71**, 2556–2562

Purchase, I.F.H., Tustin, R.C. & Van Rensburg, S.J. (1975) Biological testing of food grown in the Transkei. *Food Cosmet. Toxicol.*, **13**, 639–647

Qiu, M. & Liu, X. (2001) Determination of sphinganine, sphingosine and Sa/So ratio in urine of humans exposed to dietary fumonisin B_1. *Food Add. Contam.*, **18**, 263–269

Ramirez, M.L., Pascale, M., Chulze, S., Reynoso, M.M., March, G. & Visconti, A. (1996) Natural occurrence of fumonisins and their correlation to *Fusarium* contamination in commercial corn hybrids growth in Argentina. *Mycopathologia*, **135**, 29–34

Ramljak, D., Calvert, R.J., Wiesenfeld, P.W., Diwan, B.A., Catipovic, B., Marasas, W.F.O., Victor, T.C., Anderson, L.M. & Gelderblom, W.C.A. (2000) A potential mechanism for fumonisin B_1-mediated hepatocarcinogenesis: Cyclin D1 stabilization associated with activation of Akt and inhibition of GSK-3β activity. *Carcinogenesis*, **21**, 1537–1546

Rapior, S., Miller, J.D., Savard, M.E. & ApSimon, J.W. (1993) [Production *in vitro* of fumonisins and fusarins by European strains of *Fusarium moniliforme*]. *Microbiol. Alim. Nutr.*, **11**, 327–333 (in French)

Reddy, R.V., Johnson, G., Rottinghaus, G.E., Casteel, S.W. & Reddy, C.S. (1996) Developmental effects of fumonisin B_1 in mice. *Mycopathologia*, **134**, 161–166

Rheeder, J.P., Marasas, W.F.O., Thiel, P.G., Sydenham, E.W., Shephard, G.S. & van Schalkwyk, D.J. (1992) *Fusarium moniliforme* and fumonisins in corn in relation to human esophageal cancer in Transkei. *Phytopathology*, **82**, 353–357

Ribar, S., Mesaric, M. & Bauman, M. (2001) High-performance liquid chromatographic determination of sphinganine and sphingosine in serum and urine of subjects from an endemic nephropathy area in Croatia. *J. Chromatogr. B*, **754**, 511–519

Richard, J.L., Meeridink, G., Maragos, C.M., Tumbleson, M., Bordson, G., Rice, L.G. & Ross, P.F. (1996) Absence of detectable fumonisins in the milk of cows fed *Fusarium proliferatum* (Matsushima) Nirenberg culture material. *Mycopathologia*, **133**, 123–126

Riley, R.T., An, N.H., Showker, J.L., Yoo, H.-S., Norred, W.P., Chamberlain, W.J., Wang, E., Merrill, A.H., Jr, Motelin, G., Beasley, V.R. & Haschek, W.M. (1993) Alteration of tissue and serum sphinganine to sphingosine ratio: An early biomarker of exposure to fumonisin-containing feeds in pigs. *Toxicol. appl. Pharmacol.*, **118**, 105–112

Riley, R.T., Wang, E. & Merril, A.H., Jr (1994a) Liquid chromatographic determination of sphinganine and sphingosine: Use of the free sphinganine-to-sphingosine ratio as a biomarker for consumption of fumonisins. *J. Assoc. off. anal. Chem. int.*, **77**, 533–540

Riley, R.T., Hinton, D.M., Chamberlain, W.J., Bacon, C.W., Wang, E., Merrill, A.H., Jr & Voss, K.A. (1994b) Dietary fumonisin B_1 induces disruption of sphingolipid metabolism in Sprague-Dawley rats: A new mechanism of nephrotoxicity. *J. Nutr.*, **124**, 594–603

Riley, R.T., Showker, J.L., Owens, D.L. & Ross, P.F. (1997) Disruption of sphingolipid metabolism and induction of equine leukoencephalomalacia by *Fusarium proliferatum* culture material containing fumonisin B_2 or B_3. *Environ. Toxicol. Pharmacol.*, **3**, 221–228

Riley, R.T., Enongene, E., Voss, K.A., Norred, W.P., Meredith, F.I., Sharma, R.P., Spitsbergen, J., Williams, D.E., Carlson, D.B. & Merrill, A.H., Jr (2001) Sphingolipid perturbations as mechanisms for fumonisin carcinogenesis. *Environ. Health Perspect.*, **109** (Suppl. 2), 301–308

Ross, P.F., Nelson, P.E., Richard, J.L., Osweiler, G.D., Rice, L.G., Plattner, R.D. & Wilson, T.M. (1990) Production of fumonisins by *Fusarium moniliforme* and *Fusarium proliferatum* isolates associated with equine leukoencephalomalacia and a pulmonary edema syndrome in swine. *Appl. environ. Microbiol.*, **56**, 3225–3226

Ross, P.F., Rice, L.G., Plattner, R.D., Osweiler, G.D., Wilson, T.M., Owens, D.L., Nelson, H.A. & Richard, J.L. (1991a) Concentrations of fumonisin B_1 in feeds associated with animal health problems. *Mycopathologia*, **114**, 129–135

Ross, P.F., Rice, L.G., Reagor, J.C., Osweiler, G.D., Wilson, T.M., Nelson, H.A., Owens, D.L., Plattner, R.D., Harlin, K.A., Richard, J.L., Colvin, B.M. & Banton, M.I. (1991b) Fumonisin B_1 concentrations in feeds from 45 confirmed equine leukoencephalomalacia cases. *J. vet. diagn. Invest.*, **3**, 238–241

Ross, P.F., Ledet, A.E., Owens, D.L., Rice, L.G., Nelson, H.A., Osweiler, G.D. & Wilson, T.M. (1993) Experimental equine leukoencephalomalacia, toxic hepatosis, and encephalopathy caused by corn naturally contaminated with fumonisins. *J. vet. Diagn. Invest.*, **5**, 69–74

Ross, P.F., Nelson, P.E., Owens, D.L., Rice, L.G., Nelson, H.A., & Wilson, T.M. (1994) Fumonisin B_2 in cultured *Fusarium proliferatum*, M-6104, causes equine leukoencephalomalacia. *J. vet. Diagn. Invest.*, **6**, 263–265

Rotter, B.A., Thompson, B.K., Prelusky, D.B., Trenholm, H.L,, Stewart, B., Miller, J.D. & Savard, M.E. (1996) Response of growing swine to dietary exposure to pure fumonisin B_1 during an eight-week period: Growth and clinical parameters. *Natural Toxins*, **4**, 42–50

Rotter, B.A., Prelusky, D.B., Fortin, A., Miller, J.D. & Savard, M.E. (1997) Impact of pure fumonisin B_1 on various metabolic parameters and carcass quality of growing-finishing swine — Preliminary findings. *Can. J. Anim. Sci.*, **77**, 465–470

Rumbeiha, W.K. & Oehme, F.W. (1997) Fumonisin exposure to Kansans through consumption of corn-based market foods. *Vet. hum. Toxicol.*, **39**, 220–225

Rylova, S.N., Somova, O.G., Zubova, E.S., Dudnik, L.B., Kogtev, L.S., Kozlov, A.M., Alesenko, A.V. & Dyatlovitskaya, E.V. (1999) Content and structure of ceramide and sphingomyelin and sphingomyelinase activity in mouse hepatoma-22. *Biochemistry (Mosc.)*, **64**, 437–441

Sahu, S.C., Eppley, R.M., Page, S.W., Gray, G.C., Barton, C.N. & O'Donnell, M.W. (1998) Peroxidation of lipids and membrane oxidative DNA damage by fumonisin B_1 in isolated rat liver nuclei. *Cancer Lett.*, **125**, 117–121

Sanchis, V., Abadias, M., Oncins, L., Sala, N., Viñas, I. & Canela, R. (1994) Occurrence of fumonisins B_1 and B_2 in corn-based products from the Spanish market. *Appl. environ. Microbiol.*, **60**, 2147–2148

Sanchis, V., Abadias, M., Oncins, L., Sala, N., Viñas, I. & Canela, R. (1995) Fumonisins B_1 and B_2 and toxigenic *Fusarium* strains in feeds from the Spanish market. *Int. J. Food Microbiol.*, **27**, 37–44

Savard, M.E. & Blackwell, B.A. (1994) Spectral characteristics of secondary metabolites from *Fusarium* fungi. In: Miller, J.D. & Trenholm, H.L., eds, *Mycotoxins in Grain: Compounds Other than Aflatoxin*, St Paul, MN, Eagan Press, pp. 59–260

Scott, P.M., Delgado, T., Prelusky, D.B., Trenholm, H.L. & Miller, J.D. (1994) Determination of fumonisins in milk. *J. environ. Sci. Health*, **B29**, 989–998

Scudamore, K.A. & Chan, H.K. (1993) Occurrence of fumonisin mycotoxins in maize and millet imported into the United Kingdom. In: Scudamore, K., ed., *Proceedings of the Workshop on Occurrence and Significance of Mycotoxins*, Slough, Ministry of Agriculture, Fisheries and Food, Central Science Laboratory, pp. 186–189

Scudamore, K.A., Nawaz, S. & Hetmanski, M.T. (1998) Mycotoxins in ingredients of animal feeding stuffs: II. Determination of mycotoxins in maize and maize products. *Food. Addit. Contam.*, **15**, 30–55

Seegers, J.C., Joubert, A.M., Panzer, A., Lottering, M.L., Jordan, C.A., Joubert, F., Maree, J.-L., Bianchi, P., de Kock, M. & Gelderblom, W.C.A. (2000) Fumonisin B_1 influenced the effects of arachidonic acid, prostaglandins E2 and A2 on cell cycle progression, apoptosis induction, tyrosine- and CDC2-kinase activity in oesophageal cancer cells. *Prost. Leuk. Essen. Fatty Acids*, **62**, 75–84

Semple-Roberts, E., Hayes, M.A., Armstrong, D., Becker, R.A., Racz, W.J. & Farber, E. (1987) Alternative methods of selecting rat hepatocellular nodules resistant to 2-acetylamino-fluorene. *Int. J. Cancer*, **40**, 643–645

Sewram, V., Nair, J.J., Nieuwoudt, T.W., Gelderblom, W.C.A., Marasas, W.F.O. & Shephard, G.S. (2001) Assessing chronic exposure to fumonisin mycotoxins: The use of hair as a suitable noninvasive matrix. *J. anal. Toxicol.*, **25**, 450–455

Sharma, R.P., Dugyala, R.R. & Voss, K.A. (1997) Demonstration of in-situ apoptosis in mouse liver and kidney after short-term repeated exposure to fumonisin B_1. *J. comp. Pathol.*, **117**, 371–381

Sharma, R.P., Riley, R.T. & Voss, K.A. (2001) Cytokine involvement in fumonisin-induced cellular toxicity. In: de Koe, W.J., Samson, R.A., van Egmond, H.P., Gilbert, J. & Sabino, M., eds, *Mycotoxins and Phycotoxins in Perspective at the Turn of the Millennium* (Proceedings of the Xth International IUPAC Symposium on Mycotoxins and Phycotoxins, 21–25 May, Guarujá, Brazil), Wageningen, the Netherlands, Ponsen en Looyen, pp. 223–230

Shelby, R.A., Rottinghaus, G.E. & Minor, H.C. (1994a) Comparison of thin-layer chromatography and competitive immunoassay methods for detecting fumonisin on maize. *J. agric. Food Chem.*, **42**, 2064–2067

Shelby, R.A., White, D.G. & Bauske, E.M. (1994b) Differential fumonisin production in maize hybrids. *Plant Dis.*, **78**, 582–584

Shephard, G.S. (1998) Chromatographic determination of the fumonisin mycotoxins. *J. Chromatogr.*, **A815**, 31-39

Shephard, G.S. & Snijman, P.W. (1999) Elimination and excretion of a single dose of the mycotoxin fumonisin B_2 in a non-human primate. *Food chem. Toxicol.*, **37**, 111–116

Shephard, G.S., Sydenham, E.W., Thiel, P.G. & Gelderblom, W.C.A. (1990) Quantitative determination of fumonisins B_1 and B_2 by high-performance liquid chromatography with fluorescence detection. *J. liq. Chromatogr.*, **13**, 2077–2087

Shephard, G.S., Thiel, P.G. & Sydenham, E.W. (1992a) Determination of fumonisin B_1 in plasma and urine by high-performance liquid chromatography. *J. Chromatogr.*, **574**, 299–304

Shephard, G.S., Thiel, P.G. & Sydenham, E.W. (1992b) Initial studies on the toxicokinetics of fumonisin B_1 in rats. *Food chem. Toxicol.*, **30**, 277–279

Shephard, G.S., Thiel, P.G., Sydenham, E.W., Alberts, J.F. & Gelderblom, W.C.A. (1992c) Fate of a single dose of the ^{14}C-labelled mycotoxin fumonisin B_1 in rats. *Toxicon*, **30**, 768–770

Shephard, G.S., Thiel, P.G., Sydenham, E.W., Alberts, J.F. & Cawood, M.E. (1994a) Distribution and excretion of a single dose of the mycotoxin fumonisin B_1 in a non-human primate. *Toxicon*, **32**, 735–741

Shephard, G.S., Thiel, P.G., Sydenham, E.W., Vleggaar, R. & Alberts, J.F. (1994b) Determination of the mycotoxin fumonisin B_1 and identification of its partially hydrolyzed metabolites in the faeces of non-human primates. *Food chem. Toxicol.*, **32**, 23–29

Shephard, G.S., Thiel, P.G., Sydenham, E.W. & Alberts, J.F. (1994c) Biliary excretion of the mycotoxin fumonisn B_1 in rats. *Food chem. Toxicol.*, **32**, 489–491

Shephard, G.S., Thiel, P.G. & Sydenham, E.W. (1995a) Liquid chromatographic determination of the mycotoxin fumonisin B_2 in physiological samples. *J. Chromatogr.*, **A692**, 39–43

Shephard, G.S., Thiel, P.G., Sydenham, E.W. & Savard, M.E. (1995b) Fate of a single dose of ^{14}C-labelled fumonisin B_1 in vervet monkeys. *Natural Toxins*, **3**, 145–150

Shephard, G.S., Thiel, P.G., Sydenham, E.W. & Snijman, P.W. (1995c) Toxicokinetics of the mycotoxin fumonisin B_2 in rats. *Food chem. Toxicol.*, **33**, 591–595

Shetty, P.H. & Bhat, R.V. (1998) Sensitive method for the detection of fumonisin B_1 in human urine. *J. Chromatogr.*, **B705**, 171–173

Sheu, C.W., Rodriguez, I., Eppley, R.M. & Lee, J.K. (1996) Lack of transforming activity of fumonisin B_1 in BALB/3T3 A31-1-1 mouse embryo cells. *Food chem. Toxicol.*, **34**, 751–753

Shurtleff, M.C. (1980) *Compendium of Corn Diseases*, St Paul, MN, American Phytopathological Society

Siame, B.A., Mpuchane, S.F., Gashe, B.A., Allotey, J. & Teffera, G. (1998) Occurrence of aflatoxins, fumonisin B_1, and zearalenone in foods and feeds in Botswana. *J. Food Prot.*, **61**, 1670–1673

Smith, J.S. & Thakur, R.A. (1996) Occurrence and fate of fumonisins in beef. *Adv. exp. Med. Biol.*, **392**, 39–55

Smith, G.W., Constable, P.D., Bacon, C.W., Meredith, F.I. & Haschek, W.M. (1996) Cardiovascular effects of fumonisins in swine. *Fundam. appl. Toxicol.*, **31**, 169–172

Smith, G.W., Constable, P.D., Tumbleson, M.E., Rottinghaus, G.E. & Haschek, W.M. (1999) Sequence of cardiovascular changes leading to pulmonary edema in swine fed culture material containing fumonisin. *Am. J. vet. Res.*, **60**, 1292–1300

Smith, G.W., Constable, P.D., Eppley, R.M., Tumbleson, M.E., Gumprecht, L.A. & Haschek-Hock, W.M. (2000) Purified fumonisin B_1 decreases cardiovascular function but does not alter pulmonary capillary permeability in swine. *Toxicol. Sci.*, **56**, 240–249

Spiegel, S. (1999) Sphingosine 1-phosphate: A prototype of a new class of second messenger. *J. Leuk. Biol.*, **65**, 341–344

Stack, M.E. (1998) Analysis of fumonisin B_1 and its hydrolysis product in tortillas. *J. Assoc. off. anal. Chem. int.*, **81**, 737–740

Stack, M.E. & Eppley, R.M. (1992) Liquid chromatographic determination of fumonisin B_1 and B_2 in corn and corn products. *J. Assoc. off. anal. Chem. int.*, **75**, 834–837

Stevens, V.L. & Tang, J. (1997) Fumonisin B_1-induced sphingolipid depletion inhibits vitamin uptake via the glycosylphosphatidylinositol-anchored folate receptor. *J. Biol. Chem.*, **272**, 18020–18025

Sun, T.S.C. & Stahr, H.M. (1993) Evaluation and application of a bioluminescent bacterial genotoxicity test. *J. Assoc. off. anal. Chem. int.*, **76**, 893–898

Suprasert, D. & Chulamorakot, T. (1999) Mycotoxin contamination in detected sorghum in Thailand. *Food (Inst. Food res. Product Develop., Kasatsart Univ.)*, **29**, 187–192

Sydenham, E.W. (1994) *Fumonisins: Chromatographic Methodology and their Role in Human and Animal Health* (PhD Thesis), Cape Town, University of Cape Town

Sydenham, E.W. & Shephard, G.S. (1996) Chromatographic and allied methods of analysis for selected mycotoxins. In: Gilbert, J., ed., *Progress in Food Contaminant Analysis*, London, Blackie Publishers, pp. 65–146

Sydenham, E.W., Gelderblom, W.C.A., Thiel, P.G. & Marasas, W.F.O. (1990a) Evidence for the natural occurrence of fumonisin B_1, a mycotoxin produced by *Fusarium moniliforme*, in corn. *J. agric. Food Chem.*, **38**, 285–290

Sydenham, E.W., Thiel, P.G., Marasas, W.F.O., Shephard, G.S., Van Schalkwyk, D.J. & Koch, K.R. (1990b) Natural occurrence of some *Fusarium* mycotoxins in corn from low and high esophageal cancer prevalence areas of the Transkei, Southern Africa. *J. agric. Food Chem.*, **38**, 1900–1903

Sydenham, E.W., Shephard, G.S., Thiel, P.G., Marasas, W.F.O. & Stockenström, S. (1991) Fumonisin contamination of commercial corn-based human foodstuffs. *J. agric. Food Chem.*, **39**, 2014–2018

Sydenham, E.W., Marasas, W.F.O., Shephard, G.S., Thiel, P.G. & Hirooka, E.Y. (1992) Fumonisin concentrations in Brazilian feeds associated with field outbreaks of confirmed and suspected animal mycotoxicoses. *J. agric. Food Chem.*, **40**, 994–997

Sydenham, E.W., Shepard, G.S., Gelderblom, W.C.A., Thiel, P.G. & Marasas, W.F.O. (1993a) Fumonisins: Their implications for human and animal health. In: Scudamore, K., ed., *Proceedings of the UK Workshop on Occurrence and Significance of Mycotoxins*, Slough, Ministry of Agriculture, Fisheries and Food, Central Science Laboratory, pp. 42–48

Sydenham, E.W., Shephard, G.S., Thiel, P.G., Marasas, W.F.O., Rheeder, J.P., Sanhueza, C.E.P., González, H.H.L. & Resnick, S.L. (1993b) Fumonisins in Argentinian field-trial corn. *J. agric. Food Chem.*, **41**, 891–895

Sydenham, E.W., Shephard, G.S., Stockenström, S., Rheeder, J.P., Marasas, W.F.O. & van der Merwe, M.J. (1997) Production of fumonisin B analogues and related compounds by *Fusarium globosum*, a newly described species from corn. *J. agric. Food Chem.*, **45**, 4004–4010

Tejada-Simon, M.V., Marovatsanga, L.T. & Pestka, J.J. (1995) Comparative detection of fumonisin by HPLC, ELISA, and immunocytochemical localization in *Fusarium* cultures. *J. Food Prot.*, **58**, 666–672

Thiel, P.G., Gelderblom, W.C.A., Marasas, W.F.O., Nelson, P.E. & Wilson, T.M. (1986) Natural occurrence of moniliformin and fusarin C in corn screenings known to be hepatocarcinogenic in rats. *J. agric. Food Chem.*, **34**, 773–775

Thiel, P.G., Marasas, W.F.O., Sydenham, E.W., Shephard, G.S., Gelderblom, W.C.A. & Nieuwenhuis, J.J. (1991a) Survey of fumonisin production by *Fusarium* species. *Appl. environ. Microbiol.*, **57**, 1089–1093

Thiel, P.G., Shephard, G.S., Sydenham, E.W., Marasas, W.F.O., Nelson, P.E. & Wilson, T.M. (1991b) Levels of fumonisins B_1 and B_2 in feeds associated with confirmed cases of equine leukoencephalomalacia. *J. agric. Food Chem.*, **39**, 109–111

Thiel, P.G., Marasas, W.F.O., Sydenham, E.W., Shephard, G.S. & Gelderblom, W.C.A. (1992) The implications of naturally occurring levels of fumonisins in corn for human and animal health. *Mycopathologia*, **117**, 3–9

Tolleson, W.H., Couch, L.H., Melchior, W.B., Jr, Jenkins, G.R., Muskhelishvili, M., Muskhelishvili, L., McGarrity, L.J., Domon, O., Morris, S.M. & Howard, P.C. (1999) Fumonisin B_1 induces apoptosis in cultured human keratinocytes through sphinganine accumulation and ceramide depletion. *Int. J. Oncol.*, **14**, 823–843

Torres, M.R., Sanchis, V. & Ramos, A.J. (1998) Occurrence of fumonisins in Spanish beers analyzed by an enzyme-linked immunosorbent assay method. *Int. J. Food Microbiol.*, **39**, 139–143

Trucksess, M.W., Stack, M.E., Allen, S. & Barrion, N. (1995) Immunoaffinity column coupled with liquid chromatography for determination of fumonisin B_1 in canned and frozen sweet corn. *J. Assoc. off. anal. Chem. int.*, **78**, 705–710

Tseng, T.C. & Liu, C.Y. (1997) Natural occurrence of fumonisins in corn-based foodstuffs in Taiwan. *Cereal Res. Commun.*, **25**, 393–394

Tsunoda, M., Shanna, R.P. & Riley, R.T. (1998) Early fumonisin B_1, toxicity in relation to disrupted sphingolipid metabolism in male BALB/c mice. *J. biochem. mol. Toxicol.*, **12**, 281–289

Turner, P.C., Nikiema, P. & Wild, C.P. (1999) Fumonisin contamination of food: Progress in development of biomarkers to better assess human health risks. *Mutat. Res.*, **443**, 81–93

Ueno, Y., Aoyama, S., Sugiura, Y., Wang, D.-S., Lee, U.-S., Hirooka, E.Y., Hara, S., Karki, T., Chen, G. & Yu, S.-Z. (1993) A limited survey of fumonisins in corn and corn-based products in Asian countries. *Mycotoxin Res.*, **9**, 27–34

Ueno, Y., Iijima, K., Wang, S.-D., Sugiura, Y., Sekijima, M., Tanaka, T., Chen, C. & Yu, S.-Z. (1997) Fumonisins as a possible contributory risk factor for primary liver cancer: A 3-year study of corn harvested in Haimen, China by HPLC and ELISA. *Food chem. Toxicol.*, **35**, 1143–1150

Usleber, E., Straka, M. & Terplan, G. (1994a) Enzyme immunoassay for fumonisin B_1 applied to corn-based food. *J. agric. Food Chem.*, **42**, 1392–1396

Usleber, E., Schlichtherle, C. & Märtlbauer, E. (1994b) [Occurrence of mycotoxins in the environment. Fumonisins.] *Mag. Food Dairy Ind.*, **115**, 1220–1227 [in German]

van der Westhuizen, L., Brown, N.L., Marasas, W.F.O., Swanevelder, S. & Shephard, G.S. (1999) Sphinganine/sphingosine ratio in plasma and urine as a possible biomarker for fumonisins exposure in humans in rural areas of Africa. *Food chem. Toxicol.*, **37**, 1153–1158

van der Westhuizen, L., Shephard, G.S. & van Schalkwyk, D.J. (2001) The effect of a single gavage dose of fumonisin B_1 on the sphinganine and sphingosine levels in vervet monkeys. *Toxicon*, **39**, 273–281

Vasanthi, S. & Bhat, R.V. (1998) Mycotoxins in foods — Occurrence, health and economic significance and food control measures. *Indian J. med. Res.*, **108**, 212–224

Viljoen, J.H., Marasas, W.F.O. & Thiel, P.G. (1994) Fungal infection and mycotoxic contamination of commercial maize. In: Du Plessis, J.G., Van Rensberg, J.B.J., McLaren, N.W. &

Flett, B.C., eds, *Proceedings of the Tenth South African Maize Breeding Symposium*, Potchefstroom, Department of Agriculture, pp. 26–37

Visconti, A. & Doko, M.B. (1994) Survey of fumonisin production by *Fusarium* isolated from cereals in Europe. *J. Assoc. off. anal. Chem. int.*, **77**, 546–550

Visconti, A., Doko, M.B., Bottalico, C., Schurer, B. & Boenke, A. (1994) Stability of fumonisins (FB_1 and FB_2) in solution. *Food Addit. Contam.*, **11**, 427–431

Visconti, A., Boenke, A., Doko, M.B., Solfrizzo, M. & Pascale, M. (1995) Occurrence of fumonisins in Europe and the BCR-measurements and testing projects. *Natural Toxins*, **3**, 269–274

Visconti, A., Solfrizzo, M. & De Girolamo, A. (2001) Determination of fumonisins B_1 and B_2 in corn and corn flakes by liquid chromatography with immunoaffinity column cleanup: Collaborative study. *J. Assoc. off. anal. Chem. int.*, **84**, 1828-1837

Voss, K.A., Chamlerlain, W.J., Bacon, C.W. & Norred, W.P. (1993) A preliminary investigation on hepatic toxicity in rats fed purified fumonisin B_1. *Natural Toxins*, **1**, 222–228

Voss, K.A., Chamlerlain, W.J., Bacon, C.W., Riley, R.T. & Norred, W.P. (1995a) Subchronic toxicity of fumonisin B_1 to male and female rats. *Food Addit. Contam.*, **121**, 473–478

Voss, K.A., Chamberlain, W.J., Bacon, C.W., Herbert, R.A., Walters, D.B. & Norred, W.P. (1995b) Subchronic feeding study of the mycotoxin fumonisin B_1 in B6C3F_1 mice and Fischer 344 rats. *Fundam. appl. Toxicol.*, **24**, 102–110

Voss, K.A., Riley, R.T., Bacon, C.W., Chamberlain, W.J. & Norred, W.P. (1996a) Subchronic toxic effects of Fusarium moniliforme and fumonisin B_1 in rats and mice. *Natural Toxins*, **4**, 16–23

Voss, K. A., Bacon, C. W., Norred, W. P., Chapin, R.E., Chamberlain, W. J., Plattner, R.D. & Meredith, F.I. (1996b) Studies on the reproductive effects of *Fusarium moniliforme* culture material in rat and the biodistribution of [^{14}C]fumonisin B_1 in pregnant rats. *Natural Toxins*, **4**, 24–33

Voss, K.A., Bacon, C.W., Meredith, F.I. & Norred, W.P. (1996c) Comparative subchronic toxicity studies of nixtamalized and water-extracted *Fusarium moniliforme* culture material. *Food chem. Toxicol.*, **34**, 623–632

Voss, K.A., Riley, R.T., Bacon, C.W., Meredith, F.I. & Norred, W.P. (1998) Toxicity and sphinganine levels are correlated in rats fed fumonisin B_1 (FB_1) or hydrolyzed FB_1. *Environ. Toxicol. Pharmacol.*, **5**, 101–104

Voss, K.A., Riley, R.T., Norred, W.P., Bacon, C.W., Meredith, F.I., Howard, P.C., Plattner, R.D., Collins, T.F.X., Hansen, D.K. & Porter, J.K. (2001) An overview of rodent toxicities: Liver, and kidney effects of fumonisins and *Fusarium moniliforme*. *Environ. Health Perspect.*, **109** (Suppl. 2), 259–266

Vudathala, D.K., Prelusky, D.B., Ayroud, M., Trenholm, H.L. & Miller, J.D. (1994) Pharmacokinetic fate and pathological effects of ^{14}C-fumonisin B_1 in laying hens. *Natural Toxins*, **2**, 81–88

Wang, E., Norred, W.P., Bacon, C.W., Riley, R.T. & Merrill, A.H., Jr (1991) Inhibition of sphingolipid biosynthesis by fumonisins: Implications for diseases associated with *Fusarium moniliforme*. *J. biol. Chem.*, **266**, 14486–14490

Wang, E., Ross, P.F., Wilson, T.M., Riley, R.T. & Merrill, A.H., Jr (1992) Increases in serum sphingosine and sphinganine and decreases in complex sphingolipids in ponies given feed containing fumonisins, mycotoxins produced by *Fusarium moniliforme*. *J. Nutr.*, **122**, 1706–1716

Wang, D.-S., Sugiura, Y., Ueno, Y., Buddhanont, P. & Suttajit, M. (1993) A limited survey for the natural occurrence of fumonisins in Thailand. *Thai J. Toxicol.*, **9**, 42–46

Wang, D.-S., Liang, Y.-X., Chau, N.T., Dien, L.D., Tanaka, T. & Ueno, Y. (1995) Natural co-occurrence of *Fusarium* toxins and aflatoxin B_1 in corn for feed in North Vietnam. *Natural Toxins*, **3**, 445–449

Wang, E., Riley, R.T., Meredith, F.I. & Merrill, A.H., Jr (1999) Fumonisin B_1 consumption by rats causes reversible, dose-dependent increases in urinary sphinganine and sphingosine. *J. Nutr.*, **129**, 214–220

Weibking, T.S., Ledoux, D.R., Bermudez, A.J., Turk, J.R., Rottinghaus, G.E, Wang, E. & Merrill, A.H., Jr (1993) Effects of feeding *Fusarium moniliforme* culture material, containing known levels of fumonisin B_1, on the young broiler chick. *Poult. Sci.*, **72**, 456–466

WHO (2000) *Fumonisin B_1* (Environmental Health Criteria 219), Geneva, International Programme on Chemical Safety, World Health Organization

WHO (2002) *Evaluation of Certain Mycotoxins in Food* (WHO Technical Report Series No. 906), *Fifty-sixth Report of the Joint FAO/WHO Expert Committee on Food Additives*, Geneva, World Health Organization

Wild, C.P., Castegnaro, M., Ohgaki, H., Garren, L., Galendo, D. & Miller, J.D. (1997) Absence of a synergistic effect between fumonisin B_1 and *N*-nitrosomethylbenzylamine in the induction of oesophageal papillomas in the rat. *Natural Toxins*, **5**, 126–131

Wilson, T.M., Nelson, P.E. & Knepp, C.R. (1985) Hepatic neoplastic nodules, adenofibrosis, and cholangiocarcinomas in male Fischer 344 rats fed corn naturally contaminated with *Fusarium moniliforme*. *Carcinogenesis*, **6**, 1155–1160

Wilson, T.M., Ross, P.F., Rice, L.G., Osweiler, G.D., Nelson, H.A., Owens, D.L., Plattner, R.D., Reggiardo, C., Noon, T.H. & Pickrell, J.W. (1990) Fumonisin B_1 levels associated with an epizootic of equine leukoencephalomalacia. *J. vet. diagn. Invest.*, **2**, 213–216

Wilson, T.M., Ross, P.F., Owens, D.L., Rice, L.G., Green, S.A., Jenkins, S.J. & Nelson, H.A. (1992) Experimental reproduction of ELEM — A study to determine the minimum toxic dose in ponies. *Mycopathologia*, **117**, 115–120

Wölfle, D., Schmutte, C., Westendorf, J. & Marquardt, H. (1990) Hydroxyanthraquinones as tumor promoters: Enhancement of malignant transformation of C3H mouse fibroblasts and growth stimulation of primary rat hepatocytes. *Cancer Res.*, **50**, 6540–6544

Wu, J.M., DiPietrantonio, A.M. & Hsieh, T.-C. (2001) Mechanism of fenretinide (4-HPR)-induced cell death. *Apoptosis*, **6**, 377–388

Yamashita, A., Yoshizawa, T., Aiura, Y., Sanchez, P., Dizon, E.I., Arim, R.H. & Sardjono, (1995) Fusarium mycotoxins (fumonisins, nivalenol and zearalenone) and aflatoxins in corn from southeast Asia. *Biosci. Biotechnol. Biochem.*, **59**, 1804–1807

Yasui, Y. & Takeda, N. (1983) Identification of a mutagenic substance in *Rubia tinctorum* L. (madder) root as lucidin. *Mutat. Res.*, **121**, 185–190

Yeung, J.M., Wang, H.-Y. & Prelusky, D.B. (1996) Fumonisin B_1 induces protein kinase C translocation via direct interaction with diacylglycerol binding site. *Toxicol. appl. Pharmacol.*, **141**, 178–184

Yoshizawa, T. & Gao, H.-P. (1999) Risk assessment of mycotoxins in staple foods from the high-risk area for human esophageal cancer in China. *Mycotoxins*, **Suppl. 99**, 55-62

Yoshizawa, T., Yamashita, A. & Luo, Y. (1994) Fumonisin occurrence in corn from high- and low-risk areas for human esophageal cancer in China. *Appl. environ. Microbiol.*, **60**, 1626–1629

Yu, C.-H., Lee, Y.-M., Yun, Y.-P. & Yoo, H.-S. (2001) Differential effects of fumonisin B_1 on cell death in cultured cells: The significance of elevated sphinganine. *Arch. pharmacol. Res.*, **24**, 136–143

Zacharias, C., Van Echten-Deckert, G., Wang, E., Merrill, A.H. & Sandhoff, K. (1996) The effect of fumonisin B_1 on developing chick embryos: Correlation between de novo sphingolipid biosynthesis and gross morphological changes. *Glycoconj. J.*, **13**, 167–175

Zhong, L., Manzi, A., Skowronski, E., Notterpek, L., Fluharty, A.L., Faull, K.F., Masada, I., Rabizadeh, S., Varsanyi-Nagy, M., Ruan, Y., Oh, J.D., Butcher, L.L. & Bredesen, D.E. (2001) A monoclonal antibody that induces neuronal apoptosis binds a metastasis marker. *Cancer Res.*, **61**, 5741–5748

Zoller, O., Sager, F. & Zimmerli, B. (1994) [Occurrence of fumonisins in food.] *Mitt. Gebiete. Lebensm. Hyg.*, **85**, 81–99 (in German)

NAPHTHALENE

1. Exposure Data

1.1 Chemical and physical data

1.1.1 *Nomenclature*

Chem. Abstr. Serv. Reg. No.: 91-20-3
Chem. Abstr. Name: Naphthalene
IUPAC Systematic Name: Naphthalene
Synonyms: Naphthalin; naphthene; tar camphor; white tar

1.1.2 *Structural and molecular formulae and relative molecular mass*

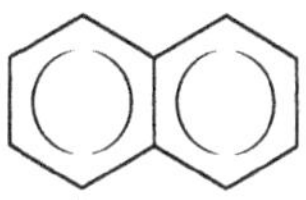

$C_{10}H_8$ Relative molecular mass: 128.17

1.1.3 *Chemical and physical properties of the pure substance*

(*a*) *Description*: White monoclinic prismatic plates (Lide & Milne, 1996; O'Neil *et al.*, 2001)
(*b*) *Boiling-point*: 217.9 °C, sublimes (Lide & Milne, 1996; Verschueren, 1996)
(*c*) *Melting-point*: 80.2 °C (Lide & Milne, 1996)
(*d*) *Density*: d_4^{20} 1.162 (O'Neil *et al.*, 2001)
(*e*) *Spectroscopy data*: Infrared (prism [865]; grating [169]), ultraviolet [265], nuclear magnetic resonance (proton [62]; ^{13}C [139]) and mass spectral data have been reported[1] (Sadtler Research Laboratories, 1980; Lide & Milne, 1996)
(*f*) *Solubility*: Slightly soluble in water (31–34 mg/L at 25 °C; Verschueren, 2001); soluble in ethanol and methanol; very soluble in acetone, benzene, carbon

[1] The numbers in brackets are referenced in Sadtler Research Laboratories (1980).

disulfide, carbon tetrachloride, chloroform and diethyl ether (Lide & Milne, 1996)

(*g*) *Volatility*: Vapour pressure, 0.011 kPa at 25 °C (Lide & Milne, 1996); relative vapour density (air = 1), 4.42 (Verschueren, 1996); flash-point, 88 °C (closed-cup) (O'Neil *et al.*, 2001)

(*h*) *Stability*: Volatilizes appreciably at room temperature; sublimes appreciably at temperatures above the melting-point (O'Neil *et al.*, 2001)

(*i*) *Octanol/water partition coefficient (P)*: log P, 3.30 (Sangster, 1989)

(*j*) *Conversion factor*[1]: $mg/m^3 = 5.24 \times ppm$

1.1.4 *Technical products and impurities*

Naphthalene is usually sold commercially according to its freezing or solidification point, because there is a correlation between the freezing-point and the naphthalene content of the product; the correlation depends on the type and relative amount of impurities that are present. Because the freezing point is changed appreciably by the presence of water, values and specifications are listed as dry, wet or as-received basis, using an appropriate method, e.g., ASTM D1493 (Mason, 1995).

Typical specifications are: for crude naphthalene-CRI, solidification point, 77.5 °C min; crude naphthalene-CRII, solidification point, 78.5 °C min.; crude naphthalene-CRIII, solidification point, 79.3 °C min; [sulfur, 0.5% max.; non-volatiles, 0.25% max.; and water content, 0.5% max.]. Typical specifications for highly refined naphthalene-RFII are: solidification point, 79.9 °C min.; refined naphthalene-RFI, solidification, 79.6 °C min; naphthalene content, 99.0% min.; [sulfur, 0.1% max.; and non-volatiles, 0.005% max.] (Recochem, 1995a,b).

The naphthalene content of the technical product is at least 95%. Virtually the sole impurity found in naphthalene obtained from coal tar is benzo[*b*]thiophene (thianaphthene). Methylindenes are essentially the only impurities found in naphthalene derived from petroleum (BUA, 1989).

Trade names for naphthalene include: Albocarbon; Dezodorator (National Toxicology Program, 2000).

1.1.5 *Analysis*

Gas–liquid chromatography is used extensively to determine the naphthalene content of mixtures. Naphthalene can be separated easily from thionaphthalene, the methyl- and dimethylnaphthalenes and other aromatics. Analysis of other impurities may require the use of high-resolution capillary columns (Mason, 1995). Selected methods for the analysis of naphthalene in various media are presented in Table 1.

---

[1] Calculated from: mg/m^3 = (relative molecular mass/24.45) × ppm, assuming a temperature of 25 °C and a pressure of 101 kPa

Table 1. Selected methods for analysis of naphthalene

Sample matrix	Sample preparation	Assay procedure[a]	Limit of detection	Reference
Air	Adsorb (charcoal or Chromosorb W); desorb (carbon disulfide)	GC/FID	1–10 μg/sample; 4 μg/sample	Eller (1994) [Method 1501]; Occupational Safety and Health Administration (1990) [Method 35]
	Adsorb (solid sorbent); desorb (organic solvent)	HPLC/UV	0.6–13 μg/sample	Eller (1994) [Method 5506]
	Adsorb (solid sorbent); desorb (organic solvent)	GC/FID	0.3–0.5 μg/sample	Eller (1994) [Method 5515]
Drinking-, ground- and surface water	Purge (inert gas); trap (Chromosorb W); desorb into capillary GC column	GC/MS	0.04 μg/L	Environmental Protection Agency (1995a) [Method 524.2]
Drinking-water and raw source water	Purge (inert gas); trap (Chromosorb W); desorb into capillary GC column	GC/PID	0.01–0.05 μg/L	Environmental Protection Agency (1995b) [Method 502.2]
Drinking-water	Extract in liquid–solid extractor; elute with dichloromethane; dry; concentrate	HPLC/UV/FD	2.20 μg/L	Environmental Protection Agency (1990a) [Method 550.1]
Wastewater, municipal and industrial	Extract with dichloromethane; dry; concentrate	HPLC/UV or GC/FID	1.8 μg/L	Environmental Protection Agency (1996a,b, 1999a) [Methods 610, 8100 & 8310]
	Extract with dichloromethane; dry; concentrate	GC/MS	1.6 μg/L	Environmental Protection Agency (1999b) [Method 625]
	Add isotope-labelled analogue; extract with dichloromethane; dry over sodium sulfate; concentrate	GC/MS	10 μg/L	Environmental Protection Agency (1999c) [Method 1625B]

Table 1 (contd)

Sample matrix	Sample preparation	Assay procedure	Limit of detection	Reference
Solid waste matrices[b]	Purge (inert gas); trap (Tenax or Chromosorb W); desorb into capillary GC column	GC/PID	0.06 μg/L	Environmental Protection Agency (1996c) [Method 8021B]
	Purge (inert gas); trap (suitable sorbent); thermal desorption *or* headspace sampling *or* direct injection	GC/MS	0. 04–0.1 μg/L	Environmental Protection Agency (1996d) [Method 8260B]
Air sampling media, water samples, solid waste matrices, soil samples	Liquid–liquid extraction *or* Soxhlet extraction *or* ultrasonic extraction *or* waste dilution *or* direct injection	GC/MS	10 μg/L (aqueous); 660 μg/kg (soil/ sediment) (EQL)[c]	Environmental Protection Agency (1996e) [Method 8270C]
Soils, sludges, solid wastes	Thermal extraction; concentrate; thermal desorption	GC/MS	0.01–0.5 mg/kg	Environmental Protection Agency (1996f) [Method 8275A]
Wastewater, soil, sediment, solid waste	Liquid–liquid extraction (water); Soxhlet or ultrasonic extraction (soil/sediment/waste)	GC/FT-IR	20 μg/L	Environmental Protection Agency (1996g) [Method 8410]

[a] Abbreviations: GC, gas chromatography; FID, flame ionization detection; FT-IR, Fourier transform infrared detection; MS, mass spectrometry; PID, photoionization detection; HPLC, high-performance liquid chromatography; UV, ultraviolet detection; FD, fluorescence detection

[b] Includes: groundwater, aqueous sludges, caustic and acid liquors, waste solvents, oily wastes, mousses, tars, fibrous wastes, polymeric emulsions, filter cakes, spent carbons, spent catalysts, soils, and sediments

[c] EQL, estimated quantitation limit

1.2 Production

Naphthalene is produced commercially from either coal tar or petroleum. Naphthalene has long been produced by the destructive distillation of high-temperature coal tars, called carbonization or coking (IARC, 1985). Coal tar was the traditional source of naphthalene until the late 1950s when it was in short supply, and the generation of naphthalene from petroleum by dealkylation of aromatics-rich fractions from reforming and catalytic cracking became commercially viable (IARC, 1989). In 1960, the first petroleum–naphthalene plant was brought on stream in the USA and, by the late 1960s, petroleum-derived naphthalene accounted for over 40% of total US naphthalene production. The availability of large quantities of *ortho*-xylene during the 1970s undercut the position of naphthalene as the prime raw material for phthalic anhydride. In 1971, 45% of phthalic anhydride capacity in the USA was based on naphthalene, as compared with only 29% in 1979 and 17% in 1990. The last dehydroalkylation plant for petroleum naphthalene was shut down late in 1991 (Mason, 1995).

World production of naphthalene in 1987 was around one million tonnes; about one-fourth came from western Europe (210 thousand tonnes), one-fifth each from Japan (175 thousand tonnes) and eastern Europe (180 thousand tonnes) and one-eighth from the USA (107 thousand tonnes). In 2000, over 90% of naphthalene in the USA was produced from coal tar; most naphthalene in western Europe was produced from coal tar; and all naphthalene produced in Japan was from coal tar (Lacson, 2000). Naphthalene supply and demand by major region in 2000 is presented in Table 2. Available information on production trends in Japan, the USA and western Europe is summarized in Table 3.

Table 2. Naphthalene supply and demand by major region in 2000 (thousand tonnes)[a]

Region	Capacity	Production	Consumption
Japan	221	179	172
USA	143	107	109
Western Europe	230	205	133
Total	594	491	414

[a] From Lacson (2000); data for Japan are from 1999.

Information available in 2001 indicated that crude naphthalene was manufactured by 36 companies in China, six companies in Japan, four companies each in Brazil and Russia, three companies each in Spain and the USA, two companies each in Argentina, India and Ukraine, and one company each in Australia, Bangladesh, Belgium, Bosnia, Canada, Colombia, Denmark, Egypt, France, Italy, Korea (Republic of), Mexico, the Netherlands, Turkey and the United Kingdom. Refined naphthalene was manufactured by 16 companies in China, five companies each in India and the USA, four companies

Table 3. Naphthalene production (thousand tonnes)[a]

		1965	1970	1975	1980	1985	1990	1995	2000
Japan	Crude	NR	NR	90.1	128.6	175.5	202.2	192.7	172.9[b]
	Refined	NR	NR	7.2	8.8	13.0	14.2	10.9	6.4[b]
USA	From coal tar	210.5	194.1	159.2	142.4	83.5	81.6	100.2	99.3
	From petroleum	157.4	132.0	50.0	61.7	24.9	22.7	7.3	7.3
	Total	367.9	326.1	209.1	204.1	108.4	104.3	107.5	106.6
Western Europe	Crude	NR	NR	NR	NR	212	210	160	166

[a] From Lacson (2000)
[b] Data reported for 1999
NR, not reported

each in Spain and Turkey, three companies in Japan, and one company each in Canada, the Czech Republic, Egypt, France, Italy, Korea (Republic of), the Netherlands, Ukraine and the United Kingdom. Naphthalene (grade unspecified) was manufactured by 39 companies in China, three companies in Ukraine, two companies each in Germany and Mexico, and one company each in Brazil, India, Japan, Russia, Turkey, the United Kingdom and the USA (Chemical Information Services, 2001).

1.3 Use

The main use for naphthalene worldwide is the production of phthalic anhydride by vapour-phase catalytic oxidation, particularly in Japan and the USA, where this accounted for 73% and 60% of naphthalene demand, respectively, in 1999. Phthalic anhydride is used as an intermediate for polyvinyl chloride plasticizers, such as di(2-ethylhexyl) phthalate. Naphthalene is also used in the manufacture of a wide variety of intermediates for the dye industry; in the manufacture of synthetic resins, celluloid, lampblack and smokeless powder; and in the manufacture of hydronaphthalenes (Tetralin (tetrahydronaphthalene), Decalin (decahydronaphthalene)) which are used as solvents, in lubricants and in motor fuels (Mason, 1995; Lacson, 2000; O'Neil *et al.*, 2001).

Naphthalene sulfonates represent a growing outlet for naphthalene. The products are used as wetting agents and dispersants in paints and coatings and in a variety of pesticides and cleaner formulations. Naphthalene is also a starting material for the manufacture of 1-naphthyl-*N*-methylcarbamate (carbaryl), an insecticide, and several other organic compounds and intermediates (Mason, 1995; Lacson, 2000).

The use of naphthalene as a moth-repellent and insecticide is decreasing due to the introduction of chlorinated compounds such a *para*-dichlorobenzene. In 2000, about 6500 tonnes of naphthalene were used (in Japan (1100 tonnes), the USA (450 tonnes) and Europe (5000 tonnes)), in moth-proofing and fumigation. Another new use for naphthalene is in production of polyethylene naphthalene for making plastic beer bottles. It has also been used in veterinary medicine in dusting powders, as an insecticide and internally as an intestinal antiseptic and vermicide (Sax & Lewis, 1987; Agency for Toxic Substances and Disease Registry, 1995a; Mason, 1995; Budavari, 1998; Lacson, 2000; O'Neil *et al.*, 2001).

Consumption of naphthalene by major region in selected years is presented in Table 4.

Table 4. Consumption of naphthalene by major region (thousand tonnes)[a]

End use	Japan		USA		Western Europe	
	1995	1999	1995	2000	1995	2000
Phthalic anhydride	137	124	66	66	42	45
Naphthalene sulfonates[b]	16	9	21	27	34	45
Pesticides[c]	2	1	17	14	15	22
Dyestuff intermediates	22	23	–	–	14	11
Other[d]	16	15	2	3	14	10
Total[e]	193	172	106	109	119	133

[a] From Lacson (2000)
[b] Includes alkylnaphthalene sulfonates and naphthalene sulfonate–formaldehyde condensates (NSF). NSF includes concrete additives and synthetic tanning agents.
[c] Includes carbaryl and moth repellents.
[d] Includes diisopropyl naphthalene, naphthalene dicarboxylic acid, tetrahydronaphthalene (Tetralin), decahydronaphthalene (Decalin) and chloronaphthalenes.
[e] Totals may not equal sums of the columns because of independent rounding.

1.4 Occurrence

1.4.1 *Natural occurrence*

Naphthalene, discovered in 1819 by A. Garden (BUA, 1989), is a natural constituent of coal tar and crude oil, which are major contributors to its presence in the environment. They contain up to 11% and 1.3% of the chemical, respectively (BUA, 1989; O'Neil *et al.*, 2001). Forest fires also contribute to the presence of naphthalene in the environment, as the chemical is a natural combustion product of wood (Agency for Toxic Substances and Disease Registry, 1995a).

1.4.2 *Occupational exposure*

From the National Occupational Exposure Survey conducted between 1981–83, the National Institute for Occupational Safety and Health (NIOSH) estimated that approximately 113 000 workers, about 4.6% females, in 31 major industrial groups were potentially exposed to naphthalene in the USA. The top six industries, by total workers, accounted for over 50% of the total potentially exposed workers. The petroleum and coal products and oil and gas extraction industries were among the top three industries and comprised about 21.4% of the workers potentially exposed to naphthalene. An estimated 1840 agricultural services workers were exposed to naphthalene; over 87% were females (National Institute for Occupational Safety and Health, 1990; National Toxicology Program, 1992).

Naphthalene has been measured in a wide variety of workplaces for many years. On the basis of the major results summarized in Table 5, the following ranking of respiratory exposure to naphthalene in the major industries can be made: creosote impregnation > coke manufacturing > asphalt industry > other industries. Exposure to naphthalene has also been extensively measured in Germany. Samples collected between 1991 and 1995 were predominantly from wood manufacturing (50%), construction (16%) and metal-working and machine construction (12%). Ninety-five per cent of the available 183 measurements from 94 factories were below the analytical limit of detection (1.0 mg/m^3 for a 2-h sampling time). Only in the manufacture of repellents and perfumed disinfectants were concentrations above the limit of detection (Bock *et al.*, 1999).

Naphthalene is the most abundant component of creosote vapour and constitutes 10–16 wt% of creosote oils (Nylund *et al.*, 1992). Dermal wipe samples were taken in a study among asphalt workers; less than 10% of the samples showed detectable concentrations ranging from 5.5 to 520 ng/cm^2 (Hicks, 1995).

Urinary naphthols have been measured as biomarkers of occupational exposure to naphthalene (Bieniek, 1997) (see Section 4.1.1(*c*)).

1.4.3 *Environmental occurrence*

The extensive use of naphthalene as an intermediate in the production of plasticizers, resins, insecticides and surface active agents, its presence as a major component of coal tar and coal-tar products such as creosote and its inclusion in a wide variety of consumer products (e.g., moth-repellents) has led to its frequent occurrence in industrial effluents and outdoor and indoor environments (Agency for Toxic Substances and Disease Registry, 1995a,b; Environmental Protection Agency, 2001).

Naphthalene has been identified in the USA by the Environmental Protection Agency (2001) and the Agency for Toxic Substances and Disease Registry (1995b) as one of the most commonly found substances at hazardous waste sites on the 'National Priorities List'. In the USA, naphthalene is listed as one of 189 hazardous air pollutants under the Clean Air Act Amendments of 1990 (Title III) of the Environmental Protection Agency which mandates reduction of its emissions (Environmental Protection Agency, 1990b; Kelly *et al.*, 1994). Naphthalene features in the Canadian Priority List of hazardous substances (Fellin & Otson, 1994).

The general population is exposed to naphthalene principally by inhalation of ambient and indoor air, with naphthalene-containing moth-repellents and tobacco smoke as the main contributors. Another source is the use of kerosene heaters (Traynor *et al.*, 1990). Assuming an urban/suburban average air concentration of 0.95 μg/m^3 and an inhalation rate of 20 m^3 per day, it has been estimated that the average daily intake of naphthalene from ambient air in the USA is 19 μg (Howard, 1989; Agency for Toxic Substances and Disease Registry, 1995a). Much lower exposure to naphthalene may occur from ingestion of drinking-water and/or food. Estimated exposure from drinking-

Table 5. Occupational exposure to naphthalene in various industries

Industry	Country	Year	No. of positive samples	TWA (µg/m³)	Range (µg/m³)	Phase	Reference
Hot mix plants	USA	–	8	2.3		Fume/vapour	Hicks (1995)
Paving	USA	–	9	6.5		Fume/vapour	Hicks (1995)
Paving, roofing, steel and silicon carbide industries	Canada	–	51	11.4		Vapour (particulate)	Lesage *et al.* (1987)
				0.1		Fume (gaseous)	
Roofing/waterproofing	USA	1985	11		0–1.9		Zey & Stephenson (1986)
Roofing manufacturing	USA	–	7	7.5		Fume/vapour	Hicks (1995)
Roofing			11	5.2		Fume/vapour	
Enhanced oil recovery	USA	1986			5.0–11.0		Daniels & Gunter (1988)
Refineries/terminals	USA	–	9	5.5		Fume/vapour	Hicks (1995)
Coke manufacturing	Belgium		16		0.7–959	Vapour	Buchet *et al.* (1992)
Coke plant	Finland	1988–90	90[b]	44–500	111–1989	Vapour	Yrjänheikki *et al.* (1995)
(modern technology)	Poland		66		0–6000	Vapour	Bieniek (1994)
Coke plant, tar distillation	Poland		69	773[a]	–	Vapour	Bieniek (1997)
Coke plant, naphthalene oil distillation	Poland		33	867[a]	–	Vapour	Bieniek (1997)
Coke plant	Poland	1997	48	170–1210	10–3280	Vapour	Bieniek (1998)
Creosote impregnation (wood)	Finland		18	2200		Vapour	Heikkilä *et al.* (1987)
Switch assembly (wood)	Finland		8	2600		Vapour	Heikkilä *et al.* (1987)
Creosote impregnation (wood)	Finland		30	1540	400–4200	Vapour	Heikkilä *et al.* (1997)
	Finland		15	1000	22–1960	Vapour	Heikkilä *et al.* (1995)
Construction	Finland		1	160		Vapour	Heikkilä *et al.* (1995)
Aluminium refinery	Canada		7	1111		Vapour	Lesage *et al.* (1987)
			7	0.5		Fume	

Table 5 (contd)

Industry	Country	Year	No. of positive samples	TWA (µg/m^3)	Range (µg/m^3)	Phase	Reference
Manufacture of graphite electrodes	Belgium		106		0.2–1212	Vapour	Buchet *et al.* (1992)
Steel industry	USA	1982	NR		< 107		Almaguer & Orris (1985)
Aluminium reduction plant	Norway				0.72–311	'Gaseous'	Bjørseth *et al.* (1978a)
Coke plant	Norway				11–1151	'Gaseous'	Bjørseth *et al.* (1978b)
Iron foundries	Denmark	1989–90	24	mean, 0.5–10.6		Vapour	Hansen *et al.* (1994)
Silicon carbide	Canada		6	75.4		Vapour	Lesage *et al.* (1987)
			6	0.0		Fume	
Refractory brick	Canada		7	16.3		Vapour	Lesage *et al.* (1987)
			7	0.0		Fume	
Printing industry	USA	1977	8		40.0–12 000		Fannick (1978)
Shoe manufacture	USA	1983	17		1–600		Gunter (1984)
	USA	1984			< 10		Albers (1984)
Forest fighting crews	USA	1991	40		< 6.1	Vapour	Kelly (1992)
Food	Finland		5[c]	[5.8] 3 h	0.2–25.6	Fume/vapour	Vainiotalo & Matveinen (1993)
Fish smokehouses	Denmark		47		2918 (max)	Vapour (gas)	Nordholm *et al.* (1986)

Abbreviations: TWA, arithmetic mean of 8-h time-weighted average personal samples; NR, not reported
Gaseous = vapour
[a] Geometric mean
[b] 2-h TWA
[c] Stationary measurements

water assuming a water concentration of 0.001–2.0 μg/L naphthalene and water consumption of 2 L per day is 0.002–4.0 μg per day (Howard, 1989).

(*a*) *Air*

Most of the naphthalene entering the environment is discharged to the air (92.2%), the largest releases (more than 50%) resulting from the combustion of wood and fossil fuels and the off-gassing of naphthalene-containing moth-repellents and deodorants. In 1989, about 12 million pounds [5.5 million kg] were released from these sources. The highest atmospheric concentrations of naphthalene have been found in the immediate vicinity of specific industrial sources and hazardous waste sites (Agency for Toxic Substances and Disease Registry, 1995a). Air emissions in the USA reported to the Environmental Protection Agency decreased from 1598 tonnes for 473 industrial facilities in 1989 to 1224 tonnes for 744 industrial facilities in 1999 (Environmental Protection Agency, 2001).

The median ambient air concentration of naphthalene determined at nine locations (84 samples) in the USA was 1.2 μg/m^3 (Kelly *et al.*, 1994). In another series of data from 1970–87, the average concentration of naphthalene in ambient air at several locations in the USA was 0.991 ppb [5.19 μg/m^3] for 67 samples, 60 of which were from source-dominated locations (Shah & Heyerdahl, 1988).

In the USA, the Environmental Protection Agency (2000) reviewed studies and calculated summary statistics for concentrations of chemicals in indoor air from selected sources, including naphthalene. Studies were selected which provided the best available estimates of 'typical concentrations' in indoor environments. These sources included the Building Assessment Survey and Evaluation (BASE) study, National Association of Energy Service Companies (NAESCO) study and School Intervention Studies (SIS). The data are reported in Table 6. [The Working group noted the unusually high outdoor air concentration in the School Intervention Studies, the fact that it is based on only three positive observations and that the original data in this review are unpublished.]

Table 6. Typical concentrations of naphthalene in indoor and outdoor air according to the Building Assessment Survey and Evaluation (BASE) study and the School Intervention Studies (SIS)

	BASE (μg/m^3)		SIS (μg/m^3)	
	Indoor	Outdoor	Indoor	Outdoor
Arithmetic mean concentration	0.95	0.31	1.3	290
95th percentile upper limit	2.6	0.81	1.7	1500
No. of buildings	70	69	10	10
No. of observations	209	69	39	10
Frequency of detection	83%	58%	21%	30%

A median naphthalene concentration of 0.18 ppb [0.94 μg/m^3] has been reported in urban air in 11 cities in the USA (Howard, 1989). An average naphthalene concentration of 170 μg/m^3 in outdoor air in a residential area of Columbus, OH (Chuang *et al.*, 1991) and a concentration of 3.3 μg/m^3 naphthalene in ambient air in Torrance, CA, have also been reported (Propper, 1988; Agency for Toxic Substances and Disease Registry, 1995a).

Average naphthalene concentrations in ambient air at five hazardous waste sites and one landfill in New Jersey ranged from 0.42 to 4.6 μg/m^3 (range of arithmetic means) (LaRegina *et al.*, 1986). Naphthalene wasfound at concentrations of less than 0.6 ng/m^3 in the air above shale-oil wastewaters in Wyoming (Hawthorne & Sievers, 1984).

Naphthalene was found in particulate matter in the atmosphere of La Plata, Argentina, at concentrations ranging from 0.18 ± 1.05 ng/m^3 to 13.2 ± 1.30 ng/m^3 when sampled on seven occasions between September 1984 and June 1986 (Catoggio *et al.*, 1989).

In a study to compare naphthalene concentrations (among 35 volatile organic compounds) in indoor and outdoor air in northern Italy in 1983–84, the mean values found in indoor and outdoor samples were 11 and 2 μg/m^3, respectively, with the median indoor/outdoor ratio being 4 for 11 samples (De Bortoli *et al.*, 1986).

Average indoor air concentrations in various residential areas in the USA ranged from 0.75 to 1600 μg/m^3 (Chuang *et al.*, 1991 (measurements done in 1986–87); Wilson & Chung, 1991). A more representative upper limit concentration of naphthalene in indoor air of 32 μg/m^3 was recorded in buildings in heavy traffic urban areas of Taiwan (Hung *et al.*, 1992).

In a study of the effect of smoking on polycyclic aromatic hydrocarbon levels, including naphthalene, in eight homes in the USA, naphthalene was found in the living room of homes of smokers (with gas heating and electric cooking) and homes of non-smokers (electric heating and cooking) at concentrations of 2.2 μg/m^3 and 1.8 μg/m^3, respectively, and the respective outdoors concentrations of naphthalene were 0.33 and 0.11 μg/m^3 (Wilson & Chuang, 1991).

In a summary of concentrations of volatile organic compounds in 230 homes in Germany, naphthalene was found at a mean concentration of 2.3 μg/m^3 with a range of < 1.0–14 μg/m^3; the 50th and 90th percentiles were 2.1 and 3.9 μg/m^3, respectively (Gold *et al.*, 1993). In a study of the relationship between climatic factors and the concentrations of 26 volatile organic compounds in Canadian homes in 1991, naphthalene concentrations in the winter, spring, summer and autumn were 3.24, 1.10, 3.82 and 8.10 μg/m^3, respectively (Fellin & Otson, 1994). In a comparison of naphthalene concentrations in the indoor air of 50 normal and 38 'sick' homes (in which people complained about the odour or had symptoms) in Finland, naphthalene was found at a median concentration of 0.31 μg/m^3 in normal homes (with 6% of normal houses having 1.5–3.1 μg/m^3), while in 'sick' homes 2.6% had concentrations of 1.5–3.1 μg/m^3, 5.3% of 3.1–15 μg/m^3 and 5.3% of 15–62 μg/m^3 (Kostiainen, 1995).

Naphthalene has been identified in the emissions of diesel light-duty vehicles (2–6 mg naphthalene released per km 'distance' driven on a chassis dynamometer)

(Scheepers & Bos, 1992). The average concentration of naphthalene reported inside automobiles in commuter traffic was about 4.5 μg/m^3 (Löfgren *et al.*, 1991).

(*b*) *Water*

Naphthalene released to the atmosphere may be transported to surface water and/or soil by wet or dry deposition. About 2–3% of naphthalene emitted to air is transported to other environmental media by dry deposition (Coons *et al.*, 1982; Agency for Toxic Substances and Disease Registry, 1995a). Naphthalene is degraded in water by photolysis and biological processes. The half-life for photolysis of naphthalene in surface water is about 71 h, but in deeper water (5 m) it is estimated to be 550 days (Agency for Toxic Substances and Disease Registry, 1995a).

Surface water discharges of naphthalene from 744 industrial facilities in the USA in 1999 amounted to 17.7 tonnes, as reported to the Toxics Release Inventory. An additional 73 tonnes of naphthalene were discharged though underground injection (Environmental Protection Agency, 2001). About 5% of all naphthalene entering the environment is released to water, mostly arising from coal tar production and distillation processes (Agency for Toxic Substances and Disease Registry, 1995a).

Naphthalene was detected in 7% of 630 ambient water samples in the USA at a median concentration of less than 10 μg/L, as shown in an analysis of 1980–82 data from the Environmental Protection Agency STORET (STOrage and RETrieval) database (Staples *et al.*, 1985; Agency for Toxic Substances and Disease Registry, 1995a). Naphthalene was also detected in 11% of 86 urban run-off samples up to 1982 at concentrations ranging from 0.8 to 2.3 μg/L (Cole *et al.*, 1984).

In the USA, naphthalene was detected in 35% of samples of groundwater at an average concentration of 3.3 mg/L at five wood-treatment facilities (Rosenfeld & Plumb, 1991) and in leachate or groundwater plume from industrial and municipal landfills at concentration ranges of < 10–19 mg/L and 0.1–19 mg/L, respectively (Brown & Donnelly, 1988). Naphthalene was detected in groundwater samples from three wells at concentrations of 380, 740 and 1800 μg/L, respectively, near an underground coal gasification site in north-western Wyoming (Stuermer *et al.*, 1982). Concentrations of naphthalene ranging from < 0.2 to 63 μg/L were detected at five out of six landfill sites in southern Ontario (Barker, 1987). Naphthalene was found at concentrations of 4.3 and 8.8 μg/L in two groundwater samples collected near an Orange County landfill site in central Florida (and not in surface water) in 1989–90, but not in 1992–93; this was believed to result from the decomposition of municipal solid waste (Chen & Zoltek, 1995).

Naphthalene has been infrequently reported in drinking-water (Agency for Toxic Substances and Disease Registry, 1995a). It was found in four samples of drinking-water extracts at concentrations ranging from 6 to 16 ng/L in Athens, GA, in 1976 (Thruston, 1978) and in another area in the USA at concentrations up to 1.4 μg/L (Coons *et al.*, 1982).

Naphthalene was measured in samples of raw river water from the Adige River, Italy, collected at 19 sampling stations in the Trento province, during two campaigns in 1989, and detected at average concentrations of 51 ng/L (range, 3–109) and 284 ng/L (range, 3–2240), respectively (Benfenati *et al.*, 1992). It was also detected in two polluted rivers, Besós and Llobregat, in Barcelona, Spain, in 1985–86 at mean concentrations of 1300 (SD, 150) ng/L and 180 (SD, 130) ng/L, respectively (Gomez-Belinchon *et al.*, 1991). Naphthalene was one of the main aromatic hydrocarbons detected (at a concentration of 0.02 μg/L) in surface waters of Admiralty Bay, King George Island, Antarctica, during the summers of 1989, 1990, 1992 and 1993 (Bícego *et al.*, 1996).

Naphthalene was found at concentrations of 0.5–35 ng/L (arithmetic mean, 12 ng/L) over 15 months at a coastal site near piers extending into Vineyard Sound, MA, USA. A dominant wintertime source considered was use of space-heating oil, with higher concentrations of naphthalene found in the winter and lowest concentrations found in the summer (Gschwend *et al.*, 1982).

(*c*) *Soil and sediments*

Releases of naphthalene to land from 744 industrial facilities in the USA in 1999 amounted to 66 tonnes (Environmental Protection Agency, 2001).

In untreated agricultural soils, naphthalene has been found at concentrations ranging from 0 to 3 μg/kg in 1942–84 (Wild *et al.*, 1990). It has been found at 6.1 (SD, 0.2) mg/kg in coal tar-contaminated soil (Yu *et al.*, 1990), at 16.7 mg/kg in soil from a former tar-oil refinery (Weissenfels *et al.*, 1992; Agency for Toxic Substances and Disease Registry, 1995a) and at up to 66 μg/kg in sludge-treated soils (Wild *et al.*, 1990).

In the USA, naphthalene was reported to be detectable in 7% of 267 sediment samples (with the median concentration for all samples of less than 500 μg/kg) entered into the Environmental Protection Agency STORET database (1980–82) (Staples *et al.*, 1985).

Naphthalene has been detected in contaminated sediments in Texas, USA, at average concentrations of 54.7 and 61.9 μg/kg at 10 m and 25 m from an oil platform and in nearby non-contaminated estuarine sediments at 2.1 μg/kg in 1982–85 (Brooks *et al.*, 1990). It was found at 200 mg/kg in a tar-contaminated sediment of the River Warnow at Schwaan near Rostock, Germany, in August 1989 (Randow *et al.*, 1996). Naphthalene was found in all four Canadian marine sediments analysed (representing varying concentrations and sources of polycyclic aromatic hydrocarbon contamination) at concentrations ranging from 0.1 to 115 mg/kg dry sediment (Simpson *et al.*, 1995).

Naphthalene concentrations ranging from < 2 to 20.2 mg/kg dry wt were reported in three out of four sediments from lakes in the Northwest Territories in Canada (Lockhart *et al.*, 1992). Primarily due to oxygen limitation, naphthalene persists in coal tar-contaminated surface sediments (Madsen *et al.*, 1996). Naphthalene concentrations in soils and sewage sludges are usually less than 1 mg/kg in the United Kingdom (Wild & Jones, 1993).

(*d*) *Biodegradation*

Studies on biodegradation of polycyclic aromatic hydrocarbons in soil suggest that absorption to organic matter significantly reduces the bioavailability and thus the biodegradability of naphthalene (Heitzer *et al.*, 1992; Weissenfels *et al.*, 1992; Agency for Toxic Substances and Disease Registry, 1995a). Reported naphthalene half-lives in soil vary considerably. The estimated half-life of naphthalene reported for a solid waste site was 3.6 months, while in typical soils more rapid biodegradation is expected to occur (Heitkamp *et al.*, 1987; Howard, 1989).

Biodegradation of naphthalene is accomplished via the action of aerobic micro-organisms and generally declines precipitously when soil conditions become anaerobic (Klecka *et al.*, 1990). Naphthalene biodegrades to carbon dioxide in aerobic soils with salicylate as an intermediate product (Heitzer *et al.*, 1992; Agency for Toxic Substances and Disease Registry, 1995a).

Although polycyclic aromatic hydrocarbons are persistent in a strictly anaerobic environment, naphthalene can be degraded anaerobically under sulfate-reducing conditions: it was oxidized to carbon dioxide in petroleum-contaminated marine harbour sediments in San Diego, CA (Coates *et al.*, 1997).

(*e*) *Food*

Naphthalene was detected in only two of 13 980 samples of foods analysed in six states of the USA in 1988–89 (Minyard & Roberts, 1991). Naphthalene is not generally reported to be present in fish, but has been detected in shellfish in the USA, with concentrations ranging from 5 to 176 μg/kg in oysters, from 4 to 10 μg/kg in mussels and from < 1 to 10 μg/kg in clams (Bender & Huggett, 1989).

Naphthalene was found in 1993 in many samples of edible portions of nine types of shrimp and fish in Kuwaiti seafood at concentrations ranging from 2 to 156 μg/kg dry wt. These elevated concentrations of naphthalene were attributed to the pollution of Kuwait's territorial waters with crude oils as a result of oil spillage during the Gulf War or the chronic pollution due to oil production, transportation or natural seepage from the seabed. Naphthalene constituted the highest burden of the 14 polycyclic aromatic hydrocarbons screened (Saeed *et al.*, 1995).

Mean concentrations of naphthalene of 19.5 μg/kg dry wt have been found in edible muscle of fish collected from the Red Sea coast of Yemen (DouAbul *et al.*, 1997). Naphthalene has been found at maximum concentrations of 27.7 μg/kg and 137 μg/kg in muscle and liver tissue, respectively, of burbot fish from lakes in the Northwest Territories in Canada (Lockhart *et al.*, 1992).

Naphthalene has also been detected in various fish species collected from the Gulf of Naples, Italy, e.g., in muscle samples of anchovy, comber and rock goby at concentrations of 63, 4 and 20 μg/kg wet wt, respectively, and in razor fish, wart venus and short-necked clams, at levels of 20, 25 and 32 μg/kg wet wt, respectively (Cocchieri *et al.*, 1990).

Naphthalene was measured in 1987–88 in six species of aquatic organisms (sea mullet, bony bream, blue catfish, mud crab, pelican and the silver gull) from the Brisbane River estuarine system in Australia. The mean concentrations (μg/kg, wet wt) (in parentheses: lipid wt basis) were: bony bream, 14.1 (306; 8 samples); blue catfish, 21.3 (433; 8 samples); sea mullet, 37.3 (773; 8 samples); mud crab, 16.5 (407; 8 samples); pelican, 21.0 (276; 3 samples) and silver gull, 31.6 (395; 3 samples) (Kayal & Connell, 1995).

Naphthalene was identified in the neutral fraction of roast beef flavour isolate (Min *et al.*, 1979).

Use of a mathematical model of naphthalene migration into milk from an atmosphere having a relatively high level of naphthalene suggested that naphthalene is first absorbed by the packaging material (low-density polyethylene). It was cautioned that when low-density polyethylene is used as the packaging material, the concentration of naphthalene vapour in the storage area should be kept low to minimize the transfer of naphthalene to milk (Lau *et al.*, 1995).

(*f*) *Miscellaneous sources*

In the USA, naphthalene was found in mainstream cigarette smoke at a concentration of 2.8 μg per cigarette and at 46 μg per cigarette in the sidestream smoke from one commercial unfiltered cigarette, and at a concentration of 1.2 μg in the smoke from a filtered 'little' cigar (Schmeltz *et al.*, 1976).

Naphthalene has been detected in ash from municipal refuse and hazardous waste incinerators. It was found in seven of eight municipal refuse ash samples at 6–28 000 μg/kg, with higher concentrations detected in bottom ash than in fly ash (Shane *et al.*, 1990) and in five of 18 ash samples from hazardous waste incinerators at 0.17–41 (mean, 4.1) mg/kg (Carroll & Oberacker, 1989).

(*g*) *Human tissues and secretions*

Naphthalene was found in 40% of human adipose tissue samples at concentrations ranging from < 9 to 63 μg/kg in a National Human Adipose Tissue Survey (NHATS) in the USA in 1982 (Stanley, 1986). Naphthalene was also detected (concentrations not reported) in six of eight selected breast milk samples from women in four cities in the USA (Pellizzari *et al.*, 1982). It was also released in expired air from three out of eight individuals at concentrations of 1.5, 2.4 and 0.12 μg/h, respectively (Conkle *et al.*, 1975).

1.5 Regulations and guidelines

Occupational exposure limits and guidelines for naphthalene are presented in Table 7.

Table 7. Occupational exposure limits and guidelines for naphthalene[a]

Country	Year	Concentration (mg/m^3)	Interpretation[b]
Argentina	1991	50	TWA
		75	STEL (15 min)
Australia	1993	50	TWA
		75	STEL (15 min)
Belgium	1993	50	TWA
		75	STEL (15 min)
Canada	1994	50	TWA
		75	STEL (15 min)
Denmark	1993	50	TWA
Finland	2002	50[c]	TWA
		100[c]	STEL (15 min)
France	1993	50	TWA
Germany	2001	50 (CAT-2, skin)	TRK
Hungary	1993	40	TWA
		80	STEL (15 min)
Ireland	1997	50	TWA
		75	STEL (15 min)
Netherlands	1999	50	TWA
Philippines	1993	50	TWA
Poland	1993	20	TWA
Russia	1989	20	STEL (15 min)
Sweden	1991	0.2 (skin)	TWA
		0.6	STEL (15 min)
Switzerland	1993	50	TWA
United Kingdom	2000	50	TWA
		75	STEL (15 min)
USA			
ACGIH[c] (TLV)	2001	10 ppm [50] (A4, skin)	TWA
		15 ppm [75]	STEL (15 min)
NIOSH (REL)	2000	50	TWA
		75	STEL (15 min)
OSHA (PEL)	2001	50	TWA

[a] From International Labour Office (1991); American Conference of Governmental Industrial Hygienists (ACGIH) (2000, 2001); Deutsche Forschungsgemeinschaft (2001); Occupational Safety and Health Administration (OSHA) (2001); Sosiaali-ja terveysministeriö (2002); United Nations Environment Programme (2002)
[b] TWA, 8-h time-weighted average; STEL, short-term exposure limit; A4, not classifiable as a human carcinogen; CAT-2, substances that are considered to be carcinogenic for man because sufficient data from long-term animal studies or limited evidence from animal studies substantiated by evidence from epidemiological studies indicate that they can make a significant contribution to cancer risk; skin, danger of cutaneous absorption; TRK, technical exposure limit; TLV, threshold limit value: REL, recommended exposure limit; PEL, permissible exposure limit
[c] Values have been rounded.

2. Studies of Cancer in Humans

Case reports

A cluster of cancer cases in a naphthalene purification plant was reported in the former East Germany (Wolf, 1976, 1978). This plant operated between 1917 and 1968 and a total of 15 employees were reported to have worked in this unit of the plant during the preceding 20–30 years. Seven employees were diagnosed with cancer, including four cases of laryngeal cancer. Diagnosis was established between 1964 and 1973 and the age at diagnosis was 60–71 years. The incidence rate for laryngeal cancer in the former East Germany in 1970 was given as 6.3 per 100 000. The four workers had been exposed for 7–31 years. The limit value for exposure to naphthalene at that time was 20 mg/m^3, with peak values of 50 mg/m^3. Concomitant exposure to various tar products was mentioned. All four cases were reported to have been smokers. [The Working Group noted that no inference on the carcinogenicity of naphthalene can be drawn from these observations.]

Ajao *et al.* (1988) reported on 23 consecutive cases of colorectal carcinoma admitted during June 1982 and May 1984 to a university college hospital in Nigeria. Eleven of these patients were 30 years or younger at diagnosis. Based on family history, procto-sigmoidoscopy, barium enema and autopsy, no indication of familial polyposis among these cases was ascertained. Half of the patients mentioned a history of taking *Kafura*, a local indigenous treatment for anorectal problems, which contains naphthalene. The other half of the patients did not know whether they had been given *Kafura* during early childhood. [The Working Group noted that no inference on the carcinogenicity of naphthalene can be drawn from these observations.]

3. Studies of Cancer in Experimental Animals

3.1 Oral administration

Rat: A group of 28 BD I and BD III rats [sex and number of each strain not specified], about 100 days old, was fed a diet [not specified] containing naphthalene (spectrographically pure) in oil [type unspecified] at a dose of 10–20 mg per day on six days per week, for 100 weeks. Animals were kept under observation until they died. The average life expectancy was 800 days, which was said to be similar to that of control rats [no details were provided regarding control animals]. All animals were subjected to necropsy with histopathological examination of abnormal tissues only. No tumours were found in any of the rats examined. (Schmähl, 1955). [The Working Group noted the small number of animals used and the incomplete reporting of this study.]

3.2 Inhalation exposure

3.2.1 *Mouse*

Groups of 70 male and 70 female B6C3F$_1$ mice, 10–11 weeks of age, were subjected to whole-body exposure to 0 or 10 ppm (0 or 52 mg/m^3) naphthalene (> 99% pure) and a group of 135 males and 135 females to 30 ppm naphthalene (157 mg/m^3) in inhalation chambers for 6 h per day, five days per week, for 104 weeks. During periods of non-exposure, animals were housed in groups of five. Mean body weight of exposed mice was slightly lower than that of the controls throughout the study. Survival rates at the end of the study were significantly lower in control male mice than in exposed males due to wound trauma and secondary infection related to fighting (survival: controls, 26/70 (37%); 10 ppm, 52/69 (75%); and 30 ppm, 118/133 (89%)). Survival in the exposed female mice was similar to that of controls: controls, 59/69 (86%); 10 ppm, 57/65 (88%); and 30 ppm, 102/135 (76%). There was a statistically significant increase in the incidence of bronchiolo-alveolar adenomas in high-dose females (controls, 5/69 (7%); 10 ppm, 2/65, (3%); 30 ppm, 28/135 (21%) [$p = 0.01$; logistic regression test]). One bronchiolo-alveolar carcinoma was noted in a high-dose female. Exposed male mice also showed an increased incidence of bronchiolo-alveolar adenomas and carcinomas but the increases were not statistically significant (adenomas: 7/70 (10%), 15/69 (22%) and 27/135 (20%); carcinomas: 0/70, 3/69 (4%) and 7/135 (5%) in controls, 10 ppm and 30 ppm dose groups, respectively). Non-neoplastic changes were seen only in the lungs and nose. A dose-related increase in bronchiolo-alveolar inflammation was seen (males: 0/70 (0%), 21/69 (30%) and 56/135 (41%); females: 3/69 (4%), 13/65 (30%) and 52/135 (39%) in the 0-, 10- and 30-ppm dose groups, respectively). Virtually all exposed animals but none of the controls had nasal chronic inflammation, respiratory epithelial hyperplasia and metaplasia of the olfactory epithelium (National Toxicology Program, 1992).

In a screening assay based on increased multiplicity and incidence of lung tumours in a strain of mice highly susceptible to the development of this neoplasm, groups of 30 female A/J mice, 8–10 weeks of age, were exposed in inhalation chambers to 0, 10 or 30 ppm [0, 52 or 157 mg/m^3] naphthalene (purity, 98–99%) for 6 h per day, on five days per week, for six months. Survival was unaffected by treatment. At the end of the experimental period, survivors were killed and examined for pulmonary adenomas. Exposure to 10 or 30 ppm did not cause a significant increase in the incidence of lung adenomas compared with concurrent controls, but histopathological evaluation of the lungs showed an increase in numbers of alveolar adenomas per tumour-bearing mouse but not in adenomas per mouse compared with the concurrent controls (controls, 21% (0.21 ± 0.39 adenomas per mouse and 1.00 ± 0.00 adenomas per adenoma-bearing mouse); 10 ppm, 29% (0.35 ± 0.55 adenomas per mouse and 1.25 ± 0.07 adenomas per adenoma-bearing mouse); 30 ppm, 30% (0.37 ± 0.55 adenomas per mouse and 1.25 ± 0.07 adenomas per adenoma-bearing mouse) (Adkins *et al.*, 1986).

3.2.2 *Rat*

Groups of 49 male and 49 female Fischer 344/N rats, six weeks of age, were exposed in inhalation chambers to 0, 10, 30 or 60 ppm [0, 52, 157 or 314 mg/m^3] naphthalene (> 99% pure) for 6 h per day, on five days per week, for 105 weeks. Mean body weights of all exposed groups of male rats were less than that of the chamber control group throughout the study, but mean body weights of exposed groups of females were similar to that of the chamber control group. Survival rates in all exposed groups were similar to that of the chamber controls. At the end of the study, 24/49, 22/49, 23/49 and 21/49 males and 28/49, 21/49, 28/49 and 24/49 females were alive in the 0, 10, 30 and 60 ppm groups, respectively. Neuroblastomas of the nasal olfactory epithelium were observed in 0/49, 0/49, 4/48 ($p = 0.056$, Poly-3 test) and 3/48 male rats and in 0/49, 2/49, 3/49 and 12/49 ($p = 0.001$, Poly-3 test) females in the 0, 10, 30 and 60 ppm groups, respectively. In addition, adenomas of the nasal respiratory epithelium were observed in 0/49, 6/49 ($p = 0.013$, Poly-3 test), 8/48 ($p = 0.003$, Poly-3 test) and 15/48 ($p < 0.001$, Poly-3 test) males and 0/49, 0/49, 4/49 ($p = 0.053$, Poly-3 test) and 2/49 females in the 0, 10, 30 and 60 ppm groups, respectively. These olfactory neuroblastomas and respiratory epithelium adenomas had not been observed in the larger database of historical controls in National Toxicology Program two-year inhalation studies in which animals were fed National Institute of Health (NIH)-07 diet or in the smaller National Toxicology Program database [all routes] in which they were fed NTP-2000 diet. In addition to the nasal neoplasms, the incidences of a variety of non-neoplastic lesions of the nasal tract in both male and female rats were significantly increased in naphthalene-exposed animals compared with controls (see Section 4.2.2(*b*)) (National Toxicology Program, 2000).

3.3 Intraperitoneal administration

3.3.1 *Mouse*

A group of 31 male and 16 female CD-1 mice received intraperitoneal injections of a 0.05-M solution of naphthalene [purity unspecified] in dimethyl sulfoxide (DMSO) on days 1, 8 and 15 of life. The total dose received was 1.75 μmol per mouse. Groups of 21 male and 21 female mice receiving DMSO alone served as vehicle controls. [The number of mice in the above four groups are reported as the effective number of mice that survived at least six months of treatment and not the starting number.] Mice were weaned at 21 days, separated by gender and maintained until termination at 52 weeks, at which time they were necropsied and gross lesions as well as liver sections were examined histologically. There was no increase in the incidence of tumours in the naphthalene-treated mice compared with the vehicle controls (LaVoie *et al.*, 1988).

3.3.2 *Rat*

Ten BD I and BD III rats, 100 days old [sex and number of each strain not specified], received weekly intraperitoneal injections of 20 mg naphthalene (spectrographically pure) as a 2% solution in 'specially purified oil' for 40 weeks and were held under observation until they died. The average age at death was 900 days, which was reported to be similar to that of controls [no details were provided regarding control animals]. All animals were necropsied with histopathological examination of abnormal tissues only. No tumours were found in any of the rats examined (Schmähl, 1955). [The Working Group noted the small number of animals used and the limited reporting of this study.]

3.4 Subcutaneous administration

Rat: Ten BD I and BD III rats, 100 days old, [sex and number of each strain unspecified] received weekly subcutaneous injections of 20 mg naphthalene (spectrographically pure) as a 2% solution in 'specially purified oil' for 40 weeks and were kept under observation until they died. The average age at death was 700 days, which was reported to be similar to that of controls [no details were provided regarding control animals]. All animals were necropsied with histopathological examination of abnormal tissues only. No tumours were found in any of the rats examined (Schmähl, 1955). [The Working Group noted the small number of animals and the limited reporting of this study.]

Groups of 38 white inbred rats [age, strain and sex unspecified] received seven subcutaneous injections of 0 or 50 mg/kg bw naphthalene (purified by chromatography) as a 15% solution in sesame oil at intervals of around 14 days extending over 3.5 months. Survival was poor due to infectious pneumonia [agent unspecified], with 5/38 treated and 11/38 vehicle controls alive at 12 months and 0/38 treated and 4/38 vehicle controls alive at the termination of the study at 18 months. In the test group, a total of five sarcomas (one uterine and four lymphosarcomas) and a single mammary fibroadenoma developed and, in the control group, a single sarcoma and a single mammary fibroadenoma (Knake, 1956). [The Working Group noted the small number of animals, the poor survival and the limited reporting of this study.]

4. Other Data Relevant to an Evaluation of Carcinogenicity and its Mechanisms

4.1 Absorption, distribution, metabolism and excretion

4.1.1 *Humans*

(*a*) *Absorption*

No studies were found that quantitatively determined the extent of absorption of naphthalene in humans following oral or inhalation exposure. Naphthalene can be absorbed through the skin. Kanikkannan *et al.* (2001a) examined the permeation of JP-8 (jet fuel) containing 0.26% (w/w) naphthalene through human skin *in vitro*. For 18 samples of human skin, the steady-state flux was 0.45 μg/cm^2 per hour and the permeability coefficient was 2.17×10^{-4} cm per hour.

(*b*) *Distribution*

In a survey, naphthalene was detected in 40% of the human adipose tissue samples tested, with concentrations up to 63 ng/g lipid (Stanley, 1986). Naphthalene has also been identified in samples of human breast milk [incidence not clear; concentrations not reported] (Pellizzari *et al.*, 1982).

(*c*) *Metabolism*

The major metabolic pathways of naphthalene are illustrated in Figure 1. Naphthalene is metabolized first to naphthalene 1,2-oxide (2, see Figure 1), which can yield 1-naphthol (3, see Figure 1) or be converted by epoxide hydrolase to *trans*-1,2-dihydro-1,2-dihydroxynaphthalene (*trans*-1,2-dihydrodiol) (5, see Figure 1). The hydroxyl group of 1-naphthol may also be sulfated or glucuronidated. The 1,2-dihydrodiol can also be converted to 2-naphthol (10, see Figure 1). The epoxide is also a substrate for glutathione *S*-transferase, yielding glutathione conjugates which are eventually eliminated as mercapturic acids. Boyland and Sims (1958) showed that trace quantities of a precursor of 1-naphthyl mercapturic acid, tentatively identified as an *N*-acetyl-L-cysteine derivative, are eliminated in human urine after oral administration of 500 mg naphthalene. Tingle *et al.* (1993) examined the metabolism of naphthalene by human and mouse liver microsomes. The ratio of the *trans*-1,2-dihydrodiol to 1-naphthol was 8.6 for human microsomes compared with 0.4 for microsomes from phenobarbital-treated mice, indicating the ready detoxification of the epoxide to the diol in humans.

Buckpitt and Bahnson (1986) measured the metabolism of naphthalene by human lung microsomes derived from two individuals and detected naphthalene dihydrodiol

Figure 1. Main metabolic pathways of naphthalene and resulting products in mammals

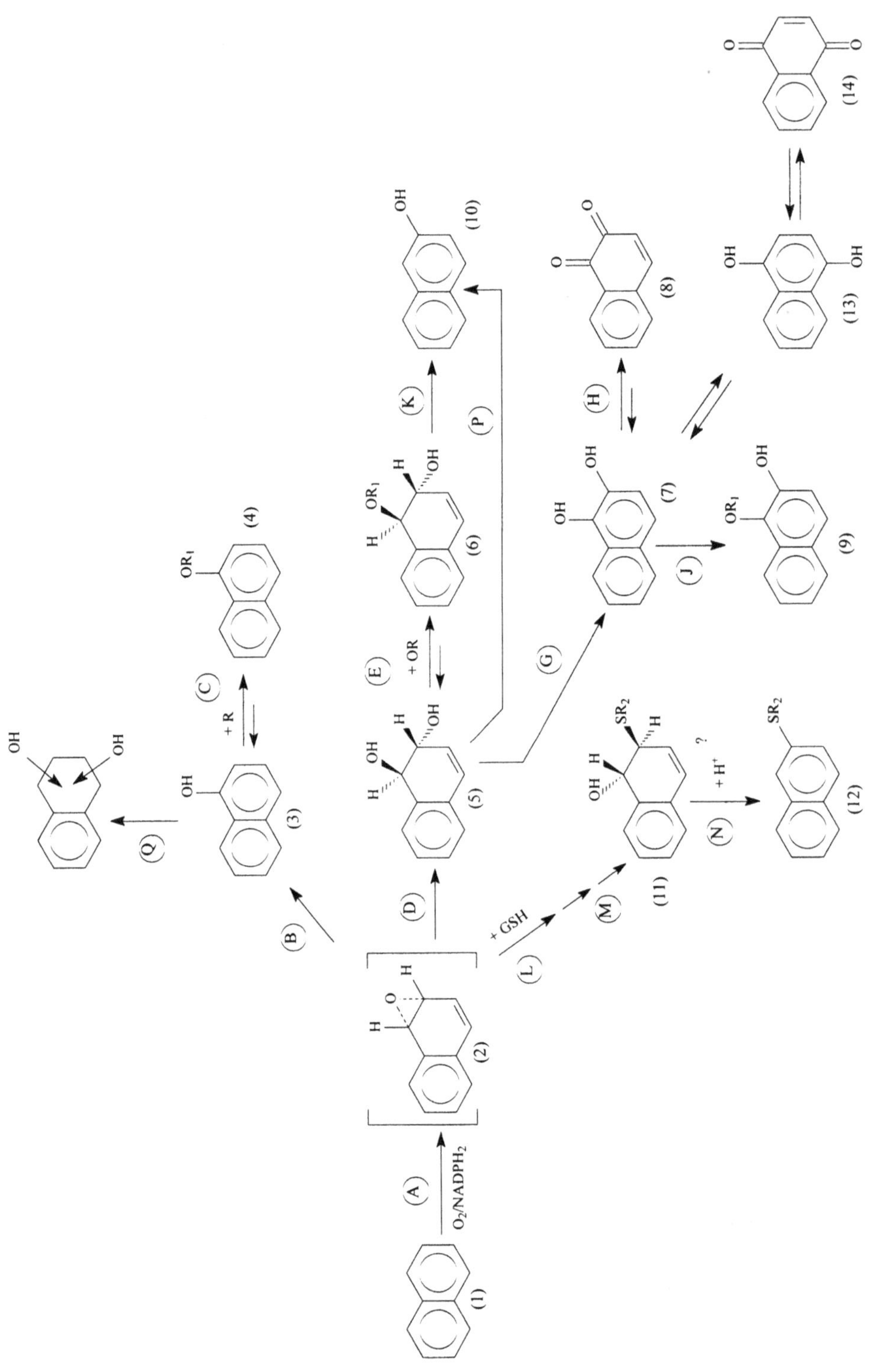

Based on BUA (1989) and Agency for Toxic Substances and Disease Registry (1995a)

(1)	Naphthalene	A,Q	=	O_2- and $NADPH_2$-dependent monooxygenase (e.g., cytochrome P450-NADP-cytochrome-c-reductase system, microsomal)
(2)	Naphthalene 1,2-oxide			
(3)	1-Naphthol (α-naphthol)	B	=	Spontaneous isomerization
(4)	1-Naphthyl glucuronide or sulfate	C,E,J	=	Conjugation reaction with sulfate (sulfotransferase, cytosolic) or with glucuronic acid (UDP-glucuronyltransferase, microsomal)
(5)	*trans*-1,2-Dihydro-1,2-dihydroxynaphthalene			
(6)	*trans*-1,2-Dihydro-2-hydroxynaphthyl-1-glucuronide	D	=	Epoxide hydrolase, synonym: epoxide hydrase (microsomal)
(7)	1,2-Dihydroxynaphthalene	F,N,P	=	Chemical dehydration
(8)	1,2-Naphthoquinone	G	=	Dihydrodiol-dehydrogenase (cystosolic); 3,5-cyclohexadiene-1,2-diol-NADP-oxidoreductase
(9)	2-Hydroxynaphthyl-1-sulfate or -glucuronide			
(10)	2-Naphthol (β-naphthol)	H	=	Chemical dehydration
(11)	*N*-Acetyl-*S*-(1,2-dihydro-1-hydroxy-2-naphthyl)-L-cysteine	K	=	Chemical hydrolysis + dehydration
(12)	*N*-Acetyl-*S*-(1-naphthyl)-L-cysteine (1-naphthyl mercapturic acid)	L	=	Enzymatic reaction with glutathione
(13)	1,4-Dihydroxynaphthalene	M	=	γ-Glutamyl transferase, peptidase, *N*-acetylase
(14)	1,4-Naphthoquinone			

GSH = Glutathione
R_1 = Sulfate or glucuronate group
R_2 = *N*-acetyl-L-cysteine residue

and three glutathione conjugates. These metabolites were also identified in animal studies, as discussed in Section 4.1.2.

Urinary metabolites of naphthalene are useful biomarkers of exposure. Seventy-five workers exposed to naphthalene while distilling naphthalene oil excreted 7.48 mg/L (4.35 mg/g creatinine) 1-naphthol (geometric mean values) at the end of the workshift. For 24 non-occupationally exposed individuals, the mean urinary concentration of 1-naphthol was 0.13 mg/L (Bieniek, 1994). 1-Naphthol, 2-naphthol and 1,4-naphthoquinone (14, see Figure 1) were identified in the urine of 69 coke-plant workers exposed to a geometric mean air concentration of naphthalene of 0.77 mg/m^3 during tar distillation. The end-of-workshift urinary concentrations of 1-naphthol and 2-naphthol were 693 and 264 µmol/mol creatinine. The correlation coefficients between the urinary excretion of naphthols and exposure to naphthalene were 0.64–0.75 for 1-naphthol and 0.70–0.82 for 2-naphthol. There was a linear relationship between the overall concentration of naphthols in urine and the naphthalene concentration in air (Bieniek, 1997). In a further study of a coke plant, Bieniek (1998) measured the concentrations of 1-naphthol and 2-naphthol in urine from eight workers in coke batteries, 11 workers in the sorting department and 29 workers in the distillation department. The mean urinary concentrations of 1-naphthol and 2-naphthol were 294 and 89 µmol/mol creatinine for the coke-battery workers, 345 and 184 µmol/mol creatinine for the sorters and 1100 and 630 µmol/mol creatinine for the distillation workers, respectively.

Andreoli *et al.* (1999) examined 15 urine samples from workers in a naphthalene-producing plant who were exposed to 0.1–0.7 mg/m^3 naphthalene. At the end of the workshift, the median urinary concentrations of 2-naphthyl sulfate, 2-naphthyl glucuronide and 1-naphthyl glucuronide were 0.030 (range, 0.014–0.121), 0.086 (range, 0.013–0.147) and 0.084 (range, 0.021–0.448) mg/L, respectively.

Since naphthalene is the most abundant component of creosote (Heikkilä *et al.*, 1987), urinary excretion of 1-naphthol was determined in three assembly workers handling creosote-impregnated wood. The average airborne concentration of naphthalene in the breathing zone was approximately 1 mg/m^3. The average end-of-shift concentration of 1-naphthol in urine changed from 254–722 (mean, 556) µmol/mol creatinine on Monday to 1820–2190 (mean, 2060) µmol/mol creatinine on Wednesday and 870–2330 (mean, 1370) µmol/mol creatinine on Friday. The same metabolite was measured in the urine of six workers exposed to creosote in a plant impregnating railroad ties (Heikkilä *et al.*, 1997). As measured by use of personal air samplers, the mean airborne concentration of naphthalene in the workers' breathing zone was 1.5 (range, 0.37–4.2) mg/m^3. The mean end-of-shift concentration of 1-naphthol was 20.5 (range, 3.5–62.1) µmol/L. There was a good correlation (r = 0.745) between concentrations of airborne naphthalene and urinary 1-naphthol. No 1-naphthol was detected (limit of detection < 0.07 µmol/L) in the urine of five non-exposed controls. Hill *et al.* (1995) measured 1-naphthol and 2-naphthol in the urine of 1000 adults without occupational exposure — a subset of the National Health and Nutrition Examination Survey III — who may have been exposed to low levels of naphthalene or pesticides that would yield these naphthols as metabolites. The frequency of detection was 86% for 1-naphthol and 81% for 2-naphthol. The mean concentrations were 15 and 5.4 μg/g creatinine, respectively. Concentrations of 1-naphthol ranged up to 1400 μg/g creatinine.

Yang *et al.* (1999) examined the relationship between certain enzyme polymorphisms and naphthalene metabolism in 119 men who were not occupationally exposed to polycyclic aromatic hydrocarbons. A polymorphism in exon 7 of the *CYP1A1* gene was not related to urinary naphthol excretion. Smokers with the *c1/c2* or *c2/c2* genotype in *CYP2E1* excreted higher concentrations of 2-naphthol in the urine than smokers with the *c1/c1* genotype. Smokers deficient in glutathione *S*-transferase M1 (GSTM1) showed higher urinary concentrations (without correction for creatinine) of both 1-naphthol and 2-naphthol.

Nan *et al.* (2001) examined the effects of occupation, lifestyle and genetic polymorphisms of *CYP1A1*, *CYP2E1* and the glutathione *S*-transferases *GSTM1* and *GSTT1* on the concentrations of 2-naphthol in the urine of 90 coke-oven workers in comparison with 128 university students. The urinary excretion of 2-naphthol was higher in the coke-oven workers (7.69 μmol/mol creatinine) than in the students (2.09 μmol/mol creatinine). In the control group, the excretion was higher in smokers (3.94 μmol/mol creatinine) than in nonsmokers (1.55 μmol/mol creatinine). Urinary 2-naphthol concentrations were higher in coke-oven workers with the *c1/c2* or *c2/c2* genotypes than in those with the more common *c1/c1* genotype of *CYP2E1*. Urinary 2-naphthol concentrations were also higher in the urine of *GSTM1*-null workers than in *GSTM1*-positive workers.

4.1.2 *Experimental systems*

(*a*) *Absorption, distribution and excretion*

Early studies indicated that in rats naphthalene is well absorbed from the gastrointestinal tract (Chang, 1943). When naphthalene was fed to white male rats (weight, about 300 g) [strain unspecified] at a concentration of 1% (w/w) in the diet, none was detected in the faeces. Similarly, when it was administered as a single dose by stomach tube (0.1 g), it was not measurable in the faeces.

Eisele (1985) examined the distribution of [^{14}C]naphthalene in laying pullets, swine and dairy cattle following oral administration. In pullets given a dose of 0.44 mg, the major site of deposition was the kidney followed by fat, lung and liver. Following acute administration of 2.46 mg in swine, the major site of deposition was fat, where the level was up to 10 times higher than that in liver. After chronic administration (0.112 mg per day for 31 days), the lung, liver and heart were major sites of accumulation. In cows, chronic exposure (5.115 mg per day for 31 days) led to deposition primarily in the liver.

When [^{14}C]naphthalene was applied dermally (3.3 μg/cm^2; total dose, 43 μg) to male Sprague-Dawley rats, the plasma half-life for absorption was 2.1 h and that for elimination was 12.8 h. The highest concentration of radioactivity 48 h after dosing was found in the skin followed by ileum, duodenum and kidney. Seventy per cent of the radioactivity was found in the urine in the first 48 h, with 3.7% appearing in the faeces and 13.6% in the expired air. The primary urinary metabolites identified were 2,7-dihydroxynaphthalene (31.1% of the total radioactivity in the first 12 h), 1,2-dihydroxynaphthalene (7, see Figure 1) (17.2%), 1,2-naphthoquinone (8, see Figure 1) (11.4%), 2-naphthol (4.3%) and 1-naphthol (3.4%). The parent compound naphthalene accounted for 0.3% of the radioactivity (Turkall *et al.*, 1994). [The Working Group noted that 2,7-dihydroxynaphthalene has not been identified as a major metabolite in other studies.]

Kilanowicz *et al.* (1999) studied the distribution, metabolism and excretion of tritiated naphthalene given intraperitoneally at a dose of 20 mg/kg bw to male IMP:Wist rats. Approximately 88% of the radioactivity was excreted in urine (68%) and faeces (20%) in the first 72 h, with maximum blood concentrations observed 2 h after dosing. The elimination of radioactivity from the blood was biphasic with half-lives of 0.8 and 99 h. [The Working Group noted that the 99-h half-life component may have been due to tritium exchange.] The highest initial tissue concentrations were found in fat, liver and kidneys. Urinary metabolites were identified as primarily the parent naphthalene, 1-naphthol and 2-naphthol with smaller amounts of 1,2-dihydro-1,2-dihydroxynaphthalene (1,2-dihydrodiol) and methylthionaphthalenes.

Sartorelli *et al.* (1999) investigated the percutaneous penetration of naphthalene from lubricating oil *in vitro* using full-thickness monkey skin. The flux for naphthalene was 0.274 nmol/cm^2 per hour, which was higher than that for acenaphthene, fluorene, anthracene, phenanthrene, pyrene and chrysene. Kanikkannan *et al.* (2001b) examined the percutaneous permeation of naphthalene in JP-8 + 100 jet fuel, which contained 0.26% (w/w) naphthalene and was spiked with [^{14}C]naphthalene, using a pig ear skin model. The steady-state flux was 0.42 μg/cm^2 per hour, similar to that of nonane but less than for tridecane.

(*b*) *Metabolism — species comparison*

Corner and Young (1954) compared the urinary metabolites of naphthalene in rats, rabbits, mice and guinea-pigs [strains unspecified] following administration of a single dose of naphthalene (500 mg/kg bw) either orally or by intraperitoneal injection. 1-Naphthol and its glucuronide and sulfate were identified in the urine of all four species (with the exception of the glucuronide in guinea-pigs). 2-Naphthol was detected in all four species but no conjugates of this metabolite were found. Although 1,2-dihydro-1,2-dihydroxynaphthalene (1,2-dihydrodiol) was found in the urine of all four species, 1,2-dihydroxynaphthalene was present only in urine of guinea-pigs. Rabbits and rats excreted more 2-naphthol than 1-naphthol, guinea-pigs excreted 1- and 2-naphthol in equal amounts and mice excreted more 1-naphthol than 2-naphthol. As in humans, a precursor of 1-naphthyl mercapturic acid has been detected as a urinary metabolite in all rodent species tested (Boyland & Sims, 1958); the amounts of this metabolite present in the urine of mice, rats and hamsters were greater than those observed in guinea-pigs, which were greater than those in humans. However, these data did not take into consideration the widely different doses given: mice, rats, hamsters, guinea-pigs and humans received total doses of 20, 100, 100, 400 and 500 mg per animal, respectively. Chen and Dorough

(1979) investigated the formation of glutathione conjugates using [^{14}C]naphthalene given to female Spague-Dawley rats. After intraperitoneal injection of 100 mg/kg bw [^{14}C]naphthalene, 65% of the water-soluble fraction of the radioactivity in urine was identified as glutathione-derived conjugate (premercapturic acid) over a 72-h period. Total recovery of radioactivity was 74% after 72 h, with 60% present in the urine and 14% in the faeces. 1,2-Dihydro-1,2-dihydroxynaphthalene (28%) and 1-naphthol (60%) were the major metabolites in the ether-extractable fraction, which accounted for 6% of the administered dose. Summer *et al.* (1979) found a dose-dependent increase in the urinary excretion of mercapturic acid conjugates in male Wistar rats given 30, 75 or 200 mg/kg bw naphthalene by stomach tube but did not find any such increase in chimpanzees (*Pan troglodytes* S.). A similar lack of a significant role for glutathione conjugation in primates was observed in rhesus monkeys (*Macaca mulatta*) (Rozman *et al.*, 1982). [The Working Group noted that standards of naphthalene mercapturates were not available in these studies and the analytical method employed may have underestimated the amounts of mercapturates present in the urine samples.]

Horning *et al.* (1980) gave naphthalene (100 mg/kg bw) intraperitoneally to male Sprague-Dawley rats and identified 21 oxygenated metabolites in the urine, all but one being generated via epoxidation. Along with those identified in other studies, the total number of known naphthalene metabolites was 31, excluding mercapturic acids, conjugates and related compounds. Bakke *et al.* (1985) gave [^{14}C]naphthalene orally to male Sprague-Dawley rats and found 4.6% of the ^{14}C dose as naphthols or their glucuronides in the urine by 24 h. In addition, they found 1,2-dihydro-1-hydroxy-2*S*-(*N*-acetyl)-cysteinylnaphthalene (11, see Figure 1) (38.1%), 1,2-dihydroxynaphthalene (7, see Figure 1) (4.9%), 1,2-dihydro-1,2-dihydroxynaphthalene glucuronide (23.9%), 1,2-dihydro-1-hydroxy-2-methylthionaphthalene glucuronide (4.6%) and uncharacterized metabolites (2.4%). Buonarati *et al.* (1990) showed that a consistent percentage of a dose of either *trans*-1*S*-hydroxy-2*S*-glutathionyl-1,2-dihydronaphthalene or *trans*-1*R*-hydroxy-2*R*-glutathionyl-1,2-dihydronaphthalene administered intravenously to male Swiss Webster mice was eliminated as the corresponding diastereomeric mercapturic acid in the urine. In contrast, a significant percentage of a dose of *trans*-1*R*-glutathionyl-2*R*-hydroxy-1,2-dihydroxynaphthalene (14–25%, depending on the dose) was metabolized to (2-hydroxy-1,2-dihydronaphthalenylthio)pyruvic acid. These observations indicate that mercapturic acids generated by conjugation at the C2 position of the napththalene nucleus can be used to assess the stereochemistry of naphthalene metabolism *in vivo*.

Pakenham *et al.* (2002) showed that 24–35% of an intraperitoneal dose of [^{14}C]-naphthalene was eliminated as mercapturates by both mice and rats at 24 h after dosing. For both species, this percentage was the same over a wide dose range (3.12–200 mg/kg bw). In contrast, after inhalation exposure, the amounts of mercapturic acid in mouse urine were approximately twice those in rat urine at the same level of exposure. Over a 24-h period, approximately 100–500 µmol/kg bw mercapturates were eliminated in urine of mice given intraperitoneal injections of 50–200 mg/kg bw naphthalene. In mice exposed by inhalation to 1–100 ppm (5.24–524 mg/m^3) naphthalene for 4 h,

1–240 μmol/kg bw total mercapturic acids were eliminated, while rats exposed to the same concentrations eliminated 0.6–67 μmol/kg bw.

Jerina *et al.* (1970) used rat [strain unspecified] liver microsomes to examine naphthalene metabolism *in vitro* and identified naphthalene 1,2-oxide (2, see Figure 1) as an intermediate in the formation of all major metabolites including glutathione conjugates. Bock *et al.* (1976) used hepatocytes from male Sprague-Dawley rats to show that 1,2-dihydro-1,2-dihydroxynaphthalene glucuronide was a major metabolite of naphthalene.

Usanov *et al.* (1982) compared the metabolism of naphthalene by microsomal preparations from rat liver and rabbit lung [strains and sex not specified] by measuring the formation of 1-naphthol. The metabolic efficiency, i.e. the rate of hydroxylation per nmol of cytochrome P450, was 7.35 times higher in rabbit lung than in rat liver microsomes.

d'Arcy Doherty *et al.* (1985) examined the metabolism of 1-naphthol by a reconstituted cytochrome P450 system from male Wistar rats and identified the products as 1,2- and 1,4-naphthoquinones (8 and 14, see Figure 1). Smithgall *et al.* (1988) examined the metabolism of *trans*-1,2-dihydro-1,2-dihydroxynaphthalene (*trans*-1,2-dihydrodiol) to the *ortho*-quinone by cytosolic dihydrodiol dehydrogenase from rat liver, and investigated the reactivity of the *ortho*-quinone with the cellular nucleophiles, cysteine and glutathione. The results showed that *ortho*-quinones formed by enzymatic oxidation of dihydrodiols may be effectively scavenged and detoxified by nucleophiles. Buckpitt and coworkers (Buckpitt & Warren, 1983; Buckpitt *et al.*, 1984, 1985) examined the relationships among the initial steps in the oxidative metabolism of naphthalene, conjugation with glutathione and the ability of reactive metabolites of naphthalene to covalently bind to protein in tissues of male Swiss Webster mice given intraperitoneal doses of [^{14}C]naphthalene. Binding of naphthalene in lung, liver and kidney was similar *in vivo*, but the rate of microsomal metabolic activation of naphthalene was much lower in the kidney than in liver or lung. Phenobarbital pretreatment increased the binding in all three tissues but only at the highest dose (400 mg/kg bw). 1-Naphthol was shown not to be an obligate intermediate in the binding process. The metabolism of naphthalene by mouse, rat and hamster pulmonary, hepatic and renal microsomal preparations was compared by Buckpitt *et al.* (1987). In all cases, glutathione adducts derived from naphthalene 1,2-oxide were formed and overall activity was particularly high in mouse lung, with a particular preference in this tissue for the formation of the naphthalene 1*R*,2*S*-oxide isomer (10:1 ratio with the 1*S*,2*R*-isomer).

Lanza *et al.* (1999) examined the ability of microsomal fractions from human lymphoblastoid cells expressing recombinant human CYP2F1 enzyme to metabolize naphthalene to glutathione adducts. The predominant conjugates formed were derived from naphthalene 1*S*,2*R*-oxide (see Table 8), in contrast to the findings in mice (Buckpitt *et al.*, 1992) (see Figure 2).

In view of the mouse lung as a target tissue, a number of investigators have examined species, tissue and cytochrome P450 (CYP) isozyme specificities in naphthalene metabolism. Nagata *et al.* (1990) identified the principal pulmonary enzyme in the mouse as

Table 8. Species comparison in the rates of conversion of naphthalene to naphthalene 1,2-oxides by recombinant enzymes

Recombinant enzyme	Species	Rate of metabolism[a]	Stereoselectivity[b]	Reference
CYP2F1	Human	35.5[c]	0.13:1	Lanza *et al.* (1999)
CYP2F2	Mouse[d]	104 000[e]	66:1	Shultz *et al.* (1999)

Modified from Buckpitt *et al.* (2002)
[a] Expressed in pmol/min/nmol enzyme
[b] Expressed as ratio of epoxide stereoisomers (1*R*,2*S*):(1*S*,2*R*)
[c] Total amount of glutathione conjugates (1 + 2 + 3) (see Figure 2)
[d] Sequence homology with human enzyme, 82%
[e] Amount of glutathione conjugate 2 (see Figure 2)

Figure 2. Metabolism of naphthalene by murine CYP2F2 to reactive epoxides and their subsequent trapping as glutathione conjugates

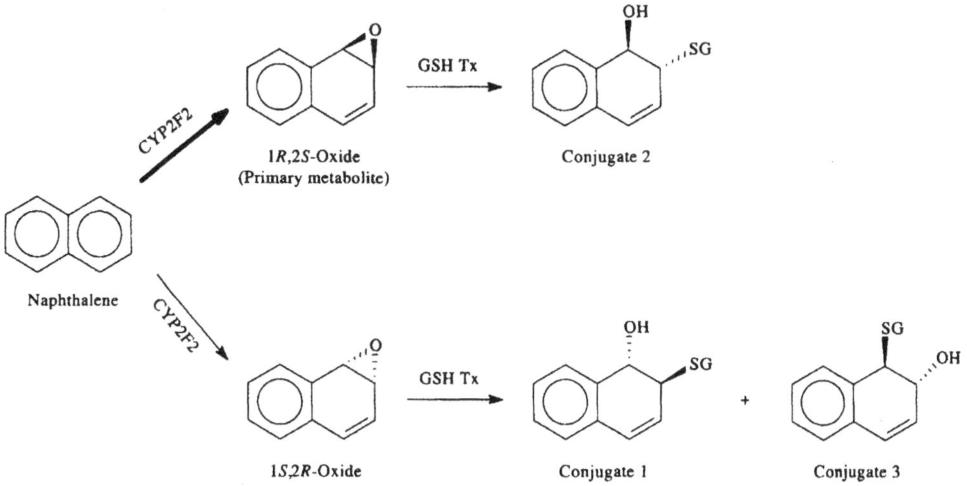

From Shultz *et al.* (1999)
Conjugates are numbered in the order of their elution after separation by reversed-phase HPLC.
GSH Tx, glutathione transferases

P450m50b [CYP2F2], which formed predominantly naphthalene 1*R*,2*S*-oxide (see Table 8 and Figure 2). Ritter *et al.* (1991) confirmed that the primary isoform responsible for naphthalene metabolism in the mouse lung was in the 2F subfamily. It was not inducible by phenobarbital, pyrazole, pregnenolone 16α-carbonitrile or 3-methylcholanthrene. Kanekal *et al.* (1991) examined the relationship between cytotoxicity and metabolism of naphthalene oxide using the isolated perfused lung of male CFW mice.

Perfusion of the lung with naphthalene 1,2-oxide reduced glutathione levels to 40–60% of control. 1,4-Naphthoquinone and naphthyl-glucuronide were the major polar metabolites, along with smaller amounts of the dihydrodiol and thioether conjugates. When lungs were perfused with naphthalene, the thioethers and the dihydrodiol predominated as metabolites. Chichester *et al.* (1991, 1994) demonstrated that Clara cells isolated from male Swiss Webster mice metabolized naphthalene to the dihydrodiol and glutathione conjugates. Microsomal preparations from Clara cells (supplemented with glutathione and glutathione *S*-transferases) metabolized naphthalene to the dihydrodiol as a minor product and formed a single glutathione adduct, derived from the 1*R*,2*S*-isomer of naphthalene oxide, as the major product, whereas the dihydrodiol predominated in intact cells. Buckpitt *et al.* (1992) determined the rates of formation and the stereochemistry of metabolites of naphthalene in postmitochondrial supernatant (S9) preparations from nasal mucosa and in microsomes from lung and liver of mice (Swiss Webster), rats (Sprague-Dawley), hamsters (Syrian golden) and rhesus monkeys (*Macaca mulatta*) (see Table 9 for details). Metabolism by mouse lung was considerably greater than that by the lungs of the rat, hamster and monkey. Using total diol and conjugates for comparison, the activity of mouse lung was two orders of magnitude higher than that of the monkey lung. In mouse lung there was preferential formation of the naphthalene 1*R*,2*S*-oxide, as judged from the stereochemistry of the glutathione conjugates. In microdissected airways, the extent of metabolism of naphthalene to the dihydrodiol and the glutathione conjugates was much higher in the airways of Swiss Webster mice compared with Sprague-Dawley rats or Syrian golden hamsters. In all three species, the rate of metabolism was higher in the distal airways than in the trachea (Buckpitt *et al.*, 1995). In mice, there was a high degree of stereoselectivity, the only glutathione conjugate being

Table 9. Species comparison in the rates of conversion of naphthalene to naphthalene oxides: pulmonary and nasal tissue

Microsome source	Species	Rate of metabolism (nmol/min/mg protein)[a]	Stereoselectivity ratio (1*R*,2*S*):(1*S*,2*R*)[b]	Dihydrodiol as % of total metabolites
Pulmonary microsomes	Mouse	13.8	11.1	7.6
	Rat	1.69	0.48	4.6
	Hamster	5.12	0.61	24.6
	Rhesus macaque	0.15	0.12	20.6
Post-mitochondrial supernatant (olfactory)	Mouse	87.1	12.7	7.4
	Rat	43.5	~ 36	4.1
	Hamster	3.9	?	7.8

Modified from Buckpitt *et al.* (2002)

[a] Total amount of dihydrodiol plus conjugates 1, 2, 3 formed (see Figure 2)

[b] The stereoselectivity varies with the concentration of substrate; the values given here are derived from incubations containing 0.5 mM naphthalene.

that derived from the 1*R*,2*S*-oxide of naphthalene. Airways of mice formed the dihydrodiol and naphthalene 1*R*,2*S*-oxide at rates substantially higher than those of rats. Immunolocalization of CYP2F2 correlated well with the sites of metabolism, in agreement with the findings of Nagata *et al.* (1990) and Ritter *et al.* (1991) as to the importance of this isoenzyme. This was confirmed in later studies (Shultz *et al.*, 1999), in which CYP2F2, expressed in *Spodoptera frugiperda* and *Trichoplusia ni* cells by use of a baculovirus expression vector system, metabolized naphthalene with a high degree of stereoselectivity to naphthalene 1*R*,2*S*-oxide (66:1 enantiomeric ratio). Substituted naphthalenes such as 1-nitronaphthalene and 2-methylnaphthalene are also substrates for purified CYP2F2 (Shultz *et al.*, 2001).

Willems *et al.* (2001) developed a physiologically based pharmacokinetic model for naphthalene administered by inhalation or by intravenous injection to Fischer 344 rats and B6C3F$_1$ mice. Model simulations for exposure by inhalation indicated that approximately 88–96% of the absorbed naphthalene was metabolized by rats and 96–98% by mice. The overall percentage of naphthalene metabolized by mice exposed to 30 ppm [157 mg/m^3] was higher than for rats exposed to 60 ppm [314 mg/m^3] because of the higher ventilation and metabolic rates in mice. The steady-state concentrations in the lungs of the mice and rats were similar at the same level of naphthalene exposure. Cumulative metabolism of naphthalene by the lung was markedly higher in the mouse than in the rat. The rates of metabolism did not increase proportionally with concentration, suggesting saturation of metabolism in this organ. The model indicated that the metabolism of naphthalene by the liver was similar in the two species.

4.2 Toxic effects

4.2.1 *Humans*

The major toxicological responses reported in humans from acute exposure to naphthalene have been haemolytic anaemia and cataracts (Zuelzer & Apt, 1949; Ghetti & Mariani, 1956; Dawson *et al.*, 1958; Zinkham & Childs, 1958). Poisoning from naphthalene has been accidental or suicidal and occurs as a result of either inhalation of fumes containing naphthalene or by ingestion of mothballs (Ojwang *et al.*, 1985; Todisco *et al.*, 1991). Accidental ingestion of household products containing naphthalene, such as mothballs or deodorant blocks, frequently occurs in children. Twelve cases of haemolytic anaemia in children secondary to the ingestion of naphthalene were reported between 1949 and 1959 (Anziulewicz *et al.*, 1959). Each child had either sucked or swallowed mothballs. Haemolytic anaemia was also observed in babies exposed to naphthalene from mothball-treated diapers, blankets or clothes (Anziulewicz *et al.*, 1959; Valaes *et al.*, 1963). In 1990, 2400 cases of accidental naphthalene ingestion were reported to 72 poison control centres in the USA. Nearly 90% of these cases occurred in children under six years of age (Woolf *et al.*, 1993). Haemolytic anaemia has also been associated with ingestion of naphthalene-containing anointing oil (Ostlere *et al.*, 1988).

Siegel and Wason (1986) reviewed a number of case studies to examine the haematological properties of naphthalene; one day after exposure, Heinz-body haemolytic anaemia leads to a sharp fall in haemoglobin, haematocrit and red blood cell counts and, in some cases, to concurrent leukocytosis. Reticulocytosis then follows with a gradual restoration of normal blood levels, except in cases of severe poisoning.

Individuals with decreased glucose-6-phosphate dehydrogenase activity in their erythrocytes are sensitive to haemolytic anaemia following exposure to naphthalene, although toxic reactions have also been observed in individuals without red cell defects. Four black patients (three male and one female) who had been exposed to naphthalene were found to have haemolytic anaemia. One of the patients was a newborn whose mother had ingested mothballs. All four patients had red blood cell glucose-6-phosphate dehydrogenase deficiency (Zinkham & Childs, 1958).

[The Working Group noted that in all of these cases the amount of naphthalene ingested was not reported.]

Additional data on the effects of exposure to naphthalene in infants and children are presented in Section 4.3.1.

4.2.2 *Experimental systems*

(*a*) *In-vivo studies with single doses*

The oral LD_{50} of naphthalene in CD-1 mice was 533 mg/kg bw for males and 710 mg/kg bw for females (Shopp *et al.*, 1984).

Naphthalene causes cataracts in rats (Rathbun *et al.*, 1990; Tao *et al.*, 1991), rabbits (van Heyningen, 1979) and mice (Wells *et al.*, 1989). Haemolytic anaemia has not been reported in experimental animals.

The biochemical pathways that modulate the cataractogenicity of naphthalene were investigated using biochemical probes and naphthalene bioactivation and detoxification in C57BL/6 and DBA/2 mice (Wells *et al.*, 1989). These mouse strains differ in susceptibility to the induction of CYP enzymes and the development of naphthalene-induced cataracts. Intraperitoneal injection of 500–2000 mg/kg bw naphthalene caused cataracts in C57BL/6 mice in a dose-dependent fashion. 1-Naphthol (56–62 mg/kg bw), 1,2-naphthoquinone and 1,4-naphthoquinone (5–250 mg/kg bw), metabolites of naphthalene, administered intraperitoneally to these mice were also found to cause cataracts. Pretreatment of the mice with SKF525A or metyrapone (CYP inhibitors), several types of antioxidants, the glutathione precursor *N*-acetylcysteine or the free radical-trapping agent, α-phenyl-*N*-*tert*-butylnitrone, decreased the incidence of cataracts. Cataracts were enhanced by pretreatment with phenobarbital and the glutathione depletor diethyl maleate. DBA/2 mice treated intraperitoneally with naphthalene (2000 mg/kg bw), 1,2- or 1,4-naphthoquinone or diethyl maleate followed by naphthalene did not develop cataracts. These results support the hypothesis that the cataractogenesis by naphthalene in C57BL/6 mice requires CYP-catalysed bioactivation to a reactive intermediate and that detoxification is dependent upon glutathione. The lack of cataract induction in DBA/2 mice after treatment with naphthalene or its metabolites 1,2- or 1,4-naphthoquinone suggests that these mice may be unable to convert the latter into free radical intermediates.

Spiro-(2,7-difluorofluorene-9,4′-imidazoline)-2′,5′-dione (ALØ1576) is an aldose reductase inhibitor that can prevent the development of naphthalene-induced cataracts in rats. This action is believed to involve inhibition of the reduction of naphthalene dihydrodiol to 1,2-dihydroxynaphthalene (Xu *et al.*, 1992; Lou *et al.*, 1996). Male rats of five strains (4–5 weeks of age), including two pigmented (Long-Evans and Brown Norway) and three albino (Sprague-Dawley, Wistar and Lewis) strains, were given naphthalene by gavage at 500 mg/kg bw per day for three days, then 1000 mg/kg bw per day for up to six weeks. Each experiment included groups treated with vehicle alone (control), naphthalene, naphthalene plus ALØ1576 or ALØ1576 alone. The aldose reductase inhibitor was given at 10 mg/kg bw per day by gavage one hour before the naphthalene feeding. Naphthalene induced cataracts in all five strains. During administration of naphthalene, whole-lens glutathione levels were 20–30% lower than those of controls. After four weeks of administration, an almost 20-fold increase in the content of protein–glutathione mixed disulfides was observed in the lenses. ALØ1576 completely prevented all morphological and chemical changes in the lenses of naphthalene-treated rats (Xu *et al.*, 1992). To determine whether pigmentation is required for naphthalene-induced cataract formation, Murano *et al.* (1993) studied the progression of cataracts in Brown Norway and Sprague-Dawley rats that had received an oral dose of naphthalene (1000 mg/kg bw) every other day for six weeks. The changes in the lens were qualitatively similar in the two strains, but the cataract progressed more uniformly and more rapidly in the Brown Norway rats.

An intraperitoneal injection of 0.05–2.0 mmol/kg bw naphthalene in corn oil in C57BL/6J mice (weighing 15–20 g) caused a specific bronchiolar lesion characterized by a dilation of Clara cells with loss of apical projections. The Clara cells later became exfoliated from large areas of the bronchioles. After loss of the Clara cells, abnormalities appeared on the surface of the ciliated cells and, within 48 h after administration of naphthalene, there was rapid division of the remaining cells. The repopulated Clara cells were distributed randomly in the bronchioles, with gradual re-establishment of the classic canal-like pattern (Mahvi *et al.*, 1977).

Intraperitoneal injection of 200–375 mg/kg bw naphthalene in male Swiss Webster mice produced highly selective necrosis of the bronchiolar epithelial cells but no necrosis in the kidney or liver. This pulmonary damage was more severe when the mice were pretreated with diethyl maleate (which depletes glutathione) and less severe after pre-treatment with piperonyl butoxide (which inhibits CYP enzymes). In contrast, treatment with SKF 525A (another CYP inhibitor) before treatment with naphthalene had no effect on naphthalene-induced pulmonary damage. Intraperitoneal injection of 25–600 mg/kg bw [^{14}C]naphthalene resulted in covalent binding of naphthalene-derived radioactivity to tissue macromolecules, with the highest levels of binding in lung, liver and kidney. The level of binding corresponded with rapid depletion of glutathione in both the lung and liver. Pretreatment with diethyl maleate increased the binding of radiolabelled material, while piperonyl butoxide and SKF 525A decreased the binding by 75% and 50%, respectively (Warren *et al.*, 1982).

O'Brien *et al.* (1985) studied the species-dependent pulmonary toxicity of naphthalene. Male Swiss T.O. mice (weighing 20–25 g) and male Wistar-derived rats (weighing 200–225 g) were given intraperitoneal doses of naphthalene at 200–600 and 400–1600 mg/kg bw, respectively. The lungs, livers, kidneys and spleen were removed 24 h after the injection and prepared for light microscopy. In mice, there was selective damage to the non-ciliated bronchiolar epithelial (Clara) cells at low doses of naphthalene. At high doses of naphthalene, vascular and hydropic degeneration of cells in the proximal convoluted tubule was observed together with protein casts in the collecting ducts. Tissue damage was not observed in the lung, liver or kidney of rats that received up to 600 mg/kg bw naphthalene. Non-protein sulfhydryl was depleted in a time-dependent manner in the lungs, liver, spleen and kidneys of naphthalene-treated mice, but only in the lung and liver of treated rats. Administration of 1-naphthol (200 mg/kg bw to mice and 200–250 mg/kg bw to rats) did not lead to depletion of non-protein sulfhydryl levels or tissue damage in the liver, lung or kidney of either species. Covalent binding and metabolism of naphthalene were approximately 10-fold greater in mouse lung microsomes than in rat lung microsomes. The authors attributed the differences in naphthalene-induced toxicity in mice and rats to differences in metabolism between the two species.

Intraperitoneal injection of 1.6–4.7 mmol/kg bw (200–600 mg/kg bw) naphthalene in male ddY mice resulted in a dose-dependent increase in lung damage mainly in the bronchiolar region, with no damage following a dose of 0.78 mmol/kg (100 mg/kg bw). The response was enhanced by diethyl maleate treatment. Increasing the dose of naphthalene from 1 to 3 mmol/kg bw resulted in a decrease in pulmonary glutathione levels. Naphthalene did not affect lipid peroxidation or phospholipid content in the lungs. In lung slice preparations, the covalent binding of naphthalene was increased or decreased when the mice had been pretreated with inducers or inhibitors of CYP enzymes, respectively (Honda *et al.*, 1990).

A single intraperitoneal injection of 2 mmol/kg bw naphthalene in ddY mice resulted in a 50% reduction of carbonyl reductase activity and microsomal mixed-function oxidase activities in the Clara cells (Matsuura *et al.*, 1990).

Naphthalene was given by intraperitoneal injection to Swiss Webster mice (0–400 mg/kg bw), Syrian golden hamsters (0–800 mg/kg bw) and Sprague-Dawley rats (0–1600 mg/kg bw). The animals were killed 24 h later for identification of the specific sites of the respiratory tract affected by the treatment (nasal cavity and tracheobronchial airway tree). In mice, the injury to the tracheobronchial epithelium was dose-dependent and Clara cell-specific. At 50 mg/kg bw, naphthalene produced swelling and vacuolation of Clara cells in terminal bronchioles. The number of terminal bronchioles with vacuolated Clara cells and the number of Clara cells within the terminal bronchioles that showed vacuolation increased after 100 mg/kg bw naphthalene. Following 200 and 300 mg/kg bw, almost all of the non-ciliated cells lining the terminal bronchioles in mice were exfoliated and necrotic. In contrast, no effect was observed on Clara cells or ciliated cells of terminal bronchioles in rats treated with up to 1600 mg/kg bw naphthalene. Only minor changes in Clara cells at some terminal bronchioles were observed in hamsters dosed with 800 mg/kg bw naphthalene. The nasal cavity showed specific injury in the olfactory epithelium in a dose- and species-dependent manner, with rats being the most sensitive species (Plopper *et al.*, 1992a). A morphometric comparison of changes in the epithelial population of the terminal bronchioles and lobar bronchi showed that Clara cells and ciliated cells in mice were affected by treatment with naphthalene, while the bronchiolar epithelium of rats and hamsters was insensitive to this treatment (Plopper *et al.*, 1992b). In a companion study, Buckpitt *et al.* (1992) demonstrated that the stereochemistry of the epoxidation of naphthalene may be important in the target tissue (lung) and species selectivity (mouse) of naphthalene toxicity (see Section 4.1.2(*b*)).

Female FVB/n mice (2–4 months of age) were given an intraperitoneal injection of 0, 50, 100 or 200 mg/kg bw naphthalene in corn oil and killed at various time points to investigate the phenotypic changes in airway epithelial cells after acute Clara cell injury. Clara cell cytotoxicity from naphthalene resulted in the exfoliation of epithelial cells containing CC10 protein, a Clara cell secretory protein. This exfoliation occurred at the same time (24 h after treatment) as a reduction in the levels of mRNA for CC10 and CYP2F monooxygenase. The mRNA for cyclin-dependent kinase 1 (CDK1), a marker of cell cycling, was detected in a large number of cells in and around the bronchioles and terminal bronchioles 48 h after treatment with naphthalene. The airways were re-populated with immature epithelial cells lacking normal levels of CC10 mRNA and over-expressing the mRNA for surfactant protein B. At 72 h after injection of naphthalene, a reduction in the number of CDK1 mRNA-positive cells was observed, except at the airway bifurcations, where increased expression of mRNA CDK1 was observed relative to the 48-h time point. The results suggest that the repair of acute airway epithelial cell injury induced by naphthalene occurs in overlapping stages, beginning with clearance of dead cells followed by the proliferative re-population of injured areas and maturation of newly re-populated regions (Stripp *et al.*, 1995).

Naphthalene (300 mg/kg bw) was administered intraperitoneally to male FVB/n mice (7–9 weeks of age) 36–72 h before intraperitoneal administration of [^{3}H]thymidine (20 Ci/mmol) at a dose of 2.5 μCi/kg bw. Lungs were removed five days after treatment. Naphthalene toxicity resulted in pulmonary neuroendocrine-cell hyperplasia, which was characterized by increased numbers of neuroepithelial bodies without significant changes in the number of isolated pulmonary neuroendocrine cells and with increased [^{3}H]thymidine labelling of cells that produce calcitonin gene-related peptide, a marker of neuroendocrine cells. These results suggest a key role of neuroendocrine cells in the reparative process of airway epithelial cell renewal after naphthalene-induced injury in mice (Stevens *et al.*, 1997). Five days after an intraperitoneal injection of naphthalene (300 mg/kg bw) to male FVB/n mice (12–14 weeks of age), an abundance of pulmonary neuroendocrine cells and neuroepithelial bodies was observed along the main axial pathway of the right middle lobe of the lung. Calcitonin gene-related peptide was used to identify the location, size and number of these cells in the airways. Neuroepithelial bodies were significantly increased in number and pulmonary neuroendocrine cells were significantly enlarged in naphthalene-treated lungs compared with controls (Peake *et al.*, 2000).

Immature Clara cells of neonatal mice are more susceptible to the toxicity of naphthalene than are mature Clara cells of adult mice. Vacuolation and exfoliation associated with cytotoxicity of Clara cells were dose-dependent when 7-day-old, 14-day-old or adult Swiss Webster mice were given a single intraperitoneal injection of 0, 25, 50 or 100 mg/kg bw naphthalene in corn oil and killed 24 h later. The range of doses at which Clara cell injury occurred varied with age, with the youngest animals being the most susceptible. The 7- and 14-day-old mice were more sensitive to the toxicity of naphthalene despite the fact that, at these ages, the airways have lower ability to activate naphthalene to its reactive intermediates compared with adult mice (Fanucchi *et al.*, 1997).

To define the repair pattern of Clara cells after massive injury, male Swiss Webster mice (2–3 months of age) were given an intraperitoneal injection of naphthalene (200 mg/kg bw) in corn oil and the lungs were evaluated at various times up to 14 days after treatment. Clara cells of terminal bronchioles were vacuolated and swollen on day 1 after the naphthalene injection, exfoliated on day 2 and resembled those of the controls on day 14. Cell proliferation was increased within the epithelium and interstitium at day 1, reached a maximum at day 2 and was close to the control level at all other time points. Markers of Clara cell differentiation were barely detectable in the terminal bronchiolar epithelium at days 1 and 2, clearly detectable at day 4 and returned to control levels between days 5 and 14. The results showed that repair of the bronchiolar epithelium after naphthalene treatment involved distinct phases of cell proliferation and differentiation, including proliferation of cells other than Clara cells, and interaction of multiple cell types including non-target cells (Van Winkle *et al.*, 1995).

Swiss-Webster mice (8–10 weeks of age) were given an intraperitoneal injection of 0 or 200 mg/kg bw naphthalene and the temporal pattern of intracellular changes was evaluated up to 6 h following treatment. Whole-lung preparations from these mice were stained with cell-permeant and -impermeant nuclear binding fluorochromes and examined by means of high-resolution light, electron and confocal fluorescence microscopy. These methods allowed the assessment of Clara cell necrosis and cell permeability on the same samples. After acute exposure to naphthalene *in vivo*, early stages of injury to bronchiolar Clara cells included swelling of the smooth endoplasmic reticulum and bleb formation, followed by increases in cell membrane permeability (Van Winkle *et al.*, 1999).

In a study in which mice were treated similarly, intracellular glutathione content was measured and compared with the degree of cytotoxicity up to 3 h after treatment. Loss of intracellular glutathione is an early event that precedes initial signs of cellular damage. Once glutathione concentration dropped below 25% of the control, injury was irreversible (Plopper *et al.*, 2001).

Male Swiss Webster mice and male Sprague-Dawley rats were exposed to naphthalene (0–110 ppm) [0–580 mg/m^3] for 4 h via whole-body inhalation (West *et al.*, 2001). Other groups of mice were given an intraperitoneal injection of 0, 50, 100, 200 or 400 mg/kg bw naphthalene. Inhalation exposure of rats to naphthalene did not result in any detectable changes in the airway epithelial cells. In mice, exposure to naphthalene at concentrations above 2 ppm [10.5 mg/m^3] resulted in a concentration-dependent increase in Clara cell injury. At low concentrations, naphthalene caused injury in the proximal airways, while at high concentrations, there was injury both in proximal airways and in the more distal conducting airways. Parenteral exposure of mice to naphthalene caused injury that was limited to the distal airways at low dose (≤ 200 mg/kg bw), while at higher doses (> 300 mg/kg bw), the injury also included the proximal conducting airways. The higher sensitivity of the distal airways was initially attributed to higher rates of naphthalene metabolism (Buckpitt *et al.*, 1995), but the results of West *et al.* (2001) suggest that cells throughout the airways are equally sensitive and that sensitivity differences between proximal and distal airways to naphthalene treatment may be due to differences in the distribution of the compound, although there are no data to support this suggestion. In conclusion, the pattern of injury after exposure to naphthalene is species-specific and highly dependent on route of exposure.

The ability of naphthalene to cause oxidative stress was assessed in female Sprague-Dawley rats (weighing 160–180 g) given vitamin E succinate for three days and then administered 1100 mg/kg bw naphthalene as a single oral dose on day 4. Another group of rats received naphthalene alone. Naphthalene induced oxidative stress as measured by increased lipid peroxidation in mitochondria in liver and brain and reduction of glutathione concentrations in these organs. The treatment also increased DNA single-strand breaks in liver tissue, and induced an increase in membrane fluidity in liver and brain microsomes, together with increases in the urinary elimination of malonaldehyde, formaldehyde, acetaldehyde and acetone. These indices of oxidative stress were less strong in the rats that had been pretreated with vitamin E succinate (Vuchetich *et al.*, 1996). [The Working Group noted the high dose used in this study.]

(*b*) *In-vivo studies with multiple doses*

Yamauchi *et al.* (1986) reported that daily administration of oral doses [number of days not given] of 1 g/kg bw naphthalene to male Wistar rats (weighing 150–170 g) resulted in increases in levels of serum and liver lipid peroxides, suggesting enhanced lipid peroxidation. Oral administration of naphthalene in dose increments up to 750 mg/kg bw over nine weeks also enhanced lipid peroxidation and decreased the activity of selenium-dependent glutathione peroxidase in the liver of male Blue-Spruce rats (Germansky & Jamall, 1988).

Male albino rats (weighing 100–125 g) were administered 1 g/kg bw naphthalene orally in refined groundnut oil daily for 10 days. Biochemical alterations in the liver, kidney and eye were evaluated. Significant changes were observed only in the liver, including increased liver weight, lipid peroxidation and aniline hydroxylase activity.

Alkaline phosphatase activity was slightly increased in the liver and eye. No significant changes were observed in the kidney (Rao & Pandya, 1981). [The Working Group noted the high dose used in this study.]

Male and female CD-1 mice were administered daily doses of 0, 27, 53 or 267 mg/kg bw naphthalene orally for 14 consecutive days or a suspension of 0, 5.3, 53 or 133 mg/kg bw naphthalene in corn oil daily for 90 consecutive days. Surviving animals were killed 24 h after the final dose. In the 14-day study, naphthalene caused a decrease in body weight and thymus weight in male mice and decreases in body and spleen weights and increases in lung weights in female mice, at the high dose only. In the 90-day study, the spleen weights were also reduced in the female mice at the high dose only. Immunotoxicity end-points (e.g., humoral immune response, response to mitogens, delayed hypersensitivity response, popliteal lymph node response, bone-marrow stem cell number and bone-marrow function) did not show any significant change from controls after either 14 days or 90 days of naphthalene administration. Although there was a slight alteration in haematological parameters, no haemolytic anaemia was observed. Serum enzymes, electrolyte levels and hexobarbital sleep times did not show consistent or dose-dependent changes after 14 or 90 days of naphthalene treatment. There was no treatment-related effect on the hepatic mixed-function oxidase system or glutathione levels after 90 days of exposure. A dose-related inhibition of hepatic aryl hydrocarbon hydroxylase activity was observed in both males and females (Shopp *et al.*, 1984).

Bronchiolar airways from male Swiss Webster mice treated intraperitoneally with naphthalene daily for seven days (50, 100 or 200 mg/kg bw) differed only slightly from those of untreated control mice, with no evidence of necrotic or exfoliated cells. In contrast, a single intraperitoneal dose of naphthalene (50, 100 or 200 mg/kg bw) caused a dose-dependent increase in the incidence and severity of bronchiolar epithelial cell necrosis. Also, when mice were treated with intraperitoneal doses of 200 mg/kg bw daily for seven days and then challenged on day 8 with a dose of 300 mg/kg bw naphthalene, the bronchiolar injury was less severe than in mice treated with a single dose of 300 mg/kg bw naphthalene. To evaluate the mechanism of this tolerance to naphthalene, the rate of formation of naphthalene 1*R*,2*S*-oxide was measured in microsomes from treated (7 × 200 mg/kg bw) and control mice. In lung microsomes from naphthalene-treated mice, there was a > 60% decrease in the rate of naphthalene metabolism compared with lung microsomes from controls. This effect was not seen in liver microsomes from these mice. However, there was no difference in the rate of formation of reactive, covalently bound naphthalene metabolites *in vivo* or *in vitro* (measured in lung microsomal preparations) between tolerant and control mice (O'Brien *et al.*, 1989).

The bronchiolar epithelium of male Swiss Webster mice treated orally daily for seven days with 200 mg/kg bw naphthalene resembled that of control mice with respect to the ciliated and non-ciliated cells and nuclear and cytoplasmic volumes. Subsequent treatment of these mice with higher doses (300 mg/kg bw) did not cause the Clara cell injury observed previously in untreated mice after a single injection of 300 mg/kg bw naphthalene. Repeated exposure to naphthalene resulted in lower activities of CYP monooxygenases in the bronchiolar epithelium. Covalent binding of reactive naphthalene metabolites to proteins in lungs of tolerant mice was similar to that in control mice (Lakritz *et al.*, 1996).

Male Swiss Webster mice were made tolerant by seven daily injections of 200 mg/kg bw naphthalene. The concentration of glutathione in the terminal airways, measured 24 h after the last injection, was 2.7-fold higher than in vehicle control mice. A challenge dose of naphthalene (300 mg/kg bw, given on day 8) did not produce injury. However, tolerant mice that were allowed to recover for 96 h after the seventh injection were again susceptible to injury induced by a challenge dose, and the concentration of glutathione in the terminal airways had declined to control values. Tolerant mice treated on day 8 simultaneously with the challenge dose of naphthalene and buthionine sulfoximine, an inhibitor of γ-glutamylcysteine synthetase, appeared as susceptible to injury as naphthalene-challenged controls. These results showed that increased rates of glutathione synthesis were critical for resistance to naphthalene toxicity in male Swiss-Webster mice (West *et al.*, 2000).

A 10% solution (w/v) of naphthalene in corn oil was administered by gavage to Brown Norway rats at a dose of 0.7 g/kg bw per day for 102 days; control rats received corn oil only. Two of the naphthalene-treated groups were given normal diet containing one of two types of aldose reductase inhibitor at concentrations known to inhibit sugar cataract formation in galactose-treated rats. The remaining naphthalene-treated groups and the controls were given unmodified diet. Gradual, progressive development of zonal opacities with decreased lens glutathione peroxidase and glutathione reductase activities was observed in rats given naphthalene or naphthalene plus a carboxylic acid aldose reductase inhibitor, but not naphthalene plus a hydantoin-type aldose reductase inhibitor. These results led the authors to suggest an oxidative mechanism in naphthalene-induced cataract formation (Tao *et al.*, 1991). Rathbun *et al.* (1990) also showed the progressive development of cataracts in Black-Hooded rats given a daily dose of 1 mL of a 10% (w/v) solution of naphthalene in corn oil by gavage for up to 79 days, together with a progressive loss of lens glutathione peroxidase and glutathione reductase activity, i.e., impairment of the defence system against oxidative damage. Holmén *et al.* (1999) later demonstrated that 0.5 and 1.0 g/kg bw naphthalene given by gavage twice a week for 10 weeks causes cataractous changes in Brown Norway rats.

Male and female B6C3F$_1$ mice (10–11 weeks of age) were exposed by inhalation to 0, 10 or 30 ppm [0, 52 or 157 mg/m^3] naphthalene for 6 h per day on five days per week for 104 weeks (Abdo *et al.*, 1992; National Toxicology Program, 1992). Naphthalene caused increased incidence and severity of chronic inflammation, metaplasia of the olfactory epithelium and hyperplasia of the respiratory nasal epithelium and chronic inflammation in the lungs of both male and female mice. In another chronic inhalation study, male and female Fischer 344 rats were exposed to 0, 10, 30 or 60 ppm [0, 52, 157 or 314 mg/m^3] naphthalene for 6 h per day on five days per week for 105 weeks (National Toxicology Program, 2000). Non-neoplastic lesions that were observed in exposed rats at incidences greater than those in the chamber controls included atypical hyperplasia, atrophy, chronic inflammation and hyaline degeneration of the olfactory epithelium; hyperplasia, squamous metaplasia, hyaline degeneration and goblet cell hyperplasia of the respiratory epithelium; and glandular hyperplasia and squamous metaplasia. The incidence and severity of these lesions increased with increasing exposure concentration.

(*c*) *In-vitro studies*

Naphthalene and its metabolite 1-naphthol (0–100 μM; 2 h, 37°C) were cytotoxic to mononuclear leukocytes after metabolic activation with human liver microsomes. Other metabolites of naphthalene, including 1,2-naphthoquinone and 1,4-naphthoquinone (0–100 μM; 2 h, 37 °C), were directly toxic to mononuclear leukocytes and depleted glutathione to 1.0% of control levels. The primary metabolite of naphthalene, the 1,2-epoxide, was not cytotoxic at concentrations up to 100 μM and did not deplete glutathione, suggesting that the quinones are responsible for the cytotoxicity of naphthalene in human mononuclear leukocytes (Wilson *et al.*, 1996).

Perfusion of the lungs of male Swiss Webster mice (4–5 weeks of age) with naphthalene (0.02–2 mM in Waymouth's medium for a one-hour period followed by 4 h in medium) resulted in swelling and vacuolation of Clara cells. This was followed by concentration-dependent losses of Clara cells from the bronchiolar epithelium. Pulmonary glutathione levels decreased over a range of 60% (at the 0.2-mM dose) to less than 10% (at the 2-mM dose) of the corresponding control level. Following perfusion with [^{14}C]naphthalene (1.67–167 μM for 30 min followed by 4.5 h in medium), reactive metabolites were covalently bound to protein in the lung and perfusate. Total binding (nanomoles bound) and specific activity (nanomoles per milligram protein) increased in lung tissue with increasing concentrations of naphthalene (Kanekal *et al.*, 1990). A subsequent study with this isolated perfused mouse lung system demonstrated that the circulating epoxides of naphthalene play a significant role in naphthalene-induced lung injury. Injury to Clara cells in lungs perfused with naphthalene or secondary metabolites such as naphthoquinones, 1-naphthol and 1,2-dihydro-1,2-dihydroxynaphthalene was less dramatic than the effects observed following exposure to naphthalene 1,2-oxide (Kanekal *et al.*, 1991).

The metabolism and cytotoxicity of naphthalene and its metabolites were investigated *in vitro* in Clara cells isolated from male Swiss Webster mice (4–5 weeks of age). The cells were incubated for 2 or 4 h with 0.1, 0.5 and 1 mM naphthalene, 1,4-naphthoquinone, 1,2-naphthoquinone, 1-naphthol, naphthalene 1,2-oxide and 1,2-dihydroxy-1,2-dihydronaphthalene. The only metabolites that were more toxic to the Clara cells than the parent compound were 1,4-naphthoquinone and naphthalene 1,2-oxide, the latter being the most potent. Piperonyl butoxide, a CYP monooxygenase inhibitor, blocked the toxic effect of naphthalene but not that of naphthalene 1,2-oxide, which suggested that the epoxide is a key participant in the process leading to the loss of cell viability in isolated Clara cells (Chichester *et al.*, 1994).

Airways microdissected from male Swiss Webster mice by filling the trachea with 1% agarose were maintained in culture for 8 h. When these explants were incubated with 0.5 mM naphthalene, the cytotoxic response of the bronchial epithelium was identical to the vacuolation and exfoliation observed *in vivo* in bronchioles of mice 24 h after intraperitoneal administration of naphthalene (100 or 300 mg/kg bw). Pre-incubation with piperonyl butoxide prevented naphthalene-induced cytotoxicity (Plopper *et al.*, 1991).

To determine whether the formation of reactive metabolites of naphthalene in defined target and non-target regions of the lung correlates with the susceptibility of these areas to naphthalene toxicity, the binding of metabolites in various cell types and in various subcompartments of the mouse lung was investigated. Binding was greater in distal bronchioles and isolated Clara cells incubated with [^{3}H]naphthalene than in explants of mouse trachea or bronchus. Binding was also greater in mouse Clara cells than in mouse hepatocytes (non-target cells) or rat trachea cells (non-susceptible species). There was a good correlation between cellular susceptibility to toxicity and the amount of reactive metabolite bound *in vitro* (Cho *et al.*, 1994a,b).

4.3 Reproductive and developmental effects

4.3.1 *Humans*

Sensorineural hearing loss was reported in an infant with neonatal hyperbilirubinaemia from haemolysis due to glucose-6-phosphate dehydrogenase deficiency and naphthalene exposure. The baby had normal hearing at 13 days of age, but had developed profound bilateral hearing loss by seven months of age. On the day of admittance to hospital, the infant had been dressed in clothes and placed on a blanket that had all been stored in naphthalene mothballs for five years (Worley *et al.*, 1996).

In a survey of neonatal jaundice in association with household drugs and chemicals in Nigeria, the overall incidence of jaundice did not differ significantly in neonates from households with or without a history of exposure to drugs or chemicals. Severe neonatal jaundice, as judged by the need for exchange blood transfusion or death of the infant, was, however, significantly more frequent among neonates from families with a history of naphthalene exposure than in those without (Familusi & Dawodu, 1985). [The Working Group noted that aflatoxin-contaminated food is a possible confounder in this populaton.]

Melzer-Lange and Walsh-Kelly (1989) reported naphthalene-induced haemolysis in a black female infant deficient in glucose-6-phosphate dehydrogenase. Fourteen of 24 children identified as having glucose-6-phosphatase deficiency were diagnosed with haemolysis associated with exposure to naphthalene-containing moth-repellents (Santucci & Shah, 2000).

Acute haemolysis with the presence of Heinz bodies and fragmented erythrocytes occurred following inhalation of naphthalene in 21 newborn Greek infants, 12 of whom had deficient glucose-6-phosphate dehydrogenase activity in the erythrocytes (Valaes *et al.*, 1963).

Two case reports of haemolytic anaemia in newborn infants secondary to maternal ingestion of mothballs or inhalation exposure to naphthalene have been reported. This indicates that naphthalene and/or its metabolites can pass the placenta (Anziulewicz *et al.*, 1959; Athanasiou *et al.*, 1997). [The Working Group noted that the exposure concentration of naphthalene was not reported.]

4.3.2 *Experimental systems*

(*a*) *Developmental toxicity studies* in vivo

In CD-1 mice given 300 mg/kg bw naphthalene per day by gavage on gestation days 7–14, maternal lethality was increased, maternal weight gain was decreased and the average number of live offspring per litter was decreased. There was no concomitant increase in the number of dead pups, suggesting that the smaller litter size was due to early embryonic resorption (Plasterer *et al.*, 1985).

In a teratogenicity study in New Zealand white rabbits treated by gavage with up to 120 mg/kg bw on gestation days 6–19, no signs of developmental or maternal toxicity were found (Navarro *et al.*, 1992).

When Sprague-Dawley rats were given intraperitoneal injections of 395 mg/kg bw naphthalene per day on days 1–15 of gestation, no evidence of maternal toxicity or developmental toxicity was found (Hardin *et al.*, 1981).

Some indications of developmental toxicity were observed when Sprague-Dawley rats were given up to 450 mg/kg bw naphthalene per day by gavage on gestational days 6–15. Maternal weight gain was reduced in the groups that received 150 and 450 mg/kg bw per day. There was a significant trend towards decreased fetal body weight and towards an increased percentage of adversely affected implants per litter (i.e., non-live or malformed). An increased incidence of visceral malformations was reported, especially enlarged ventricles of the brain. The percentage of malformed fetuses per litter seen in the 450-mg/kg bw group was 2.5 times greater than in controls, but this difference was not significant (Navarro *et al.*, 1991).

(*b*) *Developmental toxicity studies* in vitro

Preimplantation embryos of ICR mice exposed *in vivo* by intraperitoneal injection of the mother with 14 or 56 mg/kg bw naphthalene on gestational day 2 were collected on gestational day 3.5. Their subsequent in-vitro growth was markedly reduced during 72 h of culture. The viability and implantation capacity were also significantly inhibited (Iyer *et al.*, 1990).

In a study of the role of biotransformation on the rodent in-vitro preimplantation embryotoxicity of naphthalene, no toxic effects were observed in the absence of a rat S9 activation system in the culture medium. In the presence of the S9 system, naphthalene caused concentration-dependent embryolethality (approximate LC_{50}, 0.18 mM). This result indicates that the embryotoxicity of naphthalene is dependent on activation to reactive metabolites (Iyer *et al.*, 1991).

(*c*) *Reproductive toxicity in male rats*

Perturbation of glutathione levels in the testes, epididymides and liver following a single intraperitoneal dose of naphthalene (500 mg/kg bw) was studied in adult Sprague-Dawley rats. Naphthalene decreased hepatic and epididymal glutathione, but had little effect on the concentration in the testis. Chemical-induced lowering of glutathione levels in the male reproductive tract may be a mechanism for potentiation of chemically induced germ-cell mutations (Gandy *et al.*, 1990).

4.4 Genetic and related effects

4.4.1 *Humans*

No data were available to the Working Group.

4.4.2 *Experimental systems* (see Table 10 for references)

Naphthalene has consistently been found inactive in standard bacterial mutagenicity tests. However, when it was tested in the presence of nitrogen-containing reagents under photo-oxidizing or photolytic conditions, mutagenicity was observed, probably as a result of formation of nitronaphthalenes or hydroxynitronaphthalenes (Suzuki *et al.*, 1982; Arey *et al.*, 1992). Naphthalene also increased the mutagenicity of benzo[*a*]pyrene towards *Salmonella typhimurium* in the presence of an exogenous metabolic activation system (Hermann, 1981).

Naphthalene induced somatic mutations and recombination in the *Drosophila melanogaster* wing-spot test following larval feeding. It also induced sister chromatid exchange and, in the presence of exogenous metabolic activation, chromosomal aberrations in Chinese hamster ovary cells *in vitro*. Naphthalene did not induce gene mutations at the *TK* or *HPRT* locus in human MCL-5B-lymphoblastoid cells; however, an increase in the frequency of CREST-negative micronuclei, indicative of clastogenicity, was reported in this cell line. The naphthalene metabolites, 1,2-naphthoquinone and 1,4-naphthoquinone also caused an increase in sister chromatid exchange in dividing human lymphocytes, while naphthalene 1,2-oxide was inactive.

Oral exposure of mice and rats to naphthalene caused enhanced DNA fragmentation in brain and liver tissues, as judged from the presence of fragmented DNA in supernatants of homogenized tissue lysates. Mice were more sensitive to these effects, particularly p53-deficient mice. Micronuclei were not induced in bone-marrow erythrocytes of Swiss mice exposed to naphthalene *in vivo*. Adducts to haemoglobin, albumin and other proteins were found in liver, lung, kidney, brain tissue and blood cells of CFW and

Table 10. Genetic and related effects of naphthalene and its metabolites

Test system	Result[a]		Dose[b] (LED or HID)	Reference
	Without exogenous metabolic system	With exogenous metabolic system		
Escherichia coli K12 *envA*$^-$ *uvrB*$^-$, prophage induction	NT	–	500	Ho & Ho (1981)
Escherichia coli GY5027 *envA*$^-$ *uvrB*$^-$, GY40415 *amp*^{R}, prophage induction	NT	–	2000 μg/plate	Mamber *et al.* (1984)
Escherichia coli PQ37, SOS induction (chromotest)	–	NT	NR	Mersch-Sundermann *et al.* (1993)
Salmonella typhimurium TA1535/pSK1002, *umu* gene expression (SOS-inducing activity)	–	–	83 μg/mL	Nakamura *et al.* (1987)
Escherichia coli WP2/WP100 *uvr*A$^-$ *rec*A$^-$ assay, differential toxicity	NT	–	2000 μg/plate	Mamber *et al.* (1984)
Salmonella typhimurium TA100, TA1535, TA1537, TA98, reverse mutation	NT	–	100 μg/plate	McCann *et al.* (1975)
Salmonella typhimurium TA100, TA98, reverse mutation	–	–	384 μg/plate	Florin *et al.* (1980)
Salmonella typhimurium TA100, TA98, UHT8413, UHT8414, reverse mutation	–	–	2000 μg/plate	Connor *et al.* (1985)
Salmonella typhimurium TA100, TA98, TA2637, reverse mutation	–	–	500 μg/plate[c]	Nohmi *et al.* (1985)
Salmonella typhimurium TA100, TA98, TA97, reverse mutation	–	–	50 μg/plate[d]	Sakai *et al.* (1985)
Salmonella typhimurium TA100, TA1535, TA1537, TA98, reverse mutation	–	–	33 μg/plate[d]	Mortelmans *et al.* (1986)
Salmonella typhimurium TA1535, reverse mutation	NT	–	1000 μg/plate	Narbonne *et al.* (1987)
Salmonella typhimurium TA1537, reverse mutation	NT	–	100 μg/plate[d]	Gatehouse (1980)
Salmonella typhimurium TA1537, reverse mutation	NT	–	200 μg/plate	Seixas *et al.* (1982)
Salmonella typhimurium TA1538, reverse mutation	NT	–	500 μg/plate	Gatehouse (1980)
Salmonella typhimurium TA98, reverse mutation	NT	–	500 μg/plate	Ho *et al.* (1981)
Salmonella typhimurium TA98, reverse mutation	NT	–	100 μg/plate[d]	Narbonne *et al.* (1987)
Salmonella typhimurium TM677, reverse mutation	NT	–	256 μg/plate	Kaden *et al.* (1979)

Table 10 (contd)

Test system	Result[a]		Dose[b] (LED or HID)	Reference
	Without exogenous metabolic system	With exogenous metabolic system		
Drosophila melanogaster, somatic mutation and recombination	+		640 (larval feed)	Delgado-Rodriguez *et al.* (1995)
DNA strand breaks, rat hepatocytes *in vitro*, alkaline elution	–	NT	38	Sina *et al.* (1983)
Micronucleus formation, newt larvae (*Pleurodeles waltl*) erythrocytes	(+)		0.25 ppm	Djomo *et al.* (1995)
Sister chromatid exchange, Chinese hamster ovary cells *in vitro*	+	+	27	Galloway *et al.*, 1987; National Toxicology Program (2000)
Chromosomal aberrations, Chinese hamster ovary cells *in vitro*	–	+	30	Galloway *et al.*, 1987; National Toxicology Program (2000)
DNA fragmentation, macrophage J774A.1 cells *in vitro*, centrifugation	+	NT	26	Bagchi *et al.* (1998a)
Gene mutation, human MCL-5B-lymphoblastoid cells, *TK* and *HPRT* loci, *in vitro*	–	NT	40	Sasaki *et al.* (1997)
Sister chromatid exchange, human lymphocytes *in vitro*	–	–	13	Tingle *et al.* (1993); Wilson *et al.* (1995)
Micronucleus formation (CREST$^-$), human MCL-5B-lymphoblastoid cells *in vitro*	+	NT	30	Sasaki *et al.* (1997)
Cell transformation, BALB/c 3T3 cells	–	NT	150	Rundell *et al.* (1983)
Cell transformation, RLV-infected Fischer rat embryo cells	–	NT	0.5	Freeman *et al.* (1973)
DNA fragmentation, female Sprague-Dawley rat liver and brain tissue *in vivo*	+		110 po × 30[e]	Bagchi *et al.* (1998b)
DNA fragmentation, female C57BL/6NTac mouse liver and brain tissue *in vivo*	+		220 po × 1[f]	Bagchi *et al.* (2000)
DNA fragmentation, female C57BL/6TSG-p53 mouse liver and brain tissue *in vivo*	+		22 po × 1[f]	Bagchi *et al.* (2000)

Table 10 (contd)

Test system	Result[a]		Dose[b] (LED or HID)	Reference
	Without exogenous metabolic system	With exogenous metabolic system		
Micronucleus formation, male ICR Swiss mouse bone marrow erythrocytes *in vivo*	–		500 po × 1	Harper *et al.* (1984)
1,2-Naphthoquinone				
Sister chromatid exchange, human lymphocytes *in vitro*	+	NT	1.6	Wilson *et al.* (1996)
1,4-Naphthoquinone				
Sister chromatid exchange, human lymphocytes *in vitro*	+	NT	1.6	Wilson *et al.* (1996)
Naphthalene 1,2-oxide				
Sister chromatid exchange, human lymphocytes *in vitro*	–	NT	15	Wilson *et al.* (1996)

[a] +, positive; (+), weak positive; –, negative; NT, not tested; NR, not reported
[b] LED, lowest effective dose; HID, highest ineffective dose; in-vitro tests, μg/mL; in-vivo tests, mg/kg bw/day; po, oral; ip, intraperitoneal
[c] Toxicity in the absence of metabolic activation
[d] Toxicity at next higher dose
[e] Rats sacrificed on days 15, 30, 45, 60, 75, 90, 105 and 120; increased DNA fragmentation starting day 30 through day 120
[f] LED for liver, based on an oral LD_{50} = 2200 mg/kg; LED for brain is 1100 mg/kg bw.

$B6C3F_1$ mice given a single intraperitoneal injection of naphthalene (Cho *et al.*, 1994b; Tsuruda *et al.*, 1995).

4.5 Mechanistic considerations

Mechanistic studies conducted in experimental animals and tissues using a variety of approaches have attempted to determine the modes of action of naphthalene with respect to its toxicity and carcinogenicity. Such studies can provide insights into the relevance of the rodent tumours (lung tumours in female but not male mice, and nasal tumours in male and female rats) in predicting the carcinogenic response in humans.

In general, mice appear to be more susceptible to lung tumour induction by epoxides and epoxide-forming chemicals than rats (Melnick & Sills, 2001). Thus inhalation of ethylene oxide, 1,3-butadiene, isoprene and chloroprene induced lung tumours in mice but not in rats (Lynch *et al.*, 1984; National Toxicology Program, 1984; Snellings *et al.*, 1984; National Toxicology Program, 1987; Owen *et al.*, 1987; Melnick *et al.*, 1994; National Toxicology Program, 1998, 1999; Melnick & Sills, 2001). The determinants underlying the susceptibility of the mouse lung towards tumour formation may rely in part on toxicokinetic considerations, but toxicodynamic determinants probably also play a role. If naphthalene 1,2-oxide is responsible for the lung tumours observed in mice, species differences in response at this organ may be due to a combination of higher rates of naphthalene 1,2-oxide production in the Clara cells of the mouse lung, and, possibly, a greater susceptibility of the mouse lung to epoxide-induced carcinogenesis (National Toxicology Program, 2000).

4.5.1 *Interspecies differences in toxicokinetics and metabolism of naphthalene*

The initial step in naphthalene metabolism involves the formation of a 1,2-epoxide and this process is a key step in the generation of cytotoxic metabolites. Substantial differences in both the rates of epoxide formation and the stereochemistry of the epoxides formed are observed between target tissues (mouse lung and olfactory epithelium, rat olfactory epithelium) and non-target tissues (rat and hamster lung, hamster olfactory epithelium). The rates of metabolism in lung microsomes from humans and non-human primates are very similar and are 10–100-fold lower than in lung microsomes from rodents. CYP2F in mouse lung is important in the local metabolism of naphthalene and this is likely to be a critical determinant in naphthalene-induced cytotoxicity in the mouse.

4.5.2 *Interspecies differences in toxicodynamics and mode of action of naphthalene*

There is no evidence for mutagenic activity of naphthalene in the most widely used genotoxicity assays. For example, naphthalene was not mutagenic in the *Salmonella* assay with or without metabolic activation or in metabolically competent human lymphoblastoid cells at either of two loci tested. In contrast, positive results were obtained in assays for micronucleus formation, chromosomal aberrations and chromosomal recombinations *in vitro*, consistent with a potential clastogenic mechanism of action. Some, but not all, of these tests required metabolic activation for induction of genotoxicity. It is not clear, however, which reactive naphthalene metabolite is responsible for the clastogenic, and presumably carcinogenic, effects, as evidence for the reactivity of both naphthalene 1,2-oxide and naphthoquinone exists.

Exposure to naphthalene causes cellular injury and increases cell replication rates, suggesting a cytotoxic mode of action. For example, intraperitoneal administration of naphthalene produces injury (swelling, vacuolation, exfoliation, necrosis) of the tracheobronchial epithelial Clara cells of mice but not of rats (Plopper *et al.*, 1992a,b). In the same study, naphthalene was also cytotoxic to the olfactory epithelium of both rats and mice, but the effect was seen at much higher doses in mice, suggesting higher sensitivity of the rat nose. These site and species differences in toxicity correlate well with the higher rates of metabolism by mouse lung tissue and rat nasal tissue; metabolism *in vitro* in pulmonary tissue fractions from human and non-human primates is 1–2 orders of magnitude lower than that in rodents.

Overall, the proposed mechanism of action of naphthalene is that the higher rates of metabolism lead to cytotoxic metabolites in mouse lung, causing increased cell turnover and tumours. The absence of rat lung tumours is entirely consistent with this mechanism. Significantly, the maximal rates of metabolism in human lung microsomes are about two orders of magnitude lower than those in mice. The high rates of metabolism in rat nasal epithelium similarly lead to tissue damage and nasal tumours; however, the etiology of these nasal tumours, particularly the neuroblastomas, is not fully understood.

5. Summary of Data Reported and Evaluation

5.1 Exposure data

Naphthalene is a commercially important aromatic hydrocarbon which is produced from coal tar and petroleum. It is used mainly as an intermediate in the production of phthalic anhydride, naphthalene sulfonates and dyes and to a lesser extent as a moth-repellent. Human exposure to naphthalene can occur during its production, in creosote treatment of wood, in coal coking operations, during its use as an industrial intermediate, as a result of its use as a moth-repellent, and as a result of cigarette smoking.

5.2 Human carcinogenicity data

The only data available to the Working Group were two case series. No inference on the carcinogenicity of naphthalene could be drawn from these.

5.3 Animal carcinogenicity data

Naphthalene was tested for carcinogenicity by oral administration in one study in rats, by inhalation in one study in mice and one in rats and in one screening assay in mice, by intraperitoneal administration in newborn mice and in rats, and by subcutaneous administration in two studies in rats. Exposure of rats by inhalation was associated with induction of neuroblastomas of the olfactory epithelium and adenomas of the nasal respiratory epithelium in males and females. Both of these tumours were considered to be rare in untreated rats. In the screening assay study by inhalation using only female mice, there was an increase in lung adenomas per tumour-bearing mouse. In the inhalation study in mice, there was an increase in the incidence of bronchiolo-alveolar adenomas in female mice. An apparent increase in the incidence of these tumours in male mice was not statistically significant. The studies by oral administration in rats, intraperitoneal administration in mice and subcutaneous administration in rats were too limited for an evaluation of the carcinogenicity of naphthalene.

5.4 Other relevant data

Animal studies suggest that naphthalene is readily absorbed following oral or inhalation exposure. Although no data are available from human studies on absorption of naphthalene, the determination of metabolites in the urine of workers indicates that absorption does occur, and there is a good correlation between exposure to naphthalene and the amount of 1-naphthol excreted in the urine. A number of metabolites, including quinones, naphthols and conjugates (glucuronides, sulfates, glutathione) are derived from the 1,2-epoxide either directly or through multiple metabolic steps.

Naphthalene causes cataracts in humans, rats, rabbits and mice. Humans accidentally exposed to naphthalene by ingestion develop haemolytic anaemia, but there is no evidence of haemolytic anaemia in rodents. Cases of haemolytic anaemia have been reported in children and infants after oral or inhalation exposure to naphthalene or after maternal exposure during pregnancy.

Naphthalene causes lung toxicity in mice, but not rats, following either intraperitoneal injection or inhalation exposure. In mice, the injury is dose-dependent and Clara cell-specific. After repeated administration of naphthalene, mouse Clara cells become tolerant to the naphthalene-induced injury that occurs following a single dose of naphthalene. Acute and chronic exposure to naphthalene caused nasal toxicity in both mice and rats.

In isolated mouse Clara cells, 1,4-naphthoquinone and naphthalene 1,2-oxide were more toxic than naphthalene. Injury to Clara cells in perfused lungs occurred at lower concentrations of naphthalene 1,2-oxide compared with naphthalene or its other metabolites.

There is some evidence of developmental toxicity in rats and mice at dose levels that caused clear maternal toxicity. Clara cells of neonatal mice are more sensitive than those of adult mice to the cytotoxic effects of naphthalene.

There is little evidence for induction of gene mutations by naphthalene. In contrast, positive results were obtained in assays for micronucleus formation, chromosomal aberrations and chromosomal recombinations *in vitro*, which are consistent with a clastogenic potential.

Overall, the proposed mechanism of carcinogenic action is that the higher rates of metabolism of naphthalene in mice lead to cytotoxic metabolites in the lung, causing increased cell turnover and tumours. The absence of lung tumours in rats is entirely consistent with this mechanism. The maximal rates of metabolism measured in human lung microsomes are about 10–100 times lower than those in mice.

5.5 Evaluation

There is *inadequate evidence* in humans for the carcinogenicity of naphthalene.

There is *sufficient evidence* in experimental animals for the carcinogenicity of naphthalene.

Overall evaluation

Naphthalene is *possibly carcinogenic to humans (Group 2B)*.

6. References

Abdo, K.M., Eustis, S.L., McDonald, M., Jokinen, M.P., Adkins, B., Jr & Haseman, J.K. (1992) Naphthalene: A respiratory tract toxicant and carcinogen for mice. *Inhal. Toxicol.*, **4**, 393–409

Adkins, B., Jr, Van Stee, E.W., Simmons, J.E. & Eustis, S.L. (1986) Oncogenic response of strain A/J mice to inhaled chemicals. *J. Toxicol. environ. Health*, **17**, 311–322

Agency for Toxic Substances and Disease Registry (1995a) *Toxicological Profile for Naphthalene, 1-Methylnaphthalene and 2-Methylnaphthalene*, Atlanta, GA, Department of Health and Human Services

Agency for Toxic Substances and Disease Registry (1995b) *1995 Priority List of Hazardous Substances*, Atlanta, GA, United States Department of Health and Human Services

Ajao, O.G., Adenuga, M.O. & Lapido, J.K. (1988) Colorectal carcinoma in patients under the age of 30 years: A review of 11 cases. *J.R. Coll. Surg. Edinb.*, **33**, 277–279

Albers, A.T. (1984) *Butte Sheltered Workshops, Incorporated, Butte, Montana* (Health Hazard Evaluation Report No. HETA-84-116-1486), Cincinnati, OH, National Institute for Occupational Safety and Health

Almaguer, D. & Orris, P. (1985) *Inland Steel, East Chicago, Indiana* (Health Hazard Evaluation Report No. HETA-82-309-1630), Cincinnati, OH, National Institute for Occupational Safety and Health

American Conference of Governmental Industrial Hygienists (2000) *TLVs® and Other Occupational Exposure Values — 2000 CD-ROM*, Cincinnati, OH

American Conference of Governmental Industrial Hygienists (2001) *TLVs® and BEIs® — Threshold Limit Values for Chemical Substances and Physical Agents & Biological Exposure Indices*, Cincinnati, OH

Andreoli, R., Manini, P., Bergamaschi, E., Mutti, A., Franchini, I. & Niessen, W.M.A. (1999) Determination of naphthalene metabolites in human urine by liquid chromatography–mass spectrometry with electrospray ionization. *J. Chromatogr. A*, **847**, 9–17

Anziulewicz, J.A., Dick, H.J. & Chiarulli, E.E. (1959) Transplacental naphthalene poisoning. *Am. J. Obstet. Gynecol.*, 78, 519–521

Arey, J., Harger, W.P., Helmig, D. & Atkinson, R. (1992) Bioassay-directed fractionation of mutagenic PAH atmospheric photooxidation products and ambient particulate extracts. *Mutat. Res.*, **281**, 61–76

Athanasiou, M., Tsantali, C. & Trachana, M. (1997) Hemolytic anemia in a female newborn infant whose mother inhaled naphthalene before delivery. *J. Pediatr.*, **130**, 680–681

Bagchi, M., Bagchi, D., Balmoori, J., Ye, X. & Stohs, S.J. (1998a) Naphthalene-induced oxidative stress and DNA damage in cultured macrophage J774A.1 cells. *Free Radic. Biol. Med.*, **25**, 137–43

Bagchi, D., Bagchi, M., Balmoori, J., Vuchetich, P.J. & Stohs, S.J. (1998b) Induction of oxidative stress and DNA damage by chronic administration of naphthalene to rats. *Res. Commun. mol. Pathol. Pharmacol.*, **101**, 249–257

Bagchi, D., Balmoori, J., Bagchi, M., Ye, X., Williams, C.B. & Stohs, S.J. (2000) Role of *p53* tumor suppressor gene in the toxicity of TCDD, endrin, naphthalene, and chromium (VI) in liver and brain tissues of mice. *Free Radic. Biol. Med.*, **28**, 895–903

Bakke, J., Struble, C., Gustafsson, J.-Å. & Gustafsson, B. (1985) Catabolism of premercapturic acid pathway metabolites of naphthalene to naphthols and methylthio-containing metabolites in rats. *Biochemistry*, **82**, 668–671

Barker, J.F. (1987) Volatile aromatic and chlorinated organic contaminants in groundwater at 6 Ontario landfills. *Water Pollut. Res. J. Canada*, **22**, 33–48

Bender, M.E. & Huggett, R.J. (1989) Polynuclear aromatic hydrocarbon residues in shellfish: Species variations and apparent intraspecific differences. In: Kaiser, H.E., ed., *Comparative Aspects of Tumor Development*, Dordrecht, Kluwer-Academic Publishers, pp. 226–234

Benfenati, E., Di Toro, N., Fanelli, R., Lualdi, G., Tridico, R., Stella, G., Buscaini, P. & Stimilli, L. (1992) Characterization of organic and inorganic pollutants in the Adige river (Italy). *Chemosphere*, **25**, 1665–1674

Bícego, M.C., Weber, R.R. & Ito, R.G. (1996) Aromatic hydrocarbons on surface waters of Admiralty Bay, King George Island, Antarctica. *Marine Pollut. Bull.*, **32**, 549–553

Bieniek, G. (1994) The presence of 1-naphthol in the urine of industrial workers exposed to naphthalene. *Occup. environ. Med.*, **51**, 357–359

Bieniek, G. (1997) Urinary naphthols as an indicator of exposure to naphthalene. *Scand. J. Work Environ. Health*, **23**, 414–420

Bieniek, G. (1998) Aromatic and polycyclic hydrocarbons in air and their urinary metabolites in coke plant workers. *Am. J. ind. Med.*, **34**, 445–454

Bjørseth, A., Bjørseth, O. & Fjeldstad, P.E. (1978a) Polycyclic aromatic hydrocarbons in the work atmosphere I. Determination in an aluminum reduction plant. *Scand. J. Work Environ. Health*, **4**, 212–223

Bjørseth, A., Bjørseth, O. & Fjeldstad, P.E. (1978b) Polycyclic aromatic hydrocarbons in the work environment II. Determination in a coke plant. *Scand. J. Work Environ. Health*, **4**, 224–236

Bock, K.W., Van Ackeren, G., Lorch, F. & Birke, F.W. (1976) Metabolism of naphthalene to naphthalene dihydrodiol glucuronide in isolated hepatocytes and in liver microsomes. *Biochem. Pharmacol.*, **25**, 2351–2356

Bock, W., Brock, T.H., Stamm, R. & Wittneben, V. (1999) *Altstoffe-Expositionen am Arbeitsplatz* (BGAA-Report 1/99), Sankt Augustin, Hauptverband der gewerblichen Berufsgenossenschaften (HVBG)

Boyland, E. & Sims, P. (1958) Metabolism of polycyclic compounds. *Biochem. J.*, **68**, 440–447

Brooks, J.M., Kennicutt, M.C., Wade, T.L., Hart, A.D., Denoux, G.J. & McDonald, T.J. (1990) Hydrocarbon distributions around a shallow water multiwell platform. *Environ. Sci. Technol.*, **24**, 1079–1085

Brown, K.W. & Donnelly, K.C. (1988) An estimation of the risk associated with the organic constituents of hazardous and municipal waste landfill leachates. *Haz. Waste haz. Mater.*, **5**, 1–30

BUA (1989) *Naphthalene (BUA (Beratergremium für Umweltrelevante Altstoffe) Report 39, June 1989)*, Weinheim, GDCh-Advisory Committee on Existing Chemicals of Environmental Relevance

Buchet, J.P., Gennart, J.P., Mercado-Calderon, F., Delavignette, J.P., Cupers, L. & Lauwerys, R. (1992) Evaluation of exposure to polycyclic aromatic hydrocarbons in a coke production and a graphite electrode manufacturing plant: Assessment of urinary excretion of 1-hydroxypyrene as a biological indicator of exposure. *Br. J. ind. Med.*, **49**, 761–768

Buckpitt, A.R. & Bahnson, L.S. (1986) Naphthalene metabolism by human lung microsomal enzymes. *Toxicology*, **41**, 333–341

Buckpitt, A.R. & Warren, D.L. (1983) Evidence for hepatic formation, export and covalent binding of reactive naphthalene metabolites in extrahepatic tissues *in vivo*. *J. Pharmacol. exp. Ther.*, **225**, 8–16

Buckpitt, A.R., Bahnson, L.S. & Franklin, R.B. (1984) Hepatic and pulmonary microsomal metabolism of naphthalene to glutathione adducts: Factors affecting the relative rates of conjugate formation. *J. Pharmacol. exp. Ther.*, **231**, 291–300

Buckpitt, A.R., Bahnson, L.S. & Franklin, R.B. (1985) Evidence that 1-naphthol is not an obligate intermediate in the covalent binding and the pulmonary bronchiolar necrosis by naphthalene. *Biochem. biophys. Res. Comm.*, **126**, 1097–1103

Buckpitt, A.R., Castagnoli, N., Jr, Nelson, S.D., Jones, A.D. & Bahnson, L.S. (1987) Stereoselectivity of naphthalene epoxidation by mouse, rat, and hamster pulmonary, hepatic, and renal microsomal enzymes. *Drug Metab. Disp.*, **15**, 491–498

Buckpitt, A., Buonarati, M., Avey, L.B., Chang, A.M., Morin, D. & Plopper, C.G. (1992) Relationship of cytochrome P450 activity to Clara cell cytotoxicity. II. Comparison of stereoselectivity of naphthalene epoxidation in lung and nasal mucosa of mouse, hamster, rat and rhesus monkey. *J. Pharmacol. exp. Ther.*, **261**, 364–372

Buckpitt, A., Chang, A.-M., Weir, A., Van Winkle, L., Duan, X., Philpot, R. & Plopper, C.G. (1995) Relationship of cytochrome P450 activity to Clara cell cytotoxicity. IV. Metabolism of naphthalene and naphthalene oxide in microdissected airways from mice, rats, and hamsters. *Mol. Pharmacol.*, **47**, 74–81

Buckpitt, A., Boland, B., Isbell, M., Morin, D., Shultz, M., Baldwin, R., Chan, K., Karlsson, A., Lin, C., Taff, A., West, J., Fanucchi, M., Van Winkle, L. & Plopper, C. (2002) Naphthalene induced respiratory tract toxicity: Metabolic mechanisms of toxicity. *Drug. Metab. Rev.* (in press)

Budavari, S., ed. (1998) *The Merck Index*, 12th Ed., Version 12:2, Whitehouse Station, NJ, Merck & Co. [CD-ROM]

Buonarati, M., Jones, A.D. & Buckpitt, A. (1990) *In vivo* metabolism of isomeric naphthalene oxide glutathione conjugates. *Drug Metab. Disp.*, **18**, 183–189

Carroll, G.J. & Oberacker, D.A. (1989) *Characteristics of Pilot- and Full-scale Hazardous Waste Incinerator Ash* (EPA/600/D-89/232), Cincinnati, OH, Environmental Protection Agency

Catoggio, J.A., Succar, S.D. & Roca, A.E. (1989) Polynuclear aromatic hydrocarbon content of particulate matter suspended in the atmosphere of La Plata, Argentina. *Sci. total Environ.*, **79**, 43–58

Chang, L.H. (1943) The fecal excretion of polycyclic hydrocarbons following their administration to the rat. *J. biol. Chem.*, **151**, 93–99

Chemical Information Services (2001) *Directory of World Chemical Producters*, Dallas, TX

Chen, K.-C. & Dorough, H.W. (1979) Glutathione and mercapturic acid conjugations in the metabolism of naphthalene and 1-naphthyl N-methylcarbamate (carbaryl). *Drug chem. Toxicol.*, **2**, 331–354

Chen, C.S. & Zoltek, J., Jr (1995) Organic priority pollutants in wetland-treated leachates at a landfill in central Florida. *Chemosphere*, **31**, 3455–3464

Chichester, C.H., Philpot, R.M., Weir, A.J., Buckpitt, A.R. & Plopper, C.G. (1991) Characterization of the cytochrome P-450 monooxygenase system in nonciliated bronchiolar epithelial (Clara) cells isolated from mouse lung. *Am. J. respir. Cell mol. Biol.*, **4**, 179–186

Chichester, C.H., Buckpitt, A.R., Chang, A. & Plopper, C.G. (1994) Metabolism and cytotoxicity of naphthalene and its metabolites in isolated murine Clara cells. *Mol. Pharmacol.*, **45**, 664–672

Cho, M., Jedrychowski, R., Hammock, B. & Buckpitt, A. (1994a) Reactive naphthalene metabolite binding to hemoglobin and albumin. *Fundam. appl. Toxicol.*, **22**, 26–33

Cho, V., Chichester, C., Morin, D., Plopper, C. & Buckpitt, A. (1994b) Covalent interactions of reactive naphthalene metabolites with proteins. *J. Pharmacol. exp. Ther.*, **269**, 881–889

Chuang, J.C., Mack, G.A., Kuhlman, M.R. & Wilson, N.K. (1991) Polycyclic aromatic hydrocarbons and their derivatives in indoor and outdoor air in an eight-home study. *Atmos. Environ.*, **25B**, 369–380

Coates, J.D., Woodward, J., Allen, J., Philp, P. & Lovley, D.R. (1997) Anaerobic degradation of polycyclic aromatic hydrocarbons and alkanes in petroleum-contaminated marine harbor sediments. *Appl. environ. Microbiol.*, **63**, 3589–3593

Cocchieri, R.A., Arnese, A. & Minicucci, A.M. (1990) Polycyclic aromatic hydrocarbons in marine organisms from Italian central Mediterranean coasts. *Marine Pollut. Bull.*, **21**, 15–18

Cole, P.H., Frederick, R.E., Healy, R.P. & Rolan, R.G. (1984) Preliminary findings of the priority pollutant monitoring project of the nationwide urban runoff program. *J. Water Pollut. Control Fed.*, **56**, 898–908

Conkle, J.P., Camp, B.J. & Welch, B.E. (1975) Trace composition of human respiratory gas. *Arch. environ. Health*, **30**, 290–298

Connor, T.H., Theiss, J.C., Hanna, H.A., Monteith, D.K. & Matney, T.S. (1985) Genotoxicity of organic chemicals frequently found in the air of mobile homes. *Toxicol. Letters*, **25**, 33–40

Coons, S., Byrne, M. & Goyer, M. (1982) *An Exposure and Risk Assessment for Benzo(a)pyrene and Other Polycyclic Aromatic Hydrocarbons*, Vol. II, *Naphthalene. Final Draft Report (EPA-440/4-85-020)*, Washington, DC, Environmental Protection Agency, Office of Water Regulations and Standards

Corner, E.D.S. & Young, L. (1954) Biochemical studies of toxic agents. 7. The metabolism of naphthalene in animals of different species. *Biochem. J.*, **58**, 647–655

Daniels, W.J. & Gunter, B. (1988) *Surtek, Incorporated, Golden, Colorado* (Health Hazard Evaluation Report No. HETA-86-468-1875), Cincinnati, OH, National Institute for Occupational Safety and Health

d'Arcy Doherty, M., Makowski, R., Gibson, G.G. & Cohen, G.M. (1985) Cytochrome P-450 dependent metabolic activation of 1-naphthol to naphthoquinones and covalent binding species. *Biochem. Pharmacol.*, **34**, 2261–2267

Dawson, J.P., Thayer, W.W, & Desforges, J.F. (1958) Acute hemolytic anemia in the newborn infant due to naphthalene poisoning: Report of two cases, with investigations into the mechanism of the disease. *Blood*, **13**, 1113–1125

De Bortoli, M., Knöppel, H., Pecchio, E., Peil, A., Rogora, H., Schauenburg, H., Schlitt, H. & Vissers, H. (1986) Concentrations of selected organic pollutants in indoor and outdoor air in Northern Italy. *Environ. int.*, **12**, 343–350

Delgado-Rodriguez, A., Ortíz-Marttelo, R., Graf, U., Villalobos-Pietrini, R. & Gómez-Arroyo, S. (1995) Genotoxic activity of environmentally important polycyclic aromatic hydrocarbons and their nitro derivatives in the wing spot test of *Drosophila melanogaster*. *Mutat. Res.*, **341**, 235–247

Deutsche Forschungsgemeinschaft (2001) *List of MAK and BAT Values 2001* (Report No. 37), Weinheim, Wiley-VCH Verlag, pp. 80, 143

Djomo, J.E., Ferrier, V., Gauthier, L., Zoll-Moreux, C. & Marty, J. (1995) Amphibian micronucleus test *in vivo*: Evaluation of the genotoxicity of some major polycyclic aromatic hydrocarbons found in crude oil. *Mutagenesis*, **10**, 223–226

DouAbul, A.A.-Z., Heba, H.M.A. & Fareed, K.H. (1997) Polynuclear aromatic hydrocarbons (PAHs) in fish from the Red Sea coast of Yemen. *Hydrobiologia*, **352**, 251–262

Eisele, G.R. (1985) Naphthalene distribution in tissues of laying pullets, swine, and dairy cattle. *Bull. environ. Contam. Toxicol.*, **34**, 549–556

Eller, P.M., ed. (1994) *NIOSH Manual of Analytical Methods* (DHHS (NIOSH) Publ. No. 94-113), 4th Ed., Cincinnati, OH, National Institute for Occupational Safety and Health [Methods 1501, 5506, 5515]

Environmental Protection Agency (1990a) Method 550.1. Determination of polycyclic aromatic hydrocarbons in drinking water by liquid-solid extraction and HPLC with coupled ultraviolet and fluorescence detection. In: *Methods for the Determination of Organic Compounds in Drinking Water, Supplement I* (EPA Report No. EPA-600/4-90/020; US NTIS PB91-146027), Cincinnati, OH, Environmental Monitoring Systems Laboratory

Environmental Protection Agency (1990b) *Clean Air Act Amendments of 1990 Conference Report to Accompany S.1630* (Report No. 101-952),Washington DC, Office of Pollution Prevention and Toxics

Environmental Protection Agency (1995a) Method 524.2. Measurement of purgeable organic compounds in water by capillary column gas chromatography/mass spectrometry [Rev. 4.1]. In: *Methods for the Determination of Organic Compounds in Drinking Water, Supplement III* (EPA Report No. EPA-600/R-95/131; US NTIS PB-216616), Cincinnati, OH, Environmental Monitoring Systems Laboratory

Environmental Protection Agency (1995b) Method 502.2. Volatile organic compounds in water by purge and trap capillary column gas chromatography with photoionization and electrolytic detectors in series [Rev. 2.1]. In: *Methods for the Determination of Organic Compounds in Drinking Water, Supplement III* (EPA Report No. EPA-600/R-95/131; US NTIS PB95-261616), Cincinnati, OH, Environmental Monitoring Systems Laboratory

Environmental Protection Agency (1996a) Method 8100. Polynuclear aromatic hydrocarbons. In: *Test Methods for Evaluating Solid Waste — Physical/Chemical Methods* (US EPA No. SW-846), Washington DC, Office of Solid Waste

Environmental Protection Agency (1996b) Method 8310. Polynuclear aromatic hydrocarbons. In: *Test Methods for Evaluating Solid Waste — Physical/Chemical Methods* (US EPA No. SW-846), Washington DC, Office of Solid Waste

Environmental Protection Agency (1996c) Method 8021B. Aromatic and halogenated volatiles by gas chromatography using photoionization and/or electrolytic conductivity detectors [Rev. 2]. In: *Test Methods for Evaluating Solid Waste — Physical/Chemical Methods* (US EPA No. SW-846), Washington DC, Office of Solid Waste

Environmental Protection Agency (1996d) Method 8260B. Volatile organic compounds by gas chromatography/mass spectrometry (GC/MS) [Rev. 2]. In: *Test Methods for Evaluating Solid Waste — Physical/Chemical Methods* (US EPA No. SW-846), Washington DC, Office of Solid Waste

Environmental Protection Agency (1996e) Method 8270C. Semivolatile organic compounds by gas chromatography/mass spectrometry (GC/MS) [Rev. 3]. In: *Test Methods for Evaluating Solid Waste — Physical/Chemical Methods* (US EPA No. SW-846), Washington DC, Office of Solid Waste

Environmental Protection Agency (1996f) Method 8275A. Semivolatile organic compounds (PAHs and PCBs) in soils/sludges and solid wastes using thermal extraction/gas chromatography/mass spectrometry (TE/GC/MS) [Rev. 1]. In: *Test Methods for Evaluating Solid Waste — Physical/Chemical Methods* (US EPA No. SW-846), Washington DC, Office of Solid Waste

Environmental Protection Agency (1996g) Method 8410. Gas chromatography/Fourier transform infrared (GC/FT-IR) spectrometry for semivolatile organics: Capillary column. In: *Test Methods for Evaluating Solid Waste — Physical/Chemical Methods* (US EPA No. SW-846), Washington DC, Office of Solid Waste

Environmental Protection Agency (1999a) Methods for organic chemical analysis of municipal and industrial wastewater. Method 610 — Polynuclear aromatic hydrocarbons. *US Code Fed. Regul.*, **Title 40**, Part 136, App. A, pp. 145–157

Environmental Protection Agency (1999b) Methods for organic chemical analysis of municipal and industrial wastewater. Method 625 — Base/neutrals and acids. *US Code Fed. Regul.*, **Title 40**, Part 136, App. A, pp. 202–228

Environmental Protection Agency (1999c) Methods for organic chemical analysis of municipal and industrial wastewater. Method 1625B — Semivolatile organic compounds by isotope dilution GC/MS. *US Code Fed. Regul.*, **Title 40**, Part 136, App. A, pp. 286–306

Environmental Protection Agency (2000) *Ranking Air Toxics Indoors*, Washington DC, Indoor Environments Division

Environmental Protection Agency (2001) *TRI [Toxics Release Inventory] Explorer* (TRI Explorer Report (USCH), Washington DC

Familusi, J.B. & Dawodu, A.H. (1985) A survey of neonatal jaundice in association with household drugs and chemicals in Nigeria. *Ann. trop. Paediatr.*, **5**, 219–222

Fannick, N.L. (1978) *The New York Times Co., Inc., Carlstadt, New Jersey, September, 1978* (Health Hazard Evaluation Determination, Report No. HHE-78-23-530), Cincinnati, OH, National Institute for Occupational Safety and Health

Fanucchi, M.V., Buckpitt, A.R., Murphy, M.E. & Plopper, C.G. (1997) Naphthalene cytoxicity of differentiating Clara cells in neonatal mice. *Toxicol. appl. Pharmacol.*, **144**, 96–104

Fellin, P. & Otson, R. (1994) Assessment of the influence of climatic factors on concentration levels of volatile organic compounds (VOCs) in Canadian homes. *Atmos. Environ.*, **28**, 3581–3586

Florin, I., Rutberg, L., Curvall, M. & Enzell, C.R. (1980) Screening of tobacco smoke constituents for mutagenicity using the Ames' test. *Toxicology*, **18**, 219–232

Freeman, A.E., Weisburger, E.K., Weisburger, J.H., Wolford, R.G., Maryak, J.M. & Huebner, R.J. (1973) Transformation of cell cultures as an indication of the carcinogenic potential of chemicals. *J. natl Cancer Inst.*, **51**, 799–808

Galloway, S.M., Armstrong, M.J., Reuben, C., Colman, S., Brown, B., Cannon, C., Bloom, A.D., Nakamura, F., Ahmed, M., Duk, S., Rimpo, J., Margolin, B.H., Resnick, M.A., Anderson, B. & Zeiger, E. (1987) Chromosome aberrations and sister chromatid exchanges in Chinese hamster ovary cells: Evaluations of 108 chemicals. *Environ. mol. Mutagen.*, **10** (Suppl. 10), 1–175

Gandy, J. Millner, G.C., Bates, H.K., Casciano, D.A. & Harbison, R.D. (1990) Effects of selected chemicals on the glutathione status in the male reproductive system of rats. *J. Toxicol. environ. Health*, **29**, 45–57

Gatehouse, D. (1980) Mutagenicity of 1,2 ring-fused acenaphthenes against *S. typhimurium* TA1537 and TA1538: Structure-activity relationships. *Mutat. Res.*, **78**, 121–135

Germansky, M. & Jamall, I.S. (1988) Organ-specific effects of naphthalene on tissue peroxidation, glutathione peroxidases and superoxide dismutase in the rat. *Arch. Toxicol.*, **61**, 480–483

Ghetti, G. & Mariani, L. (1956) Eye changes due to naphthalene. *Med. Lav.*, **47**, 533–538

Gold, K.W., Naugle, D.F. & Perry, M.A. (1991) *Indoor Air-assessment: Indoor Air Concentrations of Environmental Carcinogens* (EPA 600/8-90/042), Research Triangle Park, NC, Environmental Protection Agency, Environmental Criteria and Assessment Office, Office of Research and Development

Gold, K.W., Naugle, D.F. & Berry, M.A. (1993) Indoor concentrations of environmental carcinogens. In: Seifert, B., Van De Wiel, H.J., Dodet, B. & O'Neil, I.K., eds, *Environmental*

Carcinogens. Methods of Analysis and Exposure Measurement, Vol. 12, *Indoor Air* (IARC Scientific Publications No. 109), Lyon, IARC, pp. 41–71

Gomez-Belinchon, J.I., Grimalt, J.O. & Albaigès, J. (1991) Volatile organic compounds in two polluted rivers in Barcelona (Catalonia, Spain). *Water Res.*, **25**, 577–589

Gschwend, P. M., Zafiriou, O.C., Mantoura, R.F.C., Schwarzenbach, R.P. & Gagosian, R.B. (1982) Volatile organic compounds at a coastal site. 1. Seasonal variations. *Environ. Sci. Technol.*, **16**, 31–38

Gunter, B.J. (1984) *Sheltered Workshop, Butte, Montana* (Health Hazard Evaluation Report No. HETA-83-421-1446), Cincinnati, OH, National Institute for Occupational Safety and Health

Hansen, A.M., Omland, Ø., Poulsen, O.M., Sherson, D., Sigsgaard, T., Christensen, J.M. & Overgaard, E. (1994) Correlation between work process-related exposure to polycyclic aromatic hydrocarbons and urinary levels of alpha-naphthol, beta-naphthylamine and 1-hydroxypyrene in iron foundry workers. *Int. Arch. occup. environ. Health*, **65**, 385–394

Hardin, B.D., Bond, G.P., Sikov, M.R., Andrew, F.D., Beliles, R.P. & Niemeier, R.W. (1981) Testing of selected workplace chemicals for teratogenic potential. *Scand. J. Work Environ. Health*, **7** (Suppl. 4), 66–75

Harper, B.L., Ramanujam, V.M.S., Gad-El-Karim, M.M. & Legator, M.S. (1984) The influence of simple aromatics on benzene clastogenicity. *Mutat. Res.*, **128**, 105–114

Hawthorne, S.B. & Sievers, R.E. (1984) Emission of organic air pollutants from shale oil waste waters. *Environ. Sci. Technol.*, **18**, 483–490

Heikkilä, P.R., Hämeilä, M., Pyy, L. & Raunu, P. (1987) Exposure to creosote in the impregnation and handling of impregnated wood. *Scand. J. Work Environ. Health*, **13**, 431–437

Heikkilä, P., Luotamo, M., Pyy, L. & Riihimäki, V. (1995) Urinary 1-naphthol and 1-pyrenol as indicators of exposure to coal tar products. *Int. Arch. occup. environ. Health*, **67**, 211–217

Heikkilä, P.R., Luotamo, M. & Riihimäki, V. (1997) Urinary 1-naphthol excretion in the assessment of exposure to creosote in an impregnation facility. *Scand. J. Work Environ. Health*, **23**, 199–205

Heitkamp, M.A., Freeman, J.P. & Cerniglia, C.E. (1987) Naphthalene biodegradation in environmental microcosms: Estimates of degradation rates and characterization of metabolites. *Appl. environ. Microbiol.*, **53**, 129–136

Heitzer, A., Webb, O.F., Thonnard, J.E. & Sayler, G.S. (1992) Specific and quantitative assessment of naphthalene and salicylate bioavailability by using a bioluminescent catabolic reporter bacterium. *Appl. environ. Microbiol.*, **58**, 1839–1846

Hermann, M. (1981) Synergistic effects of individual polycyclic aromatic hydrocarbons on the mutagenicity of their mixtures. *Mutat. Res.*, **90**, 399–409

van Heyningen, R. (1979) Naphthalene cataract in rats and rabbits: A resumé. *Exp. Eye Res.*, **28**, 435–439

Hicks, J.B. (1995) Asphalt industry cross-sectional exposure assessment study. *Appl. occup. environ. Hyg.*, **10**, 840–848

Hill, R.H., Jr, Head, S.L., Baker, S., Gregg, M., Shealy, D.B., Bailey, S.L., Williams, C.C., Sampson, E.J. & Needham, L.L. (1995) Pesticide residues in urine of adults living in the United States: Reference range concentrations. *Environ. Res.*, **71**, 99–108

Ho, Y.L. & Ho, S.K. (1981) Screening of carcinogens with the prophage λcI*ts*857 induction test. *Cancer Res.*, **41**, 532–536

Ho, C.-H., Clark, B. R., Guerin, M. R., Barkenbus, B. D., Rao, T. K. & Epler, J. L. (1981) Analytical and biological analyses of test materials from the synthetic fuel technologies. IV. Studies of chemical structure–mutagenic activity relationships of aromatic nitrogen compounds relevant to synfuels. *Mutat. Res.*, **85**, 335–345

Holmén, J.B., Ekesten, B. & Lundgren, B. (1999) Naphthalene-induced cataract model in rats: A comparative study between slit and retroillumination images, biochemical changes and naphthalene dose and duration. *Curr. Eye Res.* **19**, 418–425

Honda, T., Kiyozumi, M. & Kojima, S. (1990) Alkylnaphthalene. XI. Pulmonary toxicity of naphthalene, 2-methylnaphthalene, and isopropylnaphthalenes in mice. *Chem. pharm. Bull.*, **38**, 3130–3135

Horning, M.G., Stillwell, W.G., Griffin, G.W. & Tsang, W.-S. (1980) Epoxide intermediates in the metabolism of naphthalene by the rat. *Drug Metab. Disp.*, **8**, 404–414

Howard, P.H. (1989) *Handbook of Environmental Fate and Exposure Data for Organic Chemicals*, Vol. I, Chelsea, MI, Lewis Publishers, pp. 408–421

Hung, I.-F., Fan, H.-F. & Lee, T.-S. (1992) Aliphatic and aromatic hydrocarbons in indoor air. *Bull. environ. Contam. Toxicol.*, **48**, 579–584

IARC (1985) *IARC Monographs on the Evaluation of the Carcinogenic Risk of Chemicals to Humans*, Vol. 35, *Polynuclear Aromatic Compounds, Part 4, Bitumens, Coal-tars and Derived Products, Shale-oils and Soots*, Lyon, IARC*Press*, pp. 83–159

IARC (1989) *IARC Monographs on the Evaluation of Carcinogenic Risks to Humans*, Vol. 45, *Occupational Exposures in Petroleum Refining; Crude Oil and Major Petroleum Fuels*, Lyon, IARC*Press*, pp. 39–117

International Labour Office (1991) *Occupational Exposure Limits for Airborne Toxic Substances*, 3rd Ed. (Occupational Safety and Health Series), Geneva

Iyer, P., Gollahon, L.S., Martin, J.E. & Irvin, T. R. (1990) Evaluation of the in-vitro growth of rodent preimplantation embryos exposed to naphthalene in vivo (Abstract). *Toxicologist*, **10**, 274

Iyer, P., Martin, J.E. & Irvin, T.R. (1991) Role of biotransformation in the in vitro preimplantation embryotoxicity of naphthalene. *Toxicology*, **66**, 257–270

Jerina, D.M., Daly, J.W., Witkop, B., Zaltzman-Nirenberg, Z. & Udenfriend, S. (1970) 1,2-Naphthalene oxide as an intermediate in the microsomal hydroxylation of naphthalene. *Biochemistry*, **9**, 147–156

Kaden, D.A., Hites, R.A. & Thilly, W.G. (1979) Mutagenicity of soot and associated polycyclic aromatic hydrocarbons to *Salmonella typhimurium*. *Cancer Res.*, **39**, 4152–4159

Kanekal, S., Plopper, C., Morin, D. & Buckpitt, A. (1990) Metabolic activation and bronchiolar Clara cell necrosis from naphthalene in the isolated perfused mouse lung. *J. Pharmacol. exp. Ther.*, **252**, 428–437

Kanekal, S., Plopper, C., Morin, D. & Buckpitt, A. (1991) Metabolism and cytotoxicity of naphthalene oxide in the isolated perfused mouse lung. *J. Pharmacol. exp. Ther.*, **256**, 391–401

Kanikkannan, N., Patel, R., Jackson, T., Shaik, M.S. & Singh, M. (2001a) Percutaneous absorption and skin irritation of JP-8 (jet fuel). *Toxicology*, **161**, 1–11

Kanikkannan, N., Burton, S., Patel, R., Jackson, T., Shaik, M.S. & Singh, M. (2001b) Percutaneous permeation and skin irritation of JP-8 + 100 jet fuel in a porcine model. *Toxicol. Lett.*, **119**, 133–142

Kayal, S. & Connell, D.W. (1995) Polycyclic aromatic hydrocarbons in biota from the Brisbane River estuary, Australia. *Estuarine coastal Shelf Sci.*, **40**, 475–493

Kelly, J.E. (1992) *US Department of the Interior, National Park Service, New River Gorge National River, West Virginia* (Health Hazard Evaluation Report No. HETA-92-045-2260), Cincinnati, OH, National Institute for Occupational Safety and Health

Kelly, T.J., Mukund, R., Spicer, C.W. & Pollack, A.J. (1994) Concentrations and transformations of hazardous air pollutants. *Environ. Sci. Technol.*, **28**, 378A–387A

Kilanowicz, A., Czerski, B. & Sapota, A. (1999) The disposition and metabolism of naphthalene in rats. *Int. J. occup. Med. environ. Health*, **12**, 209–219

Klecka, G.M., Davis, J.W., Gray, D.R. & Madsen, S.S. (1990) Natural bioremediation of organic contaminants in ground water: Cliff-Dow Superfund site. *Ground Water*, **28**, 534–543

Knake, E. (1956) [Weak carcinogenic activity of naphthalene and benzene.] *Virchows Arch. Pathol. Anat. Physiol.*, **329**, 141–176 (in German)

Kostiainen, R. (1995) Volatile organic compounds in the indoor air of normal and sick houses. *Atmos. Environ.*, **29**, 693–702

Lacson, J.G. (2000) *CEH Product Review — Naphthalene*, Menlo Park, CA, Chemical Economics Handbook (CEH)-SRI International

Lakritz, J., Chang, A. Weir, A., Nishio, S., Hyde, D., Philpot, R., Buckpitt, A. & Plopper, C. (1996) Cellular and metabolic basis of Clara cell tolerance to multiple doses of cytochrome P450-activated cytotoxicants. I: Bronchiolar epithelial reorganization and expression of cytochrome P450 monooxygenases in mice exposed to multiple doses of naphthalene. *J. Pharmacol. exp. Ther.*, **278**, 1408–1418

Lanza, D.L., Code, E., Crespi, C.L., Gonzalez, F.J. & Yost, G.S. (1999) Specific dehydrogenation of 3-methylindole and epoxidation of naphthalene by recombinant human CYP2F1 expressed in lymphoblastoid cells. *Drug Metab. Disp.*, **27**, 798–803

LaRegina, J., Bozzelli, J.W., Harkov, R. & Gianti, S. (1986) Volatile organic compounds at hazardous waste sites and a sanitary landfill in New Jersey: An up-to-date review of the present situation. *Environ. Prog.*, **5**, 18–28

Lau, O.-W., Wong, S.-K. & Leung, K.-S. (1995) A mathematical model for the migration of naphthalene from the atmosphere into milk drink packaged in polyethylene bottles. *Packaging Technol. Sci.*, **8**, 261–270

LaVoie, E.J., Dolan, S., Little, P., Wang, C-X., Sugie, S. & Rivenson, A. (1988) Carcinogenicity of quinoline, 4- and 8-methylquinoline and benzoquinolines in newborn mice and rats. *Food chem. Toxicol.*, **26**, 625–629

Lesage, J., Perrault, G. & Durand, P. (1987) Evaluation of worker exposure to polycyclic aromatic hydrocarbons. *Am. ind. Hyg. Assoc. J.*, **48**, 753–759

Lide, D.R. & Milne, G.W.A. (1996) *Properties of Organic Compounds*, Version 5.0, Boca Raton, FL, CRC Press [CD-ROM]

Lockhart, W.L., Wagemann, R., Tracey, B., Sutherland, D. & Thomas, D.J. (1992) Presence and implications of chemical contaminants in the freshwaters of the Canadian Arctic. *Sci. total Environ.*, **122**, 165–243

Löfgren, L., Persson, K., Strömvall, A.M. & Petersson, G. (1991) Exposure of commuters to volatile aromatic hydrocarbons from petrol exhaust. *Sci. total Environ.*, **108**, 225–233

Lou, M.F., Xu, G.-T., Zigler, S., Jr & York, B., Jr (1996) Inhibition of naphthalene cataract in rats by aldose reductase inhibitors. *Curr. Eye Res.*, **15**, 423–432

Lynch, D.W., Lewis, T.R., Moorman, W.J., Burg, J.R., Groth, D.H., Khan, A., Ackerman, L.J. & Cockrell, B.Y. (1984) Carcinogenic and toxicologic effects of inhaled ethylene oxide and propylene oxide in F344 rats. *Toxicol. appl. Pharmacol.*, **76**, 69–84

Madsen, E.L., Mann, C.L. & Bilotta, S.E. (1996) Oxygen limitations and aging as explanations for the field persistence of naphthalene in coal tar-contaminated surface sediments. *Environ. Toxicol. Chem.*, **15**, 1876–1882

Mahvi, D., Bank, H. & Harley, R. (1977) Morphology of a naphthalene-induced bronchiolar lesion. *Am. J. Pathol.*, **86**, 559–572

Mamber, S.W., Bryson, V. & Katz, S.E. (1984) Evaluation of the *Escherichia coli* K12 inductest for detection of potential chemical carcinogens. *Mutat. Res.*, **130**, 141–151

Mason, R.T. (1995) Naphthalene. In: Kroschwitz, J.I. & Howe-Grant, M., eds, *Kirk-Othmer Encyclopedia of Chemical Technology*, 4th Ed., Vol. 16, New York, John Wiley & Sons, pp. 963–979

Matsuura, K., Hara, A., Sawada, H., Bunai, Y. & Ohya, I. (1990) Localization of pulmonary carbonyl reductase in guinea pig and mouse: Enzyme histochemical and immunohistochemical studies. *J. Histochem. Cytochem.*, **38**, 217–223

McCann, J., Choi, E., Yamasaki, E. & Ames, B. N. (1975) Detection of carcinogens as mutagens in the *Salmonella*/microsome test: Assay of 300 chemicals. *Proc. natl Acad. Sci. USA*, **72**, 5135–5139

Melnick, R.L. & Sills, R.C. (2001) Comparative carcinogenicity of 1,3-butadiene, isoprene, and chloroprene in rats and mice. *Chem.-biol. Interact.*, **135–136**, 27–42

Melnick, R.L., Sills, R.C., Roycroft, J.H., Chou, B.J., Ragan, H.A. & Miller, R.A. (1994) Isoprene, an endogenous hydrocarbon and industrial chemical, induces multiple organ neoplasia in rodents after 26 weeks of inhalation exposure. *Cancer Res.*, **54**, 5333–5339

Melzer-Lange, M. & Walsh-Kelly, C. (1989) Naphthalene-induced hemolysis in a black female toddler deficient in glucose-6-phosphate dehydrogenase. *Pediatr. Emerg. Care*, **5**, 24–26

Mersch-Sundermann, V., Mochayedi, S., Kevekordes, S., Kern, S. & Wintermann, F. (1993) The genotoxicity of unsubstituted and nitrated polycyclic aromatic hydrocarbons. *Anticancer Res.*, **13**, 2037–2044

Min, D.B.S., Ina, K., Peterson, R.J. & Chang, S.S. (1979) Preliminary identification of volatile flavor compounds in the neutral fraction of roast beef. *J. Food Sci.*, **44**, 639–642

Minyard, J.P., Jr & Roberts, W.E. (1991) Chemical contaminants monitoring — State findings on pesticide residues in foods — 1988 and 1989. *J. Assoc. off. anal. Chem.*, **74**, 438–452

Mortelmans, K., Haworth, S., Lawlor, T., Speck, W., Tainer, B. & Zeiger, E. (1986) *Salmonella* mutagenicity tests: II. Results from the testing of 270 chemicals. *Environ. Mutag.*, **8** (Suppl. 7), 1–119

Murano, H., Kojima, M. & Sasaki, K. (1993) Differences in naphthalene cataract formation between albino and pigmented rat eyes. *Ophthalm. Res.*, **25**, 16–22

Nagata, K., Martin, B.M., Gillette, J.R. & Sasame, H.A. (1990) Isozymes of cytochrome P-450 that metabolize naphthalene in liver and lung of untreated mice. *Drug Metab. Disp.*, **18**, 557–564

Nakamura, S.-I., Oda, Y., Shimada, T., Oki, I. & Sugimoto, K. (1987) SOS-inducing activity of chemical carcinogens and mutagens in *Salmonella typhimurium* TA1535/pSK1002: Examination with 15 chemicals. *Mutat. Res.*, **192**, 239–246

Nan, H.-M., Kim, H., Lim, H.-S., Choi, J.K., Kawamoto, T., Kang, J.-W., Lee, C.-H., Kim, Y.-D. & Kwon, E.H. (2001) Effects of occupation, lifestyle and genetic polymorphisms of CYP1A1, CYP2E1, GSTM1 and GSTT1 on urinary 1-hydroxypyrene and 2-naphthol concentrations. *Carcinogenesis*, **22**, 787–793

Narbonne, J.F., Cassand, P., Alzieu, P., Grolier, P., Mrlina, G. & Calmon, J.P. (1987) Structure–activity relationships of the N-methylcarbamate series in *Salmonella typhimurium*. *Mutat. Res.*, **191**, 21–27

National Institute of Occupational Safety and Health (NIOSH) (1990) *National Occupational Exposure Survey (NOES) (1981–1983)*, unpublished provisional data as of July 1, 1990 (cited by National Toxicology Program, 2000)

National Toxicology Program (1984) *Toxicology and Carcinogenesis Studies of 1,3-Butadiene (CAS No. 106-99-0) in B6C3F$_1$ Mice (Inhalation Studies)* (Technical Report Series No. 288; NIH Publ. No. 84-2544), Research Triangle Park, NC

National Toxicology Program (1987) *Toxicology and Carcinogenesis Studies of Ethylene Oxide (CAS No. 75-21-8) in B6C3F$_1$ Mice (Inhalation Studies)* (Technical Report Series No. 326; NIH Publ. No. 88-2582), Research Triangle Park, NC

National Toxicology Program (1992) *Toxicology and Carcinogenesis Studies of Naphthalene (CAS No. 91-20-3) in B6C3F$_1$ Mice (Inhalation Studies)* (NTP Technical Report No. 410; NIH Publ. No. 92-3141), Research Triangle Park, NC

National Toxicology Program (1998) *Toxicology and Carcinogenesis Studies of Chloroprene (CAS No. 126-99-8) in F344/N Rats and B6C3F$_1$ Mice (Inhalation Studies)* (Technical Report Series No. 467; NIH Publ. No. 98-3957), Research Triangle Park, NC

National Toxicology Program (1999) *Toxicology and Carcinogenesis Studies of Isoprene (CAS No. 78-79-5) in F344/N Rats (Inhalation Studies)* (Technical Report Series No. 486; NIH Publ. No. 99-3976), Research Triangle Park, NC

National Toxicology Program (2000) *Toxicology and Carcinogenesis Studies of Naphthalene (CAS No. 91-20-3) in F344/N Rats (Inhalation Studies)* (NTP Technical Report No. 500; NIH Publ. No. 01-4434), Research Triangle Park, NC

Navarro, H.A., Price, C.J., Marr, M.C., Myers, C.B., Heindel, J.J. & Schwetz, B.A. (1991) *Final Report on the Developmental Toxicity of Naphthalene (CAS No. 91-20-3) in Sprague-Dawley Rats (NTP TER-91006)*, Research Triangle Park, NC, National Institute of Environmental Health Sciences

Navarro, H.A., Price, C.J., Marr, M.C., Myers, C.B., Heindel, J.J. & Schwetz, B.A. (1992) *Final Report on the Developmental Toxicity of Naphthalene (CAS No. 91-20-3) in New Zealand White Rabbits (NTP TER-91021)*, Research Triangle Park, NC, National Institute of Environmental Health Sciences

Nohmi, T., Miyata, R., Yoshikawa, K. & Ishidate, M., Jr (1985) [Mutagenicity tests on organic chemical contaminants in city water and related compounds. I. Bacterial mutagenicity tests.] *Eisei Shikenjo Hokoku*, **103**, 60–64 (in Japanese)

Nordholm, L., Espensen, I.-M., Jensen, H.S. & Holst, E. (1986) Polycyclic aromatic hydrocarbons in smokehouses. *Scand. J. Work Environ. Health*, **12**, 614–618

Nylund, L., Heikkilä, P., Hämeilä, M., Pyy, L., Linnainmaa, K. & Sorsa M. (1992) Genotoxic effects and chemical compositions of four creosotes. *Mutat. Res.*, **265**, 223–236

O'Brien, K.A.F., Smith, L.L. & Cohen, G.M. (1985) Differences in napthalene-induced toxicity in the mouse and rat. *Chem.-biol. Interact.*, **55**, 109–122

O'Brien, K.A.F., Suverkropp, C., Kanekal, S., Plopper, C.G. & Buckpitt, A.R. (1989) Tolerance to multiple doses of the pulmonary toxicant, naphthalene. *Toxicol. appl. Pharmacol.*, **99**, 487–500

O'Neil, M.J., Smith, A. & Heckelman, P.E. (2001) *The Merck Index*, 13th Ed., Whitehouse Station, NJ, Merck & Co.

Occupational Safety and Health Administration (1990) *OSHA Analytical Methods Manual*, 2nd Ed., *Part 1 - Organic Substances*, Vol. 2, *Methods 29-54*, Salt Lake City, UT, Department of Labor [Method 35]

Occupational Safety and Health Administration (2001) Labor. *US Code. Fed. Regul.*, **Title 29**, Part 1910, Subpart 1910.1000, pp. 7–19

Ojwang, P.J., Ahmed-Jushuf, I.H. & Abdullah, M.S. (1985) Naphthalene poisoning following ingestion of moth balls: Case report. *East Afr. med. J.*, **62**, 71–73

Ostlere, L., Amos, R. & Wass, J.A.H. (1988) Haemolytic anaemia associated with ingestion of naphthalene-containing anointing oil. *Postgrad. med. J.*, **64**, 444–446

Owen, P.E., Glaister, J.R., Gaunt, I.F. & Pullinger, D.H. (1987) Inhalation toxicity studies with 1,3-butadiene. 3. Two year toxicity/carcinogenicity study in rats. *Am. ind. Hyg. Assoc. J.*, **48**, 407–413

Pakenham, G., Lango, J., Buonarati, M., Morin, D. & Buckpitt, A. (2002) Urinary naphthalene mercapturates as biomarkers of exposure and stereoselectivity of naphthalene epoxidation. *Drug Metab. Disp.*, **30**, 1–7

Peake, J.L. Reynolds, S.D., Stripp, B.R., Stephens, K.E. & Pinkerton, K.E. (2000) Alteration of pulmonary neuroendocrine cells during epithelial repair of naphthalene-induced airway injury. *Am. J. Pathol.*, **156**, 279–286

Pellizzari, E.D., Hartwell, T.D., Harris, B.S.H., III, Waddell, R.D., Whitaker, D.A. & Erickson, M.D. (1982) Purgeable organic compounds in mother's milk. *Bull. environ. Contam. Toxicol.*, **28**, 322–328

Plasterer, M.R., Bradshaw, W.S., Booth, G.M., Carter, M.W., Schuler, R.L. & Hardin, B.D. (1985) Developmental toxicity of nine selected compounds following prenatal exposure in the mouse: Naphthalene, p-nitrophenol, sodium selenite, dimethyl phthalate, ethylenethiourea, and four glycol ether derivatives. *J. Toxicol. environ. Health*, **15**, 25–38

Plopper, C.G., Chang, A.M., Pang, A. & Buckpitt, A.R. (1991) Use of microdissected airways to define metabolism and cytotoxicity in murine bronchiolar epithelium. *Exp. Lung Res.*, **17**, 197–212

Plopper, C.G., Suverkropp, C., Morin, D., Nishio, S. & Buckpitt, A. (1992a) Relationship of cytochrome P-450 activity to Clara cell cytotoxicity. I. Histopathologic comparison of the respiratory tract of mice, rats and hamsters after parenteral administration of naphthalene. *J. Pharmacol. exp. Ther.*, **261**, 353–363

Plopper, C.G., Macklin, J., Nishio, S.J., Hyde, D.M. & Buckpitt, A.R. (1992b) Relationship of cytochrome P-450 activity to Clara cell cytotoxicity. III. Morphometric comparison of changes in the epithelial populations of terminal bronchioles and lobar bronchi in mice, hamsters, and rats after parenteral administration of naphthalene. *Lab. Invest.*, **67**, 553–565

Plopper, C.G., Van Winkle, L.S., Fanucchi, M.V., Malburg, S.R.C, Nishio, S.J., Chang, A. & Buckpitt, A.R. (2001) Early events in naphthalene-induced acute Clara cell toxicity. II. Comparison of glutathione depletion and histopathology by airway location. *Am. J. respir. Cell mol. Biol.*, **24**, 272–281

Propper, R. (1988) *Polyaromatic Hydrocarbons (PAH), a Candidate Toxic Air Contaminant* (Report TR SS-88-01), Springfield, VA, National Technical Information Service, Air Resources Board

Randow, F.F.E., Hübener, T. & Merkel, G. (1996) Hazards for the Rostock water supply from a tar-contaminated sediment in the River Warnow. *Toxicol. Lett.*, **88**, 355–358

Rao, G.S. & Pandya, K.P. (1981) Biochemical changes induced by naphthalene after oral administration in albino rats. *Toxicol. Lett.*, **8**, 311–315

Rathbun, W.B., Holleschau, A.M., Murray, D.L., Buchanan, A., Sawaguchi, S. & Tao, R.V. (1990) Glutathione synthesis and glutathione redox pathways in naphthalene cataract of the rat. *Curr. Eye Res.*, **9**, 45–53

Recochem (1995a) *Technical Data Sheet No. 2: Crude Naphthalene — CR II*, Montreal, Quebec, Canada

Recochem (1995b) *Technical Data Sheet No. 5: Refined Naphthalene — RF II*, Montreal, Quebec, Canada

Ritter, J.K., Owens, I.S., Negishi, M., Nagata, K., Sheen, Y.Y., Gillette, J.R. & Sasame, H.A. (1991) Mouse pulmonary cytochrome P-450 naphthalene hydroxylase: cDNA cloning, sequence, and expression in *Saccharomyces cerevisiae*. *Biochemistry*, **30**, 11430–11437

Rosenfeld, J.K. & Plumb, R.H., Jr (1991) Ground water contamination at wood treatment facilities. *Ground Water Monit. Rev.*, **II**, 133–140

Rozman, K., Summer, K.H., Rozman, T. & Greim, H. (1982) Elimination of thioethers following administration of naphthalene and diethylmaleate to the rhesus monkey. *Drug chem. Toxicol.*, **5**, 265–275

Rundell, J.O., Guntakatta, M. & Matthews, E.J. (1983) Criterion development for the application of BALB/c-3T3 cells to routine testing for chemical carcinogenic potential. *Environ. Sci. Res.*, **27**, 309–324

Sadtler Research Laboratories (1980) *Sadtler Standard Spectra, 1980 Cumulative Index*, Philadelphia, PA, p. 361

Saeed, T., Al-Yakoob, S., Al-Hashash, H. & Al-Bahloul, M. (1995) Preliminary exposure assessment for Kuwaiti consumers to polycyclic aromatic hydrocarbons in seafood. *Environ. int.*, **21**, 255–263

Sakai, M., Yoshida, D. & Mizusaki, S. (1985) Mutagenicity of polycyclic aromatic hydrocarbons and quinones on *Salmonella typhimurium* TA97. *Mutat. Res.*, **156**, 61–67

Sangster, J. (1989) Octanol-water partition coefficients of simple organic compounds. *J. phys. chem. Ref. Data*, **18**, 1128

Santucci, K. & Shah, B. (2000) Association of naphthalene with acute hemolytic anemia. *Acad. Emerg. Med.*, **7**, 42–47

Sartorelli, P., Cenni, A., Matteucci, G., Montomoli, L., Novelli, M.T. & Palmi, S. (1999) Dermal exposure assessment of polycyclic aromatic hydrocarbons: In vitro percutaneous penetration from lubricating oil. *Int. Arch. occup. environ. Health*, **72**, 528–532

Sasaki, J.C., Arey, J., Eastmond, D.A., Parks, K.K. & Grosovsky, A.J. (1997) Genotoxicity induced in human lymphoblasts by atmospheric reaction products of naphthalene and phenanthrene. *Mutat. Res.*, **393**, 23–35

Sax, N.I. & Lewis, R.J., Sr (1987) *Hawley's Condensed Chemical Dictionary*, 11th Ed., New York, Van Nostrand Reinhold, p. 806

Scheepers, P.T.J. & Bos, R.P. (1992) Combustion of diesel fuel from a toxicological perspective. *Int. Arch. occup. environ. Health*, **64**, 149–161

Schmähl, D. (1955) [Testing of naphthalene and anthracene for carcinogenic effects in rats.] *Z. Krebsforsch.*, **60**, 697–710 (in German)

Schmeltz, I., Tosk, J. & Hoffmann, D. (1976) Formation and determination of naphthalenes in cigarette smoke. *Anal. Chem.*, **48**, 645–650

Seixas, G. M., Andon, B. M., Hollingshead, P. G. & Thilly, W. G. (1982) The aza-arenes as mutagens for *Salmonella typhimurium*. *Mutat. Res.*, **102**, 201–212

Shah, J.J. & Heyerdahl, E.K. (1988) *National Ambient Volatile Organic Compounds (VOCs) Data Base Update* (EPA/600/3-88/010a), Research Triangle Park, NC, Environmental Protection Agency, Atmospheric Sciences Research Laboratory, Office of Research and Development

Shane, B.S., Henry, C.B., Hotchkiss, J.H., Klausner, K.A., Gutenmann, W.H. & Lisk, D.J. (1990) Organic toxicants and mutagens in ashes from eighteen municipal refuse incinerators. *Arch. environ. contam. Toxicol.*, **19**, 665–673

Shopp, G.M., White, K.L., Jr, Holsapple, M.P., Barnes, D.W., Duke, S.S., Anderson, A.C., Condie, L.W., Jr, Hayes, J.R. & Borzelleca, J.F. (1984) Naphthalene toxicity in CD-1 mice: General toxicology and immunotoxicology. *Fundam. appl. Toxicol.*, **4**, 406–419

Shultz, M.A., Choudary, P.V. & Buckpitt, A.R. (1999) Role of murine cytochrome P-450 2F2 in metabolic activation of naphthalene and metabolism of other xenobiotics. *J. Pharmacol. exp. Ther.*, **290**, 281–288

Shultz, M.A., Morin, D., Chang, A.-M. & Buckpitt, A. (2001) Metabolic capabilities of CYP2F2 with various pulmonary toxicants and its relative abundance in mouse lung subcompartments. *J. Pharmacol. exp. Ther.*, **296**, 510–519

Siegel, E. & Wason, S. (1986) Mothball toxicity. *Pediatr. Toxicol.*, **33**, 369–374

Simpson, C.D., Cullen, W.R., Quinlan, K.B. & Reimer, K.J. (1995) Methodology for the determination of priority pollutant polycyclic aromatic hydrocarbons in marine sediments. *Chemosphere*, **31**, 4143–4155

Sina, J.F., Bean, C.L., Dysart, G.R., Taylor, V.I. & Bradley, M.O. (1983) Evaluation of the alkaline elution/rat hepatocyte assay as a predictor of carcinogenic/mutagenic potential. *Mutat. Res.*, **113**, 357–391

Smithgall, T.E., Harvey, R.G. & Penning, T.M. (1988) Spectroscopic identification of *ortho*-quinones as the products of polycyclic aromatic *trans*-dihydrodiol oxidation catalyzed by dihydrodiol dehydrogenase. *J. biol. Chem.*, **263**, 1814–1820

Snellings, W.M., Weil, C.S. & Maronpot, R.R. (1984) A two-year inhalation study of the carcinogenic potential of ethylene oxide in Fischer 344 rats. *Toxicol. appl. Pharmacol.*, **75**, 105–117

Sosiaali-ja terveysministeriö [Ministry of Health and Social Affairs] (2002) HTP-arvot 2002 [Values not Known to be Harmful], Tampere, Ministry of Health and Social Affairs, Department of Occupational Health, Kirjapaino Öhrling

Stanley, J.S. (1986) *Broad Scan Analysis of the FY82 National Human Adipose Tissue Survey Specimens*, Vol. I, *Executive Summary (EPA-560/5-86-035)*, Washington DC, Environmental Protection Agency, Office of Toxic Substances

Staples, C.A., Werner, A.F. & Hoogheem, T.J. (1985) Assessment of priority pollutant concentration in the United States using STORET database. *Environ. Toxicol. Chem.*, **4**, 131–142

Stevens, T.P., McBride, J.T., Peake, J.L., Pinkerton, K.E. & Stripp, B.R. (1997) Cell proliferation contributes to PNEC hyperplasia after acute airway injury. *Am. J. Physiol.*, **272**, L486–L493

Stripp, B.R., Maxson, K., Mera, R. & Singh, G. (1995) Plasticity of airway cell proliferation and gene expression after acute naphthalene injury. *Am. J. Physiol.*, **269**, L791–L799

Stuermer, D.H., Ng, D.I. & Morris, C.J. (1982) Organic contaminants in groundwater near an underground coal gasification site in northeastern Wyoming. *Environ. Sci. Technol.*, **16**, 582–587

Summer, K.H., Rozman, K., Coulston, F. & Greim, H. (1979) Urinary excretion of mercapturic acids in chimpanzees and rats. *Toxicol. appl. Pharmacol.*, **50**, 207–212

Suzuki, J., Okazaki, H., Nishi, Y. & Suzuki, S. (1982) Formation of mutagens by photolysis of aromatic compounds in aqueous nitrate solution. *Bull. environ. Contam. Toxicol.*, **29**, 511–516

Tao, R.V., Holleschau, A.M. & Rathbun, W.B. (1991) Naphthalene-induced cataract in the rat. II. Contrasting effects of two aldose reductase inhibitors on glutathione and glutathione redox enzymes. *Ophthalm. Res.*, **23**, 272–283

Thruston, A.D., Jr (1978) High pressure liquid chromatography techniques for the isolation and identification of organics in drinking water extracts. *J. chromatogr. Sci.*, **16**, 254–259

Tingle, M.D., Pirmohamed, M., Templeton, E., Wilson, A.S., Madden, S., Kitteringham, N.R. & Park, B.K. (1993) An investigation of the formation of cytotoxic, genotoxic, protein-reactive and stable metabolites from naphthalene by human liver microsomes. *Biochem. Pharmacol.*, **46**, 1529–1538

Todisco, V., Lamour, J. & Finberg, L. (1991) Hemolysis from exposure to naphthalene mothballs. *New Engl. J. Med.*, **325**, 1660–1661

Traynor, G.W., Apte, M.G., Sokol, H.A., Chuang, J.C., Tucker, W.G. & Mumford, J.L. (1990) Selected organic pollutant emissions from unvented kerosene space heaters. *Environ. Sci. Technol.*, **24**, 1265–1270

Tsuruda, L.S., Lamé, M.W. & Jones, A.D. (1995) Formation of epoxide and quinone protein adducts in $B6C3F_1$ mice treated with naphthalene, sulfate conjugate of 1,4-dihydroxy-naphthalene and 1,4-naphthoquinone. *Arch. Toxicol.*, **69**, 362–367

Turkall, R.M., Skowronski, G.A., Kadry, A.M. & Abdel-Rahman, M.S. (1994) A comparative study of the kinetics and bioavailability of pure and soil-absorbed naphthalene in dermally exposed male rats. *Arch. environ. Toxicol.*, **26**, 504–509

United Nations Environment Programme (2002) *UNEP Chemicals Data Bank Legal File*, Geneva

Usanov, S.A., Erjomin, A.N., Tishchenko, I.V. & Metelitza, D.I. (1982) Comparative study on oxidation of aromatic compounds by rat liver and rabbit lung microsomes. *Acta biol. med. germ.*, **41**, 759–769

Vainiotalo, S. & Matveinen, K. (1993) Cooking fumes as a hygienic problem in the food and catering industries. *Am. ind. Hyg. Assoc. J.*, **54**, 376–382

Valaes, T., Doxiadis, S.A., & Fessas, P. (1963) Acute hemolysis due to naphthalene inhalation. *J. Pediatr.*, **63**, 904–915.

Van Winkle, L.S., Buckpitt, A.R., Nishio, S.J., Isaac, J.M. & Plopper, C.G. (1995) Cellular response in naphthalene-induced Clara cell injury and bronchiolar epithelial repair in mice. *Am. J. Physiol.*, **269**, L800–L818

Van Winkle, L.S., Johnson, Z.A., Nishio, S.J. Brown, C.D. & Plopper, C.G. (1999) Early events in naphthalene-induced acute Clara cell toxicity. Comparison of membrane permeability and ultrastructure. *Am. J. respir. Cell mol. Biol.*, **21**, 44–53

Verschueren, K. (1996) *Handbook of Environmental Data on Organic Chemicals*, 3rd Ed., New York, Van Nostrand Reinhold, pp. 1349–1359

Vuchetich, P.J., Bagchi, D., Bagchi, M., Hassoun, E.A., Tang, L. & Stohs, S.J. (1996) Naphthalene-induced oxidative stress in rats and the protective effects of vitamin E succinate. *Free Radic. Biol. Med.*, **21**, 577–590

Warren, D.L., Brown, D.L., Jr & Buckpitt, A.R. (1982) Evidence for cytochrome P-450 mediated metabolism in the bronchiolar damage by naphthalene. *Chem.-biol. Interact.*, **40**, 287–303

Weissenfels, W.D., Klewer, H.-J. & Langhoff, J. (1992) Adsorption of polycyclic aromatic hydrocarbons (PAHs) by soil particles: Influence on biodegradability and biotoxicity. *Appl. Microbiol. Biotechnol.*, **36**, 689–696

Wells, P.G., Wilson, B. & Lubek, B.M. (1989) *In vivo* murine studies on the biochemical mechanism of naphthalene cataractogenesis. *Toxicol. appl. Pharmacol.*, **99**, 466–473

West, J.A.A., Buckpitt, A.R. & Plopper, C.G. (2000) Elevated airway GSH resynthesis confers protection to Clara cells from naphthalene injury in mice made tolerant by repeated exposures. *J. Pharmacol. exp. Ther.*, **294**, 516–523

West, J.A.A., Pakenham, G., Morin, D., Fleschner, C.A., Buckpitt, A.R. & Plopper, C.G. (2001) Inhaled naphthalene causes dose dependent Clara cell cytotoxicity in mice but not in rats. *Toxicol. appl. Pharmacol.*, **173**, 114–119

Wild, S.R. & Jones, K.C. (1993) Biological and abiotic losses of polynuclear aromatic hydrocarbons (PAHs) from soils freshly amended with sewage sludge. *Environ. Toxicol. Chem.*, **12**, 5–12

Wild, S.R., Waterhouse, K.S., McGrath, S.P. & Jones, K.C. (1990) Organic contaminants in an agricultural soil with a known history of sewage sludge amendments: Polynuclear aromatic hydrocarbons. *Environ. Sci. Technol.*, **24**, 1706–1711

Willems, B.A.T., Melnick, R.L., Kohn, M.C. & Portier, C.J. (2001) A physiologically based pharmacokinetic model for inhalation and intravenous administration of naphthalene in rats and mice. *Toxicol. appl. Pharmacol.*, **176**, 81–91

Wilson, N.K. & Chang, J.C. (1991) Indoor air levels of polynuclear aromatic hydrocarbons and related compounds in an eight-home pilot study. In: Cooke, M., Loening, K. & Merritt, J., eds, *Proceeding of the Eleventh International Symposium on Polynuclear Aromatic Hydrocarbons, Gaithersburg, MD, September 1987*, Columbus, OH, Battelle, pp. 1053–1064

Wilson, A.S., Tingle, M.D., Kelly, M.D. & Park, B.K. (1995) Evaluation of the generation of genotoxic and cytotoxic metabolites of benzo[*a*]pyrene, aflatoxin B_1, naphthalene and tamoxifen using human liver microsomes and human lymphocytes. *Hum. exp. Toxicol.*, **14**, 507–515

Wilson, A.S., Davis, C.D., Williams, D.P., Buckpitt, A.R., Pirmohamed, M. & Park, B.K. (1996). Characterization of the toxic metabolite(s) of naphthalene. *Toxicology*, **114**, 233–242

Wolf, O. (1976) [Cancers in chemical workers in a former naphthalene purification plant]. *Dt. Gesundh. Wesen*, **31**, 996–999 (in German)

Wolf, O. (1978) [Carcinoma of the larynx in naphthalene purifiers]. *Z. ges. Hyg.*, **24**, 737–739 (in German)

Woolf, A.D., Saperstein, A., Zawin, J., Cappock, R. & Sue, Y.-J. (1993) Radiopacity of household deodorizers, air fresheners, and moth repellents. *Clin. Toxicol.*, **31**, 415–428

Worley, G., Erwin, C.W., Goldstein, R.F., Provenzale, J.M. & Ware, E.R. (1996) Delayed development of sensorineural hearing loss after neonatal hyperbilirubinemia: A case report with brain magnetic resonance imaging. *Dev. Med. Child Neurol.*, **38**, 271–277

Xu, G., Zigler, J.S., Jr & Lou, M.F. (1992) The possible mechanism of naphthalene cataract in rat and its prevention by an aldose reductase inhibitor (ALØ1576). *Exp. Eye Res.*, **54**, 63–72

Yamauchi, T., Komura, S. & Yagi, K. (1986) Serum lipid peroxide levels of albino rats administered naphthalene. *Biochem. int.*, **13**, 1–6

Yang, M., Koga, M., Katoh, T. & Kawamoto, T. (1999) A study for the proper application of urinary naphthols, new biomarkers for airborne polycyclic aromatic hydrocarbons. *Arch. environ. Contam. Toxicol.*, **36**, 99–108

Yrjänheikki, E., Pyy, L., Hakala, E., Lapinlampi, T., Lisko, A. & Vähäkangas, K. (1995) Exposure to polycyclic aromatic hydrocarbons in a new coking plant. *Am. ind. Hyg. Assoc. J.*, **56**, 782–787

Yu, X., Wang, X., Bartha, R. & Rosen, J.D. (1990) Supercritical fluid extraction of coal tar contaminated soil. *Environ. Sci. Technol.*, **24**, 1732–1738

Zey, J.N. & Stephenson, R. (1986) *Roofing and Waterproofing Sites, Chicago, Illinois* (Health Hazard Evaluation Report No. HETA-85-416-1742), Cincinnati, OH, National Institute for Occupational Safety and Health

Zinkham, W.H. & Childs, B. (1958) A defect of glutathione metabolism in erythrocytes from patients with a naphthalene-induced hemolytic anemia. *Pediatrics*, **22**, 461–471

Zuelzer, W.W. & Apt, L. (1949) Acute hemolytic anemia due to naphthalene poisoning. A clinical and experimental study. *J. Am. med. Assoc.*, **141**, 185–190

STYRENE

This substance was considered by previous Working Groups, in February 1978 (IARC, 1979), in March 1987 (IARC, 1987) and in February 1994 (IARC, 1994a). Since that time, new data have become available, and these have been incorporated into the monograph and taken into consideration in the present evaluation.

1. Exposure Data

1.1 Chemical and physical data

1.1.1 *Nomenclature*

Chem. Abstr. Serv. Reg. No.: 100-42-5
Replaced CAS Reg. No.: 79637-11-9
Chem. Abstr. Name: Ethenylbenzene
IUPAC Systematic Name: Styrene
Synonyms: Cinnamene; phenethylene; phenylethene; phenylethylene; styrol; styrole; styrolene; vinylbenzene; vinylbenzol

1.1.2 *Structural and molecular formulae and relative molecular mass*

C_8H_8 Relative molecular mass: 104.15

1.1.3 *Chemical and physical properties of the pure substance*

(*a*) *Description*: Colourless, viscous liquid with a pungent odour (WHO, 1983)
(*b*) *Boiling-point*: 145 °C (Lide, 2001)
(*c*) *Melting-point*: –31 °C (Lide, 2001)
(*d*) *Density*: d_4^{20} 0.9060 (Lide, 2001)

(*e*) *Spectroscopy data*: Infrared, ultraviolet, nuclear magnetic resonance and mass spectral data have been reported (Weast & Astle, 1985; Sadtler Research Laboratories, 1991; Lide, 1996).
(*f*) *Solubility*: Insoluble in water; soluble in acetone, diethyl ether and ethanol (Lide, 2001). Very soluble in benzene and petroleum ether (WHO, 1983)
(*g*) *Volatility*: Vapour pressure, 867 Pa at 25 °C; relative vapour density (air = 1), 3.6 (WHO, 1983)
(*h*) *Stability*: Lower flammable limit, 1.1% by volume in air (Quincy, 1991); flash point, 34 °C (National Institute for Occupational Safety and Health, 1983)
(*i*) *Reactivity*: Polymerizes easily at room temperature in the presence of oxygen and oxidizes on exposure to light and air (WHO, 1983)
(*j*) *Octanol-water partition coefficient* (*P*): log P = 2.95 (Hansch *et al.*, 1995)
(*k*) *Conversion factor*: mg/m^3 = 4.26 × ppm[1]

1.1.4 *Technical products and impurities*

Styrene is available as a commercial product with the following specifications: purity, 99.6–99.9% min.; ethylbenzene, 85 ppm max.; polymer content, 10 ppm max.; *para-tert-* butylcatechol (polymerization inhibitor), 10–15 ppm or 45–55 ppm; aldehydes (as benzaldehyde), 200 ppm max.; peroxides (as hydrogen peroxide), 0.0015 wt% or 100 ppm max.; benzene, 1 ppm max.; sulfur, 1 ppm typical; chlorides (as chlorine), 1 ppm typical (James & Castor, 1994; Chevron Phillips Chemical Co., 1996; Chen, 1997; Chevron Phillips Chemical Co., 2001).

Typical analysis of a commercial styrene product reported the following components: purity, 99.93%; benzene, < 1 ppm; toluene, < 1 ppm; ethylbenzene, 50 ppm; α-methylstyrene, 150 ppm; *meta-* + *para*-xylene, 240 ppm; *ortho*-xylene, 80 ppm; cumene, 70 ppm; *n*-propylbenzene, 40 ppm; *meta-* + *para*-ethyltoluene, 20 ppm; vinyl-toluene, 10 ppm; phenylacetylene, 50 ppm; *meta-* + *para*-divinylbenzene, < 10 ppm; *ortho*-divinylbenzene, < 5 ppm; polymer, 1 ppm; *para-tert*-butylcatechol, 12 ppm; aldehydes (as benzaldehyde), 15 ppm; peroxides (as benzoyl peroxides), 5 ppm; chlorides (as chlorine), < 1 ppm; and sulfur, < 1 ppm (Chevron Phillips Chemical Co., 2001).

1.1.5 *Analysis*

(*a*) *Environmental monitoring*

Styrene in workplace air can be determined by capillary column gas chromatography (GC) with a flame ionization detector (FID). The sample is adsorbed on charcoal and desorbed with carbon disulfide. This method (NIOSH Method 1501) has an estimated limit of detection of 0.001–0.01 mg per sample (Eller, 1984).

---

[1] Calculated from: mg/m^3 = (relative molecular mass/24.45) × ppm, assuming normal temperature (25 °C) and pressure (101.3 kPa)

EPA Method 8260B can be used to determine the concentration of various volatile organic compounds, including styrene, by GC–mass spectrometry (MS), in a variety of matrices, such as groundwater, aqueous sludges, waste solvents, oily wastes, tars, soils and sediments. Samples may be analysed using direct injection, purge-and-trap, closed-system vacuum distillation, static headspace (solid samples), or desorption from trapping media (air samples) (EPA Methods 5021, 5030, 5032, 5041); the practical quantification limits are 5 μg/L for groundwater samples, 5 μg/kg (wet weight) for low-level soil and sediment samples, 250 μg/L for water-miscible liquid waste samples, 625 μg/kg for high-level soil and sludge samples and 2500 μg/L for non-water-miscible waste samples (Environmental Protection Agency, 1996).

(*b*) *Biological monitoring*

Biological methods for monitoring exposure to styrene have been reviewed (Guillemin & Berode, 1988; Lauwerys & Hoet, 1993; Pekari *et al.*, 1993; American Conference of Governmental Industrial Hygienists, 2001). Generally accepted biological markers of exposure are mandelic acid (2-hydroxy-2-phenylacetic acid) and phenylglyoxylic acid, the main metabolites of styrene (see Section 4.1.1(*c*)) in urine and styrene in blood. GC procedures have been described for the quantitative determination of urinary phenylglyoxylic and mandelic acids, which involve solvent extraction of the acids and their subsequent determination as derivatives by GC-FID on packed or capillary columns (Guillemin & Bauer, 1976; Flek & Šedivec, 1980; Bartolucci *et al.*, 1986; Dills *et al.*, 1991). High-performance liquid chromatography (HPLC) is widely used for determination of these metabolites. The acids, which may or may not be solvent-extracted, are separated on reverse-phase columns and quantified with an ultraviolet (UV) detector (Ogata & Sugihara, 1978; Ogata & Taguchi, 1987, 1988; Chua *et al.*, 1993). Styrene has been determined in blood by GC with FID or mass selective detection either after solvent extraction (Karbowski & Braun, 1978) or by head-space techniques (Pezzagno *et al.*, 1985; Bartolucci *et al.*, 1986; Brugnone *et al.*, 1993).

Measurement of adducts of styrene 7,8-oxide to the N-terminal valine in haemoglobin has been proposed for monitoring occupational exposure. After enrichment of adducted globin chains by ion-exchange chromatography, the samples are analysed by GC–MS after Edman degradation (Christakopoulos *et al.*, 1993). Conjugation of styrene 7,8-oxide with glutathione, a minor metabolic pathway in humans, leads to specific mercapturic acid products that can be measured in the urine. Ghittori *et al.* (1997) measured these mercapturic acids and mandelic and phenylglyoxylic acids and styrene in the urine of 22 workers in a reinforced plastics plant, and assessed the correlation of urinary metabolites with time-weighted-average (TWA) styrene exposures measured by personal dosimetry. Correlation coefficients of individual or total mercapturic acids (as μg/g creatinine) with TWA styrene exposure ranged up to 0.56, compared with correlation coefficients of 0.86 and 0.82 for mandelic and phenylglyoxylic acids, respectively, with styrene exposure.

Methods of isotope-dilution GC–MS have been described for determination of styrene and styrene 7,8-oxide in blood. Styrene and styrene 7,8-oxide were measured directly in pentane extracts of blood from 35 reinforced plastics workers exposed to 4.7–97 ppm [20–414 mg/m^3] styrene. Positive ion chemical ionization allowed detection of styrene at concentrations greater than 2.5 μg/L blood and styrene 7,8-oxide at concentrations greater than 0.05 μg/L blood. An alternative method for measurement of styrene 7,8-oxide used a reaction with valine followed by derivatization with pentafluorophenyl isothiocyanate and analysis via negative ion chemical ionization GC-MS-MS (styrene 7,8-oxide detection limit, 0.025 μg/L blood). The detection limits for styrene 7,8-oxide by these two methods were 10–20-fold lower than those of the GC assays reported earlier, based upon either electron impact MS or FID (Tornero-Velez *et al.*, 2001).

Elia *et al.* (1980) found an excellent correlation (correlation coefficient, 0.96) between styrene exposure and urinary mandelic acid either alone or in combination with phenylglyoxylic acid. Ikeda *et al.* (1982) found a good correlation (correlation coefficient, 0.88) in 96 male workers in glass fibre-reinforced boat production plants between styrene concentration in air by personal sampling and the combined measurements of mandelic and phenylglyoxylic acid corrected for creatinine. Droz and Guillemin (1983) developed a biological model for the absorption, distribution and elimination of styrene. They examined its validity against data for 60 workers in 10 field studies in the polyester industry and found a good correlation.

Pezzagno *et al.* (1985) exposed 14 male and six female volunteers to styrene (273–1654 μmol/m^3 for 1–3 h) and found a correlation coefficient of 0.93 between exposure and levels of unchanged styrene in urine. Ong *et al.* (1994) studied 39 male workers exposed to concentrations of styrene less than 40 ppm [170 mg/m^3] (time-weighted average) and compared end-of-shift breath levels, blood concentrations of styrene, urinary styrene levels and urinary styrene metabolite levels with exposure levels; the best correlation was observed between blood styrene and styrene in air, while correlation was poor between styrene in urine and styrene in air. There were good correlations between styrene in air and mandelic acid and phenylglyoxylic acid in urine, but the best correlation was found between exposure and the sum of the two acids in urine corrected for creatinine.

Löf and Johanson (1993) exposed two volunteers to 26, 77, 201 and 386 ppm styrene (110, 328, 856 and 1672 mg/m^3) for 2 h during light (50 W) physical exercise. A non-linear relationship between styrene in air and styrene in blood was observed which was attributed to metabolic saturation at exposure levels of 100–200 ppm.

1.2 Production

Styrene was first isolated in 1831 by distillation of storax, a natural balsam. Commercial production of styrene via dehydrogenation of ethylbenzene began in Germany in 1925 (Tossavainen, 1978; Lewis *et al.*, 1983; National Institute for Occupational Safety and Health, 1983; Chen, 1997).

Styrene is produced mainly by catalytic dehydrogenation of high-purity ethylbenzene in the vapour phase. Typical catalysts are based on ferric oxide with the additives chromia (Cr_2O_3) (stabilizer) and potassium oxide (coke retardant) (Lewis *et al.*, 1983). Fractionation of the product results in separation of high-purity styrene, unconverted ethylbenzene and minor reaction by-products such as toluene and benzene (WHO, 1983; James & Castor, 1994; Chen, 1997; Ring, 1999).

A smaller amount of styrene is produced as a co-product from a propylene oxide process. In this route, ethylbenzene is oxidized to its hydroperoxide and reacted with propylene to yield propylene oxide. The co-product methyl phenyl carbinol is then dehydrated to styrene (Mannsville Chemical Products Corp., 1987; Collins & Richey, 1992; James & Castor, 1994; Chen, 1997; Ring, 1999).

Data on the 1998 global production (and production capacity) and consumption of styrene by region are presented in Table 1.

Table 1. Worldwide supply and demand for styrene in 1998 (thousand tonnes)[a]

Region	Capacity	Production	Consumption
North America	6 763	6 095	5 241
South America	400	339	548
Western Europe	4 852	4 040	4 163
Eastern Europe	1 176	366	411
Middle East	555	518	275
Asia	7 294	6 503	7 155
Other[b]	112	84	119
Total	21 152	17 945	17 912

[a] From Ring (1999)
[b] Includes Africa and Oceania

Information available in 2001 indicated that styrene was produced by 21 companies in China, 10 in the USA, nine in Japan, six in Korea (Republic of), four each in Germany, Russia and the Ukraine, three each in Brazil, Canada, France, Mexico and the Netherlands, two each in Azerbaijan, Belgium, Singapore and the United Kingdom and one each in Argentina, Australia, Belarus, Bulgaria, Croatia, the Czech Republic, India, Iran, Poland, Saudi Arabia, Spain and Thailand (Chemical Information Services, 2001).

1.3 Use

Worldwide, styrene is one of the most important monomers for polymers and copolymers that are used in an increasingly wide range of applications. The major uses for styrene are in plastics, latex paints and coatings, synthetic rubbers, polyesters and styrene-alkyd coatings (Collins & Richey, 1992). The broad spectrum of uses of these products includes construction, packaging, automotive and household goods (Mannsville Chemical Products Corp., 1987). Packaging is the single largest application for styrene-containing resins, particularly foams, used as fillers and cushioning. Construction applications include pipes, fittings, tanks, lighting fixtures and corrosion-resistant products. Household goods include synthetic marble, flooring, disposable tableware and moulded furnishings. Transport applications range from tyres to reinforced plastics and automobile body putty (Mannsville Chemical Products Corp., 1987).

Most styrene is converted to polystyrene resins, which are readily moulded and are compatible with a range of colourants, modifiers and fillers. These are used extensively in the fabrication of plastic packaging, disposable beverage tumblers, toys and other moulded goods. Expandable polystyrene beads are used for disposable cups, containers and packaging as well as for insulation. Copolymers and adducts are the second largest family of styrene derivatives. Acrylonitrile–butadiene–styrene (ABS) and styrene–acrylonitrile (SAN) resins have a variety of applications, including in appliance, automotive, construction, pipes and electronics mouldings (Mannsville Chemical Products Corp., 1987).

A variety of special resins have the styrene functionality. Styrene–butadiene rubber (SBR), used for tyres and other elastomer applications, is the largest volume synthetic rubber produced in the USA. Styrene–butadiene latex is used for carpet backing and paper processing. Styrene is the essential co-reactant and solvent in unsaturated polyesters used in reinforced plastic fabrications, including boats, corrosion-resistant tanks and pipes and automobile body parts (Mannsville Chemical Products Corp., 1987). Typical use patterns for styrene worldwide in 1998 and in Canada, Japan, Mexico, the USA and western Europe for selected years are presented in Tables 2 and 3.

Table 2. Use patterns for styrene worldwide in 1998 (thousand tonnes)[a]

Use	North America	Western Europe	Asia	Other	Total
Polystyrene	3 203	2 649	4 409	978	11 239
Unsaturated polyester resins	323	220	161	45	749
ABS/SAN resins	396	433	1 462	43	2 334
Styrene–butadiene copolymer latexes	410	383	279	44	1 116
SBR and latexes	263	123	270	185	841
Other	619	355	574	85	1 633
Total	5 214	4 163	7 155	1 380	17 912

ABS, acrylonitrile–butadiene–styrene; SAN, styrene–acrylonitrile; SBR, styrene–butadiene rubber
[a] From Ring (1999)

Table 3. Use patterns for styrene by country/region (thousand tonnes)[a]

Use	1985	1990	1994	1998
Canada				
Polystyrene	145	183	160	192
Unsaturated polyester resins	14	14	23	24
ABS resins	21	20	11	14
Styrene–butadiene copolymer latexes	25	26	23	11
SBR and latex	14	6	1	1
SAN resins	2	2	–	–
Other	6	7	20	20
Total	227	258	238	262
Japan				
Polystyrene	1 032	1 416	1 388	1 295
Unsaturated polyester resins	72	137	113	114
ABS/SAN resins	312	373	389	372
Styrene–butadiene copolymer latexes	87	129	146	158
SBR and latexes	93	104	86	98
Other	33	130	39	100
Total	1 629	2 289	2 161	2 137
Mexico				
Polystyrene	96	126	151	248
Unsaturated polyester resins	5	5	6	7
ABS resins	4	7	11	12
SBR and latex	29	24	22	28
SAN resins	1	1	1	3
Other	10	5	3	6
Total	145	168	194	304
USA				
Polystyrene	1 844	2 271	2 657	2 876
Unsaturated polyester resins	205	205	246	284
ABS resins	252	284	411	321
Styrene–butadiene copolymer latexes	363	455	562	642
SBR and latex	133	160	192	208
SAN resins	28	43	46	36
Other	123	122	130	308
Total	2 948	3 540	4 244	4 675
Western Europe				
Polystyrene	1 970	2 518	2 513	2 649
Unsaturated polyester resins	136	224	213	220
ABS/SAN resins	301	386	411	433
Styrene–butadiene copolymer latexes	182	286	361	383
SBR and latex	177	158	143	123
Other	248	278	334	355
Total	3 014	3 850	3 975	4 163

ABS, acrylonitrile–butadiene–styrene; SAN, styrene–acrylonitrile; SBR, styrene–butadiene rubber
[a] From Ring (1999); totals may not add due to independent rounding.

1.4 Occurrence

A comprehensive review on styrene exposure and health has appeared recently (Cohen *et al.*, 2002).

1.4.1 *Natural occurrence*

Styrene has been identified in trace amounts in the gummy exudate (Storax balsam) from the damaged trunks of certain trees, probably resulting from the natural degradation of the cinnamic acid derivatives that occur in large quantities in these exudates (Furia & Bellanca, 1971; Tossavainen, 1978; Duke, 1985). Styrene has been found at very low levels in many agricultural products and foods, but it is not clear whether this styrene results from natural processes within the plant (see Section 1.4.3 (*c*)).

1.4.2 *Occupational exposure*

The National Occupational Exposure Survey conducted by the National Institute for Occupational Safety and Health between 1981 and 1983 indicated that 1 112 000 employees in the USA were potentially exposed to styrene at work (National Institute for Occupational Safety and Health, 1993). The estimate was based on a survey of companies in the USA and did not involve actual measurements of exposure.

Workers may be exposed in a number of industries and operations, including styrene production, production of polystyrene and other styrene-containing polymer resins, plastics and rubber products fabrication, fabrication of reinforced-polyester plastics composites and use of products containing styrene, such as floor waxes and polishes, paints, adhesives, putty, metal cleaners, autobody fillers and varnishes (National Institute for Occupational Safety and Health, 1983).

(*a*) *Production of styrene and polystyrene*

Average exposure to styrene in styrene production and polymerization factories has been reported rarely to exceed 20 ppm [85 mg/m^3] and is usually due to occasional bursts and leakages of reactors, tubing and other equipment (Tossavainen, 1978). Surveys conducted in plants in the USA engaged in the development or manufacture of styrene-based products between 1962 and 1976 showed that the average exposure of employees in all jobs was below 10 ppm [43 mg/m^3]. Peak concentrations of up to 50 ppm [213 mg/m^3] were measured during the drumming of styrene. Batch polymerization of styrene in 1942 produced concentrations up to 88 ppm [375 mg/m^3] during filling operations; subsequent continuous polymerization processes generally resulted in exposure levels of 1 ppm [4.3 mg/m^3] or below (Ott *et al.*, 1980). In a plant in the USA where styrene was produced and polymerized, the highest levels of styrene were found in polymerization, manufacturing and purification areas (mean, 8–35 ppm [34–149 mg/m^3]), while levels of less than 5 ppm [21 mg/m^3] occurred in maintenance, laboratory and packaging operations. Urinary mandelic acid and blood styrene were undetectable in most samples taken from these workers at the end of a shift: < 10 mg/g creatinine for mandelic acid (5 ng/mL) and < 2 ng/mL for styrene in blood. The maximal concentrations were 140 mg/g creatinine for mandelic acid and 90 ng/mL for styrene in blood (Wolff *et al.*, 1978). In a German styrene production, polymerization and processing plant, samples taken in 1975–76 in various areas of the plant contained none (< 0.01 ppm [0.04 mg/m^3]) to 6.8 ppm [29 mg/m^3], most values being below 1 ppm [4.3 mg/m^3]. In a part of the plant where polystyrene was manufactured, area samples in 1975 contained from none (< 0.01 ppm [0.04 mg/m^3]) to 47 ppm [200 mg/m^3], most values being below 1 ppm [4.3 mg/m^3]. Of 67 employees engaged in either area of the plant, six had urinary concentrations of mandelic acid above 50 mg/L (Thiess & Friedheim, 1978).

Other substances that may be found in workplace air during the manufacture of styrene and polystyrene include benzene, toluene, ethylbenzene, other alkylbenzene compounds and ethylene (Ott *et al.*, 1980; Lewis *et al.*, 1983; National Institute for Occupational Safety and Health, 1983). Exposure to benzene was previously a primary concern in these processes. In the plant in the USA described above, the TWA concentration of benzene in styrene monomer manufacture was 0.3–14.7 ppm [1–47 mg/m^3] between 1953 and 1972. Samples taken in 1942 during a washing operation in the polymerization plant contained up to 63 ppm [202 mg/m^3] benzene (Ott *et al.*, 1980).

(*b*) *Production of styrene–butadiene rubber (SBR) and other styrene-based polymers*

Concentrations of styrene in area samples and breathing-zone air measured in 1965 in various plants of a styrene–butadiene latex manufacturing company in the USA (see above) were 4–22 ppm [17–94 mg/m^3]. The initial stages of the process, including loading, operating and cleaning of polymerization reactors, involved the highest exposure, and operators in these job categories were exposed to concentrations ranging from 3.6 to 7.3 ppm [15–31 mg/m^3] in 1973 (Ott *et al.*, 1980).

In two adjacent SBR production plants in the USA, the TWA concentrations of styrene were 0.94 and 1.99 ppm [4 and 8.5 mg/m^3], with an overall range of 0.03–12.3 ppm [0.13–52.4 mg/m^3] (Meinhardt *et al.*, 1982). The mean concentrations in 159 personal air samples taken in 1979 in various departments at another SBR production plant in the USA were usually below 1 ppm [4.3 mg/m^3], except for factory service and tank farm workers, for whom the means were 1.69 and 13.7 ppm [7.2 and 58.2 mg/m^3], respectively (Checkoway & Williams, 1982). Company data provided by five of eight SBR plants in the USA for the period 1978–83 gave an average styrene level in 3649 samples from all plants of 3.53 ppm [15 mg/m^3], with a standard deviation of 14.3 ppm [61 mg/m^3] (Matanoski *et al.*, 1993). A study by Macaluso *et al.* (1996) of the same facilities as the Matanoski study used industrial hygiene data together with a series of air dispersion models to estimate how TWA styrene exposure levels in the styrene–butadiene resin industry may have changed since the 1940s. Their calculations suggest that TWA exposures declined from an average of 1.8 ppm [7.7 mg/m^3] during the 1940s to 0.1 ppm [0.4 mg/m^3] in the 1990s.

In a plant in the USA where acrylic ester–styrene copolymers [wrongly called polystyrene by the authors] were produced, concentrations in the breathing zone in 50 samples ranged from none detected (less than 1 ppb [4.3 μg/m³]) to 19.8 ppm [84 mg/m³], with an average of about 0.6 ppm [2.5 mg/m³]; the highest concentrations occurred during styrene unloading operations (Samimi & Falbo, 1982).

The numerous other substances to which workers may be exposed in these processes include 1,3-butadiene, acrylonitrile, acrylates, acrylic acid, α-methylstyrene (*meta*-vinyltoluene), 4-vinylcyclohexene, toluene, benzene, ammonia, formaldehyde, colourants and a variety of solvents (Ott *et al.*, 1980; Samimi & Falbo, 1982; National Institute for Occupational Safety and Health, 1983). Accelerators are chemical compounds that increase the rate of cure and improve the physical properties of natural and synthetic rubbers, including SBR. Thiuram sulfides and salts of dialkyldithiocarbamic acids, including dimethyldithiocarbamate, are used as vulcanization accelerators. Sodium and potassium dimethyldithiocarbamates are used as modifiers in emulsion polymerization (Schubart, 1987; Lattime, 1997).

(*c*) *Processing of styrene-based polymers*

Styrene was measured as a thermal degradation product in the air of a Finnish factory during the processing of polystyrene, impact polystyrene and acrylonitrile–butadiene–styrene (ABS) resins. The mean concentrations (6 h) were 0.4, 0.1 and 0.06 mg/m³, respectively (Pfäffli, 1982). Personal 8-h samples taken in 1978, 1979 and 1980 in companies in the USA where polystyrene and ABS moulding was performed contained [17–285 mg/m³] (Burroughs, 1979); 1.4–3.2 mg/m³ (Belanger & Elesh, 1980) and < 0.01 mg/m³ (below the limit of detection) (Ruhe & Jannerfeldt, 1980).

Styrene is one of the volatile organic compounds released during extrusion and vulcanization of SBR. Rappaport and Fraser (1977) reported styrene concentrations of 61–146 ppb [0.3–0.6 mg/m³] in the curing area of the press room of a company manufacturing passenger-car tyres. Area samples taken in the vulcanization and extrusion areas of shoe-sole, tyre retreading and electrical cable insulation plants contained styrene at concentrations of 2–500 μg/m³ (vulcanization) and 0–20 μg/m³ (extrusion) (Cocheo *et al.*, 1983). A more complete description of the work environment encountered in the rubber products manufacturing industry may be found in a previous monograph (IARC, 1982).

(*d*) *Manufacture of glass fibre-reinforced polyester products*

Occupational exposure to styrene is most extensive, with respect to number of workers and levels of exposure, in the fabrication of objects from glass fibre-reinforced polyester composite plastics, such as boats, tanks, wall panels, bath and shower units and automotive parts (National Institute for Occupational Safety and Health, 1983). Styrene serves as a solvent and a reactant for the unsaturated polyester resin, in which it constitutes about 40% by weight. In the open mould process, a releasing agent is usually applied to the mould, a first coat containing pigments (gel coat) is applied, then successive layers of chopped and/or woven fibre glass are deposited manually or with a chopper gun at the same time as the resin is sprayed or brushed on, and then the surface is rolled. During lamination and curing, about 10% of the styrene may evaporate into the workplace air (National Institute for Occupational Safety and Health, 1983; Crandall & Hartle, 1985). Exposure to styrene in this industry has been extensively documented and summarized in several reports (National Institute for Occupational Safety and Health, 1983; WHO, 1983; Pfäffli & Säämänen, 1993). Table 4 lists levels of occupational exposure to styrene (personal breathing zone samples) reported in various countries in the larger studies.

Among the biological monitoring methods available (Section 1.1.5(*b*)), measurements of mandelic acid and phenylglyoxylic acid (see Section 4.1.1(*c*)) in urine are the most commonly used biological indices of exposure to styrene. Table 5 gives the concentrations of the classical biological indicators of exposure from various studies. Symanski *et al.* (2001) examined the variation in urinary levels of mandelic acid and phenylglyoxylic acid among workers exposed to styrene in the reinforced plastics industry. Levels of phenylglyoxylic acid varied less than those of mandelic acid, as did metabolite levels expressed in terms of urinary creatinine concentration. Urinary metabolite levels were highest for laminators and for samples collected at the end of the working week.

Several factors influence the level of styrene in air. The manufacture of objects with large surface area, such as boats, truck parts, baths and showers, by the open-mould process results in the highest exposure. Data from 28 plants producing reinforced plastics products in the USA showed that the average exposure to styrene in open-mould processes was two to three times higher than that in press-mould processes: 24–82 ppm [102–350 mg/m^3] versus 11–26 ppm [47–111 mg/m^3] (Lemasters *et al.*, 1985). In a detailed survey of 12 plants making fibreglass in Washington State, USA, 40% of 8-h samples contained more than 100 ppm [430 mg/m^3]. Chopper gun operators had the highest exposure, followed by laminators and gel-coat applicators; boat-building involved higher exposures than any other sector. For 11 plants, a relationship was seen between level of exposure and the quantity of resin consumed per month per exposed employee (Schumacher *et al.*, 1981). Similar results were reported by Sullivan and Sullivan (1986) in their survey of 10 plants in Ontario, Canada, who also noted that although dilution ventilation and often auxiliary fans were used in almost all plants, there was little use of local exhaust ventilation. This was also the case for boat construction in the USA. Gel coaters have lower exposure because they generally work in ventilated booths (Crandall & Hartle, 1985). The presence of flexible exhaust ventilation hoses was reported to reduce styrene concentrations by a factor of two at a boat construction company in Japan (Ikeda *et al.*, 1982). So-called 'low-styrene emission resins' are in theory promising for reducing exposure, but their potential to do so in the workplace has not been sufficiently validated and they are not widely used (A.D. Little, Inc., 1981; Sullivan & Sullivan, 1986; Säämänen *et al.*, 1993).

An extensive data source on exposures to styrene in the styrene composites manufacturing industry is the report of a study conducted in 1986 by Cal/OSHA (1986). This

Table 4. Occupational exposure to styrene in the glass fibre-reinforced plastics industry in various countries

Country and year of survey	No. of plants	Job/task	Duration of samples	No. of samples	Air concentration in personal breathing zone (mg/m^3)		Reference
					Mean	Range	
Canada (Ontario) 1981	10	All jobs	25 min	126	< 4.3–716[a]	< 4.3–1393	Sullivan & Sullivan (1986)
		Boat laminating		59	430 GM	8.1 GSD	
		Non-boat laminating		23	124 GM	29.4 GSD	
		Chopper gun use		8	554 GM	7.7 GSD	
		Gel-coat spraying		6	298 GM	7.7 GSD	
		Filament winding		3	533 GM	6.0 GSD	
Canada (Québec) NR	3	Chopper gun use	8 h	7	564	307–938	Truchon *et al.* (1992)
		Painting (gel coat)		9	517	280–843	
		Laminating (rollers)		18	502	292–865	
		Foreman		8	97	18–279	
		Cutter		11	75	16–234	
		Warehouse work		19	35	9–187	
		Finishing		31	34	8–110	
		Mould repair		8	28	8–147	
Denmark							Jensen *et al.* (1990)
1955–70	30	NR	1–60 min	227	714	10–4700	
1971–80	97	NR		1117	274	4–1905	
1981–88	129	NR		1184	172	1–4020	
Italy 1978–90	87	Hand laminating	Variable	1028	227		Galassi *et al.* (1993)
		Spraying laminating		166	134		
		Rolling		40	163		
		Semi-automatic process		71	85		
		Non-process work		159	71 (38 GM)	3.8 GSD	
Italy NR	10	NR	8 h	64	113.6 GM	8–770.4	Gobba *et al.* (1993)
Japan NR	5	Boat fabrication:	4 h				Ikeda *et al.* (1982)
		Hull lamination		25	507 GM	145–1091	
		Hull lamination with local exhaust ventilation		9	277 GM	196–383	
		Lamination of hold walls		25	537 GM	371–916	

Table 4 (contd)

Country and year of survey	No. of plants	Job/task	Duration of samples	No. of samples	Air concentration in personal breathing zone (mg/m³)		Reference
					Mean	Range	
Switzerland	10	NA	Full shift	90	201	8–848	Guillemin *et al.* (1982)
NR							
Netherlands	4	Filament winding	4 h	18	[314 GM]	134–716	Geuskens *et al.* (1992)
NR		Spraying		62	[227 GM]	48–602	
		Hand laminating		180	[148 GM]	18–538	
USA	7	Boat fabrication	Full-shift				Crandall & Hartle (1985)
NR		Hull lamination		168	331	7–780	
		Deck lamination		114	313	52–682	
		Small parts lamination		70	193	34–554	
		Gel coating		45	202	23–439	
Europe (5 countries)[b]		Lamination	Variable				Bellander *et al.* (1994)
< 1980		Boat fabrication		1703	332		
≥ 1980				2993	234		
< 1980		Containers		437	247		
≥ 1980				1098	187		
< 1980		Panels and construction		401	213		
≥ 1980				846	145		
< 1980		Small pieces		486	251		
≥ 1980				629	158		
≥ 1980		Hand lamination		3205	281		
≥ 1980		Spray lamination		414	132		
≥ 1980		Non-manual lamination		231	68		

Table 4 (contd)

Country and year of survey	No. of plants	Job/task	Duration of samples	No. of samples	Air concentration in personal breathing zone (mg/m³)		Reference
					Mean	Range	
USA	30	Spray-up/lay-up	≥ 60 min; 8-h TWA computed	NR	256[c]	21–511	A.D. Little, Inc. (1981)
1967–78		Gel coating			192	43–256	
		Winding			170	64–362	
		Sheet-moulding compound production			170	43–341	
		Foaming			128	64–213	
		Mixing			107	9–341	
		Casting			85	21–192	
		Cut, press and weigh			64	21–341	
		Other jobs[d]			≤ 43	0–213	

GM, geometric mean; GSD, geometric standard deviation; NA, not applicable; NR, not reported; TWA, time-weighted average; concentrations originally reported in ppm were converted to mg/m³ for this table

[a] Range of arithmetic means for different plants

[b] Italian plants reported by Galassi *et al.* (1993) and Finland, Norway, Sweden and the United Kingdom

[c] Typical level

[d] Includes general and non-production, finish and assembly, store and ship, office and other, injection molding, field service, preform production and pultrusion (a continuous process for producing composite materials of constant cross-sectional area)

Table 5. Biological monitoring of occupational exposure to styrene in the glass fibre-reinforced plastics industry

Country and year of survey	No. of plants	Job/task	No. of samples	Concentrations at end of shift						Reference
				Mandelic acid in urine (mg/g creatinine)		Phenylglyoxylic acid in urine (mg/g creatinine)		Styrene in blood (mg/L)		
				Mean	SD	Mean	SD	Mean	SD	
Canada (Quebec) NR	3	Chopper gun operation	7	[980]	[980]					Truchon *et al.* (1992)
		Painting (gel coat)	9	[750]	[310]					
		Laminating (rolling)	18	[1690]	[605]					
		Foreman	8	[350]	[470]					
		Cutting	11	[320]	[380]					
		Warehouse worker	19	[70]	[70]					
		Finishing	31	[110]	[120]					
		Mould repair	8	[30]	[50]					
Germany NR	4	Laminating boats, pipes, or containers	36	210 (10–3640)[a]		190 (10–870)[a]		0.39 (0.04–4.82)[a]		Triebig *et al.* (1989)
Italy NR	4	Refrigerating containers	6	493	434	121	96	0.32	0.42	Bartolucci *et al.* (1986)
		Flooring tiles	6	428	248	72	22	0.42	0.16	
		Fibre-glass canoes	5	270	54	62	24	0.52	0.32	
		Fibre-glass tanks	3	323	129	132	41	NR		
Italy 1978–90	118	Hand lamination	2386	450 GM	2.75 GSD					Galassi *et al.* (1993)
		Spray lamination	250	211 GM	3.3 GSD					
		Rolling	63	182 GM	3.08 GSD					
		Semi-automatic process operation	121	154 GM	259 GSD					
		Non-process work	762	94 GM	3.27 GSD					
Switzerland NR	10	NR	88	1004	1207	339	360			Guillemin *et al.* (1982)
United Kingdom 1979	1	Boat industry	27	[780]	555			[0.72]	[0.43]	Cherry *et al.* (1980)

SD, standard deviation; NR, not reported; GM, geometric mean; GSD, geometric standard deviation
[a] Median (range) in mg/L

study was an in-depth industrial hygiene survey of styrene and other workplace exposures. A total of 141 workplaces with 2600 workers were inspected, and in 50 of these workplaces, a total of 379 workers were monitored over a full work shift. Exposures were sampled by a charcoal tube method with personal sampling pumps, and in addition with passive organic vapour dosimeters, and by analysis of the urine of 327 workers (85% of the study population) for mandelic acid. The focus of the study was on large open-mould spray-up/lay-up operations. Styrene exposures at these processes ranged from 0.2 to 288 ppm [0.85–971 mg/m^3] TWA for eight hours; the 8-h TWA arithmetic mean and the median for these sample results were 43.0 ppm [183 mg/m^3] and 34.0 ppm [145 mg/m^3], respectively. In a comparison of worker exposure levels by industry, the Cal/OSHA study showed that the geometric mean exposure levels were highest in tub/shower manufacturing facilities (53.6 ppm [228 mg/m^3]), followed by camper manufacturing (41.0 ppm [175 mg/m^3]), boat manufacturing (29.1 ppm [124 mg/m^3]), spa manufacturing (25.8 ppm [110 mg/m^3]), miscellaneous manufacturing (22.0 ppm [94 mg/m^3]), and tank manufacturing facilities (12.7 ppm [54 mg/m^3]). Operations ranked according to percentage of styrene exposures above 100 ppm [430 mg/m^3] as an 8-h TWA were: tub/shower manufacturing (19%), spa manufacturing (11%), camper manufacturing (6%), miscellaneous plastics manufacturing (4%) and boat and tank manufacturing (none).

Styrene exposures were measured for 82 Dutch workers at four plants where unsaturated polyester resins were applied to moulds in the production of glass fibre-reinforced plastics (Geuskens *et al.*, 1992). The study considered three resin application categories: filament winding techniques; hand laminating or spraying using more than 25 kg resin every 4 h; hand laminating using less than 25 kg resin per 4 h; workers who were in the production area but did not come into direct contact with styrene were the control group. The geometric mean 8-h TWA concentrations for the four plants ranged from 11 to 49 ppm [47–209 mg/m^3] for the workers who came into direct contact with styrene and from 5 to 16 ppm [21–68 mg/m^3] for those who were in the production area but had no contact with styrene.

Wong *et al.* (1994) followed 15 826 American workers in the reinforced plastics industry who had worked in areas with exposure to styrene for at least six months between 1948 and 1977. Using estimates of exposure based on current (around 1980) measurements and a historical assessment of work practices, plant design, process design and occupational hygiene factors, they divided their cohort into strata by cumulative styrene exposure. Approximately 23% of the workers had estimated cumulative exposures exceeding 100 ppm–years. The estimated average exposure among members of this group was 298 ppm–years.

The Norwegian National Institute of Occupational Health reviewed a collection of more than 7000 styrene concentration measurements taken in 234 reinforced-plastics production companies in Norway during the years 1972 through 1996 (Lenvik *et al.*, 1999). Of these plants, 124 (accounting for 60.2% of the measurements) produced boats, 65 (accounting for 30.2%) produced small items, five (accounting for 3.4%) produced car body parts and another 40 (accounting for 6.2%) produced miscellaneous items.

Since 1990, the long-term average measurements (sampling time more than 1 h) fell below 20 ppm [85 mg/m³] (except for a value slightly above 20 ppm in 1996). The analyses show a decrease in the median from 62 ppm [264 mg/m³] in the 1970s to 7.1 ppm [30 mg/m³] in the 1990s.

A comprehensive approach to biological monitoring of 44 workers occupationally exposed to styrene in a hand lamination plant in Europe was performed by considering several end-points, including styrene in workplace air, styrene in exhaled air, and styrene in blood. Other end-points included DNA strand breaks, chromosomal aberrations, immune parameters and genotyping of polymorphic genes (see Section 4.5). The set of workers consisted of four groups: the high-exposure group, consisting of hand lamination workers; the medium-exposure group, consisting mainly of sprayers; the low-exposure group, consisting of maintenance workers; and the control group, consisting of non-exposed clerks. The mean duration of exposure for the whole group was 13.0 years. The mean concentration of styrene in the workplace for the whole group was 101 mg/m³ and the mean blood concentration of styrene was 601 μg/L. For the high-, medium- and low-exposure groups, mean air (and blood) concentrations were 199 mg/m³ (2098 μg/L), 55 mg/m³ (81 μg/L) and 27 mg/m³ (85 μg/L) (Somorovská *et al.*, 1999).

Measurement of biological indicators of exposure complements the picture based on air concentrations because biological levels incorporate the influence of other routes of absorption and of the use of personal protective equipment (see also Section 4.1). Despite early reports that percutaneous absorption of styrene was an important route of exposure, measurement of biological indicators of the exposure of workers who did and did not wear gloves and other forms of protective clothing indicated that absorption through the skin makes a negligible contribution to overall exposure in the manufacture of glass fibre-reinforced polyester products (Brooks *et al.*, 1980; Bowman *et al.*, 1990; Truchon *et al.*, 1992). Wearing a respirator appropriate for organic vapours reduces exposure markedly, but not entirely (Brooks *et al.*, 1980; Ikeda *et al.*, 1982; Bowman *et al.*, 1990; Truchon *et al.*, 1992). Respirators are worn most often by gel-coat and chopper gun operators but not by laminators, who consider that they hinder their work (Truchon *et al.*, 1992). Single-use dust respirators, which provide no protection against styrene vapours, were often the only type of protection worn (Schumacher *et al.*, 1981; Sullivan & Sullivan, 1986).

Limasset *et al.* (1999) compared the level of styrene absorbed percutaneously with that absorbed by inhalation in a real situation in the glass fibre-reinforced polyester industry. The study protocol consisted of comparisons of the patterns of urinary excretion of styrene metabolites by four groups of workers, all of whom performed the same task at the same time in the same workshop but wore the following different protective equipment: total protection with an insulating suit and mask, respiratory equipment only, skin protection only, and no protection. The urinary excretion level of the group with total protection did not significantly differ from that of the group with respiratory protection only. Percutaneous absorption was not a particularly important pathway for styrene absorption during stratification work in the polyester industry.

Completely insulating personal protective equipment provided no greater level of protection than does a respirator at positive pressure alone.

The protection afforded by respirators to styrene-exposed workers has also been evaluated by measuring the reduction in urinary excretion of styrene (Gobba *et al.*, 2000). Seven glass fibre-reinforced plastics workers not using respiratory protection devices were studied for a week. External exposure to styrene was evaluated by personal passive sampling and the internal dose by measurement of urinary styrene. Workers then wore half-mask respirators for a week and styrene exposure and internal dose were reassessed. Mean TWA concentrations of styrene for the morning half-shift for the first week ranged from 246.5 to 261.2 mg/m^3; mean levels for the second week ranged from 228.5 to 280.8 mg/m^3. Mean TWA concentrations of styrene for the afternoon half-shift for the first week ranged from 169.2 to 330.2 mg/m^3; mean levels for the second week ranged from 207.3 to 335.7 mg/m^3. Mean urinary concentrations of styrene for the first week ranged from 80.0 to 96.1 μg/L; for the second week, the means ranged from 31.5 to 47.2 μg/L. The authors concluded that the protection afforded by negative-pressure half-mask respirators varies widely, and stressed the need to assess the effective reduction of exposure whenever these devices are introduced for styrene-exposed workers. Measurement of urinary excretion of unmodified styrene was useful for the evaluation of respirator effectiveness in exposed workers.

Other substances may be found in workplace air in plants for the production of unsaturated polyester-reinforced plastics, although at levels usually considerably lower than that of styrene. These include: solvents, mainly used to clean tools and equipment, such as ketones (e.g., acetone), chlorinated hydrocarbons (e.g., dichloromethane), aliphatic alcohols and esters and aromatic hydrocarbons; organic peroxides used as initiators (e.g., methyl ethyl ketone peroxide, benzoyl peroxide); styrene 7,8-oxide and other oxidation products resulting from the reaction of peroxides with styrene; hydroquinone and analogues used as inhibitors (e.g., hydroquinone, quinone, catechol); dusts and fibres originating mainly from filler and reinforcement materials (e.g., glass fibres, silica, asbestos); foaming agents such as isocyanates; and cobalt salts and amines used as accelerators (Pfäffli *et al.*, 1979; A.D. Little, Inc., 1981; Makhlouf, 1982; Högstedt *et al.*, 1983; National Institute for Occupational Safety and Health, 1983; Coggon *et al.*, 1987; Jensen *et al.*, 1990; Bellander *et al.*, 1994).

(*e*) *Miscellaneous operations*

In a study of exposures of firefighters, air samples taken during the 'knockdown' phase of a fire contained styrene at a concentration of 1.3 ppm [5.5 mg/m^3]; none was detected during the 'overhaul' phase (Jankovic *et al.*, 1991). During working operations at a US hazardous waste site in 1983, a mean styrene concentration of 235 μg/m^3 (maximum, 678 μg/m^3) was measured in air for a group of workers nearest the areas where chemically contaminated materials were handled (Costello, 1983). During the manufacture of polyester paints, lacquers and putties in Finland, occasional high exposure to styrene was recorded, with 5% of measurements above 20 ppm [85 mg/m^3]; use of the same products resulted in exposures below 1 ppm [4.3 mg/m^3] (Säämänen *et al.*, 1991). Application of polyester putty during cable splicing operations for a telephone company in the USA resulted in short-term levels (3–16 min) ranging from 2 to 16 ppm [8.5–68 mg/m^3] in four samples (Kingsley, 1976). In a Japanese plant where plastic buttons were manufactured from polyester resins, the 8-h TWA concentration of styrene for 34 workers was 7.1 ppm [30 mg/m^3], with a maximum of 28 ppm [119 mg/m^3] (Kawai *et al.*, 1992).

In four 100-min area air samples taken in 1982 at a college in the USA during a sculpture class in which polyester resins were used, styrene concentrations ranged from 0.8 to 1.2 ppm [3.4–5.1 mg/m^3]; two personal breathing zone air samples contained 2.8 and 3.0 ppm [11.9 and 12.8 mg/m^3]. The concentration of methyl ethyl ketone peroxide was below the detection limit (< 0.02 ppm) (Reed, 1983).

Taxidermists who used polyester resins during specimen preparation were shown to be exposed for short periods (2–34 min) to concentrations of styrene ranging from 21 to 300 mg/m^3 (12 samples) (Kronoveter & Boiano, 1984a,b).

In two cooking-ware manufacturing companies in the USA where styrene-based resins were used, the 8-h TWA concentrations of styrene ranged from 0.2 to 81 ppm [0.85–345 mg/m^3]; two short-duration samples (24 min) contained 142 and 186 ppm (605 and 792 mg/m^3] (Fleeger & Almaguer, 1988; Barsan *et al.*, 1991).

1.4.3 *Environmental occurrence*

Human exposure to styrene has been assessed on the basis of a review of data in the published literature (Tang *et al.*, 2000). The authors estimated that styrene exposure for the general population is in the range of 18.2–55.2 μg/person/day (0.3–0.8 μg/kg bw) or 6.7–20.2 mg/person/year (95.7–288 μg/kg bw), mainly resulting from inhalation and from food intake. The inhaled styrene accounts for more than 90% of the total intake. The styrene in food occurs mainly by migration from polymer packaging materials. The authors also concluded that cigarette smoking is another important source of styrene intake for smokers. The intake of styrene due to smoking 20 cigarettes was estimated to be higher than the total daily intake from food and air.

A Canadian study estimated a daily total styrene intake for the Canadian general population ranging from below 0.19 to over 0.85 μg/kg bw. Intakes from ambient air ranged from 0.004 up to 0.17 μg/kg bw and those from indoor air from 0.07 up to 0.10 μg/kg bw. Intake from food was calculated to range from below 0.11 to over 0.58 μg/kg bw. The estimated intakes from drinking water and soil were negligible. Potential exposure from cigarette smoke, on the basis of the styrene content reported for mainstream smoke (10 μg per cigarette) and a smoking rate of 20 cigarettes per day, was estimated to be 2.86 μg/kg bw per day for adults. The Canadian study estimated that styrene in food may represent a major exposure source for the general population (Health Canada, 1993; Newhook & Caldwell, 1993).

(*a*) *Air*

Styrene has been detected in the atmosphere in many locations. Its presence in air is due principally to emissions from industrial processes involving styrene and its polymers and copolymers. Other sources of styrene in the environment include vehicle exhaust, cigarette smoke and other forms of combustion and incineration of styrene polymers (WHO, 1983).

Styrene emissions reported to the European Union by member countries (Bouscaren *et al.*, 1987) are shown in Table 6. Air emissions in the USA, reported to the US Environmental Protection Agency by approximately 1500 industrial facilities, increased from 15 600 tonnes in 1988 to 24 800 tonnes in 1999 (Environmental Protection Agency, 2001a). Based on Environmental Protection Agency (1999) national emission estimates, the total styrene emissions for 1990 can be allocated as follows: on-road vehicles, 17 900 tonnes (32.9% of total emissions); reinforced plastic composites production and boat manufacturing, 21 700 tonnes (39.8%); and all other sources, 14 900 tonnes (33.9%). Ambient air levels of styrene sampled in the vicinity (< 500 m) of seven reinforced plastic processors in three states in the USA ranged from 0.29 to 2934 μg/m^3, and those in communities near the processors (500–1000 m) from 1.67 to 23.8 μg/m^3 (McKay *et al.*, 1982). Styrene levels of 1.1–6.6 μg/m^3 were measured in air samples from the Pennsylvania Turnpike Allegheny Mountain Tunnel in 1979. The mean concentration in the tunnel intake air was below 0.1 μg/m^3 (Hampton *et al.*, 1983). Air concentrations of styrene in the vicinity of five rural hazardous waste sites in New Jersey, USA, ranged up to 66 μg/m^3 (LaRegina *et al.*, 1986).

Table 6. Estimated emissions of styrene in member countries of the European Union (thousand tonnes per year)

Country	Source	
	Road traffic (gasoline)	Chemical industry
Belgium	0.5	0.75
Denmark	0.28	NR
France	2.9	3.4
Germany	2.9	3.4
Greece	0.5	NR
Ireland	0.19	NR
Italy	3.0	3.5
Luxembourg	0.02	0.03 (other sources)
Netherlands	0.7	1.45
Portugal	0.5	0.3
Spain	2.0	1.2
United Kingdom	3.0	3.7
Total	16.0	18.0

From Bouscaren *et al.* (1987); NR, not reported

Ambient air monitoring data from the USA include databases compiled and maintained by the California Air Resources Board. The 20 test stations are located in urban areas, representing the greatest portion of the California population. Styrene is measured on a 24-hour sample collected once each month. Based on data from the Board reflecting the measurements for each test station for each month from 1989 to 1995, the average reading for styrene was approximately 0.20 ppb [0.9 μg/m^3] over six years. The detection level for styrene was 0.1 ppb [0.4 μg/m^3] and the highest measurement was 2.9 ppb [12.4 μg/m^3] (Styrene Information & Research Center, 2001).

Styrene levels in ambient air were determined in a survey of 18 sites (mostly urban) in Canada in 1988–90. The mean concentrations in 586 24-h samples ranged from 0.09 to 2.35 μg/m^3. In a national survey of styrene levels in indoor air in 757 single-family dwellings and apartments, representative of the homes of the general population of Canada in 1991, the mean 24-h concentration was 0.28 μg/m^3, with values ranging from none detected (limit of detection, 0.48 μg/m^3) up to 129 μg/m^3 (Newhook & Caldwell, 1993).

For residential exposure, median concentrations obtained by personal air sampling are generally in the 1–3 μg/m^3 range (Wallace *et al.*, 1985; Wallace, 1986). Exposure to styrene is approximately six times higher for smokers than for nonsmokers, and tobacco smoke is the major source of styrene exposure for smokers (Wallace *et al.*, 1987, 1989). Measurements in homes with and without smokers revealed that average styrene concentrations in the homes of smokers were approximately 0.5 μg/m^3 higher than those in the homes of nonsmokers. Hodgson *et al.* (1996) also found that environmental tobacco smoke (ETS) can contribute significantly to indoor airborne styrene concentrations. ETS was estimated to contribute 8% to the total styrene inhalation exposure of all non-smoking Californians (Miller *et al.*, 1998).

Styrene is one of the hundreds of individual components that may be quantified in tobacco smoke (IARC, 1986; Darrall *et al.*, 1998; Health Canada, 1999; IARC, 2003). The styrene content of cigarette smoke has been reported to be 18–48 μg per cigarette (WHO, 1983). Off-gassing of styrene from some styrene-containing household products may also contribute to indoor air levels (Knöppel & Schauenburg, 1989).

In order to illustrate the relative significance of various sources of exposure to styrene, Fishbein (1992) estimated approximate exposure levels in several environments and compared nominal daily intakes from those sources (Table 7).

Thermal degradation of styrene-containing polymers also releases styrene into ambient air (Hoff *et al.*, 1982; Lai & Locke, 1983; Rutkowski & Levin, 1986). Gurman *et al.* (1987) reported that styrene monomer is the main volatile product of the thermal decomposition of polystyrene, comprising up to 100% of the volatiles in special laboratory conditions.

Volatile organic compounds found at municipal structural fires have been characterized in order to identify sources of long-term health risks to firefighters. These compounds were identified and quantified using GC–MS in selected ion-monitoring mode. The compounds with the highest levels found were benzene, toluene and naphthalene

Table 7. Estimated intake of styrene from different sources of exposure

Source	Estimated concentration	Nominal daily intake[a]
Reinforced plastics industry	200 000 μg/m³	2 g
Styrene polymerization	10 000 μg/m³	100 mg
Within 1 km of a production unit	30 μg/m³	600 μg
Polluted urban atmosphere	20 μg/m³	400 μg
Urban atmosphere	0.3 μg/m³	6 μg
Indoor air	0.3–50 μg/m³	6–1000 μg
Polluted drinking-water (2 L/day)	1 μg/L	2 μg
Cigarette smoke (20 cigarettes/day)	20–48 μg/cigarette	400–960 μg

From Fishbein (1992)
[a] Calculated on the assumption of a daily respiratory intake of 10 m³ at work or 20 m³ at home or in an urban atmosphere

(see monograph in this volume); styrene and other alkyl-substituted benzene compounds were frequently identified (Austin *et al.*, 2001).

Little is known about factors that influence blood levels of volatile organic compounds in non-occupationally exposed populations. Possible relationships were examined between recent self-reported chemical exposures and elevated blood volatile organic compound levels among 982 adult participants in the Third National Health and Nutrition Examination Survey in the USA. A strong relationship was found ($p < 0.001$) between increasing lifetime pack–years of cigarettes smoked and elevated levels of toluene, styrene, and benzene (Churchill *et al.*, 2001).

(*b*) *Water*

Although styrene has been detected occasionally in estuaries and inland waters and in drinking-water, its presence is usually traceable to an industrial source or to improper disposal (WHO, 1983; Law *et al.*, 1991). In surveys of Canadian drinking-water supplies, the frequency of detection of styrene was low; when detected, it was generally at a concentration below 1 μg/L (Newhook & Caldwell, 1993). After accidental drinking-water contamination with styrene in Spain, transient levels up to 900 μg/L were reported (Arnedo-Pena *et al.*, 2002).

(*c*) *Food*

Polystyrene and its copolymers have been used widely as food packaging materials, and residual styrene monomer can migrate into food from such packaging (WHO, 1983). Analysis of styrene in 133 plastic food containers from retail food outlets in the United Kingdom showed concentrations ranging from 16 to 1300 mg/kg; 73% of containers had styrene concentrations of 100–500 mg/kg, and only five containers had levels exceeding 1000 mg/kg. The food in the containers had levels of monomer ranging from below 1 to

200 μg/kg, although 77% of the foods had levels below 10 μg/kg and 26% had levels below 1 μg/kg (Gilbert & Startin, 1983).

Similar surveys were carried out by the Food Safety Directorate in 1992 and 1994 in the United Kingdom, with styrene concentrations similar to those found in the 1983 survey. Within each food type, higher levels of styrene were generally found for products with high fat content or packed in small containers (Ministry of Agriculture, Fisheries and Food, 1994).

Between 1991 and 1999, the Food and Drug Administration's Total Diet Study in the USA analysed 320 different foods and found styrene residues in 49 of them. In 258 samples containing styrene, the mean concentrations for individual food items varied between 10 μg/kg (eggs) and 274 μg/kg (strawberries). The median concentration for the 49 foods was 21 μg/kg (Food and Drug Administration, 2000).

Several assessments have been made of estimated daily intake (EDI) of styrene in food packaged in polymers or copolymers of styrene. In 1981, the Food and Drug Administration in the USA measured the migration of styrene over a 24-h period from foam, impact and crystal polystyrene cups into 8% ethanol and water at 49 °C. The observed migration was 0.036, 0.064 and 0.210 μg/cm^2, respectively. This corresponds to 6 μg styrene from a 2 dL (8 oz.) foam cup. Migration from foam cups into hot water, tea and coffee was one fifth of these levels: approximately 1 μg from a 2 dL (8 oz.) cup (Varner & Breder, 1981). In a 1983 study, the EDI of styrene monomer from polystyrene food packaging, including polystyrene foam cups, was in the range of 1 to 4 μg per day (Ministry of Agriculture, Fisheries and Food, 1989). More recently, Lickly *et al.* (1995) estimated total dietary intake of styrene monomer from polystyrene food-contact polymers to be 9 μg per day.

Styrene has been detected at low levels (ppb) in several foods and beverages which had not previously been in contact with styrene-containing packaging materials (Maarse, 1992a,b; Steele, 1992, Steele *et al.*, 1994). Much higher levels (around 40 mg/kg) have been measured in cinnamon, and enzymatic degradation of cinnamic acid derivatives was proposed as a possible source (Oliviero, 1906; Ducruet, 1984). The low-level occurrence of styrene in other foods has been suggested to result from enzymatic and/or microbial activity, but it is unclear to what extent these processes are, in fact, responsible for the levels detected (Steele *et al.*, 1994; Tang *et al.*, 2000).

1.5 Regulations and guidelines

1.5.1 *Exposure limits and guidelines*

Occupational exposure limits and guidelines for styrene are presented in Table 8. A tolerable daily intake (TDI) of 7.7 μg/kg bw for styrene has been established by WHO (1993), with a guideline value of 20 μg/L in drinking-water. The Environmental Protection Agency (2001b) has set a maximum contaminant level (MCL) for styrene in public water systems in the USA at 0.1 mg/L.

The Food and Drug Administration (2001) has established regulations for the use of polymers and copolymers of styrene in products in contact with food in the USA. For styrene and methyl methacrylate copolymers as components of paper and paperboard in contact with fatty foods, the monomer content in the copolymer is limited to 0.5%. For styrene–acrylic copolymers, the level of residual styrene monomer in the polymer should not exceed 0.1% by weight.

Table 8. Occupational exposure limits and guidelines for styrene

Country or region	Year	Concentration (mg/m^3)	Interpretation
Australia	1993	213	TWA
		426	STEL
Belgium	1993	213 (sk)	TWA
		426	STEL
Czech Republic	1993	200	TWA
		1000	STEL
Denmark	1993	106	TWA
Finland	2002	86	TWA
		430	STEL
France	1993	213	TWA
Germany	2001	86 (category 5)[a]	MAK; substance with systemic effects (onset < 2 h)
Hungary	1993	50 (Ca)	TWA
Ireland	1993	426	TWA
		1065	STEL
Japan	2000	85 (sk)	TWA (provisional value)
Mexico	1984	215	TWA
		425	STEL
Netherlands	1999	106	TWA
Philippines	1993	426	TWA
Poland	1998	50 (sk)	TWA
		200	STEL
Sweden	1993	106 (sk)	TWA
		320	STEL
Switzerland	1993	213	TWA
		426	STEL
Thailand	1993	426	TWA
		852	STEL
Turkey	1993	426	TWA
United Kingdom	2000	430	TWA; maximum exposure limit
		1065	STEL

Table 8 (contd)

Country or region	Year	Concentration (mg/m^3)	Interpretation
USA			
ACGIH (TLV)	2001	85 (A4, ir)	TWA
		170	STEL
OSHA (PEL)	2001	426	TWA
		852	Ceiling
NIOSH (REL)	2000	213	TWA
		426	Ceiling

From American Conference of Governmental Industrial Hygienists (ACGIH) (2000, 2001); Deutsche Forschungsgemeinschaft (2001); Occupational Safety and Health Administration (OSHA) (2001); Sosiaali-ja terveysministeriö (2002); United Nations Environment Program (UNEP) (2002)

TWA, 8-h time-weighted average; STEL, short-term exposure limit; MAK, maximum workplace concentration; TLV, threshold limit value; PEL, permissible exposure level; REL, recommended exposure level; A4, not classifiable as a human carcinogen; Ca, suspected of having carcinogenic potential; ir, irritant; sk, absorption through the skin may be a significant source of exposure

[a] Category 5, substances with carcinogenic and genotoxic potential, the potency of which is considered to be so low that, provided that the MAK value is observed, no significant contribution to human cancer risk is to be expected

1.5.2 *Reference values for biological monitoring of exposure*

The relationship between external (air concentrations) and biological measures of exposure has been studied more extensively for styrene than for most other organic compounds in the occupational environment. Various reported correlations between the concentration of styrene in air and those in venous blood and with mandelic acid and phenylglyoxylic acid levels in urine have been reviewed by Guillemin and Berode (1988), Lauwerys and Hoet (1993), Pekari *et al.* (1993) and the American Conference of Governmental Industrial Hygienists (2001).

For example, the concentration of mandelic acid in urine that corresponds to inhalation of 50 ppm styrene (213 mg/m^3) for 8 h would be approximately 800–900 mg/g creatinine at the end of a shift and 300–400 mg/g creatinine the following morning (Droz & Guillemin, 1983; Guillemin & Berode, 1988; Pekari *et al.*, 1993). The phenylglyoxylic acid concentration in urine that corresponds to an 8-h exposure to 50 ppm styrene would be expected to be 200–300 mg/g creatinine at the end of a shift and about 100 mg/g creatinine the following morning (Pekari *et al.*, 1993; American Conference of Governmental Industrial Hygienists, 2001). The styrene concentration in blood that corresponds to an 8-h exposure to 50 ppm styrene would be expected to be 0.5–1 mg/L at the end of a shift and about 0.02 mg/L in blood the following morning (Guillemin & Berode, 1988; American Conference of Governmental Industrial Hygienists, 2001).

Each year, the American Conference of Governmental Industrial Hygienists (ACGIH) (2001) and the Deutsche Forschungsgemeinschaft (2001) publish biological reference values for use in interpreting the results of biological monitoring for styrene in the workplace. The results must be interpreted in relation to the different definitions of those reference values. ACGIH biological exposure indices are reference values intended for use as guidelines for evaluating potential health hazards in the practice of industrial hygiene. The indices represent the levels of the determinants that are most likely to be observed in specimens collected from healthy workers exposed by inhalation to air concentrations at the level of the threshold limit value (American Conference of Governmental Industrial Hygienists, 2001). In Germany, the biological tolerance value (BAT) for occupational exposures is defined as the maximal permissible quantity of a chemical compound or its metabolites, or the maximum permissible deviation from the norm of biological parameters induced by those substances in exposed humans. According to current knowledge, these conditions generally do not impair the health of an employee, even if exposure is repeated and of long duration. The BAT values are conceived as ceiling values for healthy individuals (Deutsche Forschungsgemeinschaft, 2001).

Biological monitoring reference values for exposure to styrene, based on styrene metabolite levels in urine or styrene in blood, are given in Table 9.

Table 9. Reference values for biological monitoring of exposure to styrene

Determinant	Sampling time	Biological exposure index[a]	BAT[b]
Mandelic acid in urine	End of shift	800 mg/g creatinine[c]	Does not apply
	Prior to next shift	300 mg/g creatinine[c]	Does not apply
Phenylglyoxylic acid in urine	End of shift	240 mg/g creatinine[c]	Does not apply
	Prior to next shift	100 mg/g creatinine[c]	Does not apply
Mandelic acid plus phenylglyoxylic acid in urine	End of shift	Does not apply	600 mg/g creatinine
Styrene in venous blood	End of shift	0.55 mg/L[d]	Does not apply
	Prior to next shift	0.02 mg/L[d]	

[a] American Conference of Governmental Industrial Hygienists (2001)

[b] BAT, Biologischer Arbeitsstoff-Toleranz-Wert (biological tolerance value for occupational exposures) (Deutsche Forschungsgemeinschaft, 2001)

[c] Non-specific, as it is also observed after exposure to other chemicals such as ethylbenzene

[d] Semiquantitative, because of short half-life of styrene in blood

2. Studies of Cancer in Humans

2.1 Case reports

Cases of leukaemia and lymphoma were identified among workers engaged in the production of styrene–butadiene rubber (Lemen & Young, 1976), in the manufacture of styrene–butadiene (Block, 1976) and in the manufacture of styrene and polystyrene (Nicholson *et al.*, 1978). A total of 16 cases of leukaemia and 9 of lymphoma were reported in these studies. In addition to styrene, exposure to benzene, 1,3-butadiene, ethylbenzene and other chemicals could have occurred in these operations.

2.2 Cohort studies (see Table 10)

2.2.1 *Styrene manufacture and polymerization*

A study by Frentzel-Beyme *et al.* (1978) of 1960 workers engaged in the manufacture of styrene and polystyrene in Germany between 1931 and 1976 showed no significant excess mortality from cancer. The cohort had accumulated 20 138 person–years. Follow-up was 93% for the German workers but only 29% for the non-German workers and there were only 74 deaths (96.5 expected) available for analysis. Only one death from lymphatic cancer was observed. There were two deaths from pancreatic cancer (0.7 expected). In 1975 and 1976, concentrations of styrene in the plant were generally below 1 ppm [4.3 mg/m^3], but higher concentrations were occasionally recorded (Thiess & Friedheim, 1978). [The Working Group noted that insufficient information was provided to assess the risk for cause-specific deaths by exposure period or duration of exposure.]

Ott *et al.* (1980) studied a cohort of workers at four plants in the USA where styrene-based products were developed and produced. Exposure to styrene varied by process and time period. During container filling for batch polymerization in 1942, styrene concentrations ranged from 5 to 88 ppm [21–375 mg/m^3]; in continuous polymerization and extrusion units, the concentrations were below 10 ppm [43 mg/m^3] and generally below 1 ppm in 1975 and 1976. Cohorts from each plant had been exposed between 1937 and 1970 and were followed from 1940 to 1975. Other potential exposures included benzene, acrylonitrile, 1,3-butadiene, ethylbenzene, dyes and pigments. Age- and race-specific US mortality rates were used to calculate the expected numbers of deaths. A total of 2904 workers with a minimum of one year of employment were included. Bond *et al.* (1992) updated the study, adding a further 11 years of follow-up. Based on this update, the standardized mortality ratio (SMR) for all causes was 0.76 (95% confidence interval [CI], 0.70–0.82; 687 deaths), for all cancers was 0.81 (95% CI, 0.69–0.95; 162 deaths) and for lymphatic and haematopoietic malignancies was 1.4 (95% CI, 0.95–2.1; 28 deaths). The excess of lymphatic and haematopoietic neoplasms was restricted to

Table 10. Characteristics of cohort and nested case–cohort studies of incidence or mortality from neoplasms among workers exposed to styrene

Reference (country)	Type of plant; study period; number of subjects; minimal period employed; follow-up	No of deaths/ cancer deaths	Results				Comments (previous evaluation)
			No.	SMR/ SIR	95% CI	Site	
Frentzel-Beyme *et al.* (1978) (Germany)	Styrene and polystyrene manufacture facility; 1931–76; 1960 subjects; 1 month; 93% among German, 29% among non-German	74/11	1	–	–	Lymphoma	Lymphoma case was a malignant neoplasm of the spleen (IARC, 1994a). Expected number based on the statistics for the town of Ludwigshafen during 1970–75 was 0.06. Two pancreatic cancer deaths were observed, 0.7 expected, respectively; and three lung cancer deaths, 5.7 expected.
Ott *et al.* (1980) (USA)	Dow Chemical workers in development or production of styrene-based products; 1940–75; 2904 men; 1 year; 97.0%	303/58	12 21	1.6 1.6	0.84–2.8 1.0–2.5	L&H, mortality L&H, incidence	Excess incidence of lymphatic leukaemia (SIR, 4.3; 95% CI, 1.7–8.8; 7 cases); highest risks in workers exposed to styrene, ethylbenzene, other fumes, solvents and colourants (IARC, 1994a)
Hodgson & Jones (1985) (United Kingdom)	Production, polymerization and processing of styrene; 1945–78; 622 men; 1 year; 99.7% among exposed, 94.4% among referents	34/10 (exposed)	5 3 0 4	[1.2] [5.4] – [2.5]	[0.39–2.8] [1.1–16] – [0.67–6.4]	Lung, mortality Lymphomas, mortality Leukaemias, mortality L&H, incidence	Exposure to styrene among other chemicals such as 1,3-butadiene, acrylonitrile, benzene, dyestuff, and ethylene oxide (IARC, 1994a)
Okun *et al.* (1985) (USA)	Reinforced plastics boat building (two facilities); 1959–78; 5021 subjects; 1 day; 98.1%	176/36	16 0	1.4 0	[0.82–2.3] [0–0.88]	Lung L&H	From L&H, 4.2 deaths expected in the whole cohort, and about one death expected within the group with high exposure to styrene (IARC, 1994a)

Table 10 (contd)

Reference (country)	Type of plant; study period; number of subjects; minimal period employed; follow-up	No of deaths/ cancer deaths	Results				Comments (previous evaluation)
			No.	SMR/ SIR	95% CI	Site	
Coggon *et al.* (1987) (United Kingdom)	Production of glass-reinforced plastics (8 facilities); 1947–84; 7949 subjects; no minimal employment; variable, 61.9–99.7%	693/181	6	[0.40]	[0.15–0.88]	L&H, mortality	One of six deaths from L&H with high exposure to styrene. Extended follow-up of this cohort included by Kogevinas *et al.* (1994a,b) (IARC, 1994a)
Bond *et al.* (1992) (USA)	Dow Chemical workers in development or production of styrene-based products; 1940–86; 2904 men; 1 year; 96.7%	687/162	15	0.86	0.48–1.4	Large intestine	A mortality study, updating of Ott *et al.* (1980). From among L&H cancers, elevated incidences of multiple myeloma and Hodgkin disease (IARC, 1994a)
			2	[0.39]	0.04–1.4	Rectum	
			5	0.49	0.16–1.1	Pancreas	
			56	0.81	0.61–1.0	Lung	
			6	0.95	0.35–2.1	Brain/other nervous system	
			28	1.4	0.95–2.1	L&H	
Kolstad *et al.* (1993, 1994, 1995) (Denmark)	Production of glass fibre-reinforced plastics and other plastics (552 facilities); 1970–89; 53 720 men and 10 798 women; no minimal employment; 98.2%	4281/2285 incident cases	174	[1.1]	[0.93–1.3]	L&H, men/women	Significantly higher incidence of leukaemia (SIR, 1.6) among men at ≥ 10 years since first exposure; incidence of leukaemia higher among short-term male workers with estimated higher exposure to styrene; part of this cohort included by Kogevinas *et al.* (1994a,b) (IARC, 1994a)
			66	[1.1]	[0.86–1.4]	Leukaemia, men/women	
			112	1.2	0.98–1.4	L&H, men[a]	
			42	1.2	0.88–1.7	Leukaemia, men[a]	
			90	0.87	0.70–1.1	Breast, women	
			16	1.2	0.68–1.9	Brain and nervous system, women	

Table 10 (contd)

Reference (country)	Type of plant; study period; number of subjects; minimal period employed; follow-up	No of deaths/ cancer deaths	Results				Comments (previous evaluation)
			No.	SMR/ SIR	95% CI	Site	
Kogevinas *et al.* 1994a,b) (six European countries)	Production of glass fibre-reinforced plastics (660 facilities); 1945–91; varies between cohorts; 40 688 workers; 97%	2714/686	39	0.77	0.55–1.1	Colon	Risk of L&H increased with latency (p for trend = 0.012) and with average exposure (p for trend = 0.019); risk did not increase with duration of exposure or cumulative exposure. Non-significant excess risk observed for pancreas for high cumulative exposure to styrene (p for trend = 0.068) (IARC, 1994a). Results presented in this table are for the full cohort which included a non-exposed group.
			21	0.62	0.38–0.95	Rectum	
			37	1.0	0.71–1.4	Pancreas	
			235	0.99	0.87–1.1	Lung	
			18	0.62	0.37–0.98	Brain	
			60	0.93	0.71–1.2	L&H	
			28	1.0	0.69–1.5	Leukaemia	
Wong *et al.* (1994) (USA)	Reinforced plastics manufacturing plants (n = 30); 1948–89; 15 826 subjects; 6 months; 96.5%	1628/425	36	1.2	0.83–1.6	Large intestine	Higher risk of L&H among workers with cumulative exposure > 100 ppm–years at > 20 years since first exposure (SMR, 1.3; 5 deaths); SMR of lung cancer among workers with cumulative exposure > 100 ppm–years, 1.0 (34 deaths) (IARC, 1994a)
			19	1.1	0.68–1.8	Pancreas	
			162	1.4	1.20–1.6	Bronchus, trachea, lung	
			14	0.62	0.34–1.1	Breast	
			8	0.59	0.25–1.2	Central nervous system	
			31	0.82	0.56–1.2	L&H	
			11	0.74	0.37–1.3	Leukaemia and aleukaemia	
Anttila *et al.* (1998) (Finland)	Database of workers biologically monitored for urinary mandelic acid; 1973–92; 2580 subjects; no minimal employment; 98.5%	NR/48	1	0.36	0.01–2.0	Colon	Both of the L&H cases were Hodgkin lymphomas.
			6	3.1	1.14–6.8	Rectum	
			3	1.7	0.34–4.9	Pancreas	
			5	0.59	0.19–1.4	Lung, trachea	
			5	0.62	0.20–1.5	Breast	
			6	1.6	0.59–3.5	Nervous system	
			2	0.39	0.05–1.4	L&H	

Table 10 (contd)

Reference (country)	Type of plant; study period; number of subjects; minimal period employed; follow-up	No of deaths/ cancer deaths	Results				Comments (previous evaluation)
			No.	SMR/ SIR	95% CI	Site	
Sathiakumar *et al.* (1998) (USA and Canada)	Styrene–butadiene rubber plants ($n = 8$); 1943–91; 15 649 men; one year; 95%	3967/950	87	0.97	0.78-1.2	Large intestine	Among ever hourly subjects with at least ten years worked and 20 years since hire: significantly increased mortality from L&H and leukaemia (SMRs, 1.5 and 2.2; 49 and 28 deaths)
			20	0.78	0.48-1.2	Rectum	
			43	0.82	0.60-1.1	Pancreas	
			349	1.01	0.91-1.1	Lung	
			25	0.92	0.59-1.4	Central nervous system	
			101	1.1	0.88-1.3	L&H	
			48	1.3	0.97-1.7	Leukaemia	
			9	0.82	0.37-1.6	Benign neoplasms	
Loughlin *et al.* (1999) (USA)	Students attending a school adjacent to a styrene–butadiene facility; 1963–95; 15 403 students, 7882 men, 7521 women; 3 months; NR	338/44	6	[1.1]	[0.37–2.2]	Lung	No data on specific environmental exposures to chemicals among school students
			14	[1.2]	[0.66–2.0]	L&H	
			7	[1.3]	[0.51–2.6]	Leukaemia	
			6	[4.2]	[1.5–9.1]	Benign neoplasms	
Delzell *et al.* (2001) (USA and Canada)	Styrene–butadiene rubber plants ($n = 6$); 1943–91; 13 130 men; one year; vital status known for > 99%, death certificates for 98% of the decedents	3892/NR	59	3.2	1.2–8.8	Leukaemia (unadjusted)	Within-cohort comparison; the cohort follow-up reported in Sathiakumar *et al.* (1998). Relative risks for the high cumulative exposure to styrene (18 exposed cases); whether unadjusted, or adjusted for exposure to 1,3-butadiene and DMDTC; both models included also terms for age and years since hire.
			59	0.8	0.2–3.8	Leukaemia (adjusted)	

L&H, malignancies of the lymphatic and haematopoietic tissues; NR, not reported; SIR, standardized incidence ratio; SMR, standardized mortality ratio; CNS, central nervous system; DMDTC, dimethyldithiocarbamate

[a] Only men in the companies producing reinforced plastics

workers with less than five years of employment and was significantly increased among workers after 15 years of follow-up (SMR, 1.6; 95% CI, 1.0–2.4).

Hodgson and Jones (1985) reported on 622 men who had worked for at least one year in the production, polymerization and processing of styrene at a plant in the United Kingdom between 1945 and 1974 who were followed until 1978. Of these, 131 men were potentially exposed to styrene in laboratories and 491 in production of styrene monomer, polymerization of styrene or manufacture of finished products. No measurements of exposure were provided, but many other chemicals were present in the working environment. Expected numbers of deaths were calculated on the basis of national rates. There were 34 deaths (43.1 expected) among the 622 exposed workers. A significant excess of deaths from lymphoma (SMR, 5.4; [95% CI, 1.1–16]; 3 deaths) was observed. An analysis of cancer registrations for this population revealed an additional case of lymphatic leukaemia, giving a total of four incident cases of lymphatic and haematopoietic cancer, whereas 1.6 would have been expected from local cancer registration rates [standardized incidence ratio (SIR), 2.5; 95% CI, 0.67–6.4]. In addition, three incident cases of laryngeal cancer were found (0.5 expected; [SIR, 6.0; 95% CI, 1.2–18]).

2.2.2 *Use of styrene in reinforced plastics*

Okun *et al.* (1985) studied 5021 workers who had been employed in two reinforced-plastic boat-building facilities in the USA for at least one day between 1959 and 1978. On the basis of industrial hygiene surveys, 2060 individuals were classified as having had high exposure to styrene, with means in the two facilities of 42.5 and 71.7 ppm [181 and 305 mg/m^3]. Of these, 25% had worked for less than one month, 49% had worked for one month to one year and only 7% had worked for more than five years. There were 47 deaths in the high-exposure group (41.5 expected); no cases of lymphatic or haematopoietic cancer were observed in the high-exposure group (approximately one expected) or in the full cohort (4.2 expected).

Wong (1990) and Wong *et al.* (1994) reported on a cohort of 15 826 male and female employees who had worked at one of 30 reinforced-plastics plants in the USA for at least six months between 1948 and 1977. Workers were followed until 1989; vital status was determined using Social Security Administration files, the National Death Index and the records of credit agencies. A total of 307 932 person–years at risk were accumulated. Expected numbers of deaths were based on national age-, gender-, cause- and year-specific death rates for whites, as no information was available on race. Exposure to styrene was calculated using a job–exposure matrix that included work history and current and past time-weighted average exposures. A total of 1628 (10.3%) members of the cohort were found to have died, and death certificates were obtained for 97.4% of them. The overall SMR was 1.08 (95% CI, 1.03–1.13) and the SMR from all cancers was 1.16 (95% CI, 1.05–1.27). Mortality from cancers at a number of sites was increased significantly; the SMRs for these sites were: oesophagus, 1.9 (95% CI, 1.1–3.2; 14 deaths); bronchus, trachea and lung, 1.4 (95% CI, 1.2–1.6; 162 deaths); cervix uteri, 2.8

(95% CI, 1.4–5.2; 10 deaths); and other female genital organs, 2.0 (95% CI, 1.1–3.5; 13 deaths). In the cohort as a whole, no excess was observed for lymphohaematopoietic cancers (SMR, 0.82; 95% CI, 0.56–1.2; 31 deaths), nor was there a strong suggestion of excess risk for any of these cancers in any of the subgroups analysed. The data were analysed by Cox regression with age, gender, length of exposure and cumulative exposure included in the model. Neither cumulative exposure nor length of exposure was significantly related to risk in the model for lymphatic and haematopoietic cancer. No positive dose–response relationship was found for any other cancer that was analysed. [The Working Group noted that the possibility that the inclusion of two correlated indices of styrene exposure in the regression models may have artificially reduced the coefficients of both.]

Kogevinas *et al.* (1993, 1994a,b) included an extended follow-up of the study by Coggon *et al.* (1987) and parts of the material of Kolstad *et al.* (1993, 1994, 1995) and Härkönen *et al.* (1984). The study included 40 688 workers employed in 660 plants of the reinforced plastics industry and enrolled in eight subcohorts in Denmark, Finland, Italy, Norway, Sweden and the United Kingdom. Exposure to styrene was reconstructed from job and production records, environmental measurements and, in Italy, biological monitoring. An exposure database was constructed on the basis of about 16 400 personal air samples and 18 695 measurements of styrene metabolites in urine. Styrene exposure levels decreased considerably during the study period. The data from Denmark, considered to be representative of all six countries, showed exposures of about 200 ppm [852 mg/m^3] in the early 1960s, about 100 ppm [430 mg/m^3] in the late 1960s and about 20 ppm [85 mg/m^3] in the late 1980s. The 40 688 workers accumulated 539 479 person–years at risk and were followed for an average of 13 years. Workers lost to follow-up and those who emigrated constituted 3.0% of the total cohort, and in no individual cohort did this proportion exceed 8.0%; 60% of the cohort had less than two years' exposure and 9% had more than 10 years' exposure. The WHO mortality data bank was used to compute national mortality reference rates by sex, age (in five-year groups) and calendar year. No excess was observed for mortality from all causes [SMR, 0.96; 95% CI, 0.92–1.00; 2196 deaths] or from all neoplasms (SMR, 0.91; 95% CI, 0.83–0.99; 536 deaths) among styrene-exposed workers. The SMR for malignant neoplasms was 0.91 (95% CI, 0.78–1.06; 167 deaths) among laminators and 0.73 (95% CI, 0.59–0.88; 106 deaths) among unexposed workers. The mortality rate in exposed workers for neoplasms of the lymphatic and haematopoietic tissues was not elevated (SMR, 0.98; 95% CI, 0.72–1.3; 49 deaths) and was not associated with length of exposure. When the duration of exposure was two years or more and at least 20 years had elapsed since first exposure, the SMR for all lymphatic and haematopoietic cancers was 1.7 (95% CI, 0.70–3.6; seven deaths) and that for leukaemia was 1.9 (95% CI, 0.40–5.7; three deaths). Evaluation of risk by job type showed no meaningful pattern. In an analysis by country, one of the cohorts in the United Kingdom and that in Denmark had moderately increased mortality from lymphatic and haematopoietic cancer. There was no significant increase in the SMRs for other cancers; there was a nearly significant increase in risk for pancreatic cancer among workers in the highest exposure category (≥ 500 ppm-years) (SMR, 2.6; 95% CI, 0.90–7.3; 10 deaths). [The Working Group noted that the overall SMR for the unexposed group was very low, possibly reflecting problems in ascertaining mortality in some subcohorts. This may have negatively biased the SMRs in this study.] In contrast with the results of SMR analyses using the external referents, the results of internal analyses did show some statistically significant elevations in risk related to exposure to styrene. An increased risk for lymphatic and haematopoietic cancers was observed in Poisson regression models for average exposure ($p = 0.019$), culminating in a relative risk of 3.6 (95% CI, 1.0–13; $n = 8$) for the highest category, > 200 ppm. A statistically significant trend ($p = 0.012$) was also observed for time since first exposure, the relative risk reaching a peak after 20 years of 4.0 (95% CI, 1.3–12; $n = 9$). No trend was observed in the Poisson regression analysis with cumulative exposure or duration of exposure.

In the study by Kolstad *et al.* (1993, 1994), 64 529 workers (53 731 men and 10 798 women) in 552 companies engaged in production of reinforced plastics were followed up from 1970 to the end of 1989 through the Danish national cancer registry and the national mortality database. [The Working Group noted that the females were excluded from the analysis in the second paper by Kolstad *et al.* (1994) because 'the majority were not involved in the production of reinforced plastics'.] Information on the companies' activities was obtained from two dealers. Altogether, 386 companies were classified as ever producing reinforced plastics (with 36 525 male workers) and 84 were classified as never producing (with 14 254 male workers); for 82 companies, the information was unknown. All 12 837 men and 2185 women from 287 plants where the main product was reinforced plastics (involving more than 50% of the workforce in the plant) were included in the study by Kogevinas *et al.* (1993, 1994a,b). Fifty-three persons had disappeared from the follow-up and, in addition, 1104 had emigrated during the study period; the total loss to follow-up was 1.9%. A total of 584 556 person–years were accumulated until death, emigration, disappearance or the end of the study (whichever came first). The mean annual levels of styrene exposure calculated for 128 of these companies reflect the exposures measured in the industry, which ranged from 180 ppm [767 mg/m^3] in 1964–70 to 43 ppm [183 mg/m^3] in 1976–88. Duration of employment was calculated from pension fund payments made beginning in 1964–89. However, in a small validation sub-study of 671 workers, the authors found errors in the records which led to underestimation of duration of employment in the industry for about 40% of the workers and to overestimation for about 13%. There were too few women to provide statistically stable results. Among men, there were a total of 4281 deaths in the cohort and 1915 incident cases of cancer (SIR, 1.02; 95% CI, 0.97–1.07). Within companies producing reinforced plastics, there were slight increases in risk for lymphatic and haematopoietic cancers (SIR, 1.2; 95% CI, 0.98–1.4; 112 cases) and for leukaemia (SIR, 1.2; 95% CI, 0.88–1.7; 42 cases). A statistically significant increased risk for leukaemia was found after 10 years since first employment (SIR, 1.6; 95% CI, 1.1–2.2; 32 cases); however, the risk was confined to workers employed for less than one year (SIR, 2.3; 95% CI, 1.4–3.6; 20 cases). A significant increase in the incidence of leukaemia was observed for those who had been employed in 1964–70 (the period with the highest exposure to styrene) (SIR, 1.5; 95% CI, 1.0–2.2; 30 cases). [The Working Group noted that the misclassification of duration of employment that was noted by the authors would also have applied to the Danish component of the European study.]

Kolstad *et al.* (1995) also published results on solid cancers (not including lymphatic and haematopoietic cancers) among men within the reinforced plastics industry in Denmark. The study cohort was essentially the same as that in the previous studies (Kolstad *et al.*, 1993; Kogevinas *et al.*, 1994a,b; Kolstad *et al.*, 1994). There were 36 310 male workers from the reinforced plastics companies, and 14 293 workers in similar industries not producing reinforced plastics (127 more workers than in the previous studies). Altogether, 1134 solid cancers were observed during 1970–89 within the reinforced plastics industry (SIR, 0.99; 95% CI, 0.93–1.1); there were 47 cases of rectal cancer (SIR, 0.78; 95% CI, 0.58–1.0), 41 cases of pancreatic cancer (SIR, 1.2; 95% CI, 0.86–1.6) and 46 cases of tumours of the brain and nervous system (SIR, 0.97; 95% CI, 0.71–1.3). The relative risk for pancreatic cancer incidence was 2.2 (95% CI, 1.1–4.5; 17 cases) for employment with a high probability of exposure to styrene, and duration of employment at least one year, compared with no exposure.

2.2.3 *Styrene–butadiene rubber manufacture*

A number of reports have presented results on the mortality of workers in the styrene–butadiene synthetic rubber (SBR) industry (McMichael *et al.*, 1976a,b; Meinhardt *et al.*, 1982; Matanoski *et al.*, 1990; Santos-Burgoa *et al.*, 1992; Matanoski *et al.*, 1993). With some exceptions, these workers were included in the University of Alabama study. These earlier studies suggested an increased risk for lymphatic and haematopoietic malignancies in the SBR industry, but generally did not provide data on exposure to styrene *per se* and are far less informative than the more recent investigations from the University of Alabama, which are described below.

A study performed by researchers at the University of Alabama (Delzell *et al.*, 1996; Sathiakumar *et al.*, 1998) assessed the mortality experience of 15 649 male synthetic rubber workers employed for at least one year at eight SBR plants in the USA and Canada (information concerning exposure to 1,3-butadiene was summarized in IARC (1999)). A two-plant complex had been previously studied by Meinhardt *et al.* (1982) and the other six plants by Matanoski *et al.* (1990, 1993) and Santos-Burgoa *et al.* (1992). Complete work histories were available for 97% of the subjects. During 1943–91, the cohort had a total of 386 172 person–years of follow-up. Altogether 10 939 (70%) of the subjects were classified as being alive, 3976 (25%) were deceased and 734 (5%) were lost to follow-up. Information on cause of death was available for 97% of decedents. The observed total of 3976 deaths compared with 4553 deaths expected on the basis of general population mortality rates for the USA and Ontario (SMR, 0.87; 95% CI, 0.85–0.90). Cancer mortality was slightly lower than expected, with 950 deaths (SMR, 0.93; 95% CI, 0.87–0.99). Lymphopoietic cancers accounted for 101 deaths, slightly more than the number expected (SMR, 1.1; 95% CI, 0.88–1.3). There were 48 observed deaths from leukaemia in the overall cohort (SMR, 1.3; 95% CI, 0.97–1.7). There was a statistically significant excess of leukaemia among workers in polymerization (15 deaths; SMR, 2.5; 95% CI, 1.4–4.1), maintenance labour (13 deaths; SMR, 2.7; 95% CI, 1.4–4.5) and laboratories (10 deaths; SMR, 4.3; 95% CI, 2.1–7.9), which were three areas with potential for relatively high exposure to 1,3-butadiene or styrene monomers. Among the 'ever hourly-paid' workers with 10 or more years of employment and 20 or more years since hire, there was a significant excess of leukaemia deaths (28 deaths; SMR, 2.2; 95% CI, 1.5–3.2).

Delzell *et al.* (2001) and Sielken & Valdez-Flores (2001) re-analysed the University of Alabama results on leukaemia deaths within the US and Canadian cohorts. In these studies, the exposure assessments for 1,3-butadiene and styrene were revised, as compared with the earlier report (Macaluso *et al.*, 1996) and exposure estimates were developed for dimethyldithiocarbamate (DMDTC). [The Working Group noted that, unlike in the earlier report, assessment of exposure was performed after knowing the jobs and departments of the leukaemia cases.] Workers from two facilities had exposure records of insufficient quality and were dropped from the original study, leaving 13 130 men from six of the plants, and 59 deaths from leukaemia (medical records were obtained for 48 cases and one case was an acute unspecified leukaemia) during 1943–91 (in 234 416 person–years). Vital status was known for over 99% of the subjects, and death certificates were available for 98% of the decedents. In the within-cohort comparisons (Poisson regression), unlike in the cohort (SMR) study, all leukaemia deaths (leukaemia being either an underlying or a contributing cause of death) were used; 49 of the leukaemia deaths were confirmed from medical records. About 79% of the cohort subjects were exposed to 1,3-butadiene, the median cumulative exposure being 71 ppm–years [301 mg/m^3–years], and 85% to styrene, with a median cumulative exposure of 17 ppm–years [72 mg/m^3–years], and 62% to DMDTC with a median cumulative exposure[1] of 373.9 mg–years/cm. Poisson regression analyses with the individual exposures indicated a positive and monotonically increasing association between grouped cumulative exposure to styrene (relative risks of 1.0, 1.2, 2.3 and 3.2, for exposures of 0, > 0–< 20.6, 20.6–< 60.4 and ≥ 60.4 ppm–years) and leukaemia, and between exposure to 1,3-butadiene and leukaemia. For both of the exposures, a statistically significant relative risk was obtained for the highest cumulative exposure category (for styrene: relative risk, 3.2; 95% CI, 1.2–8.8; 18 deaths; for 1,3-butadiene: relative risk, 3.8; 95% CI, 1.6–9.1; 17 deaths). The exposure–response relationship for DMDTC and leukaemia did not increase monotonically, but a significantly increased relative risk was observed for each of the exposed groups. In models that included all three exposures, the exposure–response

[1] The DMDTC exposure estimation procedure yielded: (1) an estimate of the concentration of DMDTC in the solution wetting the skin of the exposed worker (in mg/cm^3); (2) an estimate of the skin surface exposed (in cm^2); and (3) an estimate of the frequency and duration of exposure. The exposure intensity unit was (mg/cm^3) × (cm^2) = mg/cm.

relationship for styrene became negative. The exposure–response relationship for 1,3-butadiene was weakly positive, and the exposure–response relationship for DMDTC remained irregular. A positive exposure–response model for styrene persisted in a model that included 1,3-butadiene (and not DMDTC). [The Working Group noted the strong correlation between exposure to styrene and 1,3-butadiene and considered this a major obstacle to assessing risks due to styrene *per se*. The Working Group noted that levels of exposure to styrene in this industry were considerably lower than in studies within reinforced plastics industries. The Working Group questioned whether there was sufficient justification for controlling for DMDTC in the analysis since there are no epidemiological or toxicological data suggesting that DMDTC is a potential carcinogen, although it does have toxic effects on haematopoietic tissues in rodents. The results from a previous analysis of this study, based on somewhat different subject pools, somewhat different diagnostic criteria and an earlier exposure assessment (Macaluso *et al.*, 1996), were slightly different from those reported by Delzel *et al.* (2001). The Working Group judged the more recent results to be the most informative.]

2.2.4 *Other cohort studies*

In addition to the studies on occupational exposures, a study was conducted on lymphatic and haematopoietic cancers among students attending a high school in Texas adjacent to a styrene–butadiene facility (Loughlin *et al.*, 1999). A cohort of 15 403 students attending this school from 1963–64 to 1992–93 was identified. Altogether, 338 deaths were identified in the 310 254 person–years in the follow-up period from 1963 up to the end of 1995 (SMR, 0.84; 95% CI, 0.74–0.95 among men; SMR, 0.89; 95% CI, 0.73–1.1 among women). There were 44 cancer deaths (SMR, 1.2; 95% CI, 0.83–1.7 among men; SMR, 0.52; 95% CI, 0.28–0.88 among women); 14 deaths were from any lymphatic or haematopoietic cancer [SMR, 1.2; 95% CI, 0.66–2.0 for both genders combined] and seven from leukaemia or aleukaemia (SMR, 1.8; 95% CI, 0.67–4.0 among men; SMR, 0.45; 95% CI, 0.01–2.5 among women; [SMR, 1.3; 95% CI, 0.51–2.6] for both genders combined). The only cause of death with a statistically significant excess was benign neoplasms (six deaths; [SMR, 4.2; 95% CI, 1.5–9.1] for both genders combined; SMR, 6.3; 95% CI, 2.0–15; five deaths among men; SMR, 1.6; 95% CI, 0.04–8.7; one death among women). All six who died of benign neoplasms had brain tumours. [It was not explicitly mentioned how many subjects (men or women) were lost to follow-up. Potential exposures to styrene or other chemical agents were not described.]

A programme of biomonitoring for styrene was undertaken by the Finnish Institute of Occupational Health among workers who were exposed to styrene from 1973. Mandelic acid was used as a biomarker of styrene exposure. Anttila *et al.* (1998) carried out a cancer follow-up study among 2580 subjects who were tested between 1973 and 1983. The overall mandelic acid level was 2.3 mmol/L (range, 0–47 mmol/L urine) during the monitoring period. Altogether 34 288 person–years were accrued in the follow-up from 1973 up to the end of 1992. In the styrene-exposed cohort, 5549 urine mandelic acid samples were collected; for 84 samples (1.5%), the personal identifier could not be traced. There was no loss to follow-up among those whose personal identifier had been traced successfully [the number of those who migrated was not mentioned.] There were 48 cases of cancer observed (SIR, 0.80; 95% CI, 0.59–1.1); six cases of rectal cancer (SIR, 3.1; 95% CI, 1.1–6.8), three cases of pancreatic cancer (SIR, 1.7; 95% CI, 0.34–4.9), six cases of nervous system cancer (SIR, 1.6; 95% CI, 0.59–3.5) and two cases of lymphatic and haematopoietic cancer (SIR, 0.39; 95% CI, 0.05–1.4; both of the cases had Hodgkin lymphoma). There was no evidence of an exposure–response relationship for any cancer site when mean lifetime mandelic acid level was used as a surrogate for exposure to styrene. There were no cases of pancreatic cancer within this highest exposure category (0.8 cases expected). [The Working Group noted that the relatively small size and low exposures of this cohort may have limited its power to detect any association with lymphatic and haematopoietic neoplasms.]

2.3 Case–control studies

Flodin *et al.* (1986) conducted a matched case–control study of 59 cases of acute myeloblastic leukaemia and 354 controls in Sweden to assess potential risk factors, which included radiation, medications and various occupational exposures. Cases were aged 20–70 years and were identified at hospitals in Sweden between 1977 and 1982. Two series of controls were drawn from a population register: one was matched to cases for sex, age (within five years) and location, and the other was a random population sample. Information on exposure was obtained through a questionnaire mailed to subjects. Of eight occupational exposures examined, styrene was reported by three cases and one control, leading to an estimated standardized odds ratio of 19 (95% CI, 1.9–357). [The Working Group noted the small numbers and the self-reported nature of the exposure estimates, which may be problematic for styrene.]

A population-based case–control study of cancer comprised 3730 histologically confirmed male cases of cancer at 15 major sites (including non-Hodgkin lymphoma and Hodgkin lymphoma) newly diagnosed between 1979 and 1986 in Montreal, Canada, aged 35–70 and ascertained in 19 major hospitals, as well as 533 population controls (Siemiatycki, 1991; Gérin *et al.*, 1998; Dumas *et al.*, 2000). The exposure of each subject to 293 occupational agents was assessed by a group of chemists on the basis of jobs held, and cases of cancer at each site were compared with those in the rest of the study population, after adjustment for age, ethnic group, family income, alcohol drinking and tobacco smoking. Two per cent of the subjects were classified as ever having been exposed to styrene. A synthetic form of cumulative exposures at low, medium or high level was also computed, based on the sum product of duration, frequency and concentration of exposure. The odds ratio for rectal cancer for any exposure to styrene was 1.7 (95% CI, 0.7–4.5; six exposed cases) (Siemiatycki, 1991; Dumas *et al.*, 2000). A significant increase in the risk for cancer of the rectum was seen for medium to high exposure level (odds ratio, 5.1; 95% CI, 1.4–19; five exposed cases in single-exposure model; odds ratio, 4.4; 95% CI, 1.1–17 in multiple-exposure models adjusting additionally for some other solvents). In the single-exposure models, the odds ratios for exposure to styrene were: 1.2 (95% CI, 0.6–2.5, 11 exposed cases for any exposure) for colon cancer, 0.3 (95% CI, 0.0–2.6; one exposed case for any exposure) for pancreatic cancer, 0.9 (95% CI, 0.2–3.3; five exposed cases for medium to high exposure) for lung cancer, 2.0 (95% CI, 0.8–4.8; eight cases for any exposure) for non-Hodgkin lymphoma and 2.4 (95% CI, 0.5–12; two exposed cases for any exposure) for Hodgkin lymphoma (Gérin *et al.*, 1998). [The Working Group noted that the exposure levels of the exposed subjects were probably lower than those in the cohort studies.]

3. Studies of Cancer in Experimental Animals

Six studies in rats and four studies in mice by oral and inhalation routes of exposure to styrene were presented in the previous monograph (IARC, 1994a). These are included here in summary form together with reviews of new inhalation studies in rats and mice.

Data on the carcinogenicity of styrene 7,8-oxide in experimental animals were evaluated as *sufficient evidence* in the same volume (IARC, 1994b).

3.1 Oral administration

3.1.1 *Mouse*

Groups of 50 male and 50 female B6C3F$_1$ mice, six weeks of age, received daily administrations of 150 or 300 mg/kg bw styrene (purity, 99.7%) in corn oil by gavage on five days per week, for 78 weeks, and the animals were killed after a further 13 weeks. Control groups of 20 male and 20 female mice received corn oil alone. The incidence of bronchiolo-alveolar carcinomas in males was 0/20, 3/44 and 5/43, while the incidence of adenomas and carcinomas combined was 0/20, 6/44 and 9/43 ($p = 0.024$) for doses of 0, 150 and 300 mg/kg, respectively. There were no bronchiolo-alveolar carcinomas in female mice. The incidence of bronchiolo-alveolar adenomas in females was 0/20, 1/43 and 3/43, respectively (National Cancer Institute, 1979a). [The Working Group noted the small number of control animals and that the incidence of both adenomas and carcinomas combined was within the historical control ranges.]

Groups of 50 male and 50 female B6C3F$_1$ mice, six weeks of age, received a commercial mixture of β-nitrostyrene ((2-nitroethenyl)benzene) (30% β-nitrostyrene, 70% styrene) in corn oil; the doses of styrene were 204 or 408 mg/kg bw, administered by gavage, on three days per week, for 78 weeks. Animals were killed after a further 14 weeks. Control groups of 20 male and 20 female mice received corn oil alone. There were no increases in the incidence of any tumours (National Cancer Institute, 1979b). [The Working Group noted that a mixture of styrene with β-nitrostyrene was tested.]

3.1.2 *Rat*

Groups of 50 male and 50 female Fischer 344 rats, six weeks of age, received daily doses of 1000 or 2000 mg/kg bw styrene (purity, 99.7%) in corn oil by gavage on five days per week, for 78 weeks; survivors were held until 105 weeks. Control groups of 20 male and 20 female rats received corn oil alone. Due to increased treatment-related mortality in the 2000-mg/kg bw group, additional groups of males and females receiving 0 and 500 mg/kg bw styrene per day in corn oil by gavage for 103 weeks were included. No treatment-related increase in the incidence of any type of tumour was observed (National Cancer Institute, 1979a).

Groups of 50 male and 50 female Fischer 344 rats, six weeks of age, received commercial β-nitrostyrene (30% β-nitrostyrene, 70% styrene) in corn oil by gavage on three days per week, for 79 weeks, at styrene doses of 175 (females), 350 (males and females) and 750 (males) mg/kg bw; survivors were observed for an additional 29 weeks. No treatment-related increase in the incidence of any type of tumour was observed (National Cancer Institute, 1979b).

Groups of 40 male and 40 female Sprague-Dawley rats, 13 weeks of age, were administered 0, 50 or 250 mg/kg bw styrene (99.8% pure) in olive oil by gavage on 4–5 days per week for 52 weeks. Survivors were held until death. In males, there was no increase in tumour incidence. In females, there was a decreased incidence of malignant and malignant plus benign mammary tumours at 250 mg/kg bw, which the authors attributed to decreased survival (Conti *et al.*, 1988). [The Working Group noted the limited reporting of this study and the short duration of treatment.]

Groups of 50 male and 70 female Charles River COBS (SD) BR rats, seven weeks of age, were administered 125 or 250 ppm (mg/L) styrene (> 98.9% pure) daily in the drinking-water for two years. Groups of 76 male and 106 female rats received drinking-water alone. Dosing was limited by the solubility of styrene in water. Styrene intake was calculated from the concentration, water consumption and body weight as being 7.7 and 14 mg/kg bw per day in males and 12 and 21 mg/kg bw per day in females. No treatment-related increase in the incidence of any type of tumour was observed (Beliles *et al.*, 1985).

3.2 Prenatal exposure followed by postnatal oral administration

3.2.1 *Mouse*

A group of 29 pregnant O20 female mice, a strain very susceptible to the formation of lung tumours, was administered 1350 mg/kg bw styrene (99% pure) by gavage in olive oil on day 17 of gestation. Pups (45 male, 39 female) were then given a high dose of 1350 mg/kg bw styrene by gavage weekly following weaning. Treatment was terminated after 16 weekly doses due to toxicity and mortality. All mice were held until spontaneous death, euthanasia due to moribund condition, or until 120 weeks. Lung tumour incidence was increased in both males (8/19 controls versus 20/23 treated) and females (14/21 controls versus 32/32 treated). No increase in tumours at other sites was observed (Ponomarkov & Tomatis, 1978). [The Working Group noted the high treatment-related toxicity and mortality early in the study.]

Using a similar study design, a group of 15 pregnant C57BL mice received 300 mg/kg bw styrene (99% pure) by gavage in olive oil on day 17 of gestation. Twenty-seven male and 27 female offspring received weekly gavage doses of 300 mg/kg bw styrene from weaning for 120 weeks. There was no increase in the incidence of treatment-related tumours (Ponomarkov & Tomatis, 1978).

3.2.2 *Rat*

A group of 21 pregnant BD IV rats was given 1350 mg/kg bw styrene (99.8% pure) by gavage in corn oil on day 17 of gestation; 10 pregnant rats received corn oil only. Beginning at weaning, 73 male and 71 female offspring received 500 mg/kg bw styrene once per week for 120 weeks; 36 male and 39 female control offspring received corn oil once a week for 120 weeks. There was no increase in any tumour incidence in the styrene-treated rats (Ponomarkov & Tomatis, 1978).

3.3 Inhalation exposure

3.3.1 *Mouse*

Groups of 50 male and 50 female mice CD-1 mice, approximately four weeks of age, were exposed by whole-body inhalation to 0, 20, 40, 80 or 160 ppm [85, 170, 341 or 682 mg/m^3] styrene vapour (> 99.5% pure) for 6 h per day, five days per week, for 104 (males) and 98 weeks (females). Due to mortality in control females (23/50 mice), the surviving females were killed six weeks earlier than originally scheduled; all four treated groups had greater survival than the controls (32, 33, 34 and 35/50 for 20, 40, 80 and 160 ppm dose groups, respectively). Other groups of 10 males and 10 females from each exposure level were killed after 52 and 78 weeks. Male and female mice exposed to 80 and 160 ppm had decreased body weights over the course of the study. Statistically significantly increased incidences of bronchiolo-alveolar adenomas were seen in males exposed to 40, 80 or 160 ppm for 24 months (control, 15/50; 20 ppm, 21/50; 40 ppm, 35/50 [$p < 0.05$]; 80 ppm, 30/50 [$p < 0.05$]; 160 ppm, 33/50 [$p < 0.05$]) with no dose–response relationship, but the incidences of bronchiolo-alveolar carcinomas were not increased (4, 5, 3, 6 and 7/50 for 0, 20, 40, 80 and 160 ppm). In females, the incidences of bronchiolo-alveolar adenomas in the groups exposed to 20, 40, and 160 ppm, but not 80 ppm, for 22.5 months were statistically significantly increased (control, 6/50; 20 ppm, 16/50 [$p < 0.05$]; 40 ppm, 16/50 [$p < 0.05$]; 80 ppm, 11/50; 160 ppm, 24/50 [$p < 0.05$]). In females, the incidences of bronchiolo-alveolar carcinoma were 0, 0, 2, 0 and 7/50 ($p < 0.05$) for 0, 20, 40, 80 and 160 ppm. The 14% incidence at 160 ppm was slightly higher than the historical control range of 0–4% for the investigating laboratory (five oral studies) and 0–13.5% for the breeder's database (nine oral studies). No increase in the incidence of lung tumours was seen in males or females at 12 or 18 months. The increase seen after 24 months was largely in small tumours, as demonstrated by a decreased average tumour size compared with controls. The fact that increased tumour incidences was seen only later than 18 months and the small size of these tumours indicated that these were late-developing tumours. No difference in tumour morphology between control and treated mice was seen. Epithelial hyperplasia of the terminal bronchioles extending into the alveolar duct was seen in a dose-related pattern at all interim and terminal necropsies. Hyperplasia was preceded by decreased eosinophilic staining of Clara cells and cellular crowding (Cruzan *et al.*, 2001).

3.3.2 *Rat*

Groups of 30 male and 30 female Sprague-Dawley rats, 12 weeks of age, were exposed by inhalation to 0, 25, 50, 100, 200 or 300 ppm [106, 213, 430, 850 or 1280 mg/m^3] styrene (99.8% pure) for 4 h per day, five days per week, for 12 months (and then held until death). In females, malignant mammary tumours were diagnosed in 6/60 (10%), 6/30 (20%), 4/30 (13%), 9/30 (30%), 12/30 (40%) and 9/30 (30%) rats inhaling 0, 25, 50, 100, 200 or 300 ppm, respectively. For total mammary tumours, the incidences were 34/60 (57%), 24/30 (80%), 21/30 (70%), 23/30 (77%), 24/30 (80%) and 25/30 (83%) for the respective exposure levels (Conti *et al.*, 1988). [The Working Group noted the short duration of treatment, the incomplete reporting of the study and the high incidence of spontaneous mammary tumours in animals of this strain.]

Groups of 60 male and 60 female Charles River CD rats, approximately four weeks of age, were exposed by whole-body inhalation to 0, 50, 200, 500 or 1000 ppm [213, 850, 2130 or 4260 mg/m^3] styrene (> 99.5% pure) for 6 h per day on five days per week for 104 weeks. Females exposed to 500 and 1000 ppm weighed less than controls throughout the study. However, there was a dose-related increase in survival; at termination, survival was 48, 47, 48, 67 and 82% in females exposed to 0, 50, 200, 500 and 1000 ppm, respectively. There were no increased incidences of tumours related to styrene exposure. A dose-dependent decrease in mammary tumours in females was reported. Mammary adenocarcinomas were diagnosed in 20/61 (33%), 13/60 (22%), 9/60 (15%), 5/60 (8%) and 2/60 (3%) female rats exposed to 0, 50, 200, 500 or 1000 ppm styrene, respectively, for two years. A decrease in benign mammary fibroadenomas (including those with epithelial atypia) was seen, the incidence being 27/61 (44%), 22/60 (37%), 18/60 (30%), 21/60 (35%) and 19/60 (32%) for the above exposure levels, respectively (Cruzan *et al.*, 1998).

3.4 Intraperitoneal administration

3.4.1 *Mouse*

In a screening assay based on multiplicity and incidence of lung tumours in a highly susceptible strain of mice (A/J), administration of 200 μmol (~100 mg/kg bw) styrene by intraperitoneal injection three times per week for 20 doses, followed by observation for 20 weeks, produced no increase in lung adenoma incidence in the styrene-treated mice compared with controls (Brunnemann *et al.*, 1992).

3.4.2 *Rat*

Groups of 40 male and 40 female Sprague-Dawley rats, 13 weeks of age, received four intraperitoneal injections of 50 mg styrene (99.8% pure) per animal in olive oil at two-month intervals. Control groups received olive oil alone. The study was terminated when the last rat died [duration unspecified]. There was no increase in tumour incidence (Conti *et al.*, 1988). [The Working Group noted the incomplete reporting of data, the short duration of treatment and the low total dose.]

3.5 Subcutaneous administration

Rat: Groups of 40 male and 40 female Sprague-Dawley rats, 13 weeks of age, received a single subcutaneous injection of 50 mg styrene (> 99% pure) per animal in olive oil. Control groups received olive oil alone. The study was terminated when the last rat died [duration unspecified]. There was no increase in tumour incidence (Conti *et al.*, 1988). [The Working Group noted the incomplete reporting of data and the single low-dose treatment.]

4. Other Data Relevant to an Evaluation of Carcinogenicity and its Mechanisms

4.1 Absorption, distribution, metabolism and excretion

Studies on the pharmacokinetics and metabolism of styrene were evaluated in 1994 as part of a previous IARC monograph (IARC, 1994a), which may be consulted for more details on the earlier studies.

4.1.1 *Humans*

(*a*) *Absorption*

As noted previously (IARC, 1994a), the pulmonary retention of styrene is 60–70% of the inhaled dose based on studies in both volunteers and workers (Stewart *et al.*,

1968; Engström *et al.*, 1978a,b; Ramsey *et al.*, 1980; Wigaeus *et al.*, 1983, 1984; Pezzagno *et al.*, 1985; Wieczorek & Piotrowski, 1985; Löf *et al.*, 1986a,b). This has been confirmed by more recent studies (Johanson *et al.*, 2000; Wenker *et al.*, 2001a,b). Wrangskog *et al.* (1996) developed a simple one-compartment model for estimation of styrene uptake based on measurements of urinary excretion of mandelic and phenylglyoxylic acids.

In human volunteers exposed by placing one hand in liquid styrene for 10–30 min, absorption was low, averaging 1 μg/cm²/min (Berode *et al.*, 1985). Limasset *et al.* (1999) carried out a field study comparing urinary excretion of metabolites of styrene in four groups of workers who performed the same task but wore different protective equipment (see Section 1.4.1(*d*)), and concluded that percutaneous absorption of styrene was not an important contribution to the body burden.

(*b*) *Distribution*

Several studies have suggested that styrene accumulates in subcutaneous fat (Wolff *et al.*, 1977; Engström *et al.*, 1978a,b). However, based on measurement of urinary metabolites, workers exposed to 37 ppm (160 mg/m³) styrene showed no accumulation during the working week (Pekari *et al.*, 1993). As noted below, pharmacokinetic analysis of the disposition of styrene supports this conclusion.

(*c*) *Metabolism*

The metabolic pathways for styrene are shown in Figure 1. Styrene is primarily metabolized to styrene 7,8-oxide by cytochrome P450 (CYP) enzymes. The oxide is metabolized by epoxide hydrolase to phenylethylene glycol and then to mandelic, phenylglyoxylic and benzoic acids. Additional routes of metabolism include ring hydroxylation, but this appears to be a minor pathway in humans. Pfäffli *et al.* (1981) identified small amounts of 4-vinylphenol in the urine of workers exposed to styrene (less than 1% of the amount of mandelic acid excreted by these workers). Another pathway is conversion of styrene to 1- and 2-phenylethanol, which is further metabolized to phenylacetaldehyde, phenylacetic acid, phenylaceturic acid and hippuric acid. Styrene 7,8-oxide may also be metabolized by conjugation with glutathione to form mercapturic acids. This appears to be a minor pathway in humans, amounting to less than 1%. Ghittori *et al.* (1997) evaluated urinary excretion in 22 workers in a reinforced-plastics factory (see Section 1.1.5(*b*)) and identified racemic mercapturates, i.e., the *R*- and *S*-diastereoisomers of *N*-acetyl-*S*-(1-phenyl-2-hydroxyethyl)cysteine and *N*-acetyl-*S*-(2-phenyl-2-hydroxyethyl)cysteine. Assuming a styrene uptake of 64%, the conversion to mercapturic acids was calculated to be between 0.021 and 0.325% of the dose.

Korn *et al.* (1994) found a linear correlation between concentrations of styrene 7,8-oxide in the blood of workers exposed to styrene (10–73 ppm [43–310 mg/m³]) and styrene in ambient air and blood. At 20 ppm [85 mg/m³] styrene, the steady-state level of styrene 7,8-oxide was about 1 μg/L blood.

Figure 1. Main metabolic pathways for styrene

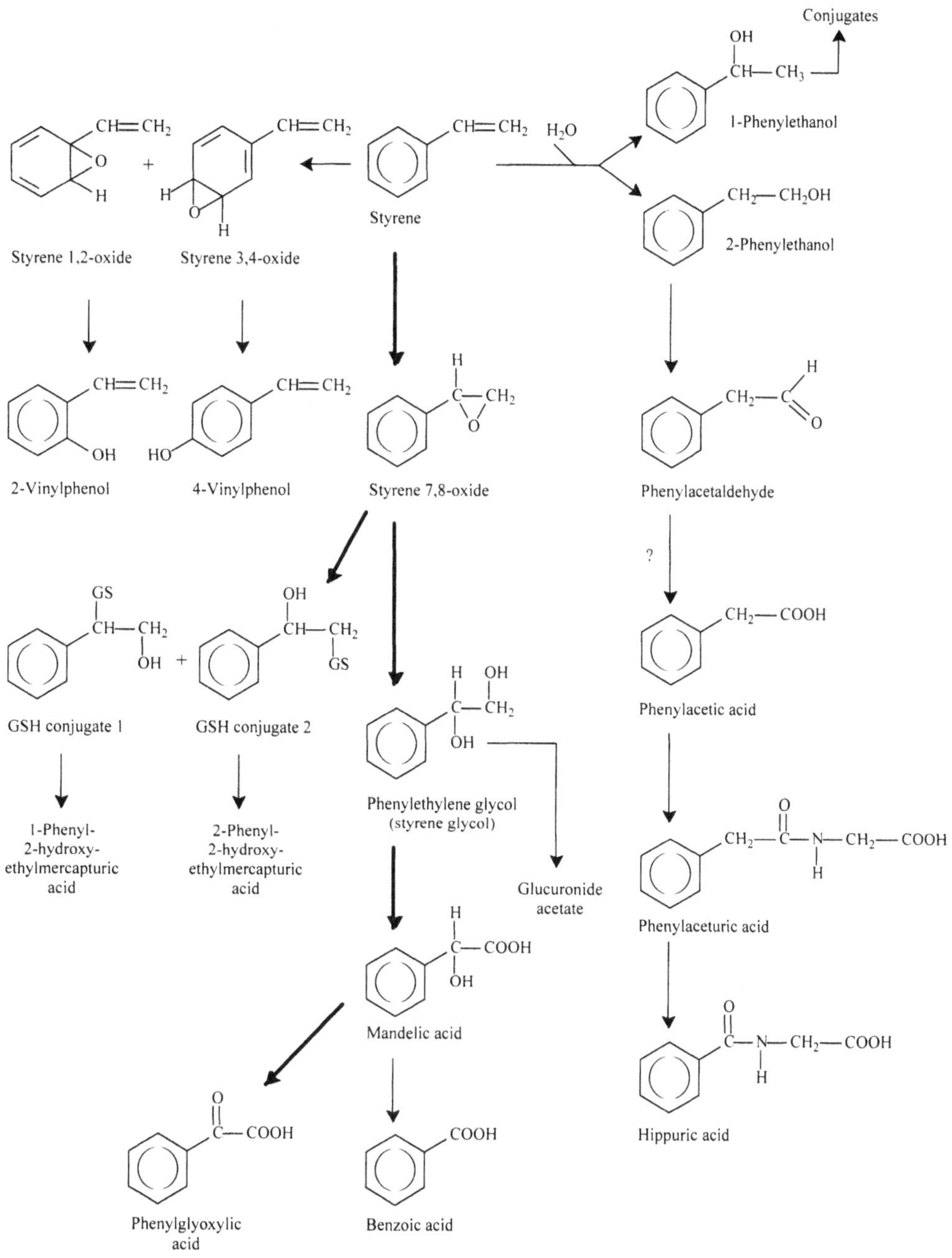

From IARC (1994a). Main pathways are indicated by thick arrows.
GSH, glutathione

Many investigators have used measurements of urinary mandelic and phenylglyoxylic acids as biomarkers of exposure in workers (reviewed by Guillemin & Berode, 1988) (see Section 1.1.5(*b*)). Stereochemical studies have been carried out on these metabolites. Korn *et al.* (1984) found the average ratio of L-mandelic acid to D-mandelic acid to be 1.62 in 11 male styrene-exposed workers and 1.44 in six female workers. In 20 male volunteers exposed to styrene (360 mg/m^3) for 1 h while performing 50-W physical exercise, the average pulmonary retention was 62%, with mandelic acid and phenylglyoxylic acid accounting for 58% of the absorbed styrene. The maximum concentration and area-under-the-curve for the free *R*-styrene glycol in blood were 1.3 and 1.7 times higher, respectively, than those of the *S*-styrene glycol. Cumulative excretion and renal clearance of conjugated styrene glycol were three and four times higher for *S*-styrene glycol than for *R*-styrene glycol, respectively (Wenker *et al.*, 2001a).

Studies in humans have attempted to assess interactions of chemicals that might alter styrene metabolism by comparing data on metabolite concentrations in urine of workers exposed to mixtures with historical information on workers exposed to styrene alone. Kawai *et al.* (1992) measured urinary metabolites of styrene in 34 male workers engaged in the production of plastic buttons and exposed to a mixture of styrene (mean concentration, 7 ppm [30 mg/m^3]), toluene and methanol. Styrene metabolism was similar to that reported previously in populations exposed to styrene alone (Ikeda *et al.*, 1982). This conclusion was supported by a subsequent study of 39 male workers co-exposed to methanol and 12.4 ppm [53 mg/m^3] styrene (Kawai *et al.*, 1995). Apostoli *et al.* (1998) examined the urinary excretion of mandelic and phenylglyoxylic acids in 50 workers involved in the production of polyester buttons who were exposed to styrene (16–439 mg/m^3) and acetone (15–700 mg/m^3). In groups with different levels of acetone co-exposure, the acetone had no effect on excretion of these metabolites. Similarly, De Rosa *et al.* (1993) monitored mandelic acid excretion in 22 workers exposed to styrene (22–522 mg/m^3) and acetone (40–1581 mg/m^3) and found no effect due to the co-exposure to acetone. Cerný *et al.* (1990) examined the metabolism of styrene (exposure concentration, 420 mg/m^3) for 8 h with and without co-administration of ethanol (four doses of 14 g given at 2-h intervals) in five volunteers and found that this acute administration decreased the excretion of mandelic and phenylglyoxylic acids. To examine whether exposure to styrene would increase styrene metabolism in subsequent exposures, Wang *et al.* (1996) used the data of Wigaeus *et al.* (1983) and Löf *et al.* (1986a) from a study in which two groups of subjects, one previously exposed to styrene and one not, were exposed to 80 ppm [341 mg/m^3] styrene for 2 h, and estimated the metabolic rate constant based on a three-compartment pharmacokinetic model. No effect on styrene metabolism was found, but the number of subjects in this study (six or seven per group) was limited.

Johanson *et al.* (2000) exposed four volunteers for 2 h to 50 ppm [213 mg/m^3] [$^{13}C_8$]styrene vapour via a face mask while they performed light physical exercise. Two hours after exposure had ended, the concentration of styrene 7,8-oxide in blood reached a maximum and averaged 6.7 nM. There was no evidence of glutathione conjugation or ring epoxidation.

Individual differences in the metabolic response to styrene were studied in 20 male volunteers exposed on separate occasions to 104 and 360 mg/m^3 styrene for 1 h while performing 50-W exercise. Urinary mandelic and phenylglyoxylic acids were measured to make correlations with individual metabolic capacities determined with enzyme-specific substrates for CYP2E1, CYP1A2 and CYP2D6. No correlation was found among individuals between the blood clearance of styrene and the metabolic capacity. Based on the high apparent blood clearance of styrene (1.4 L/min), one explanation proposed by the authors is that styrene metabolism is limited by the blood flow to the liver, which has a similar value (Wenker *et al.*, 2001b). Symanski *et al.* (2001) examined inter- and intra-individual differences in urinary concentrations of mandelic and phenylglyoxylic acids based on 1714 measurements in 331 workers over the period 1985–99. Interindividual differences were greater for post-shift urine samples than for pre-shift samples.

Hallier *et al.* (1995) measured *R*- and *S*-mandelic acids in the urine of 20 male workers exposed to 29–41 ppm styrene [124–175 mg/m^3]. The ratio of the *R*- to *S*-enantiomers of mandelic acid ranged from 0.7 to 2.2. This variation could not be explained by differences in exposure levels and was attributed to interindividual differences in styrene metabolism, probably related to enzyme polymorphisms.

The role of specific CYP isozymes in the metabolism of styrene has been examined. Nakajima *et al.* (1994a), using 12 different purified human isozymes, determined that CYP2B6 and CYP2E1 were the most active in human liver microsomes in forming styrene glycol, and CYP2F1 was the most active in lung microsomes. Guengerich *et al.* (1991) also demonstrated the significance of CYP2E1 in styrene metabolism, using diethyldithiocarbamate as a specific inhibitor of CYP2E1 and correlations with chlorzoxazone 6-hydroxylation — an indicator of CYP2E1 activity — in human liver microsomes. Using antibodies against specific human CYP isozymes and by comparing rates of styrene glycol formation by microsomes isolated from human livers, Kim *et al.* (1997) identified CYP2E1 and CYP2C8 as being the most important metabolic enzymes at low styrene concentration (0.085 mM), while CYP2B6 and CYP2C8 were most prominent at a high concentration of styrene (1.8 mM).

Wenker *et al.* (2000) examined the stereospecificity and interindividual variation in microsomal epoxide hydrolase activity in 20 human liver samples. V_{max}, K_m and V_{max}/K_m values for the substrates *R*- and *S*-styrene 7,8-oxides varied three- to five-fold between livers. The rate of hydrolysis of *S*-styrene 7,8-oxide was approximately five times slower than that of *R*-styrene 7,8-oxide, although the K_m was six times higher for *S*-styrene 7,8-oxide. In a further report (Wenker *et al.*, 2001c), an eightfold variation in V_{max} was found for styrene metabolism by 20 human liver samples (0.39–3.2 nmol/mg protein/min), and, although CYP2E1 was found to be the most important cytochrome P450 isoform, there was no correlation between the enzymic activity (V_{max} or K_m) and genetic polymorphisms of this enzyme.

Haufroid *et al.* (2001) examined possible influences of genetic polymorphisms in metabolic enzymes on the metabolism of styrene (average concentration, 18.2 ppm [77.5 mg/m^3]) in 30 workers from a glass fibre-reinforced plastics factory. The presence of the rare *CYP2E1*1B* allele was associated with increased excretion of urinary mandelic and phenylglyoxylic acids as well as mercapturic acid metabolites. Individuals deficient in glutathione *S*-transferase (GST) M1 excreted less than one-half the amount of mercapturic acids, a minor pathway in humans, compared with *GSTM1*– proficient individuals.

De Palma *et al.* (2001) examined polymorphisms in GSTM1-1 and T1-1 and microsomal epoxide hydrolase in 56 styrene-exposed workers (geometric mean exposure, 157 mg/m^3, time-weighted average over 8 h). GSTM1-1 was identified as the primary isozyme for conjugation of glutathione with styrene 7,8-oxide. Workers who were positive for GSTM1-1 excreted 5–6 times more phenylhydroxyethylmercapturic acids than did individuals who were GSTM1-1-deficient. No association was found between microsomal epoxide hydrolase polymorphisms and excretion of the mercapturic acids.

(*d*) *Excretion*

Only small amounts of styrene (0.7–4.4%) are exhaled unchanged (see IARC, 1994a for details). This has been confirmed in more recent studies, in which 0.7–2.2% of the amount of inhaled styrene was found unchanged in the exhaled breath of four subjects exposed to 50 ppm [213 mg/m^3] styrene for 2 h (Johanson *et al.*, 2000). Small amounts of styrene are also excreted unmetabolized in the urine (Pezzagno *et al.*, 1985; Gobba *et al.* 1993). Ramsey *et al.* (1980) examined the pharmacokinetics of inhaled styrene (80 ppm [341 mg/m^3]) in four human volunteers and calculated half-life values of 0.6 and 13.0 h for the two phases of elimination.

Brugnone *et al.* (1993) measured styrene concentrations in blood of 76 exposed workers at the end of the work shift and the morning after, and reported a half-life of elimination of 3.9 h.

4.1.2 *Experimental systems*

(*a*) *Absorption*

McDougal *et al.* (1990) compared the dermal absorption of vapours from several organic chemicals in male Fischer 344 rats. The animals breathed fresh air through a latex mask. For styrene (3000 ppm [12 800 mg/m^3] for 4 h), a maximal blood concentration of about 10 μg/mL and a permeability constant of 1.753 cm/h were reported. In a mixed exposure situation (inhalation and skin absorption), the skin uptake was estimated to account for 9.4% of the total uptake in these rats. Morgan *et al.* (1991) examined the dermal absorption of styrene in Fischer 344 rats. When administered neat (2 mL of styrene in a sealed dermal cell), the peak blood level of styrene, reached after 1 h, was 5.3 μg/mL. Absorption decreased when the styrene was diluted with water.

(*b*) *Distribution*

An early study in which [^{14}C]styrene labelled at the β-carbon atom was administered subcutaneously to Wistar rats (0.10 mL of a 20% solution to 100–125-g rats [20 mg per animal]) indicated that styrene was distributed rapidly (< 1 h) to tissues and that excretion of the radioactivity was primarily via the urine, with only a small fraction (2–3%) exhaled unchanged by the lung and approximately 12% appearing as $^{14}CO_2$ during 24 h after dosing (Danishefsky & Willhite, 1954).

Teramoto and Horiguchi (1979) exposed rats [strain unspecified] to 500 and 1000 ppm [2130 and 4260 mg/m^3] styrene by inhalation for 4 h and found the highest concentration of styrene in adipose tissue. The biological half-life in this tissue was 6.3 h, as opposed to 2.0–2.4 h in blood, liver, kidney, spleen, muscle and brain. No accumulation of styrene was found when rats were exposed to 700 ppm [2980 mg/m^3] for 4 h per day for five days. Withey and Collins (1979) exposed Wistar rats to styrene by inhalation (50–2000 ppm [213–8520 mg/m^3] for 5 h) and found a proportionally increasing level of styrene uptake into blood. Tissue distribution was concentration-dependent, and the concentration of styrene in the perirenal fat was higher than in any other tissue. This is probably a consequence of the high oil:blood partition coefficient of styrene (130; see Van Rees, 1974). When [^{14}C]styrene was administered orally to rats (20 mg/kg), the highest levels of radioactivity were found in the kidney followed by the liver and pancreas. Ninety per cent of the dose appeared in the urine within 24 h after dosing (Plotnick & Weigel, 1979). Löf *et al.* (1983, 1984) studied the effect of time and dose on styrene distribution and metabolism following intraperitoneal administration of styrene (1.1–4.9 mmol/kg bw) to NMRI mice. The highest concentrations of styrene were measured in adipose tissue, pancreas, liver and brain. Styrene 7,8-oxide concentrations were highest in kidney and subcutaneous adipose tissue and lowest in lung. The highest concentrations of styrene glycol were found in kidney, liver, blood and lungs. Male Wistar rats given styrene (100 mg/kg) by intubation excreted both *R*- and *S*-mandelic acids, with a preference towards excretion of the *R*-form (Drummond *et al.*, 1989).

The uptake of styrene by the upper respiratory tract of the CD-1 mouse and Sprague-Dawley rat has been determined using a surgically isolated upper respiratory tract preparation (Morris, 2000). In rats, steady-state uptake efficiency of styrene decreased with increasing exposure concentration, with 24% and 10% efficiency at 5 and 200 ppm [21 and 852 mg/m^3], respectively. In mice, the uptake efficiency — which did not maintain a steady state but declined during exposure — averaged between 42% and 10% at 5–200 ppm. Treatment with metyrapone, an inhibitor of cytochrome P450, abolished the concentration-dependence of the uptake efficiency in both species, providing evidence that inhaled styrene is metabolized by nasal tissues and that this determines the concentration-dependence.

(*c*) *Metabolism*

Watabe *et al.* (1978) clearly demonstrated the NADPH-dependent conversion of styrene to styrene 7,8-oxide in liver microsomes from male Wistar-King rats, with further conversion to styrene glycol by epoxide hydrolase.

The role of glutathione conjugation in the metabolism of styrene was recognized by Seutter-Berlage *et al.* (1978), who characterized urinary mercapturic acids amounting to approximately 11% of an intraperitoneal dose of styrene (250 mg/kg bw) in Wistar rats. The compounds were identified as *N*-acetyl-*S*-(1-phenyl-2-hydroxyethyl)cysteine, *N*-acetyl-*S*-(2-phenyl-2-hydroxyethyl)cysteine and *N*-acetyl-*S*-(phenacyl)cysteine, which were present in a 65:34:1 ratio (Seutter-Berlage *et al.*, 1978; Delbressine *et al.*, 1981). Watabe *et al.* (1982) administered styrene, racemic styrene 7,8-oxide, *R*-styrene 7,8-oxide and *S*-styrene 7,8-oxide (all at 2 mmol/kg bw) to male Wistar rats and found the two major metabolites mentioned above. For the *R*-enantiomer, the rate of metabolism to mercapturic acid was 2.5 times higher than for the *S*-enantiomer. Nakatsu *et al.* (1983) similarly identified these metabolites in the urine of male Sprague-Dawley rats after intraperitoneal administration of styrene 7,8-oxide (100 mg/kg bw). Truchon *et al.* (1990) exposed Sprague-Dawley rats to styrene by inhalation (25–200 ppm [106–852 mg/m^3], 6 h per day, five days per week for four weeks) and found dose-dependent excretion of *N*-acetyl-*S*-(1-phenyl-2-hydroxyethyl)cysteine and *N*-acetyl-*S*-(2-phenyl-2-hydroxyethyl)cysteine in urine. Recent attention has focused on stereochemical aspects of the formation of these mercapturic acids. Coccini *et al.* (1996) exposed Sprague-Dawley rats subchronically to styrene (300 ppm [1280 mg/m^3], 6 h per day, five days per week for two weeks). Urine was collected during the last 6 h of exposure. Approximately 6.5% of the dose was recovered as *N*-acetyl-*R*-(1-phenyl-2-hydroxyethyl)cysteine, 19.5% as *N*-acetyl-*S*-(1-phenyl-2-hydroxyethyl)cysteine and 10.3% as *N*-acetyl-*S*- or *R*-(2-phenyl-2-hydroxyethyl)cysteine. Linhart *et al.* (1998) dosed Wistar rats intraperitoneally with *R*-, *S*- and racemic styrene 7,8-oxide (150 mg/kg bw) and found that the regioselectivity was similar for all three treatments, yielding a 2:1 mixture of *N*-acetyl-*S*-(1-phenyl-2-hydroxyethyl)cysteine and *N*-acetyl-*S*-(2-phenyl-2-hydroxyethyl)cysteine, although the conversion to the mercapturic acids was higher with *R*-styrene 7,8-oxide (28% of the dose) than with the *S*-isomer (19%). In male B6C3F$_1$ mice given an intraperitoneal injection of 400 mg/kg bw styrene, the two major urinary metabolites were *N*-acetyl-*S*-(2-hydroxy-2-phenylethyl)cysteine and *N*-acetyl-*S*-(2-hydroxy-1-phenylethyl) cysteine, together comprising 12.5% of the dose (Linhart *et al.*, 2000).

Several studies have focused on the influence of inducers and inhibitors of xenobiotic metabolism on the biotransformation of styrene. Treatment of female Wistar rats with sodium phenobarbital (37.5 mg/kg bw injected intraperitoneally, twice daily for four days) increased the metabolism of styrene (given by intraperitoneal injection of 455 mg/kg bw on the fifth day), as determined by measurements of urinary metabolites (Ohtsuji & Ikeda, 1971). Sato and Nakajima (1985) found that treatment of male Wistar rats with phenobarbital (oral dose of 80 mg/kg bw daily for three days) enhanced the hepatic microsomal metabolism of styrene approximately six-fold. Elovaara *et al.* (1991) compared the influence of acetone, phenobarbital and 3-methylcholanthrene on the urinary metabolites of male Han/Wistar rats exposed to styrene by inhalation (2100 mg/m^3 for 24 h) and found increases of 30–50% in the excretion of phenylglyoxylic acid and mandelic acid in the rats treated with acetone (1% in drinking-water for one week) but not in the other groups. Co-administration by intraperitoneal injection of toluene (217 mg/kg bw) suppressed the metabolism of styrene (228 mg/kg bw) in Wistar rats (Ikeda *et al.*, 1972). Co-administration of ethanol (2 mM) decreased the uptake and metabolism of styrene (given as a 28.8-μM solution) in a perfused rat liver system in fed animals and increased these parameters in fasted animals. This was associated with increased levels of NADPH due to oxidation of ethanol (Sripaung *et al.*, 1995).

Salmona *et al.* (1976) compared formation of styrene 7,8-oxide and its further transformation to styrene glycol by epoxide hydrolase in microsomal preparations from different tissues from CD rats and found mono-oxygenase and epoxide hydrolase activity in liver, lung and kidney, but not in heart, spleen and brain, with the highest activities in male rat liver.

When apparent hepatic V_{max} values for the metabolism of styrene to styrene 7,8-oxide by styrene mono-oxygenase were compared among species, the order was Dunkin Hartley guinea-pig > New Zealand rabbit > Swiss mouse > Sprague-Dawley rat (Belvedere *et al.*, 1977). However, the order for the metabolism of styrene 7,8-oxide to styrene glycol by epoxide hydrolase was rat > rabbit > guinea-pig >mouse, so that the V_{max} ratios for epoxide hydrolase *vs* styrene mono-oxygenase were 3.9, 2.4, 1.6 and 1.4 for the rat, rabbit, guinea-pig and mouse, respectively. Watabe *et al.* (1981) reported that liver microsomal preparations from male Wistar rats preferentially formed *S*-styrene 7,8-oxide over *R*-styrene 7,8-oxide (*R*- to *S*-ratio, 0.77). Foureman *et al.* (1989) examined the stereoselectivity of styrene 7,8-oxide formation in the livers of male Sprague-Dawley rats and found an *R*- to *S*- ratio of 0.65. This ratio increased to 0.92 in phenobarbital-treated rats. This *R*- to *S*-ratio contrasts with results indicating preferential formation of *R*-styrene 7,8-oxide in the liver of CD-1 mice (ratio *R*/*S*, 1.2–1.8) (Carlson, 1997a; Hynes *et al.*, 1999) and in pulmonary microsomes from rabbits (ratio of 1.6) (Harris *et al.*, 1986).

Mendrala *et al.* (1993) compared the kinetics of styrene and styrene 7,8-oxide metabolism in rat, mouse and human livers *in vitro*. K_m values for styrene epoxidation by mono-oxygenase were similar for the three species. The V_{max} values, however, varied from 13 nmol/min/mg protein for B6C3F$_1$ mice, 11 nmol/min/mg protein for Fischer 344 rats, 9.3 nmol/min/mg protein for Sprague-Dawley rats to 2.1 nmol/min/mg protein for humans (average of five donors). V_{max} values for epoxide hydrolase were similar for all three species (14–15 nmol/min/mg protein).

Nakajima *et al.* (1994a) compared the rates of styrene metabolism using microsomal preparations from different sources. At a low substrate concentration (0.085 mM), the rates of hepatic metabolism decreased from mouse (2.43 nmol/mg protein/min) to rat

(1.07 nmol/mg protein/min) to human (0.73 nmol/mg protein/min), whereas at a high styrene concentration (1.85 mM), the rates were in the order of rat (4.21 nmol/mg protein/min) to mouse (2.72 nmol/mg protein/min) to human (1.91 nmol/mg protein/min). Activity in pulmonary microsomes from 38 individuals was much lower (0.006–0.0125 nmol/mg protein/min) than in liver. Carlson *et al.* (2000) also observed low activity for styrene metabolism in microsomal preparations from six human lungs. Styrene 7,8-oxide formation was detectable in only one human sample (0.088 nmol/mg protein/min). [The Working Group noted that enzyme activity measurements based upon analysis of whole organ homogenates do not reflect the substantial differences that exist between cell types, as has been demonstrated by immunohistochemical methods for CYP2E1 and CYP2F2 in lung (Buckpitt *et al.*, 1995; Green *et al.*, 1997).]

A few studies have described the formation of the ring-hydroxylated product 4-vinylphenol after exposure of rodents to styrene. Bakke and Scheline (1970) reported finding 4-vinylphenol in amounts up to 0.1% of the dose in enzymatically hydrolysed urine of male rats [strain unspecified] dosed orally with styrene (100 mg/kg bw). Similarly, Pantarotto *et al.* (1978) identified small amounts of 4-vinylphenol, 4-hydroxymandelic acid, 4-hydroxybenzoic acid and 4-hydroxyhippuric acid in the urine of male Sprague-Dawley rats administered styrene intraperitoneally. More recently, when rats and mice were exposed by inhalation to [*ring*-U-^{14}C]styrene, Boogaard *et al.* (2000a) reported finding $^{14}CO_2$, suggesting that ring hydroxylation may occur followed by ring opening. The fraction of the dose eliminated as $^{14}CO_2$ varied from 6.4–8.0% of the retained dose of styrene in mice to 2% of the retained dose in rats. Watabe *et al.* (1984) demonstrated the formation of 4-vinylphenol using ^{14}C-labelled styrene and a rat hepatic microsomal preparation fortified with NADPH after addition of a large amount of unlabelled 4-vinylphenol as a trapping agent. They suggested that 4-vinylphenol was formed via the 3,4-oxide but very rapidly metabolized to 4-hydroxystyrene-7,8-glycol. Carlson *et al.* (2001) were unable to confirm 4-vinylphenol formation when styrene was incubated with hepatic and pulmonary microsomal preparations from CD-1 mice and Sprague-Dawley rats. However, considerable 4-vinylphenol-metabolizing activity, as determined by the loss of 4-vinylphenol, was observed in both mouse and rat liver and lung microsomal preparations. This activity was three times higher in mouse liver microsomes than in rat liver microsomes and eight times higher in mouse lung microsomes than in rat lung microsomes, and it was completely absent in the absence of NADPH. Studies with cytochrome P450 inhibitors indicated the involvement of CYP2E1 and CYP2F2.

Several studies have focused on the pharmacokinetics of styrene and some species comparisons. Ramsey and Young (1978) exposed Sprague-Dawley rats to styrene by inhalation at concentrations of 80, 200, 600 and 1200 ppm [341, 852, 2560 and 5110 mg/m^3] for up to 24 h. The increase in styrene concentration in blood with increased exposure concentration was non-linear, indicating saturation of metabolism after about 6 h of exposure. Filser *et al.* (1993) compared male Sprague-Dawley rats and B6C3F$_1$ mice exposed to styrene by inhalation in a closed chamber (550–5000 ppm [2340–21 300 mg/m^3]), by intraperitoneal injection (20–340 mg/kg bw) or by oral gavage (100–350 mg/kg bw). In both species, the rate of metabolism of the inhaled styrene was concentration-dependent. At exposure concentrations above 300 ppm [1280 mg/m^3], metabolism was progressively limited by metabolic capacity and was saturated at concentrations of approximately 700 ppm [2980 mg/m^3] in rats and 800 ppm [3410 mg/m^3] in mice. Little accumulation occurred at exposures below 300 ppm [1280 mg/m^3] since uptake was rate-limiting. In rats and mice, intraperitoneal injection of styrene followed by analysis of exhaled styrene in the closed chamber resulted in concentration–time curves in agreement with the applied pharmacokinetic model. After oral administration of styrene, the concentration–time curves showed considerable inter-animal variability.

In a chronic inhalation study in Sprague-Dawley rats exposed (whole-body) to styrene vapour at 0, 50, 200, 500 or 1000 ppm [0, 213, 852, 2130 or 4260 mg/m^3] for 6 h per day, five days per week for 104 weeks, blood levels of styrene and styrene 7,8-oxide were measured at the end of a 6-h exposure period during week 95. Styrene 7,8-oxide in blood was undetectable in males or females at 0 or 50 ppm. While the styrene 7,8-oxide concentration in blood continued to increase with higher styrene exposure concentration, some saturation of metabolism was observed (Cruzan *et al.*, 1998). In CD-1 mice exposed to styrene vapour at 0, 20, 40, 80 or 160 ppm [0, 85, 170, 341 or 682 mg/m^3] for 98 weeks (females) or 104 weeks (males), blood levels of styrene and styrene 7,8-oxide were measured at the end of a 6-h exposure period during week 74. Styrene 7,8-oxide in blood was undetectable in males and females at 0 and 20 ppm. At higher exposure doses, concentration-dependent increases were observed for styrene and styrene 7,8-oxide in blood, with no evidence for saturation of metabolism (Cruzan *et al.*, 2001). Comparing the two studies, blood styrene concentrations in male and female rats reached 2780 ng/mL and 1950 ng/mL at 200 ppm styrene exposure and blood styrene 7,8-oxide concentrations reached 66 ng/mL and 28 ng/mL at this exposure level, respectively, whereas for mice the corresponding values at the highest dose (160 ppm) were 1461 ng/mL and 1743 ng/mL for styrene and 33.5 ng/mL and 20.1 ng/mL for styrene 7,8-oxide in males and females, respectively.

The effect of antibodies to CYP2C11/6, CYP2B1/2, CYP1A1/2 and CYP2E1 on the rates of styrene metabolism was studied in male Wistar rat lung and liver microsomal preparations. All four antibodies decreased styrene metabolism in microsomes from rat liver, with anti-CYP2C11/6 having the strongest effect, but only anti-CYP2B1/2 affected the metabolism in lung microsomes. This indicates that the major CYP isoform responsible for metabolism of styrene is different in these two tissues (Nakajima *et al.*, 1994b). Studies with human cytochromes P450 ectopically expressed in cultured hepatoma G2 cells indicate that CYP2B6 and CYP2F1 may also be involved in styrene metabolism (Nakajima *et al.*, 1994a). The potential contributions of various cytochromes P450 to styrene metabolism have also been examined by use of chemical inhibitors of specific isozymes in hepatic and pulmonary microsomal preparations of male CD-1 mice. Diethyldithiocarbamate inhibited the formation of both enantiomers of styrene 7,8-oxide in lung and liver, supporting the importance of CYP2E1. 5-Phenyl-1-pentyne showed a high degree of inhibition in pulmonary microsomes, but caused only a small decrease in hepatic microsomes, indicating the importance in mouse lung of CYP2F2 (Carlson, 1997a). α-Naphthoflavone, an inhibitor of CYP1A, had little effect on styrene metabolism. α-Methylbenzylaminobenzotriazole caused only a 16–19% inhibition of styrene 7,8-oxide formation at a concentration (1 μM) that caused substantial (87%) inhibition of benzyloxyresorufin metabolism, indicating that CYP2B plays a minor role in styrene metabolism (Carlson *et al.*, 1998).

Green *et al.* (2001a) examined the effects of 5-phenyl-1-pentyne (100 mg/kg bw) — an inhibitor of CYP2E1 and CYP2F2 — on the metabolism of styrene *in vitro* by pulmonary microsomes from male CD-1 mice isolated at various times up to 48 h after dosing. After six hours, styrene metabolism was reduced to 8% of control activity for the formation of the *R*-enantiomer of styrene 7,8-oxide and to 25% for the *S*-enantiomer. A single dose of 1 g/kg diethyldithiocarbamate inhibited the pulmonary microsomal metabolism of styrene to 16% of control activity for *R*- and 23% for *S*-styrene 7,8-oxide at 24 h after dosing. These studies further indicate the importance of the enzymes CYP2E1 and CYP2F2 in the pulmonary metabolism of styrene. Green *et al.* (2001b) carried out additional studies on the metabolism of styrene by nasal epithelium of CD-1 mice and Sprague-Dawley rats. The rates of formation of styrene 7,8-oxide were similar in microsomal preparations of the olfactory epithelium of mice and rats, with values of 6.59 and 7.46 nmol/min/mg protein, respectively, for the *R*-isomer and 2.24 and 2.51 nmol/min/mg protein for the *S*-isomer. Activity was approximately 50% lower in the respiratory epithelium than in the olfactory epithelium. No styrene metabolism was detected in nine samples of human nasal tissue (limit of detection: 0.04 nmol/min/mg protein). Both chlorzoxazone and 5-phenyl-1-pentyne inhibited styrene metabolism, again indicating the importance of the enzymes CYP2E1 and CYP2F2.

Recent studies have addressed the question of which pulmonary cell types are responsible for styrene metabolism and which cytochromes P450 are associated with the bioactivation of styrene. In enriched fractions of Clara cells and type II cells from male CD-1 mice and male Sprague-Dawley rats, styrene metabolism was determined with and without chemical inhibitors. Mouse Clara cells readily metabolized styrene to styrene 7,8-oxide, but type II pneumocytes had little or no activity. Styrene metabolism was several-fold higher in mouse Clara cells than in rat Clara cells. *R*-Styrene 7,8-oxide was preferentially formed in mouse lung and Clara cells, and *S*-styrene 7,8-oxide was preferentially formed in rat lung and Clara cells (Table 11) (Hynes *et al.*, 1999). Studies with both microsomes and isolated cells indicated that CYP2E1 and CYP2F2 are the main cytochromes P450 involved in pulmonary styrene metabolism. Green *et al.* (2001b) similarly found that hepatic microsomal fractions isolated from Sprague-Dawley rats preferentially formed *S*-styrene 7,8-oxide (ratio *R*:*S*, 0.72), whereas microsomal fractions prepared from CD-1 mice preferentially formed *R*-styrene 7,8-oxide (ratio *R*:*S*, 2.70). In olfactory and respiratory nasal cells, the rates of this bioactivation step were similar in both species and were higher in the respiratory epithelium than in the olfactory cells. Both chlorzoxazone (specific for CYP2E1) and 5-phenyl-1-pentyne (inhibitor of

Table 11. Metabolism of styrene to styrene 7,8-oxide enantiomers by mouse and rat lung microsomes and isolated Clara cell-enriched fraction

	R-enantiomer	*S*-enantiomer	*R*/*S*
Lung microsomes			
Mouse (n = 4)	1.50 ± 0.23[a]	0.63 ± 0.06	2.40 ± 0.36
Rat (n = 5)	0.49 ± 0.09	0.95 ± 0.18	0.52 ± 0.01
Clara cell fraction			
Mouse (n = 4)	83.3 ± 27.7	23.0 ± 8.2	3.98 ± 0.75
Rat (n = 3)	11.2 ± 3.6	11.0 ± 3.2	1.02 ± 0.09

Data from Hynes *et al.* (1999)
Values for *R*- and *S*-enantiomers are in nmol/mg protein/min for lung microsomes and in pmol/10^6 cells/min (not corrected for cell type) for Clara cell fractions. The percentages of Clara cells in these fractions were 56% for mouse and 37% for rat.
[a] Mean ± standard error

CYP2E1 and CYP2F2) inhibited nasal metabolism. Subsequent metabolism of styrene 7,8-oxide by epoxide hydrolase and GST was much higher in rat nasal tissue than in mouse nasal tissue.

(*d*) *Excretion*

Sumner *et al.* (1997) compared the metabolism of [7-^{14}C]styrene administered by inhalation (250 ppm [1060 mg/m^3], 6 h per day for 1–5 days) to B6C3F$_1$ mice, CD-1 mice and Fischer 344 rats. Rats and CD-1 mice excreted the radioactivity faster than did B6C3F$_1$ mice following a single exposure, but the rates were similar for the three groups after repeated exposures on 3 or 5 days. Boogaard *et al.* (2000a) studied the disposition of [*ring*-U-^{14}C]styrene in Sprague-Dawley rats and CD-1 mice exposed by recirculating nose-only exposure (160 ppm [682 mg/m^3] for 6 h). Urinary excretion was the primary route of elimination (amounting to 75% of the retained styrene for rats and 63% for mice), and there was little quantitative difference between rats and mice except that mice exhaled $^{14}CO_2$ which amounted to 6.4–8.0% of the retained styrene, while for rats this was only 2%. Whole-body autoradiography showed significantly higher non-specific binding of radioactivity in mouse lung and nasal passages compared with rat lung.

4.1.3 *Pharmacokinetic modelling*

Ramsey and Young (1978) and Ramsey *et al.* (1980) analysed the pharmacokinetic profile of inhaled styrene in male rats [strain unspecified] and humans. In human volunteers exposed to 80 ppm [341 mg/m^3] for 6 h, styrene was cleared from the blood according to a two-compartment linear pharmacokinetic model that was similar to that for rats. The model predicted that there would be no accumulation of styrene at this exposure concentration and upon repeated administration. Ramsey and Andersen (1984) developed a physiologically based pharmacokinetic model, which indicated that there was saturation of styrene metabolism at inhaled doses above 200 ppm [852 mg/m^3] in mice, rats and humans. At lower levels of exposure, the ratio of styrene concentration in blood to that in inhaled air is controlled by perfusion-limited metabolism.

Another physiologically based pharmacokinetic model was based on data collected by Mandrala *et al.* (1993) to predict concentration–time curves for styrene and styrene 7,8-oxide in blood and tissues (Csanády *et al.*, 1994). At low concentrations, styrene is rapidly removed from blood and the rate of metabolism is limited by the blood flow through the liver.

Cohen *et al.* (2002) developed a pharmacokinetic model viewing the lung as two compartments, the alveolar tract and the capillary bed. This model included both pulmonary and hepatic metabolism and predicted tissue or organ concentrations of styrene and styrene 7,8-oxide (both the *S*- and *R*-enantiomers, separately), under specified conditions. It was based on the Csanády *et al.* (1994) model, which only accounted for styrene metabolism by the liver. The authors indicated that the model has a substantial degree of uncertainty, however, because of inconsistencies between studies in reporting styrene 7,8-oxide levels in blood, which complicated the validation of the model.

Filser *et al.* (2002) compared the modelled concentrations of styrene 7,8-oxide in the lungs of the mouse, rat and human over a range of styrene concentrations assuming 6- or 8-h exposures. The highest concentrations of styrene 7,8-oxide were predicted for the mouse followed by the rat, with humans having by far the lowest concentration. These differences reflected the species differences in the pulmonary metabolism of styrene to styrene 7,8-oxide and subsequent detoxification of the oxide.

The most comprehensive modelling for styrene metabolism and styrene 7,8-oxide dosimetry comparing rodents and humans was undertaken by Sarangapani *et al.* (2002). It expanded upon the Csanády *et al.* (1994) model by incorporating information on the metabolic production of styrene 7,8-oxide and its decrease in the respiratory tract and was used to predict the concentrations of styrene and styrene 7,8-oxide in blood, liver and respiratory tract. This model predicts a ten-fold lower styrene 7,8-oxide concentration in the terminal bronchioles of rats than in mice exposed to identical concentrations of styrene. The model suggests that styrene 7,8-oxide concentrations in human bronchioles would be 100-fold lower than for the mouse.

Tornero-Velez and Rappaport (2001) modified the physiologically based pharmacokinetic model of Csanády *et al.* (1994) to compare the predicted contribution of styrene 7,8-oxide resulting from metabolism with that from direct exposure in the workplace. They tested their model against air and blood concentrations of styrene and styrene 7,8-oxide in 252 workers in the reinforced plastics industry. Due to efficient detoxification of the oxide formed in the liver, direct exposure via inhalation appeared to be a more important source of styrene 7,8-oxide than the bioactivation of inhaled styrene. The model indicated that absorbed styrene 7,8-oxide contributed 3640 times more of this compound to the blood than an equivalent amount of inhaled and metabolized styrene.

4.2 Toxic effects

4.2.1 *Humans*

Exposure to styrene has been reported to cause irritation of the eyes, throat and respiratory tract (Lorimer *et al.*, 1976, 1978). Subjective health complaints were not reported in the glass fibre-reinforced plastics industry with concentrations of styrene below 105 mg/m^3 (Geuskens *et al.*, 1992).

There are a number of reports of central and peripheral nervous system effects in exposed workers. Some studies have reported decreased nerve conduction velocities in workers exposed to styrene (Lilis *et al.*, 1978; Rosén *et al.*, 1978; Cherry & Gautrin, 1990; Murata *et al.*, 1991; Štetkárová *et al.*, 1993), whereas others have shown no changes at styrene concentrations below 100 ppm [430 mg/m^3] (Edling & Ekberg 1985; Triebig *et al.*, 1985). Electroencephalographic (Seppäläinen & Härkönen, 1976; Matikainen *et al.*, 1993), dopaminergic (Mutti *et al.*, 1984a; Arfini *et al.*, 1987; Checkoway *et al.*, 1992, 1994; Bergamaschi *et al.*, 1996, 1997), functional (Lindström *et al.*, 1976; Cherry *et al.*, 1980) and psychiatric anomalies (Flodin *et al.*, 1989; Edling *et al.*, 1993) were observed in styrene-exposed workers compared with controls. Most of these effects were observed at concentrations of styrene of about 100 ppm [430 mg/m^3], although memory and neurobehavioural disturbances were noted at styrene concentrations ranging from 10 to 30 ppm [43–130 mg/m^3] and above (Flodin *et al.*, 1989; Letz *et al.*, 1990; Edling *et al.*, 1993; Jégaden *et al.*, 1993; Tsai & Chen, 1996; Viaene *et al.*, 2001).

Thresholds for hearing were unchanged in workers exposed to styrene concentrations less than 150 mg/m^3. However, a comparison within the group of styrene-exposed workers (least exposed versus most exposed) showed a significant difference in hearing thresholds at high frequencies (Muijser *et al.*, 1988). In a group of 18 workers exposed for 6–15 years to styrene at levels below 110 mg/m^3, seven workers showed disturbances in the central auditory pathways (Möller *et al.*, 1990). Morioka *et al.* (1999) found a reduction in the upper limit of hearing in workers occupationally exposed to organic solvents. This effect was dose-dependent and correlated with the concentration of styrene in the breathing zone of the workers and with the amount of mandelic acid in urine. In a group of 299 workers in the glass fibre-reinforced plastics industry, there was no evidence for an effect of exposure to styrene on hearing acuity when both noise and lifetime exposures to styrene were accounted for (Sass-Kortsak *et al.*, 1995).

Colour vision was found to be impaired in styrene-exposed workers in a number of studies (Gobba *et al.*, 1991; Fallas *et al.*, 1992; Chia *et al.*, 1994; Campagna *et al.*, 1995; Eguchi *et al.*, 1995; Campagna *et al.*, 1996; Gobba, 2000). Solvent-induced loss of colour vision is indicative of changes in neural functioning along optic pathways. A positive relationship was observed between styrene exposure and early colour vision dysfunction in several of these studies, with effects detected at exposure concentrations as low as 20 ppm [85 mg/m^3]. In one study, a dose–response relationship between the concentration of mandelic acid in urine and colour vision loss was noted (Kishi *et al.*, 2001). Triebig *et al.* (2001) showed that styrene-induced colour vision dysfunction is reversible when the person is in a styrene-free environment for four weeks.

In a study of mortality among styrene-exposed workers, an increased number of deaths attributed to 'symptoms, senility and ill-defined conditions' was ascribed to a high local registration of these conditions in comparison with national statistics (Bond *et al.*, 1992). Welp *et al.* (1996a) followed a cohort of workers in the reinforced-plastics industry and found that mortality from diseases of the central nervous system, especially epilepsy, increased with exposure to styrene. Exposure was evaluated as duration (< 1–>10 years), average concentration (60–120 ppm [256–511 mg/m^3]) and cumulative exposure (< 50–> 500 ppm–years [< 213–> 2130 mg/m^3–years]). The effects of styrene on the respiratory tract of workers exposed to concentrations above 100 mg/m^3 include chronic bronchitis (Härkönen, 1977). Haematological changes (in blood coagulation and fibrinolysis) have been observed (Chmielewski & Renke, 1976). Cases of styrene-induced asthma (Moscato *et al.*, 1987; Hayes *et al.*, 1991) and one case of contact dermatitis (Sjöborg *et al.*, 1984) have also been reported. Welp *et al.* (1996b) reported that mortality from pneumonia among more than 40 000 men and women in 660 European factories manufacturing reinforced plastics appeared to be associated with exposure to styrene, but mortality from bronchitis, emphysema and asthma was not. In a more recent study, no relationship was observed between exposure to styrene and mortality from non-malignant respiratory diseases among 15 826 workers exposed to styrene (average styrene concentrations were below 10 ppm [43 mg/m^3], with 14% of the workers exposed to ≥ 60 ppm [256 mg/m^3]) in the reinforced-plastics and composites industry in the USA (Wong & Trent, 1999).

Several studies have reported signs of liver damage, as measured by liver enzyme activities in serum, but it was concluded in a review that no clear trend towards altered liver function had been demonstrated (WHO, 1983). Elevated serum bile acid concentrations were observed in one study (Edling & Tagesson, 1984) but not in another (Härkönen *et al.*, 1984).

More recently, two independent cross-sectional studies in 47 workers in the glass fibre-reinforced plastics industry and in 21 boat and tank fabricators, both with separate control groups, were carried out by Brodkin *et al.* (2001). Exposure to styrene was assessed by measurements in blood and air samples. Workers were grouped as controls, low-exposure (≤ 25.0 ppm [106 mg/m^3] styrene), or high-exposure (> 25.0 ppm styrene). Direct bilirubin levels, a marker of altered hepatic clearance, increased with styrene exposure. Hepatic transaminase activities (alanine and aspartate transaminases) increased with styrene exposure when data from both studies were pooled.

Altered kidney function in styrene-exposed workers was measured by increased urinary excretion of albumin (Askergren *et al.*, 1981). Vyskocil *et al.* (1989) studied female workers in the glass fibre-reinforced plastics industry who were exposed to styrene at 225 mg/m^3 for a mean duration of 11 years. No difference was found in the excretion of a number of urinary markers (including albumin) of renal injury compared with control female workers. Another study showed no difference in urinary markers for renal toxicity in workers exposed to styrene at ≤ 215 mg/m^3 for 1–13 years (mean, six years) compared with an unexposed control group (Viau *et al.*, 1987). In ten workers exposed to styrene (8-h time-weighted average exposure estimated to be 21–405 mg/m^3) and 15 non-exposed workers, only a weak correlation was noted between styrene exposure concentration and alterations in urinary markers for renal toxicity (Verplanke & Herber, 1998).

Early studies on the effects of styrene on the haematopoietic and immune system did not consistently reveal changes (WHO, 1983). Thiess and Friedheim (1978) found no difference in haemoglobin, erythrocyte and leukocyte concentrations in 84 workers in styrene production, polymerization and processing plants exposed to styrene (50–500 ppm [213–2130 mg/m^3] for 1–36 years) compared with a reference group. In a study of 163 workers in a styrene–butadiene synthetic rubber manufacturing plant, there was no pronounced evidence of haematological abnormality, as determined from measurements taken in peripheral blood, from exposure to styrene (< 15 ppm [64 mg/m^3]) (Checkoway & Williams, 1982). In a group of workers exposed to 56 mg/m^3 styrene in the reinforced-plastics industry, a 30% increase in the number of peripheral blood monocytes was noted (Hagmar *et al.*, 1989). In 221 styrene-exposed (1–100 ppm [4.3–426 mg/m^3] for 1–20 years) workers in the reinforced-plastics industry, an exposure-related decrease was detected in the mean corpuscular haemoglobin and neutrophil concentrations compared with the values in 104 controls (Stengel *et al.*, 1990). Mean corpuscular volume and monocytes increased with exposure to styrene. No changes were found in haemoglobin, red blood cell, white blood cell, lymphocyte or platelet concentrations in the styrene-exposed workers compared with the unexposed group. In another group of 32 workers exposed to styrene, changes in lymphocyte subpopulations were observed mainly at exposure levels above 50 ppm (Mutti *et al.*, 1992). Bergamaschi *et al.* (1995) reported a reduction in total T lymphocytes ($CD3^+$) and T helper lymphocytes ($CD4^+$) and an increase in natural killer cells in styrene-exposed workers (10–50 ppm [43–213 mg/m^3], 8-h time-weighted average). An association was observed between styrene exposure of hand-lamination workers (mean blood styrene levels of 946 μg/L) and alterations of cell-mediated immune response of T lymphocytes, with an imbalance of leukocyte subsets (increased number of monocytes and decreased number of lymphocytes) in the peripheral blood (Tulinska *et al.*, 2000).

4.2.2 *Experimental systems*

Single exposures of rats and guinea-pigs [strain not stated] to 1300 ppm [5630 mg/m^3] styrene resulted in central nervous system effects, including weakness and unsteadiness. After exposure to 2500 ppm [10.8 g/m^3] styrene for 10 h, both rats and guinea-pigs lost consciousness; exposure to 5000–10 000 ppm [21.3–42.6 g/m^3] resulted in unconsciousness and death. The principal pathological findings in these animals were severe pulmonary irritation, congestion, oedema, haemorrhage and leukocytic infiltration (Spencer *et al.*, 1942).

Early studies on the neurotoxicity of styrene gave equivocal results (WHO, 1983), but a decrease in the activity of monoamine oxidase in brain was seen in male rats after repeated oral doses of styrene (1.2 g/kg bw per day for seven days) (Zaprianov & Bainova, 1979). In subsequent studies, exposure of male rabbits to styrene vapour (0, 750 or 1500 ppm [0, 3200 or 6400 mg/m^3] for 12 h per day during three or seven days) caused a dose-dependent decrease in striatal and tuberoinfundibular dopamine content and an increase in homovanillic acid content, consistent with disturbance of the dopaminergic functions of the brain (Mutti *et al.*, 1984b). The levels of norepinephrine were unchanged. The styrene metabolites phenylglyoxylic acid and mandelic acid were shown to deplete dopamine through a direct chemical reaction, thereby rendering it inactive as a neurotransmitter (Mutti *et al.*, 1988). Chakrabarti (2000) administered styrene at 0.25 and 0.5 mg/kg bw by gavage to male Sprague-Dawley rats on seven days per week for 13 weeks; there was a decrease in dopamine and its metabolites in the corpus striatum, hypothalamus and the lateral olfactory tract regions in the high-dose group. A loss of motor function that accompanied these changes lasted for approximately a month after the last dose. Styrene (300 ppm [1280 mg/m^3], 6 h per day, five days per week for 12 weeks) also caused cell loss and dopamine depletion in retinas isolated from female Sprague-Dawley rats (Vettori *et al.*, 2000).

An increase in the concentration of glial fibrillary acidic protein was measured in the sensory motor cortex and in the hippocampus of Sprague-Dawley rats that had been continuously exposed to 320 ppm [1390 mg/m^3] styrene for three months and then kept free from exposure for another four months. This effect was taken to be an indication of solvent-induced brain damage (Rosengren & Haglid, 1989). Mild neurobehavioural disturbances were seen in rats exposed to 1400 ppm [6070 mg/m^3] styrene for 16 h per day, five days per week for 18 weeks followed by a six-week exposure-free period (Kulig, 1989) and in mice exposed to 425 ppm [1840 mg/m^3] styrene for 4 h per day, five days per week for two weeks (Teramoto *et al.*, 1988).

In dissociated primary cultures of murine spinal cord–dorsal root ganglia–skeletal muscle, styrene and styrene 7,8-oxide were cytotoxic to neuronal and non-neuronal cells at concentrations in excess of 2 and 0.2 mM, respectively (Kohn *et al.*, 1995).

Styrene is ototoxic in rats (Pryor *et al.*, 1987; Yano *et al.*, 1992; Crofton *et al.*, 1994). Exposure of male Long-Evans rats to 850 ppm and 1000 ppm [3620 and 4260 mg/m^3] styrene for 6 h per day, five days per week for four weeks, caused permanent hearing loss at mid-frequency ranges (16 kHz). At higher exposure concentrations, the increase in the auditory threshold became frequency-independent and was observed in the mid-low, mid- and high-frequency ranges (Loquet *et al.*, 1999).

Styrene suppressed the activity of mouse splenic T lymphocyte killer cells *in vitro* (Grayson & Gill, 1986). When Wistar rats were exposed to styrene (210 mg/m^3)

continuously for seven days, δ-aminolaevulinate dehydratase activity was inhibited in erythrocytes and bone marrow (Fujita *et al.*, 1987). Male Swiss mice dosed orally with 20–50 mg/kg bw styrene daily for five days showed impairment of humoral and cell-mediated immunity (Dogra *et al.*, 1989); a similar dose regimen given for four weeks decreased the resistance of mice to viral, malarial and hookworm infections (Dogra *et al.*, 1992). Male Sprague-Dawley rats given styrene by intraperitoneal injection (400 mg/kg bw per day for three days) or exposed by inhalation (300 ppm [1280 mg/m^3], 6 h per day, five days per week for two weeks) showed an increased number of erythropoietic cells. The granulocytopoietic series was not affected by injected styrene but was depressed following longer-term inhalation exposure (Nano *et al.*, 2000).

Ohashi *et al.* (1985) investigated the respiratory toxicity of styrene in rats. Epithelial changes were observed in the nose and trachea of rats exposed to 800 ppm [3410 mg/m^3] styrene for 4 h per day during eight weeks followed by a three-week period without exposure. The changes included vacuolation of epithelial cells, nuclear pyknosis and exfoliation of epithelial cells. Mild changes in the nasal mucosa occurred at exposure levels of 30 ppm [130 mg/m^3]. Morphological damage was more severe in the upper than in the lower respiratory tract.

In Swiss-Albino mice, intraperitoneal injections of a high dose of styrene (800 mg/kg bw) or its metabolite styrene 7,8-oxide (300 mg/kg bw) induced an increase in γ-glutamyl transpeptidase and lactate dehydrogenase measured in bronchoalveolar lavage fluid. These enzymes are biomarkers of pulmonary toxicity (Gadberry *et al.*, 1996). Treatment of male Sprague-Dawley rats with styrene either by intraperitoneal injection (40 or 400 mg/kg bw, daily for three days) or by inhalation exposure (300 ppm [1280 mg/m^3], 6 h per day, five days per week for two weeks) showed the same pattern of cytoplasmic changes involving bronchiolar and alveolar type II cells as observed in mice (Coccini *et al.*, 1997). Damage in the lung was accompanied by glutathione depletion. Damage was more severe following intraperitoneal injection compared with inhalation exposure. Male CD-1 mice were exposed to styrene at 0, 40 and 160 ppm [170 and 682 mg/m^3], and male Sprague-Dawley rats to 0 and 500 ppm [2130 mg/m^3] for 6 h per day for up to 10 days (Green *et al.*, 2001a). In mice, styrene exposure caused pulmonary toxicity characterized by focal loss of cytoplasm and focal crowding of non-ciliated Clara cells, particularly in the terminal bronchiolar region. The toxicity was accompanied by an increase in cell replication rates in terminal and large bronchioles of mice exposed for three days or longer. Morphological and cell-proliferative effects were not observed in the alveolar region of the mouse lung or in the lungs of rats exposed to 500 ppm styrene. 5-Phenyl-1-pentyne, an inhibitor of cytochrome P450-mediated metabolism, was given to a group of mice before styrene exposure. Cell replication rates in the lungs of these mice were similar to those in control mice, suggesting that the pulmonary effects are due to a styrene metabolite.

Khanna *et al.* (1994) examined pancreatic damage following oral administration of styrene to mice (25 or 50 mg/kg bw) and rats or guinea-pigs (160 or 320 mg/kg bw) on five days per week for four weeks. In all three species, styrene caused congestion of pancreatic lobules, inflammatory reactions around pancreatic islets (in mice only), and altered serum insulin levels but no change in blood glucose levels.

Early studies reported morphological changes in the kidney of Sprague-Dawley rats after intraperitoneal injections of styrene (2.9 and 5.8 mmol/kg bw on five days per week for six weeks) (Chakrabarti *et al.*, 1987) and in the respiratory mucosa of SD rats exposed to styrene by inhalation (150 or 1000 ppm [639–4260 mg/m^3], 4 h per day, five days per week, three weeks) (Ohashi *et al.*, 1986). Effects in kidney and lung are associated with depletion of glutathione (Chakrabarti & Tuchweber, 1987; Elovaara *et al.*, 1990) and possibly with direct toxicity of glutathione conjugates to the kidney, as was shown in experiments with a synthetic glutathione–styrene conjugate (Chakrabarti & Malick, 1991). Viau *et al.* (1987) reported that Sprague-Dawley rats given intraperitoneal injections of styrene (1 g/kg bw, which is one-fifth the oral LD_{50}) daily for 10 days showed only mild tubular damage. In female Sprague-Dawley rats exposed to styrene (300 ppm [1280 mg/m^3], 6 h per day, five days per week, for 12 weeks), a small increase was observed in the urinary excretion of plasma proteins, with minor changes in kidney histopathology consisting of interstitial fibrosis, cystic dilatations and hyaline tubules (Mutti *et al.*, 1999).

Early studies reported that styrene caused hepatotoxicity in rats concomitantly with a depletion of glutathione, which may be either a direct effect of styrene or mediated by lipid peroxidation (Srivastava *et al.*, 1983; Katoh *et al.*, 1989). Male albino rats [strain not stated] were dosed orally with 200 or 400 mg/kg bw styrene on six days per week for 100 days (Srivastava *et al.*, 1982). Liver changes were observed only after the high dose and were characterized by tiny areas of focal necrosis composed of a small number of degenerated hepatocytes and inflammatory cells.

Exposure of B6C3F$_1$ mice to 0, 125, 250 or 500 ppm [0, 530, 1060 or 2130 mg/m^3] styrene for 6 h per day for 14 days caused severe centrilobular hepatic necrosis and death after one 6-h exposure to 500 ppm or two 6-h exposures to 250 ppm styrene (Morgan *et al.*, 1993a). Mortality was higher in male than in female mice and, after 14 days of exposure, the incidence and severity of liver lesions were also greater in male mice. In a subsequent study, the greater sensitivity of male mice to styrene exposure compared with females could not be explained by differences in glutathione depletion or styrene or styrene 7,8-oxide concentrations in blood (Morgan *et al.*, 1993b). Further investigation of the role of metabolism in the toxicity of styrene in mice revealed that styrene and styrene 7,8-oxide concentrations in blood did not correlate well with strain differences in sensitivity (Morgan *et al.*, 1993c). Mortality and hepatotoxicity induced by exposure to styrene were greater in B6C3F$_1$ and C57BL/6 mice than in Swiss mice, while DBA/2 mice were the least sensitive. Styrene and styrene 7,8-oxide concentrations in blood as well as glutathione depletion were similar in B6C3F$_1$ and DBA/2 mice but greater than those in Swiss mice. Although strain and gender differences in sensitivity to styrene toxicity have been suggested to be due to differences in metabolism, styrene 7,8-oxide concentrations in blood did not correlate with toxicity in these studies. Morgan *et al.* (1995) further investigated strain and gender differences in styrene toxicity at inhalation exposure concentrations below the saturation of metabolism (150 and 200 ppm [639 and 852 mg/m³] styrene, 6 h per day, five days per week, for two weeks). Female B6C3F$_1$ mice were more susceptible than male B6C3F$_1$ and male or female Swiss mice to the hepatotoxic effects of styrene. This was suggested to be due to the fact that styrene caused greater depletion of glutathione in female B6C3F$_1$ mice, which also had a lower capacity to regenerate glutathione. Morgan *et al.* (1997) showed that hepatotoxicity in male B6C3F$_1$ mice that had been treated with phenobarbital before a 6-h exposure to 500 ppm [2130 mg/m³] styrene was more severe than in control mice treated with styrene alone. In male ddY mice pretreated with buthionine sulfoximine, a glutathione depletor, oral administration of styrene (0.96–5.76 mmol/kg) or styrene 7,8-oxide (3.84 mmol/kg) caused hepatotoxicity characterized by centrilobular necrosis and an increase in serum alanine transaminase (Mizutani *et al.*, 1994).

Cruzan *et al.* (1997) carried out several subchronic inhalation studies with styrene in Sprague-Dawley rats and CD-1 and B6C3F$_1$ mice. In rats exposed to 200–1500 ppm [852–6390 mg/m³] styrene (6 h per day, five days per week) for 13 weeks, styrene caused treatment-related lesions in the olfactory epithelium of the nasal passages at doses ≥ 500 ppm [2130 mg/m³], which were characterized by focal disorganization and hyperplasia of basal cells, single-cell necrosis and apparent cell loss. No increase in cell proliferation was observed in liver or lung following exposure to styrene for 2, 5 or 13 weeks. When CD-1 and B6C3F$_1$ mice were exposed to styrene at concentrations ranging from 15 to 500 ppm [64–2130 mg/m³] (6 h per day, five days per week) for two weeks, exposure to 250 or 500 ppm [1060 or 2130 mg/m³] caused mortality in both strains, together with hepatotoxic effects, which could be characterized microscopically as centrilobular hepatocyte necrosis. Mortality and liver changes were more severe in female mice at 250 ppm than at 500 ppm styrene. In a 13-week inhalation study, CD-1 mice were exposed to styrene (6 h per day, five days per week), with concentrations ranging from 50 to 200 ppm [213–852 mg/m³]. Liver toxicity was evident, with females generally more severely affected than males. Atrophy of the olfactory epithelium and olfactory nerve fibres occurred with or without focal respiratory metaplasia. In the lungs of a majority of exposed mice in all dose groups, decreased eosinophilia of the bronchial epithelium occurred. Cell proliferation studies revealed a highly variable labelling index in different cell types and between animals. There was a statistically significant increase in proliferation of the Clara cells of mice after two weeks of exposure to 150 and 200 ppm [639 and 852 mg/m³] styrene. These studies showed that mice are more sensitive to the toxic effects of styrene than rats. When Sprague-Dawley rats were exposed to styrene (50–1000 ppm [213–4260 mg/m³], 6 h per day, five days per week for 104 weeks), no changes in haematology, clinical chemistry, urinalysis or organ weights were observed (Cruzan *et al.*, 1998). However, the incidence of lesions of the olfactory epithelium of the nasal mucosa increased with increasing exposure concentration.

Male non-Swiss Albino mice were given single intraperitoneal injections of styrene (250, 500 or 1000 mg/kg bw), racemic styrene 7,8-oxide (100, 200, 300 or 400 mg/kg) or *R*- or *S*-styrene 7,8-oxide (300 mg/kg bw) and killed 24 h later (Gadberry *et al.*, 1996).

Styrene caused a dose-dependent increase in serum sorbitol dehydrogenase activity, an indicator of hepatotoxicity. Pretreatment of the animals with pyridine and β-naphthoflavone, inducers of CYP enzymes, increased the toxicity of styrene, but pretreatment with phenobarbital had no such effect. Trichloropropene oxide, an inhibitor of epoxide hydrolase, increased the hepatotoxicity of styrene 7,8-oxide, while buthionine sulfoximine, a glutathione depletor, did not. When given at the same dose (300 mg/kg bw), the racemic mixture of styrene 7,8-oxides and *R*-styrene 7,8-oxide caused a greater hepatotoxic effect than did *S*-styrene 7,8-oxide. However, differences in pulmonary toxicity between the enantiomers were not statistically significant. In a subsequent study to assess the susceptibility of mouse strains to the effects of styrene and styrene 7,8-oxide, male mice (A/J, C57BL/6 or CD-1) were given an intraperitoneal injection of 800 mg/kg bw styrene and killed 24 h later (Carlson, 1997b). Biomarkers of hepatotoxicity and pneumotoxicity were evaluated as in the study mentioned above (Gadberry *et al.*, 1996). CD-1 mice were not as sensitive as the other strains to the hepatotoxic or pneumotoxic effects of styrene, but were equally sensitive to the toxic effects of styrene 7,8-oxide given by intraperitoneal injection at 300 mg/kg bw.

Hepatotoxicity and cell proliferation induced in male $B6C3F_1$ mice by a single 6-h exposure to 500 ppm [2130 mg/m^3] styrene was compared with that caused by repeated exposures to 500 ppm styrene for 6 h per day for up to 14 days (Mahler *et al.*, 1999). Single or repeated inhalation exposure of mice to styrene caused severe necrosis of centrilobular hepatocytes followed by regeneration of necrotic zones. Re-exposure to styrene of mice that had received only a single dose of styrene resulted in hepatocellular necrosis, indicating that regenerated centrilobular cells were again susceptible to the cytolethal effect of styrene following a 14-day recovery. In contrast, continuous styrene exposure induced hepatocellular resistance to the cytolethal effect of styrene. New hepatocytes were produced with reduced metabolic activity. Centrilobular hepatocytes were resistant to styrene while they were actively synthesizing DNA and continued styrene exposure was required for sustained DNA synthesis.

4.3 Reproductive and developmental effects

Reproductive and developmental effects of styrene have been reviewed by IARC (1994a) and very extensively by Brown *et al.* (2000). The present evaluation is mainly based on these reviews and recent studies not included therein.

4.3.1 *Humans*

Early human studies suggested an association between occupational exposure to styrene and congenital central nervous system malformation (Holmberg, 1977, 1979), but subsequent, more extensive studies by the same and other investigators did not confirm the association (Brown *et al.*, 2000).

The frequency of spontaneous abortions among women with definite or assumed occupational exposure to styrene has been investigated. An early study suggested an association between exposure to styrene and spontaneous abortions (Hemminki *et al.*, 1980), but several later investigations, by the same group and others, of greater population size failed to confirm the association (see, e.g., Lindbohm *et al.*, 1985). It is possible to conclude that styrene is not associated with a major increase in the occurrence of spontaneous abortions, but because of the relatively small study population sizes in all of the investigations, the possibility of a small increase in risk cannot be excluded (Brown *et al.*, 2000).

There are only few studies on potential effects of exposure to styrene on birth weight. One study suggested that exposure to multiple solvents, including exposure to styrene levels above 80 ppm [340 mg/m^3], may be associated with around 4% reduced birth weight; however, the difference was not statistically significant (Lemasters *et al.*, 1989; Brown *et al.*, 2000).

Data regarding the effects of styrene on the menstrual cycle are conflicting. Abnormalities in pituitary secretion in women exposed to styrene have been suggested and may be connected with putative effects on the menstrual cycle (Brown *et al.*, 2000). In a recent study, exposure to aromatic solvents including styrene was associated with a trend towards increased frequency of oligomenorrhoea (average menstrual cycle length, > 35 days). Individually, exposure to styrene showed the greatest increase in odds ratio among the solvents (1.7; 95% CI, 1.1–2.6) and this increase was statistically significant (Cho *et al.*, 2001).

One study (Jelnes, 1988) suggested an increase in sperm abnormalities in workers exposed to high levels of styrene in the reinforced-plastics industry, but the data are weak and no firm conclusions can be drawn (Brown *et al.*, 2000). In a later study, semen samples were collected from 23 workers during the first week of employment in the reinforced-plastics industry and after six months of exposure to styrene and from 21 non-exposed farmers. A significant decline in sperm density was seen during exposure to styrene, whereas no decline was seen in the non-exposed men. No indication of an exposure–response relationship was seen when individual changes in semen quality within the group of reinforced-plastics workers were related to post-shift urinary mandelic acid concentrations, controlled for change in potential time-dependent confounders. However, the exposure gradient in the group was modest. The earlier finding of an increase in sperm abnormalities was not corroborated by this study (Kolstad *et al.*, 1999).

Among 1560 male workers in the reinforced-plastics industry in Denmark, Italy and The Netherlands, 220 styrene-exposed workers and 382 unexposed referents who had fathered a child were identified. The relationship between occupational exposure to styrene of these men and time-to-pregnancy of their partners was investigated. No consistent pattern of reduced male fecundity was found when the time to pregnancy was related to work tasks that involved higher levels of exposure to styrene or for which semiquantitative or quantitative measures of exposure to styrene were available. The workers with high exposure levels showed a fecundity ratio of 1.1 (95% CI, 0.69–1.7).

On the basis of these results, the authors concluded that it is unlikely that exposure to styrene has a strong effect on male fecundity (Kolstad *et al.*, 2000).

4.3.2 *Experimental systems*

(*a*) *Developmental toxicity studies*

Styrene has been shown to cross the placenta in rats and mice. In rats, fetal styrene concentrations of around 50% of the levels in maternal blood have been found (Withey & Karpinski, 1985; IARC, 1994a).

The potential developmental toxicity of styrene has been tested in several experimental animals including rats, mice, rabbits and hamsters. Throughout the studies, there was no evidence for an increased incidence of malformations. There have been reports of increases in embryonic, fetal and neonatal death, and possibly skeletal and kidney abnormalities at inhalation exposure levels of around 250 ppm [1060 mg/m^3] and higher during pregnancy. These dose levels also caused some effects on the dams such as decreased maternal weight gain (IARC, 1994a; Brown *et al.*, 2000).

Decreased pup weight, postnatal developmental delays as well as neurobehavioural and neurochemical abnormalities have been reported in rats exposed by inhalation to styrene prenatally and/or postnatally (Brown *et al.*, 2000). In a recent study, rats (JCL: Wistar) were exposed to 0, 50 or 300 ppm [213 or 1280 mg/m^3] styrene for 6 h per day during gestation days 6–20 and the offspring were investigated postnatally for neurochemical changes, growth and physical landmarks of development. In order to adjust for nutrient conditions, pair-fed and ad-libitum controls were included. No significant difference in maternal weight gain was observed compared with either control group, but food consumption was decreased at the 300-ppm exposure level. An increased neonatal death rate was found at 300 ppm styrene compared with the pair-fed control group. Litter size and birth weight were not affected, while postnatal development (incisor eruption, eye opening) was delayed at 300 ppm compared with both control groups. In addition, neurochemical alterations indicated by decreased 5-hydroxytryptamine turnover were detected at postnatal day 21 in offspring exposed *in utero* to 300 ppm compared with both control groups (Katakura *et al.*, 2001). The findings in this study can be attributed to exposure to styrene rather than to insufficient nutrition due to the use of a pair-fed control group and thus the results support previous findings indicating that postnatal development and the developing brain can be affected by exposure to styrene.

In mice exposed to 100 ppm [426 mg/m^3] styrene continuously for 24 h during days 0–15 of gestation, lower fetal and placental weight as well as reduced maternal body weight gain were found (Ninomiya *et al.*, 2000).

(*b*) *Reproductive toxicity studies*

No effects on fertility were found in male and female rats exposed continuously for three generations to 125 or 250 ppm [532 or 1060 mg/m^3] styrene in the drinking-water (Beliles *et al.*, 1985). Water consumption was significantly reduced in both treatment groups. Only minor general toxic effects (slightly reduced body weight in females) were observed. [The Working Group noted that the negative results obtained in this study do not provide adequate assurance of an absence of potential to impair fertility after exposure to styrene at higher dose levels and by other routes.]

No effects on sperm head morphology or sperm development were found in male mice exposed to 300 ppm [1280 mg/m^3] styrene (6 h per day, five days per week, for five weeks) (Salomaa *et al.*, 1985). General studies of the chronic and subchronic toxicity of styrene in several species have not shown testicular pathology at inhalation concentrations of up to 160 ppm [682 mg/m^3] in mice and 1000 ppm [4260 mg/m^3] or higher in other species (rat, rabbit, guinea-pig, rhesus monkey) (Brown *et al.*, 2000). One study in rats found evidence of testicular toxicity (decreased sperm count and some changes in testicular pathology) at oral styrene doses of 400 mg/kg bw per day for 60 days (Srivastava *et al.*, 1989). Srivastava *et al.* (1992) showed similar decreases in sperm count and testicular enzymes in postnatally maturing rats after treatment with styrene (200 mg/kg bw but not 100 mg/kg bw per day) by gavage for the first 60 days of life.

In peripubertal male C57BL/6 mice, free testosterone concentrations in plasma were strongly reduced after four weeks of exposure to styrene in the drinking-water (50 mg/L; this resulted in a daily intake of styrene of approximately 12 mg/kg bw). There were no effects on body weight, testis weight or plasma corticosterone and luteinizing hormone levels (Takao *et al.*, 2000).

There are only limited data on potential effects of styrene on estrous cyclicity. One study reported increased estrus cycle length and decreased body weight in rats exposed to 11.6 ppm [49 mg/m^3] styrene (Iziumova, 1972). However, the reporting of this study is confusing, the duration of estrus was poorly defined, and it is not clear if a comparable (sham) control group was used. In addition, it seems unlikely based on other studies of styrene that such a low exposure level would cause effects on body weight (see comments in Brown *et al.*, 2000).

The potential developmental toxicity of styrene and styrene 7,8-oxide was investigated in two in-vitro culture systems: micromass cell cultures of rat embryo midbrain or limb-bud cells and whole-embryo culture. Effects of styrene 7,8-oxide on markers of differentiation of limb-bud cells were evident at concentrations that had minimal effect on cell viability. Styrene alone or in the presence of an exogenous mono-oxygenase system had minimal effects even when tested at much higher concentrations than styrene 7,8-oxide. In whole-embryo culture, styrene 7,8-oxide caused growth retardation and malformations at somewhat lower levels than those producing embryo mortality (Gregotti *et al.*, 1994). In another study using postimplantation rat embryos in vitro, styrene 7,8-oxide was embryotoxic at concentrations more than 20 times lower than styrene (Brown-Woodman *et al.*, 1994).

4.4 Genetic and related effects

4.4.1 *Styrene*

(*a*) *Humans*

The genetic toxicology of styrene in humans was previously evaluated (IARC, 1994a). In styrene-exposed workers, levels of O^6-guanine adducts in lymphocyte DNA are up to five times higher than those in non-exposed controls. Variable results have been reported with regard to the association between exposure to styrene and chromosomal damage. Increased frequencies of chromosomal aberrations *in vivo* were not reported in most studies and only weak evidence was reported for induction of sister chromatid exchange. In-vitro studies consistently reported increased frequencies of chromosomal aberrations and sister chromatid exchange in human lymphocytes. Styrene did not induce mutations in bacteria, except in some studies that used exogenous metabolic activation. Most of the data published since the previous evaluation focus on the occurrence of DNA adducts, other DNA damage and cytogenetic effects associated with occupational exposures to styrene, and are summarized below.

(i) *DNA adducts*

Vodicka *et al.* (1993) studied a group of lamination workers exposed to styrene at air concentrations of 300–700 mg/m^3 (mean, 370 mg/m^3) and reported a statistically significant increase in the formation of O^6-deoxyguanosine adducts (O^6-(2-hydroxy-1-phenylethyl)-2′-deoxyguanosine, O^6-dG-styrene) in their lymphocytes, as determined by ^{32}P-postlabelling.

The persistence of these O^6-dG-styrene DNA adducts was investigated with a modified ^{32}P-postlabelling method (Hemminki & Vodicka, 1995), by comparing adduct levels in workers before vacation, after two weeks of vacation and after an additional one month of work. Air sampling showed styrene concentrations of 40–225 mg/m^3 (mean 122 mg/m^3). There was no significant difference in adduct levels in granulocytes between the control and exposed groups in any individual samplings. In lymphocytes of the laminators, the adduct levels were significantly higher (5.4 adducts/10^8 nucleotides) than those in the controls (1.0 adduct/10^8 nucleotides). A two-week interruption of exposure did not influence the total adduct level (4.9 versus 5.1 adducts/10^8 nucleotides in the first and second samplings, respectively), which indicates a very slow removal from the DNA of the specific O^6-dG-styrene adducts (Vodicka *et al.*, 1994).

Personal dosimeters were used to monitor exposure to styrene of nine workers at a hand-lamination plant in Bohemia, and the concentrations of styrene in blood and mandelic acid in urine were measured. Blood samples taken at four occasions during a seven-month period were analysed for styrene-specific O^6-dG adducts in lymphocytes and granulocytes. DNA strand breaks and hypoxanthine guanine phosphoribosyltransferase (*HPRT*) mutant frequency in T lymphocytes were also measured (see next section). Seven administrative employees in the same factory (factory controls) and eight persons in a research laboratory (laboratory controls) were used as referents. The mean concentrations of styrene in air were 122 and 91 mg/m^3, for the first and fourth samplings, respectively. DNA adduct levels were determined by the ^{32}P-postlabelling method in lymphocytes of laminators, and appeared to be remarkably constant (5–6 adducts/10^8 nucleotides) and significantly higher ($p < 0.0001$) than those in factory controls (0.3–1 adduct/10^8 nucleotides) at all four sampling times (Vodicka *et al.*, 1995).

Results of a three-year follow-up study continued to show significant increases in O^6-dG-styrene adducts in lymphocytes of lamination workers exposed to styrene (5.9 ± 4.9 adducts/10^8 nucleotides versus 0.7 ± 0.8 adducts/10^8 nucleotides in controls) (Vodicka *et al.*, 1999). In this series of studies, the numbers of exposed workers and corresponding control subjects ranged from nine to 17 individuals. The results indicate that DNA adduct levels in lamination workers correlate with styrene exposure concentration. Evidence from long-term studies, however, indicates that adducts do not continue to accumulate with exposure time, suggesting that a balance is reached between adduct formation and removal during long-term exposures. Smoking habits did not significantly affect adduct formation in the exposed workers compared to controls.

In workers with reinforced plastics in the boat-manufacturing industry, styrene exposure (1–235 mg/m^3; mean 65.6 mg/m^3) increased the level of an unidentified DNA adduct (14.2 ± 2.3 adducts/10^8 nucleotides), as well as N^2-(2-hydroxy-1-phenylethyl)-2′-deoxyguanosine (15.8 ± 3.2 adducts/10^8 nucleotides) (Horvath *et al.*, 1994; Rappaport *et al.*, 1996). Styrene exposure concentration correlated with N^2-guanine adduct levels. The increased formation of 8-hydroxy-2′-deoxyguanosine (8-OHdG) adducts in lymphocytes of styrene-exposed workers (2.23 ± 0.54 8-OHdG/10^5 dG versus 1.52 ± 0.45 8-OHdG/10^5 dG in controls) provides evidence that styrene exposure also causes oxidative DNA damage (Marczynski *et al.*, 1997a). It was suggested that such damage could alter the balance of oxidants and antioxidants in the cell. This has led to a new hypothesis for genotoxicity of styrene based on oxidative stress (Marczynski *et al.*, 2000).

(ii) *DNA single-strand breaks*

In 17 workers with low occupational exposure to styrene (time-weighted average exposure, 29.4 mg/m^3), an exposure-dependent increase was seen in single-strand breaks (IARC, 1994a). An earlier investigation of the same group of workers reported that DNA single-strand breaks — measured by the DNA unwinding method — were also induced in workers estimated to have higher exposure to styrene (average, 300 mg/m^3) calculated from post-shift urinary mandelic acid and blood concentrations of styrene glycol (Mäki-Paakkanen *et al.*, 1991). Lack of persistence of the DNA damage indicated that the strand breaks were repaired in the exposed workers. Three studies reported on DNA breakage, measured by use of the single-cell gel electrophoresis assay, in lymphocytes of lamination workers exposed to styrene in the reinforced-plastics industry (exposure range, 68–101 mg/m^3) (Vodicka *et al.*, 1995; Somorovská *et al.*, 1999; Vodicka *et al.*, 1999). Smoking habits had no effect on the level of styrene-induced DNA damage. One study reported a significant correlation between DNA single-strand breaks and O^6-dG-styrene adduct levels (Vodicka *et al.*, 1995).

(iii) *Gene mutations and cytogenetic damage*

Since the previous evaluation of styrene, results from three studies of styrene-induced gene mutations in lymphocytes from reinforced-plastics workers have been published. One study assessed glycophorin A (GPA) variant frequency in erythrocytes of 47 workers exposed for 8 h per day to a time-weighted average concentration of 37 ppm [158 mg/m^3] styrene in the breathing zone. Overall there was no significant increase in variant frequency for GPA allele duplication or allele loss, but a significant increase was found for 28 workers with exposure ≥ 85 mg/m^3 styrene (Bigbee *et al.*, 1996). Studies that investigated mutation induction at the *HPRT* locus in lymphocytes of styrene-exposed lamination workers in two different plants reported inconclusive to weak positive results (Vodicka *et al.*, 1995, 1999). Both studies reported a higher mutation frequency in exposed workers than in controls, but the increase was significant in only one of the studies (21.5 ± 9.96 versus 12.69 ± 4.56 mutants per 10^6 cells; $p = 0.039$) (Vodicka *et al.*, 1999).

The previous review of cytogenetic damage in workers occupationally exposed to styrene (IARC, 1994a) showed mainly negative results for the induction of chromosomal aberrations, micronuclei or sister chromatid exchange. Occupational exposure to styrene induced chromosomal aberrations in nine of 22 studies, micronuclei in three of 11 studies and sister chromatid exchange in three of 12 studies. The mean styrene concentration in these studies ranged from 11 to 138 ppm [47–588 mg/m^3]. Workers having no significant increase in chromosomal aberrations were exposed to styrene at concentrations ranging from 2 to 70 ppm [8.5–298 mg/m^3] (IARC, 1994a). The results from six recent cytogenetic studies of styrene-exposed workers in the plastics and boat-manufacturing industries are summarized in Table 12; chromosomal aberrations and sister chromatid exchange were increased in three of three and three of four studies, respectively, while all three studies on micronuclei were negative. The ranges of styrene concentrations to which workers were exposed are also given in Table 12.

Artuso *et al.* (1995) reported that chromosomal aberrations and sister chromatid exchange were induced in lymphocytes of boat-manufacturing workers in both a high- (20–331 ppm [85–1410 mg/m^3]) and low- (0.5–2.9 ppm [2.1–12.4 mg/m^3]) exposure group. The response in the low-exposure group, however, was weak. Workers in a glass fibre-reinforced plastics facility who were also exposed to low concentrations of styrene (0.5–26 ppm [2.1–111 mg/m^3]) did not show increased sister chromatid exchange nor increased induction of micronuclei (Van Hummelen *et al.*, 1994). One further study reported that sister chromatid exchange frequencies were significantly increased in lymphocytes of furniture manufacturers who were exposed to styrene from polyester resins at concentrations (20–300 ppm [85–1280 mg/m^3]), similar to those reported in reinforced-plastics facilities (Karakaya *et al.*, 1997).

Table 12. Cytogenetic studies in lymphocytes from individuals occupationally exposed to styrene

No. of exposed	No. of controls	Years of exposure		Styrene in air (mg/m^3)		Urinary mandelic acid (mg/g creatinine)		Cytogenetic observation			Reference
		Range	Mean	Range	Mean	Range	Mean	CA	MN	SCE	
28	20	1–26	8	125–180	157[a]		652 mg/L			+	Hallier *et al.* (1994)
52	24		2.9	2.2–111	31	11–649	102		–	–	Van Hummelen *et al.* (1994)
18	18	10–22	14.3	NR			328	+	–		Anwar & Shamy (1995)
23	51			86–1390 2–120				+ (+)		+ (+)	Artuso *et al.* (1995)
53	41		9.9	85–1280	129	14–1482[b]	207[b]		–	+	Karakaya *et al.* (1997)
44	19		14	27–199[c]	101			+			Somorovská *et al.* (1999)

CA, chromosomal aberrations; MN, micronuclei; SCE, sister chromatid exchange; NR, not reported
[a] After modifications were made to the plant to lower styrene concentrations below 85 mg/m^3, results from SCE test were negative.
[b] Mandelic acid + phenylglyoxylic acid
[c] Range of means

(iv) *Protein adducts*

Christakopoulos *et al.* (1993) compared 17 styrene-exposed reinforced-plastics workers (years of exposure: 0.2–15; mean, 6.7) with 11 non-exposed controls and found linear correlations between concentrations of haemoglobin adducts and free styrene 7,8-oxide in blood (r = 0.71), styrene glycol in blood (r = 0.94) and mandelic acid in urine (r = 0.92). [The Working Group noted that no information was available on styrene exposure concentrations for these workers.] Severi *et al.* (1994) looked for haemoglobin adducts of styrene 7,8-oxide in 52 workers having an average styrene exposure of 31 mg/m^3 and found none (detection limit of 10 pmol/g).

Yeowell-O'Connell *et al.* (1996) measured adducts in 48 workers exposed to both styrene (0.9–235 mg/m^3; mean, 64.3 mg/m^3) and styrene 7,8-oxide (13.4–525 μg/m^3; mean, 159 μg/m^3 for 20 subjects) in a boat-building plant. They found no increase in the amount of haemoglobin adducts, but levels of albumin adducts increased with exposures to both styrene and styrene 7,8-oxide. Further analysis of these data (Rappaport *et al.*, 1996) indicated that the albumin adducts correlated better with airborne styrene 7,8-oxide than with styrene in those individuals for whom exposure to both chemicals was measured. Fustinoni *et al.* (1998) compared 22 workers selected on the basis of high levels of exposure to styrene (estimated to be about 100 mg/m^3) with 15 controls. The mean levels of 2-phenylethanol and 1-phenylethanol — obtained by chemical cleavage of styrene 7,8-oxide–cysteine adducts — were respectively 2.84 and 0.60 nmol/g albumin and 5.44 and 0.43 nmol/g haemoglobin for the workers and 2.74 and 0.50 nmol/g albumin and 5.27 and 0.39 nmol/g haemoglobin for controls. Differences between workers and controls were significant only when the data for the workers were stratified for high exposure. Johanson *et al.* (2000) exposed four male volunteers to 50 ppm [213 mg/m^3] [$^{13}C_8$]styrene for 2 h. Maximal concentrations of styrene 7,8-oxide in blood (average, 6.7 nM) were seen at 2 h. Hydroxyphenylethylvaline was estimated at 0.3 pmol/g globin. The major (> 95%) portion of the [$^{13}C_8$]styrene metabolites in urine was derived from hydrolysis of styrene 7,8-oxide, with less than 5% coming from metabolism via the phenylacetaldehyde pathway.

(*b*) *Experimental systems* (see Table 13 for references)

Results from studies published since the previous evaluation of styrene are summarized in Table 13.

(i) *DNA adducts*

Both *N*7- and *O*^{6}-deoxyguanosine adducts were formed in NMRI mouse liver, lung and spleen evaluated 3 h after a single intraperitoneal injection of styrene at doses ranging from 0.28 to 4.35 mmol/kg bw (Pauwels *et al.*, 1996). A dose–response relationship was reported for both adducts in all three tissues. The *N*7-alkyldeoxyguanosine was more abundant than the *O*^{6}-deoxyguanosine adduct and the levels of both adducts were highest in lung tissue. Samples of liver and lung DNA, taken from CD-1 mice and Sprague-Dawley rats 42 h after a 6-h exposure to 160 ppm [682 mg/m^3] [^{14}C]styrene,

Table 13. Genetic and related effects of styrene

Test system	Result[a]		Dose[b] (LED or HID)	Reference
	Without exogenous metabolic system	With exogenous metabolic system		
Drosophila melanogaster, somatic mutation (*w/w*+ mosaic assay)	–[c]		1040 μg/mL, feed	Rodriguez-Arnaiz (1998)
Sister chromatid exchange, human lymphocytes *in vitro*	+		1	Chakrabarti *et al.* (1993)
Sister chromatid exchange, human lymphocytes *in vitro*	+		52	Lee & Norppa (1995)
DNA strand breaks (single-cell gel electrophoresis assay), C57BL/6 mouse liver, kidney, bone marrow and lymphocytes *in vivo*	+[d]		250 ip × 1	Vaghef & Hellman (1998)
Sister chromatid exchange, Fischer 344 rat lymphocytes *in vivo*	–		4260 mg/m^3 (inh.), 6 h/d, 5 d/w, 4 w	Preston & Abernethy (1993)
Chromosomal aberrations, Fischer 344 rat lymphocytes *in vivo*	–		4260 mg/m^3 (inh.), 6 h/d, 5 d/w, 4 w	Preston & Abernethy (1993)
DNA adducts in NMRI mouse liver, lung, spleen and blood, *in vivo*	+		30 ip × 1	Pauwels *et al.* (1996)
DNA adducts in Sprague Dawley rat liver and lung, *in vivo*[e]	+		682 mg/m^3 (inh.), 6 h	Boogaard *et al.* (2000b)
DNA adducts in CD-1 mouse liver and lung, *in vivo*[f]	+		682 mg/m^3 (inh.), 6 h	Boogaard *et al.* (2000b)
DNA adducts (O^6-guanine) (^{32}P-postlabelling) in female CRL CD-1 mouse lung *in vivo*	–		682 mg/m^3 (inh.), 6 h/d, 5 d/w, 2 w	Otteneder *et al.* (2002)
DNA adducts (O^6-guanine) (^{32}P-postlabelling) in female CRL CD rat lung *in vivo*	–		2130 mg/m^3 (inh.), 6 h/d, 5 d/w, 2 w	Otteneder *et al.* (2002)
DNA adducts (O^6-guanine) (^{32}P-postlabelling) in male and female CRL CD rat liver *in vivo*	+		4260 mg/m^3 (inh.), 6 h/d, 5 d/w, 2 y	Otteneder *et al.* (2002)
Chromosomal aberrations, human lymphocytes *in vivo*	?		27 mg/m^3	Somarovská *et al.* (1999)

[a] +, positive;–, negative

[b] LED, lowest effective dose; HID, highest ineffective dose; in-vitro tests, μg/mL; in-vivo tests, mg/kg bw/day (unless otherwise specified); d, day; ip, intraperitoneal; inh., inhalation; w, week; y, year; h, hour

[c] A positive response was seen at this dose in insecticide-resistant strains, which have high bioactivation capacities.

[d] Positive response in bone marrow at 350 mg/kg bw

[e] DNA adduct level was 3-fold higher in lung Type-II cells than in total lung.

[f] DNA adduct levels in Clara and non-Clara cells were similar to that in total lung.

showed increased levels of *N*7-deoxyguanosine adducts for both species (Boogaard *et al.*, 2000b). The adduct level in rat lung fractions enriched in type II cells was approximately three times higher than that in total lung, whereas the adduct level in mouse lung fractions enriched in Clara cells was similar to that of total lung. In these studies, the covalent binding indices (CBI) were all below 0.5.

Otteneder *et al.* (2002) did not detect O^6-deoxyguanosine adducts in DNA from lung tissue of CD-1 mice or CD rats exposed to 160 ppm or 500 ppm [682 or 2130 mg/m^3] styrene, respectively for two weeks (6 h per day, five days per week). The limit of detection in the ^{32}P-postlabelling method used in this study was one adduct per 10^7 nucleotides. However, both α and β isomers of O^6-dG-styrene 7,8-oxide adducts were detected in DNA from CD rats that were exposed by inhalation to 1000 ppm [4260 mg/m^3] styrene for two years (6 h per day, five days per week). The authors noted that DNA adduct formation in tissue homogenates did not reflect either the species difference in carcinogenicity nor the target organ specificity of styrene.

(ii) *Mutation and allied effects*

Positive and negative results have previously been reported for styrene-induced mutations in *Salmonella typhimurium* (IARC, 1994a). Positive results were reported in studies using strains TA1530 or TA1535 with the addition of an exogenous metabolic activation system. No new data on mutations in *S. typhimurium* have been reported.

Results from one study of somatic mutations in *Drosophila melanogaster* were negative. This study used both insecticide-susceptible (IS) and -resistant (IR) strains, which differ in their activities of drug-metabolizing enzymes, to evaluate mutations in the *w/w+* mosaic assay. Females were exposed to styrene at a concentration of 1040 μg/mL in the feed medium. Although negative results were reported for the IS strain, positive results were seen in the IR strain at this dose. Two studies of the induction of sister chromatid exchange in human lymphocytes exposed to styrene *in vitro* also gave positive results. These observations are consistent with those from earlier studies.

DNA strand breaks were induced in liver, kidney, bone marrow and lymphocytes of C57BL/6 mice given a single intraperitoneal injection of styrene. On the other hand, neither chromosomal aberrations nor sister chromatid exchange were seen in lymphocytes of Fischer 344 rats exposed to styrene by inhalation at a concentration of 1000 ppm [4260 mg/m^3] for 6 h per day on five days per week for four weeks.

4.4.2 *Styrene 7,8-oxide*

(*a*) *Humans*

No data on effects of exposure of humans to styrene 7,8-oxide alone were available to the Working Group.

(*b*) *Experimental systems* (see Table 14 for references)

(i) *DNA adducts*

A detailed description of *N*7, *N*[2] and *O*[6]-deoxyguanosine adducts of styrene 7,8-oxide formed *in vivo* and *in vitro* was previously presented (IARC, 1994b). Subsequent studies have reported that *N*7-deoxyguanosine adducts are formed in human whole blood and embryonic lung fibroblasts *in vitro*, as well as in salmon testis DNA, upon exposure to styrene 7,8-oxide. In addition, *O*[6]-deoxyguanosine adducts were induced in human lymphocytes exposed to 24 μg/mL styrene 7,8-oxide *in vitro* and in rat liver DNA following inhalation exposure to 1000 ppm [4.9 mg/m^3] styrene 7,8-oxide for 6 h per day on five days per week for 104 weeks. DNA adducts were not clearly induced in other studies in which rats were exposed to styrene 7,8-oxide by intraperitoneal injection or oral exposure.

(ii) *Mutation and allied effects*

Styrene 7,8-oxide, a monofunctional alkylating agent, has given positive results in virtually every short-term genetic assay *in vitro* and, to a lesser degree, *in vivo*. Positive results have previously been reported for gene mutations in bacteria and mammalian cells *in vitro*. DNA strand breaks, alkali-labile sites, and cytogenetic damage also have been reported in mammalian cells *in vitro* and, to some degree, *in vivo*.

It was noted in the previous evaluation (IARC, 1994b) that the *R*-enantiomer of styrene 7,8-oxide may be slightly more mutagenic in *S. typhimurium* TA100 than the *S*-enantiomer (Pagano *et al.*, 1982; Seiler, 1990; Sinsheimer *et al.*, 1993). In contrast, Sinsheimer *et al.* (1993) reported that the *S*- but not the *R*-enantiomer induced chromosomal aberrations and sister chromatid exchange in mouse bone marrow.

[The Working Group noted that, except when the *R*- or *S*- enantiomer of styrene 7,8-oxide was specifically tested, exposures to styrene 7,8-oxide were probably mixtures of optical isomers.]

Results from studies on genetic and related effects of styrene 7,8-oxide published since the earlier evaluation are essentially all positive. In fact, only one study reported negative results, the *w/w+* mosaic assay for induction of somatic mutations in insecticide-sensitive *Drosophila melanogaster*. However, in insecticide-resistant *Drosophila* strains, which have an active bioactivation system, styrene 7,8-oxide induced somatic mutations.

DNA single-strand breaks and alkali-labile sites were induced in Wistar rat and human testicular and Chinese hamster V-79 cells *in vitro*. Four studies reported that DNA strand breaks and/or fragmentation were induced in human lymphocytes and human embryonic lung fibroblasts exposed to styrene 7,8-oxide *in vitro*. Similarly, positive results were reported for the induction of gene mutations at the *HPRT* locus in human lymphocytes and B-cell lymphoblastoid cells treated *in vitro*. Six studies reported that exposure to styrene 7,8-oxide *in vitro* induced a significant increase in sister chromatid exchange frequencies in human lymphocytes. One study also reported a significant

Table 14. Genetic and related effects of styrene 7,8-oxide

Test system	Result[a]		Dose[b] (LED or HID)	Reference
	Without exogenous metabolic system	With exogenous metabolic system		
Drosophila melanogaster, somatic mutation (*w/w+* mosaic assay)	–		1200 μg/mL[c] in feed	Rodriguez-Arnaiz (1998)
DNA single-strand breaks and alkali-labile sites (alkaline elution), Wistar rat testicular cells *in vitro*	+	NT	12	Bjørge *et al.* (1996)
DNA single-strand breaks and alkali-labile sites (alkaline elution), Chinese hamster V-79 cells, *in vitro*	+	NT	6	Herrero *et al.* (1997)
DNA strand breaks, human lymphocytes *in vitro*	+	NT	6	Bastlová *et al.* (1995)
DNA single-strand breaks and alkali-labile sites (alkaline elution), human testicular cells *in vitro*	+	NT	12	Bjørge *et al.* (1996)
DNA single-strand breaks human embryonic lung fibroblasts *in vitro*	+	NT	1.2	Vodicka *et al.* (1996)
DNA fragmentation, human whole blood *in vitro*	+	NT	1	Marczynski *et al.* (1997a)
DNA damage (alkaline single-cell gel electrophoresis assay), human lymphocytes *in vitro*	+	NT	2.4	Laffon *et al.* (2001)
Gene mutation, human lymphocytes, *HPRT* locus *in vitro*	(+)	NT	24	Bastlová *et al.* (1995)
Gene mutation, human T-lymphocytes, *HPRT* locus, *in vitro*	+	NT	24	Bastlová & Podlutsky (1996)
Gene mutation, human B-cell lymphoblastoid cells, *HPRT* locus, *in vitro*	(+)[d]	NT	72	Shield & Sanderson (2001)
Sister chromatid exchange, human lymphocytes *in vitro*	+	NT	6	Lee & Norppa (1995)
Sister chromatid exchange, human lymphocytes *in vitro*	+	NT	6	Ollikainen *et al.* (1998)
Sister chromatid exchange, human lymphocytes *in vitro*	+	NT	6	Uusküla *et al.* (1995)
Sister chromatid exchange, human lymphocytes *in vitro*	+	NT	6	Laffon *et al.* (2001)

Table 14 (contd)

Test system	Result[a]		Dose[b] (LED or HID)	Reference
	Without exogenous metabolic system	With exogenous metabolic system		
Sister chromatid exchange, human lymphocytes *in vitro*	+	NT	12	Chakrabarti *et al.* (1997)
Sister chromatid exchange, human lymphocytes *in vitro*	+	NT	12	Zhang *et al.* (1993)
Micronucleus test, human lymphocytes *in vitro*	+	NT	12	Laffon *et al.* (2001)
DNA alkali-labile sites (alkaline single-cell gel electrophoresis assay), C57BL/6 mouse liver, kidney, bone marrow and lymphocytes *in vivo*	+		100 ip × 1	Vaghef & Hellman (1998)
DNA alkali-labile sites (alkaline single-cell gel electrophoresis assay), ddY mouse organs *in vivo*	+		400 ip × 1	Tsuda *et al.* (2000)
DNA alkali-labile sites (alkaline single-cell gel electrophoresis assay), CD-1 mouse organs *in vivo*	+		400 ip × 1	Sasaki *et al.* (1997)
DNA adducts (*N*7-guanine), salmon testis DNA *in vitro*	+	NT	11	Koskinen *et al.* (2000)
DNA adducts (*N*7-guanine) in human embryonic lung fibroblasts *in vitro*	+	NT	1.2	Vodicka *et al.* (1996)
DNA adducts (*N*7-guanine) in human whole blood, *in vitro*	+	NT	NR	Pauwels & Veulemans (1998)
DNA adducts (O^6-guanine) in human lymphocytes *in vitro*	(+)	NT	24	Bastlová *et al.* (1995)
DNA adducts (O^6-guanine), rat liver *in vivo*	+		4260 mg/m³ (inh.), 6 h/d, 5 d/w, 104 w	Otteneder *et al.* (1999)

[a] +, positive; (+), weak positive; –, negative; NT, not tested; NR, not reported
[b] LED, lowest effective dose; HID, highest ineffective dose; in-vitro tests, μg/mL; in-vivo tests, mg/kg bw/day; d, day; ip, intraperitoneal; inh., inhalation
[c] Positive results reported in insecticide-resistant (IR) strains at 600 μg/mL (5mM)
[d] GSTM1-null more sensitive

increase in the induction of micronuclei in human lymphocytes treated *in vitro*. These results are consistent with those previously reviewed.

Three recent studies have reported that DNA strand breaks and alkali-labile sites are induced in multiple tissues (lung, liver, kidney, bone marrow and lymphocytes) of mice exposed to styrene 7,8-oxide by a single intraperitoneal injection at doses of 100–400 mg/kg bw. Results from previously evaluated studies of induction of cytogenetic damage in rodents exposed to styrene 7,8-oxide *in vivo* were predominately negative. Two of three studies reported some evidence of exposure-related sister chromatid exchange induction and two of four studies reported that styrene 7,8-oxide induced chromosomal aberrations. Neither of the two micronucleus studies previously evaluated reported increases in the frequency of micronuclei following exposures to styrene 7,8-oxide via intraperitoneal injection (IARC, 1994b).

4.5 Mechanistic considerations

Following chronic exposure to styrene, a carcinogenic response has been observed only in mouse lung, but not in tissues of rats.

Studies in humans and experimental animals and with tissues from both rodents and humans *in vitro* using a variety of experimental approaches have attempted to determine the mode of action for the styrene toxicity and carcinogenicity observed in experimental animals. In general, the objectives of many of these studies have been (1) to explore the basis for the observed interspecies differences in tumour response between rats and mice, (2) to determine the critical steps in styrene carcinogenicity and (3) to identify the most likely chemical species responsible for the initiation of tumour development.

At least two plausible modes of action, which are not mutually exclusive, for styrene carcinogenicity are suggested by the existing data. One involves a DNA-reactive mode of action initiated by the metabolic conversion of styrene to styrene 7,8-oxide, a genotoxic metabolite of styrene, and the subsequent induction of DNA damage in target tissues; the other involves cytotoxic effects in lungs of mice exposed to styrene. For either mode of action, interspecies differences in the metabolism and toxicokinetics of styrene and styrene 7,8-oxide that exist between rats and mice are likely to play a key role.

4.5.1 *Interspecies differences in toxicokinetics and metabolism*

A number of studies in human volunteers and in workers occupationally exposed to styrene by inhalation (summarized by Bond, 1989) have shown that styrene toxicokinetics and metabolic pathways are qualitatively similar in humans and experimental animals, although there are quantitative differences. In humans, between 60 and 70% of inhaled styrene is absorbed and more than 85% of the absorbed styrene is eliminated in the urine as mandelic and phenylglyoxylic acids. These metabolites are formed following hydrolysis of styrene 7,8-oxide, which indicates that humans metabolize styrene to styrene 7,8-oxide. Johanson *et al.* (2000) have quantified blood levels of styrene 7,8-oxide in human volunteers exposed to styrene.

In all experimental animal species studied, styrene is rapidly metabolized to styrene 7,8-oxide following absorption by oral, dermal or inhalation exposure. Styrene 7,8-oxide is sufficiently stable to be readily detected in the blood of both rats and mice exposed to styrene. Systemic styrene 7,8-oxide concentrations do not, however, explain interspecies differences in the carcinogenic response to styrene. At styrene exposure concentrations of 1000 ppm [4260 mg/m^3] (the highest concentration tested), styrene 7,8-oxide concentrations in the blood of rats were two orders of magnitude higher than those in the blood of mice exposed to 20–40 ppm [85–170 mg/m^3] styrene (the lowest concentrations causing tumours), although mice, but not rats, develop lung tumours (Cruzan *et al.*, 1998, 2001).

The enzymatic oxidation of styrene to styrene 7,8-oxide is mediated by several isoforms of cytochrome P450 (CYP) enzymes including CYP2E1, CYP2B6 and CYP2F1 in humans. In experimental animals, CYP2E1 appears to be the major isoform responsible for styrene oxidation in the liver, and CYP2F2 is the major form in the mouse lung, as determined by in-vitro studies (Carlson, 1997a). Interspecies differences exist in the rates of metabolism of styrene by liver and lung tissue. In particular, human lung tissue produces less styrene 7,8-oxide than that of rats and considerably less than that of mice (Nakajima *et al.*, 1994a; Carlson *et al.*, 2000; Filser *et al.*, 2002).

The metabolism of styrene to styrene 7,8-oxide in mouse lung occurs almost exclusively in the Clara cells and the rate of styrene oxidation in these cells is threefold faster than in Clara cells of rats (Hynes *et al.*, 1999). The mouse Clara cell may be at greater risk than the rat Clara cell if locally generated, rather than extrapulmonary, styrene 7,8-oxide is critical for cytotoxic or genotoxic damage.

Metabolism of styrene 7,8-oxide can proceed through hydrolysis of styrene 7,8-oxide by epoxide hydrolase or through conjugation with glutathione mediated by glutathione *S*-transferase. Conversion to phenylacetaldehyde and other ring-opened metabolites provides additional pathways for styrene and styrene 7,8-oxide metabolism (Sumner & Fennell, 1994; Johanson *et al.*, 2000). Stable products representing each of these pathways are eliminated in the urine. Interspecies differences exist in the proportion of styrene 7,8-oxide that is eliminated via each of these pathways, and these differences may play a role in the species sensitivity towards the toxic and carcinogenic effects of styrene exposure. For example, the capacity of epoxide hydrolase in human tissues to detoxify styrene 7,8-oxide exceeds that of rat or mouse tissues. In contrast, the activity of glutathione *S*-transferase in human tissues is much lower than in rodent tissues (summarized in Cohen *et al.*, 2002). Mouse tissues show significantly higher rates of activation of styrene to DNA-reactive epoxides than do rat or human tissues. While mouse, rat and human tissues all use epoxide hydrolase to detoxify styrene 7,8-oxide, the mouse also uses glutathione *S*-transferase to a significant extent, with less activity in the rat and very little in humans.

The role of glutathione in the detoxication of styrene 7,8-oxide introduces the possibility that high concentrations of styrene 7,8-oxide significantly deplete the cellular concentrations of glutathione, resulting in substantial cellular damage. Significant reductions in pulmonary glutathione were noted in mice exposed repeatedly to 80–300 ppm [341–1280 mg/m^3] of styrene (Filser *et al.*, 2002). Rats were significantly less susceptible to glutathione depletion than mice. However, while glutathione depletion in mouse lung has not been demonstrated at styrene concentrations below 80 ppm [341 mg/m^3], evidence of cancer induction exists at exposure concentrations as low as 20 ppm [85 mg/m^3].

In summary, quantitative, but not qualitative, differences exist in the toxicokinetics of styrene and styrene 7,8-oxide in mice, rats and humans. These quantitative differences alone do not appear to be sufficient to account for the development of lung tumours in mice at low styrene exposures and the resistance of rats to tumour formation at high exposures. Thus, factors of a toxicodynamic nature may play a role in the interspecies differences in response to styrene.

4.5.2 *Interspecies differences in toxicodynamics and mode of action*

In general, mice appear to be more susceptible to lung tumour induction by epoxides and epoxide-forming chemicals than rats (Melnick & Sills, 2001). Inhalation of ethylene oxide, 1,3-butadiene, isoprene and chloroprene induced lung tumours in mice but not rats (Lynch *et al.*, 1984; National Toxicology Program, 1984; Snellings *et al.*, 1984; National Toxicology Program, 1987; Owen *et al.*, 1987; Melnick *et al.*, 1994; National Toxicology Program, 1998, 1999; Melnick & Sills, 2001). The determinants underlying the susceptibility of the mouse lung towards tumour formation may rely in part on toxicokinetic considerations, but toxicodynamic determinants are probably also at play.

One potential mode of action for styrene involves the cytotoxicity of styrene 7,8-oxide. Repeated exposure to styrene progressively results in focal crowding of bronchiolar cells, bronchiolar epithelial hyperplasia and bronchiolo-alveolar hyperplasia in the lungs of mice but not in those of rats (Cruzan *et al.*, 1997, 1998; Cohen *et al.*, 2002; Cruzan *et al.*, 2001). A major factor that may play a role in the onset of hyperplasia in mice after exposure to styrene is pulmonary formation of styrene 7,8-oxide. Styrene 7,8-oxide administered to mice by intraperitoneal injection results in pulmonary toxicity as measured by release of enzymes into bronchoalveolar lavage fluid, suggesting that styrene 7,8-oxide is cytotoxic to mouse lung cells. This cytotoxicity stimulates cell replication and proliferation. The pulmonary toxicity induced by high doses of injected styrene 7,8-oxide also clearly demonstrates that systemically available styrene 7,8-oxide (i.e., that produced in extrapulmonary tissues and released into the circulation) can enter lung cells to induce a toxic response.

A second mode of action for styrene-induced carcinogenicity invokes DNA-reactivity and subsequent genotoxicity. Although styrene itself is not DNA-reactive, styrene 7,8-oxide binds covalently to macromolecules including proteins and nucleic acids.

Styrene 7,8-oxide can bind to DNA with the formation of stable N^2 and O^6 adducts of deoxyguanosine in human lymphocytes and cultured mammalian cells (Horvath *et al.*, 1994; Bastlová *et al.*, 1995; Vodicka *et al.*, 1999). In particular, the O^6 adducts are relatively persistent. The stability of the O^6 adducts is supported by studies of styrene 7,8-oxide adducts in the lymphocytes of workers exposed to styrene (Vodicka *et al.*, 1994, 1995, 1999).

The genotoxicity of styrene and styrene 7,8-oxide has been studied extensively in both experimental systems and humans. Styrene 7,8-oxide has been shown to cause mutations (with and without metabolic activation) in the Ames *Salmonella* assay. In contrast, styrene generally gives negative results for mutagenic activity in the Ames *Salmonella* assay; however, it was noted previously (IARC, 1994a) that positive results in *Salmonella typhimurium* were seen in some studies in the presence of exogenous metabolic activation. The evidence for mutations in humans occupationally exposed to styrene is equivocal.

Exposure to styrene results in an increase in sister chromatid exchange frequency and, to a limited extent, the frequency of chromosomal aberrations and micronucleated cells in experimental animals. Cytogenetic studies in exposed workers are difficult to interpret due to potential confounders and co-exposures to other genotoxic agents (Scott & Preston, 1994). Assessing quantitative relationships across human studies has shown associations between styrene exposure and the frequency of chromosomal abnormalities but not sister chromatid exchange or micronuclei (Bonassi *et al.*, 1996; Cohen *et al.*, 2002). Rappaport *et al.* (1996) reported a positive correlation between airborne styrene 7,8-oxide concentrations, but not styrene concentrations, and sister chromatid exchange frequency in styrene workers.

An association has been reported between styrene exposure and DNA strand breaks in workers employed in various styrene-related industries (Mäki-Paakkanen *et al.*, 1991; Somorovská *et al.*, 1999). However, the role of these strand breaks in disease etiology has been questioned since this type of damage is quickly and efficiently repaired in both animals and humans (Walles *et al.*, 1993; Bastlová *et al.*, 1995). As an example, studies by Walles *et al.* (1993) indicate that end-of-shift DNA damage in subjects occupationally exposed to styrene is repaired by the next morning. The DNA strand breaks are probably caused by the much less persistent *N*7 adducts (Vodicka *et al.*, 1996).

4.5.3 *Conclusion*

In summary, data from both laboratory (*in vitro* and *in vivo*) and human studies indicate that styrene exposure can result in low levels of DNA adducts and DNA damage in individuals who possess the capacity to activate styrene metabolically to its epoxide metabolite, styrene 7,8-oxide. However, as noted above, mice, but not rats, develop lung tumours following styrene exposure, even though both species form DNA adducts. DNA adducts are also found in organs other than the lung.

Circulating styrene 7,8-oxide may also play a role. However, the concentration of this metabolite in rat blood is two orders of magnitude higher than in the mouse. The lung tumours in mice probably develop as a result of in-situ formation of styrene 7,8-oxide which leads to cytotoxicity and increased cell proliferation, but a role of circulating styrene 7,8-oxide and of DNA adducts cannot be discounted.

Based on metabolic considerations, it is likely that the proposed mechanism involving metabolism of styrene to styrene 7,8-oxide in mouse Clara cells is not operative in human lungs to a biologically significant extent. However, based on the observations in human workers regarding blood styrene 7,8-oxide, DNA adducts and chromosomal damage, it cannot be excluded that this and other mechanisms are important for other organs.

5. Summary of Data Reported and Evaluation

5.1 Exposure data

Styrene is a commercially important monomer which is used extensively in the manufacture of polystyrene resins (plastic packaging, disposable cups and containers, insulation) and in copolymers with acrylonitrile and/or 1,3-butadiene (synthetic rubber and latex, reinforced plastics). Human exposure occurs at levels of milligrams per day during its production and industrial use and at much higher levels in the glass fibre-reinforced plastics industry. Exposure to the general population occurs at levels of micrograms per day due mainly to inhalation of ambient air and cigarette smoke and intake of food that has been in contact with styrene-containing polymers.

5.2 Human carcinogenicity data

Retrospective cohort studies of styrene have been conducted in three types of industry: production of styrene monomer and styrene polymers, production of glass fibre-reinforced plastic products, and production of styrene–butadiene rubber.

In a European multinational cohort study of workers in the glass fibre-reinforced plastics industry (the largest component of which was Danish), there was no excess mortality from lymphatic and haematopoietic neoplasms in the entire cohort in comparison with the general population, but the results may have been biased by problems with mortality ascertainment in some of the sub-cohorts. In internal analyses, using unexposed cohort members as the comparison group, the risk of lymphatic and haematopoietic neoplasms was significantly increased among exposed workers after more than 20 years since their first exposure to styrene, and increased with increasing intensity of exposure but not with increasing cumulative exposure to styrene.

Another study of cancer incidence in the reinforced-plastics industry conducted in Denmark involved many workers who had been included in the European study. Overall, a small and non-significant excess of leukaemia was observed. A significant excess of leukaemia was observed among workers employed before 1971 (the period with the highest styrene exposures). There was also a significant increase in the incidence of leukaemia when attention was restricted to the follow-up period after 10 years since first exposure to styrene, but only among workers with very short employment (less than 1 year). There was evidence that the available data on duration underestimated the true duration of exposure for many of the workers. (This also applies to the European cohort since the Danish cohort constituted a large fraction of this cohort).

A large study of workers exposed to styrene in the reinforced-plastics industry in the USA found no overall excess of lymphatic and haematopoietic neoplasms.

Two studies of chemical workers in the USA and the United Kingdom involved in the production of styrene and styrene derivatives found a weak association between exposure to styrene and lymphatic and haematopoietic cancers. Styrene exposures were poorly documented in these studies, and exposures to several other chemicals may have occurred.

A follow-up study of cancer incidence in Finnish workers biologically monitored for occupational exposures to styrene during the 1970s and early 1980s did not show any increase in risk for lymphatic and haematopoietic neoplasms. The relatively small size of this study and the low exposures of workers detracted from its power to detect an effect of the magnitude found in some of the other studies.

A small excess of leukaemia mortality has been reported in studies of styrene–butadiene workers in the USA. This excess increased with cumulative exposure to styrene in analyses that only considered this exposure; however, in analyses that included 1,3-butadiene, the exposure–response relationship became non-monotonic. Interpretation of the findings from this study is hampered by the high correlation between styrene and 1,3-butadiene exposures, which makes it difficult to disentangle the effects of these two exposures.

There have also been reports of increased risks of rectal, pancreatic and nervous system cancers in some of the cohort and case–control studies. The numbers of cases were quite small in these studies, and most of the larger cohort studies have not yielded similar findings. Many of the cohort studies did not examine these sites in detail.

The studies of glass fibre-reinforced plastics workers are the most informative with regard to the hypothesis that styrene exposure is associated with an increased risk of cancer in humans. This is because these workers had higher styrene exposures and less potential for exposure to other substances than the other cohorts studied. On the other hand, they are hampered by the high mobility of this workforce. In the overlapping European and Danish studies, a small excess of lymphatic and haematopoietic neoplasms was found, particularly in subgroup analyses of workers with relatively high exposures and a sufficiently long time (e.g., > 10 years) since first exposure. There was no relationship between lymphatic and haematopoietic neoplasms and cumulative styrene exposure, but these studies had problems in accurately estimating duration of employment and hence cumulative exposures.

The increased risks for lymphatic and haematopoietic neoplasms observed in some of the studies are generally small, statistically unstable and often based on subgroup analyses. These findings are not very robust and the possibility that the observations are the results of chance, bias or confounding by other occupational exposures cannot be ruled out.

5.3 Animal carcinogenicity data

Styrene was tested for carcinogenicity in mice in one inhalation study and four oral gavage studies. In the inhalation study, in male mice there was an increase in the incidence of pulmonary adenomas and in female mice, there was an increase in the incidence of pulmonary adenomas, and only an increase in that of carcinomas in the high-dose group. Two of the gavage studies were negative. The other two were considered inadequate for an evaluation of the carcinogenicity of styrene. A screening study by intraperitoneal administration also did not find an increase in tumour incidence or multiplicity in mice.

Styrene was tested for carcinogenicity in rats in four gavage studies, one drinking-water study and two inhalation studies. Overall, there was no reliable evidence for an increase in tumour incidence in rats.

Styrene 7,8-oxide is a major metabolite of styrene and has been evaluated previously (IARC, 1994b). The evaluation at that time was that there was *sufficient evidence* in experimental animals for the carcinogenicity of styrene 7,8-oxide.

5.4 Other relevant data

Styrene is absorbed following exposure via inhalation, dermal contact and orally in humans and laboratory animals. In humans, approximately 70% of the inhaled dose is absorbed. Styrene is distributed throughout the body, with the highest concentration generally found in adipose tissue. There are both quantitative and qualitative interspecies differences in styrene metabolism. In humans, styrene is metabolized primarily via the styrene 7,8-oxide pathway to be excreted in the urine as mandelic and phenylglyoxylic acids. In rodents, but not in humans, glutathione conjugation of styrene 7,8-oxide to form mercapturic acids is an important metabolic pathway. Metabolism of styrene to 1- and 2-phenylethanol and then to phenylacetaldehyde and finally to phenylacetic, phenylaceturic and hippuric acids is more important in animals than in humans.

CYP2E1 and CYP2F are the most important cytochrome P450 enzymes in rodents and humans responsible for the metabolism of styrene to styrene 7,8-oxide. In addition, CYP2B6 may be important in humans. *In vitro*, the rates of metabolism of styrene to styrene 7,8-oxide are much higher in mouse lung than in rat or human lung.

Occupational styrene exposure causes central and peripheral nervous system effects in humans. It causes a reversible decrease in colour discrimination and in some studies effects on hearing have been reported. Studies of effects of styrene on the haematopoietic and immune systems, liver and kidney in exposed workers did not reveal consistent changes.

Central nervous system effects of styrene were reported in rats, guinea-pigs and rabbits. Styrene exposure causes liver and lung toxicity in mice and nasal toxicity in rats and mice.

In humans, there is no evidence for an association between workplace exposure to styrene and spontaneous abortions, malformations or decreased male fecundity.

In rats, there is some evidence for reduced sperm count and peripubertal animals may be more sensitive than adult animals. Styrene crosses the placenta in rats and mice. It increases prenatal death at dose levels causing decreased maternal weight gain. Decreased pup weight, postnatal developmental delays as well as neurobehavioural and neurochemical abnormalities have been reported in rats exposed to styrene during pre- or postnatal development. In-vitro studies indicate that the potential for developmental toxicity is much higher for styrene 7,8-oxide than for styrene.

Occupational exposure to styrene leads to formation of O^6-deoxyguanosine (O^6-(2-hydroxy-1-phenylethyl)-2′-deoxyguanosine-3′-monophosphate) and *N*7-deoxyguanosine adducts in DNA. Low levels of these two adducts were also detected in liver of mice and rats exposed to styrene.

Inconsistent results have been reported for chromosomal aberrations, micronuclei and sister chromatid exchange in approximately 30 studies of workers exposed to styrene in various industries. These studies were predominantly from the reinforced-plastics industry where styrene exposure is high, but there was no indication of a dose–response relationship in any of the studies reporting positive results. Induction of chromosomal aberrations was reported in 12 of 25 studies, sister chromatid exchange in six of 16 studies and micronuclei in three of 14 studies.

Sister chromatid exchange and to a lesser degree chromosomal aberrations were induced in rodents *in vivo* and consistently in human lymphocytes *in vitro*.

Styrene was predominantly inactive in assays for gene mutations in bacteria, although some studies reported mutations in the presence of a metabolic activation system.

Data from both laboratory (*in vitro* and *in vivo*) and human studies indicate that styrene exposure can result in low levels of DNA adducts and DNA damage in individuals who possess the capacity to activate styrene metabolically to styrene 7,8-oxide. However, as noted above, mice, but not rats, develop lung tumours following styrene exposure, even though both species form DNA adducts. DNA adducts are also found in organs other than the lung. Circulating styrene 7,8-oxide may also play a role. However, the concentration in rat blood is two orders of magnitude higher than in the mouse.

The lung tumours in mice probably develop as a result of in-situ formation of styrene 7,8-oxide which causes cytotoxicity and increased cell proliferation, but the roles of circulating styrene 7,8-oxide and of DNA adducts cannot be discounted. Based on metabolic considerations, it is likely that the proposed mechanism involving metabolism of styrene to styrene 7,8-oxide in mouse Clara cells is not operative in human lungs to a biologically significant extent. However, based on the observations in human workers regarding blood styrene 7,8-oxide, DNA adducts and chromosomal damage, it cannot be excluded that this and other mechanisms are important for other organs.

5.5 Evaluation[1]

There is *limited evidence* in humans for the carcinogenicity of styrene.

There is *limited evidence* in experimental animals for the carcinogenicity of styrene.

Overall evaluation

Styrene is *possibly carcinogenic to humans (Group 2B).*

6. References

A.D. Little, Inc. (1981) *Industrial Hygiene Evaluation of Retrospective Mortality Study Plants*, Boston

American Conference of Governmental Industrial Hygienists (2000) *TLVs® and Other Occupational Exposure Values — 2000 CD-ROM*, Cincinnati, OH

American Conference of Governmental Industrial Hygienists (2001) *2001 TLVs® and BEIs® — Threshold Limit Values for Chemical Substances and Physical Agents and Biological Exposure Indices*, Cincinnati, OH, pp. 53, 91

Anttila, A., Pukkala, E., Riala, R., Sallmen, M. & Hemminki, K. (1998) Cancer incidence among Finnish workers exposed to aromatic hydrocarbons. *Int. Arch. occup. environ. Health*, **71**, 187–193

Anwar, W.A. & Shamy, M.Y. (1995) Chromosomal aberrations and micronuclei in reinforced plastics workers exposed to styrene. *Mutat. Res.*, **327**, 41–47

Apostoli, P., Alessandro, G., Placidi, D. & Alessio, L. (1998) Metabolic interferences in subjects occupationally exposed to binary styrene-acetone mixtures. *Int. Arch. occup. environ. Health*, **71**, 445–452

Arfini, G., Mutti, A., Vescovi, P., Ferroni, C., Ferrarri, M., Giaroli, C., Passeri, M. & Franchini, I. (1987) Impaired dopaminergic modulation of pituitary secretion in workers occupationally exposed to styrene: Further evidence for PRL response to TRH stimulation. *J. occup. Med.*, **29**, 826–830

Arnedo-Pena, A., Bellido-Blasco, J.B., Villamarin-Vazquez, J.L., Aranda-Mares, J.L., Font-Cardona, N., Gobba, F. & Kogevinas, M. (2002) Acute health effects after accidental exposure to styrene from drinking water in Spain. *Scand. J. Work Environ. Health* (in press)

[1] Two Working Group members, Dr Carlson and Dr Cruzan, recused themselves from the final discussion and the evaluation of styrene.

Artuso, M., Angotzi, G., Bonassi, S., Bonatti, S., De Ferrari, M., Gargano, D., Lastrucci, L., Miligi, L., Sbrana, C. & Abbondandolo, A. (1995) Cytogenetic biomonitoring of styrene-exposed plastic boat builders. *Arch. environ. Contam. Toxicol.*, **29**, 270–274

Askergren, A., Allgén, L.-G., Karlsson, C., Lundberg. I. & Nyberg, E. (1981) Studies on kidney function in subjects exposed to organic solvents. I. Excretion of albumin and β-2-micro-globulin in the urine. *Acta med. scand.*, **209**, 479–483

Austin, C.C., Wang, D., Ecobichon, D.J. & Dussault, G. (2001) Characterization of volatile organic compounds in smoke at municipal structural fires. *J. Toxicol. environ. Health*, **A63**, 437–458

Bakke, O.M. & Scheline, R.R. (1970) Hydroxylation of aromatic hydrocarbons in the rat. *Toxicol. appl. Pharmacol.*, **16**, 691–700

Barsan, M.E., Kinnes, G.M. & Blade, L. (1991) *Health Hazard Evaluation Report, Rubbermaid, Inc., Reynolds, IN* (Report No. 90-261-2124), Cincinnati, OH, National Institute for Occupational Safety and Health

Bartolucci, G.B., De Rosa, E., Gori, G.P., Chiesura Corona, P., Perbellini, L. & Brugnone, F. (1986) Biomonitoring of occupational exposure to styrene. *Appl. ind. Hyg.*, **1**, 125–131

Bastlová, T. & Podlutsky, A. (1996) Molecular analysis of styrene oxide-induced hprt mutation in human T-lymphocytes. *Mutagenesis*, **11**, 581–591

Bastlová, T., Vodicka, P., Peterková, K., Hemminki, K. & Lambert, B. (1995) Styrene oxide-induced HPRT mutations, DNA adducts and DNA strand breaks in cultured human lymphocytes. *Carcinogenesis*, **16**, 2357–2362

Belanger, P.L. & Elesh, E. (1980) *Health Hazard Evaluation Determination, Bell Helmets, Inc., Norwalk, CA* (Report No. 79-36-656). Cincinnati, OH, National Institute for Occupational Safety and Health

Beliles, R.P., Butala, J.H., Stack, C.R. & Makris, S. (1985) Chronic toxicity and three-generation reproduction study of styrene monomer in the drinking water of rats. *Fundam. appl. Toxicol.*, **5**, 855–868

Bellander, T., Kogevinas, M., Pannett, B., Bjerk, J.E., Breum, N.O., Ferro, S., Fontana, V., Galassi, C., Jensen, A.A. & Pfäffli, P. (1994) Part II: Historical exposure to styrene in the manufacture of glass-reinforced polyester products in six European countries. In: Kogevinas, M., Ferro, G., Saracci, R., Andersen, A., Bellander, T., Biocca, M., Bjerk, J.E., Breum, N.O., Coggon, D., Fontana, V., Ferro, S., Galassi, C., Gennaro, V., Hutchings, S., Jensen, A.A., Kolstad, H., Lundberg, I., Lynge, E., Pannett, B., Partanen, T. & Pfäffli, P., *IARC Historical Multicentric Cohort Study of Workers Exposed to Styrene. Report of the Epidemiological Study and the Industrial Hygiene Investigation* (IARC Internal Technical Report 94/002), Lyon, IARC, pp. 63–127

Belvedere, G., Cantoni, L., Facchinetti, T. & Salmona, M. (1977) Kinetic behaviour of microsomal styrene monooxygenase and styrene epoxide hydratase in different animal species. *Experientia*, **33**, 708–709

Bergamaschi, E., Smargiassi, A., Mutti, A., Franchini, I. & Lucchini R. (1995) Immunological changes among workers occupationally exposed to styrene. *Int. Arch. occup. environ. Health*, **67**, 165–171

Bergamaschi, E., Mutti, A., Cavazzini, S., Vettori, M.V., Renzulli, F.S. & Franchini, I. (1996) Peripheral markers of neurochemical effects among styrene-exposed workers. *Neurotoxicology*, **17**, 753–759

Bergamaschi, E., Smargiassi, A., Mutti, A., Cavazzini, S., Vettori, M.V., Alinovi, R., Fanchini, I. & Mergler, D. (1997) Peripheral markers of catecholaminergic dysfunction and symptoms of neurotoxicity among styrene-exposed workers. *Int. Arch. occup. environ. Health*, **69**, 209–214

Berode, M., Droz, P.-O. & Guillemin, M. (1985) Human exposure to styrene. VI. Percutaneous absorption in human volunteers. *Int. Arch. occup. environ. Health,* **55**, 331–336

Bigbee, W.L., Grant, S.G., Langlois, R.G., Jensen, R.H., Anttila, A., Pfäffli, P., Pekari, K. & Norppa, H. (1996) Glycophorin A somatic cell mutation frequencies in Finnish reinforced plastics workers exposed to styrene. Cancer Epidemiol. *Biomarkers Prev.*, **5**, 801–810

Bjørge, C., Brunborg, G., Wiger, R., Holme, J.A., Scholz, T., Dybing, E. & Søderlund, E.J. (1996) A comparative study of chemically induced DNA damage in isolated human and rat testicular cells. *Reprod. Toxicol.*, **10**, 509–519

Block, J.B. (1976) A Kentucky study: 1950–1975. In: Ede, L., ed., *Proceedings of NIOSH Styrene-Butadiene Rubber Briefing, Covington, Kentucky, April 30, 1976* (HEW Publ. No. (NIOSH) 77-129), Cincinnati, OH, National Institute for Occupational Safety and Health, pp. 28–32

Bonassi, S., Montanaro, F., Ceppi, M. & Abbondandolo, A. (1996) Is human exposure to styrene a cause of cytogenetic damage? A re-analysis of the available evidence. *Biomarkers*, **1**, 217–225

Bond, J.A. (1989) Review of the toxicology of styrene. *Crit. Rev. Toxicol.*, **19**, 227–249

Bond, G.G., Bodner, K.M., Olsen, G.W. & Cook, R.R. (1992) Mortality among workers engaged in the development or manufacture of styrene-based products — An update. *Scand. J. Work Environ. Health*, **18**, 145–154

Boogaard, P.J., de Kloe, K.P., Sumner, S.C.J., van Elburg, P.A. & Wong, B.A. (2000a) Disposition of [ring-U-^{14}C]styrene in rats and mice exposed by recirculating nose-only inhalation. *Toxicol. Sci.*, **58**, 161–172

Boogaard, P.J., de Kloe, K.P., Wong, B.A., Sumner, S.C.J., Watson, W.P. & van Sittert, N.J. (2000b) Quantification of DNA adducts formed in liver, lungs, and isolated lung cells of rats and mice exposed to ^{14}C-styrene by nose-only inhalation. *Toxicol. Sci.*, **57**, 203–216

Bouscaren, R., Frank, R. & Veldt, C. (1987) *Hydrocarbons. Identification of Air Quality Problems in Member States of the European Communities* (Report No. EUR-10646-EN; US NTIS PB88-187992), Luxembourg, European Commission

Bowman, J.D., Held, J.L. & Factor, D.R. (1990) A field evaluation of mandelic acid in urine as a compliance monitor for styrene exposure. *Appl. occup. environ. Hyg.*, **5**, 526–535

Brodkin, C.A., Moon, J.-D., Camp, J., Echeverria, D., Redlich, C.A., Willson, R.A. & Checkoway, H. (2001) Serum hepatic biochemical activity in two populations of workers exposed to styrene. *Occup. environ. Med.*, **58**, 95–102

Brooks, S.M., Anderson, L., Emmett, E., Carson, A., Tsay, J.-Y., Elia, V., Buncher, R. & Karbowsky, R. (1980) The effects of protective equipment on styrene exposure in workers in the reinforced plastics industry. *Arch. environ. Health*, **35**, 287–294

Brown, N.A., Lamb, J.C., Brown, S.M. & Neal, B.H. (2000) A review of the developmental and reproductive toxicity of styrene. *Regul. Toxicol. Pharmacol.*, **32**, 228–247

Brown-Woodman, P.D., Webster, W.S., Picker, K. & Huq, F. (1994) In vitro assessment of individual and interactive effects of aromatic hydrocarbons on embryonic development of the rat. *Reprod. Toxicol.*, **8**, 121–135

Brugnone, F., Perbellini, L., Wang, G.Z., Maranelli, G., Raineri, E., De Rosa, E., Saletti, C., Soave, C. & Romeo, L. (1993) Blood styrene concentrations in a 'normal' population and in exposed workers 16 hours after the end of the workshift. *Int. Arch. occup. environ. Health*, **65**, 125–130

Brunnemann, K.D., Rivenson, A., Cheng, S.C., Saa, V. & Hoffmann, D. (1992) A study of tobacco carcinogenesis. XLVII. Bioassays of vinylpyridines for genotoxicity and for tumorigenicity in A/J mice. *Cancer Lett.*, **65**, 107–113

Buckpitt, A., Chang, A.M., Weir, A., Van Winkle, L., Duan, X., Philpot, R. & Plopper, C. (1995) Relationship of cytochrome P450 activity to Clara cell cytotoxicity. IV. Metabolism of naphthalene and naphthalene oxide in microdissected airways from mice, rats, and hamsters. *Mol. Pharmacol.*, **47**, 74–81

Burroughs, G.E. (1979) *Health Hazard Evaluation Determination. Piper Aircraft Corporation, Vero Beach, FL* (Report No. 78-110-585), Cincinnati, OH, National Institute for Occupational Safety and Health

Cal/OSHA (1986) *Occupational Exposure to Styrene and Other Health Hazards in the Fiberglass Reinforced Plastics Industry*, State of California, Sacramento, Division of Occupational Safety and Health

Campagna, D., Mergler, D., Huel, G., Bélanger, S., Truchon, G., Ostiguy, C. & Drolet, D. (1995) Visual dysfunction among styrene-exposed workers. *Scand. J. Work Environ. Health*, **21**, 382–390

Campagna, D., Gobba, F., Mergler, D., Moreau, T., Galassi, C., Cavalleri, A. & Huel, G. (1996) Color vision loss among styrene-exposed workers neurotoxicological threshold assessment. *Neurotoxicology*, **17**, 367–373

Carlson, G.P. (1997a) Effects of inducers and inhibitors on the microsomal metabolism of styrene to styrene oxide in mice. *J. Toxicol. environ. Health*, **51**, 477–488

Carlson, G.P. (1997b) Comparison of mouse strains for susceptibility to styrene-induced hepatotoxicity and pneumotoxicity. *J. Toxicol. environ. Health*, **51**, 177–187

Carlson, G.P., Hynes, D.E. & Mantick, N.A. (1998) Effects of inhibitors of CYP1A and CYP2B on styrene metabolism in mouse liver and lung microsomes. *Toxicol. Lett.*, **98**, 131–137

Carlson, G.P., Mantick, N.A. & Powley, M.W. (2000) Metabolism of styrene by human liver and lung. *J. Toxicol. environ. Health*, **A59**, 591–595

Carlson, G.P., Perez Rivera, A.A. & Mantick, N.A. (2001) Metabolism of the styrene metabolite 4-vinylphenol by rat and mouse liver and lung. *J. Toxicol. environ. Health A.*, **63**, 541–551

Cerný, S., Mráz, J., Flek, J. & Tichý, M. (1990) Effect of ethanol on the urinary excretion of mandelic and phenylglyoxylic acids after human exposure to styrene. *Int. Arch. occup. environ. Health*, **62**, 243–247

Chakrabarti, S.K. (2000) Altered regulation of dopaminergic activity and impairment in motor function in rats after subchronic exposure to styrene. *Pharmacol. Biochem. Behav.*, **66**, 523–532

Chakrabarti, S. & Malick, M.A. (1991) In vivo nephrotoxic action of an isomeric mixture of *S*-(1-phenyl-2-hydroxyethyl) glutathione and *S*-(2-phenyl-2-hydroxyethyl) glutathione in Fischer-344 rats. *Toxicology*, **67**, 15–27

Chakrabarti, S.K. & Tuchweber, B. (1987) Effects of various pretreatments on the acute nephrotoxic potential of styrene in Fischer-344 rats. *Toxicology*, **46**, 343–356

Chakrabarti, S.K., Labelle, L. & Tuchweber, B. (1987) Studies on the subchronic nephrotoxic potential of styrene in Sprague-Dawley rats. *Toxicology*, **44**, 355–365

Chakrabarti, S., Duhr, M.-A., Senécal-Quevillon, M. & Richer, C.-L. (1993) Dose-dependent genotoxic effects of styrene on human blood lymphocytes and the relationship to its oxidative and metabolic effects. *Environ. mol. Mutag.*, **22**, 85–92

Chakrabarti, S., Zhang, X.-X. & Richer, C.L. (1997) Influence of duration of exposure to styrene oxide on sister chromatid exchanges and cell-cycle kinetics in cultured human blood lymphocytes in vitro. *Mutat. Res.*, **395**, 37–45

Checkoway, H. & Williams, T.M. (1982) A hematology survey of workers at a styrene-butadiene synthetic rubber manufacturing plant. *Am. ind. Hyg. Assoc. J.*, **43**, 164–169

Checkoway, H., Costa, L.G., Camp, J., Coccini, T., Daniell, W.E. & Dills, R.L. (1992) Peripheral markers of neurochemical function among workers exposed to styrene. *Br. J. ind. Med.*, **49**, 560–565

Checkoway, H., Echeverria, D., Moon, J.-D., Heyer, N. & Costa, L.G. (1994) Platelet monoamine oxidase B activity in workers exposed to styrene. *Int. Arch. occup. environ. Health*, **66**, 359–362

Chemical Information Services (2001) *Directory of World Chemical Producers 2001.2 Edition*, Dallas, TX [CD-ROM]

Chen, S.S. (1997) Styrene. In: Kroschwitz, J.I. & Howe-Grant, M., eds, *Kirk-Othmer Encyclopedia of Chemical Technology*, 4th Ed., Vol. 22, New York, John Wiley & Sons, pp. 956–994

Cherry, N. & Gautrin, D. (1990) Neurotoxic effects of styrene: Further evidence. *Br. J. ind. Med.*, **47**, 29–37

Cherry, N., Waldron, H.A., Wells, G.G., Wilkinson, R.T., Wilson, H.K. & Jones, S. (1980) An investigation of the acute behavioural effects of styrene on factory workers. *Br. J. ind. Med.*, **37**, 234–240

Chevron Phillips Chemical Co. (1996) *Technical Data Sheet: Styrene Monomer*, Houston, TX

Chevron Phillips Chemical Co. (2001) *Technical Bulletin: Styrene Monomer*, Houston, TX

Chia, S.-E., Jeyaratnam, J., Ong, C.-N., Ng., T.-P. & Lee, H.-S. (1994) Impairment of color vision among workers exposed to low concentrations of styrene. *Am. J. ind. Med.*, **26**, 481–488

Chmielewski, J. & Renke, W. (1976) Clinical and experimental research into the pathogenesis of toxic effects of styrene. III. Morphology, coagulation and fibronolysis systems of the blood in persons exposed to the action of styrene during their work. *Bull. Inst. mar. trop. Med. Gdynia*, **27**, 63–68

Cho, S.-I., Damokosh, A.I., Ryan, L.M., Chen, D., Hu, Y.A., Smith, T.J., Christiani. D.C. & Xu, X. (2001) Effects of exposure to organic solvents on menstrual cycle length. *J. occup. environ. Med.*, **43**, 567–575

Christakopoulos, A., Bergmark, E., Zorcec, V., Norppa, H., Mäki-Paakkanen, J. & Osterman-Golkar, S. (1993) Monitoring occupational exposure to styrene from hemoglobin adducts and metabolites in blood. *Scand. J. Work Environ. Health*, **19**, 255–263

Chua, S.C., Lee, B.L., Liau, L.S. & Ong, C.N. (1993) Determination of mandelic acid and phenylglyoxylic acid in the urine and its use in monitoring of styrene exposure. *J. anal. Toxicol.*, **17**, 129–132

Churchill, J.E., Ashley, D.L. & Kaye, W.E. (2001) Recent chemical exposures and blood volatile organic compound levels in a large population-based sample. *Arch. environ. Health*, **56**, 157–166

Coccini, T., Maestri, L., Robustelli della Cuna, F.S., Bin, L., Costa, L.G. & Manzo, L. (1996) Urinary mercapturic acid diastereoisomers in rats subchronically exposed to styrene and ethanol. *Arch. Toxicol.*, **70**, 736–741

Coccini, T., Fenoglio, C., Nano, R., De Piceis Polver, P., Moscato, G. & Manzo, L. (1997) Styrene-induced alterations in the respiratory tract of rats treated by inhalation or intraperitoneally. *J. Toxicol. environ. Health*, **52**, 63–77

Cocheo, V., Bellomo, M.L. & Bombi, G.G. (1983) Rubber manufacture: Sampling and identification of volatile pollutants. *Am. ind. Hyg. Assoc. J.*, **44**, 521–527

Coggon, D., Osmond, C., Pannett, B., Simmonds, S., Winter, P.D. & Acheson, E.D. (1987) Mortality of workers exposed to styrene in the manufacture of glass-reinforced plastics. *Scand. J. Work Environ. Health*, **13**, 94–99

Cohen, J.T., Carlson, G., Charnley, G., Coggon, D., Delzell, E., Graham, J.D., Greim, H., Krewski, D., Medinsky, M., Monson, R., Paustenbach, D., Petersen, B., Rappaport, S., Rhomberg, L., Ryan, P.B. & Thompson, K. (2002) A comprehensive evaluation of the potential health risks associated with occupational and environmental exposure to styrene. *J. Toxicol. environ. Health*, **5**, 1–265

Cole, P., Delzell, E. & Acquavella, J. (1993) Exposure to butadiene and lymphatic and hematopoietic cancer. *Epidemiology*, **4**, 96–103

Collins, D.E. & Richey, F.A., Jr (1992) Synthetic organic chemicals. In: Kent, J.A., ed., *Riegel's Handbook of Industrial Chemistry*, 9th Ed., New York, Van Nostrand Reinhold, pp. 800–862

Conti, B., Maltoni, C., Perino, G. & Ciliberti, A. (1988) Long-term carcinogenicity bioassays on styrene administered by inhalation, ingestion and injection and styrene oxide administered by ingestion in Sprague-Dawley rats and *para*-methylstyrene administered by ingestion in Sprague-Dawley rats and Swiss mice. *Ann. N.Y. Acad. Sci.*, **534**, 203–234

Costello, R.J. (1983) *Health Hazard Evaluation Report, Chem-Dyne Hazardous Waste Site, Hamilton, OH* (Report No. 83-307-1561), Cincinnati, OH, National Institute for Occupational Safety and Health

Crandall, M.S. & Hartle, R.W. (1985) An analysis of exposure to styrene in the reinforced plastic boat-making industry. *Am. J. ind. Med.*, **8**, 183–192

Crofton, K.M., Lassiter, T.L. & Rebert, C.S. (1994) Solvent-induced ototoxicity in rats: An atypical selective mid-frequency hearing deficit. *Hear. Res.*, **80**, 25–30

Cruzan, G., Cushman, J.R., Andrews, L.S., Granville, G.C., Miller, R.R., Hardy, C.J., Coombs, D.W. & Mullins, P.A. (1997) Subchronic inhalation studies of styrene in CD rats and CD-1 mice. *Fundam. appl. Toxicol.*, **35**, 152–165

Cruzan, G., Cushman, J.R., Andrews, L.S., Granville, G.C., Johnson, K.A., Hardy, C.J., Coombs, D.W., Mullins, P.A. & Brown, W.R. (1998) Chronic toxicity/oncogenicity study of styrene in CD rats by inhalation exposure for 104 weeks. *Toxicol. Sci.*, **46**, 266–281

Cruzan, G., Cushman, J.R., Andrews, L.S., Granville, G.C., Johnson, K.A., Bevan, C., Hardy, C.J., Coombs, D.W., Mullins, P.A. & Brown, W.R. (2001) Chronic toxicity/oncogenicity study of styrene in CD-1 mice by inhalation exposure for 104 weeks. *J. appl. Toxicol.*, **21**, 185–198

Csanády, G.A., Mendrala, A.L., Nolan, R.J. & Filser, J.G. (1994) A physiologic pharmacokinetic model for styrene and styrene-7,8-oxide in mouse, rat and man. *Arch. Toxicol.*, **68**, 143–157

Danishefsky, I. & Willhite, M. (1954) The metabolism of styrene in the rat. *J. biol. Chem.*, **211**, 549–553

Darrall, K.G., Figgins, J.A., Brown, R.D. & Phillips, G.F. (1998) Determination of benzene and associated volatile compounds in mainstream cigarette smoke. *Analyst*, **123**, 1095–1101

De Flora, S. (1979) Metabolic activation and deactivation of mutagens and carcinogens. *Ital. J. Biochem.*, **28**, 81–103

De Flora, S. (1981) Study of 106 organic and inorganic compounds in the Salmonella/microsome test. *Carcinogenesis*, **2**, 283–298

Delbressine, L.P.C., Van Bladeren, P.J., Smeets, F.L.M. & Seutter-Berlage, F. (1981) Stereoselective oxidation of styrene to styrene oxide in rats as measured by mercapturic acid excretion. *Xenobiotica*, **11**, 589–594

Delzell, E., Sathiakumar, N., Hovinga, M., Macaluso, M., Julian, J., Larson, R., Cole, P. & Muir, D.C. (1996) A follow-up study of synthetic rubber workers. *Toxicology*, **113**, 182–189

Delzell, E., Macaluso, M., Sathiakumar, N. & Matthews, R. (2001) Leukemia and exposure to 1,3-butadiene, styrene and dimethyldithiocarbamate among workers in the synthetic rubber industry. *Chem.-biol. Interact.*, **135–136**, 515–534

De Palma, G., Manini, P., Mozzoni, P., Andreoli, R., Bergamaschi, E., Cavazzini, S., Franchini, I., & Mutti, A. (2001) Polymorphism of xenobiotic-metabolizing enzymes and excretion of styrene-specific mercapturic acids. *Chem. Res. Toxicol.*, **14**, 1393–1400

De Rosa, E., Cellini, M., Sessa, G., Saletti, C., Rausa, G., Marcuzzo, G. & Bartolucci, G.B. (1993) Biological monitoring of workers exposed to styrene and acetone. *Int. Arch. occup. environ. Health*, **65**, S107–S110

Deutsche Forschungsgemeinschaft (2001) *MAK and BAT Values 2001* (Report No. 37), Weinheim, VCH Verlagsgesellschaft, pp. 98, 187

Dills, R.L., Wu, R.L., Checkoway, H. & Kalman, D.A. (1991) Capillary gas chromatographic method for mandelic and phenylglyoxylic acids in urine. *Int. Arch. occup. environ. Health*, **62**, 603–606

Dogra, R.K.S., Khanna, S., Srivastava, S.N., Shukla, L.J. & Shanker, R. (1989) Styrene-induced immunomodulation in mice. *Int. J. Immunopharmacol.*, **11**, 577–586

Dogra, R.K.S., Chandra, K., Chandra, S., Gupta, S., Khanna, S., Srivastava, S.N., Shukla, L.J., Katiyar, J.C. & Shanker, R. (1992) Host resistance assays as predicitive models in styrene immunomodulation. *Int. J. Immunopharmacol.*, **14**, 1003–1009

Droz, P.O. & Guillemin, M.P. (1983) Human styrene exposure. V. Development of a model for biological monitoring. Int. *Arch. occup. environ. Health*, **53**, 19–36

Drummond, L., Caldwell, J. & Wilson, H.K. (1989) The metabolism of ethylbenzene and styrene to mandelic acid: Stereochemical considerations. *Xenobiotica*, **19**, 199–207

Ducruet, V. (1984) Comparison of the headspace volatiles of carbonic maceration and traditional wine. *Lebensmittel. Wissenschaft. Technol.*, **17**, 217–221

Duke, J.A. (1985) *CRC Handbook of Medical Herbs*, Boca Raton, FL, CRC Press, p. 323

Dumas, S., Parent, M.E., Siemiatycki, J. & Brisson, J. (2000) Rectal cancer and occupational risk factors: A hypothesis-generating, exposure-based case-control study. *Int. J. Cancer*, **87**, 874–879

Edling, C. & Ekberg, K. (1985) No acute behavioural effects of exposure to styrene: A safe level of exposure? Br. J. ind. Med., **42**, 301–304

Edling, C. & Tagesson, C. (1984) Raised serum bile acid concentrations after occupational exposure to styrene: A possible sign of hepatotoxicity? *Br. J. ind. Med.*, **41**, 257–259

Edling, C., Anundi, H., Johanson, G. & Nilsson, K. (1993) Increase in neuropsychiatric symptoms after occupational exposure to low levels of styrene. *Br. J. ind. Med.*, **50**, 843–850

Eguchi, T., Kishi, R., Harabuchi, I., Yuasa, J., Arata, Y., Katakura, Y. & Miyake, H. (1995) Impaired colour discrimination among workers exposed to styrene: Relevance of a urinary metabolite. *Occup. environ. Med.*, **52**, 534–538

Elia, V.J., Anderson, L.A., MacDonald, T.J., Carson, A., Buncher, C.R. & Brooks, S.M. (1980) Determination of urinary mandelic and phenylglyoxylic acids in styrene-exposed workers and a control population. *Am. ind. Hyg. Assoc. J.*, **41**, 922–926

Eller, P.M., ed. (1984) *NIOSH Manual of Analytical Methods*, 3rd Ed., Vol. 1 (DHHS (NIOSH) Publ. No. 84-100), Washington DC, US Government Printing Office, pp. 1501-1–1501-7

Elovaara, E., Vainio, H. & Aitio, A. (1990) Pulmonary toxicity of inhaled styrene in acetone-phenobarbital- and 3-methylcholanthrene-treated rats. *Arch. Toxicol.*, **64**, 365–369

Elovaara, E., Engström, K., Nakajima, T., Park, S.S., Gelboin, H.V. & Vainio, H. (1991) Metabolism of inhaled styrene in acetone-, phenobarbital- and 3-methylcholanthrene-pretreated rats: Stimulation and stereochemical effects by induction of cytochromes P450IIE1, P450IIB and P450IA. *Xenobiotica*, **21**, 651–661

Engström, J., Åstrand, I. & Wigaeus, E. (1978a) Exposure to styrene in a polymerization plant. Uptake in the organism and concentration in subcutaneous adipose tissue. *Scand. J. Work Environ. Health*, **4**, 324–239

Engström, J., Bjurström, R., Åstrand, I. & Övrum, P. (1978b) Uptake, distribution and elimination of styrene in man — Concentration in subcutaneous adipose tissue. *Scand. J. Work Environ. Health*, **4**, 315–323

Environmental Protection Agency (1996) *Method 8260B. Volatile Organic Compounds by Gas Chromatography/Mass Spectrometry (GC/MS)*, Revision 2, December 1996 [CD-ROM], http://www.epa.gov/epaoswer/hazwaste/test/8260b.pdf

Environmental Protection Agency (1999) *1990 Emissions Inventory of Forty Potential Section 112(k) Pollutants, Final Report*, Research Triangle Park, North Carolina, p. 6-150

Environmental Protection Agency (2001a) *1999 Toxics Release Inventory — Public Data Release (EPA-260/R-01-001)*, Washington, DC, Office of Environmental Information, p. A-134

Environmental Protection Agency (2001b) National primary drinking water regulations. *US Code fed. Regul.*, **Title 40**, Subpart D, Parts 141.32, 141.61, pp. 401, 430–431

Fallas, C., Fallas, J., Maslard, P. & Dally, S. (1992) Subclinical impairment of colour vision among workers exposed to styrene. *Br. J. ind. Med.*, **49**, 679–682

Filser, J.G., Schwegler, U., Csanády, G.A., Greim, H., Kreuzer, P.E. & Kesler, W. (1993) Species-specific pharmacokinetics of styrene in rat and mouse. *Arch. Toxicol.*, **67**, 517–530

Filser, J.G., Kessler, W. & Csanády, G.A. (2002) Estimation of a possible tumorigenic risk of styrene from daily intake via food and ambient air. *Toxicol. Lett.*, **126**, 1–18

Fishbein, L. (1992) Exposure from occupational versus other sources. *Scand. J. Work Environ. Health*, **18** (Suppl. 1), 5–16

Fleeger, A.K. & Almaguer, D. (1988) *Health Hazard Evaluation Report, Polymer Engineering, Inc., Reynolds, IN* (Report No. 87-384-1895), Cincinnati, OH, National Institute for Occupational Safety and Health

Flek, J. & Šedivec, V. (1980) Simultaneous gas chromatographic determination of urinary mandelic and phenylglyoxylic acids using diazomethane derivatization. *Int. Arch. occup. environ. Health*, **45**, 181–188

Flodin, U., Fredriksson, M., Persson, B., Hardell, L. & Axelson, O. (1986) Background radiation, electrical work, and some other exposures associated with acute myeloid leukemia in a case-referent study. *Arch. environ. Health*, **41**, 77–84

Flodin U., Ekberg, K. & Andersson, L. (1989) Neuropsychiatric effects of low exposure to styrene. *Br. J. ind. Med.*, **46**, 805–808

Food and Drug Administration (2000) Total Diet Study. Summary of Residues Found. Ordered by Pesticide. Market Baskets 91-3–991, http://www.cfsan.fda.gov/~acrobat/TDS1byps.pdf

Food and Drug Administration (2001) Food and drugs. *US Code fed. Regul.*, **Title 21**, Parts 172.515, 172.615, 173.20, 173.25, 173.70, 175.105, 175.125, 175.300, 175.320, 176.170, 176.180, 176.300, 177.1010, 177.1020, 177.1030, 177.1040, 177.1050, 177.1200, 177.1210, 177.1430, 177.1630, 177.1635, 177.1640, 177.1810, 177.1820, 177.1830, 177.2260, 177.2420, 177.2600, 177.2710, 177.2800, 178.1005, 178.2010, 178.3480, 178.3610, 178.3790, 181.30, 181.32, pp. 26–432, 443–447

Foureman, G.L., Harris, C., Guengerich, F.P. & Bend, J.R. (1989) Stereoselectivity of styrene oxidation in microsomes and in purified cytochrome P-450 enzymes from rat liver. *J. Pharmacol. exp. Ther.*, **248**, 492–497

Frentzel-Beyme, R., Thiess, A.M. & Wieland, R. (1978) Survey of mortality among employees engaged in the manufacture of styrene and polystyrene at the BASF Ludwigshafen works. *Scand. J. Work Environ. Health*, **4** (Suppl.), 2231–2239

Fujita, H., Koizumi, A., Furusawa, T. & Ikeda, M. (1987) Decreased erythrocyte δ-aminolaevulinate dehydratase activity after styrene exposure. *Biochem. Pharmacol.*, **36**, 711–716

Furia, T.E. & Bellanca, N., eds (1971) *Fenaroli's Handbook of Flavor Ingredients*, Cleveland, OH, Chemical Rubber Co.

Fustinoni, S., Colosio, C., Colombi, A., Lastrucci, L., Yeowell-O'Connell, K. & Rappaport, S.M. (1998) Albumin and hemoglobin adducts as biomarkers of exposure to styrene in fiberglass-reinforced-plastics workers. *Int. Arch. occup. environ. Health*, **71**, 35–41

Gadberry, M.G., DeNicola, D.B. & Carlson, G.P. (1996) Pneumotoxicity and hepatotoxicity of styrene and styrene oxide. *J. Toxicol. environ. Health*, **48**, 273–294

Galassi, C., Kogevinas, M., Ferro, G. & Biocca, M. (1993) Biological monitoring of styrene in the reinforced plastics industry in Emilia Romagna, Italy. *Int. Arch. occup. environ. Health*, **65**, 89–95

Gérin, M., Siemiatycki, J., Desy, M. & Krewski, D. (1998) Associations between several sites of cancer and occupational exposure to benzene, toluene, xylene, and styrene: Results of a case-control study in Montreal. *Am. J. ind. Med.*, **34**, 144–156

Geuskens, R.B., van der Klaauw, M.M., van der Tuin, J. & van Hemmen, J.J. (1992) Exposure to styrene and health complaints in the Dutch glass-reinforced plastics industry. *Ann. occup. Hyg.*, **36**, 47–57

Ghittori, S., Maestri, L., Imbriani, M., Capodaglio, E. & Cavalleri, A. (1997) Urinary excretion of specific mercapturic acids in workers exposed to styrene. *Am. J. ind. Med.*, **31**, 636–644

Gilbert, J. & Startin, J.R. (1983) A survey of styrene monomer levels in foods and plastic packaging by coupled mass spectrometry-automatic headspace gas chromatography. *J. Sci. Food Agric.*, **34**, 647–652

Gobba, F. (2000) Color vision: A sensitive indicator of exposure to neurotoxins. *Neurotoxicology*, **21**, 857–862

Gobba, F., Galassi, C., Imbriani, M., Ghittori, S., Candela, S. & Cavalleri, A. (1991) Acquired dyschromatopsia among styrene-exposed workers. *J. occup. Med.*, **33**, 761–765

Gobba, F., Galassi, C., Ghittori, S., Imbriani, M., Pugliese, F. & Cavalleri, A. (1993) Urinary styrene in the biological monitoring of styrene exposure. *Scand. J. Work Environ. Health*, **19**, 175–182

Gobba, F., Ghittori, S., Imbriani, M. & Cavalleri, A. (2000) Evaluation of half-mask respirator protection in styrene-exposed workers. *Int. Arch. occup. environ. Health*, **73**, 56–60

Grayson, M.H. & Gill, S.S. (1986) Effect of in vitro exposure to styrene, styrene oxide, and other structurally related compounds on murine cell-mediated immunity. *Immunopharmacology*, **11**, 165–173

Green, T., Mainwaring, G.W. & Foster, J.R. (1997) Trichloroethylene-induced mouse lung tumors: Studies of the mode of action and comparisons between species. *Fundam. appl. Toxicol.*, **37**, 125–130

Green, T., Toghill, A. & Foster, J.R. (2001a) The role of cytochromes P-450 in styrene induced pulmonary toxicity and carcinogenicity. *Toxicology*, **169**, 107–117

Green, T., Lee, R., Toghill, A., Meadowcroft, S., Lund, V. & Foster, J. (2001b) The toxicity of styrene to the nasal epithelium of mice and rats: Studies on the mode of action and relevance to humans. *Chem.-biol. Interact.*, **137**, 185–202

Gregotti, C.F., Kirby, Z., Manzo, L., Costa, L.G. & Faustman, E.M. (1994) Effects of styrene oxide on differentiation and viability of rodent embryo cultures. *Toxicol. appl. Pharmacol.*, **128**, 25–35

Guengerich, F.P., Kim, D.-H. & Iwasaki, M. (1991) Role of human cytochrome P-450 IIE1 in the oxidation of many low molecular weight cancer suspects. *Chem. Res. Toxicol.*, **4**, 168–179

Guillemin, M.P. & Bauer, D. (1976) Human exposure to styrene. II. Quantitative and specific gas chromatographic analysis of urinary mandelic and phenylglyoxylic acids as an index of styrene exposure. *Int. Arch. occup. environ. Health*, **37**, 57–64

Guillemin, M.P. & Berode, M. (1988) Biological monitoring of styrene: A review. *Am. ind. Hyg. Assoc. J.*, **49**, 497–505

Guillemin, M.P., Bauer, D., Martin, B. & Marazzi, A. (1982) Human exposure to styrene. IV. Industrial hygiene investigations and biological monitoring in the polyester industry. *Int. Arch. occup. environ. Health*, **51**, 139–150

Gurman, J.L., Baier, L. & Levin, B.C. (1987) Polystyrenes: A review of the literature on the products of thermal decomposition and toxicity. *Fire Mater.*, **11**, 109–130

Hagmar, L., Högstedt, B., Welinder, H., Karlsson, A. & Rassner, F. (1989) Cytogenetic and hematological effects in plastics workers exposed to styrene. *Scand. J. Work Environ. Health*, **15**, 136–141

Hallier, E., Goergens, H.W., Hallier, K. & Bolt, H.M. (1994) Intervention study on the influence of reduction of occupational exposure to styrene on sister chromatid exchanges in lymphocytes. *Int. Arch. occup. environ. Health*, **66**, 167–172

Hallier, E., Goergens, H.W., Karels, H. & Golka, K. (1995) A note on individual differences in the urinary excretion of optical enantiomers of styrene metabolites and of styrene-derived mercapturic acids in humans. *Arch. Toxicol.*, **69**, 300–305

Hampton, C.V., Pierson, W.R., Schuetzle, D. & Harvey, T.M. (1983) Hydrocarbon gases emitted from vehicles on the road. 2. Determination of emission rates from diesel and spark-ignition vehicles. *Environ. Sci. Technol.*, **17**, 699–708

Hansch, C., Leo, A. & Hoekman, D. (1995) *Exploring QSAR — Hydrophobic, Electronic, and Steric Constants*, Washington, DC, Americam Chemical Society, p. 40

Härkönen, H. (1977) Relationship of symptoms to occupational styrene exposure and to the findings of electroencephalographic and psychological examination. *Int. Arch. occup. environ. Health*, **40**, 231–239

Härkönen, H., Lehtniemi, A. & Aitio, A. (1984) Styrene exposure and the liver. *Scand. J. Work Environ. Health*, **10**, 59–61

Harris, C., Philpot, R.M., Hernandez, O. & Bend, J.R. (1986) Rabbit pulmonary cytochrome P-450 monooxygenase system: Isozyme differences in the rate and stereoselectivity of styrene oxidation. *J. Pharmacol. exp. Ther.*, **236**, 144–149

Haufroid, V., Buchet, J.-P., Gardinal, S., Ghittori, S., Imbriani, M., Hirvonen, A. & Lison, D. (2001) Importance of genetic polymorphisms of drug-metabolizing enzymes for the interpretation of biomarkers of exposure to styrene. *Biomarkers*, **6**, 236–249

Hayes, J.P., Lambourn, L., Hopkirk, J.A.C., Durham, S.R. & Newman Taylor, A.J. (1991) Occupational asthma due to styrene. *Thorax*, **46**, 396–397

Health Canada (1993) *Priority Substances List Assessment Report: Styrene*, Ottawa, Minister of Supply and Services Canada

Health Canada (1999) *Determination of Pyridine, Quinoline and Styrene in Mainstream Tobacco Smoke, Health Canada — Official Method*, Ottawa

Hemminki, K. & Vodicka, P. (1995) Styrene: From characterisation of DNA adducts to application in styrene-exposed lamination workers. *Toxicol. Lett.*, **77**, 153–161

Hemminki, K., Franssila, E. & Vainio, H. (1980) Spontaneous abortions among female chemical workers in Finland. *Int. Arch. Occup. Environ. Health*, **45**, 123–126

Herrero, M.E., Arand, M., Hengstler, J.G. & Oesch, F. (1997) Recombinant expression of human microsomal epoxide hydrolase protects V79 Chinese hamster cells from styrene oxide- but not from ethylene oxide-induced DNA strand breaks. *Environ. mol. Mutag.*, **30**, 429–439

Hodgson, J.T. & Jones, R.D. (1985) Mortality of styrene production, polymerization and processing workers at a site in northwest England. *Scand. J. Work Environ. Health*, **11**, 347–352

Hodgson, A.T., Daisey, J.M., Mahanama, K.R.R. & Ten Brinke, J. (1996) Use of volatile tracers to determine the contribution of environmental tobacco smoke to concentrations of volatile organic compounds in smoking environments. *Environ. int.*, **3**, 295–307

Hoff, A., Jacobsson, S., Pfäffli, P., Zitting, A. & Frostling, H. (1982) Degradation products of plastics. Polyethylene and styrene-containing thermoplastics — Analytical, occupational, and toxicologic aspects. *Scand. J. Work Environ. Health*, **8** (Suppl. 2), 1–60

Högstedt, B., Akesson, B., Axell, K., Gullberg, B., Mitelman, F., Pero, R. W., Skerfving, S. & Welinder, H. (1983) Increased frequency of lymphocyte micronuclei in workers producing reinforced polyester resin with low exposure to styrene. *Scand. J. Work Environ. Health*, **9**, 241–246

Holmberg, P.C. (1977) Central nervous defects in two children of mothers exposed to chemicals in the reinforced plastics industry. Chance or a causal relation? *Scand. J. Work Environ. Health*, **3**, 212–214

Holmberg, P.C. (1979) Central-nervous-system defects in children born to mothers exposed to organic solvents during pregnancy. *Lancet*, **ii**, 177–179

Horvath, E., Pongracz, K., Rappaport, S. & Bodell, W.J. (1994) ^{32}P-Post-labeling detection of DNA adducts in monomuclear cells of workers occupationally exposed to styrene. *Carcinogenesis*, **15**, 1309–1315

Hynes, D.E., DeNicola, D.B. & Carlson, G.P. (1999) Metabolism of styrene by mouse and rat isolated lung cells. *Toxicol. Sci.*, **51**, 195–201

IARC (1979) *IARC Monographs on the Evaluation of the Carcinogenic Risk of Chemicals to Humans*, Vol. 19, *Some Monomers, Plastics and Synthetic Elastomers, and Acrolein*, Lyon, IARC*Press*, pp. 231–274

IARC (1982) *IARC Monographs on the Evaluation of the Carcinogenic Risk of Chemicals to Humans*, Vol. 28, *The Rubber Industry*, Lyon, IARC*Press*

IARC (1986) *IARC Monographs on the Evaluation of the Carcinogenic Risk of Chemicals to Humans*, Vol. 38, *Tobacco Smoking*, Lyon, IARC*Press*, pp. 83–126

IARC (1987) *IARC Monographs on the Evaluation of Carcinogenic Risks to Humans*, Suppl. 7, *An Updating of* IARC *Monographs Volumes 1 to 42*, Lyon, IARC*Press*, pp. 345–347

IARC (1994a) *IARC Monographs on the Evaluation of Carcinogenic Risks to Humans*, Vol. 60, *Some Industrial Chemicals*, Lyon, IARC*Press*, pp. 233–320

IARC (1994b) IARC Monographs on the Evaluation of Carcinogenic Risks to Humans, Vol. 60, Some Industrial Chemicals, Lyon, IARCPress, pp. 321–346

IARC (1999) *IARC Monographs on the Evaluation of Carcinogenic Risks to Humans*, Vol. 71, *Re-evaluation of Some Organic Chemicals, Hydrazine and Hydrogen Peroxide (Part One)*, Lyon, IARC*Press*, pp. 109–225

IARC (2003) *IARC Monographs on the Evaluation of Carcinogenic Risks to Humans*, Vol. 83, *Tobacco Smoke and Involuntary Smoking* (in press)

Ikeda, M., Ohtsuji, H. & Imamura, T. (1972) In vivo suppression of benzene and styrene oxidation by co-administered toluene in rats and effects of phenobarbital. *Xenobiotica*, **2**, 101–106

Ikeda, M., Koizumi, A., Miyasaka, M. & Watanabe, T. (1982) Styrene exposure and biologic monitoring in FRP boat production plants. *Int. Arch. occup. environ. Health*, **49**, 325–339

Iziumova, A.S. (1972) [Effect of small concentrations of styrol on the sexual function of female albino rats.] *Gig. Sanit.*, **37**, 29–30 (in Russian)

James, D.H. & Castor, W.M. (1994) Styrene. In: Elvers, B., Hawkins, S. & Russey, W., eds., *Ullmann's Encyclopedia of Industrial Chemistry*, 5th rev. Ed., Vol. A25, New York, VCH Publishers, pp. 329–344

Jankovic, J., Jones, W., Burkhart, J. & Noonan, G. (1991) Environmental study of firefighters. *Ann. occup. Hyg.*, **35**, 581–602

Jégaden, D., Amann, D., Simon, J.F., Habault, M., Legoux, B. & Galopin, P. (1993) Study of the neurobehavioural toxicity of styrene at low levels of exposure. *Int. Arch. occup. environ. Health*, **64**, 527–531

Jelnes, J.E. (1988) Semen quality in workers producing reinforced plastic. *Reprod. Toxicol.*, **2**, 209–212

Jensen, A.A., Breum, N.O., Bacher, J. & Lynge, E. (1990) Occupational exposures to styrene in Denmark 1955-88. *Am. J. ind. Med.*, **17**, 593–606

Johanson, G., Ernstgård, I., Gullstrand, E., Löf, A., Osterman-Golkar, S., Williams, C.C. & Sumner, S.C.J. (2000) Styrene oxide in blood, hemoglobin adducts, and urinary metabolites in human volunteers exposed to $^{13}C_8$-styrene vapors. *Toxicol. appl. Pharmacol.*, **168**, 36–49

Karakaya, A.E., Karahalil, B., Yilmazer, M., Aygün, N., Sardas, S. & Burgaz, S. (1997) Evaluation of genotoxic potential of styrene in furniture workers using unsaturated polyester resins. *Mutat. Res.*, **392**, 261–268

Karbowski, R.J. & Braun, W.H. (1978) Quantitative determination of styrene in biological samples and expired air by gas chromatography–mass spectrometry (selected ion monitoring). *J. Chromatogr.*, **160**, 141–145

Katakura, Y., Kishi, R., Ikeda, T. & Miyake, H. (2001) Effects of prenatal styrene exposure on postnatal development and brain serotonin and catecholamine levels in rats. *Environ. Res.*, **85**, 41–47

Katoh, T., Higashi, K. & Inoue, N. (1989) Sub-chronic effects of styrene and styrene oxide on lipid peroxidation and the metabolism of glutathione in rat liver and brain. *J. toxicol. Sci.*, **14**, 1–9

Kawai, T., Yasugi, T., Mizunuma, K., Horiguchi, S., Morioka, I., Miyashita, K., Uchida, Y. & Ikeda, M. (1992) Monitoring of workers exposed to a mixture of toluene, styrene and methanol vapours by means of diffusive air sampling, blood analysis and urine analysis. *Int. Arch. occup. environ. Health*, **63**, 429–435

Kawai, T., Mizunuma, K., Yasugi, T., Horiguchi, S., Moon, C.-S., Zhang, Z.-W., Miyashita, K., Takeda, S. & Ikeda, M. (1995) Effects of methanol on styrene metabolism among workers occupationally exposed at low concentrations. *Arch. environ. Contam. Toxicol.*, **28**, 543–546

Khanna, S., Rao, G.S., Dogra, R.K.S., Shukla, L.J., Srivastava, S.N., Dhruv, S.P. & Shanker, R. (1994) Styrene induced pancreatic changes in rodents. *Indian J. exp. Biol.*, **32**, 68–71

Kim, H., Wang, R.S., Elovaara, E., Raunio, H., Pelkonen, O., Aoyama, T., Vainio, H. & Nakajima, T. (1997) Cytochrome P450 isozymes responsible for the metabolism of toluene and styrene in human liver microsomes. *Xenobiotica*, **27**, 657–665

Kingsley, I. (1976) *Health Hazard Evaluation Determination, New York Telephone and Telegraph Company, NY* (Report No. 75-178-295), Cincinnati, OH, National Institute for Occupational Safety and Health

Kishi, R., Eguchi, T., Yuasa, J., Katakura, Y., Arata,Y., Harabuchi, I., Kawai, T. & Masuchi, A. (2001). Effects of low-level occupational exposure to styrene on color vision: Dose relation and urinary metabolite. *Environ. Res.*, **A85**, 25–30

Knöppel, H. & Schauenburg, H. (1989) Screening of household products for the emission of volatile organic compounds. *Environ. int.*, **15**, 413–418

Kogevinas, M., Ferro, G., Saracci, R., Andersen, A., Lynge, E. & Partanen, T., Biocca, M., Coggon, D., Gennaro, V., Hutchings, S., Kolstad, H., Lundberg, I., Lynge, E. & Partanen, T. (1993) Cancer mortality in an international cohort of workers exposed to styrene. In: Sorsa, M., Peltonen, K., Vainio, H. & Hemminki, K. eds, *Butadiene and Styrene: Assessment of Health Hazards* (IARC Scientific Publications No. 127), Lyon, IARCPress, pp. 289–300

Kogevinas, M., Ferro, G., Andersen, A., Bellander, T., Biocca, M., Coggon, D., Gennaro, V., Hutchings, S., Kolstad, H., Lundberg, I., Lynge, E., Partanen, T. & Saracci, R. (1994a) Cancer mortality in a historical cohort study of workers exposed to styrene. *Scand. J. Work Environ. Health*, **20**, 249–259

Kogevinas, M., Ferro, G., Saracci, R., Andersen, A., Bellander, T., Biocca, M., Bjerk, J.E., Breum, N.O., Coggon, D., Fontana, V., Ferro, S., Galassi, C., Gennaro, V., Hutchings, S., Jensen, A.A., Kolstad, H., Lundberg, I., Lyne, E., panett, B., partanen, T. & Pfäffli, P. (1994b) *IARC Historical Multicentric Cohort Study of Workers Exposed to Styrene. Report of the Epidemio-*

logical Study and the Industrial Hygiene Investigation (IARC internal Technical Report 94/002), Lyon, IARC

Kohn, J., Minotti, S. & Durham, H. (1995) Assessment of the neurotoxicity of styrene, styrene oxide, and styrene glycol in primary cultures of motor and sensory neurons. *Toxicol. Lett.*, **75**, 29–37

Kolstad, H.A., Lynge, E. & Olsen, J. (1993) Cancer incidence in the Danish reinforced plastics industry. In: Sorsa, M., Peltonen, K., Vainio, H. & Hemminki, K., eds, *Butadiene and Styrene: Assessment of Health Hazards* (IARC Scientific Publications No. 127), Lyon, IARCPress, pp. 301–308

Kolstad, H.A., Lynge, E., Olsen, J. & Breum, N. (1994) Incidence of lymphohematopoietic malignancies among styrene-exposed workers of the reinforced plastics industry. *Scand. J. Work Environ. Health*, **20**, 272–278

Kolstad, H.A., Juel, K., Olsen, J. & Lynge, E. (1995) Exposure to styrene and chronic health effects: Mortality and incidence of solid cancers in the Danish reinforced plastics industry. *Occup. environ. Med.*, **52**, 320–327

Kolstad, H.A., Bonde, J.P., Spano, M., Giwercman, A., Zschiesche, W., Kaae, D., Larsen, S.B. & Roeleveld, N. (1999) Change in semen quality and sperm chromatin structure following occupational styrene exposure. *Int. Arch. occup. environ. Health*, **72**, 135–141

Kolstad, H.A., Bisanti, L., Roeleveld, N., Baldi, R., Bonde, J.P. & Joffe, M. (2000) Time to pregnancy among male workers of the reinforced plastics industry in Denmark, Italy and The Netherlands. *Scand. J. Work. Environ. Health*, **26**, 353–358

Korn, M., Wodarz, R., Schoknecht, W., Weichardt, H.. & Bayer, E. (1984) Styrene metabolism in man: Gas chromatographic separation of mandelic acid enantiomers in the urine of exposed persons. *Arch. Toxicol.*, **55**, 59–63

Korn, M., Gfrörer, W., Filser, J.G. & Kessler, W. (1994) Styrene-7,8-oxide in blood of workers exposed to styrene. *Arch. Toxicol.*, **68**, 524–527

Koskinen, M., Vodicka, P. & Hemminki, K. (2000) Adenine N3 is a main alkylation site of styrene oxide in double-stranded DNA. *Chem.-biol. Interact.*, **124**, 13–27

Kronoveter, K.J. & Boiano, J.M. (1984a) *Health Hazard Evaluation Report, Charlie's Taxidermy and Gifts, Fleetwood, PA* (Report No. 83-276-1499), Cincinnati, OH, National Institute for Occupational Safety and Health

Kronoveter, K.J. & Boiano, J.M. (1984b) *Health Hazard Evaluation Report, Pennsylvania Institute of Taxidermy, Ebensburg, PA* (Report No. 84-322-1502), Cincinnati, OH, National Institute of Occupational Safety and Health

Kulig, B.M. (1989) The neurobehavioral effects of chronic styrene exposure in the rat. *Neurotoxicol. Teratol.*, **10**, 511–517

Laffon, B., Pásaro, E. & Méndez, J. (2001) Genotoxic effects of styrene-7,8-oxide in human white blood cells: Comet assay in relation to the induction of sister-chromatid exchange and micronuclei. *Mutat. Res.*, **491**, 163–172

Lai, S.-T. & Locke, D.C. (1983) Stepwise pyrolysis-liquid chromatography and pyrolysis-gas chromatography of polystyrene. *J. Chromatogr.*, **255**, 511–527

LaRegina, J., Bozzelli, J.W., Harkov, J. & Gianti, S. (1986) Volatile organic compounds at hazardous waste sites and a sanitary landfill in New Jersey. An up-to-date review of the present situation. *Environ. Progress*, **5**, 18–27

Lattime, R.R. (1997) Styrene–butadiene rubber. In: Kroschwitz, J.I. & Howe-Grant, M., eds, *Kirk-Othmer Encyclopedia of Chemical Technology*, 4th Ed., Vol. 22, New York, John Wiley & Sons, pp. 994–1014

Lauwerys, R.R. & Hoet, P. (1993) *Industrial Chemical Exposure: Guidelines for Biological Monitoring*, 2nd Ed., Boca Raton, FL, Lewis Publishers, pp. 143–159

Law, R.J., Fileman, T.W. & Matthiessen, P. (1991) Phthalate esters and other industrial organic chemicals in the North and Irish seas. *Water Sci. Technol.*, **24**, 127–134

Lee, S.-H. & Norppa, H. (1995) Effects of indomethacin and arachidonic acid on sister chromatid exchange induction by styrene and styrene-7,8-oxide. *Mutat. Res.*, **348**, 175–181

Lemasters, G.K., Carson, A. & Samuels, S.J. (1985) Occupational styrene exposure for twelve product categories in the reinforced plastics industry. *Am. ind. Hyg. Assoc. J.*, **46**, 434–441

Lemasters, G.K., Samuels, S.J., Morrison, J.A. & Brooks, S.M. (1989) Reproductive outcomes of pregnant workers employed at 36 reinforced plastics companies. II. Lowered birth weight. *J. occup. Med.*, **31**, 115–120

Lemen, R.A. & Young, R. (1976) Investigation of health hazards in styrene-butadiene rubber facilities. In: Ede, L., ed., *Proceedings of NIOSH Styrene-Butadiene Rubber Briefing, Covington, Kentucky, April 30, 1976* (HEW Publ. No. (NIOSH) 77-129), Cincinnati, OH, National Institute for Occupational Safety and Health, pp. 3–8

Lenvik, K., Osvoll, P.O. & Woldbaek, T. (1999) Occupational exposure to styrene in Norway, 1972–1996. *Appl. occup. environ. Hyg.*, **14**, 165–170

Letz, R., Mahoney, F.C., Hershman, D.L., Woskie, S. & Smith, T.J. (1990) Neurobehavioral effects of acute styrene exposure in fiberglass boatbuilders. *Neurotoxicol. Teratol.*, **12**, 665–668

Lewis, P.J., Hagopian, C. & Koch, P. (1983) Styrene. In: Mark, H.F., Othmer, D.F., Overberger, C.G., Seaborg, G.T. & Grayson, M., eds, *Kirk-Othmer Encyclopedia of Chemical Technology*, Vol. 21, 3rd Ed., New York, John Wiley & Sons, pp. 770–801

Lickly, T.D., Breder C.V. & Rainey, M.L. (1995) A model for estimating the daily dietary intake of a substance from food-contact articles: Styrene from polystyrene food-contact polymers. *Regul. Toxicol. Pharmacol.*, **21**, 406–417

Lide, D.R., ed. (1996) *Properties of Organic Compounds*, Version 5.0, Boca Raton, FL, CRC Press Inc. [CD-ROM]

Lide, D.R., ed. (2001) *CRC Handbook of Chemistry and Physics*, 82nd Ed., Boca Raton, FL, CRC Press, pp. 3-46, 6-58

Lilis, R., Lorimer, W.V., Diamond, S. & Selikoff, I.J. (1978) Neurotoxicity of styrene in production and polymerization workers. *Environ. Res.*, **15**, 133–138

Limasset, J.C., Simon, P., Poirot, P., Subra, I. & Grzebyk, M. (1999) Estimation of the percutaneous absorption of styrene in an industrial situation. *Int. Arch. occup. environ. Health*, **72**, 46–51

Lindbohm, M.L., Hemminki, K. & Kyyronen, P. (1985) Spontaneous abortions among women employed in the plastics industry. *Am. J. ind. Med.*, **8**, 579–596

Lindström, K., Härkönen, H. & Hernberg, S. (1976) Disturbances in psychological functions of workers occupationally exposed to styrene. *Scand. J. Work Environ. Health*, **3**, 129–139

Linhart, I., Šmejkal, J. & Mládková, I. (1998) Stereochemical aspects of styrene biotransformation. *Toxicol. Lett.*, **94**, 127–135

Linhart, I., Gut, I., Šmejkal, J. & Novák, J. (2000) Biotransformation of styrene in mice. Stereochemical aspects. *Chem. Res. Toxicol.*, **13**, 36–44

Löf, A. & Johanson, G. (1993) Dose-dependent kinetics of inhaled styrene in man. In: Sorsa, M., Peltonen, K., Vainio, H. & Hemminki, K., eds, *Butadiene and Styrene: Assessment of Health Hazards* (IARC Scientific Publications No. 127, Lyon, IARC*Press*, pp. 89–99

Löf, A., Gullstrand, E. & Byfält Nordqvist, M. (1983) Tissue distribution of styrene, styrene glycol and more polar styrene metabolites in the mouse. *Scand. J. Work Environ. Health*, **9**, 419–430

Löf, A., Gullstrand, E., Lundgren, E. & Byfält Nordqvist, M. (1984) Occurrence of styrene-7,8-oxide and styrene glycol in mouse after the administration of styrene. *Scand. J. Work Environ. Health*, **10**, 179–187

Löf, A., Lundgren, E. & Byfält Nordqvist, M. (1986a) Kinetics of styrene in workers from a plastics industry after controlled exposure: A comparison with subjects not previously exposed. *Br. J. ind. Med.*, **43**, 537–543

Löf, A., Lundgren, E., Nydahl, E.-M. & Byfält Nordqvist, M. (1986b) Biological monitoring of styrene metabolites in blood. *Scand. J. Work Environ. Health*, **12**, 70–74

Loquet, G., Campo, P. & Lataye, R. (1999) Comparison of toluene-induced and styrene-induced hearing losses. *Neurotoxicol. Teratol.*, **21**, 689–697

Lorimer, W.V., Lilis, R., Nicholson, W.J., Anderson, H., Fischbein, A., Daum, S., Rom, W., Rice, C. & Selikoff, I.J. (1976) Clinical studies of styrene workers: Initial findings. *Environ. Health Perspect.*, **17**, 171–181

Lorimer, W.V., Lilis, R., Fischbein, A., Daum, S., Anderson, H., Wolff, M.S. & Selikoff, I.J. (1978) Health status of styrene-polystyrene polymerization workers. *Scand. J. Work Environ. Health*, **4** (Suppl. 2), 220–226

Loughlin, J.E., Rothman, K.J. & Dreyer, N.A. (1999) Lymphatic and haematopoietic cancer mortality in a population attending school adjacent to styrene-butadiene facilities, 1963–1993. *J. Epidemiol. Community Health*, **53**, 283–287

Lynch, D.W., Lewis, T.R., Moorman, W.J., Burg, J.R., Groth, D.H., Khan, A., Ackerman, L.J. & Cockrell, B.Y. (1984) Carcinogenic and toxicologic effects of inhaled ethylene oxide and propylene oxide in F344 rats. *Toxicol. appl. Pharmacol.*, **76**, 69–84

Maarse, H. (1992a) *Natural Occurrence and Routes of Formation of Styrene in Food* (TNO Report No. B92.084), Zeist, TNO Nutrition and Food Research

Maarse, H. (1992b) *Quantities of Styrene in 6 Selected Food Products* (TNO Report No. B92.553), Zeist, TNO Nutrition and Food Research

Macaluso, M., Larson, R., Delzell, E., Sathiakumar, N., Hovinga, M., Julian, J., Muir, D. & Cole, P. (1996) Leukemia and cumulative exposure to butadiene, styrene and benzene among workers in the synthetic rubber industry. *Toxicology*, **113**, 190–202

Mahler, J.F., Price, H.C., Jr, O'Connor, R.W., Wilson, R.E., Eldridge, S.R., Moorman, M.P. & Morgan, D.L. (1999) Characterization of hepatocellular resistance and susceptibility to styrene toxicity in B6C3F1 mice. *Toxicol. Sci.*, **48**, 123–133

Makhlouf, J. (1982) Polyesters, unsaturated. In: Mark, H.F., Othmer, D.F., Overberger, C.G. & Seaborg, G.T., eds, *Kirk-Othmer Encyclopedia of Chemical Technology*, 3rd Ed., Vol. 18, New York, John Wiley & Sons, pp. 575–594

Mäki-Paakkanen, J. (1987) Chromosome aberrations, micronuclei and sister-chromatid exchanges in blood lymphocytes after occupational exposure to low levels of styrene. *Mutat. Res.*, **189**, 399–406

Mäki-Paakkanen, J., Walles, S., Osterman-Golkar, S. & Norppa, H. (1991) Single-strand breaks, chromosome aberrations, sister-chromatid exchanges, and micronuclei in blood lymphocytes of workers exposed to styrene during the production of reinforced plastics. *Environ. mol. Mutag.*, **17**, 27–31

Mannsville Chemical Products Corp. (1987) *Chemical Products Synopsis: Styrene*, Cortland, NY

Marczynski, B., Rozynek, P., Elliehausen, H.-J., Korn, M. & Baur, X. (1997a) Detection of 8-hydroxydeoxyguanosine, a marker of oxidative DNA damage, in white blood cells of workers occupationally exposed to styrene. *Arch. Toxicol.*, **71**, 496–500

Marczynski, B., Peel, M. & Baur, X. (2000) New aspects in genotoxic risk assessment of styrene exposure — A working hypothesis. *Med. Hypotheses*, **54**, 619–623

Matanoski, G.M., Santos-Burgoa, C. & Schwartz, L. (1990) Mortality of a cohort of workers in the styrene-butadiene polymer manufacturing industry (1943–1982). *Environ. Health Perspect.*, **86**, 107–117

Matanoski, G., Francis, M., Correa-Villasenor, A., Elliott, E., Santos-Burgoa, C. & Schwartz, L. (1993) Cancer epidemiology among styrene-butadiene rubber workers. In: Sorsa, M., Peltonen, K., Vainio, H. & Hemminki, K., eds, *Butadiene and Styrene: Assessment of Health Hazards* (IARC Scientific Publications No. 127), Lyon, IARC*Press*, pp. 363–374

Matikainen, E., Forsman-Grönholm, L., Pfäffli, P. & Juntunen, J. (1993) Nervous system effects of occupational exposure to styrene: A clinical and neurophysiological study. *Environ. Res.*, **61**, 84–92

McDougal, J.N., Jepson, G.W., Clewell III, H.J., Gargas, M.L. & Andersen, M.E. (1990) Dermal absorption of organic chemical vapors in rats and humans. *Fundam. appl. Toxicol.*, **14**, 299–308

McKay, R.T., Lemasters, G.K. & Elia, V.J. (1982) Ambient air styrene levels in communities near reinforced plastic processors. *Environ. Pollut.* (Series B), **4**, 135–141

McMichael, A.J., Spirtas, R., Gamble, J.F. & Tousey, P.M. (1976a) Mortality among rubber workers: Relationship to specific jobs. *J. occup. Med.*, **18**, 178–185

McMichael, A.J., Andjelkovic, D.A. & Tyroler, H.A. (1976b) Cancer mortality among rubber workers: An epidemiologic study. *Ann. N.Y. Acad. Sci.*, **271**, 125–137

Meinhardt, T.J., Lemen, R.A., Crandall, M.S. & Young, R.J. (1982) Environmental epidemiologic investigation of the styrene-butadiene rubber industry. Mortality patterns with discussion of the hematopoietic and lymphatic malignancies. *Scand. J. Work Environ. Health*, **8**, 250–259

Melnick, R.L. & Sills, R.C. (2001) Comparative carcinogenicity of 1,3-butadiene, isoprene, and chloroprene in rats and mice. *Chem.-biol. Interact.*, **135–136**, 27–42

Melnick, R.L., Sills, R.C., Roycroft, J.H., Chou, B.J., Ragan, H.A. & Miller, R.A. (1994) Isoprene, an endogenous hydrocarbon and industrial chemical, induces multiple organ neoplasia in rodents after 26 weeks of inhalation exposure. *Cancer Res.*, **54**, 5333–5339

Mendrala, A.L., Langvardt, P.W., Nitschke, K.D., Quast, J.F. & Nolan, R.J. (1993) In vitro kinetics of styrene and styrene oxide metabolism in rat, mouse, and human. *Arch. Toxicol.*, **67**, 18–27

Miller, S.L., Branoff, S. & Nazaroff, W.W. (1998) Exposure to toxic air contaminants in environmental tobacco smoke: An assessment for California based on personal monitoring data. *J. Expo. anal. environ. Epidemiol.*, **8**, 287–311

Ministry of Agriculture, Fisheries and Food (MAFF) (1989) *Survey of Styrene Levels in Food Contact Materials and in Foods.* The Eleventh Report of the Steering Group on Food Surveillance. The Working Party on Styrene (Food Surveillance Paper No. 11), 2nd Ed., London, Her Majesty's Stationery Office

Ministry of Agriculture, Fisheries and Food (MAFF) (1994) *Joint Food Safety and Standards Group. Food Surveillance Information Sheet Number 38.* http://archive.food.gov.uk/maff/archive/food/infsheet/1994/no38/38styr.htm

Mizutani, T., Irie, Y. & Nakanishi, K. (1994) Styrene-induced hepatotoxicity in mice depleted of glutathione. *Res. Comm. mol. Pathol. Pharmacol.*, **86**, 361–374

Möller, C., Ödkvist, L., Larsby, B., Tham, R., Ledin, T. & Bergholtz, L. (1990) Otoneurological findings in workers exposed to styrene. *Scand. J. Work Environ. Health*, **16**, 189–194

Morgan, D.L., Cooper, S.W., Carlock, D.L., Sykora, J.J., Sutton, B., Mattie, D.R. & McDougal, J.N. (1991) Dermal absorption of neat and aqueous volatile organic chemicals in the Fischer 344 rat. *Environ. Res.*, **55**, 51–63

Morgan, D.L., Mahler, J.F., O'Connor, R.W., Price, H.C., Jr & Adkins, B., Jr (1993a) Styrene inhalation toxicity studies in mice. I. Hepatotoxicity in B6C3F1 mice. *Fundam. appl. Toxicol.*, **20**, 325–335

Morgan, D.L., Mahler, J.F., Dill, J.A., Price, H.C., Jr, O'Connor, R.W. & Adkins, B., Jr (1993b) Styrene inhalation toxicity studies in mice. II. Sex differences in susceptibility of B6C3F1 mice. *Fundam. appl. Toxicol.*, **21**, 317–325

Morgan, D.L., Mahler, J.F., Dill, J.A., Price, H.C., Jr, O'Connor, R.W. & Adkins, B., Jr (1993c) Styrene inhalation toxicity studies in mice. III. Strain differences in susceptibility. *Fundam. appl. Toxicol.*, **21**, 326–333

Morgan, D.L., Mahler, J.F., Moorman, M.P., Wilson, R.E., Price, H.C., Jr, Richards, J.H. & O'Connor, R.W. (1995) Comparison of styrene hepatotoxicity in B6C3F1 and Swiss mice. *Fundam. appl. Toxicol.*, **27**, 217–222

Morgan, D.L., Mahler, J.F., Wilson, R.E., Moorman, M.P., Price, H.C., Jr, Patrick, K.R., Richards, J.H. & O'Connor, R.W. (1997) Effects of various pretreatments on the hepatotoxicity of inhaled styrene in the B6C3F1 mouse. *Xenobiotica*, **27**, 401–411

Morioka, I., Kuroda, M., Miyashita, K. & Takeda, S. (1999) Evaluation of organic solvent ototoxicity by the upper limit of hearing. *Arch. environ. Health*, **54**, 341–346

Morris, J.B. (2000) Uptake of styrene in the upper respiratory tract of the CD mouse and Sprague-Dawley rat. *Toxicol. Sci.*, **54**, 222–228

Moscato, G., Biscaldi, G., Cottica, D., Pugliese, F., Candura, S. & Candura, F. (1987) Occupational asthma due to styrene: Two case reports. *J. occup. Med.*, **29**, 957–960

Muijser, H., Hoogendijk, E.M.G. & Hooisma, J. (1988) The effects of occupational exposure to styrene on high-frequency hearing thresholds. *Toxicology*, **49**, 331–340

Murata, K., Araki, S. & Yokoyama, K. (1991) Assessment of the peripheral, central, and autonomic nervous system function in styrene workers. *Am. J. ind. Med.*, **20**, 775–784

Mutti, A., Vescovi, P.P., Falzoi, M., Arfini, G., Valenti, G. & Franchini, I. (1984a) Neuroendocrine effects of styrene on occupationally exposed workers. *Scand. J. Work Environ. Health*, **10**, 225–228

Mutti, A., Falzoi, M., Romanelli, A. & Franchini, I. (1984b) Regional alterations of brain catecholamines by styrene exposure in rabbits. *Arch. Toxicol.*, **55**, 173–177

Mutti, A., Falzoi, M., Romanelli, A., Bocchi, M.C., Ferroni, C. & Franchini, I. (1988) Brain dopamine as a target for solvent toxicity: Effects of some monocyclic aromatic hydrocarbons. *Toxicology*, **49**, 77–82

Mutti, A., Buzio, C., Perazzoli, F., Bergamaschi, E., Bocchi, M.C., Selis, L., Mineo, F. & Franchini, I. (1992) [Lymphocyte subsets in workers occupationally exposed to styrene.] *Med. Lav.*, **83**, 167–177 (in Italian)

Mutti, A., Coccini, T., Alinovi, R., Toubeau, G., Broeckaert, F., Bergamaschi, E., Mozzoni, P., Nonclercq, D., Bernard, A. & Manzo, L. (1999) Exposure to hydrocarbons and renal disease: An experimental animal model. *Ren. Fail.*, **21**, 369–385

Nakajima, T., Elovaara, E., Gonzalez, F.J., Gelboin, H.V., Raunio, H., Pelkonen, O., Vainio, H. & Aoyama, T. (1994a) Styrene metabolism by cDNA-expressed human hepatic and pulmonary cytochromes P450. *Chem. Res. Toxicol.*, **7**, 891–896

Nakajima, T., Wang, R-S., Elovaara, E., Gonzalez, F.J., Gelboin, H.V., Vainio, H. & Aoyama, T. (1994b) CYP2C11 and CYP2B1 are major cytochrome P450 forms involved in styrene oxidation in liver and lung microsomes from untreated rats, respectively. *Biochem. Pharmacol.*, **48**, 637–642

Nakatsu, K., Hugenroth, S., Sheng, L.-S., Horning, E.C. & Horning, M.G. (1983) Metabolism of styrene oxide in the rat and guinea pig. *Drug Metab. Disp.*, **11**, 463–470

Nano, R., Rossi, A., Fenoglio, C. & De Piceis Polver, P. (2000) Evaluation of a possible styrene-induced damage to the haematopoietic tissues in the rat. *Anticancer Res.*, **20**, 1615–1619

National Cancer Institute (1979a) *Bioassay of Styrene for Possible Carcinogenicity (CAS No. 100-42-5)* (Tech. Rep. Ser. No. 185; DHEW Publ. (NIH) 79-1741), Washington DC, US Government Printing Office

National Cancer Institute (1979b) *Bioassay of a Solution of β-Nitrostyrene and Styrene for Possible Carcinogenicity (CAS No. 102-96-5, CAS No. 100-42-5)* (Tech. Rep. Ser. No. 170; DHEW Publ. (NIH) 79-1726), Washington DC, US Government Printing Office

National Institute for Occupational Safety and Health (1983) *Criteria for a Recommended Standard. Occupational Exposure to Styrene* (NIOSH Publication No. 83-119), Cincinnati, OH

National Institute for Occupational Safety and Health (1993) *National Occupational Exposure Survey (1981–1983)*, Cincinnati, OH

National Toxicology Program (1984) *Toxicology and Carcinogenesis Studies of 1,3-Butadiene (CAS No. 106-99-0) in B6C3F$_1$ Mice (Inhalation Studies)* (Technical Report Series No. 288; NIH Publ. No. 84-2544), Research Triangle Park, NC

National Toxicology Program (1987) *Toxicology and Carcinogenesis Studies of Ethylene Oxide (CAS No. 75-21-8) in B6C3F$_1$ Mice (Inhalation Studies)* (Technical Report Series No. 326; NIH Publ. No. 88-2582), Research Triangle Park, NC

National Toxicology Program (1998) *Toxicology and Carcinogenesis Studies of Chloroprene (CAS No. 126-99-8) in F344/N Rats and B6C3F$_1$ Mice (Inhalation Studies)* (Technical Report Series No. 467; NIH Publ. No. 98-3957), Research Triangle Park, NC

National Toxicology Program (1999) *Toxicology and Carcinogenesis Studies of Isoprene (CAS No. 78-79-5) in F344/N Rats (Inhalation Studies)* (Technical Report Series No. 486; NIH Publ. No. 99-3976), Research Triangle Park, NC

Newhook, R. & Caldwell, I. (1993) Exposure to styrene in the general Canadian population. In: Sorsa, M., Peltonen, K., Vainio, H. & Hemminki, K., eds, *Butadiene and Styrene: Assessment of Health Hazards* (IARC Scientific Publications No. 127), Lyon, IARC*Press*, pp. 27–33

Nicholson, W.J., Selikoff, I.J. & Seidman, H. (1978) Mortality experience of styrene-polystyrene polymerization workers. Initial findings. *Scand. J. Work Environ. Health*, **4** (Suppl.), 2247–2252

Ninomiya, R., Hirokawa, Y., Yamamoto, R., Masui, H., Koizumi, N. & Kubota, A. (2000) [Effects of low concentration of styrene monomer vapour on pregnancy.] *Jpn. J. Hyg.*, **55**, 547–551 (in Japanese)

Occupational Safety and Health Administration (2001) Labor. *US Code. Fed. Regul.*, **Title 29**, Part 1910, Subpart 1910.1000, pp. 7–19

Ogata, M. & Sugihara, R. (1978) High performance liquid chromatographic procedure for quantitative determination of urinary phenylglyoxylic, mandelic and hippuric acids as indices of styrene exposure. *Int. Arch. occup. environ. Health*, **42**, 11–19

Ogata, M. & Taguchi, T. (1987) Quantitation of urinary metabolites of toluene, xylene, styrene, ethylbenzene, benzene and phenol by automated high performance liquid chromatography. *Int. Arch. occup. environ. Health*, **59**, 263–272

Ogata, M. & Taguchi, T. (1988) Simultaneous determination of urinary creatinine and metabolites of toluene, xylene, styrene, ethylbenzene and phenol by automated high performance liquid chromatography. *Int. Arch. occup. environ. Health*, **61**, 131–140

Ohashi, Y., Nakai, Y., Ikeoka, H., Koshimo., H., Esaki, Y., Horiguchi, S. & Teramoto, K. (1985) Electron microscopic study of the respiratory toxicity of styrene. *Osaka City med. J.*, **31**, 11–21

Ohashi, Y., Nakai, Y., Ikeoka, H., Koshimo, H., Nakata, J., Esaki, Y., Horiguchi, S. & Teramoto, K. (1986) Degeneration and regeneration of respiratory mucosa of rats after exposure to styrene. *J. appl. Toxicol.*, **6**, 405–412

Ohtsuji, H. & Ikeda, M. (1971) The metabolism of styrene in the rat and the stimulatory effect of phenobarbital. *Toxicol. appl. Pharmacol.*, **18**, 321–328

Okun, A.H., Beaumont, J.J., Meinhardt, T.J. & Crandall, M.S. (1985) Mortality patterns among styrene-exposed boatbuilders. *Am. J. ind. Med.*, **8**, 193–205

Oliviero, M. (1906) [Reduction of cinnamic acid to cinnamene by moulds.] *J. Pharmacol. chim.*, **24**, 62–64 (in French)

Ollikainen, T., Hirvonen, A. & Norppa, H. (1998) Influence of GSTT1 genotype on sister chromatid exchange induction by styrene-7,8-oxide in cultured human lymphocytes. *Environ. mol. Mutag.*, **31**, 311–315

Ong, C.N., Shi, C.Y., Chia, S.E., Chua, S.C., Ong, H.Y., Lee, B.L., Ng, T.P. & Teramoto, K. (1994) Biological monitoring of exposure to low concentrations of styrene. *Am. J. ind. Med.*, **25**, 719–730

Ott, M.G., Kolesar, R.C., Scharnweber, H.C., Schneider, E.J. & Venable, J.R. (1980) A mortality survey of employees engaged in the development or manufacture of styrene-based products. *J. occup. Med.*, **22**, 445–460

Otteneder, M., Eder, E. & Lutz, W.K. (1999) ^{32}P-Postlabeling analysis of DNA adducts of styrene 7,8-oxide at the O^6-position of guanine. *Chem. Res. Toxicol.*, **12**, 93–99

Otteneder, M., Lutz, U. & Lutz, W.K. (2002) DNA addducts of styrene-7,8-oxide in target and non-target organs for tumor induction in rat and mouse after repeated inhalation exposure to styrene. *Mutat. Res.*, **500**, 111–116

Owen, P.E., Glaister, J.R., Gaunt, I.F. & Pullinger, D.H. (1987) Inhalation toxicity studies with 1,3-butadiene. 3. Two year toxicity/carcinogenicity study in rats. *Am. ind. Hyg. Assoc. J.*, **48**, 407–413

Pagano, D. A., Yagen, B., Hernandez, O., Bend, J. R. & Zeiger, E. (1982) Mutagenicity of (R) and (S) styrene 7,8-oxide and the intermediary mercapturic acid metabolites formed from styrene 7,8-oxide. *Environ. Mutag.*, **4**, 575–584

Pantarotto, C., Fanelli, R., Bidoli, F., Morazzoni, P., Salmona, M. & Szczawinska, K. (1978) Arene oxides in styrene metabolism, a new perspective in styrene toxicity? *Scand. J. Work Environ. Health*, **4**, 67–77

Parent, M.E., Hua, Y. & Siemiatycki, J. (2000) Occupational risk factors for renal cell carcinoma in Montreal. *Am. J. ind. Med.*, **38**, 609–618

Pauwels, W. & Veulemans, H. (1998) Comparison of ethylene, propylene and styrene-7,8-oxide in vitro adduct formation on N-terminal valine in human haemoglobin and on N-7-guanine in human DNA. *Mutat. Res.*, **418**, 21–33

Pauwels, W., Vodiceka, P., Severi, M., Plná, K., Veulemans, H. & Hemminki, K. (1996) Adduct formation on DNA and haemoglobin in mice intraperitoneally administered with styrene. *Carcinogenesis*, **17**, 2673–2680

Pekari, K., Nylander-French, L., Pfäffli, P., Sorsa, M. & Aitio, A. (1993) Biological monitoring of exposure to styrene — Assessment of different approaches. *J. occup. Med. Toxicol.*, **2**, 115–126

Pezzagno, G., Ghittori, S., Imbriani, M. & Capodaglio, E. (1985) Urinary elimination of styrene in experimental and occupational exposure. *Scand. J. Work Environ. Health*, **11**, 371–379

Pfäffli, P. (1982) Degradation products of plastics. III. Industrial hygiene measurements. *Scand. J. Work Environ. Health*, **8** (Suppl. 2), 27–43

Pfäffli, P. & Säämänen, A. (1993) The occupational scene of styrene. In: Sorsa, M., Peltonen, K., Vainio, H. & Hemminki, K., eds, *Butadiene and Styrene: Assessment of Health Hazards* (IARC Scientific Publications No. 127), Lyon, IARC*Press*, pp. 15–26

Pfäffli, P., Vainio, H. & Hesso, A. (1979) Styrene and styrene oxide concentrations in the air during the lamination process in the reinforced plastics industry. *Scand. J. Work Environ. Health*, **5**, 158–161

Pfäffli, P., Hesso, A., Vainio, H. & Hyvönen, M. (1981) 4-Vinylphenol excretion suggestive of arene oxide formation in workers occupationally exposed to styrene. *Toxicol. appl. Pharmacol.*, **60**, 85–90

Plotnick, H.B. & Weigel, W.W. (1979) Tissue distribution and excretion of ^{14}C-styrene in male and female rats. *Res. Comm. chem. Pathol. Pharmacol.*, **24**, 515–524

Ponomarkov, V. & Tomatis, L. (1978) Effects of long-term oral administration of styrene to mice and rats. *Scand. J. Work Environ. Health*, **4**, 127–135

Preston, R.J. & Abernethy, D.J. (1993) Studies of the induction of chromosomal aberration and sister chromatid exchange in rats exposed to styrene by inhalation. In: Sorsa, M., Peltonen, K., Vainio, H. & Hemminki, K., eds, *Butadiene and Styrene: Assessment of Health Hazards* (IARC Scientific Publications No. 127), Lyon, IARC*Press*, pp. 225–233

Pryor, G.T., Rebert, C.S. & Howd, R.A. (1987) Hearing loss in rats caused by inhalation of mixed xylenes and styrene. *J. appl. Toxicol.*, **7**, 55–61

Quincy, M.A., ed. (1991) *National Fire Protection Guide, Fire Protection Guide on Hazardous Materials*, 10th Ed., Boston, National Fire Protection Association, pp. 49–162

Ramsey, J.C. & Andersen, M.E. (1984) A physiologically based description of the inhalation pharmacokinetics of styrene in rats and humans. *Toxicol. appl. Pharmacol.*, **73**, 159–175

Ramsey, J.C. & Young, J.D. (1978) Pharmacokinetics of inhaled styrene in rats and humans. *Scand. J. Work Environ. Health*, **4**, 84–91

Ramsey, J.C., Young, J.D., Karbowski, R.J., Chenoweth, M.B., McCarty, L.P. & Braun, W.H. (1980) Pharmacokinetics of inhaled styrene in human volunteers. *Toxicol. appl. Pharmacol.*, **53**, 54–63

Rappaport, S.M. & Fraser, D.A. (1977) Air sampling and analysis in a rubber vulcanization area. *Am. ind. Hyg. Assoc. J.*, **38**, 205–210

Rappaport, S.M., Yeowell-O'Connell, K., Bodell, W., Yager, J.W. & Symanski, E. (1996) An investigation of multiple biomarkers among workers exposed to styrene-7,8-oxide. *Cancer Res.*, **56**, 5410–5416

Reed, L.D. (1983) *Health Hazard Evaluation Report, Columbia College, Columbia, MO* (Report No. 81-326-1247), Cincinnati, OH, National Institute for Occupational Safety and Health

Ring, K.L. (1999) *CEH Marketing Research Report — Styrene*, Chemical Economics Handbook (CEH)-SRI International, Menlo Park, CA

Rodriguez-Arnaiz, R. (1998) Biotransformation of several structurally related 2B compounds to reactive metabolites in the somatic *w/w+* assay of *Drosophila melanogaster*. *Environ. mol. Mutag.*, **31**, 390–401

Rosén, I., Haeger-Aronsen, B., Rehnström, S. & Welinder, H. (1978) Neurophysiological observations after chronic styrene exposure. *Scand. J. Work Environ. Health*, **4** (Suppl. 2), 184–194

Rosengren, L.E. & Haglid, K.G. (1989) Long term neurotoxicity of styrene. A quantitative study of glial fibrillary acidic protein (GFA) and S-100. *Br. J. ind. Med.*, **46**, 316–320

Ruhe, R.L. & Jannerfeldt, E.R. (1980) *Health Hazard Evaluation Report, Metamora Products Corporation, Elkland, PA* (Report No. HE-80-188-797), Cincinnati, OH, National Institute for Occupational Safety and Health

Rutkowski, J.V. & Levin, B.C. (1986) Acrylonitrile-butadiene-styrene copolymers (ABS): Pyrolysis and combustion products and their toxicity — A review of the literature. *Fire Mater.*, **10**, 93–105

Säämänen, A., Anttila, A. & Pfäffli, P. (1991) *Styreeni* [Styrene] [Exposures at Work 9], Helsinki, Työterveyslaitos, Työsuojelurahasto [National Institute of Occupational Health, Work Environment Fund] (in Finnish)

Säämänen, A., Pfäffli, P. & Niemelä, R. (1993) *Altistumishuippujen Vähentäminen Polyester-ilujitemuo-vityössä* [Reduction of Peak Exposures in the Reinforced Plastics Industry] (Työsuojelurahaston hanke No. 90201), Helsinki, Työsuojelurahasto Työterveyslaitos [Work Environment Fund] (in Finnish)

Sadtler Research Laboratories (1991) *Sadtler Standard Spectra. 1981–1991 Supplementary Index*, Philadelphia, PA

Salmona, M., Pachecka, J., Cantoni, L., Belvedere, G., Mussini, E. & Garattini, S. (1976) Microsomal styrene mono-oxygenase and styrene epoxide hydrase activities in rats. *Xenobiotica*, **6**, 585–591

Salomaa, S., Donner, M. & Norppa, H. (1985) Inactivity of styrene in the mouse sperm morphology test. *Toxicol. Lett.*, **24**, 151–155

Samimi, B. & Falbo, L. (1982) Monitoring of workers exposure to low levels of airborne monomers in an acrylic ester-styrene copolymer production plant. *Am. ind. Hyg. Assoc. J.*, **43**, 858–862

Santos-Burgoa, C., Matanoski, G.M., Zeger, S. & Schwartz, L. (1992) Lymphohematopoietic cancer in styrene-butadiene polymerization workers. *Am. J. Epidemiol.*, **136**, 843–854

Sarangapani, R., Teeguarden, J.G., Cruzan, G., Clewell, H.J. & Andersen, M.E. (2002) Physiologically based pharmacokinetic modeling of styrene and styrene oxide respiratory tract dosimetry in rodents and humans. *Inhal. Toxicol.*, **14**, 789–834

Sasaki, Y.F., Izumiyama, F., Nishidate, E., Matsusaka, N. & Tsuda, S. (1997) Detection of rodent liver carcinogen genotoxicity by the alkaline single-cell gel electrophoresis (Comet) assay in multiple mouse organs (liver, lung, spleen, kidney, and bone marrow). *Mutat. Res.*, **391**, 201–214

Sass-Kortsak, A.M., Corey, P.N. & Robertson, J. McD. (1995) An investigation of the association between exposure to styrene and hearing loss. *Ann. Epidemiol.*, **5**, 15–24

Sathiakumar, N., Delzell, E., Hovinga, M., Macaluso, M., Julian, J.A., Larson, R., Cole, P. & Muir, D.C. (1998) Mortality from cancer and other causes of death among synthetic rubber workers. *Occup. environ. Med.*, **55**, 230–235

Sato, A. & Nakajima, T. (1985) Enhanced metabolism of volatile hydrocarbons in rat liver following food deprivation, restricted carbohydrate intake, and administration of ethanol, phenobarbital, polychlorinated biphenyl and 3-methylcholanthrene: A comparative study. *Xenobiotica*, **15**, 67–75

Schubart, R. (1987) Dithiocarbamic acid and derivatives. In: Gerhartz, W., Yamamoto, Y.S., Kaudy, L., Rounsaville, J.F. & Schulz, G., eds, *Ullmann's Encyclopedia of Industrial Chemistry*, 5th Rev. Ed., Vol. A9, New York, VCH Publishers, pp. 1–27

Schumacher, R.L., Breysse, P.A., Carlyon, W.R., Hibbard, R.P. & Kleinman, G.D. (1981) Styrene exposure in the fiberglass fabrication industry in Washington State. *Am. ind. Hyg. Assoc. J.*, **42**, 143–149

Scott, D. & Preston, R.J. (1994) A re-evaluation of the cytogenetic effects of styrene. *Mutat. Res.*, **318**, 175–203

Seiler, J.P. (1990) Chirality-dependent DNA reactivity as the possible cause of the differential mutagenicity of the two components in an enantiomeric pair of epoxides. *Mutat. Res.*, **245**, 165–169

Seppäläinen, A.M. & Härkönen, H. (1976). Neurophysiological findings among workers occupationally exposed to styrene. *Scand. J. Work Environ. Health*, **3**, 140–146

Seutter-Berlage, F., Delbressine, L.P.C., Smeets, F.L.M. & Ketelaars, H.C.J. (1978) Identification of three sulphur-containing urinary metabolites of styrene in the rat. *Xenobiotica*, **8**, 413–418

Severi, M., Pauwels, W., Van Hummelen, P., Roosels, D., Kirsch-Volders, M. & Veulemans, H. (1994) Urinary mandelic acid and hemoglobin adducts in fiberglass-reinforced plastics workers exposed to styrene. *Scand. J. Work Environ. Health*, **20**, 451–458

Shield, A.J. & Sanderson, B.J. (2001) Role of glutathione S-transferase mu (GSTM1) in styrene-7,8-oxide toxicity and mutagenicity. *Environ. mol. Mutag.*, **37**, 285–289

Sielken, R.L. & Valdez-Flores, C. (2001) Dose–response implications of the University of Alabama study of lymphohematopoietic cancer among workers exposed to 1,3-butadiene and styrene in the synthetic rubber industry. *Chem.-biol. Interact.*, **135–136**, 637–651

Siemiatycki J., ed. (1991) *Risk Factors for Cancer in the Workplace*, Boca Raton, FL, CRC Press

Sinsheimer, J.E., Chen, R., Das, S.K., Hooberman, B.H., Osorio, S. & You, Z. (1993) The genotoxicity of enantiomeric aliphatic epoxides. *Mutat. Res.*, **298**, 197–206

Sjöborg, S., Fregert, S. & Trulsson, L. (1984) Contact allergy to styrene and related chemicals. *Contact Derm.*, **10**, 94–96

Snellings, W.M., Weil, C.S. & Maronpot, R.R. (1984) A two-year inhalation study of the carcinogenic potential of ethylene oxide in Fischer 344 rats. *Toxicol. appl. Pharmacol.*, **75**, 105–117

Somorovská, M., Jahnová, E., Tulinská, J., Zamecníková, M., Sarmanová, J., Terenová, A., Vodicková, L., Lísková, A., Vallová, B., Soucek, P., Hemminki, K., Norppa, H., Hirvonen, A., Tates, A.D., Fuortes, L., Dusinská, M. & Vodicka, P. (1999) Biomonitoring of occupational exposure to styrene in a plastics lamination plant. *Mutat. Res.*, **428**, 255–269

Sosiaali- ja terveysministeriö [Ministry of Health and Social Affairs] (2002) *HTP-arvot 2002* [Values Known to be Harmful], Tampere, Ministry of Health and Social Affairs, Department of Occupational Health, Kirjapaino Öhrling

Spencer, H.C., Irish, D.D., Adams, E.M. & Rowe, V.K. (1942) The response of laboratory animals to monomeric styrene. *J. ind. Hyg. Toxicol.*, **24**, 295–301

Sripaung, N., Motohashi, Y., Nakata, K., Nakamura, K. & Takano, T. (1995) Effects of simultaneous administration of ethanol on styrene metabolism under fed and fasted conditions in the perfused rat liver. *J. Toxicol. environ. Health*, **45**, 439–451

Srivastava, S.P., Das, M., Mushtaq, M., Chandra, S.V. & Seth, P.K. (1982) Hepatic effects of orally administered styrene in rats. *J. appl. Toxicol.*, **2**, 219–222

Srivastava, S.P., Das, M. & Seth, P.K. (1983) Enhancement of lipid peroxidation in rat liver on acute exposure to styrene and acrylamide a consequence of glutathione depletion. *Chem-biol. Interact.*, **45**, 373–380

Srivastava, S., Seth, P.K. & Srivastava, S.P. (1989) Effect of styrene administration on rat testis. *Arch. Toxicol.*, **63**, 43–46

Srivastava, S., Seth, P.K., & Srivastava, S.P. (1992) Effect of styrene on testicular enzymes of growing rat. *Indian J. exp. Biol.*, **30**, 399–401

Steele, D.H. (1992) *The Determination of Styrene and Benzene in Selected Foods* (MRI Project No. 6450), Kansas City, MO, Midwest Research Institute

Steele, D.H., Thornburg, M.J., Stanley, J.S., Miller, R.R., Brooke, R., Cushman, J.R. & Cruzan, G. (1994) Determination of styrene in selected foods. *J. agric. Food. Chem.*, **42**, 1661–1665

Stengel, B., Touranchet, A., Boiteau, H.L., Harousseau, H., Mandereau, L. & Hémon, D. (1990) Hematological findings among styrene-exposed workers in the reinforced plastics industry. *Int. Arch. Occup. Env. Health*, **62**, 11–18

Štetkárová, I., Urban, P., Procházka, B. & Lukáš, E. (1993) Somatosensory evoked potentials in workers exposed to toluene and styrene. *Br. J. ind. Med.*, **50**, 520–527

Stewart, R.D., Dodd, H.C., Baretta, E.D. & Schaffer, A.W. (1968) Human exposure to styrene vapor. *Arch. environ. Health*, **16**, 656–662

Styrene Information & Research Center (2001) *Styrene Sources and Exposure Scenarios*, Arlington, VA

Sullivan, E.A. & Sullivan, J.L. (1986) Styrene exposure in the reinforced plastics industry in Ontario. *Occup. Health Ontario*, **7**, 38–55

Sumner, S.J. & Fennell, T.R. (1994) Review of the metabolic fate of styrene. *Crit. Rev. Toxicol.*, **24**, S11–S33

Sumner, S.C.J., Cattley, R.C., Asgharian, B., Janszen, D.B. & Fennell, T.R. (1997) Evaluation of the metabolism and hepatotoxicity of styrene in F344 rats, B6C3F1 mice, and CD-1 mice following single and repeated inhalation exposures. *Chem.-biol. Interact.*, **106**, 47–65

Symanski, E., Bergamaschi, E. & Mutti, A. (2001) Inter- and intra-individual sources of variation in levels of urinary styrene metabolites. *Int. Arch. occup. environ. Health*, **74**, 336–344

Takao, T., Nanamiya, W., Nazarloo, H.P., Asaba, K. & Hashimoto, K. (2000) Possible reproductive toxicity of styrene in peripubertal male mice. *Endocr. J.*, **47**, 343–347

Tang, W., Hemm, I. & Eisenbrand, G. (2000) Estimation of human exposure to styrene and ethylbenzene. *Toxicology*, **144**, 39–50

Teramoto, K. & Horiguchi, S. (1979) Absorption, distribution and elimination of styrene in man and experimental animals. *Arh. Hig. Rada Toksikol.*, **30**, 431–437

Teramoto, K., Horiguchi, S., Wakitani, F., Tojyo, F., Tokimoto, T. & Kuribara, H. (1988) Effects of styrene on wheel-running and ambulatory activities in mice. *J. toxicol. Sci.*, **13**, 133–139

Thiess, A.M. & Friedheim, M. (1978) Morbidity among persons employed in styrene production, polymerization and processing plants. *Scand. J. Work Environ. Health*, **4** (Suppl. 2), 203–214

Tornero-Velez, R. & Rappaport, S.M. (2001) Physiological modeling of the relative contributions of styrene-7,8-oxide derived from direct inhalation and from styrene metabolism to the systemic dose in humans. *Toxicol. Sci.*, **64**, 151–161

Tornero-Velez, R., Waidyanatha, S., Perez, H.L., Osterman-Golkar, S., Echeverria, D. & Rappaport, S.M. (2001) Determination of styrene and styrene-7,8-oxide in human blood by gas chromatography-mass spectrometry. *J. Chromatogr. B Biomed. Sci. Appl.*, **757**, 59–68

Tossavainen, A. (1978) Styrene use and occupational exposure in the plastics industry. *Scand. J. Work Environ. Health*, **4** (Suppl. 2), 7–13

Triebig, G., Schaller, K.-H., & Valentin, H. (1985) Investigations on neurotoxicity of chemical substances at the workplace. VII. Longitudinal study with determination of nerve conduction velocities in persons occupationally exposed to styrene. *Int. Arch. occup. environ. Health*, **56**, 239–247

Triebig, G., Lehrl, S., Weltle, D., Schaller, K.H. & Valentin, H. (1989) Clinical and neurobehavioural study of the acute and chronic neurotoxicity of styrene. *Br. J. ind. Med.*, **46**, 799–804

Triebig, G., Stark, T., Ihrig, A. & Dietz, M.C. (2001) Intervention study on acquired color vision deficiencies in styrene-exposed workers. *J. occup. environ. Med.*, **43**, 494–500

Truchon, G., Gérin, M. & Brodeur, J. (1990) Urinary excretion of mandelic, phenylglyoxylic, and specific mercapturic acids in rats exposed repeatedly by inhalation to various concentrations of styrene vapors. *Can. J. Physiol. Pharmacol.*, **68**, 556–561

Truchon, G., Ostiguy, C., Drolet, D., Mergler, D., Campagna, D., Bélanger, S., Larribe, F. & Huel, G. (1992) Neurotoxic effects among styrene-exposed workers. I. Environmental and biological monitoring of exposure. *Trav. Santé*, **8**, S11–S14 (in French)

Tsai, S.-Y. & Chen, J.-D. (1996) Neurobehavioral effects of occupational exposure to low-level styrene. *Neurotoxicol. Teratol.*, **18**, 463–469

Tsuda, S., Matsusaka, N., Madarame, H., Miyamae, Y., Ishida, K., Satoh, M., Sekihashi, K. & Sasaki, Y.F. (2000) The alkaline single cell electrophoresis assay with eight mouse organs: Results with 22 mono-functional alkylating agents (including 9 dialkyl *N*-nitrosamines) and 10 DNA crosslinkers. *Mutat. Res.*, **467**, 83–98

Tulinska, J., Dusinska, M., Jahnova, E., Liskova, A., Kuricova, M., Vodicka, P., Vodickova, L., Sulcova, M. & Fuortes, L. (2000) Changes in cellular immunity among workers occupationally exposed to styrene in a plastics lamination plant. *Am. J. ind. Med.*, **38**, 576–583

United Nations Environment Programme (UNEP) (2002) *IRPTC Data Profiles on Styrene*, Geneva

Uusküla, M., Järventaus, H., Hirvonen, A., Sorsa, M. & Norppa, H. (1995) Influence of GSTM1 genotype on sister chromatid exchange induction by styrene-7,8-oxide and 1,2-epoxy-3-butene in cultured human lymphocytes. *Carcinogenesis*, **16**, 947–950

Vaghef, H. & Hellman, B. (1998) Detection of styrene and styrene oxide-induced DNA damage in various organs of mice using the comet assay. *Pharmacol. Toxicol.*, **83**, 69–74

Van Hummelen, P., Severi, M., Pauwels, W., Roosels, D., Veulemans, H. & Kirsch-Volders, M. (1994) Cytogenetic analysis of lymphocytes from fiberglass-reinforced plastics workers occupationally exposed to styrene. *Mutat. Res.*, **310**, 157–165

Van Rees, H. (1974) The partition coefficients of styrene between blood and air and between oil and blood. *Int. Arch. Arbeitsmed.*, **12**, 39–47

Varner, S. & Breder, C. (1981) Headspace sampling and gas chromatographic determination of styrene migration from food-contact polystyrene cups into beverages and food stimulants. *J. Assoc. off. anal. Chem.*, **64**, 1122–1130

Verplanke, A.J.W. & Herber, R.F.M. (1998) Effects on the kidney of occupational exposure to styrene. *Int. Arch. occup. environ. Health*, **71**, 47–52

Vettori, M.V., Corradi, D., Coccini, T., Carta, A., Cavazzini, S., Manzo, L. & Mutti, A. (2000) Styrene-induced changes in amacrine retinal cells: An experimental study in the rat. *Neurotoxicology*, **21**, 607–614

Viaene, M.K., Pauwels, W., Veulemans, H., Roels, H.A. & Masschelein, R. (2001) Neurobehavioural changes and persistence of complaints in workers exposed to styrene in a polyester boat building plant: Influence of exposure characteristics and microsomal epoxide hydrolase phenotype. *Occup. environ. Med.*, **58**, 103–112

Viau, C., Bernard, A., De Russis, R., Ouled, A., Maldague, P. & Lauwerys, R. (1987) Evaluation of the nephrotoxic potential of styrene in man and in rat. *J. appl. Toxicol.*, **7**, 313–316

Vodicka, P., Vodicková, L. & Hemminki, K. (1993) ^{32}P-Postlabeling of DNA adducts of styrene-exposed lamination workers. *Carcinogenesis*, **14**, 2059–2061

Vodicka, P., Vodicková, L., Trejbalová, K., Srám, R.J. & Hemminki, K. (1994) Persistence of O^6-guanine DNA adducts in styrene-exposed lamination workers determined by ^{32}P-postlabelling. *Carcinogenesis*, **15**, 1949–1953

Vodicka, P., Bastlová, T., Vodicková, L., Peterková, K., Lambert, B. & Hemminki, K. (1995) Biomarkers of styrene exposure in lamination workers: Levels of O^6-guanine DNA adducts, DNA strand breaks and mutant frequencies in the hypoxanthine guanine phosphoribosyltransferase gene in T-lymphocytes. *Carcinogenesis*, **16**, 1473–1481

Vodicka, P., Stetina, R., Kumar, R., Plna, K. & Hemminki, K. (1996) 7-Alkylguanine adducts of styrene oxide determined by ^{32}P-postlabelling in DNA and human embryonal lung fibroblasts (HEL). *Carcinogenesis*, **17**, 801–808

Vodicka, P., Tvrdik, T., Osterman-Golkar, S., Vodicková, L., Peterková, K., Soucek, P., Sarmanová, J., Farmer, P.B., Granath, F., Lambert, B. & Hemminki, K. (1999) An evaluation of styrene genotoxicity using several biomarkers in a 3-year follow-up study of hand-lamination workers. *Mutat. Res.*, **445**, 205–224

Vyskocil, A., Emminger, S., Malir, F., Fiala, Z., Tusl, M., Ettlerova, E. & Bernard, A. (1989) Lack of nephrotoxicity of styrene at current TLV level (50 ppm). *Int. Arch. occup. environ. Health*, **61**, 409–411

Wallace, L.A. (1986) Personal exposures, indoor and outdoor air concentrations, and exhaled breath concentrations of selected volatile organic compounds measured for 600 residents of New Jersey, North Dakota, North Carolina, and California. *Toxicol. environ. Chem.*, **12**, 215–236

Wallace, L.A., Pellizzari, E.D., Hartwell, T.D., Sparacino, C.M., Sheldon, L.S. & Zelon, H. (1985) Personal exposures, indoor-outdoor relationships, and breath levels of toxic air pollutants measured for 355 persons in New Jersey, USA. *Atmos. Environ.*, **19**, 1651–1661

Wallace, L.A., Pellizzari, E.D., Hartwell, T.D., Perritt, R. & Ziegenfus, R. (1987) Exposures to benzene and other volatile compounds from active and passive smoking. *Arch. environ. Health*, **42**, 272–279

Wallace, L.A., Pellizzari, E.D., Hartwell, T.D., Davis, V., Michael, L.C. & Whitmore, R.W. (1989) The influence of personal activities on exposure to volatile organic compounds. *Environ. Res.*, **50**, 37–55

Walles, S.A.S., Edling, C., Anundi, H. & Johanson, G. (1993) Exposure dependent increase in DNA single strand breaks in leucocytes from workers exposed to low concentrations of styrene. *Br. J. ind. Med.*, **50**, 570–574

Wang, Y., Kupper, L.L., Löf, A. & Rappaport, S.M. (1996) Comparison of average estimated metabolic rates for styrene in previously exposed and unexposed groups with pharmacokinetic modelling. *Occup. environ. Med.*, **53**, 601–605

Watabe, T., Isobe, M., Yoshikawa, K. & Takabatake, E. (1978) Studies on metabolism and toxicity of styrene. I. Biotransformation of styrene to styrene glycol via styrene oxide by rat liver microsomes. *J. pharm. Dyn.*, **1**, 98–104

Watabe, T., Ozawa, N. & Yoshikawa, K. (1981) Stereochemistry in the oxidative metabolism of styrene by hepatic microsomes. *Biochem. Pharmacol.*, **30**, 1695–1698

Watabe, T., Ozawa, N. & Yoshikawa, K. (1982) Studies on metabolism and toxicity of styrene. V. The metabolism of styrene, racemic, (*R*)-(+)-, and (*S*)-(–)-phenyloxiranes in the rat. *J. pharm. Dyn.*, **5**, 129–133

Watabe, T., Hiratsuka, A., Sone, T., Ishihama, T. & Endoh, K. (1984) Hepatic microsomal oxidation of styrene to 4-hydroxystyrene 7,8-glycol via 4-hydroxystyrene and its 7,8-oxide as short-lived intermediates. *Biochem. Pharmacol.*, **33**, 3101–3103

Weast, R.C. & Astle, M.J. (1985) *CRC Handbook of Data on Organic Compounds*, Volumes I and II, Boca Raton, FL, CRC Press Inc., p. V2 304

Welp, E., Kogevinas, M., Andersen, A., Bellander, T., Biocca, M., Coggon, D., Esteve, J., Gennaro, V., Kolstad, H., Lundberg, I., Lynge, E., Partanen, T., Spence, A., Boffetta, P. Ferro, G. & Saracci, R. (1996a) Exposure to styrene and mortality from nervous system diseases and mental disorders. *Am. J. Epidemiol.*, **144**, 623–633

Welp, E., Partanen, T., Kogevinas, M., Andersen, A., Bellander, T., Biocca, M., Coggon, D., Gennaro, V., Kolstad, H., Lundberg, I., Lynge, E., Spence, A., Ferro, G., Saracci, R. &

Boffetta, P. (1996b) Exposure to styrene and mortality from non-malignant respiratory disease. *Occup. environ. Med.*, **53**, 499–501

Wenker, M.A.M., Kežic, S., Monster, A.C. & de Wolff, F.A. (2000) Metabolism of styrene-7,8-oxide in human liver in vitro: Interindividual variation and stereochemistry. *Toxicol. appl. Pharmacol.*, **169**, 52–58

Wenker, M.A.M., Kežic, S., Monster, A.C. & de Wolff, F.A. (2001a) Stereochemical metabolism of styrene in volunteers. *Int. Arch. occup. environ. Health*, **74**, 359–365

Wenker, M.A.M., Kežic, S., Monster, A.C. & de Wolff, F.A. (2001b) Metabolic capacity and interindividual variation in toxicokinetics of styrene in volunteers. *Hum. exp. Toxicol.*, **20**, 221–228

Wenker, M.A.M., Kežic, S., Monster, A.C. & de Wolff, F.A. (2001c) Metabolism of styrene in the human liver in vitro: Interindividual variation and enantioselectivity. *Xenobiotica*, **31**, 61–72

WHO (1983) *Styrene* (Environmental Health Criteria 26), Geneva, International Programme of Chemical Safety

WHO (1993) *Guidelines for Drinking-water Quality*, 2nd Ed., Vol. 1, Geneva, p. 67

Wieczorek, H. & Piotrowski, J.K. (1985) Evaluation of low exposure to styrene. I. Absorption of styrene vapours by inhalation under experimental conditions. *Int. Arch. occup. environ. Health*, **57**, 57–69

Wigaeus, E., Löf, A., Bjurström, R. & Byfält Nordqvist, M. (1983) Exposure to styrene — Uptake, distribution, metabolism and elimination in man. *Scand. J. Work Environ. Health*, **9**, 479–488

Wigaeus, E., Löf, A. & Byfält Nordqvist, M. (1984) Uptake, distribution, metabolism, and elimination of styrene in man. A comparison between single exposure and co-exposure with acetone. *Br. J. ind. Med.*, **41**, 539–546

Withey, J.R. & Collins, P.G. (1979) The distribution and pharmacokinetics of styrene monomer in rats by the pulmonary route. *J. environ. Pathol. Toxicol.*, **2**, 1329–1342

Withey, J.R. & Karpinski, K. (1985) Fetal distribution of styrene in rats after vapor phase exposures. *Biol. Res. Pregnancy Perinatol.*, **6**, 59–64

Wolff, M.S., Daum, S.M., Lorimer, W.V., Selikoff, I.J. & Aubrey, B.B. (1977) Styrene and related hydrocarbons in subcutaneous fat from polymerization workers. *J. Toxicol. environ. Health*, **2**, 997–1005

Wolff, M.S., Lilis, R., Lorimer, W.V. & Selikoff, I.J. (1978) Biological indicators of exposure in styrene polymerisation workers. Styrene in blood and adipose tissue and mandelic and phenylglyoxylic acids in urine. *Scand. J. Work Environ. Health*, **4** (Suppl. 2), 114–118

Wong, O. (1990) A cohort mortality study and a case-control study of workers potentially exposed to styrene in the reinforced plastics and composites industry. *Br. J. ind. Med.*, **47**, 753–762

Wong, O. & Trent, L.S. (1999) Mortality from nonmalignant diseases of the respiratory, genitourinary and nervous systems among workers exposed to styrene in the reinforced plastics and composites industry in the United States. *Scand. J. Work Environ. Health*, **25**, 317–325

Wong, O., Trent, L.S. & Whorton, M.D. (1994) An updated cohort mortality study of workers exposed to styrene in the reinforced plastics and composites industry. *Occup. environ. Med.*, **51**, 386–396

Wrangskog, K., Sollenberg, J. & Söderman, E. (1996) Application of a single-compartment model for estimation of styrene uptake from measurements of urinary excretion of mandelic and phenylglyoxylic acids after occupational exposure. *Int. Arch. occup. environ. Health*, **68**, 337–341

Yano, B.L., Dittenber, D.A., Albee, R.R. & Mattsson, J.L. (1992) Abnormal auditory brain stem responses and cochlear pathology in rats induced by an exaggerated styrene exposure regimen. *Toxicol. Pathol.*, **20**, 1–6

Yeowell-O'Connell, K., Jin, Z. & Rappaport, S.M. (1996) Determination of albumin and hemoglobin adducts in workers exposed to styrene and styrene oxide. *Cancer Epidemol. Biomark. Prevent.*, **5**, 205–215

Zaprianov, Z. & Bainova, A. (1979) Changes in monoamine oxidase activity (MOA) after styrene and ethanol combined treatment of rats. *Activ. nerv. Sup. (Praha)*, **21**, 262–265

Zhang, X.-X., Chakrabarti, S., Malick, A.M. & Richer, C.-L. (1993) Effects of different styrene metabolites on cytotoxicity, sister-chromatid exchanges and cell-cycle kinetics in human whole blood lymphocytes in vitro. *Mutat. Res.*, **302**, 213–218

SUMMARY OF FINAL EVALUATIONS

Agent	Degree of evidence of carcinogenicity		Overall evaluation of carcinogenicity to humans
	Human	Animal	
Herbal remedies containing plant species of the genus *Aristolochia*	S	ND	1
Naturally occurring mixtures of aristolochic acids	L		2A
Aristolochic acids		S	
Laxatives containing anthraquinone derivatives	I		
1-Hydroxyanthraquinone		S	2B
Madder root (*Rubia tinctorum)*		L	3
Riddelliine	ND	S	2B
Fumonisins, naturally occurring mixtures of	I		
Fumonisin B_1		S	2B
Aflatoxins, naturally occurring mixtures of			(1, reaffirmed)
Naphthalene	I	S	2B
Styrene	L	L	2B

S, sufficient evidence of carcinogenicity; L, limited evidence of carcinogenicity; I, inadequate evidence of carcinogenicity; ND, no data; Group 1, carcinogenic to humans; Group 2A, probably carcinogenic to humans; Group 2B, possibly carcinogenic to humans; Group 3, cannot be classified as to carcinogenicity to humans. For definitions of criteria for degrees of evidence, see Preamble.

LIST OF ABBREVIATIONS USED IN THIS VOLUME

AA-I: aristolochic acid I
AA-II: aristolochic acid II
ABS: acrylonitrile–butadiene–styrene
ACGIH: American Conference of Governmental Industrial Hygienists
AESGP: Association of the European Self-Medication Industry
AFAR: aflatoxin aldehyde reductase
AFB: B aflatoxins
AFB$_1$: aflatoxin B_1
AFG: G aflatoxins
AFM$_1$: aflatoxin M_1
AFP$_1$: aflatoxin P_1
AIN: American Institute of Nutrition
anti-HBc: antibody to hepatitis B core antigen
anti-HBs: antibody to HBsAg
AOAC: Association of Official Analytical Chemists
AP/AT: atmospheric pressure/ambient temperature
APCI: atmospheric pressure chemical ionization
AR: aldehyde reductase
AUC: area under the curve
a$_w$: water activity
BASE: Building Assessment Survey and Evaluation
CAM: complementary and alternative medicine
CBI: covalent binding index
CDSL: Commission on Dietary Supplement Labels
CHN: Chinese herb nephropathy
CI: confidence interval
CNS: central nervous system
CPA: cyclopiazonic acid
CZE: capillary zone electrophoresis
dA: deoxyadenosine
dAMP: deoxyadenosine 5′-monophosphate
dTMP: deoxythymidine 5′-monophosphate
dCMP: deoxycytidine 5′-monophosphate
dG: deoxyguanosine

DGAL: Direction Générale de l'Alimentation
DGCCRF: Direction Générale de la Concurrence, de la Consommation et de la Répression des Fraudes
DMDTC: dimethyldithiocarbamate
DMSO: dimethyl sulfoxide
DSHEA: Dietary Supplements Health and Education Act
EC: European Commission
ELEM: equine leukoencephalomalacia
ELISA: enzyme-linked immunosorbent assay
EMEA: European Agency for the Evaluation of Medicinal Products
EN: European norm
EQL: estimated quantitation limit
ESCOP: European Scientific Cooperative on Phytotherapy
ETS: environmental tobacco smoke
EU: European Union
FAO: Food and Agriculture Organization
FB_1: fumonisin B_1
FD: fluorescence detection
FDA: Food and Drug Administration
FID: flame ionization detection
FT–IR: Fourier transform infrared detection
GAP: good agriculture practice
GC: gas chromatography
GEMS: Global Environment Monitoring System
GF: growth factor
GPA: glycophorin A
Gpt: guanine phosphoribosyl transferase
GRAS: generally recognized as safe
GSH: (reduced) glutathione
GST: glutathione *S*-transferase
GSTP: placental glutathione *S*-transferase
HACCP: hazard analysis: critical control point
HBsAg: hepatitis B (virus) surface antigen
HBV: hepatitis B virus
HCC: hepatocellular carcinoma
HCV: hepatitis C virus
HFB_1: hydrolysed fumonisin B_1
HFB_2: hydrolysed fumonisin B_2
HGF: hepatocyte growth factor
HP/HT: high pressure/high temperature
HPB: Health Protection Branch
HPLC: high-performance liquid chromatography

HPRT: hypoxanthine guanine phosphoribosyl-transferase
IC: immunoaffinity column
IR: insecticide-resistant
IS: insecticide-susceptible
ISO: International Organization for Standardization
JECFA: Joint FAO/WHO Expert Committee on Food Additives
L&H: malignancies of the lymphatic and haematopoietic tissues
LC: liquid chromatography
LOH: loss of heterozygosity
MAFF: Ministry of Agriculture, Fisheries and Food
MAK: maximum workplace concentration (German)
MC: microcolumn
MCC: Medicines Control Council
mf: *Macaca fascicularis*
MFC: multifunctional column
MHV: Woodchuck (*Marmota*) hepatitis virus
MHW: Ministry of Health and Welfare
ML: maximum level
MS: mass spectrometry
NADPH: nicotinamide-adenine dinucleotide phosphate (reduced form)
NAESCO: National Association of Energy Service Companies
NAPRALERT: NAtural PRoducts ALERT
NCI: National Cancer Institute
NHATS: National Human Adipose Tissue Survey
NHI: National Health Insurance
NIH: National Institutes of Health
OR: odds ratio
OSHA: Occupational Safety and Health Administration
OTC: over-the-counter
OTD: Oregon Test Diet
OV: *Opisthorchis viverrini*
PC: phosphatidylcholine
PCR: polymerase chain reaction
PE: phosphatidylethanolamine
PEL: permissible exposure level
PGE_1: prostaglandin E_1
PGE_2: prostaglandin E_2
PHC: primary hepatocellular carcinoma
PHP: piperidinohydroxypropyl
PID: photoionization detection
PLA_2: phospholipase A_2
PLC: primary liver cancer

PMTDI: provisional maximum tolerable daily intake
PTWI: provisional tolerable weekly intake
p-yr: person–years
REL: recommended exposure level
RFLP: restriction fragment length polymorphism
RR: relative risk
Sa/So: free sphinganine/free sphingosine
Sa: free sphinganine
SAN: styrene–acrylonitrile
SBR: styrene–butadiene rubber
SIS: school intervention studies
SM: sphingomyelin
SMase: sphingomyelinase
SMR: standardized mortality ratio
So: free sphingosine
SPIR: standardized proportionate mortality ratio
SPRIA: serum plasma radioimmunoassay
SSCP: single-strand conformation polymorphism
STEL: short-term exposure limit
STORET: STOrage and RETrieval
TGA: Therapeutic Goods Administration (Australia)
TGF: transforming growth factor
TLC: thin-layer chromatography
TLV: threshold limit value
TNF: tumour necrosis factor
UNEP: United Nations Environment Programme
UV: ultraviolet detection
w/v: weight/volume
w/w: weight/weight
WHO: World Health Organization

CUMULATIVE CROSS INDEX TO *IARC MONOGRAPHS ON THE EVALUATION OF CARCINOGENIC RISKS TO HUMANS*

The volume, page and year of publication are given. References to corrigenda are given in parentheses.

B

E

F

G

H

L

M

N

Q

R

U

Y

Z

List of IARC Monographs on the Evaluation of Carcinogenic Risks to Humans*

Volume 1
Some Inorganic Substances, Chlorinated Hydrocarbons, Aromatic Amines, *N*-Nitroso Compounds, and Natural Products
1972; 184 pages (out-of-print)

Volume 2
Some Inorganic and Organometallic Compounds
1973; 181 pages (out-of-print)

Volume 3
Certain Polycyclic Aromatic Hydrocarbons and Heterocyclic Compounds
1973; 271 pages (out-of-print)

Volume 4
Some Aromatic Amines, Hydrazine and Related Substances, *N*-Nitroso Compounds and Miscellaneous Alkylating Agents
1974; 286 pages (out-of-print)

Volume 5
Some Organochlorine Pesticides
1974; 241 pages (out-of-print)

Volume 6
Sex Hormones
1974; 243 pages (out-of-print)

Volume 7
Some Anti-Thyroid and Related Substances, Nitrofurans and Industrial Chemicals
1974; 326 pages (out-of-print)

Volume 8
Some Aromatic Azo Compounds
1975; 357 pages

Volume 9
Some Aziridines, *N*-, *S*- and *O*-Mustards and Selenium
1975; 268 pages

Volume 10
Some Naturally Occurring Substances
1976; 353 pages (out-of-print)

Volume 11
Cadmium, Nickel, Some Epoxides, Miscellaneous Industrial Chemicals and General Considerations on Volatile Anaesthetics
1976; 306 pages (out-of-print)

Volume 12
Some Carbamates, Thiocarbamates and Carbazides
1976; 282 pages (out-of-print)

Volume 13
Some Miscellaneous Pharmaceutical Substances
1977; 255 pages

Volume 14
Asbestos
1977; 106 pages (out-of-print)

Volume 15
Some Fumigants, the Herbicides 2,4-D and 2,4,5-T, Chlorinated Dibenzodioxins and Miscellaneous Industrial Chemicals
1977; 354 pages (out-of-print)

Volume 16
Some Aromatic Amines and Related Nitro Compounds—Hair Dyes, Colouring Agents and Miscellaneous Industrial Chemicals
1978; 400 pages

Volume 17
Some *N*-Nitroso Compounds
1978; 365 pages

Volume 18
Polychlorinated Biphenyls and Polybrominated Biphenyls
1978; 140 pages (out-of-print)

Volume 19
Some Monomers, Plastics and Synthetic Elastomers, and Acrolein
1979; 513 pages (out-of-print)

Volume 20
Some Halogenated Hydrocarbons
1979; 609 pages (out-of-print)

Volume 21
Sex Hormones (II)
1979; 583 pages

Volume 22
Some Non-Nutritive Sweetening Agents
1980; 208 pages

Volume 23
Some Metals and Metallic Compounds
1980; 438 pages (out-of-print)

Volume 24
Some Pharmaceutical Drugs
1980; 337 pages

Volume 25
Wood, Leather and Some Associated Industries
1981; 412 pages

Volume 26
Some Antineoplastic and Immunosuppressive Agents
1981; 411 pages

Volume 27
Some Aromatic Amines, Anthraquinones and Nitroso Compounds, and Inorganic Fluorides Used in Drinking-water and Dental Preparations
1982; 341 pages

Volume 28
The Rubber Industry
1982; 486 pages

Volume 29
Some Industrial Chemicals and Dyestuffs
1982; 416 pages

Volume 30
Miscellaneous Pesticides
1983; 424 pages

---

*Certain older volumes, marked out-of-print, are still available directly from IARCPress. Further, high-quality photocopies of all out-of-print volumes may be purchased from University Microfilms International, 300 North Zeeb Road, Ann Arbor, MI 48106-1346, USA (Tel.: 313-761-4700, 800-521-0600).

Volume 31
Some Food Additives, Feed Additives and Naturally Occurring Substances
1983; 314 pages (out-of-print)

Volume 32
Polynuclear Aromatic Compounds, Part 1: Chemical, Environmental and Experimental Data
1983; 477 pages (out-of-print)

Volume 33
Polynuclear Aromatic Compounds, Part 2: Carbon Blacks, Mineral Oils and Some Nitroarenes
1984; 245 pages (out-of-print)

Volume 34
Polynuclear Aromatic Compounds, Part 3: Industrial Exposures in Aluminium Production, Coal Gasification, Coke Production, and Iron and Steel Founding
1984; 219 pages

Volume 35
Polynuclear Aromatic Compounds, Part 4: Bitumens, Coal-tars and Derived Products, Shale-oils and Soots
1985; 271 pages

Volume 36
Allyl Compounds, Aldehydes, Epoxides and Peroxides
1985; 369 pages

Volume 37
Tobacco Habits Other than Smoking; Betel-Quid and Areca-Nut Chewing; and Some Related Nitrosamines
1985; 291 pages

Volume 38
Tobacco Smoking
1986; 421 pages

Volume 39
Some Chemicals Used in Plastics and Elastomers
1986; 403 pages

Volume 40
Some Naturally Occurring and Synthetic Food Components, Furocoumarins and Ultraviolet Radiation
1986; 444 pages

Volume 41
Some Halogenated Hydrocarbons and Pesticide Exposures
1986; 434 pages

Volume 42
Silica and Some Silicates
1987; 289 pages

Volume 43
Man-Made Mineral Fibres and Radon
1988; 300 pages

Volume 44
Alcohol Drinking
1988; 416 pages

Volume 45
Occupational Exposures in Petroleum Refining; Crude Oil and Major Petroleum Fuels
1989; 322 pages

Volume 46
Diesel and Gasoline Engine Exhausts and Some Nitroarenes
1989; 458 pages

Volume 47
Some Organic Solvents, Resin Monomers and Related Compounds, Pigments and Occupational Exposures in Paint Manufacture and Painting
1989; 535 pages

Volume 48
Some Flame Retardants and Textile Chemicals, and Exposures in the Textile Manufacturing Industry
1990; 345 pages

Volume 49
Chromium, Nickel and Welding
1990; 677 pages

Volume 50
Pharmaceutical Drugs
1990; 415 pages

Volume 51
Coffee, Tea, Mate, Methyl-xanthines and Methylglyoxal
1991; 513 pages

Volume 52
Chlorinated Drinking-water; Chlorination By-products; Some Other Halogenated Compounds; Cobalt and Cobalt Compounds
1991; 544 pages

Volume 53
Occupational Exposures in Insecticide Application, and Some Pesticides
1991; 612 pages

Volume 54
Occupational Exposures to Mists and Vapours from Strong Inorganic Acids; and Other Industrial Chemicals
1992; 336 pages

Volume 55
Solar and Ultraviolet Radiation
1992; 316 pages

Volume 56
Some Naturally Occurring Substances: Food Items and Constituents, Heterocyclic Aromatic Amines and Mycotoxins
1993; 599 pages

Volume 57
Occupational Exposures of Hairdressers and Barbers and Personal Use of Hair Colourants; Some Hair Dyes, Cosmetic Colourants, Industrial Dyestuffs and Aromatic Amines
1993; 428 pages

Volume 58
Beryllium, Cadmium, Mercury, and Exposures in the Glass Manufacturing Industry
1993; 444 pages

Volume 59
Hepatitis Viruses
1994; 286 pages

Volume 60
Some Industrial Chemicals
1994; 560 pages

Volume 61
Schistosomes, Liver Flukes and *Helicobacter pylori*
1994; 270 pages

Volume 62
Wood Dust and Formaldehyde
1995; 405 pages

Volume 63
Dry Cleaning, Some Chlorinated Solvents and Other Industrial Chemicals
1995; 551 pages

Volume 64
Human Papillomaviruses
1995; 409 pages

Volume 65
Printing Processes and Printing Inks, Carbon Black and Some Nitro Compounds
1996; 578 pages

Volume 66
Some Pharmaceutical Drugs
1996; 514 pages

Volume 67
Human Immunodeficiency Viruses and Human T-Cell Lymphotropic Viruses
1996; 424 pages

Volume 68
Silica, Some Silicates, Coal Dust and *para*-Aramid Fibrils
1997; 506 pages

Volume 69
Polychlorinated Dibenzo-*para*-Dioxins and Polychlorinated Dibenzofurans
1997; 666 pages

Volume 70
Epstein-Barr Virus and Kaposi's Sarcoma Herpesvirus/Human Herpesvirus 8
1997; 524 pages

Volume 71
Re-evaluation of Some Organic Chemicals, Hydrazine and Hydrogen Peroxide
1999; 1586 pages

Volume 72
Hormonal Contraception and Post-menopausal Hormonal Therapy
1999; 660 pages

Volume 73
Some Chemicals that Cause Tumours of the Kidney or Urinary Bladder in Rodents and Some Other Substances
1999; 674 pages

Volume 74
Surgical Implants and Other Foreign Bodies
1999; 409 pages

Volume 75
Ionizing Radiation, Part 1, X-Radiation and γ-Radiation, and Neutrons
2000; 492 pages

Volume 76
Some Antiviral and Antineoplastic Drugs, and Other Pharmaceutical Agents
2000; 522 pages

Volume 77
Some Industrial Chemicals
2000; 563 pages

Volume 78
Ionizing Radiation, Part 2, Some Internally Deposited Radionuclides
2001; 595 pages

Volume 79
Some Thyrotropic Agents
2001; 763 pages

Volume 80
Non-Ionizing Radiation, Part 1: Static and Extremely Low-Frequency (ELF) Electric and Magnetic Fields
2002; 429 pages

Volume 81
Man-made Vitreous Fibres
2002; 418 pages

Volume 82
Some Traditional Herbal Medicines, Some Mycotoxins, Naphthalene and Styrene
2002; 590 pages

Supplement No. 1
Chemicals and Industrial Processes Associated with Cancer in Humans (*IARC Monographs*, Volumes 1 to 20)
1979; 71 pages (out-of-print)

Supplement No. 2
Long-term and Short-term Screening Assays for Carcinogens: A Critical Appraisal
1980; 426 pages (out-of-print)
(updated as IARC Scientific Publications No. 83, 1986)

Supplement No. 3
Cross Index of Synonyms and Trade Names in Volumes 1 to 26 of the *IARC Monographs*
1982; 199 pages (out-of-print)

Supplement No. 4
Chemicals, Industrial Processes and Industries Associated with Cancer in Humans (*IARC Monographs*, Volumes 1 to 29)
1982; 292 pages (out-of-print)

Supplement No. 5
Cross Index of Synonyms and Trade Names in Volumes 1 to 36 of the *IARC Monographs*
1985; 259 pages (out-of-print)

Supplement No. 6
Genetic and Related Effects: An Updating of Selected *IARC Monographs* from Volumes 1 to 42
1987; 729 pages

Supplement No. 7
Overall Evaluations of Carcinogenicity: An Updating of *IARC Monographs* Volumes 1–42
1987; 440 pages

Supplement No. 8
Cross Index of Synonyms and Trade Names in Volumes 1 to 46 of the *IARC Monographs*
1990; 346 pages (out-of-print)

All IARC publications are available directly from
IARCPress, 150 Cours Albert Thomas, 69372 Lyon cedex 08, France
(Fax: +33 4 72 73 83 02; E-mail: press@iarc.fr).

IARC Monographs and Technical Reports are also available from the
World Health Organization Marketing and Dissemination, 1211 Geneva 27, Switzerland
(Fax: +41 22 791 4857; E-mail: publications@who.int)
and from WHO Sales Agents worldwide.

IARC Scientific Publications, IARC Handbooks and IARC CancerBases are also available from
Oxford University Press, Walton Street, Oxford, UK OX2 6DP (Fax: +44 1865 267782).

IARC Monographs are also available in an electronic edition,
both on-line by internet and on CD-ROM, from GMA Industries, Inc.,
20 Ridgely Avenue, Suite 301, Annapolis, Maryland, USA
(Fax: +01 410 267 6602; internet: https//www.gmai.com/Order_Form.htm)

Impression : Imprimerie des Deux-Ponts - 38610 Gières
Dépôt légal n°628

www.ingramcontent.com/pod-product-compliance
Lightning Source LLC
Chambersburg PA
CBHW081152020426
42333CB00020B/2479